Polar Organometallic Reagents

Polar Organometallic Reagents

Synthesis, Structure, Properties and Applications

Edited by

Andrew E. H. Wheatley
University of Cambridge
Cambridge, UK

Masanobu Uchiyama
The University of Tokyo
Tokyo, Japan

WILEY

This edition first published 2022

Registered Office(s)
John Wiley & Sons, Inc., 111 River Street, Hoboken, NJ 07030, USA
John Wiley & Sons Ltd, The Atrium, Southern Gate, Chichester, West Sussex, PO19 8SQ, UK

Editorial Office
The Atrium, Southern Gate, Chichester, West Sussex, PO19 8SQ, UK

For details of our global editorial offices, customer services, and more information about Wiley products visit us at www.wiley.com.

Wiley also publishes its books in a variety of electronic formats and by print-on-demand. Some content that appears in standard print versions of this book may not be available in other formats.

Library of Congress Cataloging-in-Publication Data

Names: Wheatley, Andrew E. H, editor. | Uchiyama, Masanobu, editor.
Title: Polar organometallic reagents : synthesis, structure, properties and applications / edited by Andrew E. H. Wheatley, University of Cambridge, Cambridge, UK; Masanobu Uchiyama, The University of Tokyo, Tokyo, JP.
Description: First edition. | Hoboken, NJ : Wiley, 2022. | Includes bibliographical references and index.
Identifiers: LCCN 2021037022 (print) | LCCN 2021037023 (ebook) | ISBN 9781119448822 (hardback) | ISBN 9781119448860 (adobe pdf) | ISBN 9781119448846 (epub)
Subjects: LCSH: Organometallic chemistry.
Classification: LCC QD411 .P63 2022 (print) | LCC QD411 (ebook) | DDC 547/.05–dc23
LC record available at https://lccn.loc.gov/2021037022
LC ebook record available at https://lccn.loc.gov/2021037023

Cover Design: Wiley
Cover Image: © Piotr Zajc/Shutterstock, ANDREW E. H. WHEATLEY

Set in 9.5/12.5pt STIXTwoText by Strive, Pondicherry, India
Printed and bound by CPI Group (UK) Ltd, Croydon, CR0 4YY

C9781119448822_220222

Contents

Preface

Just as it has done for many years, organometallic and metalloorganic chemistry continues to play a vital role for synthetic chemists in the twenty-first century. It offers some of the most effective ways of regiospecifically elaborating organic systems or of harnessing the potential of small molecules. Nevertheless, chemists constantly encounter challenges and target new systems which are not amenable to existing reagents. It is issues such as aggressive nucleophilicity or the implications of temperature instability or solvent sensitivity of many traditional organometallic bases, which have driven much of the research discussed in this volume. A major emergent theme covered is the potential of new, heterobimetallic systems. In particular, the problems of understanding just how reagents predicated on the action of two different metals operate. Synergy vs. cooperativity will be looked at in depth.

The new and often unique reactivities of these more complex reagents (or reagent mixtures) and their ability to achieve more effective chemical transformations under increasingly mild conditions are discussed. To do this, a broad view of organometallic chemistry is adopted, taking in organooxides, amides and the like as appropriate. Mostly, combinations of main group metals will be looked at but, in particular when considering environmental applications of bimetallic systems, coverage will extend to the d-block. Divided into nine chapters, the volume broadly covers three main fields; structural chemistry in the solid state, understanding often catalytic processes by monitoring reaction pathways, and synthetic applications.

The first three chapters are dominated by the role of crystallography in understanding organometallic reagents and, in particular, heterobimetallics. Throughout the book, a range of crystal structures are reproduced. For ease, these omit minor disorder, lattice solvent and all but the most chemically relevant hydrogens. Chapter 1 focuses on the work in the laboratory where a new generation of so-called ate or synergic bases were originally developed. The advent of Kondo's non-nucleophilic zincates is considered and the implications for selective deprotonation are described. The major emphasis then switches to some of the most recent work in this area, where lithium cuprates have been studied and the idea of harnessing the oxidative flexibility of copper in coupling chemistry has been developed. If the first chapter shows anything it is that truly understanding the nature of many heterobimetallic reagents is vital to predicting or rationalizing their behaviour, and this frequently necessitates crystallographic determination. Chapter 2 focuses on this through the prism of synergistic main group combinations. In particular, a more expansive view of the less strongly polarizing metal component of these systems is considered, enabling the demonstration of some mechanistic variation in formally synergistic systems. This introduces competition between concerted transformations by individually heterobimetallic reagents and stepwise conversions, where new reactivity is enabled in spite of individual steps proceeding through the action of much more traditional reagents that cooperate. Though Chapter 2 tends towards the applications of these new systems, the theme is drawn out further in Chapter 3, where magnesiate chemistry begins to explore the seminal advances made by Mulvey and latterly O'Hara in what was originally termed 'inverse crown ether' chemistry but is now more broadly thought of in terms of a methodology for templating selective metalation. This has empowered chemists to achieve selectivities that contrast with those associated with established directing effects. Thereafter, looking elsewhere in group 2 allows an exploration of recent advances in calciate, strontiate, and bariate complexes.

Chapter 4 is the first of two chapters dominated by a catalytic emphasis on our consideration of applied chemistry. It does this, though, through the lens of understanding chemical processes. It directs the reader towards the role of computation in understanding how heterobimetallic complexes that now take in d-block as well as main group elements can work, and also the applications of these systems in environmentally relevant polymerization processes dependent on the activation of small molecules. As is pointed out, proving positive cooperative effects is easily done, though understanding the nature and origin of a cooperative mechanism remains difficult, with concerted or consecutive actions of different metals potentially competing, and then with variance between formal metal–metal bonding or metals tethered by bridging ligands. In all this it is important to bear in mind that limited direct observation of structures will typically be possible on catalytic species *in operando*. Catalysis remains the dominant theme in Chapter 5, but more through the prism of p-block chemistry, and of group 13 in particular. Noting that the domain of catalysis has been dominated by d-block metals, this chapter records that the beginning of the twenty-first century has seen a shift towards main group element alternatives. This is the context then in which the remarkable reactivity of group 13 compounds have entered the catalytic arena. While tuning of their Lewis acidity and electrophilicity is possible by varying metal coordination number, it is the conversion of covalent complexes into cationic derivatives that will dominate. The chapter provides an overview of recent developments in the synthetic, characterization, computation and reactivity studies and applications in catalysis of cationic complexes. It will cover boron for completeness, but focus on aluminium and the higher group 13 elements.

Following the introduction of convenient spectroscopic handles in Chapter 5, multinuclear NMR studies allow the interrogation of solution processes and intermediates in Chapter 6. This picks up ideas hinted at but not fully explored in Chapters 1–3. The chapter revisits early studies in the NMR spectroscopic analysis of recognizably traditional organolithium and lithium amide systems, but then brings that work up-to-date, looking at the recent elegant advances in our understanding of the solution dynamics of enolates. The main focus of this chapter, though, relates to recent developments in diffusion-ordered spectroscopy. Again, this is contextualized though reference to the early work of Williard, but then shifts to modern lithium amide chemistry capable of being done under aerobic conditions. Most particularly, though, modern diffusion methods have allowed us to better understand the real solution chemistries of heterobimetallic formulations. This has been greatly enabled by the work of Stalke, raising the bar where the accuracy of molecular weight determination in solution is concerned. In consequence, diffusion analysis and multinuclear spectroscopy has prompted a much truer understanding of the aggregation and solvation of highly applied systems—for example the turbo reagents of Knochel.

The final set of chapters deal with up-to-date aspects of (mostly main group) synthetic applications of organometallic and metalloorganic chemistry. Chapter 7 is focused on the chemistry of boryl anions. In these, the nucleophilicity at the group 13 centre contrasts with the more common Lewis acid chemistry of boron, enabling a range of boryl-substituted organic, main group, and organometallic compounds to be prepared. Chapter 8 deals very much with synthetic organic chemistry as accessed by ate complexes. A range of ate complexes are discussed, but the chapter opens, appropriately enough, with work done in Sendai in the early 1990s. Initial discussion of sterically demanding tetramethylzincate in a range of transformations eventually shifts to the advent of organoamidozincates that are structurally covered in earlier chapters. From there, the evolution of aluminate and cyanocuprate congeners follows naturally. Again, structural aspects of these species feature elsewhere, but the discussion here focuses on their potential in aromatic derivatization. The chapter also continues a theme from the previous one. Efforts to exert control over the reactivity of boryl anions through ate complexes are presented in the context of novel nucleophilic boration. Lastly, Chapter 9 finishes the book as it started, looking at organocopper compounds. Copper-mediated processes are well established, prompting much organocopper structural chemistry. Whereas efforts have often focused on alkynyl- and arylorganocopper complexes, the emphasis here is instead on traditionally unstable mono-alkenyl and butadienylcopper compounds. The potential of steric, coordinative and synergistic strategies for preventing decomposition are examined.

Overall, this book seeks to bring the reader up-to-date with developments in the field of modern polar organometallic chemistry, particularly in the context of the emergent areas of synergic and cooperative species. The recent advances in our understanding of the operation of these systems are presented, but the evolution of the area is described, offering a context and explaining the need and rationale for the advent of new reagents. The hope is that this is done in a way that makes the text accessible to students as well as academics and industrial scientists, arming them with a broad understanding of how these systems present new opportunities to the synthetic chemist.

A. E. H. Wheatley
Cambridge, UK

M. Uchiyama
Tokyo, Japan

List of Contributors

Alistair M. Broughton
Yusuf Hamied Department of Chemistry
University of Cambridge
Cambridge, UK

Andrew E. H. Wheatley
Yusuf Hamied Department of Chemistry
University of Cambridge
Cambridge, UK

Andrew J. Peel
Yusuf Hamied Department of Chemistry
University of Cambridge
Cambridge, UK

Andrew S. S. Wilson
Department of Chemistry
University of Bath
Claverton Down, UK

Anne-Frédérique Pécharman
Department of Chemistry
University of Bath
Claverton Down, UK

Antoine Buchard
Department of Chemistry
University of Bath
Claverton Down, UK

Chao Wang
Graduate School of Pharmaceutical Sciences
The University of Tokyo
Tokyo, Japan

Eva Hevia
Department für Chemie, Biochemie und Pharmazie
Universität Bern
Bern, Switzerland

Frances N. Singer
Department of Chemistry
University of Bath
Claverton Down, UK

Keiichi Hirano
Graduate School of Pharmaceutical Sciences
The University of Tokyo
Tokyo, Japan

Leonie J. Bole
Department für Chemie, Biochemie und Pharmazie
Universität Bern
Bern, Switzerland

Liang Liu
College of Chemistry
Peking University
Beijing, China

Makoto Yamashita
Department of Molecular and Macromolecular Chemistry
Graduate School of Engineering
Nagoya University
Nagoya, Japan

Mamta Bhandari
Department of Chemical Sciences
Indian Institute of Science Education and Research Mohali
India

Masanobu Uchiyama
Graduate School of Pharmaceutical Sciences
The University of Tokyo
Tokyo, Japan

Research Initiative for Supra-Materials (RISM)
Shinshu University
Nagano, Japan

Masanori Shigeno
Graduate School of Pharmaceutical Sciences
Tohoku University
Sendai, Japan

Michael S. Hill
Department of Chemistry
University of Bath
Claverton Down, UK

Robert E. Mulvey
WestCHEM, Department of Pure and Applied Chemistry
University of Strathclyde
Glasgow, UK

Sandeep Rawat
Department of Chemical Sciences
Indian Institute of Science Education and Research Mohali
India

Sanjay Singh
Department of Chemical Sciences
Indian Institute of Science Education and Research Mohali
India

Sharanappa Nembenna
School of Chemical Sciences
National Institute of Science Education and Research Bhubaneswar
India

Stuart D. Robertson
WestCHEM, Department of Pure and Applied Chemistry
University of Strathclyde
Glasgow, UK

Yoshinori Kondo
Graduate School of Pharmaceutical Sciences
Tohoku University
Sendai, Japan

Zhenfeng Xi
College of Chemistry
Peking University
Beijing, China

Acknowledgements

The editors wish to thank John Wiley & Sons for the opportunity to publish this work. Our sincere thanks go to Sarah Higginbotham for proposing and supporting the idea of a book covering the recent advances in the field of polar organometallics. We are also grateful to Emma Strickland and the handling editors, particularly Katrina Maceda, for their constant support and advice. Finally, we would thank the chapter authors for contributing their expertise and time, without which this book could not have happened.

1

The Road to Aromatic Functionalization by Mixed-metal Ate Chemistry

Masanori Shigeno[a], Andrew J. Peel[b], Andrew E. H. Wheatley[b], and Yoshinori Kondo[a]

[a] *Graduate School of Pharmaceutical Sciences, Tohoku University, Sendai, Japan*
[b] *Yusuf Hamied Department of Chemistry, University of Cambridge, Cambridge, UK*

1.1 Introduction

Multifunctionalized organometallic species have played a vital role as versatile intermediates in modern synthetic organic chemistry for many years [1, 2]. Identifying a suitable metal and controlling reaction conditions have been crucial to underpinning the handling of functionalized organometallics because the reactivity and stability of these species and the effectiveness, selectivity, and safety of the process(es) by which they display metalation are strongly dependent upon the identity of the metal. This chapter describes the advent of controlled and selective metalation using traditional monometallic reagents but also outlines its limitations and the recognition that reactivity can be tuned by introducing a second metal centre. It was this view that eventually gave rise to a major theme of the current book; the concept of ate complexation of organometallics containing a very electropositive (highly reactive but perhaps unselective) metal using a more electronegative (less reactive, more selective) metal (Figure 1.1a). This idea led, in the late 1990s, to the use of not only organozinc halides or di-organozincs but also of the development of organozincates and their extensive use in a range of organic transformations including nucleophilic addition, halogen–metal exchange, metal carbenoid synthesis, de-protonative metalation, ring-opening, and cross-coupling [3]. The prolific ability of organozinc compounds and complexes to effect efficient yet highly selective transformations in tandem with inherent functional group compatibility resulted in great interest from synthetic chemists. In particular, because the latter characteristic feeds into applications in directed aromatic metalation, whereby a pre-existing substituent on an aromatic ring acts as a directing group (DMG) to promote selective reaction. This is a methodology that has come to represent one of the most effective ways of regiospecifically elaborating functionalized aromatic systems and which has, over a number of years, focused on the strategy of achieving *ortho* reaction of said ring. Though this technique has come to the fore in the elaboration of both aromatic and heteroaromatic systems, the low acidity of the hydrogen atoms on aromatic rings has led to a dependence on organolithium reagents as the base of choice for effecting deprotonation [4–8]. However, the high nucleophilicity of these same organolithium reagents has brought with it the associated risk of competing reaction at the DMG itself. It was the need to overcome this synthetic limitation that led to the development of the heterobimetallic complexes alluded to

Polar Organometallic Reagents: Synthesis, Structure, Properties and Applications, First Edition. Edited by Andrew E. H. Wheatley and Masanobu Uchiyama.

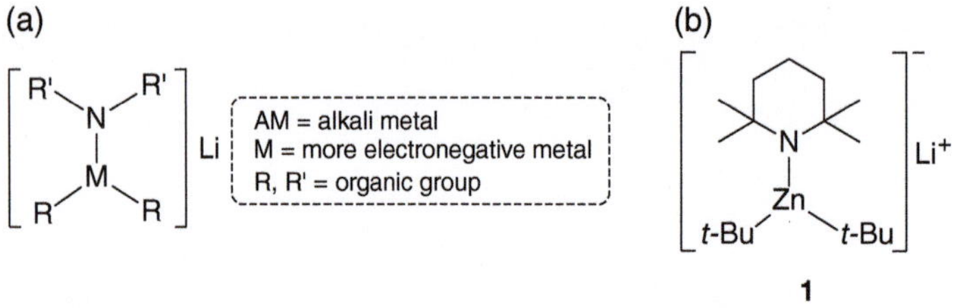

Figure 1.1 The general principles demonstrated by (a) a heterobimetallic ate complex designed for selective aromatic metalation, (b) as exemplified by the first reported example of such a complex, t-$Bu_2Zn(TMP)Li$ **1**.

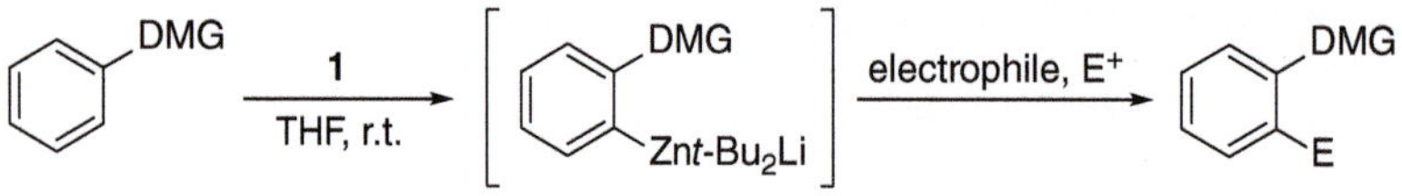

Scheme 1.1 A generic directed *ortho*-deprotometalation strategy incorporating **1**.

above, designed to promote chemoselective directed metalation reactions under mild conditions on the grounds of reduced nucleophilicity. This has now seen a range of zincates and other metal ate complexes incorporating the sterically bulky (non-nucleophilic) 2,2,6,6-tetramethylpiperidide ligand (TMP) successfully used to elaborate functionalized aromatics incorporating DMGs normally susceptible to competing nucleophilic degradation. The first work in this field appeared in 1999 when Kondo and coworkers reported the directed zincation of a functionalized aromatic compound. This was achieved using the, at that stage, putative base t-$Bu_2Zn(TMP)Li$ **1** (Figure 1.1b and Scheme 1.1) and avoided the normal (for organolithium reagents) requirement of temperatures below room temperature (r.t.) [9]. This key advance led to an explosion of interest in developing the concept of tunable ate complex reagents in directed metalation chemistry.

1.2 Deprotonation of Aromatics

1.2.1 Monometallic Bases

Since the pioneering work of Gilman [10] and Wittig [11], directed *ortho*-metalation has become widely used as a powerful and efficient method for the regioselective functionalization of aromatic compounds. A range of DMGs have been employed to facilitate the selective deprotonation of arenes, and various strong bases such as alkyllithiums (RLi) and lithium dialkylamides (R_2NLi) have been used (Scheme 1.2) [6], though often with the drawbacks outlined in Section 1.1.

Among group 1 organometallics, alkyllithium reagents are the most convenient to work with for synthetic chemists because of their solubility in ethers and/or frequently in alkanes. Moreover, or perhaps because of this, many of them are commercially available. Therefore, it has become of great practical importance to define the scope and limitations of alkyllithium-promoted deprotonation reactions. Generally, only substrates with high C–H acidity enhanced by a DMG are amenable to deprotonative lithiation [6]. In this capacity, ester and cyano groups have long been regarded as potentially important and attractive directors of metalation. However, their use has been limited because the deprotonation requires strictly controlled reaction conditions owing to the instability of intermediary aryllithium species. For example, lithium 2,2,6,6-tetramethylpiperidide (LTMP) has been used for directing the *ortho*-lithiation of aryl

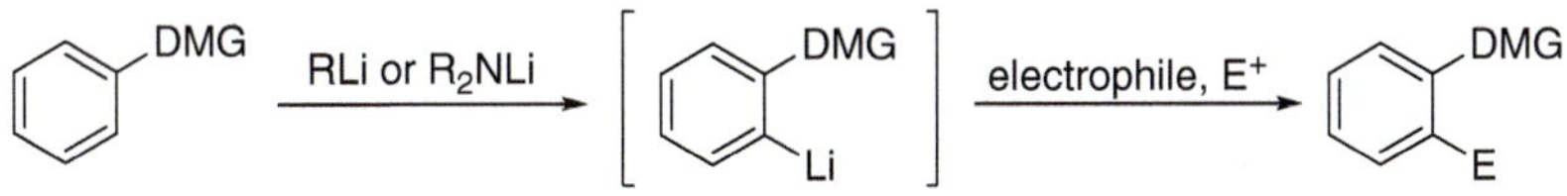

Scheme 1.2 Representation of a generic directed *ortho*-lithiation strategy.

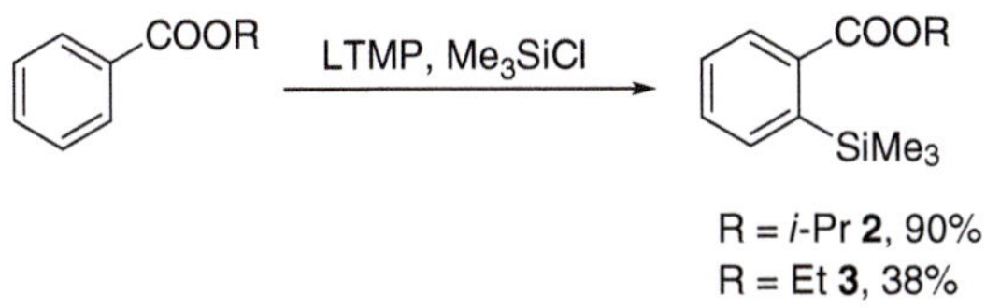

Scheme 1.3 Steric effects on the directed *ortho*-lithiation of aromatic esters.

carboxylic esters. However, the accompanying problems of unwanted condensation reactions between the aryllithium and electrophilic directing group have been well documented [12]. In spite of this, the *in situ* trapping of aryllithium species by electrophiles during the deprotonation of aryl carboxylic esters has been reported. However, the use of bulky ester groups under restricted reaction conditions is essential (Scheme 1.3) [13, 14].

The dominance of organolithiums in the deprotonation of aromatics by traditional organometallic reagents is reflected also at the structural level, and the actions of the many DMGs capable of controlling metalation have been summarized [15]. However, in spite of the long-established importance of DMGs such as carboxylic amides in directed *ortho*-metalation, it was only in 2001 that the first full structural evidence was provided for the nature of an aromatic deprotonated by an organolithium base under the influence of this directing agent [16]. Single crystal data established how it was that the simple amides $i\text{-}Pr_2NC(O)Ar$ (Ar = aryl) efficiently direct *ortho*-lithiation [17] given the well documented, sterically induced amide-arene twist-angle of ~90° in the substrate [18, 19]. Crystalline $i\text{-}Pr_2NC(O)C_6H_4Li(OEt_2)\text{-}2$ **4** was generated using a sterically congested DMG and *t*-BuLi to avoid unwanted competition from the nucleophilic reaction. The result was a dimer where Li was stabilized by both diethyl ether and DMG groups and in which the amide-arene twist-angle was been modulated to 47°. More complicated behaviour was seen for 2-alkyl-*N*,*N*-diisopropylbenzamides, with strongly solvent-dependent competition observed [20]. The dimer of $i\text{-}Pr_2NC(O)\text{-}2\text{-}Et\text{-}C_6H_3Li(THF)$ **5** (Figure 1.2a) proved to be a kinetic product, isolable from limited THF but rearranging in the presence of an excess of THF or of a stronger Lewis base [21] to give a thermodynamic benzylic metalate, presumably of the type seen in the THF-solvate of α-lithio-2-ethyl-*N*,*N*-diisopropyl-1-naphthamide [22, 23]. Meanwhile, the *ortho*-lithiate of 2-isopropyl-*N*,*N*-diisopropylbenzamide could be obtained from diethyl ether as a hemi-solvate **6** (Figure 1.2b) [24].

A further area of complexity associated with the deprotonative chemistry of organolithiums is that of anionic rearrangements, such as the long known [25] and synthetically useful [26] Fries formation of *ortho*-phenols from aromatic carbamates (see Section 1.4.2). According to rate studies, substituents at the meta position of the arene and the dialkylamino moiety of the carbamate DMG significantly influenced the relative rates of *ortho*-lithiation and subsequent rearrangement. Solution structural studies by ^6Li and ^{15}N NMR spectroscopies revealed a dimer of the lithiating agent (LDA), LDA dimer-arene complexes, an aryllithium monomer, LDA-aryllithium mixed dimers, an LDA-lithium phenolate mixed dimer, and lithium phenolate aggregates. While

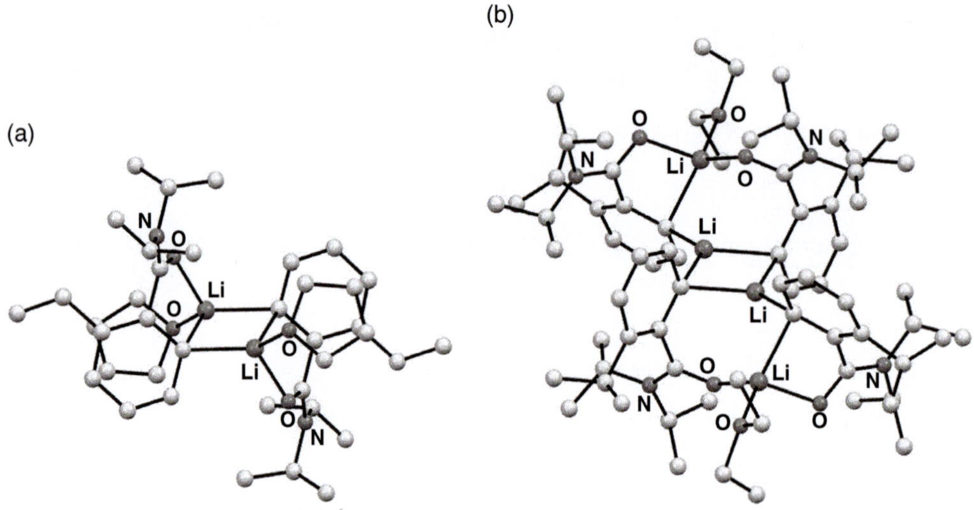

Figure 1.2 Dimer structures of (a) i-$Pr_2NC(O)$-2-Et-$C_6H_3Li(THF)$ **5** and (b) hemi-solvate [i-$Pr_2NC(O)$-2-(i-Pr)-$C_6H_3Li]_2(OEt_2)$ **6**. *Sources*: Adapted from Armstrong et al. [20]; Campbell Smith et al. [24].

monomer- and dimer-based *ortho*-lithiations and also monomer- and mixed dimer-based Fries rearrangements were identified, only the very insoluble phenolate dimer resulting from rearrangement could be isolated [27].

Though anionic Fries rearrangements compromise the regioselectivity of directed deprotonation, they are well documented. In contrast, an unusual post *ortho*-lithiation rearrangement was noted more recently for 1-lithio-naphthyllithium compounds with an *ortho*-directing 2-(dimethylamino)-methyl group [28]. The lithiation of 2-[(dimethylamino)methyl]naphthalene (dman) allowed elucidation of the 3-lithio regioisomer as a tetramer of 2-(Me_2NCH_2)$C_{10}H_6$Li-3 **7** in both the solid and solution states. In contrast, the 1-lithio regioisomer proved insoluble except in the presence of additional coordinating solvents, which gave [2-(Me_2NCH_2)$C_{10}H_6$Li-1]$_2$L (L = Et_2O **8**, dman **9**) in apolar solution. In the case of **9**, heating to 90 °C in toluene induced quantitative 1-lithio to 3-lithio rearrangement (**10**). Isotope labelling experiments suggested a rearrangement mechanism catalytic in dman and proceeding via heteroleptic intermediate [{2-(Me_2NCH_2)$C_{10}H_6$-1}{2-(Me_2NCH_2)$C_{10}H_6$-3}Li_2](dman) **11**; the dman is lithiated at its 3-position, while the formerly 1-lithio-naphthalene fragment is converted into new *N*-donor amine (Scheme 1.4).

Prior to the advent of ate chemistry, attempts to overcome the problem of organometallic nucleophilicity focused on the deployment of other metals. For example, in 1989, Eaton et al. reported the selective magnesiation of alkyl benzoates using sterically demanding magnesium amide **12** (Scheme 1.5) [29], suggesting the possibility of highly chemoselective conversion to e.g. **13** in the presence of ester and amide moieties. This protocol was used to *ortho*-carboxylate methyl benzoate, giving **14** in 81% yield.

Using a similar thesis, 1-substituted indole derivatives have been deprotonated using a magnesium diamide to give magnesioindoles, which were then successfully reacted with electrophiles. The compatibility of the magnesiated intermediates with a range of electrophilic functional groups was examined. For example, methyl 1-phenylsulfonylindole-3-carboxylate was treated with (i-$Pr_2N)_2Mg$ **15** followed by iodine or benzaldehyde to give 2-iodo derivative **16** and alcohol **17** in the respective yields 85 and 93% (Scheme 1.6) [30]. 1-Phenylsulfonylindole-3-carbonitrile was also tested in this iodination using i-Pr_2NMgBr **18** at the outset. In a similar vein, ethyl

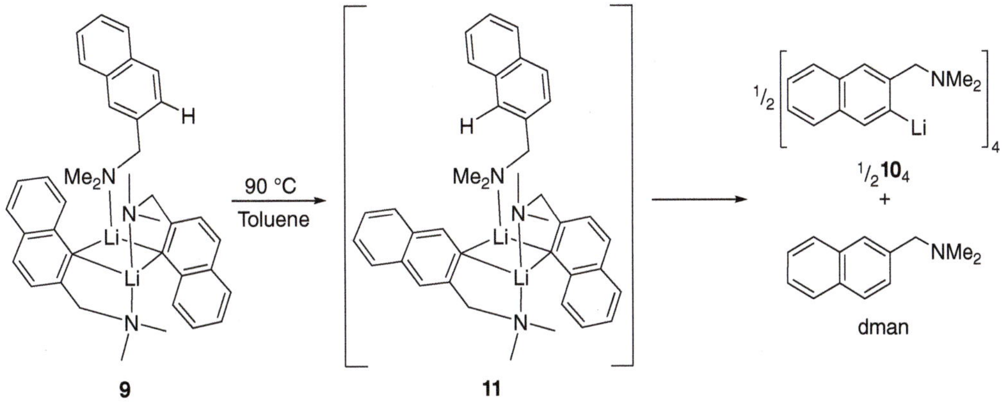

Scheme 1.4 The rearrangement of $[2\text{-}(Me_2NCH_2)C_{10}H_6Li\text{-}1]_2$(dman) **9** in hot toluene.

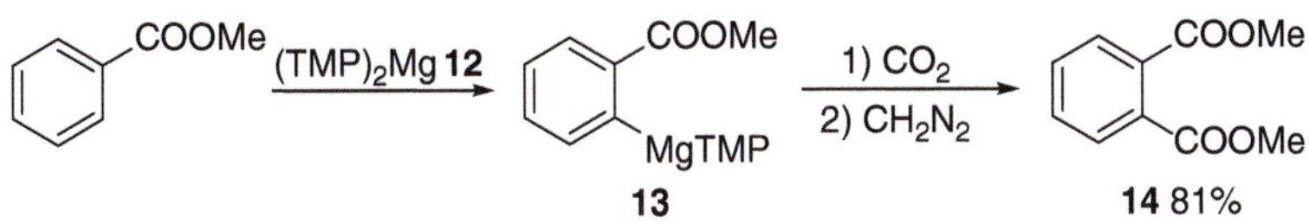

Scheme 1.5 Selective magnesiation of an alkyl benzoate using magnesium amide **12**.

Scheme 1.6 Treatment of methyl 1-phenylsulfonylindole-3-carboxylate with $(i\text{-}Pr_2N)_2Mg$ **15** en route to 2-iodinated **16** and alcohol **17**.

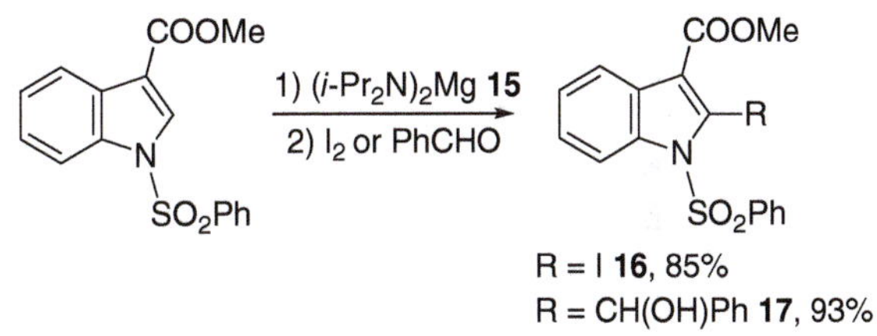

n-thiophenecarboxylate ($n = 2, 3$) has been selectively deprotonated with retention of the ester group, using $i\text{-}Pr_2NMgCl$ **19** to give 2,5- and 2,3-disubstituted thiophenes, respectively [31]. Meanwhile, the selective deprotonation of pyridine carboxamides and carbamates in conjunction with the more sterically congested magnesium amide TMPMgCl **20** has been reported [32]. Deprotonative magnesiation has been further investigated by Knochel and its scope and limitations have been the subject of review [33].

Detailed elucidation of the products of directed aromatic magnesiation has been enabled using air-sensitive crystallographic techniques. The same is true of precursors to deprotonation, with for example, the constitution of the deprotonating agents Grignard and Hauser bases probed. In these contexts, upon exposure to THF, MeMgCl **21** was found to form $MeMg_2(\mu\text{-}Cl)_3(THF)_{4\text{-}6}$ **22**, but to dimerize to give $Me_2Mg_4Cl_6(THF)_6$ **23** in the solid-state [34]. In contrast, externally solvated alkyl- and halo(amido)magnesiums formed straightforward monomers and dimers [35]. Moving to *ortho*-magnesiation, the bisamide **12** has been used to smoothly react boron-substituted benzenes. This work evolved from the advent of TMP-bases as *ortho*-metalating agents and enabled the *ortho*-magnesiation of borylbenzenes via internal reaction of an *N*-magnesiated intermediate **24**. The resulting *C,N*-magnesiate **25** was then electrophilically quenched using Me_2SO_4 to provide methylated product **26** that could be converted to a pinacolate (Scheme 1.7). The structure of a

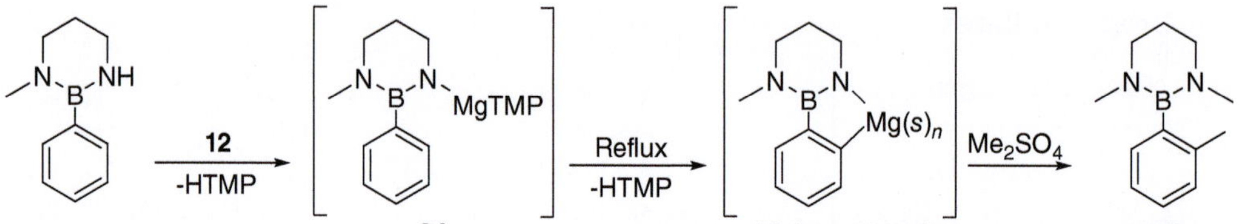

24
26

Scheme 1.7 *Ortho*-magnesiation of a borylbenzene via internal reaction of *N*-magnesiated intermediate **24** to give *C*,*N*-magnesiate **25**.

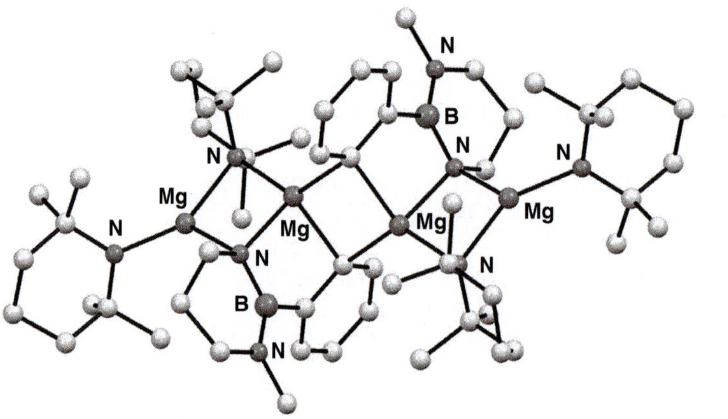

Figure 1.3 Structure of the *ortho C*,*N*-magnesiated dimer of **25(12)**. *Source*: Adapted from Kawachi et al. [36].

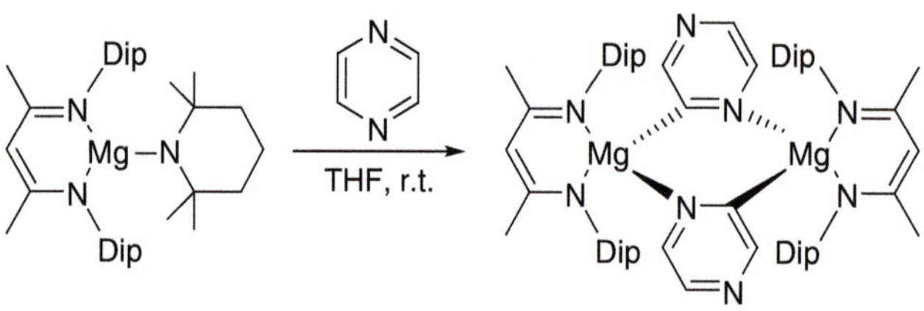

27
28_2

Scheme 1.8 The use of (^{Dip}Nacnac)MgTMP **27** in pyrazine deprotometalation.

representative *ortho C*,*N*-magnesiated intermediate proved to be a TMP-intercepted spirocycle (in effect, a dimer of **25**(**12**); Figure 1.3) [36].

β-Diketiminate magnesium complexes have been developed very recently and applied to the regioselective magnesiation of aromatics. Initially, (^{Dip}Nacnac)MgTMP **27** (^{Dip}Nacnac = DipNC(Me)CHC(Me)NDip; Dip = 2,6-(*i*-Pr)$_2$-C_6H_3), which combines kinetic base TMP with a sterically demanding spectator β-diketiminate, was reacted with pyrazine at room temperature to quantitatively yield (^{Dip}Nacnac)Mg($C_4H_3N_2$) **28** (Scheme 1.8) [37]. This work was then extended to using the kinetic TMP base to trap sensitive fluoroaryl anions for deployment in Negishi cross-coupling. This trapping behavior contrasted with that of the kinetic-amide-lacking organyl analogue (^{Dip}Nacnac)Mg(R)(THF) **29** (R = *n*-Bu, Ph, benzofuryl), which was shown to be effective in the chemically interesting [38, 39] magnesiation of perfluorinated aromatics by C–F bond alkylation/arylation [40].

1.2.2 Bimetallic Bases

1.2.2.1 Group 1/1 Reagents

An alternative approach to directed aromatic metalation has focused not upon replacing the alkyllithium base but on activating it. Two methods by which to achieve this were developed some time ago. The first was centered on TMEDA-activation (TMEDA = *N*,*N*,*N*′,*N*′-tetramethylethylenediamine) and the second involved the use of *tert*-butoxide-complexed alkyllithium reagents in the form of LICKOR superbases. The former route was employed to achieve site-selective deprotonation with different selectivity to that achieved using the alkyllithium alone. Meanwhile, whereas unimetallic superbases are known [41], 1966–1967 saw the introduction by Lochmann [42] and Schlosser [43] of heterobimetallic (Li–Na/K) superbases. Subsequently extended to incorporate a range of alkyllithium adducts of potassium alkoxides [44], the most widely known example deploys traditional organolithium reagents in tandem with KO*t*-Bu. Such heterobimetallic systems have shown enormous reactivity toward deprotonative metalation [45, 46]. As such, they enable the smooth deprotometalation of low acidity hydrocarbons [47] and weakly activated or nonactivated benzene derivatives [48] with, in some cases, unique regioselectivity [49] and also the facility for multideprotonation [50]. Whilst the synthetic importance of heterobimetallic superbases was quickly established, the characterization of such air-sensitive materials lagged behind.

Structural information on superbases has been gathered from a number of areas. Concerning the activation of organolithium bases, early evidence for the existence of organolithium-alkoxylithium aggregates of the type $n\text{-Bu}_x\text{Li}_4(\text{OR})_{4\text{-}x}$ (R = *n*-Bu or *t*-Bu; $x = 1$–4) in solution [51] was later substantiated by the crystallographic characterization of $(n\text{-BuLi})_4(\text{LiO}t\text{-Bu})_4$ **30** [52]. This sparked a search for similar complexes capable of acting as structurally well-defined models for superbases. Hence, alkoxy- and/or amido(alkali metal) combinations, now demonstrating the at least partial replacement of lithium with a higher group 1 congener, were investigated. These revealed not only a number of heterobimetallic structures [53] but also ternary alkali metal aggregates such as the twelve-vertex cage $[\{\text{PhN(H)}\}_2(\text{O}t\text{-Bu})\text{LiNaK(TMEDA)}_2]_2$ $\mathbf{31}_2$ (Figure 1.4a) [54]. However, whilst providing fascinating insights into alkali metal structural chemistry, their study has yet to allow a relationship to synthetically useful Li–K superbases to be established. A significant advance towards this was made with the study of the tetralithium–tetrapotassium amide-alkoxide $[\{t\text{-BuN(H)}\}_4(\text{O}t\text{-Bu})_4\text{Li}_4\text{K}_4(\text{C}_6\text{H}_6)](\text{C}_6\text{H}_6)$ **32** (Figure 1.4b). Upon dissolution, crystals of this material achieved the smooth metalation of toluene [55]. Meanwhile, the combination of three alkali metal alkoxides yielded similar reactivity towards toluene through forming the fully characterizable ternary complex $[\text{Li}_4\text{Na}_2\text{K}_2(\text{O}t\text{-Bu})_8(\mu\text{-L})]$ **33** (L = $\eta^6 : \eta^6$-C_6H_6, TMEDA) at ambient temperature. In contrast, none of the individual alkali metal alkoxides MO*t*-Bu (M = Li, Na, K) or their binary combinations detectably reacted with toluene under the same conditions (Scheme 1.9) [56].

Attempts to structurally investigate organo/alkoxy alkali metal aggregates, which might better represent the more frequently employed types of superbasic reagent already introduced [42, 43] have been hampered by the tendency of *in situ* preparations to precipitate microcrystalline (often organopotassium) products [57, 58]. Accordingly, a strategy was devised to overcome this problem through the *in situ* formation of a ligand providing both alkoxy and alkyl functionalities. To this end, sodium 2,4,6-trimethylphenoxide was treated with *n*-BuLi to result in lateral sodiation, and this enabled the crystallization of $[\{4,6\text{-Me}_2\text{C}_6\text{H}_2(\text{O})(\text{CH}_2)\}\text{LiNa(TMEDA)}]_4$ $\mathbf{34}_4$ (Figure 1.5), in which each phenoxide ligand had undergone a single benzylic deprotonation [59]. Consistent with previous reports that Li–O bonding takes precedent over Na–O interactions (logically a reflection of hard/soft interactions) [54, 55], X-ray diffraction revealed a structure based upon a $(\text{LiO})_4$ pseudocubic core, with the Na centres peripheral and coordinated perpendicular to the plane of the benzyl group.

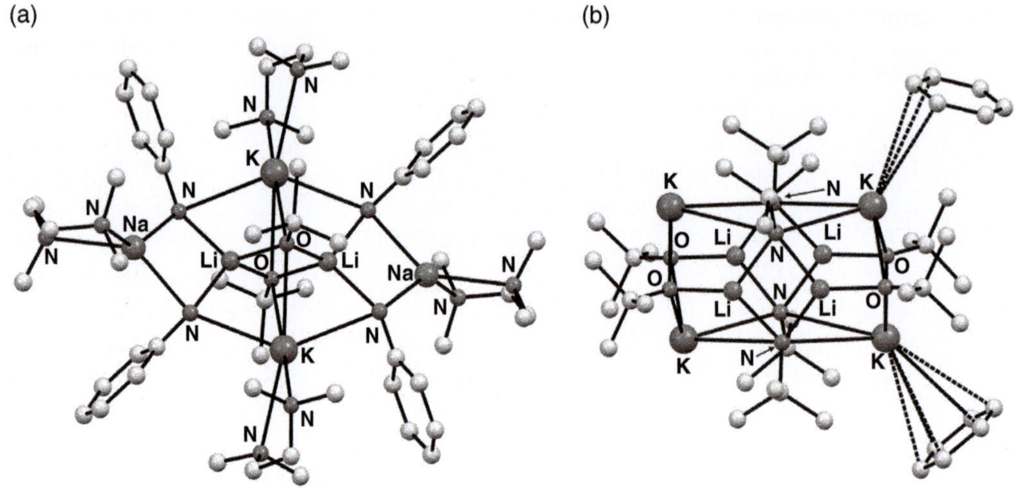

Figure 1.4 Molecular structures of (a) $[\{PhN(H)\}_2(t\text{-BuO})LiNaK(TMEDA)_2]_2$ **31**$_2$ and (b) $[\{t\text{-BuN(H)}\}_4(t\text{-BuO})_4Li_4K_4(C_6H_6)](C_6H_6)$ **32**. *Sources*: Adapted from Mackenzie et al. [54]; Kennedy et al. [55].

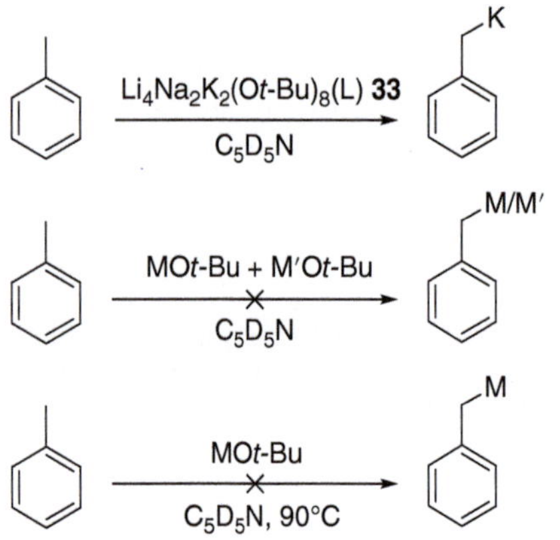

Scheme 1.9 Reactivity of metal alkoxides towards toluene in C_5D_5N. L = $\eta^6:\eta^6$-C_6H_6, TMEDA; M, M' = Li, Na, K.

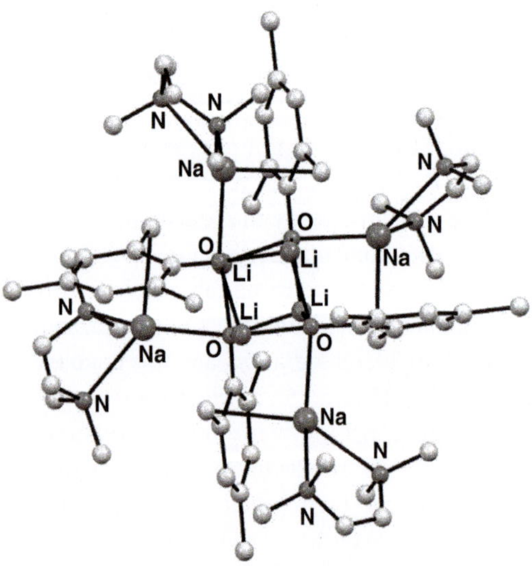

Figure 1.5 Molecular structure of $[\{4,6\text{-}Me_2C_6H_2(O)(CH_2)\}LiNa(TMEDA)]_4$ **34**$_4$. *Source*: Adapted from Harder and Streitwieser [59].

Moving forward a decade, the first organo/alkoxy Li–K species to be isolated from a LICKOR mixture was reported. To achieve this, *n*-BuLi-KO*t*-Bu was reacted with C_6H_6 in THF, yielding crystalline $(PhK)_4(PhLi)(t\text{-}BuOLi)(THF)_6(C_6H_6)_2$ **35**, an aggregate containing all the expected constituents of a superbase. Furthermore, its place in this family of reagents was confirmed by its ability to metalate toluene. From a structural perspective, in spite of the substoichiometric Li content of **35**, interactions with the lighter alkali metal dominate: bonds from *tert*-butoxide and phenyl moieties are both shorter and more strongly directed towards Li^+ compared to peripheral K^+, again likely reflecting a competing preference for hard/soft stabilization (Figure 1.6) [60].

Whilst providing encouraging evidence for the reactivity of mixed Li–K species, the high potassium content of **35** leaves some doubt surrounding its validity as a model superbase since pure PhK can achieve the same reactivity towards toluene [61]. However, further evidence for Li–K cooperativity has been gathered by the judicious choice of neopentyllithium (NpLi, Np = $CH_2C(CH_3)_3$) as a stable and hydrocarbon-soluble alkyl equivalent. When partnered with *t*-BuOK, to form a LICKOR mixture, its use has allowed for the isolation of mixed-metal aggregate $Li_4K_4Np_{2.75}(t\text{-}BuO)_{5.25}$ **36** [62]. Though detailed analysis of the structure was complicated by the positional disorder of the ligands and metal, the fundamental motif – a central, square planar arrangement of K centres, bicapped by $[RLi(Ot\text{-}Bu)_2LiR]^{2-}$ (R = Np/*t*-BuO) – was clear (Figure 1.7a). Meanwhile, the NMR spectroscopic monitoring of experiments in which the NpLi : LiO*t*-Bu : KO*t*-Bu ratio was varied revealed evidence for equilibria involving Li-rich mixed Np/O*t*-Bu species (represented by the series $Li_4K_3Np_x(Ot\text{-}Bu)_{7-x}$) in solution. In this work, the excess of Li could be rationalized by the tendency of poorly soluble NpK to precipitate. Consistent with this, cooling solutions rich in Np^- led to crystallization of the Li-rich mixed metal species, $Li_4K_3Np_{3.16}(Ot\text{-}Bu)_{3.84}$ **37**. X-ray diffraction exposed in this the coordination of alkyl moieties to both K and Li centres (Figure 1.7b). This suggested hybrid Li/K–C bond polarity, which was in turn postulated to be responsible for superbasic activity.

Recently, this work has been extended to examine other Li–K compounds of the type described above, whose compositions are dominated by either metal. Thus, the structure of $Li_4KNp_2(Ot\text{-}Bu)_3$ **38** has been elucidated in the solid-state, revealing a square pyramidal metal architecture wherein interaction of apical K with the CH_3 component of an Np unit in an adjacent monomer results in a solid-state dimer (Figure 1.8) [63]. Meanwhile, exploring K-rich end members has lately provided

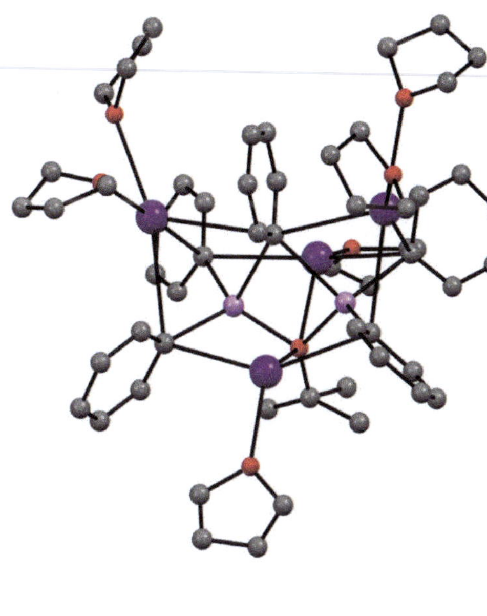

Figure 1.6 Molecular structure of the core of $(PhK)_4(PhLi)(t\text{-}BuOLi)(THF)_6(C_6H_6)_2$ **35** (K = dark purple, Li = pink, O = red). *Source*: Adapted from Unkelbach et al. [60].

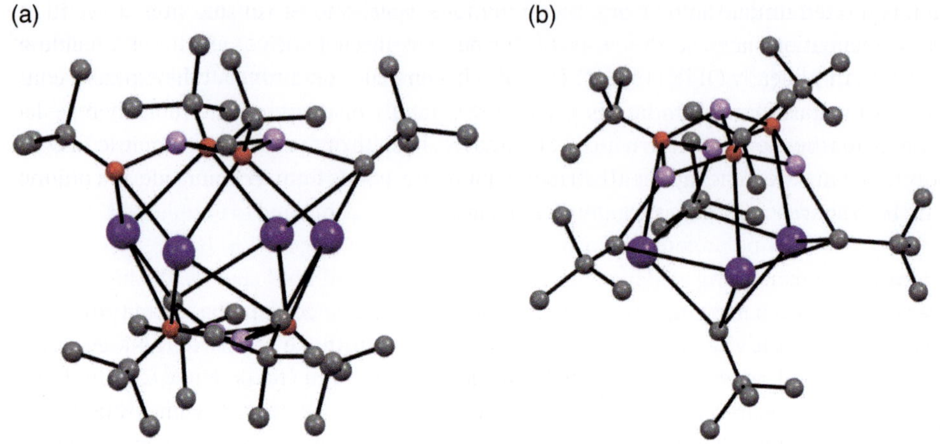

Figure 1.7 Molecular structures of (a) $Li_4K_4Np_{2.75}$(*t*-BuO)$_{5.25}$ **36** and (b) $Li_4K_3Np_{3.16}$(O*t*-Bu)$_{3.84}$ **37**. Minor ligand disorder omitted (K = dark purple, Li = pink, O = red). *Source*: Adapted from Benrath et al. [62].

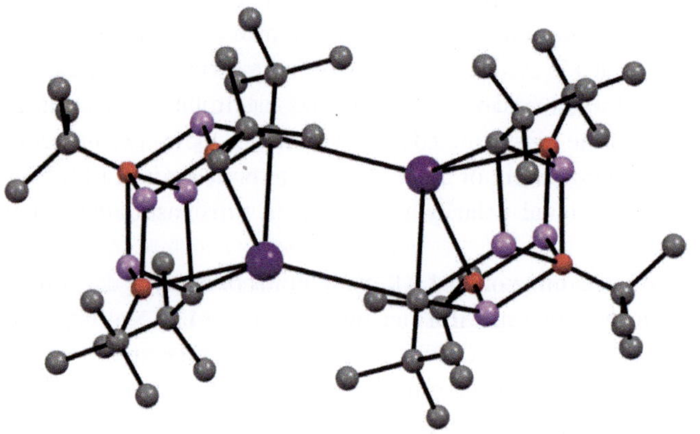

Figure 1.8 Molecular structure of [Li_4KNp_2(O*t*-Bu)$_3$]$_2$ **38**$_2$ (K = dark purple, Li = pink, O = red). *Source*: Adapted from Jennewein et al. [63].

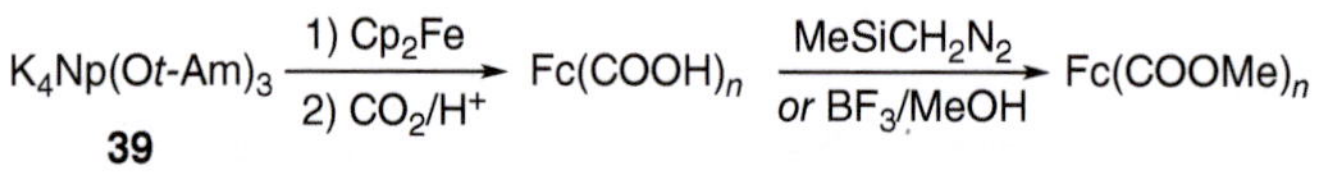

Scheme 1.10 The elaboration of ferrocene employing **39**. Fc = ferrocenyl, n = 2–4.

the first crystallographic evidence for the existence of a mixed organo/alkoxypotassium species, K_4Np(O*t*-Am)$_3$ **39** (*t*-Am = $CH_2C(CH_3)_2CH_2CH_3$). Based on its favourable solubility profile and donor solvent-free constitution, reactivity comparable with or superior to organopotassium reagents was anticipated. In the event, this was realized in the polymetalation of ferrocene. Though isolation of metalated intermediates was not possible, the regioselectivity of metalation could be competently assessed through the study of carboxylated intermediates and the crystallization of selected ferrocene methyl ester derivatives (Scheme 1.10).

The issues of reactivity that plagued the use of organolithium reagents have then, in many cases, been overcome by the advent of LICKOR superbases. However, the introduction of higher alkali metal reagents has led to new issues that have limited the applicability of heterometallic reagents except in the hands of specialists in air-sensitive and nonstandard techniques. So though they offer significant, and often unique advantages, the limitations expressed here of the bases conventionally used for metalation have led to the search for new organometallic combinations capable of similar functionality, ideally under mild conditions. This has led researchers to utilize alkali metals in tandem with metals from elsewhere in the periodic table. This search has ultimately revealed new chemistry for the metalation of aromatic compounds and the advent of ate complexes as synthetic tools [64]. However, first it is worth briefly exploring the use of group 2 elements in conjunction with alkali metals–covered at more length in Chapter 3.

1.2.2.2 Group 1/2 Reagents

In 2006 Knochel et al. augmented the synthetic toolbox available to chemists by improving on Hauser bases through the introduction of a lithium salt. The resultant family of $R_2NMgCl(LiCl)$ reagents has become known as turbo-Hauser reagents. They were prepared by combining *i*-PrMgCl(LiCl) **40** with *i*-Pr_2NH or HTMP and displayed a range of benefits. It was quickly perceived that, unlike many straightforward magnesium amide compounds, they show excellent THF solubility at e.g. room temperature, avoid the need of traditional bases like organolithiums for the use of depressed temperatures (of −78 °C or so), which complicates scale-up, and exhibit kinetic basicity so enabling the regioselective magnesiation of heterocyclic substrates [65]. For example, the selective deprotonation of pyrimidines has typically proved challenging on account of competing addition reactions exhibited by many organometallics. However, pyrimidine derivatives reacted smoothly with TMPMgCl(LiCl) **41** in THF at 55 °C to afford putative selectively magnesiated intermediates **42** (Scheme 1.11, top) that could be worked-up with a range of electrophiles – a strategy that subsequently underpinned the successive regio- and chemoselective elaboration of halogenated pyrimidines [66]. Similar reactivity was reported for pyridines, quinolones, and isoquinolines, as well as for heterocycles like thiazole, thiophene, furan, benzothiophene, and benzothiazole heterocycles that bear relatively acidic hydrogens. This work led, later the same year, to the deployment of turbo-Hauser reagents in directed *ortho*-magnesiation [67]. Impressively, the direct magnesiation of highly functionalized aromatics bearing esters, nitriles, or ketones was achieved through the mediation of an easily introduced/removed OBoc (Boc = *tert*-butoxycarbonyl) directing group. The compatibility of this approach with, for example, keto groups enabled the fabrication of a putative *meta*-magnesiated benzophenone **43** in high yield at −20 °C, ready for subsequent electrophilic quenching (Scheme 1.11, bottom).

In spite of the significant advantages to be had by using $R_2NMgCl(LiCl)$-type reagents, it was quickly seen that some moderately activated aromatics, e.g. *tert*-butyl benzoate, reacted only modestly. This observation led to the development of a new class of mixed Li/Mg base.

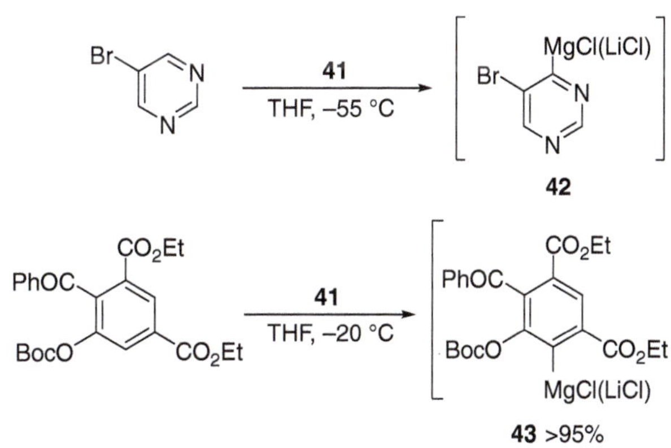

Scheme 1.11 Reactions of TMPMgCl(LiCl) **41** with 5-bromopyrimidine (top) and a highly functionalized aromatic bearing ester groups (bottom).

The reaction of R_2NLi and $MgCl_2$ in THF at 0 °C afforded $(R_2N)_2Mg(LiCl)_2$ **44**. Whereas **41** failed to react significantly (7% *ortho*-iodination after quenching with I_2) with *t*-BuO_2CPh, **44** [amide = TMP, *t*-Bu(*c*-C_6H_{11})N, *t*-Bu(*i*-Pr)N] performed excellently (77–90% *ortho*-iodination). The tolerance of e.g. $\mathbf{44}^{TMP}$ towards ketone, carbonate, and bis(dimethylamino)phosphonate groups allowed intriguing new applications in *ortho* functionalization. For example, the $OP(O)(NMe)_2$ group selectively directed 4-deprotonation via assumed **45** to give **46** on quenching. Moreover, the magnesiation of electron-rich aromatics was made significantly more accessible than had hitherto been the case. This has been utilized to *ortho*-magnesiate dimethyl-1,3-benzodioxan-4-one to give **47** in just 10 min at −40 °C en route to the preparation of 6-hexylsalicylic acid, a natural product found in *Pelargonium sidoides* DC (Scheme 1.12) [68].

Alkali metal magnesiate reagents have been developed recently. Their chemistry will be discussed at more length elsewhere. However, their often unique reactivity merits a summary here. TMEDA-solvated sodium magnesiates, which have a lot in common with the structural chemistry of THF-solvated mixed lithium-magnesium amide complexes [69], initially demonstrated similar structural motifs to those seen in, for example, zincate chemistry (see Section 1.4.2) and elucidated in depth in Chapter 3 [70–72]. Briefly, convoluted chemistry of TMPMg(TMP)(*n*-Bu)Na(TMEDA) **48** has been explored with respect to furan, which underwent α-metalation via the formation of a transient intermediate that disproportionated to the inverse crown disodium dimagnesium hexafuryl tri(THF) complex $(TMEDA)Mg_2(2\text{-}C_4H_3O)_6Na_2(THF)_3$ **49** alongside TMEDA-solvated $Mg(TMP)_3Na$ **50**, the ion-separated analogue of which ($[(TMP)_3Mg][Na(PMDETA)_2]$ **51**, PMDETA = *N,N,N′,N′′,N′′*-pentamethyldiethylenetriamine) proved isolable [71].

Although the chemistries of Grignard and Hauser reagents have long formed a staple of synthetic chemistry, recent interest has focused on their so-called LiCl-complexed 'turbo' variants. Their excellent solubility and the extensive new array of aromatic functionalizations allowed by these complexes have led to an interest in reagent structure as a means of understanding reactivity. However, it is only lately that the first solid-state evidence has emerged (see Chapter 2). The turbo-Hauser monomer $(THF)(TMP)Mg(\mu\text{-}Cl)_2Li(THF)_2$ **52** was reported in 2008 to *ortho*-magnesiate ethyl 3-chlorobenzoate [73]. Meanwhile, the different reactivities of spirocyclic $TMPMg(\mu\text{-}Cl)_2Li(THF)_2$ **53** were explained shortly thereafter [74]. Such structural work on turbo-Hauser bases has been extended to the solution phase using Diffusion Ordered SpectroscopY (DOSY) [75, 76]. This approach has allowed the dissolution of **52** and **53** with retention of

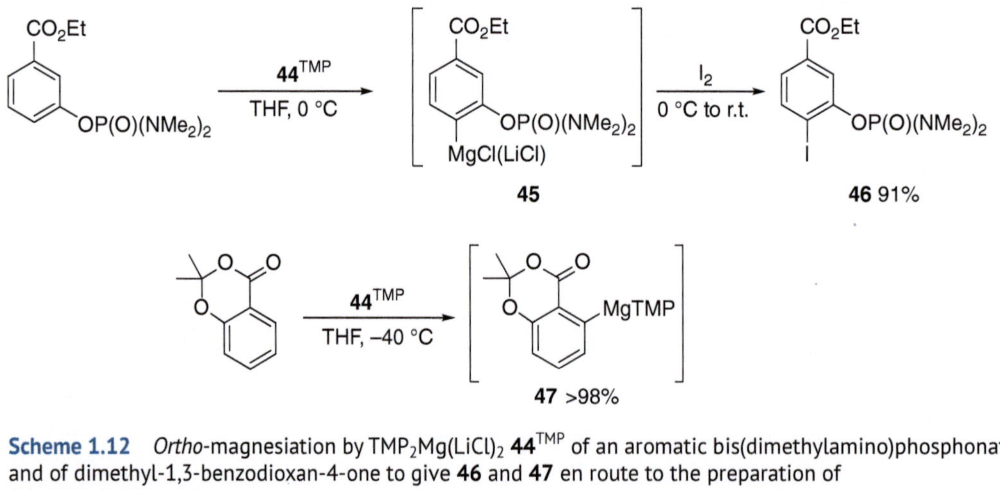

Scheme 1.12 *Ortho*-magnesiation by $TMP_2Mg(LiCl)_2$ $\mathbf{44}^{TMP}$ of an aromatic bis(dimethylamino)phosphonate and of dimethyl-1,3-benzodioxan-4-one to give **46** and **47** en route to the preparation of 6-hexylsalicylic acid.

aggregation. Elaboration of the DOSY technique allowed the corroboration of ideas posited by the authors of the solid-state work insofar as turbo reagent reactivity was attributed to the unsaturation of Mg after the loss of THF [77].

1.3 Aromatic Ate Complex Chemistry: Metal/Halogen Exchange

1.3.1 Introduction

The halogen–metal exchange reaction for the direct formation of functionalized aryl metalates from aryl halides was studied for many years prior to the development of directed deprotonation using ate complexes. Indeed, it has proved one of the most useful processes for the preparation of aromatic lithio- and magnesio-derivatives. However, despite the great potential of these species in synthetic applications, the wider use of organometals (beyond organolithiums and organomagnesiums) in this context has remained rather limited.

1.3.2 Zincates

Extending the earlier studies on forming aromatic and heteroaromatic zinc derivatives, the 1990s saw the development of halogen–zinc exchange reactions of aromatic halides using organozinc reagents to encompass the use of heterobimetallic ones. Lithium trialkylzincates (R_3ZnLi) have long been established as versatile reagents for the 1,4-addition of alkyl groups to α,β-unsaturated ketones [78, 79]. However, investigation of the potential of less reactive lithium aryldimethylzincates for smoothly effecting 1,4-addition represents a more recent development. Hence, a novel preparation of lithium aryldimethylzincates using the halogen–zinc exchange reaction of a range of aromatic halides with lithium trimethylzincate **54**, followed by reaction of the resulting intermediates with electrophiles has been reported [80]. In the first step, iodobenzene was treated with **54** in THF at −78 °C for 1 h to give assumed **55** (R = H). The introduction of benzaldehyde (R′ = Ph) gave 1,2-adduct **56** in 65% yield. Aromatic iodides with a *para* substituent were examined for the same reaction, and various functional groups were found to tolerate the halogen–zinc exchange reaction – the corresponding 1,2-adducts being obtained in good yields (Scheme 1.13). Especially noteworthy was the observed compatibility of a nitro group with this reaction sequence. Thus, the reaction of the arylzincate with propionaldehyde proceeded smoothly with various substituents; the arylzincate from *p*-iodoanisole gave the 1,2-adduct in 66% yield, with this level of effectiveness maintained using the corresponding nitro substrate (68%).

In seeking to extend the portfolio of this early class of ate chemical reactivity, and with the aim of preparing a new class of indolylzinc derivatives, the direct halogen–zinc exchange reaction of 2- and 3-iodoindoles with **54** was next studied [81]. Preparation of indolylzincate by the treatment of the 3-iodoindole with lithium trimethylzincate at −78 °C in THF was followed by the introduction of benzaldehyde to give the desired alcohol in 51% yield. A similar reaction with the indolylzincate

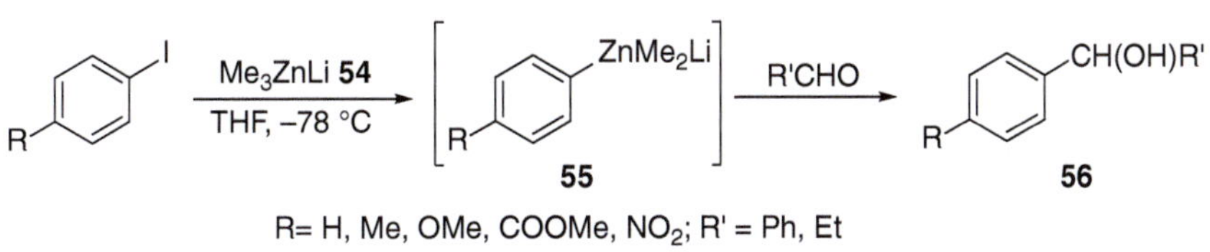

Scheme 1.13 Halogen–zinc exchange in a *para*-substituted aromatic iodide.

and allyl bromide gave 3-allylindole in 44% yield. The metalation gave the expected products in slightly higher yields when the reaction was instead conducted with a combination of lithium trimethylzincate and TMEDA. Importantly, although 3-lithio-1-phenylsulfonylindole has long been known to isomerize easily to the thermodynamically stable 2-lithioindole **57**, the alternative use of lithium zincates avoided the formation of 2-substituted isomers (Scheme 1.14). The heterobimetallic method, therefore, became considered to be advantageous for the selective metalation of the indole 3-position, giving **58**.

The outer shell of the zinc centre in a lithium trialkylzincate is occupied by 16 electrons and, as will be explored more fully in Chapter 2, this leaves room for additional ligand coordination to form an 18 electron state [82]. With this in mind, reports on tetraalkylzincates have presented X-ray studies that have disclosed a tetrahedral arrangement about the zinc [83–86]. However, the reactivities of tetraalkylzincates have not been well studied. Intramolecular carbometalation has been examined using the iodine–zinc exchange of allyl 2-iodophenyl ether. When the iodo ether was treated with **54**, halogen–metal exchange proceeded smoothly, but no intramolecular carbometalation was observed, yielding **59**. In contrast, the reaction of the iodo ether with Me_4ZnLi_2 **60** was followed by hydrolysis to give 3-methyldihydrobenzo[b]furan **61** in 42% yield, which behaviour was explained by invoking an intramolecular carbozincation (Scheme 1.15) [87, 88].

As they emerged as a practically applicable new subset of organozinc reagent, organozincates rapidly became regarded as attractive candidates for carbon–carbon bond formation. To evaluate the non-transferability of alkyl groups, the migratory aptitude of various alkyls from a range of lithium aryldialkylzincates was investigated [89]. *Tert*-butyl turned out to be the best non-transferable group and lithium tri(*tert*-butyl)zincate was therefore found to represent a highly effective reagent for the chemoselective halogen–zinc exchange of functionalized organic halides. As an exemplar, the migratory aptitude of lithium dialkylphenylzincates was investigated as illustrated in Scheme 1.16. Phenyllithium was reacted with a range of dialkylzinc substrates to form the corresponding ate complexes **62**, and these were then treated with 0.5 equivalent benzaldehyde. Spectroscopic analysis of the resulting crude mixture of products then elucidated the ratio of the two alcoholic products.

The status of *tert*-butyl as the most effective non-transfer group led to the performance of lithium tri(*tert*-butyl)zincate **65** also being examined in the halogen–zinc exchange of functionalized

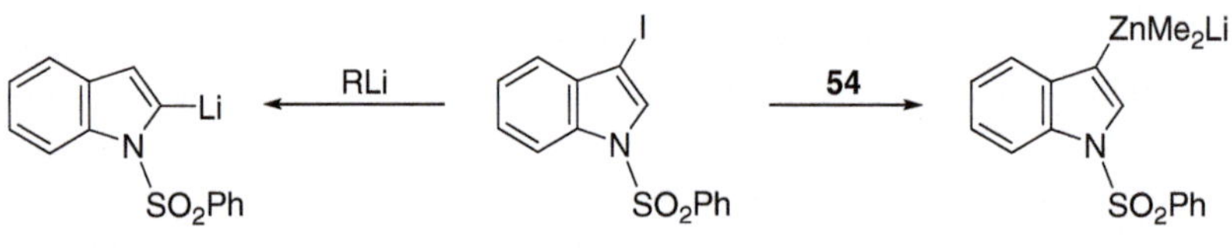

57 **58**

Scheme 1.14 Use of a lithium zincate to avoid intermediate rearrangement of the type undergone by the 3-lithio-1-phenylsulfonylindole.

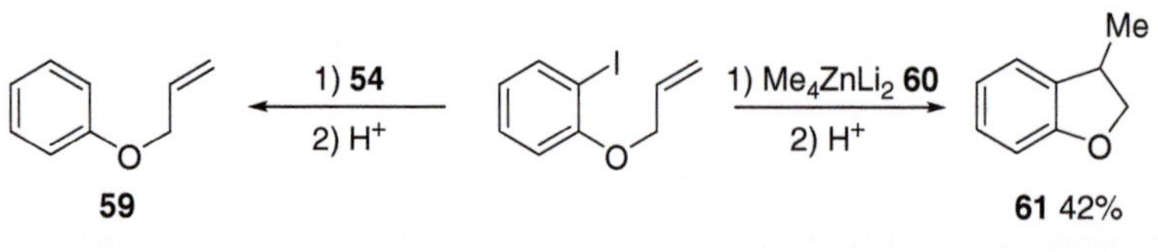

59 **61** 42%

Scheme 1.15 The contrasting reactivity of an allyl 2-iodophenyl ether with Me_3ZnLi **54** and Me_4ZnLi_2 **60**.

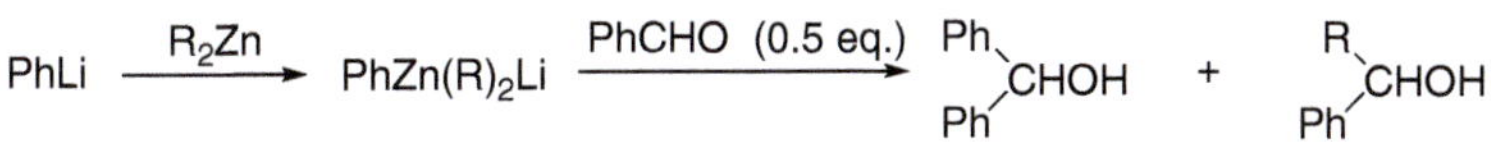

62 63 64

R	63:64
Me	93:7
n-Bu	21:79
t-Bu	100:0

Scheme 1.16 Investigation of the migratory aptitude of lithium dialkylphenylzincates using phenyllithium, a dialkylzinc and benzaldehyde.

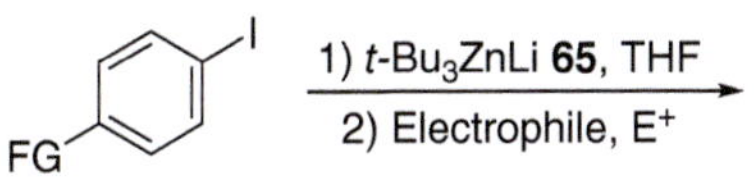

	FG	Electrophile	Yield (%)
66	H	PhCHO	70
67	OMe	PhCHO	67
68	CO_2Me	PhCHO	53
69	CO_2Me	MeI	41

Scheme 1.17 *Tert*-butyl as a nontransfer group; **65** incurs the halogen–zinc exchange of functionalized aryl halides.

aryl halides. It was observed that the exchange reaction generally proceeded smoothly and that electrophilic quenching of the arylzincate intermediates also completed, giving **66–69**, without transferring the *tert*-butyl group (Scheme 1.17).

To demonstrate the compatibility of this reaction with diverse electrophilic substituents, methyl 4-iodobenzoate was treated with **65** and the resulting assumed arylzincate **70** was then treated with a range of electrophiles. Attempts at 1,2-addition with benzaldehyde, alkylation with methyl iodide and allylation with allyl iodide all gave the expected products (e.g. **71**, Scheme 1.18), and acylation with benzoyl chloride proceeded smoothly without a palladium catalyst to give the corresponding ketone **72**. This avoidance of the use of a noble metal catalyst was advantageous not only on cost grounds but also on account of the strict guidelines that regulate trace amounts of transition metals in pharmaceuticals. In contrast to the reaction of 2-iodoanisole with lithium trimethylzincate **54**, which gave the corresponding alcohol in a low yield of 29% upon introduction of benzaldehyde, the use of tri(*tert*-butyl)zincate **65** enabled the corresponding reaction in the much higher yield of 83%.

To further investigate the synthetic potential of this halogen–zinc exchange reaction of **65**, transmetalation of the putative lithium di(*tert*-butyl)phenylzincate **73** with thienylcyanocuprate was surveyed (Scheme 1.19, Th = thienyl) [89]. The phenyl group was found to smoothly undergo 1,4-addition to cyclohex-2-enone in the presence of lithium thienylcyanocuprate **74** to give **75**, while the reaction in the absence of **74** was very sluggish.

From a structural perspective, the first fully characterized lithium alkylzincate appeared in the 1960s [83]. The structure of **60** revealed tetravalent zinc, the completion of a full outer electron shell in which has more recently characterized the reactivity of alkali metal zincates in

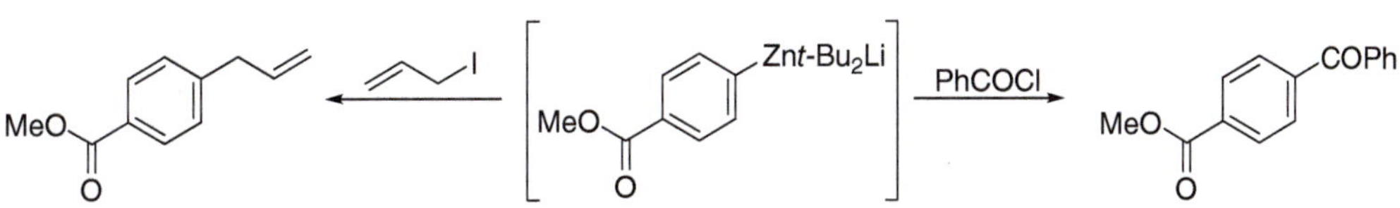

71 54% **70** **72** 56%

Scheme 1.18 Treatment of methyl 4-iodobenzoate with **65** preceded allylation (with allyl iodide) and acylation (with catalyst-free benzoyl chloride).

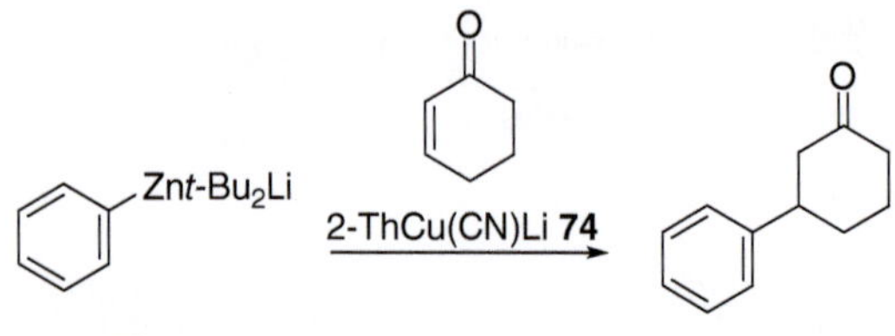

Scheme 1.19 Addition reaction involving the transmetalation of putative lithium di(*tert*-butyl) phenylzincate **73** with thienylcyanocuprate.

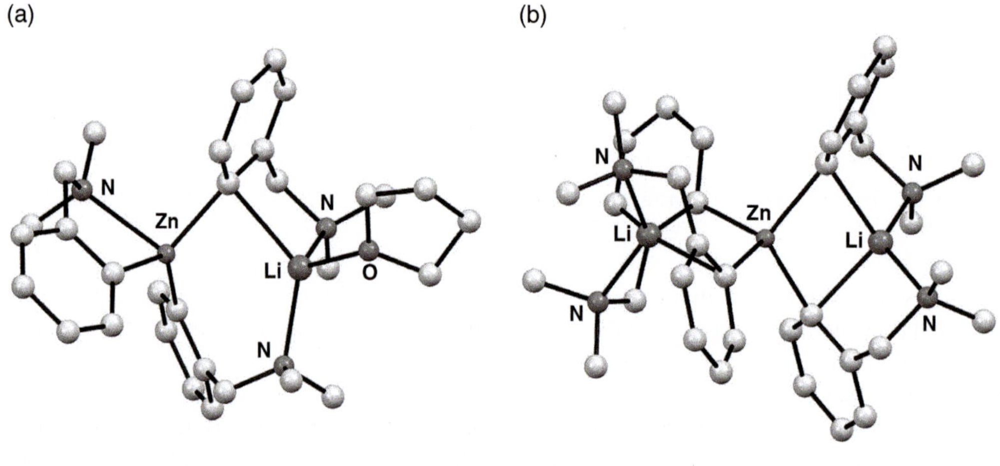

Figure 1.9 Molecular structures of (a) solvated $(DMBA)_3ZnLi$ **76** and (b) $(DMBA)_4ZnLi_2$ **77**, which can be selectively targeted by modulating the $(DMBA)_2Zn$:DMBALi ratio in reaction. *Sources*: Adapted from Wyrwa et al. [91]; Rijnberg et al. [92].

deprotometalation, of which more later. Moving on several decades, the imperative to complete zinc's outer shell was similarly revealed by external solvation in simple amine adducts in spite of the deployment of sterically demanding ligands in tri[di(trimethylsilyl)methyl]zincates [90]. Competition between tri- and tetra(organyl)zincates was further elucidated at about this time through the use of the monoanionic, potentially *C,N*-chelating $C_6H_4CH_2NMe_2$-2 (DMBA) ligand. This allowed the observation of 18 electron Zn centres in both $(DMBA)_3ZnLi$ **76** (following 4:1 $(DMBA)_2Zn$:DMBALi reaction; Figure 1.9a) and spirocyclic [91] $(DMBA)_4ZnLi_2$ **77** (2 : 1 reaction; Figure 1.9b) [92]. Interestingly though, reactions of these zincates with cyclohexanone gave a very early indication of disproportionation, a phenomenon seen more recently in alkyl, amido, and mixed alkyl–amido zincates [86]. This issue has been revisited in ate chemistry discussed elsewhere in this book that finds applications in directed aromatic deprotometalation [93, 94]. Further advances towards the isolation and full characterization of simple, trivalent, 16 electron Zn were made by the expansion of steric demands in silylmethylzincates [90, 95, 96] in combination with the introduction of Lewis bases capable of abstracting the alkali metal. Accordingly, with TMEDA present it proved possible to observe ion-separated $[Me_{3-n}Zn\{CH(SiMe_3)Ph\}_n][Li(TMEDA)_2]$ **78** (n = 1–3) [97]. Recently, remarkable insights into competing solvent-separated ion-pair (SIP, obtained in the presence of diglyme) and contact ion-pair (CIP, obtained in the presence of PMDETA) formation saw the subject go full-cycle, returning to simple lithium methylzincates but now probing 16 electron metal centres in the trimethylate (Scheme 1.20). The same work used the relatively new technique of DOSY to shed light on the solution behaviour of these species. Hence, though ^{1}H NMR spectroscopy revealed single methyl resonances for both PMDETA **79** and diglyme **80** systems in solution, DOSY was able to establish that this was due to exchange

$$\underset{\mathbf{79}}{\mathrm{MeZn(\mu\text{-}Me)_2Li(PMDETA)}} \xleftarrow{\text{PMDETA}} \mathrm{MeLi} + \mathrm{ZnMe_2} \xrightarrow{\text{diglyme}} \underset{\mathbf{80}}{\mathrm{[Me_3Zn][Li(diglyme)_2]}}$$

Scheme 1.20 Competing SIP and CIP formation in Me_3ZnLi chemistry.

in the former case and not SIP formation. The power of DOSY to advance our understanding of the potentially elaborate solution chemistry of polar organometallics is the subject of detailed discussion elsewhere in this book.

1.3.3 Cuprates

While lithium cuprates have very recently undergone major development as versatile reagents for selective carbon–carbon bond forming reactions (see below), little attention has been paid to halogen–copper exchange using ate complexes. That said, with the aim of developing a new and facile method for the preparation of arylcuprates, the halogen–copper exchange reaction of aromatic halides using lithium cuprates was investigated in the mid-1990s [98]. It was in this context that the suggested complex $Me_2Cu(CN)Li_2$ **81** was found to be an excellent metalating reagent, while organocoppers were not. The application of this protocol to the high-enantiomeric purity preparation of precursors to the CC-1065 [99–101]/duocarmycin [102–104] pharmacophore was also conducted as outlined in Scheme 1.23.

Copper-based organometallic complexes are the organotransition metal reagents most widely used as soft nucleophiles in organic synthesis. Hence, both organocopper and organocuprate reagents are employed for carbon–carbon bond formation owning to their characteristic reactivities in conjugate addition to α,β-unsaturated carbonyl compounds, in substitution reactions, and in the carbometalation of carbon–carbon triple bonds. Although both organocopper and organocuprate reagents are well established as tolerating a wide range of electrophilic functional groups, the formation of functionalized organocopper reagents has not proved promising. This has largely been because transmetalation of nucleophilic organolithium or Grignard reagents has typically been required and this has been limited by functional group tolerance. In a similar vein, functionalized organocopper reagents have been prepared by the transmetalation of functionalized organozinc compounds and by direct oxidative addition of active copper, prepared from $CuI(PBu_3)$ and lithium naphthalenide, to organic halides [105]. A number of so-called Gilman reagents – lithiocuprates of general formula R_2CuLi – have been used in organic syntheses. Mixed cuprates, $R_2Cu(CN)Li_2$, have also been reported to show high reactivity towards a variety of organic substrates. Though the halogen–metal exchange reaction is one of the most useful processes for the preparation of metalated arenes, as noted above, examples have tended to be limited to the use of lithium and magnesium compounds. The rather limited coverage of other halogen–metal exchange systems is true also of copper; although the possibility of halogen–copper exchange has been suggested in the coupling reaction of aryl halides with cuprates [106, 107], the reactions of halogen-exchange-generated organocopper intermediates with electrophiles are still unexplored from the viewpoint of synthetic chemistry.

Moving to discuss organic transformations based on organocopper/cuprate chemistry in more detail, arylcuprates have been prepared by reaction of iodobenzene with **81** in THF at −40 °C. Subsequent reaction with benzaldehyde at −78 °C gave benzhydrol **82** in 89% yield [98]. Meanwhile, mixed cuprates Me_2CuLi **83**, $Me_2Cu(SCN)_2Li_2$ **84**, $MeCuTh(CN)Li_2$ **85** were found to be less reactive. The halogen–metal exchange reaction of *p*-iodoanisole with this latter cuprate proved slower than the corresponding reaction of iodobenzene. However, satisfactory results could be obtained when the metalation was conducted at −20 °C. The *p*-methoxy and ester groups were tolerated in the halogen–copper exchange reaction, and the intermediary copper reagents reacted with benzaldehyde to give alcohols **86** and **87**. The *p*-methoxy result contrasted starkly with that obtained using an organolithium intermediate, where self-condensation was seen. However, better yields were obtained when the metalation was conducted at −78 °C (Scheme 1.21).

Interesting conjugate addition reactions have been enabled using cuprate chemistry, obviating the traditional need for additives to promote conversion. For example, the phenylcuprate presumed to be generated from iodobenzene and **81**, has been added to 2-cyclohexenone to afford 3-phenylcyclohexanone in 61% yield without additional reagents. Meanwhile, reaction of the same phenylcuprate with 1,2,-epoxycyclohexane gave *trans*-2-phenylcyclohexanol in 53% yield in the absence of (normally required) additives such as Lewis acids. To obtain post mortem information about the structure of the arylcopper intermediate in these addition processes, the putative arylcuprate **88** obtained by the halogen–copper exchange reaction of methyl *p*-iodobenzoate was tested against both hydrolysis and oxidation with oxygen. Hydrolysis of the cuprate in aqueous NH_4Cl at −40 °C gave methyl benzoate in 85% yield with no observed formation of methyl *p*-methylbenzoate, verifying the non-transferability of the Me ligand, with MeI presumably produced during halogen exchange and giving a cuprate less prone to react with electrophiles than is $MeCuAr(CN)Li_2$ (**88**). Meanwhile, oxidation of the arylcuprate by bubbling oxygen through the reaction mixture at −78 °C gave the coupling product methyl *p*-methylbenzoate **89** in 76% yield (Scheme 1.22). Overall, these data pointed towards the incorporation of LiCN alongside a Me-ligand in the arylcuprate intermediate (the structural implications of this are discussed in Section 1.4) [98].

In connection with studies into the synthesis of the CC-1065/duocarmycin pharmacophore, the syntheses of 3-hydroxymethyl-2,3-dihydroindole and 3-hydroxy-1,2,3,4-tetrahydroquinoline was investigated [98]. The availability of these precursors in enantiomerically pure form is fundamental to the straightforward asymmetric synthesis of the pharmacophore. In particular, the intramolecular ring-opening of epoxyorganometallic compounds is of interest with regard to the regioselectivity of subsequent cyclization. This led to a detailed examination of the synthesis of a precursor to CC-1065/duocarmycin pharmacophore by intramolecular ring-opening of epoxyarylmetal ate complexes. This precursor – a chiral epoxide – was treated with *n*-BuLi at −90 °C, resulting in the formation of 5-exo cyclization product **90** in 43% yield and without any detectable loss of enantiopurity. Meanwhile, the reaction of the epoxide with lithium trimethylzincate at −50 °C gave the 5-exo product in 40% and the 6-endo product in 57% yield,

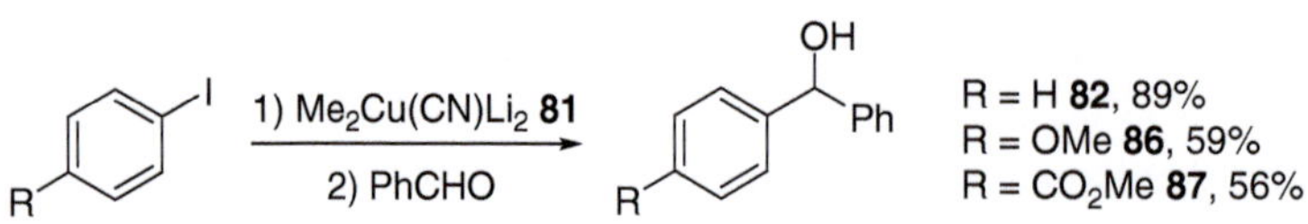

Scheme 1.21 Halogen–metal exchange of *p*-iodoanisole with cuprate **81** at −78 °C.

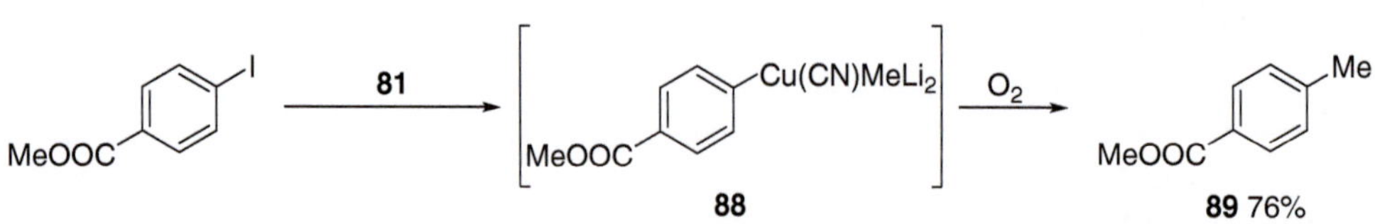

Scheme 1.22 Reaction of methyl *p*-iodobenzoate and **81**, with subsequent oxidation at −78 °C giving coupling product **89**.

respectively. In contrast, the reaction of the epoxide with cuprates showed reverse regioselectivity, and the 6-endo product was dominant when **81** was used as the metalating reagent (yield 62% 6-endo **91** versus 6% 5-exo). This was further improved using $Me_3Cu(CN)Li_3$ **92**, which gave a uniquely 6-endo reaction in 73% yield. In general, enantiomeric purity was unchanged after ring opening (Scheme 1.23) [98].

Though synthetic work outlined above is dominated by cuprate chemistry, the structures of organocopper(I) reagents continue to capture the interest of chemists in their own right. However, their thermal instability and their sensitivity towards oxygen and moisture have posed serious obstacles to the characterization of organocopper(I) species. The stability of RCu is known to depend strongly on the nature of the organic ligand, with stability increasing in the order [108] R = alkyl [109] < alkenyl [110–113] ≈ aryl [114–119] < alkynyl [120, 121]. Crystallographic studies have revealed cyclic aggregates based on (typically) two-coordinate copper centres with, in many cases, some degree of aggregation retained in solution [116, 122, 123]. For alkylcopper compounds, crystallographic data are limited to examples featuring stabilized ligands or stabilizing additives. Hence, Me_3SiCH_2Cu **93** afforded a metallacyclic tetramer [109]. Meanwhile, attempts to prepare $MeCu(PPh_3)_2$ afforded unusual heterodimer $MeCu(\mu\text{-}Me)Cu(PPh_3)_2$ **94**, best viewed as contact ion pair (CIP) $[Me_2Cu][Cu(PPh_3)]$ (Figure 1.10a) [124]. Indeed, a comparable ion-separated structure (SIP) has been reported; $[Me_2Cu][Cu(PMe_3)_4]$ **95** was based on a linear coordinate anion and tetrahedral cation (Figure 1.10b) [125].

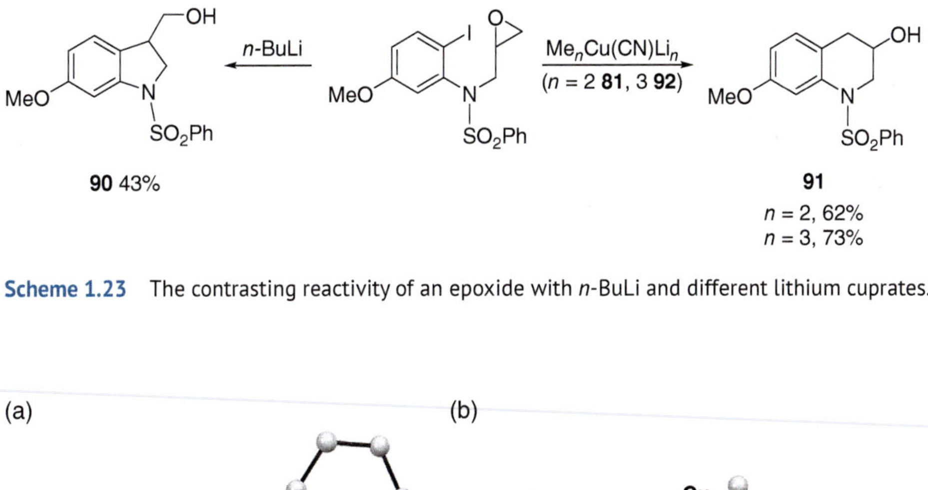

Scheme 1.23 The contrasting reactivity of an epoxide with *n*-BuLi and different lithium cuprates.

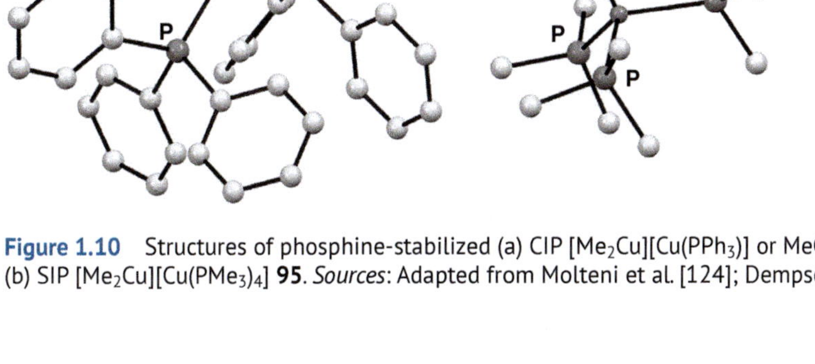

Figure 1.10 Structures of phosphine-stabilized (a) CIP $[Me_2Cu][Cu(PPh_3)]$ or $MeCu(\mu\text{-}Me)Cu(PPh_3)_2$ **94** and (b) SIP $[Me_2Cu][Cu(PMe_3)_4]$ **95**. *Sources*: Adapted from Molteni et al. [124]; Dempsey et al. [125].

Whilst homometallic organocopper compounds continue to evolve new interest in areas such as photoluminescence [126], the most extensively studied synthetically useful class of organocopper reagents are the heterobimetallic lithium cuprates. As first reported by Gilman [127], lithium cuprates differ from typical organocopper compounds in forming homogenous solutions in ethereal solvent, a property which is not only essential for their usability and reactivity but which also underpins their amenity towards structural characterization by enabling crystallization (Scheme 1.24).

It has been recognized for some time that the Cu : Li stoichiometry employed in cuprate formation offers a profound structural impact upon the resulting complex. Two fundamentally different types of cuprate have been recognized in consequence; so-called 'lower-order' and 'higher-order' forms. The former are characterized by two-coordinate Cu, the latter by Cu bearing a higher coordination number. Early structural data was gathered largely in the solution-state, where evidence from vapour pressure depression, ^{1}H NMR spectroscopy and solution X-ray scattering all lent weight to the dominance of cyclic dimers [128]. It was not until the 1980s that the first reports on the X-ray structures of lithium cuprates appeared. However, these revealed atypical copper-rich anions. The synthetic utility of phenylcuprates has been alluded to above, and the first clusters of these species to be characterized, $[Ph_6Cu_5][Li(THF)_4]$ **96** [129], $[Ph_6CuLi_4][Li(Et_2O)_4]$ **97** [130] and $[Ph_6Cu_3Li_2]_2[Li_4Cl_2(Et_2O)_{10}]$ **98** [131], were obtained by reacting PhLi with CuBr and CuCN, respectively. The cuprate moieties in these SIPs revealed the same fundamental architecture, based upon a compressed trigonal bipyramidal arrangement of metal atoms in which the apical sites could be considered to bridge three $[Ph_2Cu]^-$ units (Figure 1.11a) [132]. The subsequent isolation of the neutral phenylcuprate dimer of $Ph_2CuLi(Et_2O)$ **99** (whose structure could be derived from $[Ph_6Cu_3Li_2]^-$ by the formal replacement of one $[Ph_2Cu]^-$ unit by Et_2O (Figure 1.11b)) [133] lent support to this interpretation. Similar aggregates have also been reported where dimethyl sulfide replaces Et_2O [134].

$$\mathrm{CuI} + 2\,\mathrm{MeLi} \xrightarrow{\mathrm{Et_2O}} \underset{\mathbf{83}}{\mathrm{Me_2CuLi}} + \mathrm{LiI}$$

Scheme 1.24 Synthesis of lithium dimethylcuprate **83**.

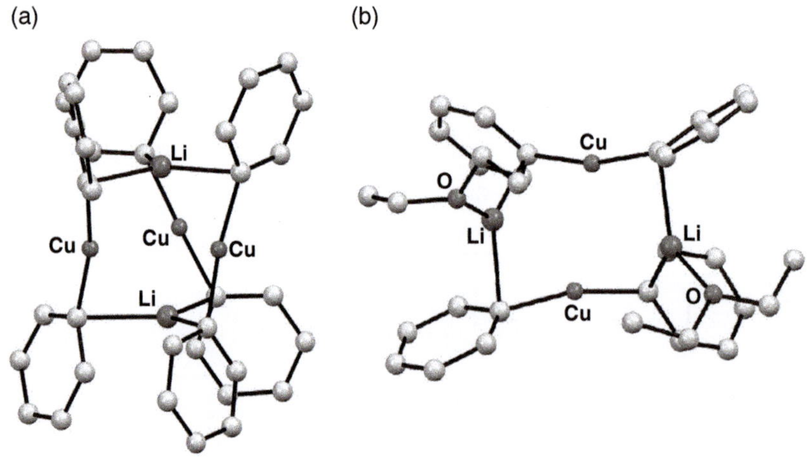

Figure 1.11 Structures of phenylcuprate species (a) $[Ph_6Cu_3Li_2]^-$ in **98** and (b) $[Ph_2CuLi(Et_2O)]_2$ $\mathbf{99}_2$. *Sources*: Adapted from Hope et al. [131]; Lorenzen and Weiss [133].

Structures of complexes such as **96** [129] and **97** [130] exhibit unusual 'higher-order' copper centres that act as bridges towards lower-order $[Ph_2Cu]^-$ units. On the other hand, compounds like $Ph_9Cu_4Li_5(SMe_2)_4$ **100** [134] and $Ph_5Cu_2Li_3(SMe_2)_4$ **101** [135] incorporate $[Ph_3Cu]^{2-}$ units as one of the primary cuprate moieties, making a straightforward higher-order description more appropriate (Figure 1.12a). Indeed, recent work has shown that a higher Cu coordination number is attainable in spirocuprate $(biph)_2CuLi_3(THF)_6$ **102** (biph = 2,2′-biphenyl, Figure 1.12b), with Cu now displaying a remarkable distorted tetrahedral geometry [dihedral angle between cuprocycles = 84.1(1)°] [136].

Moving from simple phenylcuprates, the use of aminoaryl ligands capable of providing internal coordination has enabled the isolation of neutral $(DMBA)_2CuLi$ **103**, whose dimeric structure revealed a near-planar arrangement of alternating Cu and Li atoms, bridged by aryl ligands and with only the Li centres interacting with the pendant amine functions (Figure 1.13a) [137]. In this case, the bridging mode of the aryl ligand differed from earlier reports on the tetramer of similar organocopper species $[2\text{-}(Me_2NCH_2)C_6H_4\text{-}Me\text{-}5]Cu$ **104** [138], its asymmetric nature suggesting primary σ-type interaction of the C-based sp^2 lone pair with Cu (C–Cu 1.942(3) and C–Li 2.385(6)

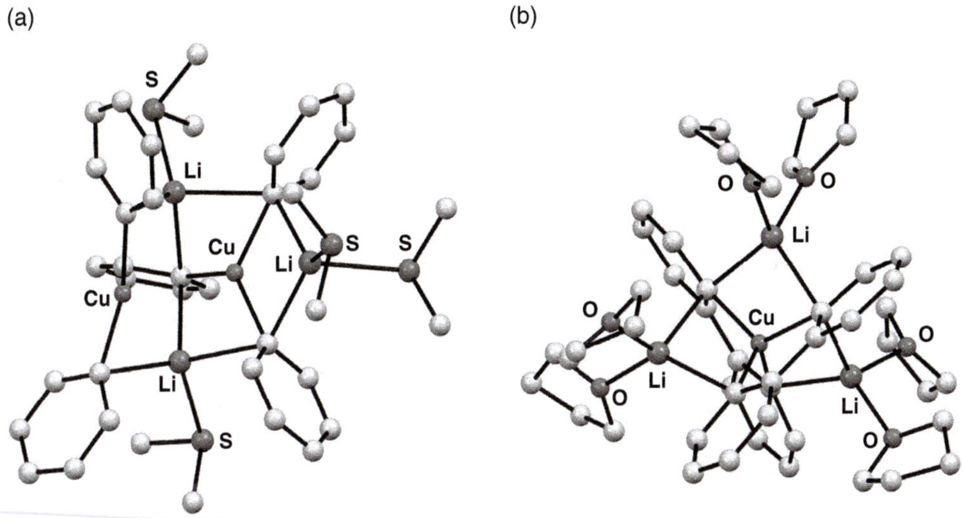

Figure 1.12 Selected higher-order cuprates (a) $Ph_5Cu_2Li_3(SMe_2)_4$ **101** and (b) $(biph)_2CuLi_3(THF)_6$ **102** (biph = 2,2′-biphenyl). *Sources*: Adapted from Olmstead et al. [135]; Liu et al. [136].

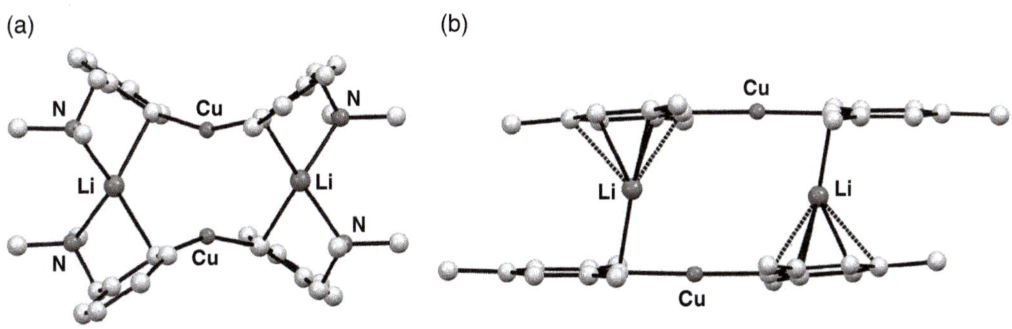

Figure 1.13 Molecular structures of the dimers of (a) $(DMBA)_2CuLi$ **103** and (b) $(Mes)_2CuLi$ **105**. *Sources*: Adapted from Van Koten et al. [137]; Davies et al. [139].

Å, respectively). A much more pronounced contrast in σ/π-bonding has been reported in $(Mes)_2CuLi$ **105**, where dimerization results in each Li centre adopting both η^1 and η^6-coordination towards mesityl groups, leaving Cu free to adopt a preferred linear geometry (C–Cu–C = 178.34(7)°, Figure 1.13b) [139]. The dominance of Li. . .π interaction here can be attributed to the absence of donor solvent in the structure.

Simple alkylcuprates of the type routinely used in synthesis have proved difficult to study due to their relatively low thermal stability. In 1984, the crystal structure of SIP $[\{(Me_3Si)_3C\}_2Cu][Li(THF)_4]$ **106** was described, providing the first solid-state evidence for the structure of a dialkylcuprate (Figure 1.14a) [140]. Crystallography revealed linear, two-coordinate copper, though the possibility that these features were imposed by the steric bulk of the anion could not be excluded. Other breakthroughs in the alkylcuprate field have included characterization of the polymer of $(Me_3SiCH_2)_2CuLi(SMe_2)$ **107** [141], a structure consisting of dimeric units (similar to those seen in the structure of **99** joined by SMe_2 ligands. Meanwhile, only two lithium dimethylcuprate structures have been reported for reagents; SIPs $[Me_2Cu][Li(12\text{-crown-}4)_2]$ **108** [142] and $[Me_2Cu][Li(DME)_3]$ **109** [143] (DME = 1,2-dimethoxyethane). Recently, these have been added to by a possible pre-reaction π-complex $(fluorenone)CuMe_2Li(THF)_3$ **110** (Figure 1.14b) [144]. In contrast to the linear cuprate ion geometry observed in the first two cases, the C=O π-complex in Figure 1.14b reveals a C–Cu–C angle of 104°. Ion separation would appear to be induced by strongly coordinating Lewis base additive. On the other hand, in the more weakly coordinating ethereal solvents in which lithium dimethylcuprate is typically used, it is believed that CIPs dominate and that these forms of reagent are responsible for observed reactivity [143].

A number of solution-state studies on the lithium methylcuprate species $(Me_{m+n}Cu_mLi_n)$ have been undertaken. Detailed solution work is covered elsewhere in this volume. However, briefly, ^1H and ^7Li NMR spectroscopies have revealed that the addition of MeLi to MeCu in THF/Et_2O results in an equilibrium between Me_2CuLi **83** and Me_3Cu_2Li **111** plus MeLi, though the existence of this equilibrium has proved to be strongly dependent upon both the solvent (it does not occur in Et_2O) and the presence of LiI (which promotes the formation of a different discrete entity). This work highlighted the fact that reagents presumed to be either 'lower-order' or 'higher-order' according to the stoichiometry of their preparation were in fact composed of varying quantities of the same species [145]. More recent DOSY studies using PFG (pulsed field gradient) NMR spectroscopy have

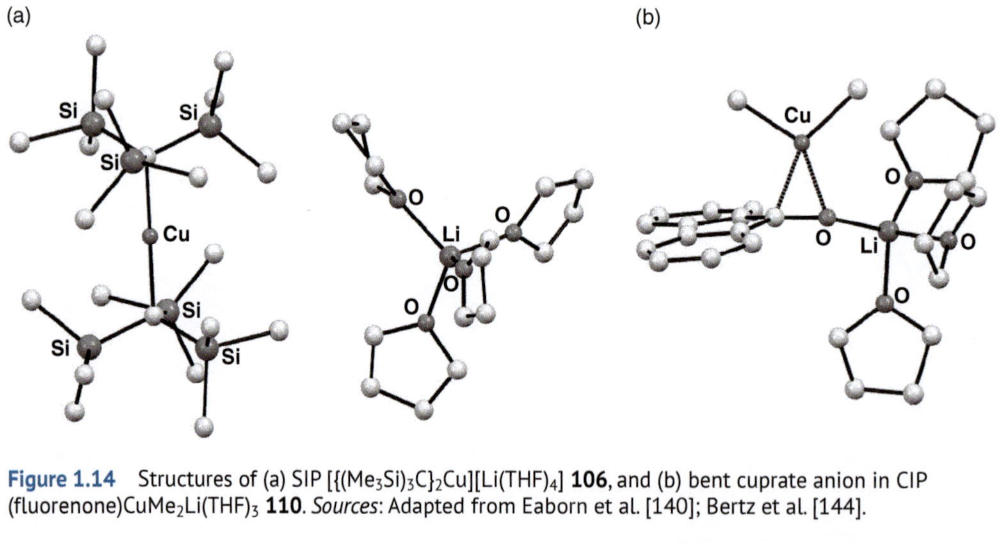

Figure 1.14 Structures of (a) SIP $[\{(Me_3Si)_3C\}_2Cu][Li(THF)_4]$ **106**, and (b) bent cuprate anion in CIP $(fluorenone)CuMe_2Li(THF)_3$ **110**. *Sources*: Adapted from Eaborn et al. [140]; Bertz et al. [144].

indicated the existence of dimethylcuprate aggregates (based on homodimeric cores) larger than dimers in Et_2O, though these exhibited depleted reactivity [146]. Likewise, multi-dimensional NMR spectroscopy showed that the addition of small amounts of THF to Et_2O-based cuprate preparations had surprisingly different effects on aggregation depending on the identity of inorganic Li salts present [147].

The dramatic effects on reactivity of incorporating a Li salt with a polar organometallic reagent have already been discussed in the context of magnesium amides and will be returned to in the context of copper amides. The importance of this notion in terms of cuprate chemistry derives from the fact that in *in situ* cuprate syntheses a Li salt by-product is often formed and is rarely separated prior to application of the cuprate. Whilst the influence upon and/or inclusion of LiX (X = an inorganic anion, often a halide) in cuprate structures has been subject to several investigations (most recently by in-depth NMR spectroscopy), solid-state evidence for association remains rare (X = CN constitutes a special class of cuprate, which is discussed separately). However, the reaction of *ortho* diamine-chelated aryllithium reagents with CuBr has given cuprates of formula $Ar_2Cu(Br)Li_2$ (Ar = $C_6H_4\{CH_2N(Me)CH_2CH_2NMe_2\}$-2 **112** and 1-$C_{10}H_6\{CH_2N(Me)CH_2CH_2NMe_2\}$-2 **113**, Figure 1.15) [148]. Interestingly, it was noted that the benzylic nitrogen centres became stereogenic upon coordination to Li, though only *R,R* and *S,S* pairs were observed in the solid-state, implying selectivity during assembly.

As mentioned above, cyanocuprates are often considered as a distinct class of cuprates. They have been recognized as highly reactive and robust reagents that offer advantages over traditional Gilman reagents in substitution [149] and addition [150] reactions. However, the structures of cyanocuprates proved controversial for many years. Unlike halides, the ability of cyanide to act as a strongly coordinating ligand raised the possibility that it might remain bonded to Cu during cuprate formation. This behaviour has been plainly evidenced over a number of years in a range of lower-order cyanocuprates (obtained from the stoichiometric reaction of CuCN and RLi) both in the solid state [151–153] and in solution [154]. However, upon adding two equivalents of RLi to CuCN (to give a *Lipshutz* cuprate, a species of the type $R_2Cu(CN)Li_2$), the outcome became less straightforward to predict. Two possibilities arose: (i) expulsion of cyanide as LiCN (or else retention by the cuprate but without a direct Cu–CN interaction) or (ii) retention of a Cu–CN bond to form a higher-order cuprate. Although ^{13}C NMR spectroscopy initially suggested CN^- to be bound to copper [155], subsequent work demonstrated that the ^{13}C NMR chemical shift of CN was indifferent to the organic R groups, arguing against a direct Cu–CN bond [156]. This latter scenario was subsequently supported by extended X-ray absorption fine structure (EXAFS) measurements [157, 158], IR spectroscopy [159], and calculations [160, 161]. However, the most conclusive evidence disfavouring higher-order structures arrived with the crystal structures of $(DMBA)_2Cu(CN)Li_2(THF)_4$ **114** [162] and [*t*-Bu_2Cu][CN{Li(THF)(PMDETA)}$_2$] **115** [163]. These structures contrasted; displaying CIP and SIP structures, respectively. However, the lack of

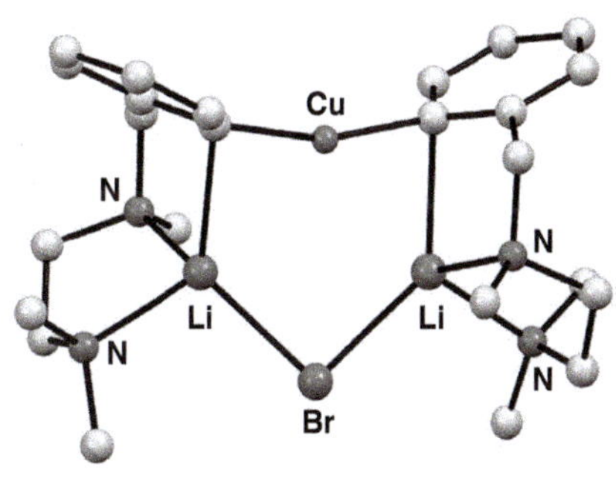

Figure 1.15 Molecular structure of $[C_6H_4\{CH_2N(Me)CH_2CH_2NMe_2\}\text{-}2]_2Cu(Br)Li_2$ **112**. *Source*: Adapted from Kronenburg et al. [148].

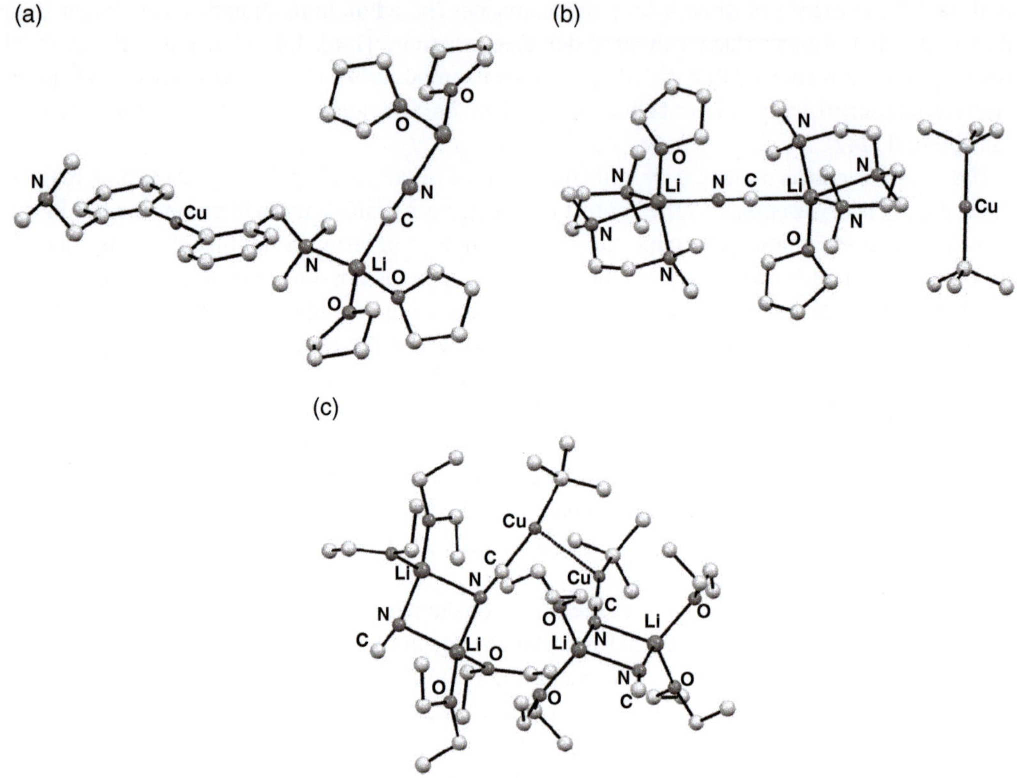

Figure 1.16 Molecular structures of (a) polymeric $(DMBA)_2Cu(CN)Li_2(THF)_4$ **114**, (b) [t-Bu_2Cu][CN{Li(THF)(PMDETA)}$_2$] **115** and (c) a cuprophilic aggregate of t-$BuCu(CN)Li_2(Et_2O)_2$ **116**. *Sources*: Adapted from Kronenburg et al. [162]; Boche et al. [163].

Cu–CN bonding in either case was obvious. This was particularly noteworthy in **115**, where the absence of a Cu–CN bond contrasted with its presence in the product of the 1 : 1 reaction of t-BuLi with CuCN; t-$BuCu(CN)Li_2(Et_2O)_2$ **116** (Figure 1.16).

Several explanations have been posited for the apparently higher reactivity of Lipshutz cuprates, though the idea has also been contested [164]. Indeed, it has been suggested that the differing solubility of organic groups may be a contributory factor to observed variations in reactivity. For example, NMR spectroscopy uncovered the possibility that unreactive Cu-rich cuprates may form in the presence of LiI when the organic groups were solubilizing [165], whereas lower-order cyanocuprates (which did not interfere with unconsumed reactant) were the preferred sink for organocopper by-product in the presence of cyanide. Differences in reactivity could then be understood in terms of the ability of the organocopper by-product to sequester otherwise reactive cuprate. However, while these ideas have been considered in the context of applied conjugate addition, they have yet to be applied in detail to directed deprotonation or to copper-halogen exchange reactions.

1.3.4 Solid-phase Synthesis

Solid-phase synthesis has become a recognized and attractive methodology for constructing libraries of biologically active small molecules in connection with combinatorial chemistry and automated synthesis oriented towards drug discovery research [166]. Various synthetic methodologies

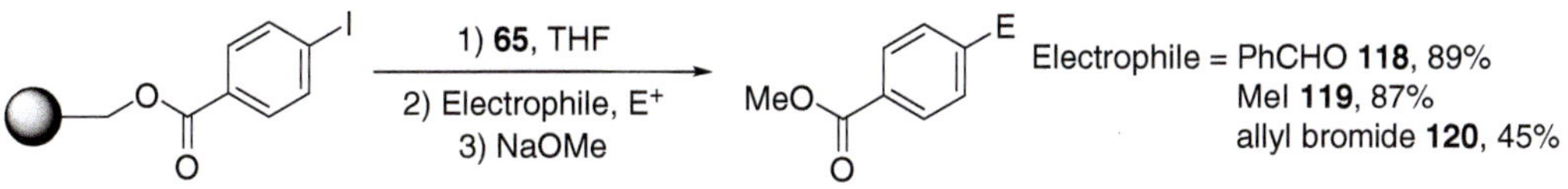

Scheme 1.25 Use of ate complex *t*-Bu_3ZnLi **65** as a solid-phase metalating agent.

have been applied to solid-phase synthesis [167]; however, organometallic chemistry has not yet been well explored in this area due to the lack of effective preparative methods of immobilized organometallic compounds for reaction with electrophilic linkages such as esters. That said, organometallic compounds have played an important role in solution-phase chemistry for selective carbon–carbon bond forming reactions directed toward construction of complex molecules. This has led to the development of new solid phase carbon–carbon bond forming processes, with the halogen–metal exchange reaction of organic halides supported on polymer with an electrophilic linkage being the target of investigation [168]. The ate complexes *t*-Bu_3ZnLi **65** and $(Me_3SiCH_2)_2Cu(CN)Li_2$ **117** turned out to be very effective as the solid phase metalating agents in this context (Scheme 1.25). Various modes of carbon–carbon bond formation were achieved using immobilized arylmetal ate complexes, including 1,2-addition to benzaldehyde, 1,4-addition to 2-cyclohexenone, and alkylations with methyl iodide and allyl iodide (**118–120**). Meanwhile, the metalation of insoluble polymers such as cross-linked polystyrene has been investigated in connection with the preparation of materials for polymer-assisted chemistry, with lithiation routinely being used for the functionalization of polystyrene resin. However, this approach has typically been restricted to examples where electrophilic functional groups are not present, and only a few reports of the solid phase lithiation of immobilized small molecules bearing electrophilic functional groups have appeared in the literature. In spite of these limitations, solid-state synthesis has retained its appeal where functional groups not compatible with solution derivatization are concerned. For example, organometallic compounds containing an alkoxycarbonyl group are typically unstable at elevated temperatures due to self-condensation or homocoupling. Immobilized organometallic compounds, on the other hand, have a track record of demonstrating enhanced stability, even when they bear highly reactive electrophilic groups, because of the diminished chance of self-condensation and homocoupling by pseudo-dilution effects. In the face, however, of the intrinsic limitations of traditional monometallic reagents, there has been great interest in developing new ways for the chemoselective metalation of small molecules on polystyrene, and this has focused on the use of metalating agents capable of being used in tandem with the presence of electrophilic functional groups. It is in this context that chemoselective solid-phase halogen–metal exchange reactions, focusing principally on the preparation and transformation of immobilized intermediary organozincate and organocuprate species, have been investigated by Kondo and coworkers. Cahiez, Knochel, and coworkers also reported that the relevant halogen–metal exchange takes place using a Grignard reagent [169].

1.4 Deprotonation Using Ate Complexes

1.4.1 Introduction

Lithium amides have been extensively studied [170] and the potential of metal amides in synthesis is well understood [171]. However, the susceptibility of many DMGs to nucleophilic attack, even in the presence of amides [172, 173], has meant that sterically demanding reagents have necessarily come to the fore. Indeed, sterically cumbersome LTMP has been used for the directed *ortho*-lithiation of arylcarboxylic esters. However, unwanted condensation between the aryllithium intermediate and electrophilic directing groups has still been known to occur during metalation [12]. These formative ideas were, however, used in attempts to develop new chemoselective metalating agents capable of the deprotonative reaction – zincation in the first instance – of functionalized aromatics and heteroaromatics. The early research in this area led to TMP-zincate reagents being developed as outlined below [174, 175].

1.4.2 Zincates

The first reported TMP-zincate, *t*-$Bu_2Zn(TMP)Li$ **1**, was targeted on account of the nontransferability of its *tert*-butyl groups [89]. It was prepared by adding *t*-Bu_2Zn to a solution of LTMP in THF at −78 °C, whereupon the complex solution was allowed to warm to room temperature (Scheme 1.26). ^{13}C NMR spectroscopy revealed a new set of signals that could not be attributed to either *t*-Bu_2Zn or LTMP, suggesting the formation of an ate complex. No spectroscopic indications of decomposition were detected after several hours at room temperature, suggesting the reagent to be suitably robust for further application.

The *ortho*-metalation of arenes with different DMGs was examined using a solution of the TMP-zincate complex. First, the reaction of alkyl benzoates with this reagent was investigated, with metalation found to proceed smoothly at room temperature. The putative arylzincates thus prepared were treated with I_2 to give iodobenzoates in excellent yields. Substantiation of arylzincate intermediate formation came from the stepwise treatment of alkyl benzoates with LTMP followed by the addition of *t*-Bu_2Zn. This was found to be an ineffective route to the formation of the arylzincates; pre-complexation of the Li and Zn reagents was evidently essential for successful zincation. *N*,*N*-diisopropylbenzamide was also metalated by the TMP-zincate, and subsequent treatment with I_2 gave the iodide. The cyano group also functioned as an excellent DMG, and metalation proceeded smoothly. The arylzincate could be trapped with I_2 or benzaldehyde to give excellent yields of the iodide or alcohol, respectively. This arylzincate preparation was applied to biaryl synthesis, employing palladium-catalyzed cross-coupling with aryl iodides. The arylzincate derived from ethyl benzoate and TMP-zincate was reacted with iodobenzene and 3-iodopyridine in the presence of $Pd(PPh_3)_4$ at room temperature for 24 h to give biarylcarboxylates (Scheme 1.27).

The metalation of a range of heteroaromatic compounds at diverse ring positions was examined using **1**. For example, just as ethyl 3-thiophenecarboxylate has been metalated at the 2-position

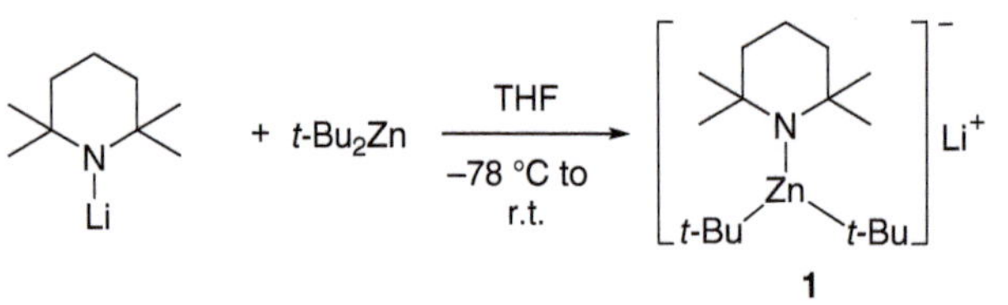

Scheme 1.26 Synthesis of *t*-$Bu_2Zn(TMP)Li$ **1**.

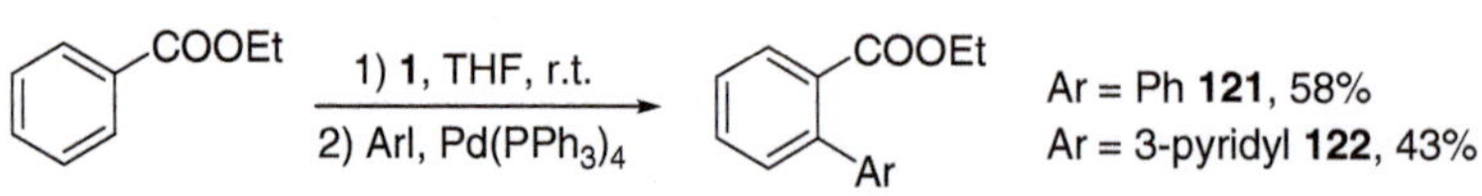

Scheme 1.27 Application of **1** in biaryl synthesis.

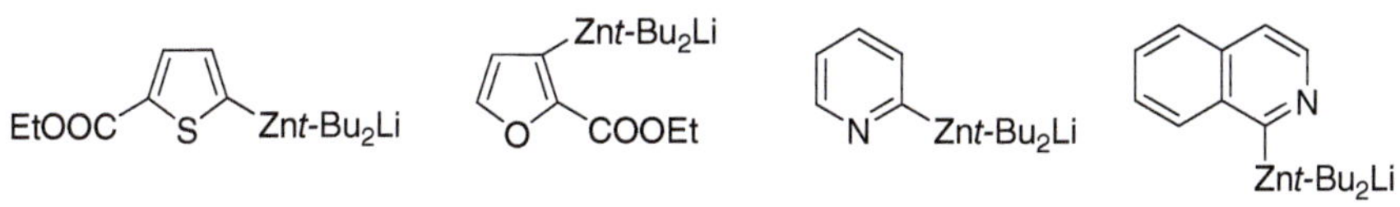

123 **124** **125** **126**

Figure 1.17 Proposed intermediates in the metalation of selected heteroaromatics.

using magnesium amides [38, 39], so too was this deprotonation achieved at room temperature using a zincate base, after which treatment with I_2 gave the 2-iodo derivative in 89% yield [174]. A similar reaction using ethyl 2-thiophenecarboxylate gave presumed **123** and thence the 5-iododerivative in 62% yield (Figure 1.17). Ethyl 2-furancarboxylate showed different regioselectivity (viz. **124**) from that of ethyl 2-thiophenecarboxylate, and 3-iodo derivative was obtained in 71% yield. α-Metalation of π-deficient heteroaromatic compounds – considered to be a challenging target – has been investigated as part of the search for more efficient, direct methods for introducing functionalities into heteroaromatic rings. Controlling the reactivity and selectivity of the metalating species has been one of the most important and essential issues in developing this approach, and this work acted to demonstrate the versatility of the TMP-zincate. The α-metalation of pyridine was found to proceed smoothly at room temperature, and the putative pyridinylzincate **125** was treated with I_2 to give 2-iodopyridine in 76% yield. Interestingly, quinoline was metalated preferentially at the 8-position, and treatment with I_2 gave 8-iodoquinoline in 61% yield (together with α-metalated 2-iodoquinoline in 26% yield). Isoquinoline was also easily metalated at the 1-position (putatively **126**), and 1-iodoisoquinoline was obtained in 93% yield (Figure 1.17). This contrasted with the directed 1-lithiation of isoquinoline, which is difficult to accomplish on account of the formation of isoquinoline dimers [176].

Since the advent of directed aromatic deprotometalation using **1** [174], efforts have been ongoing to establish in detail the nature of potential intermediates in this process. This has led to extensive structural studies that will be visited throughout this book. To summarize, however, the earliest work focused on the nature of so-called synergic bases themselves and rapidly established the predominance of a metallacyclic core based on the ability of the amido ligand and one of the two alkyl groups (in the case of Zn) to chelate the alkali metal irrespective of the presence of coordinating solvents or reagents [177]. The alkali metal can be a higher group 1 element such as sodium [178] or even potassium [179, 180], though the use of lithium has dominated synthetically applied work [82]. Representative examples that have been fully elucidated crystallographically include $Et_2Zn(TMP)Li$ **127** and t-$Bu_2Zn(TMP)Li(TMEDA)$ **128**. They are based on 4-membered NZnCLi metallacyclic cores deriving from the intermetal bridging ability of TMP and one of the two alkyl groups (e.g. Figure 1.18). Theirs and closely related single-crystal structures are explored in detail in Chapter 2, Figures 2.13 and 2.14.

More instructive in terms of understanding deprotometalation, have been investigations into model intermediates that have incorporated simple aromatics. An early example of this involved

Figure 1.18 A generalized alkali metal zincate.

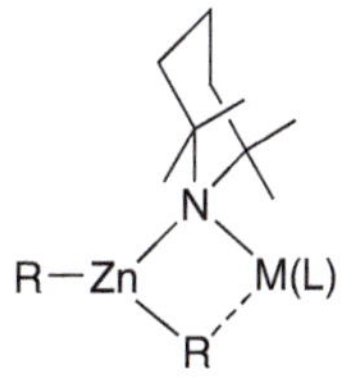

R = alkyl; M = Li, Na, K;
L = THF, TMEDA, etc.

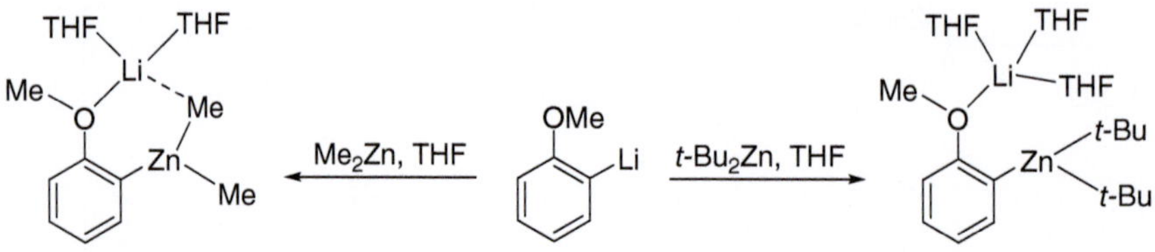

129 **130**

Scheme 1.28 Transmetalation of lithioanisole to variously solvated *ortho*-zincates **129** and **130**.

the reaction of anisole, with the corresponding *ortho*-zincates prepared under a variety of solvent conditions by transmetalation. Isolation revealed solvated aryl(dialkyl)zincates of the type $R_2Zn(C_6H_4OMe\text{-}2)Li(THF)_n$ (R = Me, n = 2 **129**; R = *t*-Bu, n = 3 **130**), in which the zincated alkyl groups stabilized the alkali metal or not depending on the level of solvent inclusion in the structure (Scheme 1.28) [181].

Moving away from transmetalation reactions, the isolation and characterization of intermediates in deprotonative zincation reactions at the aromatic *ortho* position have been the subject of study. Spectroscopy quickly suggested a structural basis for the creation alongside dominant *ortho*-metalates of *meta* and even a small amount of *para*-metalates [182]. Directed *meta* reactivity [183, 184] will be dealt with alongside more recent developments in the polydeprotonation of model aromatics at length in Chapter 2. However, immediately synthetically useful *ortho* reaction formed the basis of a range of synthetic studies. One practical advantage of *ortho*-zincation that quickly emerged surrounded the observation of apparent ambibasicity. That is to say, it rapidly became apparent that nonstoichiometric zincate activity was evidenced. This was exemplified by the observation that reaction of *N*,*N*-diisopropylbenzamide led to the isolation of $t\text{-BuZn(TMP)}_m\{C_6H_4C(O)N\textit{i}\text{-Pr}_2\text{-2}\}_n\text{M(TMEDA)}$ (m = 0, n = 2, M = Li **131**; m = 1, n = 1, M = Na **132**) [185]. Similar work afforded both $EtZn\{C_{10}H_6C(O)N\textit{i}\text{-Pr}_2\text{-2}\}_2Li(THF)_2$ **133** and $Zn\{C_6H_4C(O)N\textit{i}\text{-Pr}_2\text{-2}\}_3Li(THF)$ **134** (the solid-state structures of which are explored in Chapter 2, Scheme 2.20), with spectroscopic [186] and theoretical [187] analysis ultimately elucidating the basis for all of these observations; TMP was acting as a kinetic base whose conjugate acid could then be quenched by intermediate *ortho*-zincates up to three times before HTMP was finally being liberated once a tri(aryl)zincate had been obtained [187]. Other advantages of this realm of chemistry manifested themselves, most particularly through the recognition that traditionally underused DMGs were compatible with relatively (wrt organolithium bases) nonnucleophilic zincates. In possibly the most dramatic example of the newfound ability to sidestep unwanted nucleophilicity, **128** was reacted with aromatic nitriles to give *ortho*-zincated products instead of addition products [188]. A further manifestation of this new selectivity for *ortho* reaction saw the quantitative avoidance [189] of anionic Fries rearrangement by phenyl *N*,*N*-dialkylcarbamates of *ortho*-lithiates e.g. **135** to *ortho*-phenoxides e.g. **136**, enabling the isolation of **137** (Scheme 1.29) [27] whilst elsewhere the *ortho* reaction of benzyl methyl ester was recorded in place of the expected abstraction of a thermodynamic α-hydrogen [190]. The isolation and full characterization of the intermediate mono(aryl) zincate was investigated theoretically [189], with results suggesting that factors such as external solvation [191] and aggregation [189] interfered with the ability of zincate intermediates to undergo polybasic reactions. Applications in the deprotonation of a range of *N*-heteroaromatics (pyrazine, pyrazidine, pyrimidine, quinoxaline) to give both mono- and disubstituted products have also been demonstrated using a variant on $R_2Zn(TMP)Li$; namely $Zn(TMP)_3Li$ **138**, created *in situ* from

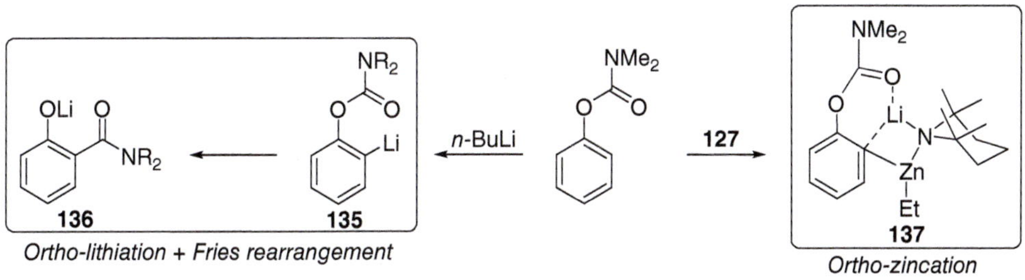

Scheme 1.29 The anionic Fries rearrangement and its avoidance using lithium zincate **127**.

$ZnCl_2$ and LTMP [192]. This work was then rapidly extended to the preparation of functionalized pyrroles [193] and finally O- and S-containing five-membered aromatic heterocycles [194].

1.4.3 Cadmates

Limited synthesis using ate complexes has been achieved with the higher elements of group 12. However, in recent years limited reports of the deprotometalation of functionalized aromatics and heterocycles have appeared. These have sought to extend the role of organocadmiums as soft nucleophilic reagents in organic synthesis [195, 196]. Whereas Ph_3CdLi **139** was first reported almost 70 years ago [197], the combination of $CdCl_2$(TMEDA) and LTMP was reported just a decade ago. This work evolved from studies suggesting the cooperative action of LTMP and $(TMP)_2Zn$ [194]. In the current case, neither $CdCl_2$(TMEDA) nor LTMP satisfactorily deprotonated anisole, whereas a 1 : 3 mixture of the two did so in 74% yield according to trapping with I_2 [198]. At the same time, ^{13}C NMR spectroscopy suggested the absence of LTMP in solution, pointing to lithiocadmate formation. Subsequently, a range of aromatic amides, esters, nitriles, and even ketones were shown to undergo selective *ortho* reaction. Aromatic halides revealed chemoselective iodination remote to the halogen (Scheme 1.30). Efforts next turned to the deprotometalation of aromatic heterocycles (both π-deficient and π-excessive), where impressive results were obtained, beating those reported for Zn (see above) [192, 194]. The reactions of a wide range of N-, O-, and S-heterocycles were subsequently probed [199]. Particularly interesting, very sensitive reagents such as diazines proved amenable to reaction (Scheme 1.30).

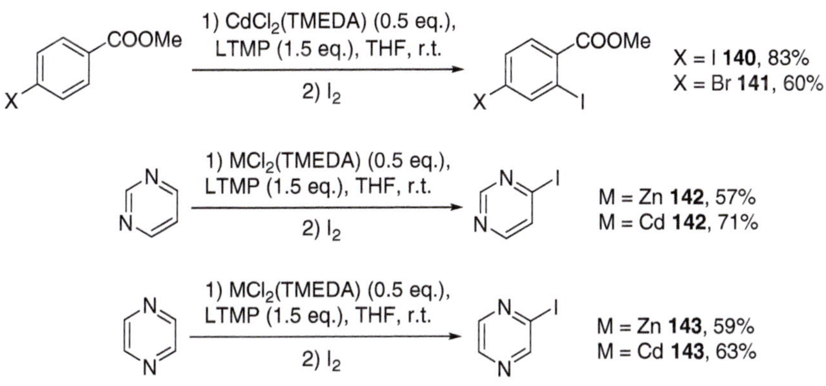

Scheme 1.30 Comparing the performance of Cd and Zn reagents in aromatic halogenation.

1.4.4 Aluminates

Aromatic aluminium compounds are potentially attractive as functional materials and synthetic building blocks [200–202]. However, aromatic aluminium chemistry has not been well developed, mainly due to the lack of efficient preparative methods compatible with the presence of ancillary functional groups. A common synthetic route to aliphatic aluminium compounds is transmetalation of organolithiums or Grignard reagents to the corresponding aluminium compounds [203]. Unfortunately, these metalating reagents or the intermediary aromatic lithium or Grignard species they form are often too reactive to coexist with electrophilic functional groups like halogens, amides, nitriles, and π-deficient heterocycles [12]. Neither insertion of aluminium into carbon–halogen bonds nor halogen–metal exchange reactions of aluminium on aromatic rings have been realized to date and this led to an interest in the development of new Al-based ate reagents for effecting direct deprotometalation. Hence, a regio- and chemoselective direct alumination of functionalized aromatics using a newly designed aluminium ate base was investigated [204]. To develop this area, halogen–metal exchange and deprotonative metalation were both investigated. First, the halogen–aluminium exchange reactions of haloaromatics using various kinds of organoaluminum reagents were examined, but all attempts were unsuccessful. Then the deprotonative alumination of functionalized benzenes was investigated, a method which would be more advantageous to generate multifunctionalized aromatic aluminium compounds from the viewpoint of precursor availability. After extensive experiments using anisole and benzonitrile as model substrates, the complex lithium tri(isobutyl)(tetramethylpiperidido)aluminate i-Bu$_3$Al(TMP)Li **144**, prepared by mixing i-Bu$_3$Al and LTMP in THF (Scheme 1.31) was found to be better than either Me$_3$Al(TMP)Li **145** or Et$_3$Al(TMP)Li **146** for achieving direct alumination under mild conditions (Scheme 1.32). The resulting aryl aluminate was treated with D_2O or I_2 to give the desired *o*-dueterio- or *o*-iodoanisole, respectively.

In the same way as the principles of structural chemistry have been applied to alkali metal zincates, allowing the interrogation of basic reagents and reaction intermediates and the development of an improved understanding of reaction processes, so have they helped elucidate the activity of aluminates. Four-membered metallacycles akin to those reported for dialkylamidozincate bases are seen for trialkylamidoaluminates such as i-Bu$_3$Al(TMP)M(L) (M = Li, L = THF **147**; M = Na, L = TMEDA **148**; the crystal structure of the first is exemplified in Chapter 2, Figure 2.21) [205, 206], though a dialkyldiamidoaluminate has also been suggested, based on two metal-bridging amides (see below) [207]. As with zincate chemistry, the alkali metal proved straightforwardly variable; though while Li and Na were relatively simple to work with [205, 206], K proved more problematic and required PMDETA as a stabilizing additive to prevent the TMP ligand from degrading [208]. As for zincates, reactivity was more complex and subtle than might at first have been expected. The polybasicity reported for zincates was not observed for **144**, a fact rationalized theoretically in terms of the contrast between closed-shell Al in a tetraorganoaluminate and 16 e$^-$ Zn in a triorganozincate [209].

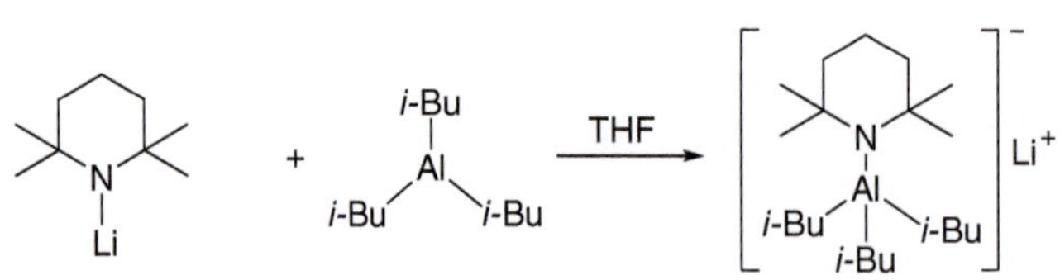

Scheme 1.31 Synthesis of i-Bu$_3$Al(TMP)Li **144**.

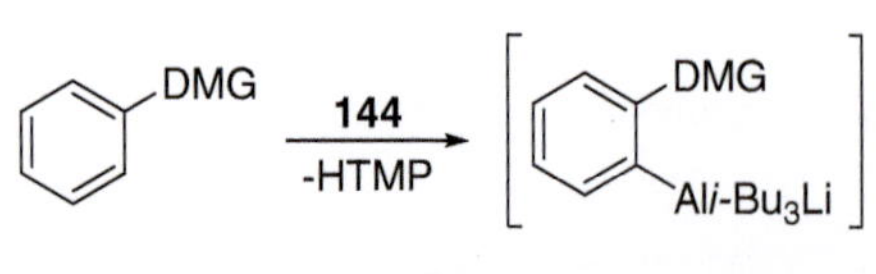

Scheme 1.32 *Ortho*-alumination of a functionalized aromatic ring.

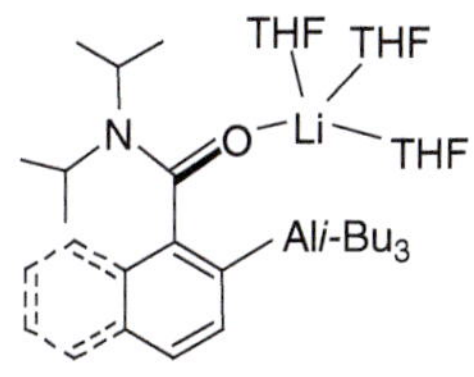

149

Figure 1.19 Model aluminate **149**, obtained by sequentially treating ArC(O)N*i*-Pr_2 with *t*-BuLi and *i*-Bu_3Al in THF.

Moreover, as spectroscopic studies have subsequently made clear, the heterobimetallic nature of isolated alkali metal aluminates belies the possibility of different reactive pathways. Instead of synergic reactivity and direct alumination of organic reagents, cooperative cleave-and-capture chemistry could occur whereby the deprotonative alkali metalation of an organic can instead precede the formation of an Al–C bond, so 'capturing' the reactive anion [93]. In either case, amide abstraction of a proton from the *ortho* position of the aromatic reagent is favoured. The intermediates expected for such processes were structurally modelled by sequentially treating ArC(O)N*i*-Pr_2 with *t*-BuLi and *i*-Bu_3Al in THF to obtain diffraction quality crystals of **149** (Figure 1.19). Their stability in solution was also studied, with partial desolvation observed in hydrocarbon media. In spite of this, exposure to excess HTMP failed (in accordance with theory) to affect the reaction (i.e. quenching of the amine) [209]. The complete passivity of either *ortho*-aluminate towards HTMP has been further evidenced by ^{1}H and ^{13}C NMR spectroscopic analysis using (non-polar) benzene or (Lewis basic) THF media and introducing excess amine by injection. In either solvent, chemical shifts attributable to *ortho*-aluminate were unchanged upon addition of amine, even after heating to reflux.

The synthetic utilization of kinetic basicity was a fundamental driver for developing the aluminate systems described here. This mode of action has been nicely demonstrated, albeit in a nonaromatic aside; regioselectively functionalizing unsymmetrical ketones. Methyl isopropyl ketone reacted with a mixture of *i*-Bu_3Al and LTMP at 0 °C in THF to selectively give the kinetic enolate with near complete selectivity. Subsequent electrophilic trapping with benzaldehyde then proceeded smoothly to give only the corresponding regioisomer. A range of other unsymmetrical acyclic and cyclic ketones then demonstrated similar kinetic regioselectivity in their functionalization, even under harsh conditions; heating to reflux in THF for 18 h [209]. In an example of interesting functional group tolerance by aluminates, use of the dialkyldiamidoaluminate [207] *i*-Bu_2Al(TMP)$_2$Li **150** has enabled the highly efficient, facile conversion of 4-halo-anisoles to synthetically important triheterohalogenated anisoles (**151–153**) through the sequential reaction of a 4-halo reagent with sulfuryl chloride, *N*-bromosuccinimide and/or iodine (Scheme 1.33) [210]. Of course, the impressive compatibility of the base with halogenated reagents underpins this reactivity. This contrasts strongly with that of many traditional organometallics and it extended to the ability to isolate and fully elucidate *ortho*-aluminated intermediates in both the first (**154**) and second (**155**) halogenation processes (Figure 1.20).

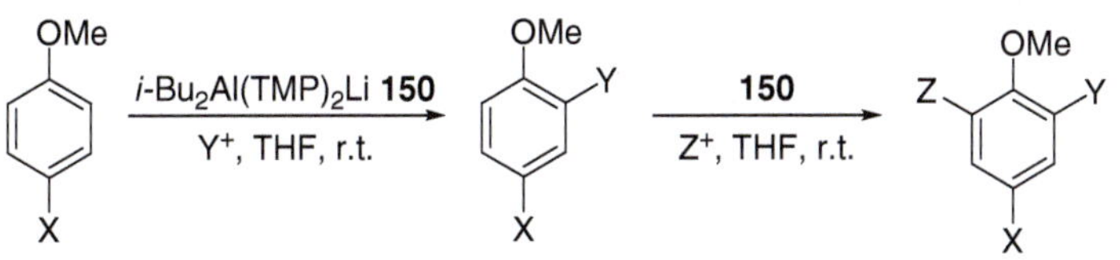

X = I, Y = Br, Z = Cl **151**
X = Br, Y = I, Z = Cl **152**
X = Cl, Y = I, Z = Br **153**

Scheme 1.33 *i*-Bu_2Al(TMP)$_2$Li **150** has enabled the conversion of 4-halo-anisoles to triheterohalogenated anisoles **151–153**.

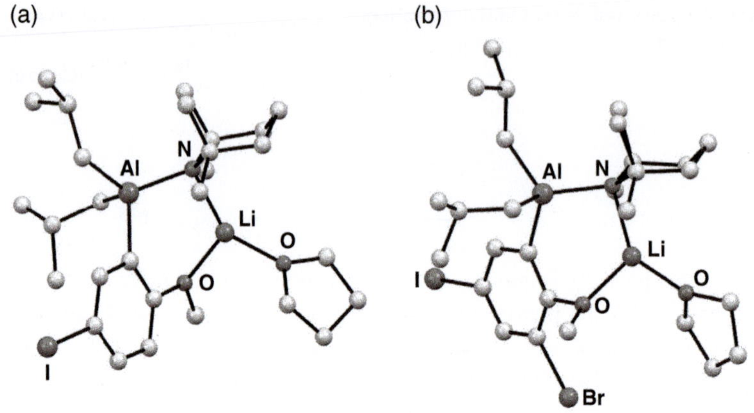

Figure 1.20 Molecular structures of aluminated precursors **154** and **155** to (a) di-, and (b) triheterohalogenated anisole derivatives, respectively. *Source*: Adapted from Conway et al. [210].

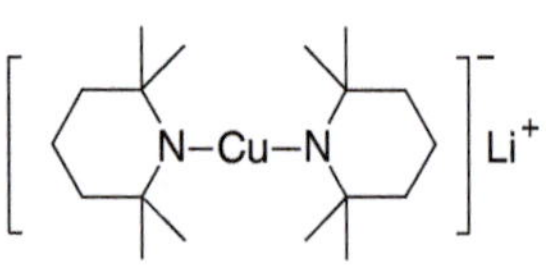

156

Figure 1.21 Representation of the Gilman amidocuprate $(TMP)_2CuLi$.

1.4.5 Cuprates

Whilst lithium organocuprates (Section 1.3.3) have established dominance amongst the transition metal organometallics used in synthesis, several different flavours of reagents have evolved to suit various applications. These include heteroleptic cuprates, which normally combine organyl and heteroatom-based ligands as a means of increasing organyl transfer efficiency; organoamidocuprates are one such well-studied member of the heterocuprate family. They have become known by virtue of their unique features and reactivities. They have many potential applications in organic transformations, especially in stereoselective synthesis because the amido ligand can act not only as a dummy (nontransferable) group but also as a chiral auxiliary [211]. The non-transferability of amido (heteroatom) ligands on cuprates in carbon–carbon bond forming reactions has also been theoretically clarified by DFT calculations [212]. A new use for amidocuprates was investigated, wherein the amido ligand transfers (reacts) first as a base for chemoselective directed *ortho*-cupration and then as a switch in successive C–C, C–O, and C–N bond formation processes [213, 214]. To develop new applications of amidocuprates, the deprotonative metalation of functionalized benzenes was investigated [215]. Initial studies used benzonitrile as a model aromatic compound with an electron-withdrawing group to identify favourable reaction conditions. These indicated that a TMP group as the amido moiety and THF as solvent were suitable starting points for the optimization of directed metalation reaction conditions (Figure 1.21). Attempts to use the Gilman amidocuprate $(TMP)_2CuLi$ **156** prepared from CuI proved unsuccessful in terms of reactivity and directed metalation selectivity. On the other hand, the use of CuCN was presumed to result in the incorporation of cyanide and the formation of so-called Lipshutz amidocuprates. Two examples, putatively $(TMP)_2Cu(CN)Li_2$ **157** and $MeCu(TMP)(CN)Li_2$ **158**, have been proposed to exemplify this, with reaction mixtures incorporating the appropriate amounts of the necessary components (CuCN, LTMP and, for **158**, MeLi) enabling metalation without any catalyst, in good yields at 0 °C (summarized in Scheme 1.34 and explored in depth in Chapter 8). It was found that although the

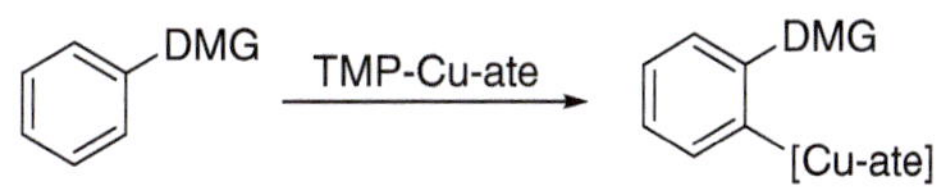

Scheme 1.34 A generalized view of directed *ortho*-cupration.

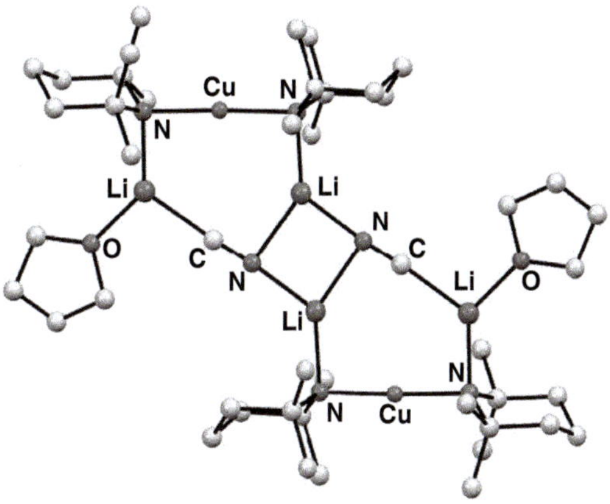

Figure 1.22 Molecular structure of Lipshutz cuprate dimer $[(TMP)_2Cu(CN)Li_2(THF)]_2$ **159**$_2$. *Source*: Adapted from Usui et al. [213].

latter case employed a CuMe-containing reagent, at least one TMP ligand, one of the bulkiest available amido ligands, was crucial for good yield and chemoselectivity.

Structural work has been integral to the evolution of directed cupration. The concept of organo(amido)cuprate bases can be viewed as emerging from the confluence of organocuprate chemistry and the (at the time) relatively new and evolving field of synergic base chemistry. From the perspective of cuprate chemistry, it had been recognized for some time that the replacement of an organyl group with a non-transferable ligand [212], to give a *heterocuprate* could (i) reduce wastage of valuable organic groups and (ii) significantly improve reagent stability. Several classes of non-transferable ligand have been investigated [216]. However, amido groups have offered the most compelling combination of excellent stabilizing properties and relative ease of access from commercial reagents [217]. Of course, the ability of the sterically congested amido ligand TMP to act as a potent kinetic base in directed metalation has already been discussed in this chapter in the context of highly successful synergic metalating reagents such as **1** [174]. The combination of these factors led to organo(TMP)cuprates being conceived as viable bases for directed cupration. Their successful application to the elaboration of functionalized aromatics is described in the preceding section [213] and is related in detail in Chapter 8. Most relevant to the work of an applied vein, the combination of LTMP with CuCN in THF was argued to form cyanide-containing Lipshutz bis(amido)cuprate complex $(TMP)_2Cu(CN)Li_2(THF)$ **159**, this system demonstrating excellent reactivity in directed *ortho*-cupration. As a part of the same study, 2 : 1 reaction of LTMP with CuCN in the presence of THF indeed enabled the isolation of, and X-ray diffraction studies on, this Lipshutz cuprate. Data revealed a dimer composed of individually 7-membered metallacycles that associate by forming a central Li_2N_2 ring (Figure 1.22). The inclusion of cyanide as a bridge between multiple Li centres without any discernable interaction with Cu was significant since it provided further important evidence (backing up prior theoretical work) [160, 161] for the absence of higher-order structures in cyanocuprates. This motif was subsequently found to apply to structures incorporating a range of inorganic anions that led to the development of the expression 'Lipshutz-type' cuprates for the recently reviewed family of complexes $(TMP)_2Cu(X)Li_2$ (X = inorganic anion ≠ CN, see below for specific examples) [218].

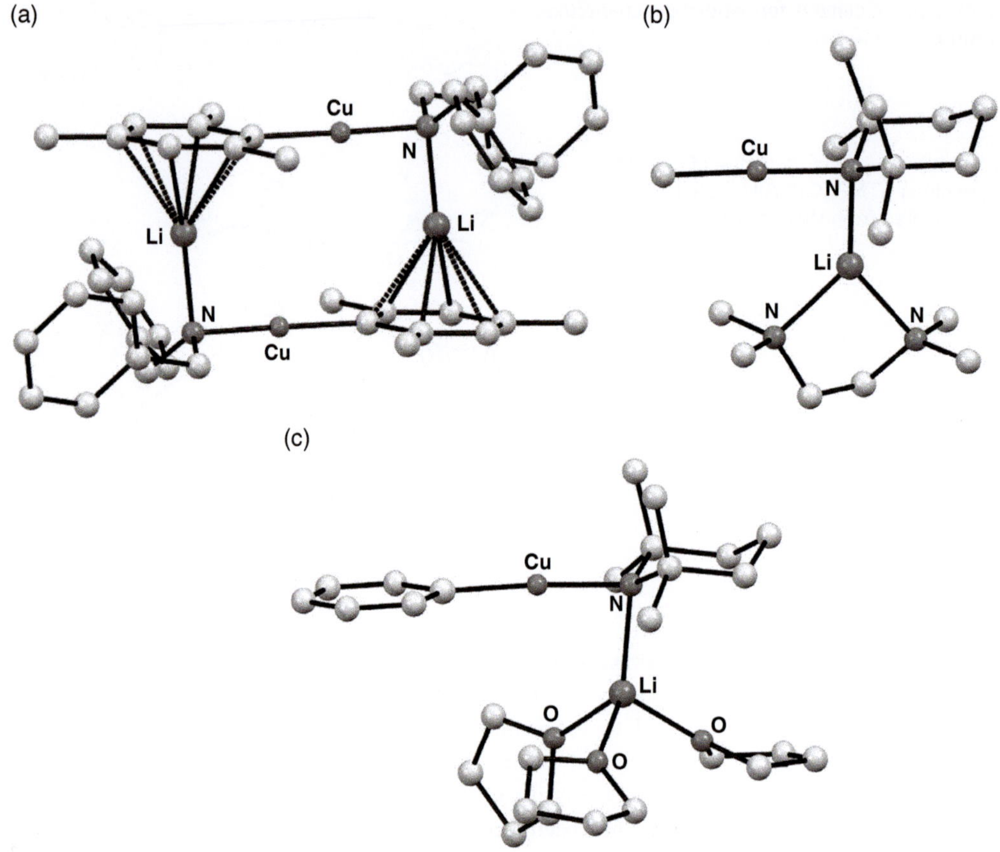

Figure 1.23 Molecular structures of organoamidocuprates (a) $[MesCu(NBn_2)Li]_2$ **160**$_2$, (b) MeCu(TMP)Li(TMEDA) **161** and (c) $PhCu(TMP)Li(THF)_3$ **162**. *Sources*: Adapted from Davies et al. [219]; Haywood et al. [220].

Reports on the structures of organo(amido)cuprates emerged around the same time as the inception of directed *ortho*-cupration. The combination of mesitylcopper with dibenzylamidolithium in toluene led to the isolation of Gilman heterocuprate $MesCu(NBn_2)Li$ **160** (Bn = benzyl), which revealed a head-to-tail dimer in the solid state (Figure 1.23) [219]. In this case, Li salts and donor solvents were excluded during cuprate formation. On the other hand, *in situ* cuprate synthesis using mixtures of organolithium and amidolithium reagents in combination with CuCN offered the possibility of LiCN inclusion in the solution and/or solid-state structures. In such cases, however, *Gilman* organo(amido)cuprates MeCu(TMP)Li(TMEDA) **161** and $PhCu(TMP)Li(THF)_3$ **162** (Figure 1.23) that did not include LiCN were isolated and analyzed in the solid-state. Both species preferred to form solvated monomers [220]. However, the inactivity of these isolated cuprates in directed *ortho*-metalation was in stark contrast to the behaviour of the *in situ* preparations, suggesting LiCN involvement to be likely in solution. This hypothesis was supported by DFT calculations, which suggested a facile equilibrium between Lipshutz and Gilman organo(amido)cuprates in solution.

Similar patterns of reactivity have been seen for bis(amido)cuprate preparations based on CuI, whereby Gilman cuprate **156** (actually a dimer – see below) proved an ineffective base, whereas Lipshutz-type $(TMP)_2Cu(I)Li_2(THF)$ **163** performed much better (Scheme 1.35 and Figure 1.24) [221]. Importantly, these results pointed towards a more general role for Li salts in disrupting unreactive Gilman aggregates, spawning a search for other Lipshutz-type reagents that could offer safer alternatives to cyanide-based preparations.

Scheme 1.35 Selective formation of Gilman and Lipshutz-type cuprates from CuI.

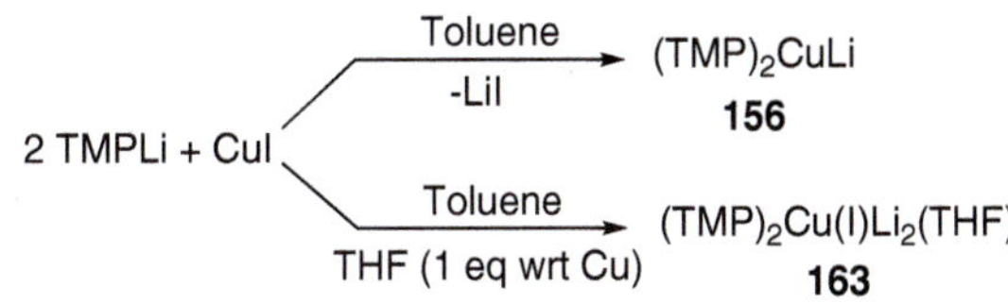

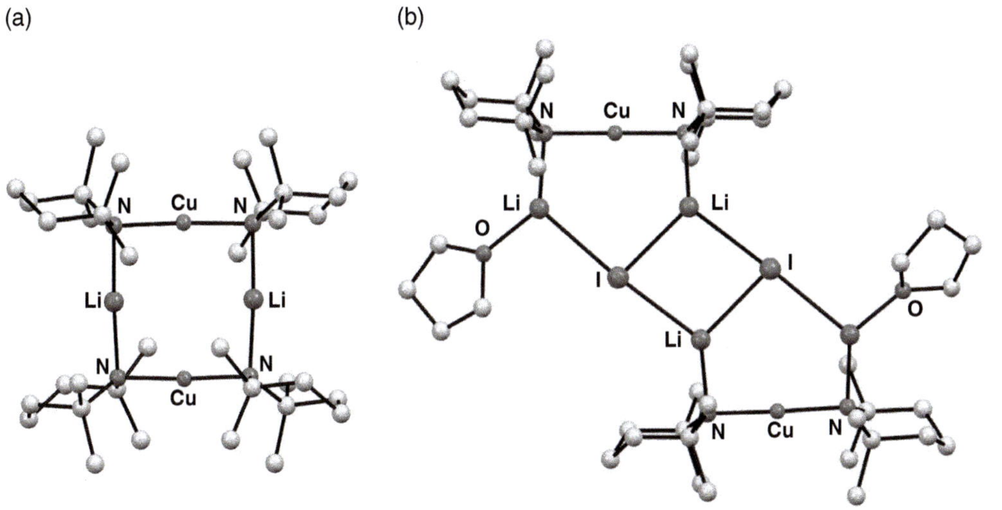

Figure 1.24 Molecular structures of (a) Gilman amidocuprate dimer of $(TMP)_2CuLi$ **156** and (b) Lipshutz-type dimer of $(TMP)_2Cu(I)Li_2(THF)$ **163**. *Source*: Adapted from Komagawa et al. [221].

Readily available copper(I) halides CuCl [222] and CuBr [223] have been investigated as sources of lithium salts for Lipshutz-type cuprates, the structures of which have been found to be very similar to that of the dimeric iodide shown in Figure 1.24. Synthetically, *in situ* preparations using the putative cuprate $(TMP)_2Cu(Cl)Li_2$ **164** were found to be excellent reagents for the directed cupration of heterocycles, opening up a new route to the synthesis of pharmacologically interesting azafluorenones [222].

In an attempt to decrease the costs associated with amidocuprate preparation [224], copper(I) halides have been employed extensively in the creation of DMP- rather than relatively expensive TMP-cuprates (DMP = *cis*-2,6-dimethylpiperidide) [225]. The 2 : 1 reaction of amidolithium LDMP with CuX (X = Cl, Br, I) was therefore attempted as a route to more economical Lipshutz-type cuprates. Remarkably, the formal replacement of two methyl groups from TMP with H-atoms led to an entirely different structure-type that could be viewed as an *adduct* of Gilman and Lipshutz-type monomers (Figure 1.25 shows X = Br **165**). In the pentametallic species seen, differences in Li–X (X = Cl, Br, I) and Li–N bond lengths were rationalized in terms of competing stabilization by hard/soft donors. Experiments in which TMP-cuprates and DMP-cuprates were both prepared in the presence of THF or Et_2O confirmed that the difference in structure-types was attributable to the amido ligand rather than the Lewis base. Importantly, the inclusion of LiX (X = Cl, Br, I) in adduct cuprates was consistent with their observed reactivity in directed *ortho*-cupration. DFT calculations reinforced this view that adducts could affect *ortho*-metalation by showing that adduct cuprates represented an energetically feasible source of reactive Gilman monomers – which prior work had already suggested to represent the active species in directed *ortho*-cupration [221].

The switch in structure-type apparently enforced by the amido ligands has led to a search for other potential replacements for HTMP that might also influence structure-type. 2-Methylpiperidide (MP) was quickly identified as an interesting target, in view of the low cost of its conjugate acid and its chirality. The

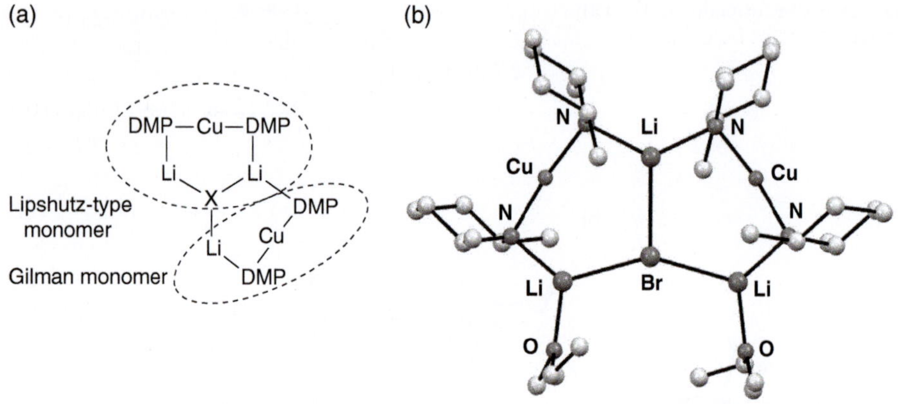

Figure 1.25 (a) Schematic of an adduct cuprate structure-type and (b) molecular structure of $\{(DMP)_2CuLi(Et_2O)\}_2LiBr$ **165**. *Source*: Adapted from Peel et al. [226].

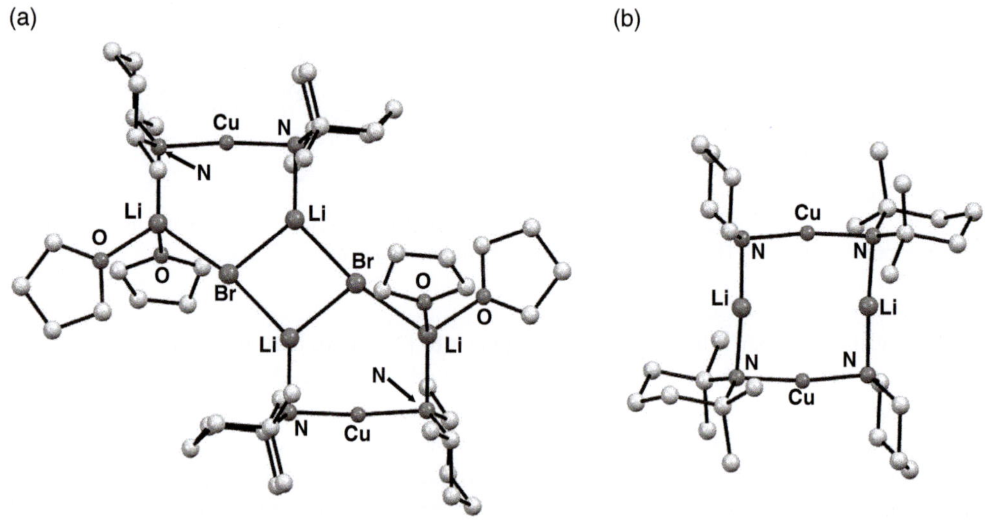

Figure 1.26 Molecular structures of heteroleptic cuprates (a) $[(TMP)(DMP)Cu(Br)Li_2(THF)_2]_2$ **167**$_2$ and (b) $[(PIP)(TMP)CuLi]_2$ **168**$_2$. *Source*: Adapted from Peel et al. [226].

reaction of racemic LMP with CuBr in a 2 : 1 ratio yielded $\{(MP)_2CuLi(THF)_2\}_2LiBr$ **166** – evidenced by X-ray diffraction to be an adduct cuprate in the solid-state [226]. In spite of a precedent from organocuprate chemistry [227] stereoselective assembly was not observed in this case, with X-ray diffraction suggesting a multi-component crystal involving permutations of *R*- and *S*-MP. Meanwhile, combining the use of either DMP or MP *and* TMP demonstrated the ability to produce heteroleptic Lipshutz-type structure **167**. Partnering TMP with piperidide (PIP) then suggested competition between Lipshutz-type and Gilman structures in heterodiamide chemistry by producing Gilman cuprate paddlewheel **168**$_2$ (Figure 1.26).

The structural influence of inorganic anions beyond halides capable of replacing cyanide in the creation of Lipshutz-type cuprates was investigated through the reaction of CuSCN with an amidolithium reagent. In the event, CuSCN provided straightforward access to a range of differently solvated Lipshutz-type cuprates $(TMP)_2Cu(SCN)Li_2(L)$ (L = THF **169**, Et_2O **170** and THP **171**; THP = tetrahydropyran) when introduced to LTMP in a 1 : 2 ratio in the presence of donor solvent [228].

A strong dependence of the geometry of the solid-state dimers of these thiocyanatocuprates on the Lewis base additives was uncovered, apparently resulting from the ability of the metallacyclic $(LiSCN)_2$ core to adopt boat-like, chair-like or planar conformations (Figure 1.27). The influence of the donor solvent was not limited to the solid-state either: Lipshutz-type thiocyanato(amido)cuprates were found to convert to Gilman cuprate in benzene solution, with the degree of conversion being strongly influenced by the identity of the donor solvent incorporated in the cuprate (and being most pronounced for Et_2O). In a synthetic setting, thiocyanatocuprates performed competitively with CuCl-derived bases in the directed *ortho*-cupration of halopyridines.

Cyanato(amido)cuprate analogues of the thiocyanate systems described above have also been investigated. Attempts to prepare these cuprates from the direct reaction of CuOCN with LTMP were not successful, with spectroscopy suggesting multiple products and crystallography indicating Cu/Li substitution in the solid-state in some of these [229]. Nonetheless, Lipshutz-type $(TMP)_2Cu(OCN)Li_2(THF)$ **172** proved accessible by inserting LiOCN into **156** in THF. The resulting solid-state dimer offered a geometry that differed substantially from those of the known THF-solvated thiocyanatocuprates, but otherwise retained all the features now established to be typical for Lipshutz-type cuprates (Figure 1.28). Curiously, when reacted with CuOCN, less sterically

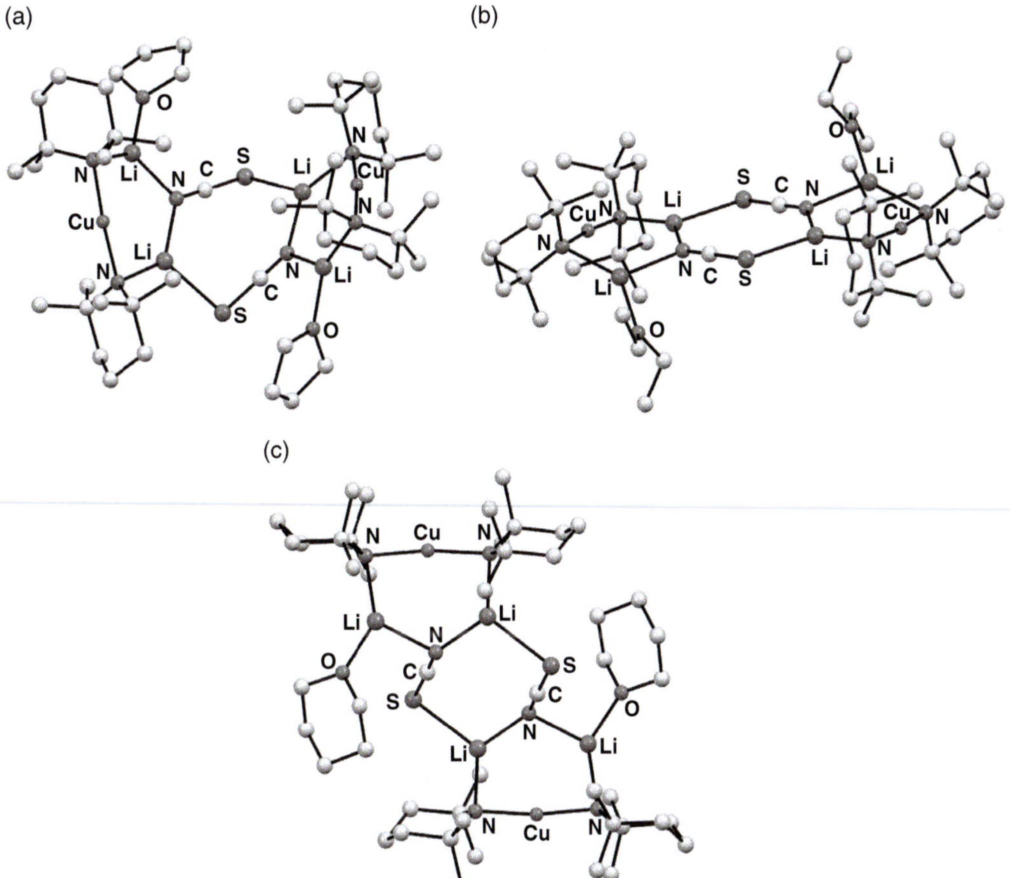

Figure 1.27 Molecular structures of the dimers of thiocyananto(amido)cuprates (a) $(TMP)_2Cu(SCN)Li_2(THF)$ **169**, (b) $(TMP)_2Cu(SCN)Li_2(Et_2O)$ **170**, and (c) $(TMP)_2Cu(SCN)Li_2(THP)$ **171**. *Source*: Adapted from Peel et al. [228].

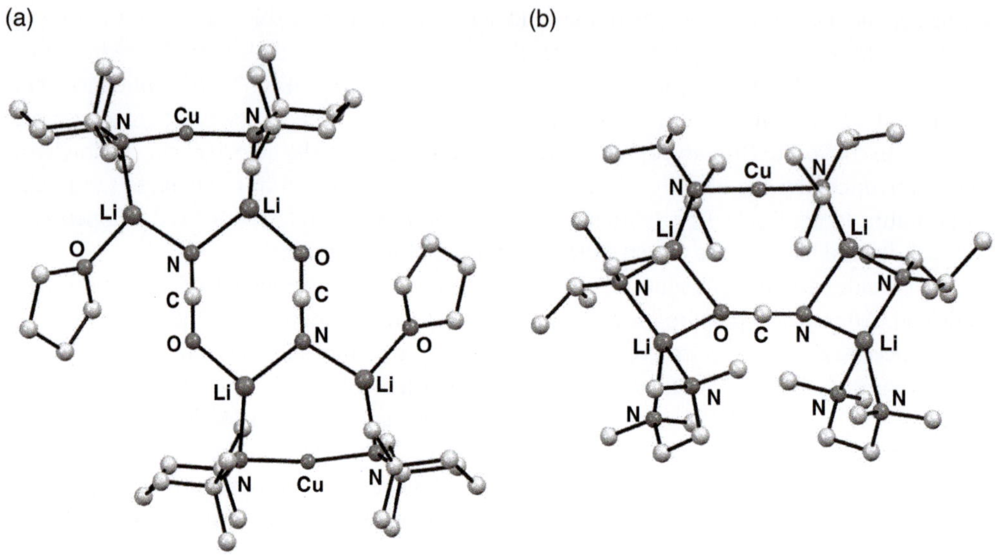

Figure 1.28 Examples of cyanatocuprates (a) $[(TMP)_2Cu(OCN)Li_2(THF)]_2$ **172**$_2$ and (b) $(DA)_4Cu(OCN)Li_4(TMEDA)_2$ **173**. *Source*: Adapted from Peel et al. [229].

encumbered lithium diisopropylamide (LDA) furnished a novel amidocuprate-amidolithium adduct $(DA)_4Cu(OCN)Li_4(TMEDA)_2$ **173**, which could be viewed as arising from the attachment of units of LDA(TMEDA) to a Lipshutz-type monomer. Meanwhile, spectroscopic investigations on CuOCN/LTMP reaction mixtures suggested the production of a new amidocopper-amidolithium aggregate $(CuTMP)_2(LTMP)_2$ **174** (Figure 1.29) – a structural isomer of previously reported Gilman dimer $[(TMP)_2CuLi]_2$ **156**$_2$ (Figure 1.24a). The reactivity of the Gilman dimer was previously established to be low. However, the knowledge that monomeric Gilman amidocuprates are reactive led to an interest in the solution behaviour and reactivity of this new adduct.

The recent realization that **174** could be accessed in a pure form by reaction of CuCl with LTMP made possible investigations into its synthetic utility. This proved both highly unexpected and potentially useful. Notably, the adduct demonstrated an uncanny ability to smoothly deprotonate benzene. As such it suggests the use of currently underutilized chemical feedstocks in aromatic elaboration. Physical mixtures of CuTMP and LTMP replicated this reactivity towards benzene, in so doing leading to the production of a series of organoamidocuprates $Ph(TMP)_3Cu_nLi_{4-n}$ ($n = 1$–4 **175–178**; Figure 1.30) [230]. This contrasted with conventional Gilman cuprate **156**$_2$, which was unreactive under the same conditions. Crystallography revealed metallacyclic structures for these product organoamidocuprates in the solid-state; the amido ligand adopted the by now usual metal-bridging mode, whilst coordination of the Ph-group varied substantially depending on metal content. In all cases, where Ph bridged Cu and Li, Cu–C σ-bonding took precedence, the C···Li interactions being π-type and suggesting an increase in hapticity with increasing Li content (a conclusion which was also supported by spectroscopy). Detailed spectroscopic discussions are

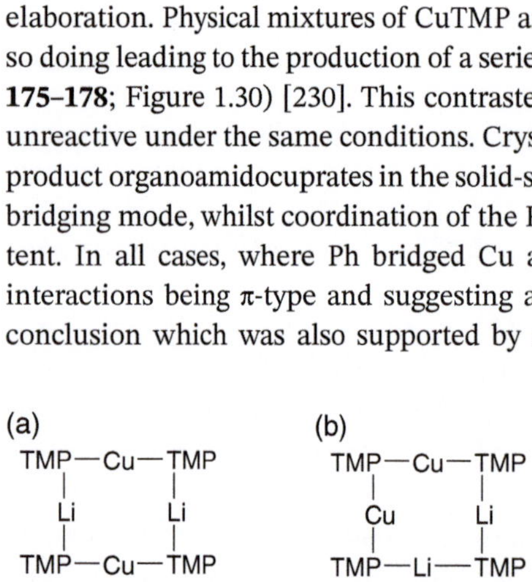

Figure 1.29 Isomers of $(TMP)_4Cu_2Li_2$; (a) dimer of conventional Gilman cuprate **156** (see Figure 1.24a) and, (b) isomeric adduct $(CuTMP)_2(LTMP)_2$ **174**.

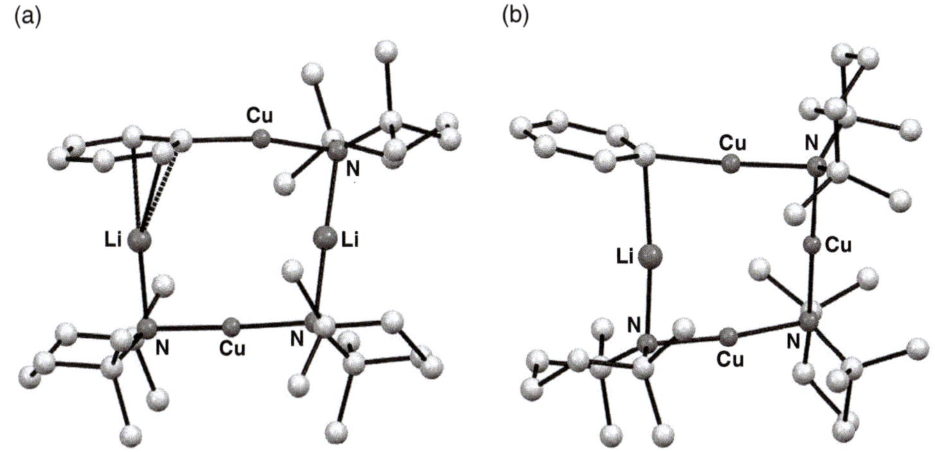

Figure 1.30 Structurally characterized organoamidocuprate aggregates (a) $Ph(TMP)_3Cu_2Li_2$ **176** and (b) $Ph(TMP)_3Cu_3Li$ **177**, illustrating changes to the hapticity of the Ph moiety in the solid-state as a function of metal composition. *Source*: Adapted from Peel et al. [230].

reserved for Chapter 5. However, briefly, $^1H,^1H$-NOESY/EXSY revealed that in solution, $Ph(TMP)_3Cu_2Li_2$ **176** was conformationally fluxional but otherwise retains its integrity. More dramatically, $^7Li,^7Li$-EXSY established a dissociative-associative equilibrium for the Li-rich species $Ph(TMP)_3CuLi_3$ **175**.

1.4.6 Argentates

To bring the current narrative right up to date, the similarities between copper and silver, vertically related in the periodic table, have led to a recent interest in lithium argentates. However, whereas applications of organocopper compounds have proved extensive [231–233], those of their silver congeners have been much more limited due to the difficulties posed by their preparation. Hence, the Ag(0 → I) redox potential [234] of 0.8 V means that, unlike organocopper(I) compounds, organosilver(I) species cannot be made by oxidatively metalating organic halides [235]. Meanwhile, neither hydro (or carbo)-argentation nor halogen–silver exchange have become well established, though isolated examples of borylargentation [236], fluoroargentation [237], and carboargentation [238] have been described. Meanwhile, transmetalation has been limited by the high reactivity of e.g. s-block organometallics towards silver salts [239–241]. This has led to efforts to expand the field of DoM [242] to encompass silver chemistry. To engender development, the argentate analogue of Lipshutz cuprate **159** [213], $(TMP)_2Ag(CN)Li_2(THF)$ **179**, was prepared using AgCN and found to form a comparable, isolable dimer (Figure 1.31). At the same time, variation of the amido component was probed, with $(HMDS)_2Ag(CN)Li_2(THF)$ (**180**, HMDS = hexamethyldisilamide), $(DA)_2Ag(CN)Li_2(THF)$ **181**, and $(Cy_2N)_2Ag(CN)Li_2(THF)$ (**182**, Cy = cyclohexyl) all being formed but only **179** and **182** proving effective in proof-of-concept iodinations of *N,N*-diisopropylbenzamide. Thereafter, **179** was deployed in a range of directed *ortho*-deprotometalations using various DMG and ancillary functional group (FG) permutations (Scheme 1.36). Work demonstrated the ability to use tertiary carboxylic amide DMGs of variable lability (yielding **183–185**). The success of nitrile (giving **186**) and methyl ester (giving **187**) DMGs was remarkable given their incompatibility with e.g. organolithiums. Likewise, in contrast to the difficulty of metalating nitroarenes using traditional strong bases [243], and the fact

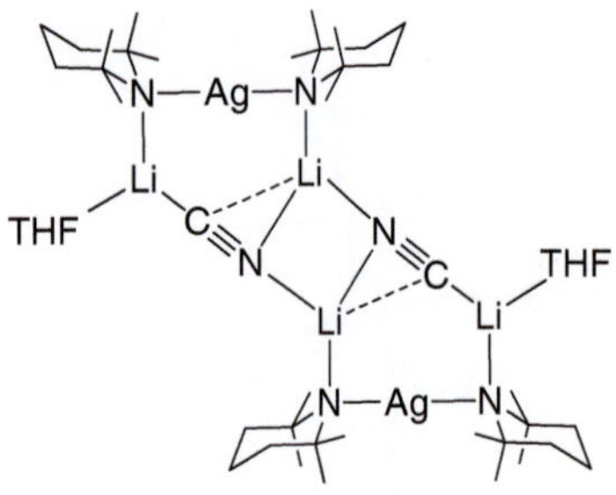

179_2

Figure 1.31 The dimer of lithium argentate $(TMP)_2Ag(CN)Li_2(THF)$ **179**.

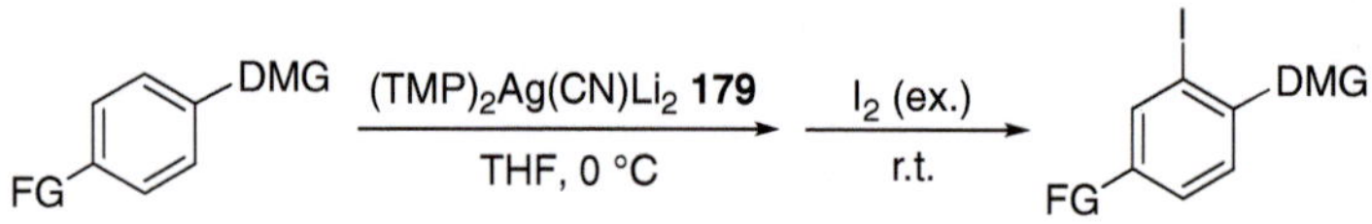

DMG	FG	Product	Yield (%)
C(O)N*i*-Pr$_2$	H	**183**	92
C(O)NEt$_2$	H	**184**	98
C(O)N(CH$_2$CH$_2$)$_2$O	H	**185**	48[a]
CN	H	**186**	95
C(O)OMe	H	**187**	86[a]
NO$_2$	OMe	**188**	80[a]
NO$_2$	NMe$_2$	**189**	81[a]
C(O)N*i*-Pr$_2$	CH=CH$_2$	**190**	86[a]
C(O)N*i*-Pr$_2$	Cl	**191**	95[a,b]
C(O)N*i*-Pr$_2$	OTf	**192**	92[a,c]

Scheme 1.36 Examples of directed deprotometalation using lithium argentate **179**. [a]Argentation at −40 °C. [b]*ortho* : *meta* >29 : 1. [c]*ortho* : *meta* >16 : 1 (Tf = trifluoromethanesulfonyl).

that this has hitherto been achievable using only specific substrates [244], **188** and **189** were now smoothly generated by **179**. Notably, the yield of **188** (80%) contrasted with just 6% obtained when using **159**. Elsewhere (forming **190–192**), ancillary styrene-type FGs could be employed without detectable polymerization, while halides and pseudohalides were tolerated, enabling the demonstration that OTf [(trifluoromethanesulfonyl)oxy] could survive a deprotonative metalation sequence [245].

The compatibility of traditionally problematic DMGs with **179** suggested the formation of *ortho*-argentate intermediates. This was further hinted at by the isolation of aryl argentates such as the crystalline dimer of $[2\text{-}(i\text{-}Pr)_2NC(O)\text{-}C_6H_4]_2AgLi(THF)$ **193** (Figure 1.32), the preliminary characterization of which points to further avenues of investigation in this area.

Having established the accessibility of a range of *ortho*-iodides, the replacement of I_2 with various disulfides was explored. Again, the use of **179** overcame the yield variability and limited FG compatibility encountered when attempting sulfide installation by treating established aryllithium [246, 247], or -magnesium [248] reagents with disulfides. Moreover, this could be combined with the DMG tolerance of **179** in preparing sulfides such as **194** (Scheme 1.37, top). Lastly, the oxidative inertness of silver meant that, unlike the corresponding cuprate, **179** avoided biaryl production when preparing azo compounds from diazonium salts [forming e.g. **195** in good yield (cf. 23% using **159**), Scheme 1.37, bottom]. This reactivity overcame the S_NAr or SET processes typically seen when arylmetals react with diazoniums [249] whilst achieving uniquely heterocoupled products [250].

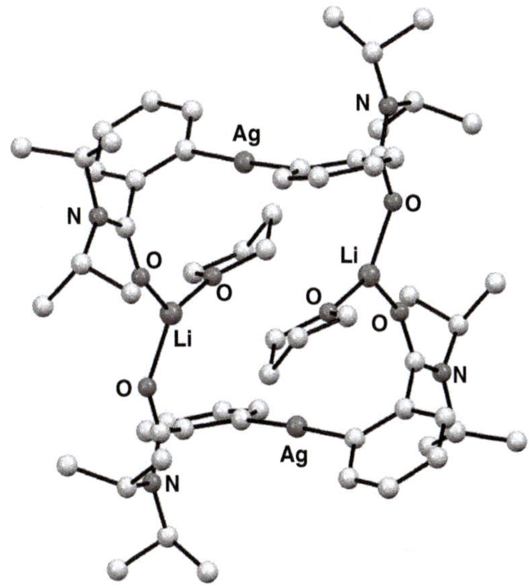

Figure 1.32 Molecular structure of the dimer of $[2\text{-}(i\text{-Pr})_2NC(O)\text{-}C_6H_4]_2AgLi(THF)$ **193**. *Source*: Adapted from Tezuka et al. [245].

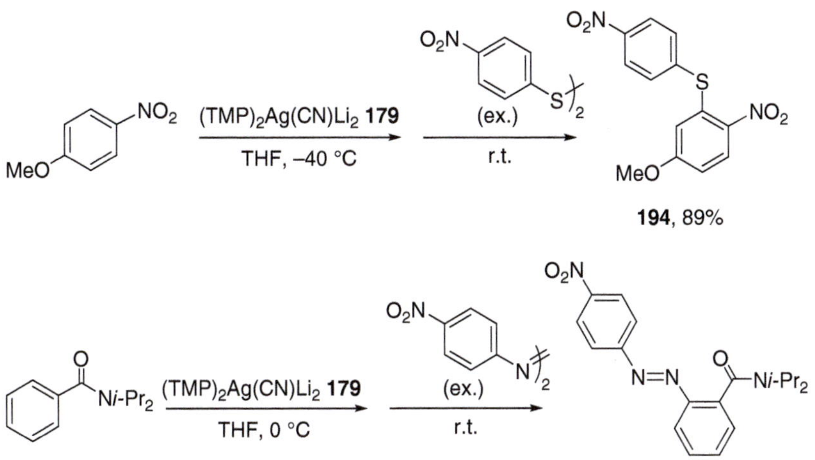

Scheme 1.37 Lithium argentate **179** shows good functional group tolerance when making sulfides while the oxidative inertness of silver avoids biaryl production when making azo compounds.

1.5 Concluding Remarks

Metalation of functionalized (hetero)aromatic compounds using a new breed of polymetallic reagent that operates according to the principles of inter-metal synergy has enabled access to otherwise challenging molecular targets that have previously resisted approach using simpler reagents with unprecedented levels of selectivity. Specifically, metal–halogen exchange and then deprotometalation reactions have been developed to provide access to versatile organometallic species. Whilst these reactions traditionally employed highly polar (often unselective) s-block organometallics (Li and Mg), the emergence of heterobimetallic systems, based on the combination of an alkali metal with a more electronegative metal has circumvented many of the selectivity and reactivity problems that limited the usefulness of monometallic reagents. Magnesiates, zincates, and cuprates have been deployed in metal–halogen exchange with high levels of functional group tolerance and have even proved compatible with solid-phase synthesis. Meanwhile, avoidance of unwanted nucleophilic reactivity by the use of the sterically hindered amido base 2,2,6,6-tetramethylpiperidide (TMP) has transformed deprotometalation chemistry, offering access to magnesium, zinc, aluminium, copper, silver, and cadmium ate complexes, each of which have found their place in (hetero)aromatic derivatization. Crystallographic work has proved vital for understanding the nature of heterobimetallic species, including reaction intermediates, while recent advances in spectroscopy have shed unprecedented light on the behaviour of selected reagents in solution. However, few systems have been probed in such depth yet. This area will no doubt prove essential to providing an understanding that will enable the rational design of the next generation of synergic reagents. These will offer new reactivity and cost-benefit ratios based upon new amide ligands with differing steric properties and potentially inclined to encourage the nonstoichiometric or even catalytic action of synergic bases reported to date in only a very few, selected examples. It will also seek to take advantage of the vast array of bimetallic combinations available to the modern chemist, seeking new and more easily handled alkali metal components – for example predicating ate complex formation on alkali metal amides rather than less stable alkali metal–carbon bonded organometallics – whilst also introducing other possibilities for the less electropositive metal.

References

1 Schlosser, M. (**2005**). *Angew. Chem. Int. Ed. 44*: 376–393.
2 Boudier, A., Bromm, L.O., Lotz, M., and Knochel, P. (**2000**). *Angew. Chem. Int. Ed. 39*: 4414–4435.
3 Knochel, P. and Jones, P. (**1998**). *Organozinc Reagents*. Oxford University Press.
4 Gschwend, H.W. and Rodriguez, H.R. (**1979**). *Org. React. 26*: 1–360.
5 Beak, P. and Snieckus, V. (**1982**). *Acc. Chem. Res. 15*: 306–312.
6 Snieckus, V. (**1990**). *Chem. Rev. 90*: 879–933.
7 Gant, T.G. and Meyers, A.I. (**1994**). *Tetrahedron 50*: 2297–2360.
8 Schlosser, M., (**2002**). *Organometallics in Synthesis*, 2nd ed. (Ed.: M. Schlosser), Chapter 1. New York: Wiley.
9 Clayden, J. (**2002**). *Organolithiums: Selectivity for Synthesis*. Oxford: Pergamon.
10 Gilman, H. and Bebb, R.L. (**1939**). *J. Am. Chem. Soc. 61*: 109–112.
11 Wittig, G., Pieper, G., and Fuhrmann, G. (**1940**). *Ber. Dtsch. Chem. Ges. B 73*: 1193–1197.
12 Upton, C. and Beak, P. (**1975**). *J. Org. Chem. 40*: 1094–1098.
13 Krizan, T. D., Martin, J. C. (**1983**). *J. Am. Chem. Soc. 105*, 6155–6157.
14 Caron, S., Hawkins, J. M. (**1998**). *J. Org. Chem. 63*, 2054–2055.
15 Wheatley, A.E.H. (**2003**). *Eur. J. Inorg. Chem.*: 3291–3303.
16 Clayden, J., Davies, R.P., Hendy, M.A. et al. (**2001**). *Angew. Chem. Int. Ed. 40*: 1238–1240.
17 Clayden, J., Frampton, C.S., McCarthy, C., and Westlund, N. (**1999**). *Tetrahedron 55*: 14161–14184.
18 Beak, P., Kerrick, S.T., and Gallagher, D.J. (**1993**). *J. Am. Chem. Soc. 115*: 10628–10636.
19 Bowles, P., Clayden, J., Helliwell, M. et al. (**1997**). *J. Chem. Soc., Perkin Trans. 1*: 2607–2616.
20 Armstrong, D.R., Boss, S.R., Clayden, J. et al. (**2004**). *Angew. Chem. Int. Ed. 43*: 2135–2138.
21 Wheatley, A.E.H., Clayden, J., Hillier, I.H. et al. (**2012**). *Beilstein J. Org. Chem. 8*: 50–60.
22 Armstrong, D.R., Clayden, J., Haigh, R. et al. (**2003**). *Chem. Commun.*: 1694–1695.
23 Clayden, J., Stimson, C.C., Keenan, M., and Wheatley, A.E.H. (**2004**). *Chem. Commun.*: 228–229.

24 Campbell Smith, A., Donnard, M., Haywood, J. et al. (**2011**). *Chem. Eur. J. 17*: 8078–8084.
25 Fries, K. and Finck, G. (**1908**). *Chem. Ber. 41*: 4271–4284.
26 MacNeil, S.L., Wilson, B.J., and Snieckus, V. (**2006**). *Org. Lett. 8*: 1133–1136.
27 Singh, K.J. and Collum, D.B. (**2006**). *J. Am. Chem. Soc. 128*: 13753–13760.
28 Jastrzebski, J.T.B.H., Arink, A.M., Kleijn, H. et al. (**2013**). *J. Am. Chem. Soc. 135*: 13371–13378.
29 Eaton, P.E., Lee, C.-H., and Xiong, Y. (**1989**). *J. Am. Chem. Soc. 111*: 8016–8018.
30 Kondo, Y., Yoshida, A., and Sakamoto, T. (**1996**). *J. Chem. Soc., Perkin Trans. 1*: 1331–2332.
31 Shilai, M., Kondo, Y., and Sakamoto, T. (**2001**). *J. Chem. Soc., Perkin Trans. 1*: 442–444.
32 Schlecker, W., Huth, A., Ottow, E., and Mulzer, J. (**1995**). *J. Org. Chem. 60*: 8414–8416.
33 Haag, B., Mosrin, M., Ila, H. et al. (**2011**). *Angew. Chem. Int. Ed. 50*: 9794–9824.
34 Sakamoto, S., Imamoto, T., and Yamaguchi, K. (**2001**). *Org. Lett. 3*: 1793–1795.
35 Yang, K.-C., Chang, C.-C., Huang, J.-Y. et al. (**2002**). *J. Organomet. Chem. 648*: 176–187.
36 Kawachi, A., Nagae, S., Onoue, Y. et al. (**2011**). *Chem. Eur. J. 17*: 8005–8008.
37 Davin, L., McLellan, R., Hernán-Gómez, A. et al. (**2017**). *Chem. Commun. 53*: 3653–3656.
38 Bakewell, C., White, A.J.P., and Crimmin, M.R. (**2016**). *J. Am. Chem. Soc. 138*: 12763–12766.
39 Bakewell, C., Ward, B.J., White, A.J.P., and Crimmin, M.R. (**2018**). *Chem. Sci. 9*: 2348–2356.
40 Davin, L., McLellan, R., Kennedy, A.R., and Hevia, E. (**2017**). *Chem. Commun. 53*: 11650–11653.
41 Caubère, P. (**1993**). *Chem. Rev. 93*: 2317–2334.
42 Lochmann, L., Pospíšil, J., and Lím, D. (**1966**). *Tetrahedron Lett. 2*: 257–262.
43 Schlosser, M. (**1967**). *J. Organomet. Chem. 8*: 9–16.
44 Schlosser, M. and Strunk, S. (**1984**). *Tetrahedron Lett. 25*: 741–744.
45 Schlosser, M., Chi, J.H., and Takagishi, S. (**1990**). *Tetrahedron 46*: 5633–5648.
46 Katsoulos, G., Takagishi, S., and Schlosser, M. (**1991**). *Synlett*: 731–732.
47 Schlosser, M. (**1988**). *Pure Appl. Chem. 60*: 1627–1634.
48 Lochmann, L. (**2000**). *Eur. J. Inorg. Chem.*: 1115–1126.
49 Schlosser, M. (**2005**). *Angew. Chem.Int. Ed. 44*: 376–393.
50 Schlosser, M., Choi, J.H., and Takagishi, S. (**1990**). *Tetrahedron 46*: 5633–5648.
51 McGarrity, J.F. and Ogle, C.A. (**1985**). *J. Am. Chem. Soc. 107*: 1805–1810.
52 Marsch, M., Harms, K., Lochmann, L., and Boche, G. (**1990**). *Angew. Chem. Int. Ed. Engl. 29*: 308–309.
53 e.g. Clegg, W., Drummond, A. M., Liddle, S. T., Mulvey, R. E., Roberston, A. (**1999**). *Chem. Commun.* 1569–1570.
54 Mackenzie, F.M., Mulvey, R.E., Clegg, W., and Horsburgh, L. (**1996**). *J. Am. Chem. Soc. 118*: 4721–4722.
55 Kennedy, A.R., MacLellan, J.G., and Mulvey, R.E. (**2001**). *Angew. Chem. Int. Ed. 40*: 3245–3247.
56 Wei, X., Dong, Q., Tong, H. et al. (**2008**). *Angew. Chem. Int. Ed. 47*: 3976–3978.
57 Lochmann, L. and Lím, D. (**1971**). *J. Organomet. Chem. 28*: 153–158.
58 Pi, R., Bauer, W., Brix, B. et al. (**1986**). *J. Organomet. Chem. 306*: C1–C4.
59 Harder, S. and Streitwieser, A. (**1993**). *Angew. Chem. Int. Ed. Engl. 32*: 1066–1068.
60 Unkelbach, C., O'Shea, D.F., and Strohmann, C. (**2014**). *Angew. Chem. Int. Ed. 53*: 553–556.
61 Gau, G. (**1976**). *J. Organomet. Chem. 121*: 1–6.
62 Benrath, P., Kaiser, M., Limbach, T. et al. (**2016**). *Angew. Chem. Int. Ed. 55*: 10886–10889.
63 Jennewein, B., Kimpel, S., Thalheim, D., and Klett, J. (**2018**). *Chem. Eur. J. 24*: 7605–7609.
64 Mulvey, R.E., Mongin, F., Uchiyama, M., and Kondo, Y. (**2007**). *Angew. Chem. Int. Ed. 46*: 3802–3824.
65 Krasovskiy, A., Krasovskaya, V., and Knochel, P. (**2006**). *Angew. Chem., Int. Ed. 45*: 2958–2961.
66 Mosrin, M. and Knochel, P. (**2008**). *Org. Lett. 10*: 2497–2500.

67 Lin, W., Baron, O., and Knochel, P. (**2006**). *Org. Lett. 8*: 5673–5676.
68 Clososki, G.C., Rohbogner, C.J., and Knochel, P. (**2007**). *Angew. Chem. Int. Ed. 46*: 7681–7684.
69 Forbes, G.C., Kennedy, A.R., Mulvey, R.E. et al. (**2001**). *J. Chem. Soc., Dalton Trans.*: 1477–1484.
70 Hevia, E., Gallagher, D.J., Kennedy, A.R. et al. (**2004**). *Chem. Commun.*: 2422–2423.
71 Graham, D.V., Hevia, E., Kennedy, A.R. et al. (**2006**). *Chem. Commun.*: 417–419.
72 Blair, V.L., Kennedy, A.R., Klett, J., and Mulvey, R.E. (**2008**). *Chem. Commun.*: 5426–5428.
73 García-Álvarez, P., Graham, D.V., Hevia, E. et al. (**2008**). *Angew. Chem. Int. Ed. 47*: 8079–8081.
74 Armstrong, D.R., García-Álvarez, P., Kennedy, A.R. et al. (**2010**). *Angew. Chem. Int. Ed. 49*: 3185–3188.
75 Li, D., Keresztes, I., Hopson, R., Williard, P. G. (**2008**). *Acc. Chem. Res. 41*, 270–280.
76 Guang, J., Hopson, R., Williard, P. G. (**2015**). *J. Org. Chem. 80*, 9102–9107.
77 Neufeld, R. and Stalke, D. (**2016**). *Chem. Eur. J. 22*: 12624–12628.
78 Tuckmantel, W., Oshima, K., and Nozaki, H. (**1986**). *Chem. Ber. 119*: 1581–1593.
79 Isobe, M., Kondo, S., Nagasawa, N., and Goto, T. (**1977**). *Chem. Lett.*: 679–682.
80 Kondo, Y., Takazawa, N., Yamazaki, C., and Sakamoto, T. (**1994**). *J. Org. Chem. 59*: 4717–4718.
81 Kondo, Y., Takazawa, N., Yoshida, A., and Sakamoto, T. (**1995**). *J. Chem. Soc., Perkin Trans. 1*: 1207–1208.
82 Kondo, Y., Morey, J.V., Morgan, J.C. et al. (**2007**). *J. Am. Chem. Soc. 129*: 12734–12738.
83 Weiss, E. and Wolfrum, R. (**1968**). *Chem. Ber. 101*: 35–40.
84 Fröhlich, H.-O., Kosan, B., Müller, B., and Hiller, W. (**1992**). *J. Organomet. Chem. 441*: 177–184.
85 Fröhlich, H.-O., Kosan, B., Undeutsch, B., and Görls, H. (**1994**). *J. Organomet. Chem. 472*: 1–14.
86 Armstrong, D.R., Dougan, C., Graham, D.V. et al. (**2008**). *Organometallics 27*: 6063–6070.
87 Uchiyama, M., Koike, M., Kameda, M. et al. (**1996**). *J. Am. Chem. Soc. 118*: 8733–8734.
88 Uchiyama, M., Kameda, M., Mishima, O. et al. (**1998**). *J. Am. Chem. Soc. 120*: 4934–4946.
89 Kondo, Y., Fujinami, M., Uchiyama, M., and Sakamoto, T. (**1997**). *J. Chem. Soc., Perkin Trans. 1*: 799–800.
90 Westerhausen, M., Rademacher, B., Schwarz, W., and Anorg, Z. (**1993**). *Allg. Chem. 619*: 675–689.
91 Wyrwa, R., Fröhlich, H.-O., and Görls, H. (**1996**). *Organometallics 15*: 2833–2835.
92 Rijnberg, E., Jastrzebski, J.T.B.H., Boersma, J. et al. (**1997**). *Organometallics 16*: 2239–2245.
93 Armstrong, D.R., Kennedy, A.R., Mulvey, R.E. et al. (**2012**). *Chem. Sci. 3*: 2700–2707.
94 Armstrong, D.R., Crosbie, E., Hevia, E. et al. (**2014**). *Chem. Sci. 5*: 3031–3045.
95 Armstrong, D.R., Emerson, H.S., Hernán-Gómez, A. et al. (**2014**). *Dalton Trans. 43*: 14229–14238.
96 Robert, A.J., Kennedy, A.R., McLellan, R. et al. (**2016**). *Eur. J. Inorg. Chem.*: 4752–4760.
97 Westerhausen, M., Wieneke, M., Ponikwar, W. et al. (**1998**). *Organometallics 17*: 1438–1441.
98 Kondo, Y., Matsudaira, T., Sato, J. et al. (**1996**). *Angew. Chem. Int. Ed. Engl. 35*: 736–738.
99 Boger, D.L. and Coleman, R.S. (**1988**). *J. Am. Chem. Soc. 110*: 1321–1323.
100 Boger, D.L. and Coleman, R.S. (**1988**). *J. Am. Chem. Soc. 110*: 4796–4807.
101 Kelly, R.C., Gehhard, I., Wicnienski, N. et al. (**1987**). *J. Am. Chem. Soc. 109*: 6837–6838.
102 Boger, D.L. and Machiya, K. (**1992**). *J. Am. Chem. Soc. 114*: 10056–10058.
103 Boger, D.L., Machiya, K., Hertog, D.L. et al. (**1993**). *J. Am. Chem. Soc. 115*: 9025–9036.
104 Muratake, H., Abe, I., and Natsume, M. (**1994**). *Tetrahedron Lett. 35*: 2573–2576.
105 Wehmeyer, G.W. and Rieke, R.D. (**1987**). *J. Org. Chem. 52*: 5056–5057.
106 House, H.O., Koepsell, D.G., and Campbell, W.J. (**1972**). *J. Org. Chem 37*: 1003–1011.
107 Hiyama, T., Yamamoto, H., Nishio, K. et al. (**1979**). *Bull. Chem. Soc. Jpn. 52*: 3632–36e37.
108 van Koten, G. (**2012**). *Organometallics 31*: 7634–7646.
109 Jarvis, J.A.J., Pearce, R., and Lappert, M.F. (**1977**). *J. Chem. Soc., Dalton Trans.*: 999–1003.
110 Noltes, J.G., ten Hoedt, R.W.M., van Koten, G. et al. (**1982**). *J. Organomet. Chem. 225*: 365–376.

111 Schulte, P., Behrens, U., and Anorg, Z. (**2000**). *Allg. Chem. 626*: 1692–1696.
112 Geng, W., Wei, J., Zhang, W.-X., and Xi, Z. (**2014**). *J. Am. Chem. Soc. 136*: 610–613.
113 Liu, L., Zhu, M., Yu, H.-T. et al. (**2018**). *Organometallics 37*: 845–847.
114 Camus, A. and Marsich, N. (**1968**). *J. Organomet. Chem. 14*: 441–446.
115 Cairncross, A. and Shappard, W. (**1971**). *J. Am. Chem. Soc. 93*: 247–248.
116 Cairncross, A., Omura, H., and Sheppard, W.A. (**1971**). *J. Am. Chem. Soc. 93*: 248–249.
117 Sundaraman, A., Lalancette, R.A., Zakharov, L.N. et al. (**2003**). *Organometallics 22*: 3526–3532.
118 van Koten, G., Leusink, A.J., and Noltes, J.G. (**1970**). *J. Chem. Soc. Chem. D, Chem. Commun*: 1107–1108.
119 Bomparola, R., Davies, R.P., Lal, S., and White, A.J.P. (**2012**). *Organometallics 31*: 7877–7883.
120 Lang, H., Jacob, A., and Milde, B. (**2012**). *Organometallics 31*: 7661–7693.
121 Buschbech, R., Low, P.J., and Lang, H. (**2011**). *Coord. Chem. Rev. 255*: 241–272.
122 Gambarotta, S., Floriani, C., Chiesi-Villa, A., and Guastini, C. (**1983**). *J. Chem. Soc., Chem. Commun.*: 1156–1158.
123 Jassen, M., Corsten, M.A., Spek, A.L. et al. (**1996**). *Organometallics 15*: 2810–2820.
124 Molteni, R., Bertermann, R., Edkins, K., and Steffen, A. (**2016**). *Chem. Commun. 52*: 5019–5022.
125 Dempsey, D.F. and Girolami, G.S. (**1988**). *Organometallics 7*: 1208–1213.
126 Jäkle, F. (**2007**). *Dalton Trans.*: 2851–2858.
127 Gilman, H., Jones, R.G., and Woods, L.A. (**1952**). *J. Org. Chem. 17*: 1630–1634.
128 Pearson, R.G. and Gregory, C.G. (**1976**). *J. Am. Chem. Soc. 98*: 4098–4104.
129 Edwards, P.G., Gellert, R.W., Marks, M.W., and Bau, R. (**1982**). *J. Am. Chem. Soc. 104*: 2072–2073.
130 Khan, S., Edwards, P.G., Yuan, H.S.H., and Bau, R. (**1985**). *J. Am. Chem. Soc. 107*: 1682–1684.
131 Hope, H., Oram, D., and Power, P. (**1984**). *J. Am. Chem. Soc. 106*: 1149–1150.
132 Davies, R.P. (**2011**). *Coord. Chem. Rev. 255*: 1226–1251.
133 Lorenzen, P. and Weiss, E. (**1990**). *Angew. Chem. Int. Ed. Engl. 29*: 300–302.
134 Olmstead, M.M. and Power, P.P. (**1990**). *J. Am. Chem. Soc. 112*: 8008–8014.
135 Olmstead, M.M. and Power, P.P. (**1989**). *J. Am. Chem. Soc. 111*: 4135–4136.
136 Liu, L., Zhu, M., Yu, H.-T. et al. (**2017**). *J. Am. Chem. Soc. 139*: 13688–13691.
137 van Koten, G., Jastrzebski, J.T.B.H., Muller, F., and Stam, C.H. (**1985**). *J. Am. Chem. Soc. 107*: 697–698.
138 Guss, J.M., Mason, R., Søtofte, I. et al. (**1972**). *J. Chem. Soc., Chem. Commun.*: 446–447.
139 Davies, R.P., Hornaur, S., and White, A.J.P. (**2007**). *Chem. Commun.*: 304–306.
140 Eaborn, C., Hitchcock, P.B., Smith, J.D., and Sullivan, A.C. (**1984**). *J. Organomet. Chem. 236*: c23–c25.
141 Olmstead, M.M. and Power, P.P. (**1990**). *Organometallics 9*: 1720–1722.
142 Hope, H., Olmstead, M.M., Power, P.P. et al. (**1985**). *J. Am. Chem. Soc. 107*: 4337–4338.
143 John, M., Auel, C., Behrens, C. et al. (**2000**). *Chem. Eur. J. 16*: 3060–3068.
144 Bertz, S.H., Hardin, R.A., Heavey, T.J., and Ogle, C.A. (**2013**). *Angew. Chem. Int. Ed. 52*: 10250–10252.
145 Lipshutz, B., Kozlowski, J.A., and Breneman, C. (**1985**). *J. Am. Chem. Soc. 107*: 3197–3204.
146 Xie, X., Auel, C., Henze, W., and Gschwind, R.M. (**2003**). *J. Am. Chem. Soc. 125*: 1595–1601.
147 Henze, W., Vyater, A., Krause, N., and Gschwind, R.M. (**2005**). *J. Am. Chem. Soc. 127*: 17335–17342.
148 Kronenburg, C.M.P., Amijs, C.H.M., Jastrzebski, J.T.B.H. et al. (**2002**). *Organometallics 21*: 4662–4671.
149 Liphsutz, B.H., Wilhelm, R.S., Kozlowski, J.A., and Parker, D. (**1984**). *J. Org. Chem. 49*: 3928–3938.
150 Lipshutz, B.H., Wilhelm, R.S., and Kozlowski, J.A. (**1984**). *J. Org. Chem. 49*: 3938–3942.

151 Boche, G., Bosold, F., Marsch, M., Harms, K. (**1998**). *Angew. Chem. Int. Ed. 37*, 1684–1686.

152 Hwang, C.-S., Power, P. P. (**1998**). *J. Am. Chem. Soc. 120*, 6409–6410.

153 Eaborn, A., El-Hamruni, S. M., Hill, M. S., Hitchcock, P. B., Smith, J. D. (**2002**). *J. Chem. Soc., Dalton Trans.* 3975–3979.

154 Bertz, S.H. (**1991**). *J. Am. Chem. Soc. 113*: 5470–5471.

155 Lipshutz, B.H., Kozlowski, J.A., and Wilhelm, R.S. (**1984**). *J. Org. Chem. 49*: 3943–3949.

156 Bertz, S.H. (**1990**). *J. Am. Chem. Soc. 112*: 4031–4032.

157 Stemmler, T., Penner-Hahn, J. E., Knochel, P *J. Am. Chem. Soc.* (**1993**). *115*, 348–350.

158 Barnhart, T. M., Huang, H., Penner-Hahn, J. E. (**1995**). *J. Org. Chem. 60*, 4310–4311.

159 Huang, H., Alvarez, K., Lui, Q. et al. (**1996**). *J. Am. Chem. Soc. 118*: 8808–8816.

160 Snyder, J., Spangler, D., Behling, J., and Rossiter, B. (**1994**). *J. Org. Chem. 59*: 2665–2667.

161 Snyder, J. and Bertz, S. (**1995**). *J. Org. Chem. 60*: 4312–4313.

162 Kronenburg, C.M.P., Jastrzebski, J.T.H., Spek, A.L., and van Koten, G. (**1998**). *J. Am. Chem. Soc. 120*: 9968–9688.

163 Boche, G., Bosold, F., Marsch, M., and Harms, K. (**1998**). *Angew. Chem. Int. Ed. 37*: 1684–1686.

164 Bertz, S.H., Chopra, A., Eriksson, M. et al. (**1999**). *Chem. Eur. J. 5*: 2680–2691.

165 Neumeier, N. and Gschwind, R.M. (**2014**). *J. Am. Chem. Soc. 136*: 5765–5772.

166 Wilson, S.R. and Szarnik, A.W. (**1997**). *Combinatorial Chemistry*. John Wiley & Sons.

167 Obrecht, D. and Villalgordo, J.M. (**1998**). *Solid-Supported Combinatorial and Parallel Synthesi of Small-Molecular-Weight Compound Libraries*. Pergamon.

168 Kondo, Y., Komine, T., Fujinami, M. et al. (**1999**). *J. Comb. Chem. 1*: 123–126.

169 Boymond, L., Rottländer, M., Cahiez, G., and Knochel, P. (**1998**). *Angew. Chem. Int. Ed. 37*: 1701–1703.

170 Gregory, K., Schleyer, P.v.R., and Snaith, R. (**1991**). *Adv. Inorg. Chem. 37*: 47–142.

171 Lappert, M., Protchenko, A., Power, P., and Seeber, A. (**2008**). *Metal Amide Chemistry*. Chichester: John Wiley & Sons, Ltd.

172 Armstrong, D. R., Davies, J. E., Davies, R. P., Raithby, P. R., Snaith, R., Wheatley, A. E. H. (**1999**). *New J. Chem. 23*, 35–41.

173 Armstrong, D. R., Davies, R. P., Raithby, P. R., Snaith, R., Wheatley, A. E. H. (**1999**). *New J. Chem. 23*, 499–507.

174 Kondo, Y., Shilai, M., Uchiyama, M., and Sakamoto, T. (**1999**). *J. Am. Chem. Soc. 121*: 3539–3540.

175 Imahori, T., Uchiyama, M., Sakamoto, T., and Kondo, Y. (**2001**). *Chem. Commun.*: 2450–2451.

176 Clarke, A.J., McNamara, S., and Meth-Cohn, O. (**1974**). *Tetrahedron Lett.*: 2373–2376.

177 Clegg, W., Dale, S.H., Hevia, E. et al. (**2006**). *Angew. Chem. Int. Ed. 45*: 2370–2374.

178 Andrikopoulos, P.C., Armstrong, D.R., Barley, H.R.L. et al. (**2005**). *J. Am. Chem. Soc. 127*: 6184–6185.

179 Conway, B., Graham, D.V., Hevia, E. et al. (**2008**). *Chem. Commun.*: 2638–2640.

180 Clegg, W., Conway, B., Graham, D.V. et al. (**2009**). *Chem. Eur. J. 15*: 7074–7082.

181 Clegg, W., Conway, B., Hevia, E. et al. (**2009**). *J. Am. Chem. Soc. 131*: 2375–2384.

182 Armstrong, D.R., Blair, V.L., Clegg, W. et al. (**2010**). *J. Am. Chem. Soc. 132*: 9480–9487.

183 Armstrong, D.R., Clegg, W., Dale, S.H. et al. (**2006**). *Angew. Chem. Int. Ed. 45*: 3775–3778.

184 Armstrong, D.R., Balloch, L., Hevia, E. et al. (**2011**). *Beilstein J. Org. Chem. 7*: 1234–1248.

185 Clegg, W., Dale, S.H., Harrington, R.W. et al. (**2006**). *Angew. Chem. Int. Ed. 45*: 2374–2377.

186 Clegg, W., Conway, B., Hevia, E. et al. (**2009**). *J. Am. Chem. Soc. 131*: 2375–2384.

187 Kondo, Y., Morey, J.V., Morgan, J.M. et al. (**2007**). *J. Am. Chem. Soc. 129*: 12734–12738.

188 Clegg, W., Dale, S.H., Hevia, E. et al. (**2008**). *Angew. Chem. Int. Ed. 47*: 731–734.

189 García, F., McPartlin, M., Morey, J.V. et al. (**2008**). *Eur. J. Org. Chem.*: 644–647.

190 Balloch, L., Kennedy, A.R., Klett, J. et al. (**2010**). *Chem. Commun. 46*: 2319–2321.
191 Balloch, L., Kennedy, A.R., Mulvey, R.E. et al. (**2011**). *Organometallics 30*: 145–152.
192 Seggio, A., Chevallier, F., Vaultier, M., and Mongin, F. (**2007**). *J. Org. Chem. 72*: 6602–6605.
193 Seggio, A., Lannou, M.-I., Chevallier, F. et al. (**2007**). *Chem. Eur. J. 13*: 9982–9989.
194 L'Helgoual'ch, J.-M., Seggio, A., Chevallier, F. et al. (**2008**). *J. Org. Chem. 73*: 177–183.
195 Jones, P.R. and Desio, P.J. (**1978**). *Chem. Rev. 78*: 491–516.
196 O'Brien, P. and Malik, M.A. (**2004**). *Sci. Synth. 3*: 91–131.
197 Wittig, G., Meyer, F.J., and Lange, G. (**1951**). *Liebigs Ann. Chem. 571*: 167–201.
198 L'Helgoual'ch, J.-M., Bentabed-Ababsa, G., Chevallier, F. et al. (**2008**). *Chem. Commun.*: 5375–5377.
199 Snégaroff, K., L'Helgoual'ch, J.-M., Bentabed-Ababsa, G. et al. (**2009**). *Chem. Eur. J. 15*: 10280–10290.
200 Mole, T., Jeffrey, E. A., (**1972**). *Organoaluminum Compounds*, Elsevier: Amsterdam.
201 Negishi, E., (**1976**). *J. Organomet. Chem. Libr. 1*, 93–125.
202 Saito, S., (**2004**). *Aluminum in Organic Synthesis. In Main Group Metals in Organic Synthesis*, Vol. 1, (ed. H. Yamamoto, K. Oshima), Chap. 6, Wiley-VCH: Weinheim.
203 Eisch, J. J., (**1982**). *Comprehensive Organometallic Chemistry*, Vol. 6, (ed. G. Wilkinson, F. G. A. Stone, E. W. Abel), Chap. 6, Pergamon Press: Oxford.
204 Uchiyama, M., Naka, H., Matsumoto, Y., and Ohwada, T. (**2004**). *J. Am. Chem. Soc. 126*: 10526–10527.
205 García-Álvarez, J., Graham, D.V., Kennedy, A.R. et al. (**2006**). *Chem. Commun.*: 3208–3210.
206 Naka, H., Uchiyama, M., Matsumoto, Y. et al. (**2007**). *J. Am. Chem. Soc. 129*: 1921–1930.
207 Mulvey, R.E., Armstrong, D.R., Conway, B. et al. (**2011**). *Inorg. Chem. 50*: 12241–12251.
208 Conway, B., García-Álvarez, P., Kennedy, A.R. et al. (**2010**). *New J. Chem. 34*: 1707–1712.
209 Naka, H., Morey, J.V., Haywood, J. et al. (**2008**). *J. Am. Chem. Soc. 130*: 16193–16200.
210 Conway, B., Crosbie, E., Kennedy, A.R. et al. (**2012**). *Chem. Commun. 48*: 4674–4676.
211 Rossiter, B.E. and Swingle, N.M. (**1992**). *Chem. Rev. 92*: 771–806.
212 Yamanaka, M. and Nakamura, E. (**2005**). *J. Am. Chem. Soc. 127*: 4697–4706.
213 Usui, S., Hashimoto, Y., Morey, J.V. et al. (**2007**). *J. Am. Chem. Soc. 129*: 15102–15103.
214 Tezuka, N., Shimojo, K., Hirano, K. et al. (**2016**). *J. Am. Chem. Soc. 138*: 9166–9171.
215 Nguyen, T.T., Marquise, N., Chevallier, F., and Mongin, F. (**2011**). *Chem. Eur. J. 17*: 10405–10416.
216 Krause, N. and Gerold, A. (**1997**). *Angew. Chem. Int. Ed. Engl. 36*: 186–204.
217 Bertz, S.H. and Dabbagh, G. (**1982**). *J. Chem. Soc., Chem. Commun.*: 1030–1032.
218 Harford, P.J., Peel, A.J., Chevallier, F. et al. (**2014**). *Dalton Trans. 43*: 14181–14203.
219 Davies, R.P., Hornauer, S., and Hitchcock, P.B. (**2007**). *Angew. Chem. Int. Ed. 46*: 5191–5194.
220 Haywood, J., Morey, J.V., Wheatley, A.E.H. et al. (**2009**). *Organometallics 28*: 38–41.
221 Komagawa, S., Usui, S., Haywood, J. et al. (**2012**). *Angew. Chem. Int. Ed. 51*: 12081–12085.
222 Marquise, N., Harford, P.J., Chevallier, F. et al. (**2013**). *Tetrahedron Lett. 54*: 3154–3157.
223 Marquise, N., Harford, P.J., Chevallier, F. et al. (**2013**). *Tetrahedron 69*: 10123–10133.
224 Armstrong, D.R., Garden, J.A., Kennedy, A.R. et al. (**2013**). *Chem. Eur. J. 19*: 13492–13503.
225 Harford, P.J., Peel, A.J., Taylor, J.P. et al. (**2014**). *Chem. Eur. J. 20*: 3908–3912.
226 Peel, A.J., Slaughter, J., and Wheatley, A.E.H. (**2016**). *J. Organomet. Chem. 812*: 259–267.
227 Kronenburg, C.M.P., Jastrzebski, J.T.H., Boersma, J. et al. (**2002**). *J. Am. Chem. Soc. 124*: 11675–11683.
228 Peel, A.J., Hedidi, M., Bentabed-Ababsa, G. et al. (**2016**). *Dalton Trans. 45*: 6094–6104.
229 Peel, A.J., Ackroyd, R., and Wheatley, A.E.H. (**2017**). *Chem. Sci. 8*: 4904–4916.
230 Peel, A.J., Tezuka, N., D'Rozario, J.M. et al. (**2019**). *Chem. Sci. 10*: 3385–3400.
231 Yoshikai, N., Nakamura, E. (**2012**). *Chem. Rev. 112*, 2339–2372.

232 Shimizu, Y., Kanai, M. (**2014**). *Tetrahedron Lett.* *55*, 3727–3737.
233 Yoshida, H. (**2016**). *ACS Catal.* *6*, 1799–1811.
234 Vanýsek, P., (**2005**). in *CRC Handbook of Chemistry and Physics*, (Ed. D. W. H. Lide), (8-20)–(8-29), 89th Edn. Boca Raton: CRC Press.
235 Rieke, R.D., Stack, D.E., Dawson, B.T., and Wu, T.-C. (**1993**). *J. Org. Chem.* *58*: 2483–2491.
236 Yoshida, H., Kageyuki, I., and Takaki, K. (**2014**). *Org. Lett.* *16*: 3512–3515.
237 Tyrra, W. and Naumann, D. (**2004**). *J. Fluor. Chem.* *125*: 823–830.
238 Kleijn, H., Tigchelaar, M., Meijer, J., and Vermeer, P. (**1981**). *Recl. Des Trav. Chim. Des Pays-Bas* *100*: 337–341.
239 Martínez de Salinas, S., Mudarra, Á.L., Benet-Buchholz, J. et al. (**2018**). *Chem. Eur. J.* *24*: 11895–11898.
240 Joven-Sancho, D., Baya, M., Martín, A., and Menjón, B. (**2018**). *Chem. Eur. J.* *24*: 13098–13101.
241 Weske, S., Hardin, R.A., Auth, T. et al. (**2018**). *Chem. Commun.* *54*: 5086–5089.
242 Chevallier, F., Mongin, F., Takita, R., and Uchiyama, M. (**2015**). *Arene Chemistry* (ed. J. Mortier), 777–812. Hoboken, New Jersey: John Wiley & Sons, Inc.
243 Bartoli, G., Dalpozzo, R., and Nardi, M. (**2014**). *Chem. Soc. Rev.* *43*: 4728–4750.
244 Wunderlich, S.H. and Knochel, P. (**2007**). *Angew. Chem. Int. Ed.* *46*: 7685–7688.
245 Tezuka, N., Hirano, K., Peel, A.J. et al. (**2020**). *Chem. Sci.* *11*: 1855–1861.
246 Figuly, G.D. and Martin, J.C. (**1980**). *J. Org. Chem.* *45*: 3728–3729.
247 Muchowski, J.M. and Venuti, M.C. (**1980**). *J. Org. Chem.* *45*: 4798–4801.
248 Du, B.-X., Quan, Z.-J., Da, Y.-X. et al. (**2015**). *Adv. Synth. Catal.* *357*: 1270–1276.
249 Zollinger, H. (ed.) (**1994**). *Diazo Chemistry I.* Weinheim: VCH Verlagsgesellschaft mbH.
250 Takeda, Y., Okumura, S., and Minakata, S. (**2013**). *Synthesis* *45*: 1029–1033.

2

Structural Evidence for Synergistic Bimetallic Main Group Bases

Robert E. Mulvey and Stuart D. Robertson

WestCHEM, Department of Pure and Applied Chemistry, University of Strathclyde, Glasgow, UK

2.1 General Introduction

The conversion of a carbon–hydrogen bond into a carbon–carbon or carbon–heteroatom bond is a fundamental chemical reaction of utmost importance in the synthesis of value-added chemicals such as pharmaceuticals and agrochemicals from cheaper starting materials. Accordingly, the chemical community is constantly striving to refine existing methodologies or to develop new methodologies to make such reactions easier and more economical. Selectivity is a key consideration in the development of such a process as most organic molecules contain many different carbon–hydrogen bonds, so the targeted C–H bond must be broken exclusively when the starting material presents more than one such possibility. A second consideration is functional group tolerance, whereby the reagent being used to effect the desired transformation does not also react with other functional groups present at alternative regions of the substrate.

One of the best ways to facilitate these transformations is to initially convert the relatively inert low polarity C–H bond into a more reactive higher polarity carbon–metal bond which is then primed to react onwards to ultimately provide the desired product. This has historically been achieved by a polar organometallic reagent which can broadly be defined as a highly electropositive main group metal coupled with anionic ligand(s) based on carbon, nitrogen, or oxygen (most commonly, alkyl/aryl, amido, and alkoxide, respectively). While there will be variations depending on the exact groups bound to the formally anionic atom, it can be generalized that carbanions are the strongest Brønsted bases (that is have the highest pK_a values and are the 'best' deprotometallating agents), followed by amides and then alkoxides. It is therefore logical that the type of inert C–H bond which is being targeted for conversion to a C–M bond requires a strong base, and so carbon-based reagents as the most reactive are also the most common of the organometallic reagents for general synthetic purposes.

At the vanguard of the polar organometallic reagents lie the alkali metals (group 1), which are so synthetically useful due to the highly electropositive nature of the metal, which confers high polarity and thus reactivity upon the resulting M–C (or M–N/M–O) bond [1, 2]. However, this high reactivity brings with it potential problems, with the possibility of nucleophilic addition of organoalkali metal reagents to unsaturated functional groups (e.g. C=O, C=C) occurring preferentially over the desired metallation reaction. Operating at sub-ambient temperatures can moderate the side-reactivity of these highly reactive reagents. Group 1 organometallic reagents themselves are

Polar Organometallic Reagents: Synthesis, Structure, Properties and Applications, First Edition. Edited by Andrew E. H. Wheatley and Masanobu Uchiyama.

dominated by their lightest member, lithium. This can be attributed in large part to their superior solubility in organic solvents compared to that of their heavier sodium or potassium congeners as well as their slightly diminished reactivity. Having more ionic character, organosodium and organopotassium reagents are more prone to attack certain solvents or additives, while they can even react with themselves (autodestruction) [3–5]. Consequently, organolithium reagents have dominated the landscape of metallation chemistry for 100 years since their discovery by Wilhelm Schlenk and Johanna Holtz [6]. It is thus somewhat ironic that Schlenk was nominated but never received the Nobel Prize for Chemistry due to the perception at the time that organolithium reagents would never prove useful [7]! Seemingly, minor modifications to an organolithium reagent can result in small but significant changes to structure and thus reactivity. For example, the weakly electron donating properties of alkyl groups means that the basicity increases with the number of alkyl substituents attached to the anionic carbon atom. This is demonstrated by the pK_a values for MeH (48), *n*-BuH (50), *s*-BuH (51), *t*-BuH (53); that is with 0, 1, 2, and 3 alkyl groups bound to the relevant carbon centre, respectively. Care must be taken here however as the basicity of the resulting carbanion will be highly dependent on a variety of other factors such as the identity of the metal to which it is bound, the aggregation state, and the presence of any aggregation altering Lewis donor molecules.

Moving away from group 1, it is found that groups 2 [8] or 13 [9] organometallic reagents are also capable of participating in direct metal-hydrogen exchange, although they require carefully selected ligands and/or substrates which contain acidic hydrogen atoms or are pre-activated towards this reactivity, although this is not the predominant use of such organometallic compounds. The dominant metals in this context are magnesium and aluminium respectively while the group 12 metal zinc also plays an important role. Whilst not strictly a main group metal, it is appropriate to consider zinc alongside the main group as it is dominated by the +2 oxidation-state with a full d-shell and thus displays many similarities. These formally greater positively charged metals require more than one anionic ligand (which could be the same or different) to satisfy their valency requirements and consequently this, combined with electronic factors, causes noticeable differences in the structures that they adopt. Their diminished actual electropositivity with respect to the alkali–metals results in their M–C, M–N and M–O bonds displaying more covalent character, which leads to their diminished reactivity. This does bring with it the advantage of increased selectivity, since these reagents are much less likely to engage in the undesirable side reactions described earlier and are considerably more tolerant of sensitive functional groups. The diminished reactivity can be enhanced through utilization at elevated temperatures if necessary.

An alternative methodology for metal–hydrogen exchange is for two organometallic reagents to work together [10]. This approach typically brings together one of each class of compound mentioned above, that is an organo-alkali–metal reagent with a multivalent post-alkali–metal-based reagent. In the best cases, the resulting bimetallic complex will display the most desirable aspects of each component part, namely, combining the high reactivity of the group 1 component with the high selectivity of the second component. The advantage here is that many of these mixed-metal compounds can operate at ambient temperature which is a considerable economic advantage especially in large-scale processes. This best-of-both-worlds approach can be considered as using a less reactive metal to tame the reactive alkali metal or perhaps more appropriately exporting the high reactivity of the alkali metal onto the selective but sluggish secondary metal, since structural evidence shows that it is in fact the secondary metal which almost invariably takes the place of the hydrogen (Section 2.3).

A second significant advantage of these bimetallic complexes is that they can also display alternative selectivity patterns whereby they will react with a C–H bond not considered to be the most susceptible to attack by a conventional organometallic reagent. This opens up new avenues in the synthesis of value-added chemicals since it increases the diversity of targets through alternative reaction pathways.

Early understanding of the reactivity of organometallic species came from analysis of organic products after a thorough work-up rather than through identification of metallated intermediates. One of the keys to understanding reactivity and selectivity of mono- or bimetallic reagents is to have knowledge of their chemical structure, either in the solid state (typically through the determination of molecular structure through the technique of single-crystal X-ray diffraction) or in the solution state (which can be approached in a variety of ways of which nuclear magnetic resonance spectroscopy is the most popular). The following sections will now look at the structural evidence for these synergic main group bases. To fully grasp the implications of structure on reactivity it is important to first consider the most important homometallic bases.

2.2 Homometallic Bases

A wealth of information has been gathered on homometallic bases and these have been reviewed in many publications which can be referred to for greater detail [11–18]. However, due to the importance of alkali-metal reagents in the field of bimetallic main group bases some brief key aspects of structure and reactivity of the most common examples will be touched on briefly here as a prelude to the commentary on bimetallic complexes. For brevity, we limit discussion here to the common alkali-metals lithium, sodium, and potassium. Lithium compounds dominate this field.

2.2.1 Carbanionic Lithium Reagents

Alkyl- and aryllithium reagents are often referred to by their empirical abbreviation RLi although in reality they exist as aggregated molecules $(RLi)_n$ with the degree of aggregation n influenced by the steric profiles of the carbanion. These aggregates form as a result of electrostatic interactions between oppositely charged Li and C atoms which can be partially or fully disrupted by competing interactions with solvents or Lewis donating additives, leading to structural (and consequently reactivity) alterations. For lithium in particular, unsolvated (donor-free) aggregates are often discrete oligomers (the most common types are shown in Figure 2.1) which has a direct bearing on their reactivity by permitting solubility in non-polar, hydrocarbon solvents.

The relationship between solvation and reactivity has long been recognized, for example Morton proposed in 1944 that a Lewis donor–Lewis acceptor adduct between triphenylamine and

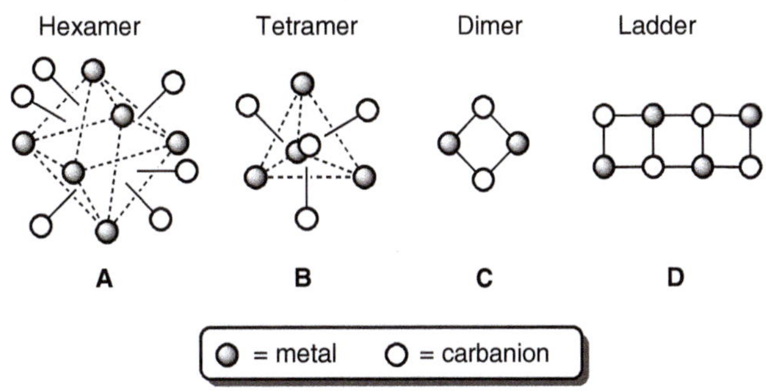

Figure 2.1 Common oligomeric structures adopted by the organolithium reagents.

butyllithium preluded the *meta*-deprotonation of the amine aromatic ring [19]. There is no set rule on the aggregation state of alkyllithium compounds in the presence of a certain Lewis donor, with electronic and steric factors both contributing. That notwithstanding, it is often the case the presence of a Lewis donor will lead to a lower aggregation state and that lower aggregation states are nearly always more reactive than higher aggregation states for a given alkyllithium species. Consequently, it is important to consider both the absence and the presence of such structure-disrupting additives to gain a fuller insight into the structure–reactivity relationship of these commonly used reagents.

As mentioned, organolithium reagents with simple alkyl or aryl ligands are typically soluble in nonpolar solvents such as pentane or hexane. Common exceptions which are not hydrocarbon soluble are the simplest alkyllithium, methyllithium **1** and the simplest aryllithium, phenyllithium **2**, with their lack of solubility attributable to their polymeric structures. The central feature of MeLi is a tetrahedral Li_4 core with the four (Li_3) faces each capped by a carbanion (type **B**) [20]. The lack of steric bulk around this carbanion facilitates longer range interactions between a methyl group and a lithium cation on an adjacent tetramer, propagating a 'polymer of tetramers' type structure. This then propagates into a three-dimensional body-centred cubic network of tetrameric units and is the underlying reason for the insolubility in non-donating hydrocarbon solvents [21]. Phenyllithium adopts a lower primary aggregate, forming dimeric **C** type $[LiC]_2$ units which propagate laterally into an infinite stepladder through interactions of the π-system of the aromatic group with the lithium atom of an adjacent dimeric unit (Figure 2.2) [22]. It is a classic example of the ring-laddering principle that was postulated by Snaith [23–26] prior to the elaboration of this structure and applicable to many other facets of structural chemistry [27, 28].

Many common Lewis donors will coordinate through their lone pairs to the lithium centres of MeLi but without actually disrupting its tetrahedral core. This is the case with the monodentate oxygen donors THF or Et_2O and the potentially bidentate donor TMEDA (*N,N,N′,N′*-tetramethylethylenediamine). Specifically, each Li atom is coordinated by a molecule of THF [29] or Et_2O [30], which disrupts the inter-tetrameric interactions resulting in a discrete tetrameric structure of formula $[MeLi(donor)]_4$ (donor = THF, **3**; Et_2O, **4**). Rather than chelate a single Li atom via both of its N atoms, here TMEDA rather mimics the ethers by binding via one N atom, with each molecule of TMEDA bridging between two tetramers to give a Lewis donor propagated polymer of tetramers with a formal formula of $[(MeLi)_4(TMEDA)_2]$ **5** [31]. The inflexibility of TMCDA (*N,N,N′,N′*-tetramethyl-1,2-diaminocyclohexane) enforced by the cyclohexyl backbone means that it cannot twist like TMEDA to bridge between metal centres, consequently it cleaves the Li_4 tetramer to give a dimeric $[MeLi(TMCDA)]_2$ complex (**6**) with a four-coordinate distorted tetrahedral lithium centre and a distorted central $[CLi]_2$ core (Scheme 2.1) [32].

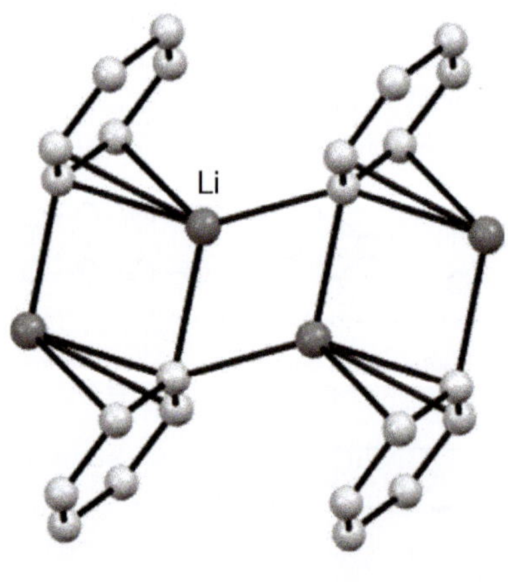

Figure 2.2 Part of the infinite solid-state ladder structure of PhLi **2**. *Source*: Adapted from Dinnebier et al. [22].

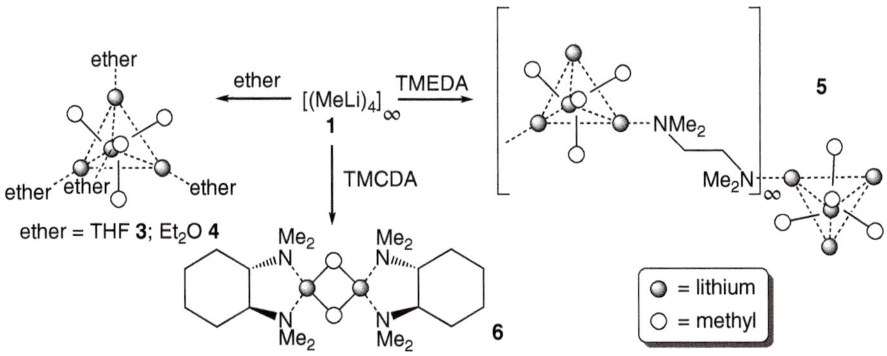

Scheme 2.1 Disruption of the polymerization of MeLi tetramers **1** by common Lewis donors to give complexes **3–6**.

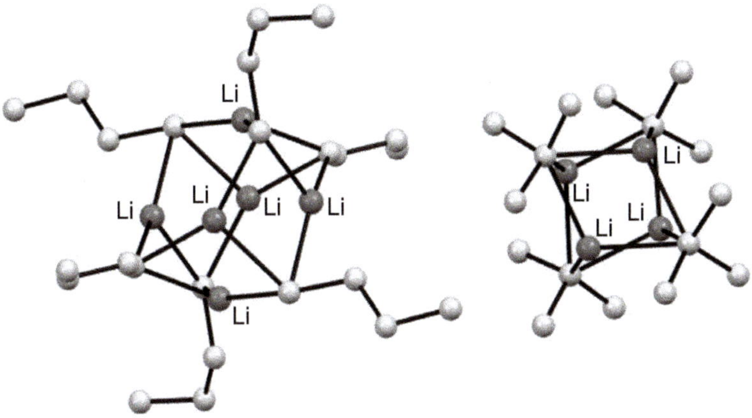

Figure 2.3 Molecular structures of *n*-BuLi **7** (left) and *t*-BuLi **8** (right). *Source*: Adapted from Kottke and Stalke [40].

The central dimeric type **C** core of phenyllithium is typically maintained in the presence of donors, with the donor ligand simply replacing the adjacent phenyl ring as the lithium-solvating group due to the stronger σ-interaction it can provide. This is indeed the case for tetrahydropyran (two equivalents per lithium) [33], bidentate TMCDA (one equivalent per lithium) [34], and cyclic tridentate TACN (1,4,7-triazacyclononane, one equivalent per lithium) [35]. Interestingly, this final example results in the unusually high-coordination number of 5 for lithium. When the acyclic tridentate donor PMDETA (*N,N,N′,N″,N″*-pentamethyldiethylenetriamine) is used, the dimeric core is broken and a monomeric PhLi unit is stabilised, with a more conventional 4-coordinate lithium [36]. Unusually, for organolithium complexes a larger oligomer, specifically a type **B** tetramer, occurs upon solvation with diethyl ether [37] or the thioether dimethyl sulfide [38]. However, NMR studies on phenyllithium suggest that dimeric and tetrameric structures coexist in diethyl ether solution [39].

Oligomeric, hydrocarbon soluble alkyllithium reagents are dominated by butyllithium. Structural differences between *n*- and *t*-butyllithium are often attributed as influencing the difference in their deprotonating power. The less bulky *n*-isomer adopts a hexameric structure in **7**; while the bulk of the *t*-butyl group enforces the same tetrameric structure seen for methyllithium, but with the steric bulk preventing the intermolecular polymerizing interactions in **8** (Figure 2.3) [40]. The six lithium atoms of *n*-butyllithium lie at the points of an octahedron with the butyl groups again capping the triangular Li_3 faces. However, there are eight such faces and only six butyl groups meaning that two of the faces are uncapped. These are diametrically opposed with their Li···Li separation distances elongated as a result of the absence of the carbanion, which would otherwise hold them in closer proximity through electrostatic interactions, allowing repulsive cation–cation forces to have a greater influence.

Despite the close proximity of the lithium centres to one another, NMR studies on ^{6}Li enriched *t*-BuLi demonstrated no ^{6}Li–^{7}Li scalar coupling and thus concluded that there is no lithium–lithium bonding [41].

In the presence of THF, *n*-BuLi is deaggregated from hexameric to tetrameric, with the structure resembling that of THF-solvated methyllithium in [*n*-BuLi(THF)]$_4$ **9** [42]. Over time, however, *n*-BuLi will attack THF by metallating α- to the heteroatom which then results in cleavage of the ring into ethene and the lithium enolate of acetaldehyde via a [3 + 2] cycloreversion [43].

The same tetrameric core is also witnessed with DME (dimethoxyethane), with the polyether coordinating via only a solitary oxygen atom but bridging between adjacent tetrameric units giving an empirical formula of [(*n*-BuLi)$_4$(DME)$_2$] **10** [42]. With regard to poly-*N*-donors, TMEDA also bridges between tetrameric units in [(*n*-BuLi)$_4$(TMEDA)]$_\infty$ **11**, however, the greater steric profile of the *n*-Bu group in comparison to methyl group means that this time only half of the lithium atoms are solvated to give two distinct lithium environments; one which is formally three-coordinate to α–C atom of three butyl arms while the other is four-coordinate due to the additional TMEDA *N*-interaction, resulting in a 1-dimensional zig-zag polymer [42, 44]. It transpires that this structure is stoichiometry dependent, with an excess of TMEDA in hexane solution resulting instead in the crystallization of the discrete dimeric complex [*n*-BuLi(TMEDA)]$_2$ **12** which displays the familiar folding of the central [CLi]$_2$ ring from planarity [42]. Likewise, the complexes formed by the chiral Lewis base TMCDA displays a stoichiometry dependency [45], with a deficit resulting in an end-capped ladder motif in **13** (type **D**) which can be considered as the joining together of two hemi-solvated dimers and results in different coordination geometries; 3 or 4 coordinate lithium centres and 2 or 3 coordinate butyl anions. Excess TMCDA results in a similar solvated dimer **14** to that seen with its achiral analogue TMEDA. Finally, PMDETA gives a similar ladder structure **15** but the tridentate nature of this ligand forces two of the edges of the ladder to break, allowing the solvated lithium atoms to maintain a coordination number of 4 but consequently leaving all *n*-butyl groups to bridge between two lithium atoms only (Scheme 2.2) [46]. This structure was considered an intermediate *en route* to a PMDETA solvated monomer of *n*-BuLi, with monomerization enhancing the reactivity of the organolithium such that it deprotometallates one terminal N–Me group of PMDETA to generate [$Me_2NCH_2CH_2N(Me)CH_2CH_2N(Me)CH_2Li$]$_2$ **16** (Scheme 2.3).

Diethyl ether does cleave the tetrameric unit in *t*-BuLi to give a dimeric four-membered [CLi]$_2$ central unit, with the steric bulk of the *t*-butyl group permitting only a single ether molecule to coordinate to each lithium for a total coordination number of 3 in [*t*-BuLi(OEt_2)]$_2$ **17** [40]. Despite this lower coordination number than in the TMCDA coordinated MeLi dimer [MeLi(TMCDA)]$_2$ **18**, the distortion to the central ring is actually greater as shown in Figure 2.4.

The aggregation state influence on reactivity is clearly demonstrated through the reaction of *n*-butyllithium with the aromatic species benzene and toluene. This champion alkyllithium reagent is essentially unreactive when stirred in either solvent but the addition of DABCO (1,4-diazabicyclo[2.2.2]octane) to a benzene solution, or TMEDA to a toluene solution, results in deprotonation of the aromatic ring, or the methyl arm respectively to give phenyl- and benzyl-lithium (**19** and **20**, Scheme 2.4) [47].

Lewis donors do not only have the capacity to enhance deprotonation reactions, they can also influence the type of reactivity witnessed. This is demonstrated by the reaction of 2-methylpyridine (2-picoline) with butyllithium; in hydrocarbon solution the highest pK_a isomer *t*-BuLi **8** adds

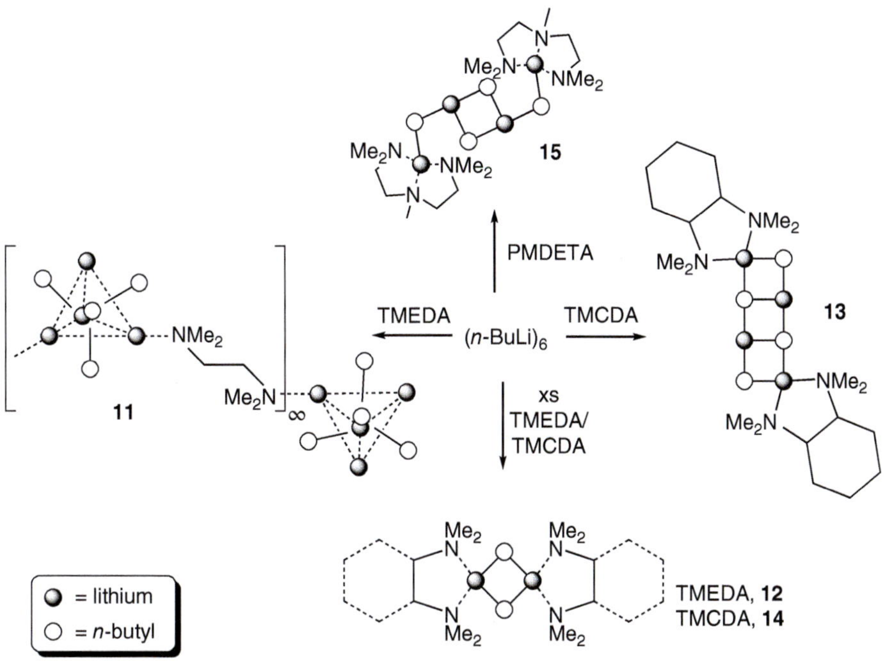

Scheme 2.2 Disruption of the n-BuLi hexamer by polydentate Lewis donors to yield complexes **11–15** including stoichiometry dependence of the TMEDA and TMCDA complexes.

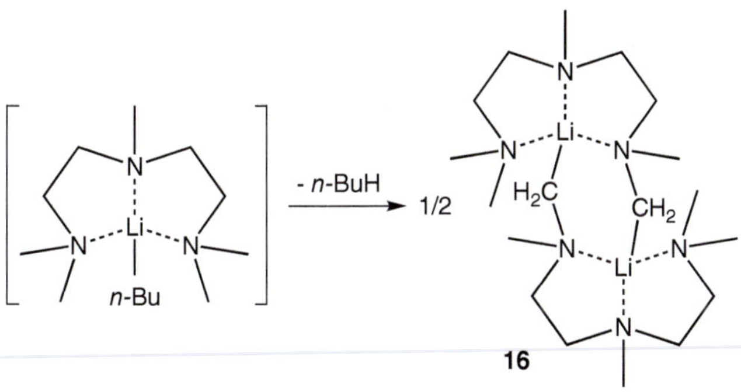

Scheme 2.3 Deprotonation of PMDETA by coordinated n-BuLi producing **16**.

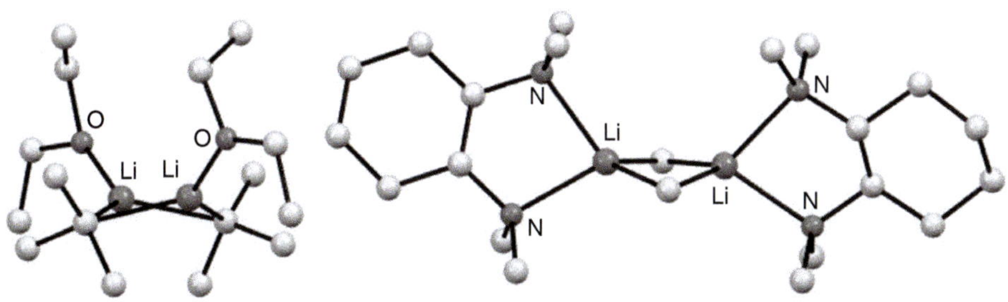

Figure 2.4 Distortion to the central Li_2C_2 rings of dimeric t-BuLi and MeLi. *Sources*: Adapted from Kottke and Stalke [40]; Strohmann and Gessner [32].

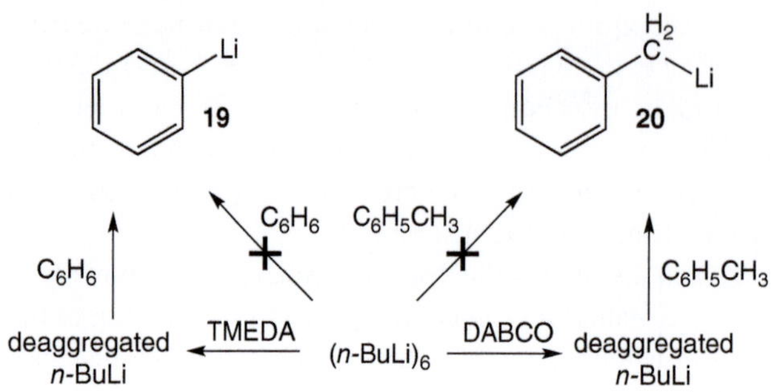

Scheme 2.4 Donor dependence of n-BuLi reactivity towards benzene and toluene.

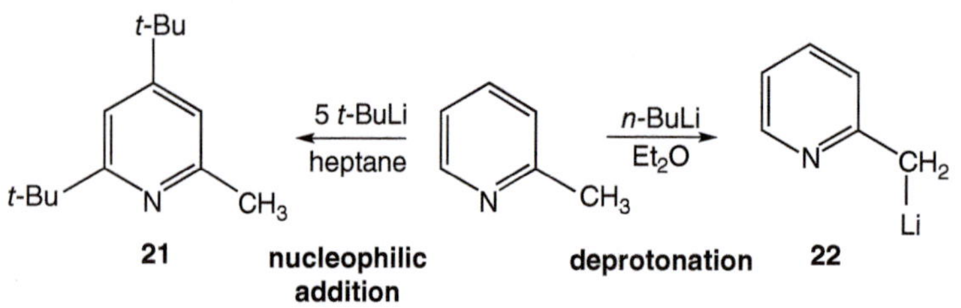

Scheme 2.5 Contrasting solvent-dependent reactivity of BuLi with the heterocycle picoline.

nucleophilically to give a twofold *t*-butyl substituted picoline **21** [48] whereas alternatively deprotonation of the methyl arm results (**22**) when applying *n*-BuLi **7** in the Lewis donor solvent diethyl ether (Scheme 2.5) [49].

2.2.2 Heavier Carbanionic Alkali Metal Reagents

In comparison to the aforementioned organolithium reagents, much less is known structurally regarding organosodium and organopotassium reagents on account of their more common insolubility in hydrocarbon solvents and their often overly aggressive reactivity with the Lewis-donating solvents or additives, which would be required to initially solubilize them for crystal growth. With a larger electronegativity difference between the heavier alkali metals and carbon (Li, 0.98; Na, 0.93; K, 0.82; C, 2.55), these compounds are more polar and thus more reactive on account of the greater carbanionic character of their anionic moieties. Consequently, given the huge popularity and applicability of the more soluble organolithium compounds discussed in Section 2.1 it is perhaps understandable that less information is known about these heavier derivatives. A rare example of a solvent-free alkylsodium structure is methylsodium **23**, which was determined via powder X-ray diffraction [50]. The structure was determined to be isostructural to that of methyllithium—a polymer of tetramers formed via long-range Me–Na interactions. However, the synthetic method used was metathesis of MeLi by adding NaO*t*-Bu, which left residual MeLi (in solvent-dependent amounts varying from ~3 to 25%) that could occupy the relatively larger cavities formed between tetramers due to the longer Me–Na interactions. Modification of the synthetic method so that MeLi was added to NaO*t*-Bu gave pure methylsodium (note deuterated methyllithium was used so that neutron diffraction characterization could be employed) since there was no residual MeLi for the newly formed MeNa to template around [51].

The presence of trimethylsilyl groups within alkylsodium and alkylpotassium reagents has been advantageous for facilitating their stability and thus enabling their structural characterization, emphasized by the carbanions CH_2SiMe_3 and $CH(SiMe_3)_2$. The sodium salt of the former (**24**) adopts a central four-membered tetrahedral (type **B**) core with the triangular faces capped by CH_2SiMe_3 ligands [52]. The presence of silicon induces a slight $Si\delta^+$–$C\delta^-$ polarity in the trimethylsilyl groups resulting in two of these groups forming interactions with neighbouring sodium tetrahedra to propagate a polymeric chain (Figure 2.5). Bulkier $NaCH(SiMe_3)_2$ **25** also forms a polymer but the increased bulk inhibits the ability to form a tetrahedral core and so it rather consists of alternating chains of sodium cations and bis(trimethylsilyl)methyl anions, resulting in formally two-coordinate (bent) sodium atoms although long range interactions with these δ- methyl groups also play a role in stabilization of the structure [53].

These polymers are influenced by TMEDA in distinctly different ways. $NaCH_2SiMe_3$ converts to a chain polymer in **26** of alternating Na and C atoms with bidentate TMEDA coordinating to the metal centres in place of any long range Na···Me interactions [54], while $NaCH(SiMe_3)_2$ is deaggregated to an unusual discrete dinuclear species **27** with each sodium centre coordinated by a bidentate TMEDA molecule and a further TMEDA molecule bridging between the two sodium atoms to complete their coordination sphere (Scheme 2.6) [55].

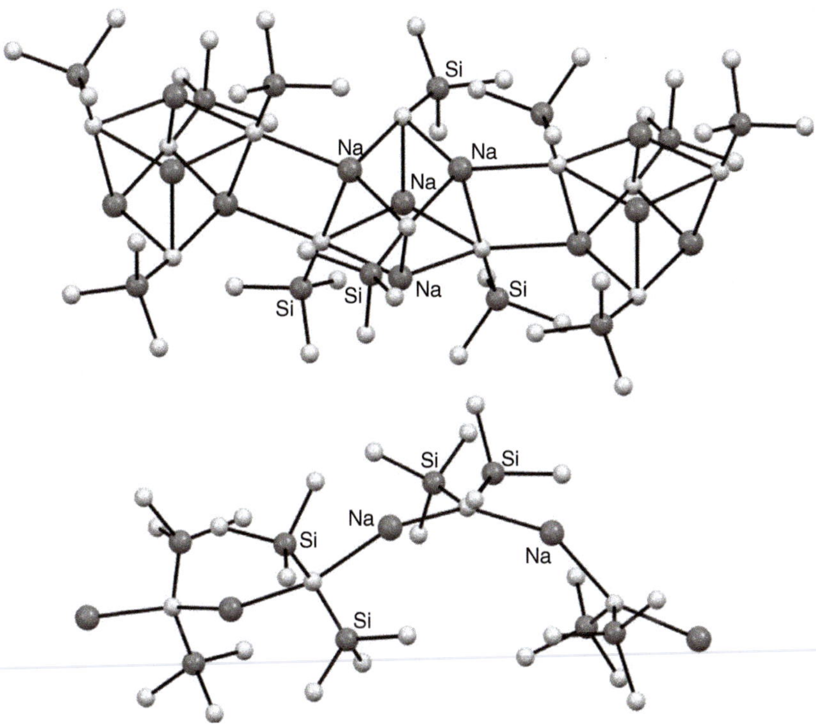

Figure 2.5 Polymeric structures of trimethylsilylmethylsodium **24** (top) and bis(trimethylsilyl) methylsodium **25** (bottom). *Sources*: Adapted from Baillie et al. [52]; Hitchcock et al. [53].

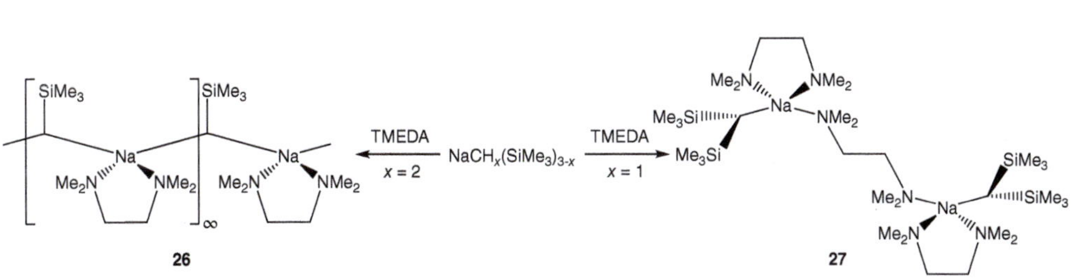

26 **27**

Scheme 2.6 Effect of TMEDA on structures of trimethylsilylmethylsodium and bis(trimethylsilyl)-methylsodium.

Trimethylsilylmethylpotassium **28** adopts a chain polymer structure when solvated by tridentate PMDETA **29**, mirroring the structure of the TMEDA solvated sodium derivative, with the larger potassium centre benefiting from the greater number of donor atoms to gain a coordination number of 5 [54]. TMEDA enforces an alternative structure on KCH_2SiMe_3, where the metals and anions form the type B tetrahedral structure of **30**. Three of the four potassium atoms are solvated by TMEDA in a bidentate manner with the fourth rather coordinated by the trimethylsilyl unit of an adjacent tetramer, giving a donor-deficient polymer of tetramers. The larger steric bulk in $KCH(SiMe_3)_2$ means that THF solvates a polymeric chain structure in **31** but with only one donor molecule per metal for a coordination number of 3 [56]. Addition of PMDETA results in the 'broken ladder' type structure of **32** witnessed previously with PMDETA solvated *n*-BuLi (Section 2.1) [57].

Phenylsodium **2** was shown by Weiss to adopt a cyclodimeric (type **C**) structure when solvated by PMDETA in **33** [58]. This is in contrast to monomeric PhLi(PMDETA) **34** and reflects the propensity of sodium to adopt larger coordination numbers, in this case 5 as opposed to 4 for lithium.

2.2.3 Alkali Metal Amides

Derived from secondary amines, the alkali metal amides (MNR_2) constitute the other main collection of commonly used metallating agents. The key difference versus carbanionic alkali metal reagents is that they are non-nucleophilic, meaning that they tolerate a wider range of substrates and functional groups. However, they tend to have a lower pK_a (indeed they are typically prepared by deprotonating the parent secondary amine with a stronger organoalkali metal base) although in reality they display strong enough Brønsted basicity to be suitable for most common deprotonation reactions.

There are a huge range of secondary amides which would be appropriate for the task at hand, with the preference normally being simply for a parent amine which is cheap, leading to the prevalence of both diisopropylamide (DA) and 1,1,1,3,3,3-hexamethyldisilazide (HMDS). A third amide, 2,2,6,6-tetramethylpiperidide (TMP) also finds sustained use as, despite its greater cost, it has a noticeable increase in steric bulk (Figure 2.6) which gives it the advantage of being a stronger base (i.e. it possesses a higher pK_a). This is due to the presence of the ring which then prevents the flanking methyl groups from relaxing away from the M–N bond as would be the case in untethered diisopropylamide. Their structural chemistry has been addressed in a variety of publications and so only the most pertinent examples will be looked at here. Like that of the carbanionic alkali metal reagents, this structural chemistry has been dominated by lithium due to their increased solubility and less aggressive nature towards certain solvents.

As for the alkyl/aryl alkali metal reagents, it is traditional to refer to these complexes formulaically as if they were monomeric (e.g. $LiNR_2$) for simplicity, however, again this is not accurate as they too have a strong propensity to oligomerize. The presence of a nitrogen anion does though result in different structural motifs being attained due to the nitrogen's lone pair of electrons. This lone pair can donate to the Lewis acidic metal centre of an adjacent molecule resulting in nitrogen

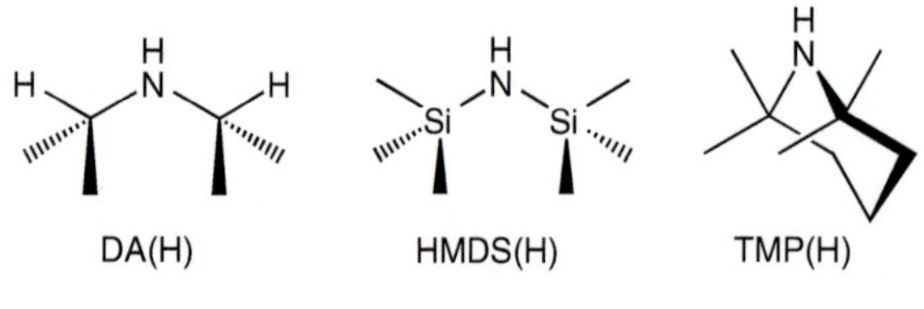

Figure 2.6 Common secondary amines employed for metallation chemistry.

having two different local interactions; an electrostatic interaction with its charge-balancing metal cation partner and a donor/acceptor interaction with the neighbouring metal. In reality, these different interactions are often indistinguishable in the solid state due to resonance between the two resulting in an averaged M–N bond distance (Scheme 2.7).

This difference in bonding means that such complexes tend not to adopt alkali metal polyhedra with face-capping anions but rather form simpler [M–N] oligomers which either alternate indefinitely to give a polymer or cyclize to give a discrete oligomer. LiDA **35** adopts the (helical) polymeric chain structure [59] while LiHMDS **36** rather forms a cyclotrimer [60] with the δ^+ nature of the silicon centre polarizing the Si–C bonds and giving the methyl groups sufficient negative charge to engage in agostic interactions with the metal centres. Interestingly, LiTMP **37** adopts two different cyclic oligomers, a tetramer (**37a**) [61] and a trimer (**37b**) (Figure 2.7) [62]. In solution, these oligomers exist in equilibrium, with a temperature dependence on the crystallization process with higher temperatures favouring the larger oligomer and sub-ambient temperatures favouring the smaller oligomer [62].

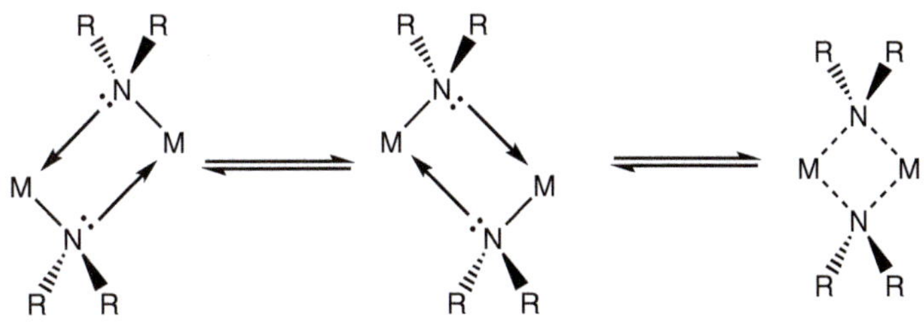

Scheme 2.7 Simplified bonding in metal-amide oligomers, using a cyclodimer as a representative example.

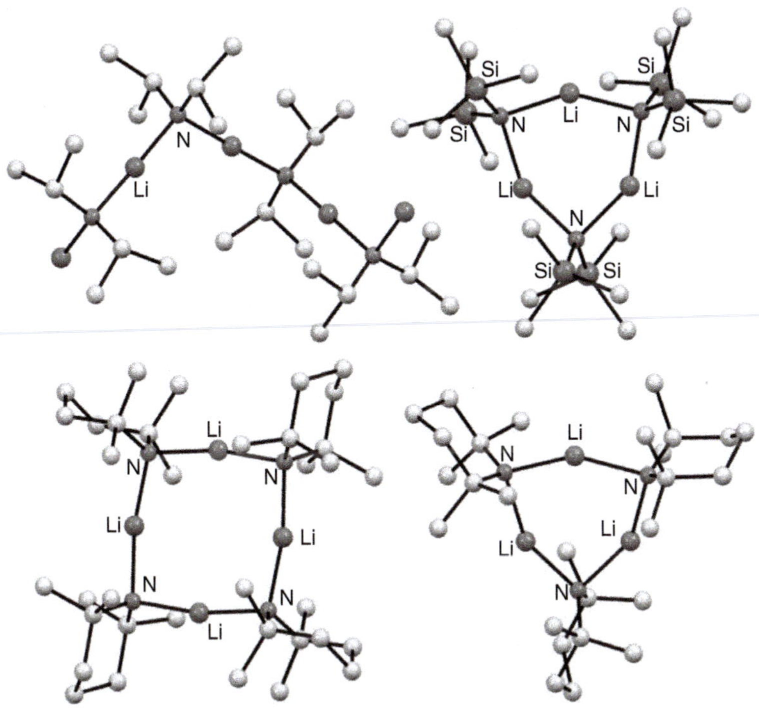

Figure 2.7 Molecular structures of LiDA **35**, LiHMDS **36**, and LiTMP **37a/37b**. *Sources*: Adapted from Barnett et al. [59]; Rogers et al. [60]; Lappert et al. [61]; Hevia et al. [62].

Despite these different structural motifs, all three lithium amides form mono-solvated cyclodimers in the presence of THF. The relative steric bulk of the amides is reflected structurally in their central Li_2N_2 rings with the least bulky (HMDS) [63] displaying a symmetric ring as described in the right-hand side of Scheme 2.7, while bulkier DA [64] and TMP [65] show a shorter and longer Li–N interaction as described in the left-hand side of Scheme 2.7, with the bulkiest TMP complex showing the largest distinction.

2.3 Heterometallic Bases

Like homometallic species, a structural understanding of bimetallic complexes provides considerable insight into their reactivity. At the forefront of X-ray crystallography of bimetallic species was the late Erwin Weiss [13], who identified a great number of such complexes in the solid state, distinguishing lower-order ates (those with an alkali metal: secondary metal ratio of unity) from the higher-order ates (those with a ratio greater than 1: 1). These 'Weiss motifs' (see Figure 2.8 for representative examples: $[(TMEDA)Li(\mu\text{-}Me)_2]_2Mg$ **38** [66] and $[(TMPDA)Li(\mu\text{-}n\text{-}Bu)_2Mg(\mu\text{-}n\text{-}Bu)]_2$ **39** where TMPDA = *N,N,N′,N′*-tetramethylpropanediamine [67] provided a great insight into ate chemistry and are central to our understanding of structure/reactivity relationships.

We now discuss some of the most important and instructive examples of bimetallic complexes, focusing on our structural understanding of them and how this relates to their reactivity.

2.3.1 Heteroalkali Metal Bases

As mentioned in Section 2.2, methylsodium **23** was characterized with varying amounts of methyllithium present within its inter-tetrameric cavities. This was achieved serendipitously as methyllithium was utilized as a starting material. Heteroalkali metallic complexes have however been deliberately targeted and shown to display reactivity different to the homometallic constituent parts. For example, Wittig prepared the mixed lithium–sodium aryl complex $[(PhLi)(PhNa)(OEt_2)]$ **40** and noted its diminished reactivity compared to that of homometallic PhNa on account that it was stable in diethyl ether for days; whereas, PhNa readily metallates the same solvent [68]. Remarkably, the presence of PhLi (**2**) was actually found to stabilize up to 24 molar equivalents of

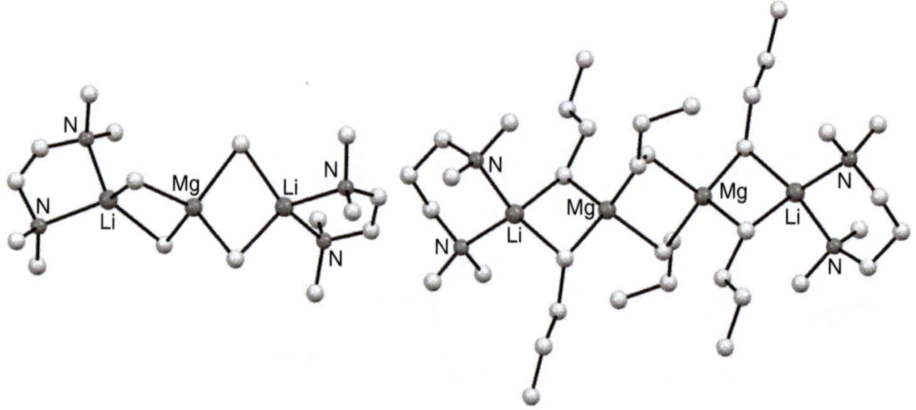

Figure 2.8 Representative example of higher-order and lower-order structures **38** and **39**. *Sources:* Adapted from Greiser et al. [66]; Zaragoza-Calero et al. [67].

PhNa towards diethyl ether! Structural enlightenment on this system came from Schümann and Weiss who crystallographically characterized a TMEDA-solvated mixed lithium–sodium structure in the monolithium-trisodium complex $[(PhLi)(PhNa)_3(TMEDA)_3]$ **41** [69]. This displays a type **B** tetrahedral structure (Figure 2.1) but is distorted on account of the shorter Li–C bonds with respect to the Na–C bonds. This has the effect of drawing the lithium atom further into the structure, leaving it inaccessible for Lewis donor coordination meaning that only the three Na atoms are coordinated by TMEDA.

Some common heteroalkali metallic secondary amides also display a wide variety in their structures. Despite Li(HMDS) and K(HMDS) both adopting cyclic structures, the heterometallic complex $LiK(HMDS)_2$ **42** is a chain polymer [70], reflecting the different M–N bond lengths which likely preclude formation of the ring as this would involve movement of the HMDS groups away from one of the metals and prevent the agostic type [71] $M\cdots CH_3Si$ interactions which stabilize such rings (Figure 2.9), that is interactions between a metal and the electrons of a C–H bond. That notwithstanding, this ring can be stabilized through coordination by THF, with a single molecule solvating the lighter alkali metal and two molecules solvating the more sterically demanding heavier alkali metal [72]. A bidentate donor ligand such as TMEDA deaggregates the heterometallic structure to its homometallic component moieties [73]. The chain polymer motif is also adopted by $LiNa(TMP)_2$ **43** [74] despite both homometallic species in isolation adopting cyclic structures (see Section 2.2.3).

Other than in their stabilities with respect to selected solvents, little has been reported regarding modification of reactivity when two alkali metals have been paired within a homo-anionic complex (i.e., $MM'R_2$ at the empirical level). This is perhaps unsurprising since the reactivity of the homo-alkali metallic complexes is already both high and reasonably similar and consequently there is not much 'middle ground' waiting to be exploited. In contrast, heteroanionic mixed-metal synergistic effects have been widely recognized. The best-known practical example of such chemistry in action is the LiCKOR base, whose name is derived from its component parts (an alkyllithium reagent – 'LiC'; and a potassium alkoxide – 'KOR'). Also known as the Lochmann–Schlosser superbase after the primary researchers who independently discovered it [75], this mixed-metal reagent clearly displays cooperativity in for example the deprotonation of toluene at its methyl arm to generate benzyl potassium **44** (Figure 2.10). *n*-BuLi and KO*t*-Bu can both be stirred individually in toluene with no obvious reaction occurring yet when stirred together immediate precipitation of orange benzyl potassium takes place which can be attributed to a cooperative effect in the potassium alkoxide synergistically boosting the deprotonating power of the alkyllithium reagent. These

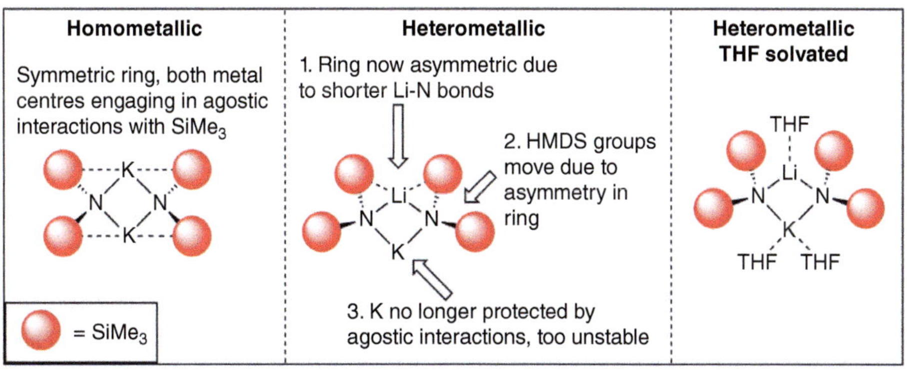

Figure 2.9 Influence of the agostic interactions on homo-, hetero- and solvated HMDS structures.

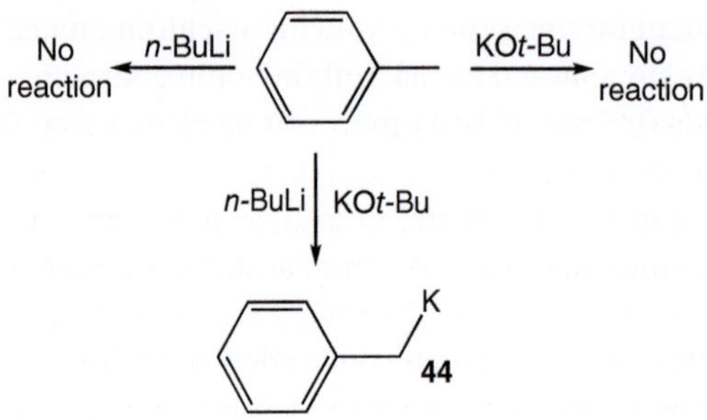

Figure 2.10 LiCKOR metallation of toluene.

Table 2.1 Effect of identity of alkali metal alkoxide on yield and *meta/para* ratio of product.

t-Pentoxide	Yield	*m/p* ratio
Li	3	15.0
Na	29	3.4
K	58	1.4
None	26	3.0

bimetallic mixtures were actually originally recognized for their alternative metallation patterns by Morton in the 1950s who noticed a change in the ratio of *meta/para* dimetallated benzene when introducing alkali–metal alkoxides to amylsodium [76]. Interestingly, the absolute yields of dimetallated product were increased when the heavier alkali metal alkoxides were used (Table 2.1).

Despite its continuous usage in organic synthesis, the molecular structure of the active LiCKOR superbase has remained thus far elusive. There have, however, been a number of closely related surrogate structures divulged which all support the position that some sort of heteroleptic bimetallic complex or complexes is/are responsible for the enhanced reactivity rather than a highly reactive alkyl potassium reagent generated via alkyl/alkoxide metathesis. These include structures containing three of the four key components (that is of general empirical formula M_2LL' in $Li_8n\text{-}Bu_4(Ot\text{-}Bu)_4$ **45** [77] and $MM'L_2$ in $Li_4M_4(Ot\text{-}Bu)_8$, [M = Na **46**; K **47**; Rb **48**; Cs **49**] [78, 79], a family of structures with a central $Li_4M_4O_8$ 'breastplate' motif (see an example in Chapter 1, Figure 1.7a) as well as those with all four components. Of those structures reported, there are currently none containing the constituent LiCKOR components in the synthetically relevant 1:1:1:1 stoichiometry, for example in $Li_4K_3Np_3(Ot\text{-}Bu)_4$ [80], $Li_4K_3Np_4(Ot\text{-}Bu)_3$ [80], $Li_4K_4Np_2(Ot\text{-}Bu)_6$ [80], $Li_4K_4Np_3(Ot\text{-}Bu)_5$ [80], $[Li_4KNp_2(Ot\text{-}Bu)_3]_2$ [81] (Np = neopentyl) and $Li_2K_4Ph_5(Ot\text{-}Bu)$ [82] (**50–55**). A stoichiometrically exact $Li_4K_4(Ot\text{-}Bu)_4(NHt\text{-}Bu)_4$ structure (**56**) was also revealed with primary amide ligands substituting for the carbanionic ligands of the experimental reagent (illustrated in Chapter 1, Figure 1.4b) [83]. Another structure relevant to the Lochmann–Schlosser superbase family is $Li_4Na_4(CH_2C_6H_2Me_2O)_4$ **57** where the carbanion and alkoxy anions co-exist within the same (dianionic) ligand. This octanuclear structure was formed by a straightforward metallation of sodium 2,4,6-trimethylphenoxide by BuLi (Scheme 2.8) [84].

Going beyond these MM′LL′ bimetallic compositions, there have also been reports of ternary (trimetallic) alkali metal complexes, spearheaded by $Li_2Na_2K_2(NHPh)_4(Ot\text{-}Bu)_2$ **58** [85] and $Li_4Na_2K_2(Ot\text{-}Bu)_8$ **59** [86]. This homoleptic alkoxide complex **59** appears to display a trimetallic cooperative effect since it abstracts a hydrogen from the methyl arm of toluene at room temperature, whereas studied bimetallic compositions are inert towards toluene at room temperature and monometallic compositions are inert towards toluene even at 90 °C. This is likely facilitated through K–arene π interactions manoeuvring the toluene into place and allowing the deprotonation to occur, a hypothesis supported by the elucidation of the structure of the benzene solvate, which shows such metal-substrate interactions (Figure 2.11).

Introducing the TMP ligand to a LiCKOR superbase can have a profound effect on the pattern of metallation. For example, LiCKOR metallates *ortho-* and *meta-* OMOM-substituted toluenes on the ring, in the position adjacent to the heteroatom containing directing group according to the well-defined principles of DoM (directed *ortho*-metallation) chemistry and the complex-induced proximity effect (CIPE) [87]. However, the deprotonation event switches to a lateral position when the secondary amide is included in the bimetallic mixture (Scheme 2.9) [88]. This can be achieved even with only a catalytic (10%) amount of added TMP(H) and is reliant on the secondary amine (which is likely deprotonated itself to give an amido anion) since neutral tertiary amines do not induce the same effect. This suggests anion migration since the initially formed *ortho*-metallated product soon diminishes in favour of the laterally metallated isomer.

Scheme 2.8 Lithiation of sodium 2,4,6-trimethylphenoxide to yield the hetero alkali metal superbase model **57**.

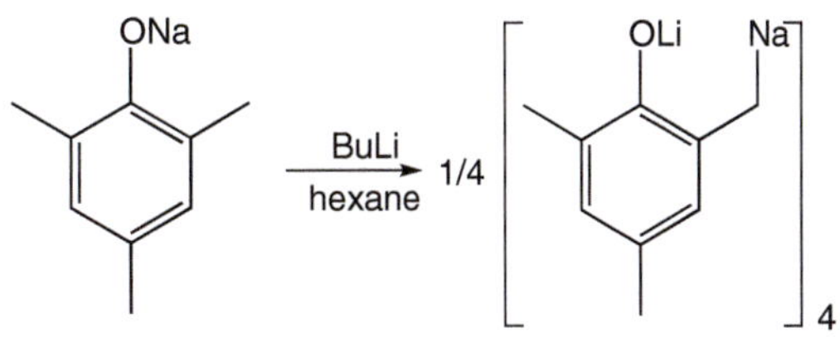

57

Figure 2.11 Molecular structure of trimetallic alkoxide complex **59** (K = dark purple, Li = pink, Na = light purple, O = red). *Source*: Adapted from Wei et al. [86].

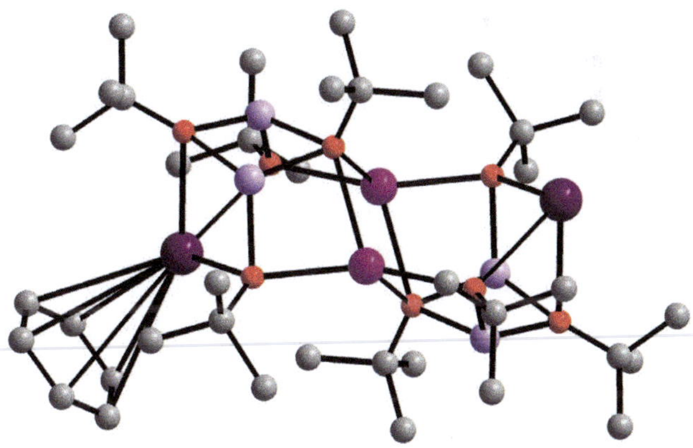

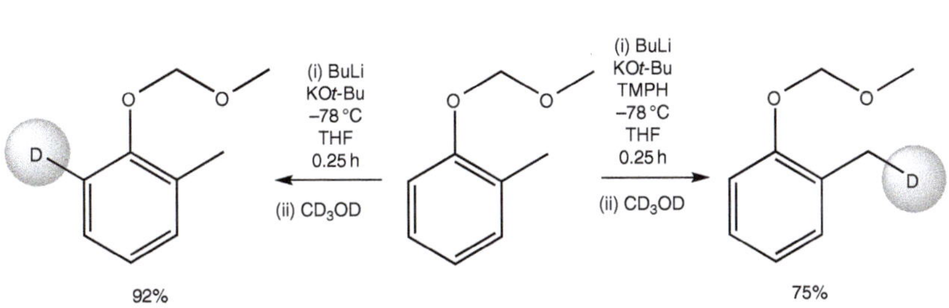

Scheme 2.9 Contrasting reactivity of LiCKOR superbase in the presence and absence of catalytic TMP(H).

While this 'LiNK' chemistry has been exploited effectively for the generation of substituted benzyl compounds [89, 90] and [2.2]metacyclophanes [91, 92], similar to the case for LiCKOR and perhaps even more so, a structural understanding of the active bimetallic reagent is lacking.

2.3.2 Alkali Metal Magnesiate Chemistry

Since the alkali metal magnesium pairing is one of the most well studied in bimetallic ate chemistry it has been granted a chapter in its own right (see Chapter 3 for full details). To avoid unnecessary repetition this chemistry is not covered here, except to mention that its most fruitful area, developed by Knochel over the past 15 years, is in its turbo reagents. These reactivity-enhanced organometallic reagents combine an alkali-metal in the form of lithium chloride with Grignard (R–Mg–Cl) or Hauser (R_2N–Mg–Cl) reagents [93]. While the synthetic utility of these reagents is extremely impressive and broad-scoped, structural and mechanistic knowledge of them remains scant. Only a small handful of crystallographic studies revealing structural evidence to help explain observed reactivity has been successful thus far. One such key study uncovered a difference in aggregation state between the more reactive (monomeric) [$(THF)_2Li(\mu\text{-}Cl)_2Mg(THF)TMP$] **60** [94] and the less reactive (dimeric) [$(THF)_2Li(\mu\text{-}Cl)_2MgN\textit{i}\text{-}Pr_2]_2$ **61** [95] (Figure 2.12; for solution behaviour see Chapter 6). One of the most exciting developments in this area of bimetallic chemistry, organomagnesium turbo-reagents could make an even more substantial impact in synthesis and catalysis in the future if greater structural and mechanistic clarity can be obtained through further study.

2.3.3 Early Signs of Synergistic Behaviour in Zincate Chemistry

Interestingly, though its development remained static for decades, alkali metal synergistic chemistry actually predates the birth of conventional organolithium chemistry in 1917 by Schlenk and Holtz [6] with Wanklyn's 1858 pioneering synthesis of the alkali metal homoleptic zincates $MZnEt_3$ (M = Na, K) [96]. Seyferth's illuminating essay on "*Zinc Alkyls, Edward Frankland, and the Beginnings of Main-Group Organometallic Chemistry*" records this epochal period in the history of main group organometallic chemistry [97]. Wanklyn interpreted sodium triethylzincate ($NaZnEt_3$) as a 'double compound between sodium–ethyl and zinc–ethyl'. In modern parlance, this would be regarded as a co-complex between the two homometallic alkyl compounds, though Wanklyn's original preparation was not a co-complexation reaction but a redox reaction between elemental

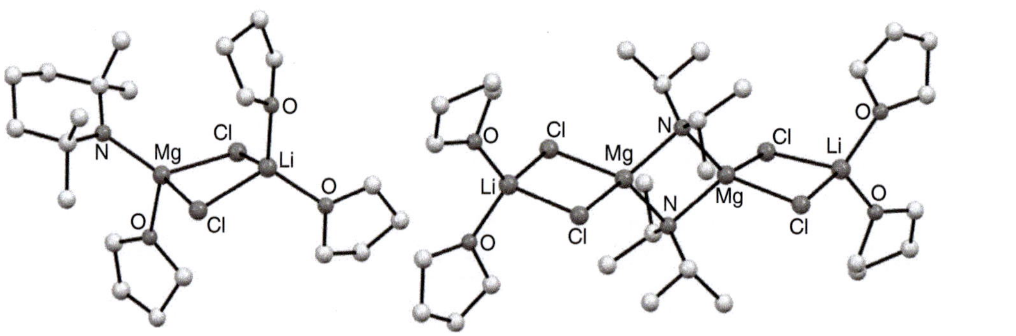

Figure 2.12 Molecular structures of [$(THF)_2Li(\mu\text{-}Cl)_2Mg(THF)TMP$] **60** (left) and [$(THF)_2Li(\mu\text{-}Cl)_2MgN\textit{i}\text{-}Pr_2]_2$ **61** (right) showing steric influence of secondary amide on aggregation state. *Sources*: Adapted from García-Álvarez et al. [94]; Armstrong et al. [95].

sodium and a large excess of diethylzinc. Surprisingly, as this article is written, the molecular structures of these parent alkali metal triethylzincates have still not been determined. Crystallization of $NaZnEt_3$ samples, in the absence or presence of any additional solvating ligands, of a suitable quality for X-ray crystallographic study is exceptionally challenging as several research groups have experienced. In a recent attempt to form crystals of $NaZnEt_3$ in benzene solution, the heteroleptic alkyl–aryl zincate $[NaZnEt_2Ph]_\infty$ **62** was obtained fortuitously [98]. Amenable to X-ray crystallographic study, $[NaZnEt_2Ph]_\infty$ forms a coordination network composed of contacted Na^+ cations and zincate $[ZnEt_2Ph]^-$ anions, where contact is through each Na cation interacting with three ethyl groups and two phenyl groups belonging to three $[ZnEt_2Ph]^-$ anions (Scheme 2.10). This formation of $[NaZnEt_2Ph]_\infty$ through metallation (zincation, Zn–H exchange) of benzene is proof positive of the synergistic nature of the chemistry of Wanklyn's $NaZnEt_3$ complex, here manifested in strong Brønsted basicity, since on its own neutral $ZnEt_2$ is a weak base incapable of abstracting a hydrogen from a C–H bond in benzene. Wanklyn's original paper also mentioned evidence of this synergistic behaviour reporting that $NaZnEt_3$ reacts with carbon dioxide to form propionic acid; whereas neutral diethylzinc is inert towards carbon dioxide.

This nineteenth century zincate work pioneered what has grown over time to be a vast class of compounds referred to collectively as 'metallates' (often abbreviated simply to "ates") [99], the term being introduced to chemistry parlance by the Nobel laureate Wittig [100]. Bimetallic ates such as the triarylzincate complex $LiZnPh_3$ **63** and its magnesium congener $LiMgPh_3$ **64**, both prepared by Wittig via co-complexation (mixing of the homometallic components), contain a metal centre in their complex anionic moiety balanced by a cationic electropositive metal partner, which may be a single metal cation (often having donor ligands attached for coordinative saturation) or

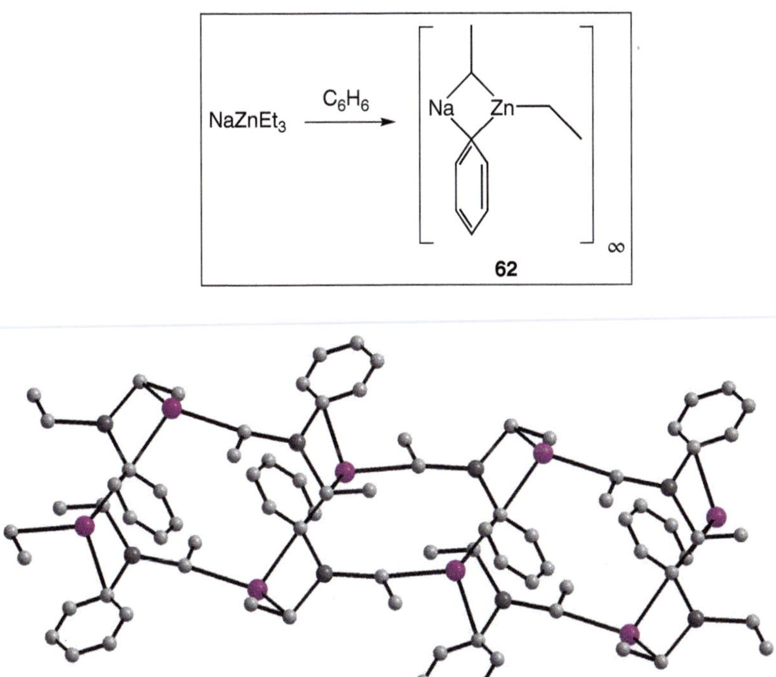

Scheme 2.10 Synthesis of heteroleptic sodium zincate **62** by metallation of benzene and molecular structure of resulting complex (Na = light purple, Zn = dark gray). *Source:* Adapted from Hedström et al. [98].

some sort of complex derivative. These features define ate compounds. Discovering that for example the triarylmagnesiate complex $LiMgPh_3$ reacts with benzalacetophenone to afford a product mainly of 1,4-addition, whereas its neutral monometallic component PhLi **2** affords mainly the alternative 1,2-addition product, Wittig also recognized the synergistic potential of ates. Wittig interpreted this synergistic behaviour as 'anionic activation', with the ligands coordinating to the zinc or magnesium metal centre activated "anionically" through an inductive effect [101]. Tochtermann stressed this anionic activation in an early review of "structures and reactions of organic ate-complexes" [102].

Synergistic properties have also been described in terms of one component acting as a promoter. For example, while di-*n*-butylzinc is inert to carbon monoxide at atmospheric pressure in the absence of a promoter, adding a stoichiometric amount of potassium *t*-butoxide promotes a reaction that produces *n*-valeroin (Scheme 2.11) [103].

Adding more perplexity to this synergistic phenomenon, the normal reactivity orders of bimetallic ate complexes can be inverted compared to those of monometallic complexes. Illustrating this point, though organomagnesium compounds are generally regarded as more reactive than organozinc compounds in monometallic chemistry, in bimetallic chemistry the potassium zinc silylamide '$KZn(HMDS)_3$' can deprotonate toluene to generate the polymeric benzyl product $[\{KZn(HMDS)_2(CH_2Ph)\}_\infty]$ **65** (Scheme 2.12), in contrast to the magnesium analogue '$KMg(HMDS)_3$' which is inert to toluene under the same conditions [104].

This polymeric potassium zincate **65** has been utilized as an effective catalyst for the benzylic C–H bond addition of diarylmethanes to alkenes and conjugated dienes. The bimetallic complex is key to the mechanism with the terminal benzyl group deprotonating the diarylmethane followed by addition of the new Zn–C bond across the unsaturated unit, with the catalytic cycle closing through a deprotonation of the diarylmethane substrate with this new bimetallic complex (Scheme 2.13).

2.3.4 Lithium TMP–Zincate Chemistry

The introduction of the sterically demanding secondary amide TMP was the watershed event that transformed organozincate synergistic reactivity from the odd noted curiosity to the synthetically useful methodology it is today. TMP(H), the free amine, was actually first reported in 1885 albeit in an impure form [105], though its synthesis was improved 20 years later [106]. Starting modestly in 1972 through two independent papers, one reporting rearrangement reactions of epoxides and the other alpha lithiation adjacent to a boron centre, the lithium derivative LiTMP **37** has since developed into a popular deprotonating reagent. Together with LiHMDS and LDA (lithium diisopropylamide), LiTMP is one of the utility amides (derived from secondary amines) [17] that find

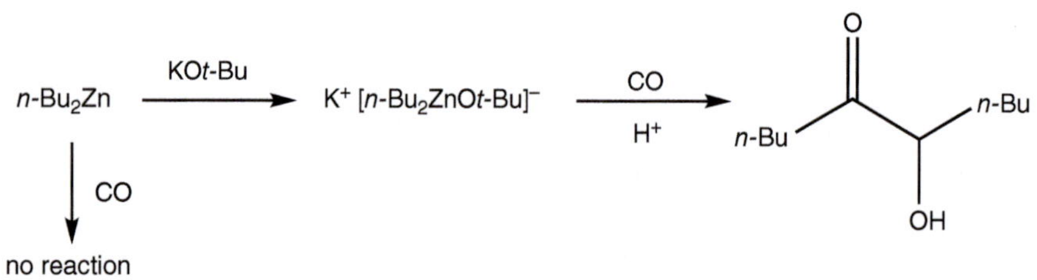

Scheme 2.11 Contrasting reactivity of a homometallic zinc reagent and a bimetallic zincate with carbon monoxide.

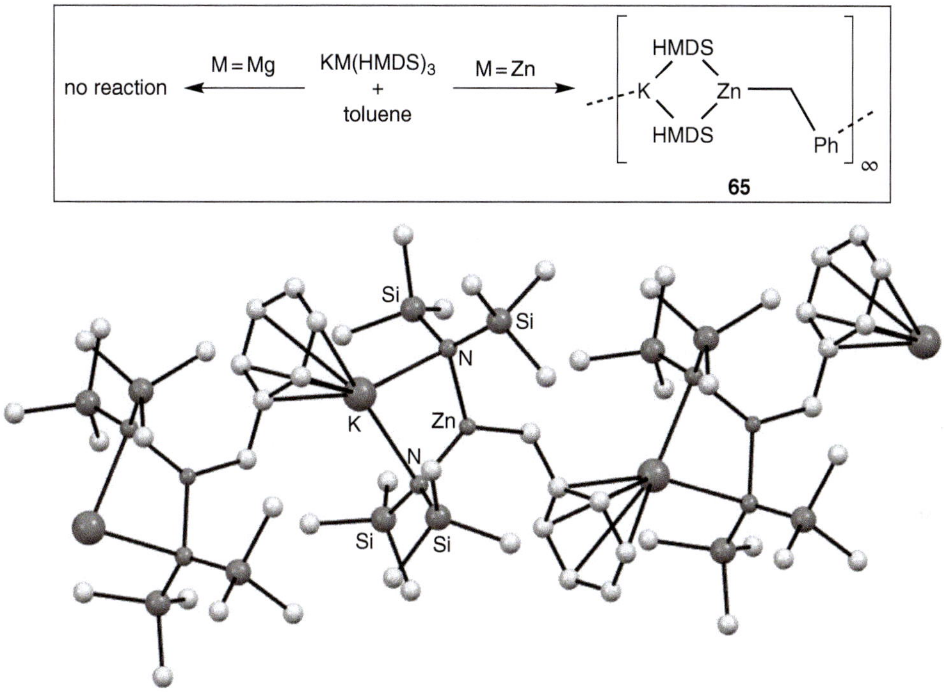

Scheme 2.12 Contrasting reactivity of homoleptic bimetallic HMDS complexes with toluene and section of polymeric structure. *Source:* Adapted from Clegg et al. [104].

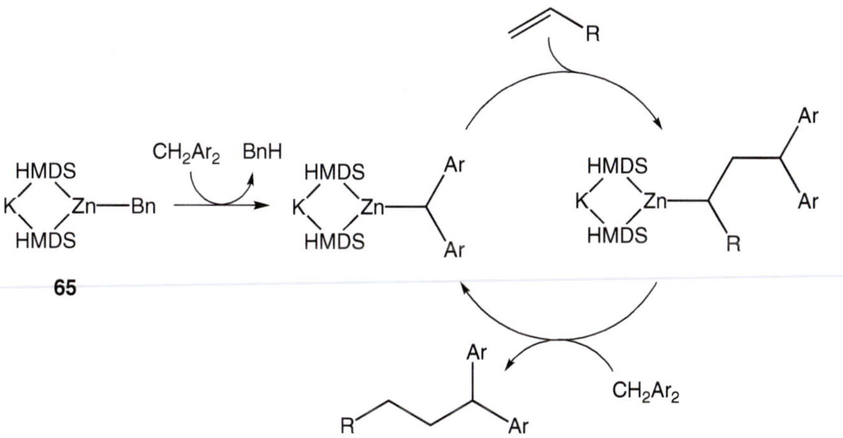

Scheme 2.13 Proposed mechanism of addition of diarylmethanes to alkenes catalyzed by potassium zincate **65**.

application in a plethora of C–H deprotonation reactions. Steric bulk limits the nucleophilicity of these utility amides so that they engage in less side reactions than alkyl lithium reagents when confronting substrates in which Brønsted basicity and nucleophilic addition are competitive.

Recently reviewed [17], the structural chemistry of LiTMP is remarkably diverse. However, it is worth highlighting here that the classic structure is the unsolvated cyclotetrameric structure

$(LiTMP)_4$, and since the publication of the review a cyclotrimeric variant $(LiTMP)_3$ has also been discovered (Figure 2.7) [62], though both forms were already known in solution [107]. These structures are archetypal of the aggregation phenomena that trademark organolithium structural chemistry, both in the solid state and in the solution, with organo here not just applying to organoC type ligands but defined more broadly to include organoN, organoO, and organoS ligands etc. that is organic ligands with electronegative heteroatomic heads.

In general, aggregation phenomena are not so common in alkali metal zincate chemistry though examples do exist. Often these zincates can be interpreted as co-complexes. In terms of proven synthetic usefulness, the most eminent zincate is the lithium TMP–zincate 'LiZn(*t*-Bu)$_2$(TMP)', which can be viewed as a co-complex between LiTMP and di-*t*-butylzinc. Some authors refer to such co-complexes as mixed aggregates, but co-complexes seem more apt given that both the lithium component and the zinc component are monomers, not aggregates. In the original paper Kondo and Uchiyama state that "precomplexation of lithium TMP and *t*-Bu$_2$Zn is essential for successful metallation" [108]. The metallation power of LiZn(*t*-Bu)$_2$(TMP) is manifested in strong metallating (zincating) efficiency that achieve high levels of chemoselectivity and regioselectivity in directed *ortho*-deprotonation reactions with a variety of functionalized aromatic (e.g. alkyl benzoates and dialkyl benzamides) and heteroaromatic (e.g. pyridines, furans and thiophenes) substrates (Scheme 2.14) [108, 109]. Since these organic substrates cannot be metallated by di-*t*-butylzinc when unassisted by the alkali metal amide, while unaccompanied LiTMP is so aggressively basic it often displays poor functional-group tolerance and generally requires sub-ambient temperatures to avoid side reactions, these metallation reactions of LiZn(*t*-Bu)$_2$(TMP) have a genuine synergistic origin. As the reactions of LiZn(*t*-Bu)$_2$(TMP) have been confined mainly to polar THF solution from which crystallization of organometallic compounds is challenging, the structural basis behind this synergistic behaviour was initially not clear. However, preparing this

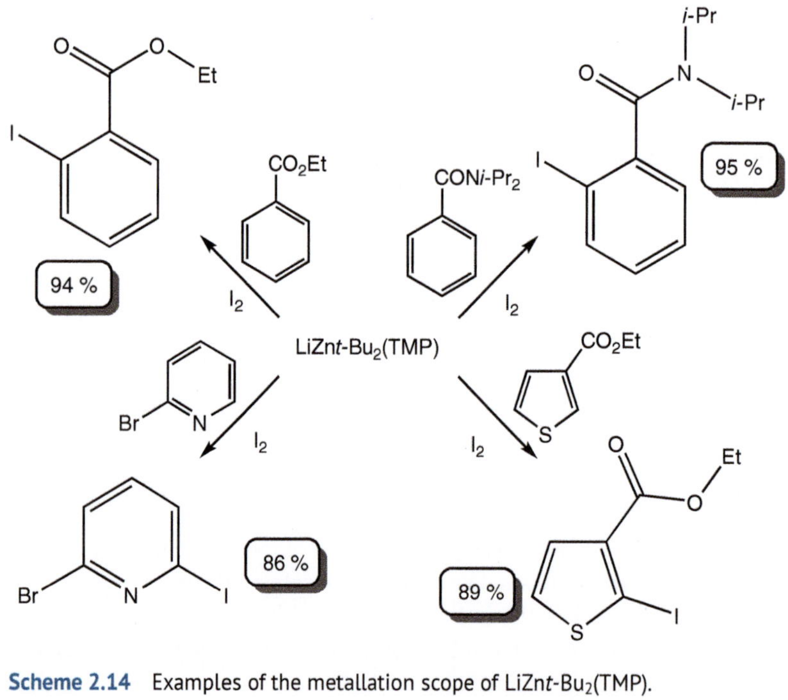

Scheme 2.14 Examples of the metallation scope of LiZn*t*-Bu$_2$(TMP).

reagent in bulk hexane solution containing a stoichiometric quantity of THF did the trick affording crystals of the mono-THF solvate [(THF)Li(TMP)(*t*-Bu)Zn(*t*-Bu)] **66**. TMP is well known to be a bridge connecting Li centres in the aforementioned classical LiTMP structures, in crystalline [(THF)Li(TMP)(*t*-Bu)Zn(*t*-Bu)] (Figure 2.13), TMP connects the Li and Zn centres within the same molecule, the key enabling feature behind the synergistic behaviour [110, 111].

This TMP bridging and the other features of this ion-contacted arrangement, a trigonal planar anionic zinc moiety and a donor (here THF)-solvated lithium cation have now become commonplace in zincate structural chemistry though departures from these features are known depending on stoichiometry, steric factors, as well as the number and denticity of donor molecules. DFT calculations (B3LYP/631SVP) on model systems [$Me_2(Me_2N)ZnLi$ and $(MeLi)_2$] provided further confirmation of the synergistically operative nature of this TMP zincate reagent with both lithium and zinc involved in the transition states of deprotonation (specifically DoM) reactions with benzonitrile and methyl benzoate [112]. Important distinctions between these lithium zincate reagents and dimeric alkyl lithium reagents have been brought to light by this theoretical study. The former heterobimetallic systems have adaptable transition state structures that make them able to function as efficient deprotonating agents with different substrates, while their unequal spread of Lewis acidity across the two metal atoms (the LUMO components are localized on the Li atom) means that the zinc atom cannot activate multiple bonds and hence competitive addition reactions are disfavoured. In contrast, the latter homometallic system having equal Lewis acidity across its two Li atoms, each able to activate multiple bonds favour addition reactions over DoM reactions (Scheme 2.15).

Having an alkali metal and zinc in the same compound means that multiple ligands also need to be present to satisfy valency requirements. Moreover, in heteroleptic systems such as [(THF)Li(TMP)(*t*-Bu)Zn(*t*-Bu)] different types of Brønsted basic ligand (here alkyl and amido) may also be present. Unsurprisingly, this diversity can lead to more complicated reaction chemistry as revealed in a case study of anisole, a benchmark aromatic substrate in the history of organolithium DoM chemistry [113, 114]. The crystallographically characterized products obtained from reactions of LiZn(t-Bu)$_2$(TMP) with anisole, namely [(THF)Li(TMP)(o-C_6H_4OMe)Zn(*t*-Bu)], [(THF)Li(TMP)(o-C_6H_4OMe)Zn(o-C_6H_4OMe)], and [(C_6H_5(Me)O)Li(TMP)(o-C_6H_4OMe)Zn(*t*-Bu)] **67–69** indicate that the zincate acted formally as an alkyl base, a dialkyl base, and a combined alkyl base/Lewis acid respectively (Scheme 2.16) [115].

The true picture is in fact more mosaic. TMP–zincates seemingly operate via a two-step mechanism with TMP the kinetically active base that generates the conjugate acid TMP(H) on

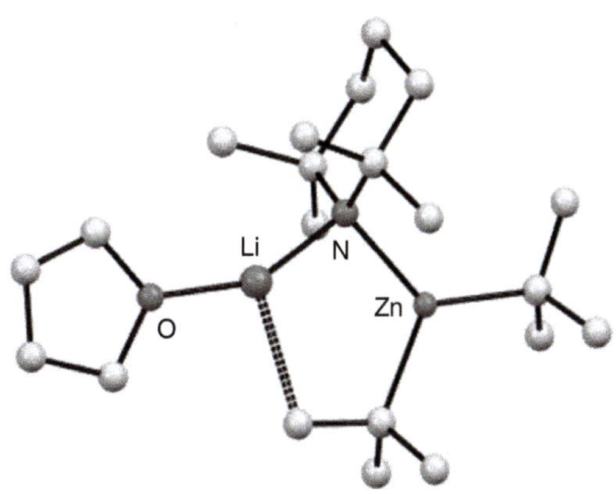

Figure 2.13 Molecular structure of [(THF)Li(TMP)(*t*-Bu)Zn(*t*-Bu)] **66**. *Source*: Adapted from Uchiyama et al. [111].

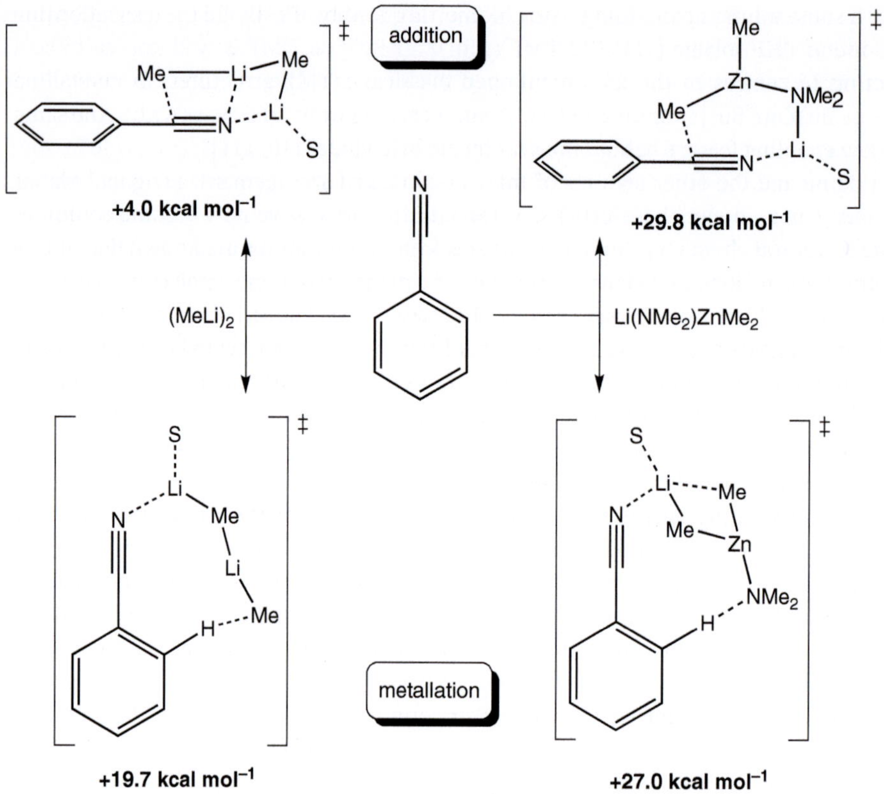

Scheme 2.15 Computed transition states for addition versus metallation reactions of homometallic MeLi and bimetallic $Li(NMe_2)ZnMe_2$ with benzonitrile.

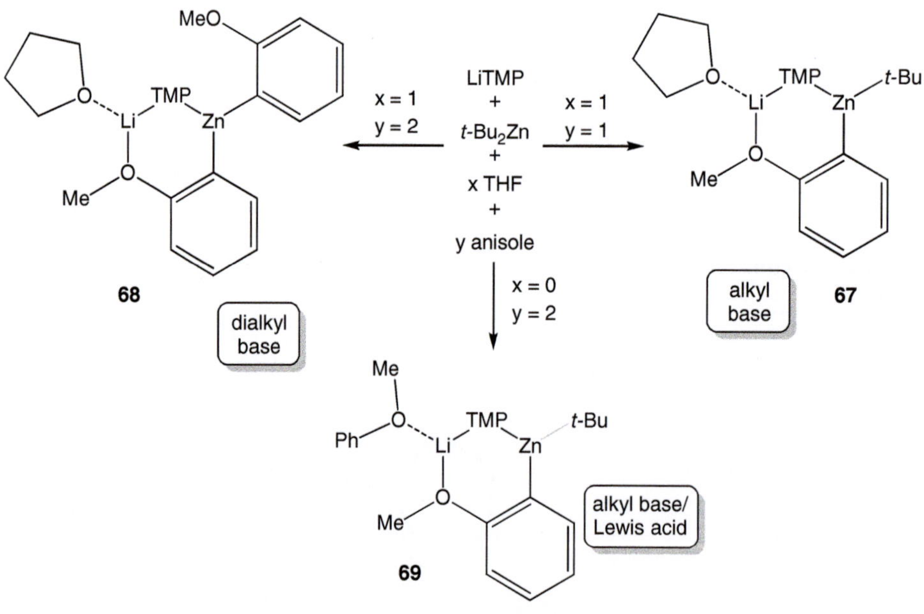

Scheme 2.16 Stoichiometry dependent variable reactivity of $LiZn(t\text{-}Bu)_2(TMP)$ with THF/anisole.

deprotonating a substrate. In turn, TMP(H) can be deprotonated by the alkyl ligand to regenerate TMP with release of alkane (Scheme 2.17).

Evidence for this two-step process is compelling. In the aforementioned theoretical study, probing the reaction of the model zincate $[(Me_2O)Li(NMe_2)(Me)Zn(Me)]$ with anisole, Uchiyama and Morokuma concluded that the amide is the kinetically preferred base to the alkyl base, so generating a metallated intermediate $[(Me_2O)Li(C_6H_4\text{–}OMe)(Me)Zn(Me)(NHMe_2)]$. Experimentally this was confirmed by Hevia, who reacted the dialkylanisolyl zincates $[(THF)_2Li(C_6H_4\text{–}OMe)(Me)Zn(Me)]$ **70**, and $[(THF)_3Li(C_6H_4\text{–}OMe)(t\text{-Bu})Zn(t\text{-Bu})]$ **71**, made by co-complexation of lithiated anisole and the appropriate dialkylzinc reagent in THF, with the amine TMP(H). In the former case the main product was [(THF)Li(TMP)(Me)Zn(Me)] **72**, with TMP substituting for the anisolyl ligand with liberation of free anisole, thus establishing an alkyl effect whereby the anisolyl ligand is a better base than the methyl ligand [116]. In the latter case, the main product was $[(THF)Li(TMP)(C_6H_4\text{–}OMe)Zn(t\text{-Bu})]$ **73**, with TMP substituting for a *t*-butyl ligand and liberating butane, thus mimicking the first step of the theorized two-step mechanism (Scheme 2.18). A solvent dependence was also noted since adding hexane to THF solutions of $[(THF)_2Li(C_6H_4\text{–}OMe)(Me)Zn(Me)]$ and $[(THF)_3Li(C_6H_4\text{–}OMe)(t\text{-Bu})Zn(t\text{-Bu})]$, produced the higher order (2:1, Li:Zn stoichiometry) tetraanisolylzincate $[(THF)_2Li_2Zn(C_6H_4\text{–}OMe)_4]$ **74**, indicative of a disproportionation event (Scheme 2.19). The reduction in steric bulk allows the Zn centre in **74** to expand its coordination from three to four, comprising a tetrahedron of *ortho*-deprotonated C atoms. Zinc lies almost coplanar with the anisole rings indicating high Zn–C σ-bonding character, while each lithium lies near perpendicular to them indicative of greater π-character. This σ/π bonding distinction is a signature feature of many mixed-metal ate structures.

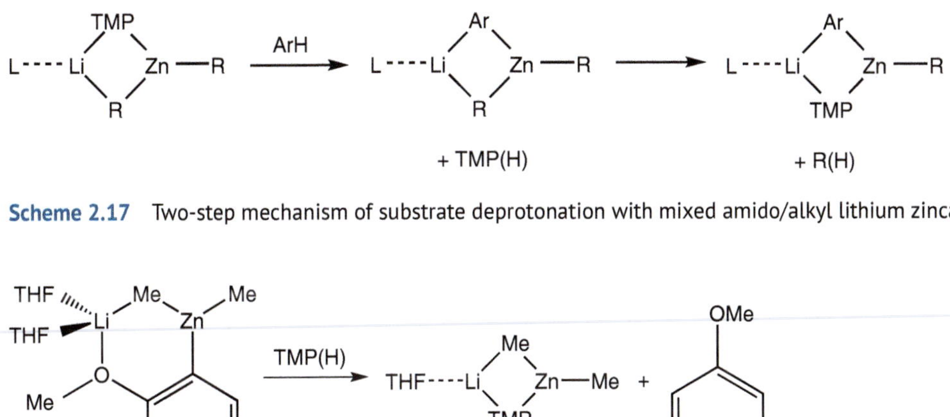

Scheme 2.17 Two-step mechanism of substrate deprotonation with mixed amido/alkyl lithium zincate.

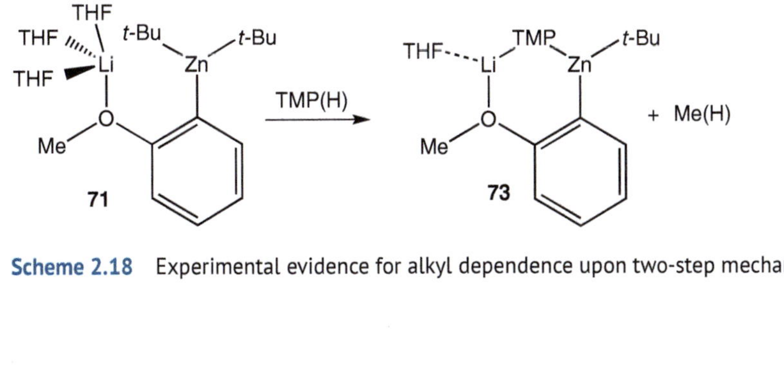

Scheme 2.18 Experimental evidence for alkyl dependence upon two-step mechanism.

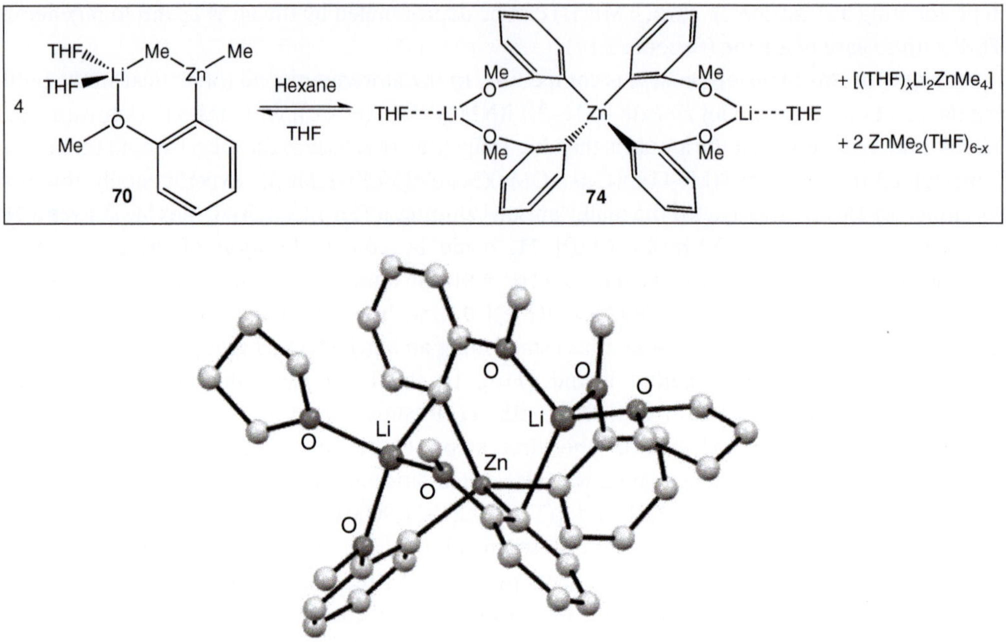

Scheme 2.19 Disproportionation of anisolyl lithium zincate and the molecular structure of the homoleptic product **74**. *Source:* Adapted from Clegg et al. [116].

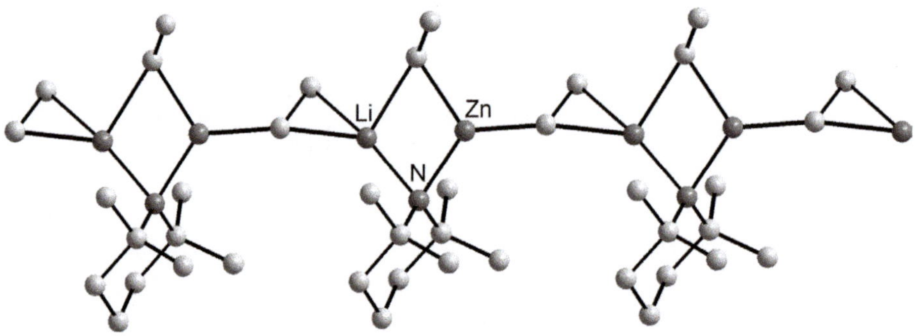

Figure 2.14 Propagation of EtZn(Et)(TMP)Li **75** into a polymer. *Source*: Adapted from Kondo et al. [117].

As already alluded to through examples above, donor solvents strongly influence the structures of lithium di(alkyl)amidozincate reagents. Their absence in certain cases can even prevent co-complexation of the lithium and zinc components as found with LiTMP and Zn(t-Bu)$_2$ [115]. Smaller dialkylzinc components do not require donor solvent participation for co-complexation but aggregation comes into play as demonstrated in polymeric [{EtZn(Et)(TMP)Li}$_\infty$] **75**, a co-complex between diethylzinc and LiTMP. Here, zinc displays its common trigonal planar coordination comprising two Et and one TMP groups, the latter participating in a Zn–N–Li bridge. One Et group also bridges to Li through its α–C atom, while the other Et group propagates the polymer through interaction of both its C atoms with Li (Figure 2.14) [117]. This polymeric structure breaks up in THF solution deprotonating the aromatic tertiary amides *N,N*-diisopropylnaphthamide and *N,N*-diisopropylbenzamide to generate [EtZn{$C_{10}H_6C(=O)N$$i$-$Pr_2$-2}$_2$Li(THF)$_2$] and [Zn{$C_6H_4C(=O)N$$i$-$Pr_2$-2}$_3$Li(THF)] both of which were crystallographically characterized (**76** and **77**, Scheme 2.20). Two and three carboxylic amide molecules respectively have been deprotonated at the *ortho* position in these structures in compliance with the DoM concept [118–122]. Though these isolated crystalline products suggest dibasic and tribasic behaviour of [EtZn(Et)(TMP)Li(THF)$_n$], respectively, ^{1}H NMR spectroscopic studies and calculations signify the solution picture is more intricate.

Lithium TMP–zincate bases have also found application in metallocene chemistry. Though reactions with ferrocene are not simple, monodeprotonation of a single Cp (C_5H_5) ring is a common feature with di(*n*-butyl)TMP–zincate, [(TMEDA)Li(TMP)(*n*-Bu)Zn(*n*-Bu)] **78** [123]. Two crystallographically characterized products obtained from hexane solution, the neutral zinc complex [{(C_5H_5)Fe(C_5H_4)}$_2$Zn(TMEDA)] **79** and the tris-ferrocenyl anionic zincate [Li(THF)$_4$][{(C_5H_5)Fe(C_5H_4)}$_3$Zn] **80** (Figure 2.15) together with NMR studies of the reaction mixture, imply a complicated sequence of reactions dependent on stoichiometry and mixing time (Scheme 2.21).

2.3.5 Sodium TMP–Zincate Chemistry

Demands for lithium, a comparatively rare metal, are surging at present due principally to its excellent energy storage properties [124, 125]. To avoid over-exploitation and plan for any future availability/supply problems of lithium, activity in organosodium chemistry is anticipated to

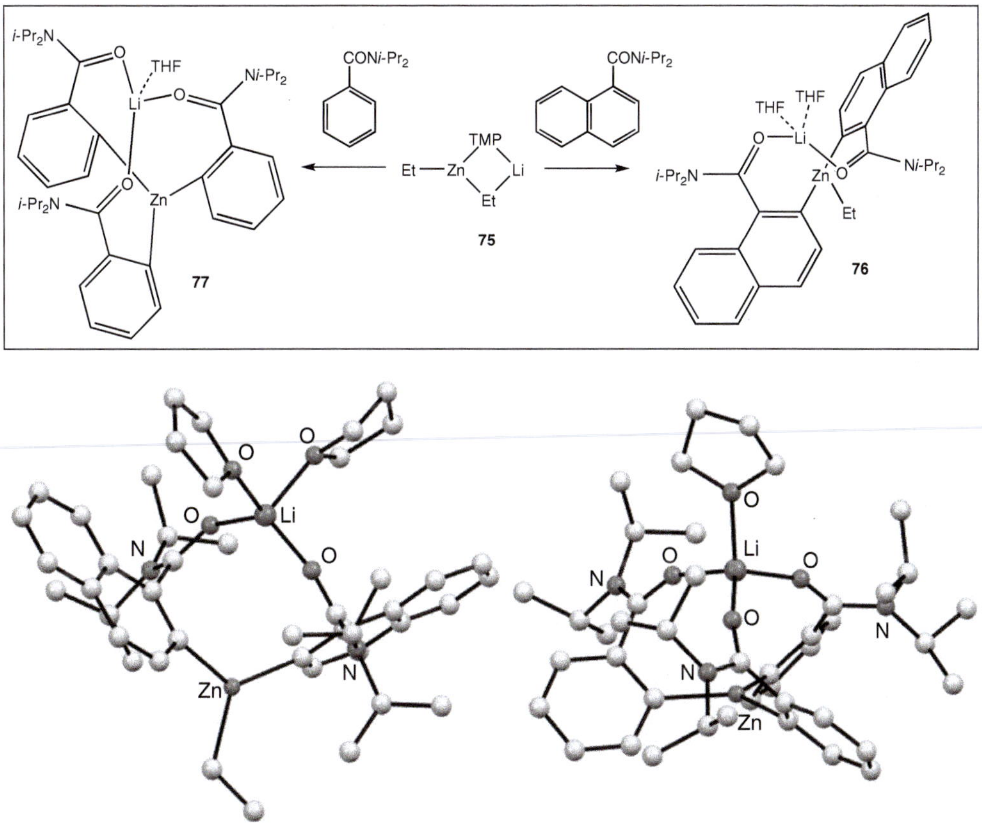

Scheme 2.20 Synthetic approach and molecular structures of [EtZn{$C_{10}H_6C(=O)Ni\text{-}Pr_2\text{-}2$}$_2$Li(THF)$_2$] **76** and [Zn{$C_6H_4C(=O)Ni\text{-}Pr_2\text{-}2$}$_3$Li(THF)] **77**. *Source:* Adapted from Kondo et al. [117].

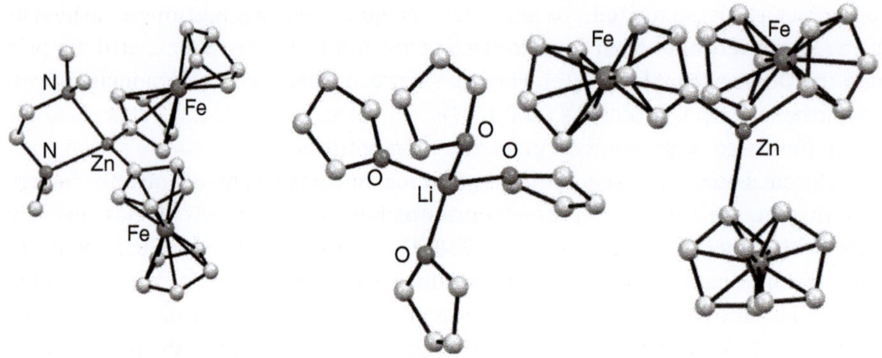

Figure 2.15 Molecular structures of [{(C_5H_5)Fe(C_5H_4)}$_2$Zn(TMEDA)] **79** (left) and [Li(THF)$_4$][{(C_5H_5)Fe(C_5H_4)}$_3$Zn] **80** (right). *Source*: Adapted from Barley et al. [123].

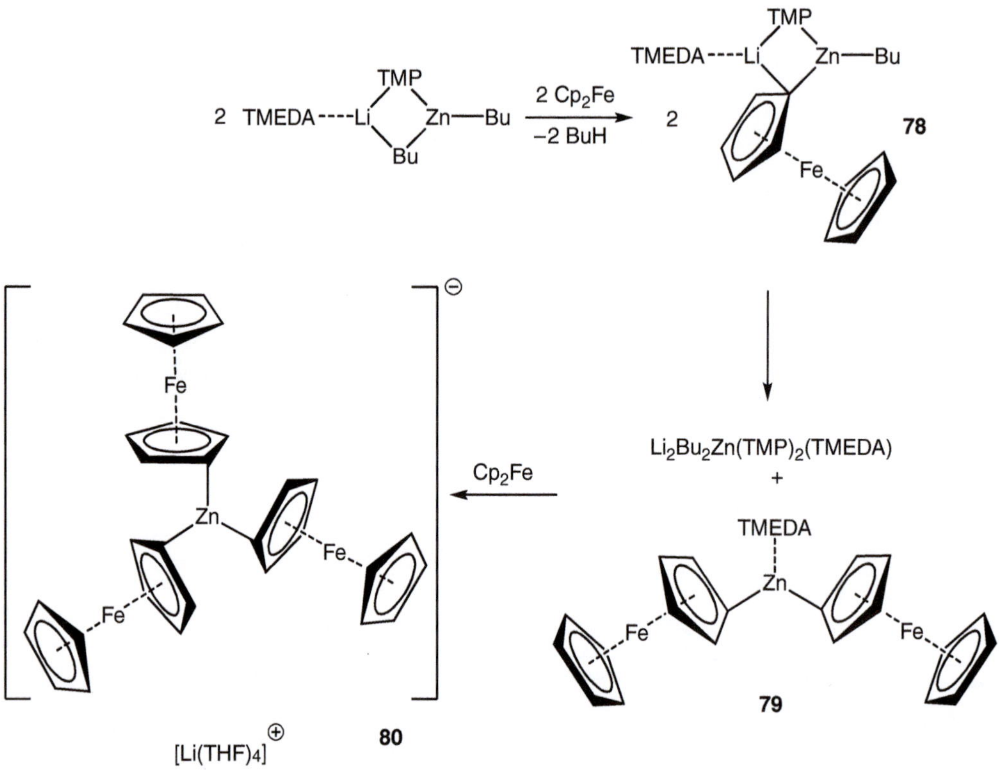

Scheme 2.21 Cascade of reactions upon deprotonating ferrocene with dialkyl-amido lithium zincate complex.

increase in attempts to develop a comparable chemistry to that of the ubiquitous organolithium reagents, since there are vast natural resources of sodium, the most abundant alkali metal on earth and in the oceans [126, 127]. That said, sodium TMP–zincate chemistry is already relatively well developed. Stimulated by the success of [(THF)Li(TMP)(*t*-Bu)Zn(*t*-Bu)], the closely related sodium analogue [(TMEDA)Na(TMP)(*t*-Bu)Zn(*t*-Bu)] (sodium TMP–zincate **81**) was introduced in 2005 [128] with its strong Brønsted basicity inspiring the term 'alkali-metal-mediated zincation'

(AMM*Zn*) to describe C–H deprotonation reactions that generate C–Zn bonds through the mediation of the also present sodium in the base molecule. Having a similar charge-contacted ion pair structure to lithium TMP–zincate, this sodium version (Figure 2.16) performs deprotonation reactions in apolar hexane solution as opposed to polar THF.

A representative sample of these reactions in which the deprotonated products have been crystallographically defined is now presented. Benzene [128, 129] and naphthalene [130], two benchmark molecules for assessing the efficiency of organometallic bases, can be regioselectively monodeprotonated and dideprotonated by sodium TMP–zincate (**82–85**, Scheme 2.22). The number of deprotonations in these reactions are controlled stoichiometrically though the dizincated products require the absence of TMEDA (prior to dideprotonation) in the case of [1,4-{(TMEDA)Na(μ-TMP)Zn(*t*-Bu)}$_2$C$_6$H$_4$] **84** and more forceful reflux conditions in the case of [(TMEDA)$_2$Na$_2$(μ-2,6-C$_{10}$H$_6$)Zn$_2$(*t*-Bu)$_2$] **85**. Highlighting the superiority of this synergistically operative zincate reagent, when naphthalene is metallated by *n*-BuLi [131] or superbasic

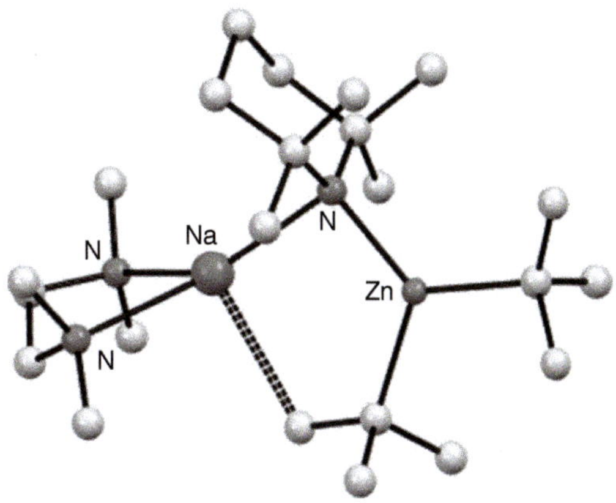

Figure 2.16 Molecular structure of [(TMEDA)Na(TMP)(*t*-Bu)Zn(*t*-Bu)] **81**. *Source*: Adapted from Andrikopoulos et al. [128].

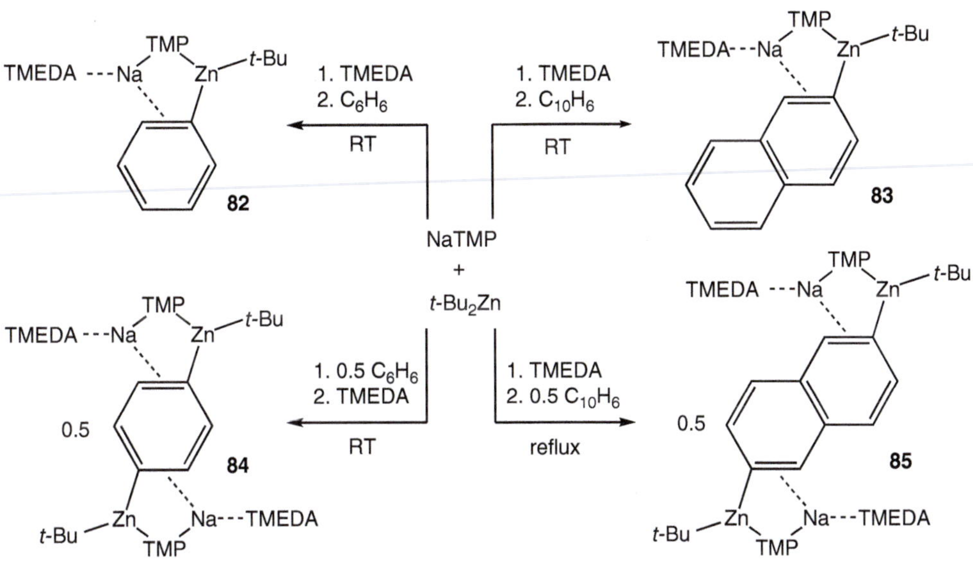

Scheme 2.22 Stoichiometry dependent mono- and di-deprotonation of aromatic substrates with sodium TMP–zincate-yielding complexes **82–85**.

LiCKOR [132] non-selective mixtures of 1- and 2-monosubstituted isomers as well as all ten possible disubstituted isomers are formed. The synergistic peculiarity of sodium TMP–zincate can also yield crystalline products in which *N,N*-dimethylaniline substrate is deprotonated in the *meta* position yielding [(TMEDA)Na(μ-Ar*)(μ-TMP)Zn(*t*-Bu)] (Ar* = 3-$C_6H_4NMe_2$) **86** (Scheme 2.23) [133] in contravention of the *ortho* (DoM) directionality usually obtained with conventional bases. Theoretical calculations on this system back up the experimental results that the *meta* isomer is the minimum-energy structure with electrostatic Na···(π-Ar*) contacts a factor in the *meta* selectivity.

With *m*-tolunitrile, sodium TMP–zincate executes the more expected *ortho* selectivity, but this is manifested in a remarkable 'sodium sodiumdizincate' $[(3\text{-Me-}C_6H_4CN)_2Na(TMEDA)_2]^+$ $[\{6\text{-Zn}(t\text{-Bu})\text{-3-Me-}C_6H_3CN\}_2Na(TMEDA)_2]^-$ **87** (Figure 2.17) [134]. The anionic moiety within this charge-separated structure has two *ortho*-zincated nitrile ligands, whereas the cationic moiety contains two neutral, non-zincated nitriles, with both moieties founded on an octahedral sodium platform completed by two equatorial TMEDA ligands. With 1-cyanonaphthalene, ortho-deprotonation again occurs but in the molecular product $[(TMEDA)_2Na\{2\text{-Zn}(t\text{-Bu})_2\text{-1-NC-}C_{10}H_6\}]$ **88**, while with *t*-BuCN, where acidic C–H bonds are absent, the trialkylzincate product $[\{(t\text{-BuCN})_2Na(TMEDA)_2\}^+(t\text{-Bu}_3Zn)^-]$ **89** indicates a dismutation process is operating.

In the absence of X-ray crystallographic studies, the above reactions with nitriles would not be well defined. However, to draw the full structural map of chemical changes occurring between the limits of the starting zinc reagent and deprotonated zincated product, other evidence such as DFT calculations are essential. For example, from DFT calculations using $[(TMEDA)Na(\mu\text{-}NMe_2)(\mu\text{-Me})Zn(Me)]$ and $[(TMEDA)Na(\mu\text{-}NMe_2)(\mu\text{-Ph})Zn(Me)]$ to mimic the

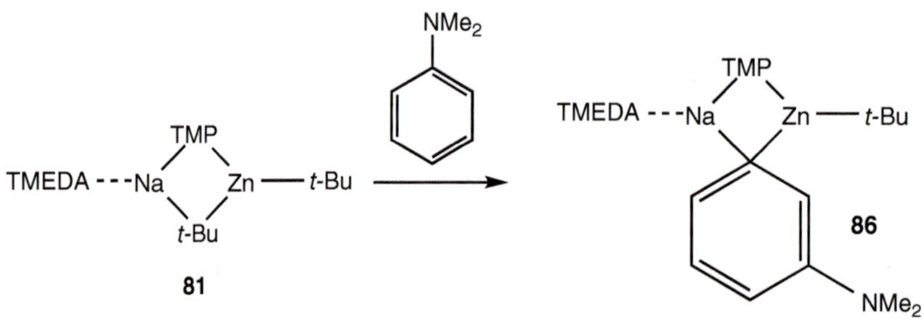

Scheme 2.23 *Meta*-deprotonation of *N,N*-dimethylaniline using sodium TMP–zincate.

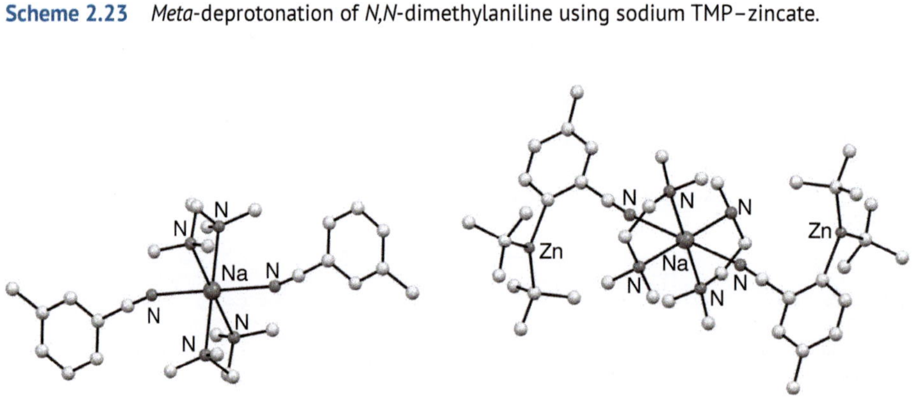

Figure 2.17 Molecular structure of $[(3\text{-Me-}C_6H_4CN)_2Na(TMEDA)_2]^+$ $[\{6\text{-Zn}(t\text{-Bu})\text{-3-Me-}C_6H_3CN\}_2Na(TMEDA)_2]^-$ **87**. *Source*: Adapted from Clegg et al. [134].

deprotonation of benzene by sodium TMP–zincate, it has been postulated that, as aforementioned with lithium TMP–zincate, the amide is the kinetic base abstracting a hydrogen from benzene to generate a bisalkyl-monoaryl intermediate [{(TMEDA)Na(μ-Me)(μ-Ph)Zn(Me)], which in turn reacts with liberated Me_2NH to afford the final products of the reaction, [{(TMEDA)Na(μ-NMe_2)(μ-Ph)Zn(Me)] and methane (Scheme 2.24) [135].

Crystallographic characterization was also essential in revealing the product of reacting *N,N*-diethyl-thiophene-2-carboxamide with sodium TMP–zincate in a 1 : 1 stoichiometric ratio in benzene. Surprisingly, the reaction was not a single-site C—H to C—Zn deprotonation as anticipated, but a 3,5 twofold deprotonation manifested in a molecular cage with a 16-crown-4 zincocyclic core in [{Na[μ-3,5-[2-C(O)NEt_2]-C_4H_1S]Zn(*t*-Bu)$_4$] **90** (Scheme 2.25) [136].

Though primarily a base, sodium TMP–zincate can also exhibit dual functionality, exemplified through its reactions with benzoylferrocene [137]. While in isolation *t*-Bu_2Zn does not react with this substituted metallocene, the synergistic reagent generates [(TMEDA)Na(μ-TMP)Zn{OC(*t*-Bu)(Ph)(η^5-C_5H_3)Fe(η^5-C_5H_5)}] **91** formed via deprotonation of the α-position of the Cp ring and a 1,2-addition of an alkyl substituent across the carbonyl function (Scheme 2.26). Quenching the reaction mixture with iodine in THF solution affords a mixture of

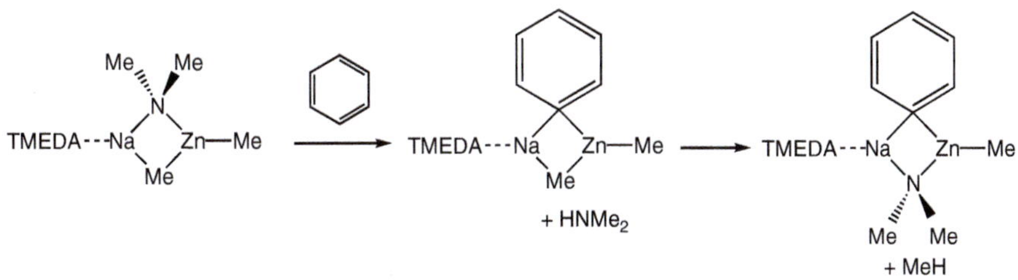

Scheme 2.24 Calculated two-step mechanism for zincation of benzene with bisalkyl-amido sodium zincate.

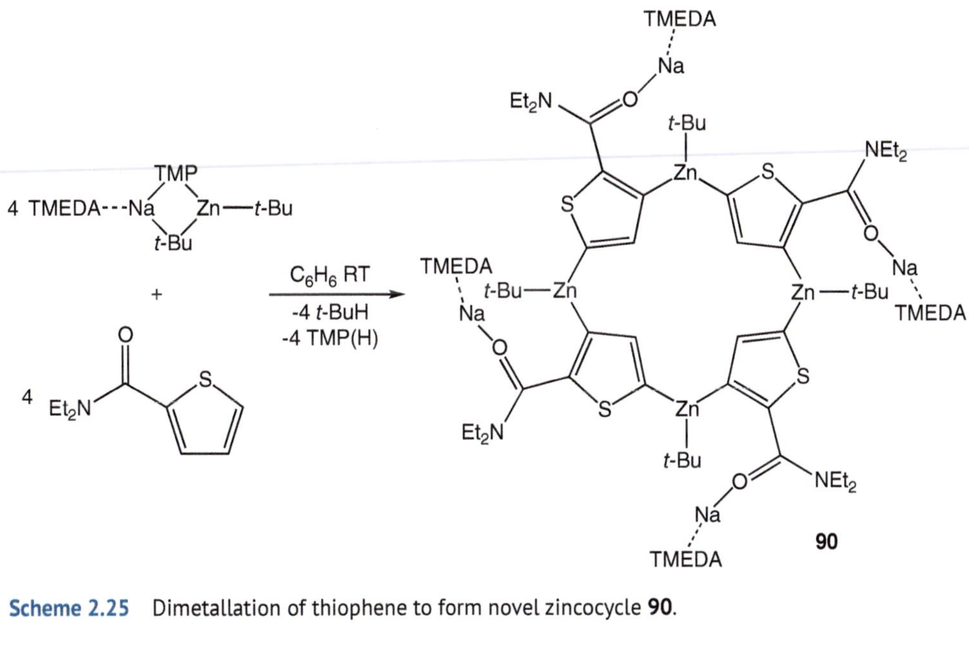

Scheme 2.25 Dimetallation of thiophene to form novel zincocycle **90**.

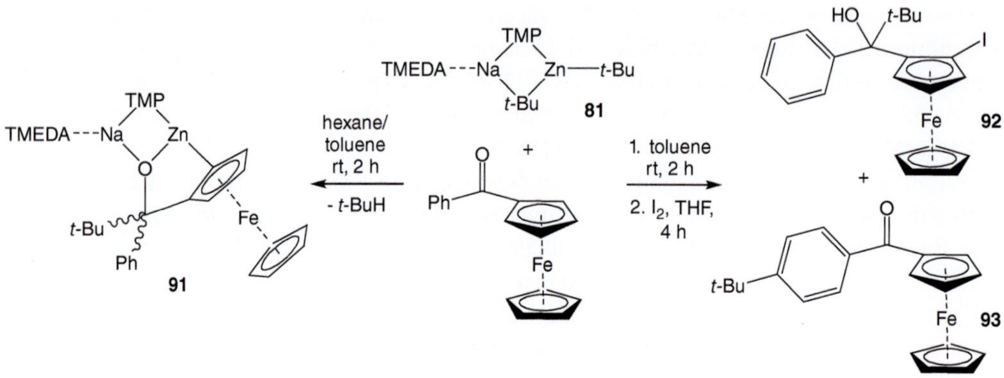

Scheme 2.26 Zincation of benzoylferrocene showing metallated intermediate **91** and products after iodine quenching.

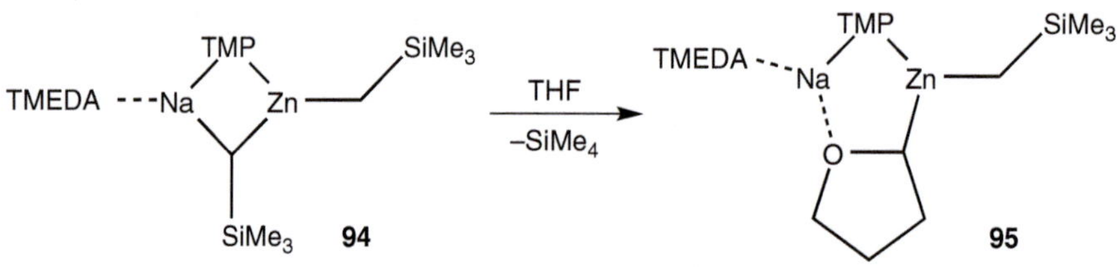

Scheme 2.27 Zincation of THF with sodium zincate **94**.

the anticipated product [PhC(OH)(*t*-Bu)(η^5-C_5H_3I)Fe(η^5-C_5H_5)] **92** and the unanticipated product [4-*t*-Bu-C_6H_4C(=O)(η^5-C_5H_4)Fe(η^5-C_5H_5)] **93**. The second product implies a remote 1,6-addition across the phenyl ring, a reaction precedented with benzophenone [138].

'Cleave and capture chemistry' is an expression that has been used to describe the deprotonative action of TMP–zincates and related synergistic bases [139]. It views such reactions as the sequential cleavage of selected C–H bonds and the capturing of the emergent anion by the multi (Lewis acid)–(Lewis base) character of the residue of the bimetallic base. THF provides a striking example. Ring opening of this ubiquitous cyclic ether liberating ethene can generally take place even at sub-ambient temperatures when deprotonated by a conventional base such as *n*-butyllithium. In contrast, the trimethylsilylmethyl-based TMP–zincate [(TMEDA)Na(TMP)(CH_2SiMe_3)Zn(CH_2SiMe_3)] **94** cleaves one α-C–H bond then captures the THF anion ($C_4H_7O^-$) intact in a crystalline complex (**95**, Scheme 2.27) [140].

2.3.6 Lithium Chloride (Turbo Charged) TMP–Zinc Chemistry

The most popular class of TMP–zinc reagent in terms of its widespread scope in organic synthesis is referred to as turbo reagents, typified by TMPZnCl·LiCl **96** [141] and $TMP_2Zn\cdot 2MgCl_2\cdot LiCl$ **97** [142]. While formally a turbo is a turbine-driven forced induction device that increases an internal combustion engine's efficiency and power output, in this colloquial chemical context it refers to the presence of lithium chloride increasing the efficiency and power of zinc amides (magnesium amides are boosted likewise) in deprotonation reactions. Good solubility in THF, enhanced functional group tolerance and fast kinetics (in comparison to conventional zinc amides) are general core attributes of these synergistic mixed lithium–zinc reagents. As mentioned previously, growing

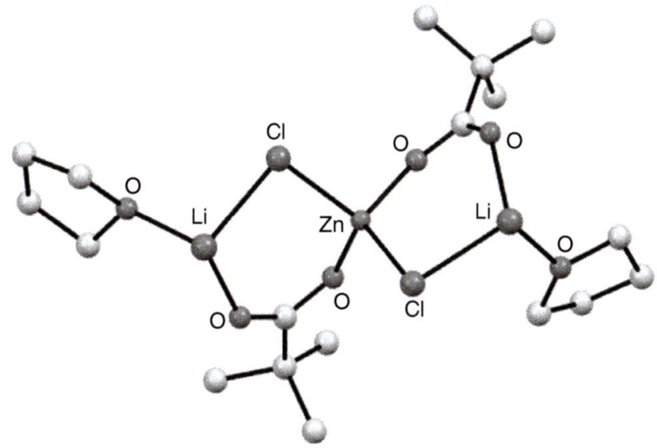

Figure 2.18 Molecular structure of Zn[(μ-OPiv)(μ-Cl)Li(THF)]$_2$ **98**. *Source:* Adapted from Hernán-Gómez et al. [143].

single crystals from neat THF solutions can be exceptionally challenging so as yet there has been no definitive crystallographic authentication of a turbo-zinc reagent structure. The dual Lewis acid (Li)–Lewis base (Cl) character of lithium chloride makes a molecular form of the salt an efficient co complexing entity so it is highly likely that these species have molecular structures akin to that of the structurally defined magnesium turbo reagent [(THF)$_2$Li(μ-Cl)$_2$Mg(THF)TMP] **60**, in which monomeric LiCl and Mg(TMP)Cl units co-complex via bridging Cl ligands with other sites filled by THF (Figure 2.12) [94].

Germane to the area of lithium–zinc, structurally induced synergistic chemistry, it has been noted that lithium chloride imparts a solubilizing effect on zinc pivalate Zn(OPiv)$_2$ in THF solution. In the absence of the lithium salt zinc pivalate is essentially insoluble in THF. A structural origin was found for this effect in the crystallographic determination of the molecular complex Zn[(μ-OPiv)(μ-Cl)Li(THF)]$_2$ **98** (Figure 2.18) [143]. Both anionic ligands, Cl and pivalate bridge the Li and Zn centres in this structure.

2.3.7 Indirect TMP Zincation

The aforementioned Zn–H exchange reactions can be classified as direct synchronized zincations involving a base containing either Li or Na in combination with Zn, where Zn attaches directly to the carbon atom of the organic substrate that has underwent deprotonation. Indirect zincations can also occur in bimetallic systems. The distinction is that this is a two-step sequential process where the metallation reaction has already been carried out by a more polar reagent (usually involving lithium) to generate an anionic intermediate that is then trapped by a zinc compound. Termed trans-metal-trapping (TMT) [144] as far as structural evidence is concerned this indirect metallation method is most well developed with lithium–aluminium systems (see Section 2.3.8). Trans-metal-trapping is characterized by retention of both metals in the final organometallic product. It has been described as a 'crossover complex' that stops before a full transmetallation is reached [145]. A structurally defined example with zinc is the formation of heteroleptic [(THF)$_3$Li(μ-Cl)ClZn(μ-Dip-dabqdi)ZnCl(μ-Cl)Li(THF)$_3$] **99** by insertion of zinc chloride into the precursor lithiated Dip-dabqdi compound (Dip-dabqdiH$_2$ = 2,5-diamino-1,4-benzoquinonediimine) (Scheme 2.28) [146]. Synthetically, trans-metal-trapping is well developed with LiTMP as the base and Zn(TMP)$_2$ as the trap as well as with ZnCl$_2$, ZnCl$_2$·2LiCl, and ZnCl$_2$(TMEDA) traps, though no structural characterization has been forthcoming [147]. However, the structures of intermediates in which both Li and Zn are present have been predicted in LiTMP-mediated metallations of methoxy-substituted arenes via DFT calculations (Figure 2.19) [148, 149].

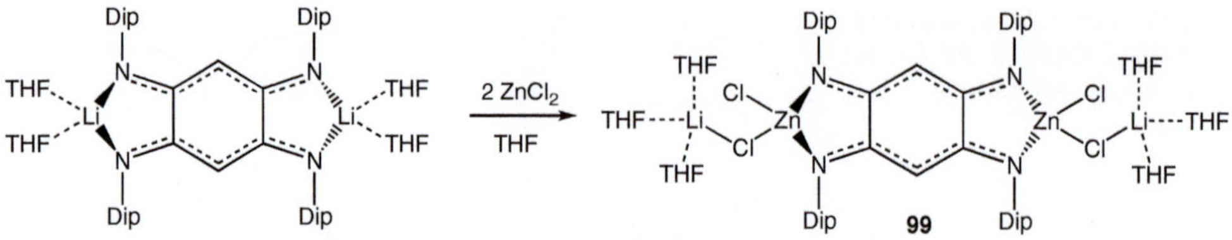

Scheme 2.28 $ZnCl_2$ insertion into lithiated Dip-dabqdi complex to yield bimetallic lithium zincate **99**.

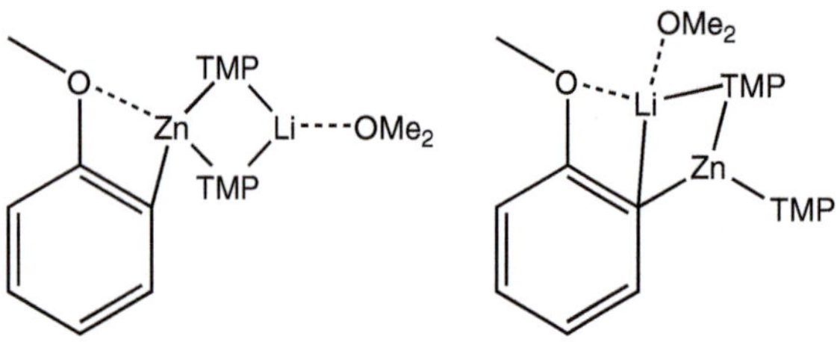

Figure 2.19 Proposed anisolyl lithium zincate intermediates from theoretical study.

2.3.8 Alkali Metal Group 13 Ates

The development of new strategies to promote alumination (C–H to C–Al exchange), including by pairing compounds of the group 13 metal with alkali-metal reagents, is a particularly attractive and timely focus as a consequence of aluminium's high natural abundance (and associated low cost) and the increasing importance of sustainability in modern chemistry. An added incentive is aluminium's relatively low toxicity in comparison to that of the precious metals commonly used in synthesis and catalysis such as those of the platinum group.

In terms of their preparation, heterobimetallic aluminates benefit from access to commercially available $LiAlH_4$ which can operate as a base towards substrates such as alcohols and secondary amides with relatively acidic X–H bonds, providing an alternative methodology for preparation of tailored aluminates beyond the co-complexation approach commonplace for the synthesis of magnesiates, zincates, etc. An exemplar is the reaction of $LiAlH_4$ with HMDS(H) in which two hydride anions react to yield heteroleptic $LiAlH_2(HMDS)_2$ **100** and liberating hydrogen gas. This heterobimetallic heteroleptic complex is an effective catalyst for aldehyde and ketone hydroboration with the rationale behind this effectiveness being revealed by knowledge of its structure [150]. Specifically, this displays that the two metals work in unison with the Lewis acidic alkali metal holding the substrate at a Lewis basic site and presenting it to the aluminium centre to aid the hydride transfer from metal to substrate (Figure 2.20).

The ligating efficiency of the Lewis base donor is crucial also, since three equivalents of monodentate THF facilitate a superior catalytic performance to that of one equivalent of tridentate PMDETA as the chelating nature of PMDETA means the lithium is not sufficiently exposed upon decoordination of one donor atom to allow substrate coordination to occur.

Co-complexation is also a viable route for the preparation of hydridoaluminates via the commercially available reagent *i*-Bu_2AlH (DIBAL). Pairing DIBAL with LiTMP affords the heteroleptic aluminate *i*-$Bu_2Al(TMP)HLi$ **101** which can catalyze hydroboration of C=O and C≡C bonds [151]. The versatility of this bimetallic complex was displayed by its capacity to also act as a TMP base in

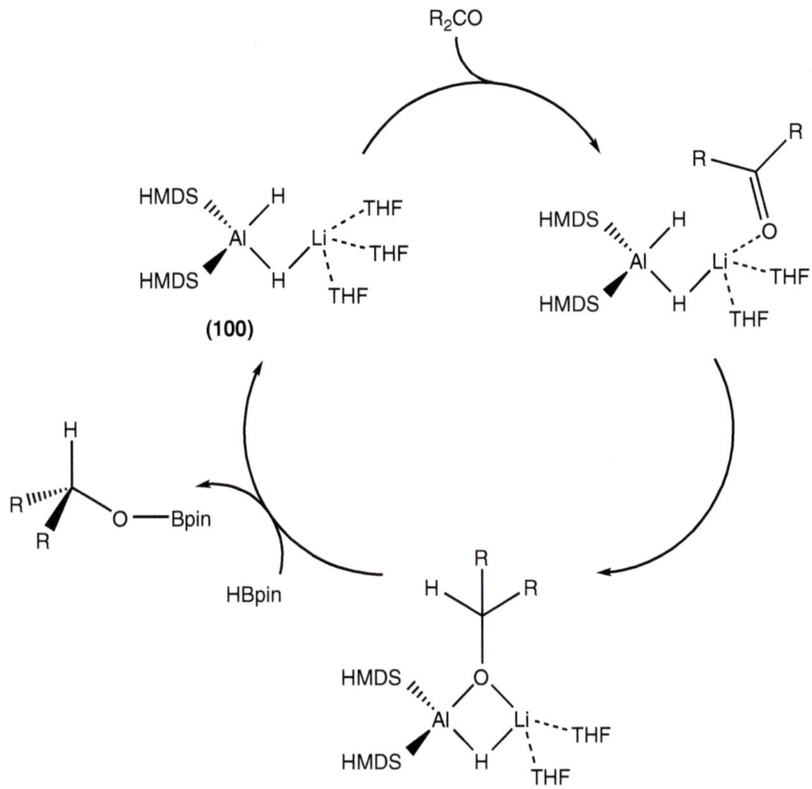

Figure 2.20 Postulated mechanism for catalytic hydroboration of carbonyl functionality by lithium aluminate complex **100**.

deprotonation of a triazole; but as a nucleophile in the Al–H addition across a C=N bond of pyrazine (Scheme 2.29).

Early structural work in synergistic lithium–aluminium deprotonations hinted at the same process occurring as described previously for synergistic zincation, that is, an alkali metal-mediated alumination process occurring whereby $LiAlR_4$ would deprotonate the substrate R′H to yield $LiAlR_3R'$, liberating RH. This was typified by the complex generated through the co-complexation of *i*-Bu_3Al and LiTMP [152], which displayed the anticipated contacted ion pair structure **102** upon recrystallization from hexane in the presence of THF (Figure 2.21) [153].

However, the putative hydride-free bis-amide complex $LiAl(TMP)_2$*i*-Bu_2, made similarly through co-complexation of *i*-$Bu_2Al(TMP)$ with LiTMP, could not be crystallographically characterized, with the crystals obtained on adding stoichiometric THF to a hexane solution instead proving to have resulted from a TMP-induced deprotonation of a THF molecule adjacent to the heteroatom in (THF)Li(μ-TMP)(μ-C_4H_7O)Al*i*-Bu_2 **103** (Figure 2.22) [154]. Using different bidentate and tridentate donors likewise resulted in deprotonation adjacent to a Lewis donor heteroatom and yielded structures with similar motifs to (THF)Li(μ-TMP)(μ-C_4H_7O)Al*i*-Bu_2 [155]. The same donors solvate a contacted ion pair type complex with the less bulky LiTMP–*i*-Bu_3Al partnership.

DOSY NMR and computational studies established that LiTMP and *i*-$Bu_2Al(TMP)$ could not co-complex due to the steric clashing that would result in a putative four-membered $LiN_{TMP}AlN_{TMP}$ ring [156]. Furthermore, it appears that the preformed aluminate (THF)Li(μ-TMP)(μ-*i*-Bu)Al*i*-Bu_2

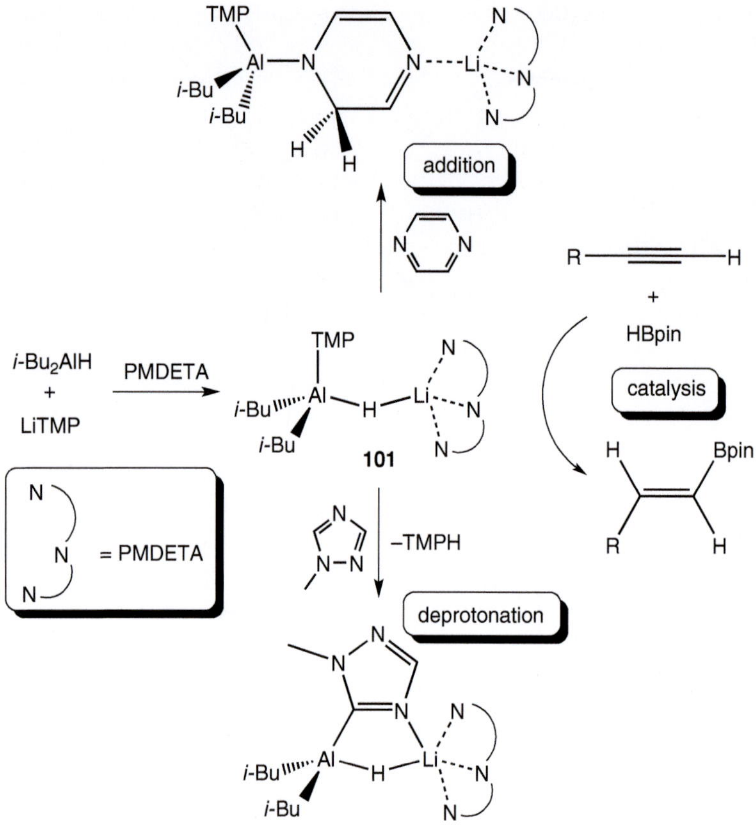

Scheme 2.29 Synthesis and versatile reactivity of *i*-Bu$_2$Al(TMP)HLi **101**.

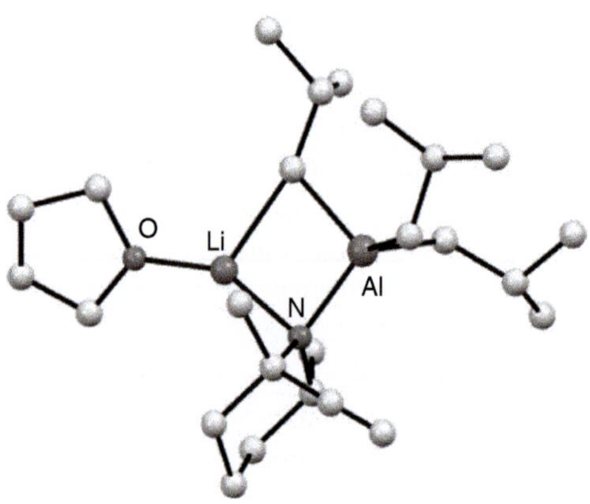

Figure 2.21 Molecular structure of (THF)Li(μ-TMP)(μ-*i*-Bu)Al*i*-Bu$_2$ **102**. *Source*: Adapted from Naka et al. [153].

is not an active bimetallic base and that in bulk THF solution it consists as a complicated mixture of five components, namely [Li(THF)$_4$]$^+$[*i*-Bu$_3$Al(TMP)]$^-$, [Li(THF)$_4$]$^+$[*i*-Bu$_4$Al]$^-$, LiTMP(THF) [65], *i*-Bu$_3$Al(THF), and *i*-Bu$_2$Al(TMP)(THF) [157], obtained through a redistribution process. The only active Brønsted base in this mixture with regard to the sp^2-C–H bonds of aromatic substrates is LiTMP. Thus, the initial reaction of these 'aluminates' is deprotonation of a substrate actioned by LiTMP, while the subsequent step involves trapping of the emergent complex anion by an organoaluminium reagent, for example, triisobutylaluminium (i-Bu$_3$Al) or [i-Bu$_2$Al(TMP)] by insertion into the Li–C bond. The origin of this trans-metal-trapping (TMT) concept lies in the complementarity of the metal reagents involved. Highly polar LiTMP is a strong Brønsted base; whereas these moderately polar organoaluminium complexes are more carbophilic and so can stabilize the resulting carbanions. The question of why these reactions are labelled trans-metal-trapping and not transmetallation is sketched out below and considered from a NMR perspective in detail in Chapter 6.

In the latter case, the metals generally exchange and produce a lithium product and an aluminium product; whereas in the former case, there is only one product, which contains both lithium and aluminium. The clear benefit of this base-trap protocol is illustrated when the base–trap, LiTMP–i-Bu$_2$Al(TMP) synergistic partnership, is administered to anisole (Scheme 2.30) [156]. The yield of the *ortho*-metallated (aluminated) product is essentially quantitative; in contrast to only 10% of *ortho*-lithiated product when the trap is omitted. The problem in the latter case is that there is an equilibrium between the lithiated product and the LiTMP starting material due to the

Figure 2.22 Molecular structure of (THF)Li(μ-TMP)(μ-C$_4$H$_7$O)Ali-Bu$_2$ **103**, featuring an intact deprotonated THF ring. *Source*: Adapted from Crosbie et al. [154].

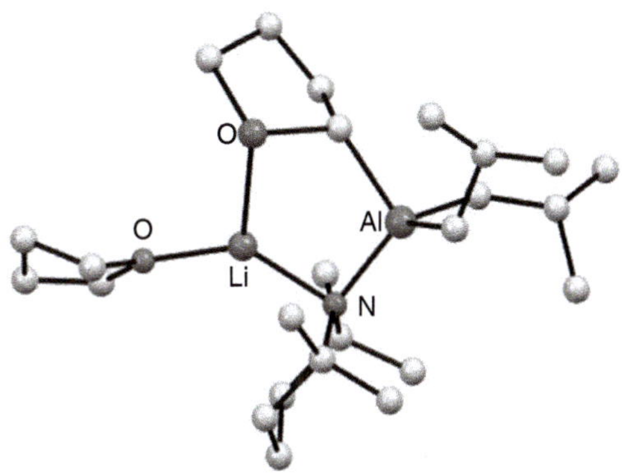

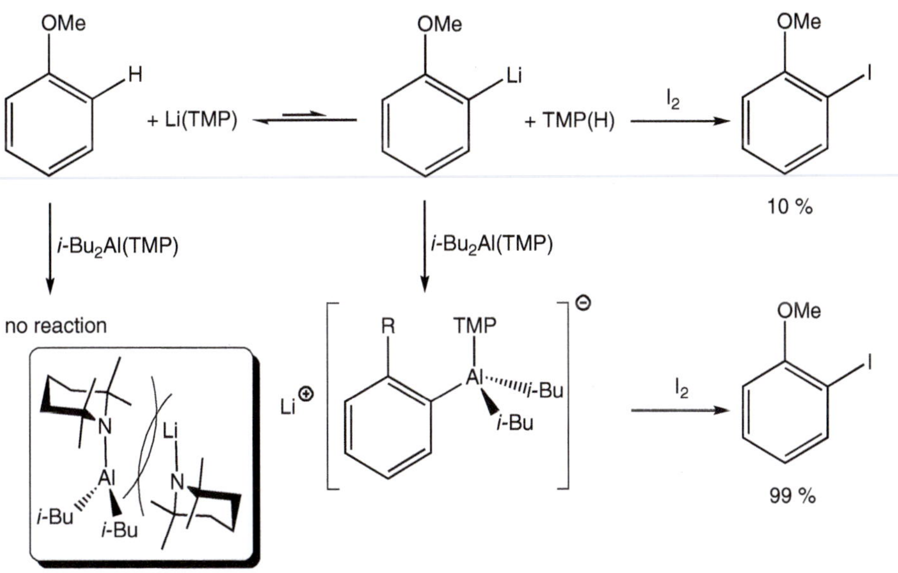

Scheme 2.30 Basis of trans-metal-trapping using a lithium base and sterically hindered aluminium partner which cannot co-complex (inset).

relatively high acidity of its conjugate acid TMP(H). The trapping action by the organoaluminium reagent removes the deprotonated anisole from the equilibrium and drives the process forward to the desired metallated substrate.

This TMT approach has been demonstrated as an effective tactic for deprotonating a mesityl arm of the common *N*-heterocyclic carbene SIMes [1,3-bis(2,4,6-trimethylphenyl)-4,5-dihydroimidazol-2-ylidene] [158]. Related unsaturated NHCs are susceptible to attack at the olefinic backbone by strong monometallic lithium bases (e.g. in **104**) [159], but this unsaturated example is alternatively attacked at a mesityl arm leading to the opening of the five-membered imidazole ring to generate new indole ring system **105** (Figure 2.23). Performing a TMT reaction by introducing a *i*-Bu_2Al(TMP) trap prevents this ring opening, enabling the aluminium centre to stabilize the newly formed carbanion in an ate arrangement (**106**) with the lithium counter-cation being solvated by the Lewis donating normal carbene centre and additional THF molecules.

The chemical character of the base can play a key structural role in the outcome of any polymetallation reaction, as evidenced by the metallation of ferrocene. When using a sodium–zinc pairing (see Section 2.3.5) tetrametallation of ferrocene can be achieved (akin to that previously recorded utilizing a sodium magnesiate), due in part to the fact that volatile and nondonating butane is generated via *t*-Bu basicity. However, when turning to the LiTMP–*i*-Bu_2Al(TMP) partnership, the lithiation reaction generates the poorly volatile Lewis basic molecule TMP(H), which can in turn act as a Lewis base donor to lithium within the TMT product and thus prevent formation of the inverse crown ring around the organometallic sandwich complex [160].

Structural information can also shed light on the possible degradations of substrates on the metallation pathway. This is seen for example in the metallation of *N*,*N*-dimethylphenylethylamine, with the most acidic benzylic position being the site of preferential attack. However, the lithiated intermediate subsequently undergoes β-elimination to liberate styrene, concomitantly producing a molecule of $LiNMe_2$, which is trapped in a bimetallic clamp by *i*-Bu_2Al(TMP) to yield (THF)Li(μ-TMP)(μ-NMe_2)Al*i*-Bu_2 **107**.

Trans-metal-trapping is not limited to aluminium in group 13 but can also be executed with gallium(III) reagents, thanks in part to the d-block contraction which means the heavier group 13 element has a smaller atomic radius and thus forms shorter metal–ligand bonds. Again, steric clashing is the obstacle here to prevent the two monometallic reagents from combining into a single bimetallic structure with diminished deprotonative power. For example, $LiCH_2SiMe_3$ and $Ga(CH_2SiMe_3)_3$ are known to co-complex into a polymer [161] which reacts as a nucleophile rather than a base with pyrazine yielding **108** (Scheme 2.31). However, NMR evidence shows that the

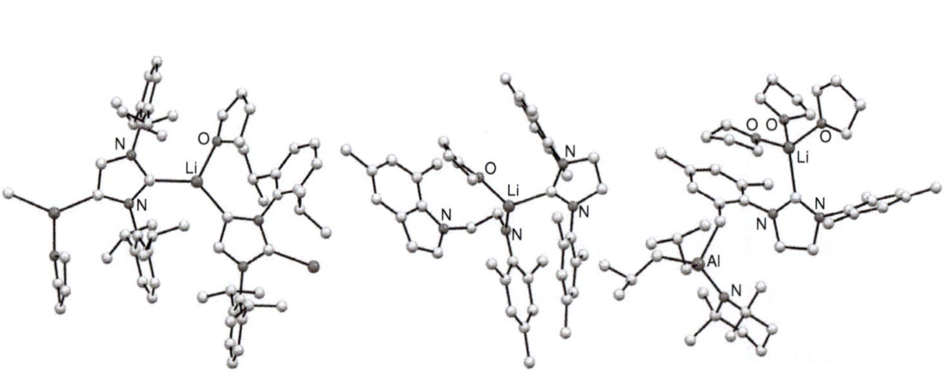

Figure 2.23 Molecular structures of polymeric lithiated NHC **104** (left), ring-opened saturated NHC **105** (centre) and metallated NHC which has undergone trans-metal-trapping to capture the intact anion **106** (right). *Sources*: Adapted from Hernan-Gomez et al. [158]; Wang et al. [159].

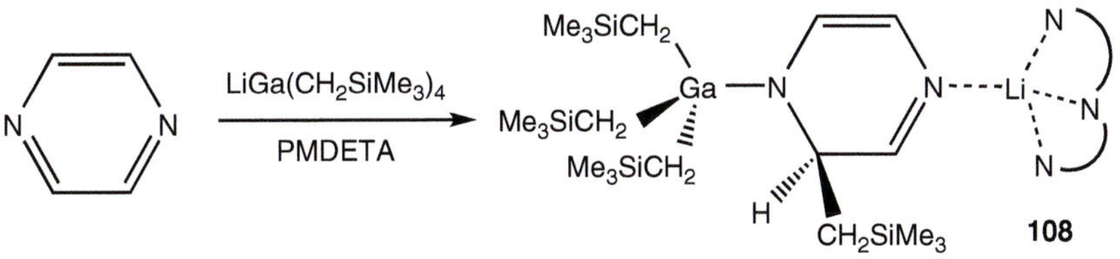

108

Scheme 2.31 Nucleophilic addition of homoleptic lithium gallate to pyrazine.

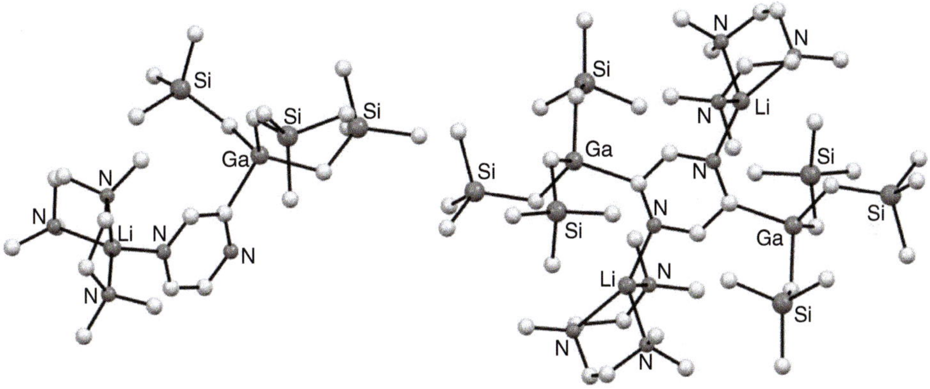

Figure 2.24 Molecular structures of mono-metallated **109** and di-metallated pyrazine **110** formed from reaction with LiTMP/Ga(CH_2SiMe_3)$_3$. *Source*: Adapted from Uzelac et al. [162].

bulkier LiTMP/Ga(CH_2SiMe_3)$_3$ partnership remains as separated monometallic species in solution which, operating synergistically, can deprotonate pyrazine, with the gallium centre ultimately bonding to the aryl anion to give gallate **109** with the remote ring nitrogen functioning as a Lewis donor to lithium (Figure 2.24) [162]. Furthermore, this reaction is stoichiometry dependent with two molar equivalents of the organometallic reagents resulting in *para*-dimetallation of the substrate and yielding **110**.

Enhanced stability is an additional benefit of utilizing gallium when metallating sensitive fluoroaromatic substrates [163]. Dehydrometallation of 3-fluoroanisole by LiTMP in the position mutually *ortho* to the two substituents clearly demonstrates this effect. The resulting anion can be trapped and stabilized by *i*-Bu_2Al(TMP) in **111** but this product quickly eliminates fluoroaluminate **112** with concomitant formation of a benzyne intermediate which itself is trapped by addition of the *in situ* generated TMP(H). Homoleptic Ga(CH_2SiMe_3)$_3$ also serves as a suitable trapping agent for metallated 1,3,5-trifluorobenzene but with no fluoride elimination/benzyne formation (giving **113**), meaning that onward electrophilic quenching provides substantially higher yields than with an unstable aluminate intermediate (Figure 2.25).

2.3.9 Bimetallic Complexes Without an Alkali Metal Component

Although this article has focused on alkali metal-induced synergistic effects, it is apparent that other metal partnerships not involving an alkali metal might also exhibit synergistic behaviour in their reactivity. Here, a few examples are spotlighted. Logically, one would perhaps look first for such behaviour in their closest relatives, namely, the alkaline-earth metals. *In situ* metathesis (or transmetallation) is a common reaction type where Grignard reagents come face-to-face with a compound of another metal. Here, the Grignard reagent RMgX is reacted with a metal salt of a less

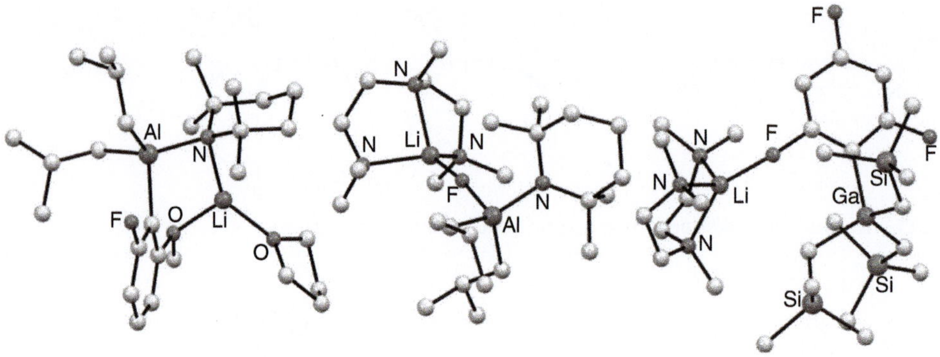

Figure 2.25 Molecular structures of aluminated fluoroanisole **111**, lithium fluoroaluminate formed from MF elimination **112** and gallated trifluorobenzene **113**. *Source*: Adapted from McLellan et al. [163].

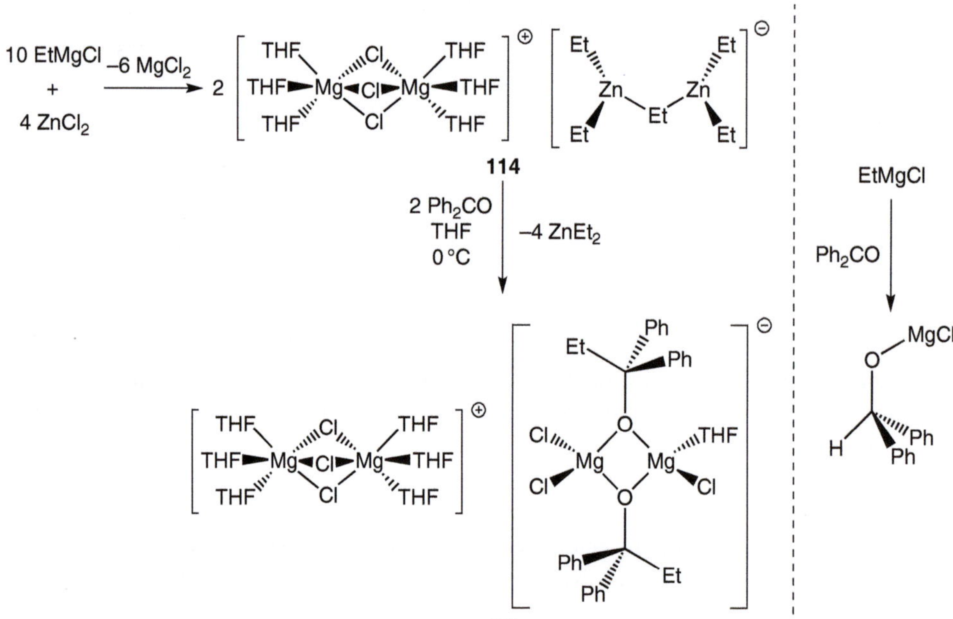

Scheme 2.32 Contrasting reactivity of benzophenone with magnesium zincate **114** and Grignard reagent.

electropositive metal such as a zinc dihalide ZnX_2 to access lower polarity organometallic reagents ZnR_2. These reactions are generally favoured thermodynamically by the concomitant formation of MgX_2 ionic salts. Interestingly, when carried out *in situ*, such reaction mixtures can operate synergistically without complete transmetallation taking place. For example, reaction of EtMgCl with $ZnCl_2$ in THF affords the magnesium dinuclear zincate $[\{(THF)_6Mg_2Cl_3\}^+\{Zn_2Et_5\}^-]$ **114** (Scheme 2.32), whose solution behaviour is looked at in Chapter 6. Its reaction with benzophenone producing $[\{(THF)_6Mg_2Cl_3\}^+ \{Mg_2(OCEtPh_2)_2Cl_3(THF)\}^-]$ **115** and $ZnEt_2$ (Scheme 2.32, left) has been rationalized in terms of a magnesium–zinc synergistic effect. This contrasts with the outcome obtained when benzophenone was allowed to react with EtMgCl in the absence of $ZnCl_2$, which preferentially leads to reduction of the ketone (Scheme 2.32, right) [164].

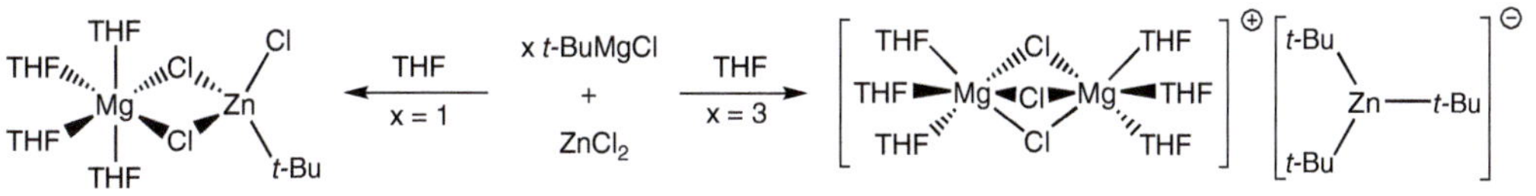

Scheme 2.33 Stoichiometric dependence on preparation of bimetallic Mg/Zn species.

Magnesium–zinc hybrid chemistry has also been illustrated nicely in reactions between *t*-BuMgCl and $ZnCl_2$ in THF solution, where an equivalent or excess of the Grignard reagent gives rise to different isolable products in [$(THF)_4MgCl_2Zn$(*t*-Bu)Cl] and [$\{Mg_2Cl_3(THF)_6\}^+$\{Zn*t*-$Bu_3\}^-$], respectively (**116** and **117**, Scheme 2.33) [165].

These reactions have been interpreted as co-complexation processes whereby the two expected products, for example $MgCl_2$ and *t*-BuZnCl have co-complexed in the presence of THF supports to give hybrid structures where magnesium and zinc are intimately linked through two chlorine bridges giving rise to a planar [Mg(μ-Cl)$_2$Zn] 4-atom ring (**116**, Figure 2.26). A terminal chlorine and a tert-butyl group complete the zinc tetrahedral geometry [166]. In contrast, [$\{Mg_2Cl_3(THF)_6\}^+$\{Zn*t*-$Bu_3\}^-$] adopts a charge-separated ion pair arrangement in which the excess *t*-Bu ligands migrate to the Zn centre in the anionic moiety. Reactivity studies on **117** have elaborated its effectiveness at promoting direct Zn–I exchange reactions with aromatic halides in THF solution at room temperature. For a detailed exploration of the solution interrogation of these processes, see Scheme 2.33 and also Chapter 6. However, briefly, it can exchange either two or three alkyl ligands as demonstrated by the crystallographically characterized products [$\{Mg(THF)_6\}^{2+}$ 2\{Zn(*o*-C_6H_4–OMe)$_3\}^-$] and [$(THF)_4$MgCl(NC-*o*-C_6H_4)ZnI(*o*-C_6H_4-CN)(THF)] obtained via reactions with 2-iodoanisole and 2-iodobenzonitrile, respectively (**118** and **119**, Figure 2.27).

Similarly, simple looking 1:1 metathesis reactions introducing $ZnCl_2$ to a series of alkyl and aryl Grignard reagents (EtMgCl, *t*-BuMgCl, *n*-BuMgCl,and *o*-OMe-C_6H_4MgBr) in THF solution have also produced novel crystalline Mg–Zn hybrid complexes of formula [$\{(THF)_2Mg(\mu\text{-}Cl)_3ZnR\}_2$] (R = Et, *t*-Bu, *n*-Bu or *o*-OMe-C_6H_4, **120–123**, respectively, see also Chapter 3, Figure 3.8). Interestingly, addition of LiCl enables the chemoselective alkylation of the sensitive ketone trifluoroacetophenone (Scheme 2.34), suggesting the synergistic chemistry at play could have a three-component Li/Mg/Zn origin [167].

Figure 2.26 Molecular structure of [$(THF)_4MgCl_2Zn$(*t*-Bu)Cl] **116**. *Source*: Adapted from Hevia et al. [166].

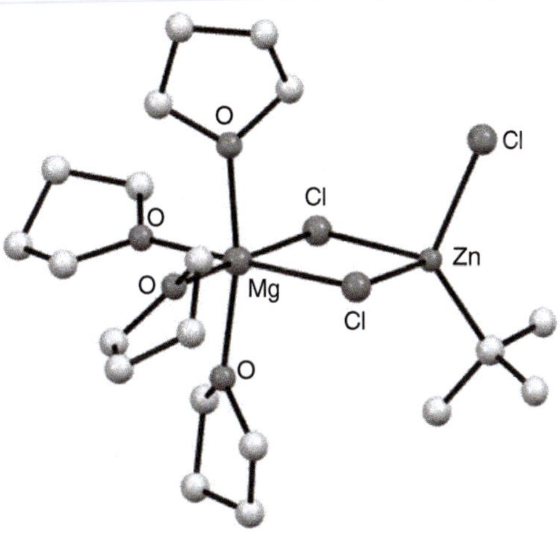

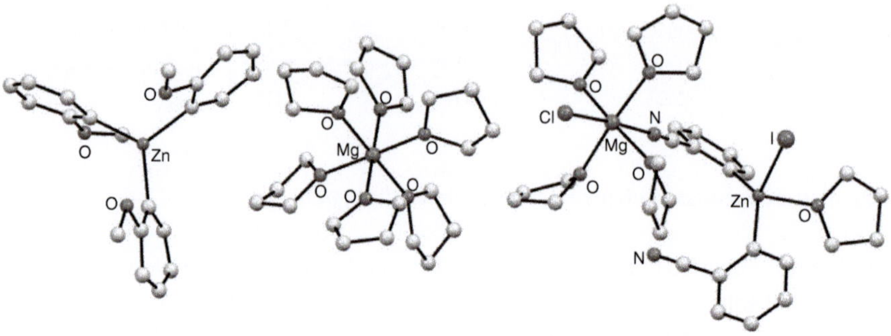

Figure 2.27 Molecular structures of $[\{Mg(THF)_6\}^{2+}\ 2\{Zn(o\text{-}C_6H_4\text{-}OMe)_3\}^-]$ **118** (only one of the zincate anions shown, left) and $[(THF)_4MgCl(NC\text{-}o\text{-}C_6H_4)ZnI(o\text{-}C_6H_4\text{-}CN)(THF)]$ **119** (right). *Source*: Adapted from Bluemke et al. [165].

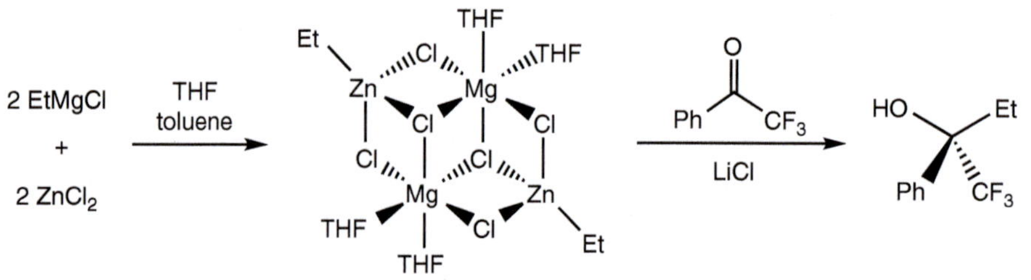

120

Scheme 2.34 Co-complexation of EtMgCl with $ZnCl_2$ (**120**) and LiCl assisted addition to trifluoroacetophenone.

Mixed magnesium–zinc complexes have also started to make an impact in polymerization chemistry. A heterodinuclear Mg(II)/Zn(II) complex of symmetrical bis-phenolate based macrocycle **124** (Scheme 2.35, left) exhibits high polymerization selectivity, control, and substantially greater activity in CO_2/epoxide and anhydride/epoxide copolymerization reactions in comparison to either the homodimagnesium or homodizinc analogues or indeed any combinations of them. This copolymerization operates via a 'chain shuttling' mechanism (Scheme 2.35, right).

Elucidation of the molecular structure of the heterodinuclear Mg(II)/Zn(II) complex implies epoxide coordination is likely to occur at the magnesium centre and is followed by attack from the zinc carbonate (Figure 2.28) [168]. This work chimes perfectly with the synergistic theme of the article, as can be evidenced from the title of the paper 'greater than the sum of its parts: a heterodinuclear polymerization catalyst' [169]. The first Mg(II)/Zn(II) heterodinuclear catalysts contained a halide co-ligand on each metal centre in addition to the macrocyclic diphenolate-tetra-amine ligand. By switching these halide ligands for carboxylate groups, this class of synergistic Mg(II)/Zn(II) complex has been extended and found to be highly active catalysts for CO_2/epoxide alternating copolymerization [170].

The cooperation of magnesium and aluminium can be exploited to trap complexes that unusually contain a carbon dioxide unit in its linear conformation. CO_2 and other related heterocumulenes typically undergo insertion into metal–carbon or metal–nitrogen bonds to yield carboxylate or carbamate complexes, respectively, yet in this example the neutral CO_2 unit remains intact, operating instead as a Lewis donor and demonstrating how metal–metal cooperativity can induce unique chemistry [171]. Commencing with $Mg(HMDS)_2$ and R_3Al (R = Me **125**; Et **126**), the final

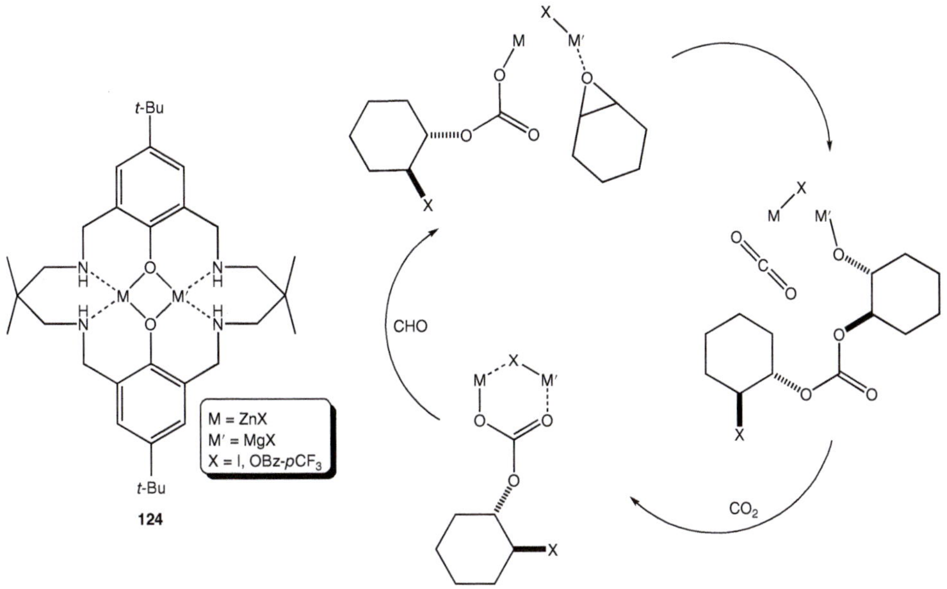

Scheme 2.35 Heteronuclear complex **124** used in polymerization alongside proposed 'chain-shuttling' mechanism (CHO = cyclohexene oxide).

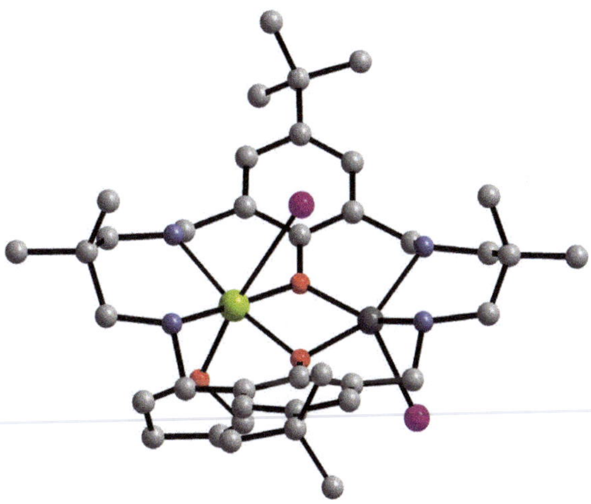

Figure 2.28 Molecular structure of Mg/Zn heterometallic polymerization catalyst **124** (I = dark pink, Mg = green, N = blue, O = red, Zn = dark gray). *Source*: Adapted from Deacy et al. [168].

product is a trimeric ring formed of Mg corners bridged with O=C=O units. The first step is reaction of the homometallic Mg bisamide with CO_2 through oxo-transfer and loss of an isocyanate to yield heteroleptic $(Me_3Si)_2NMgOSiMe_3$, followed by co-complexation with AlR_3, coordination of CO_2 and finally trimerization and loss of an R group (Scheme 2.36).

Sub-valent main group metals can also induce synergistic effects in partnership with another metal. This was demonstrated by pairing of a cationic calcium complex [as its $B(C_6F_5)_4$ salt] with an Al(I) β-diketiminate (Nacnac) complex in an attempt to utilize this low-valent Al fragment as a Lewis donor towards the calcium centre. This pairing can trap and reduce a molecule of benzene in the form of $C_6H_6^{2-}$ between the metal centres in **127** [172]. The reduction of the benzene molecule is clear within the crystal structure (Figure 2.29) due to the loss of planarity and the disparity

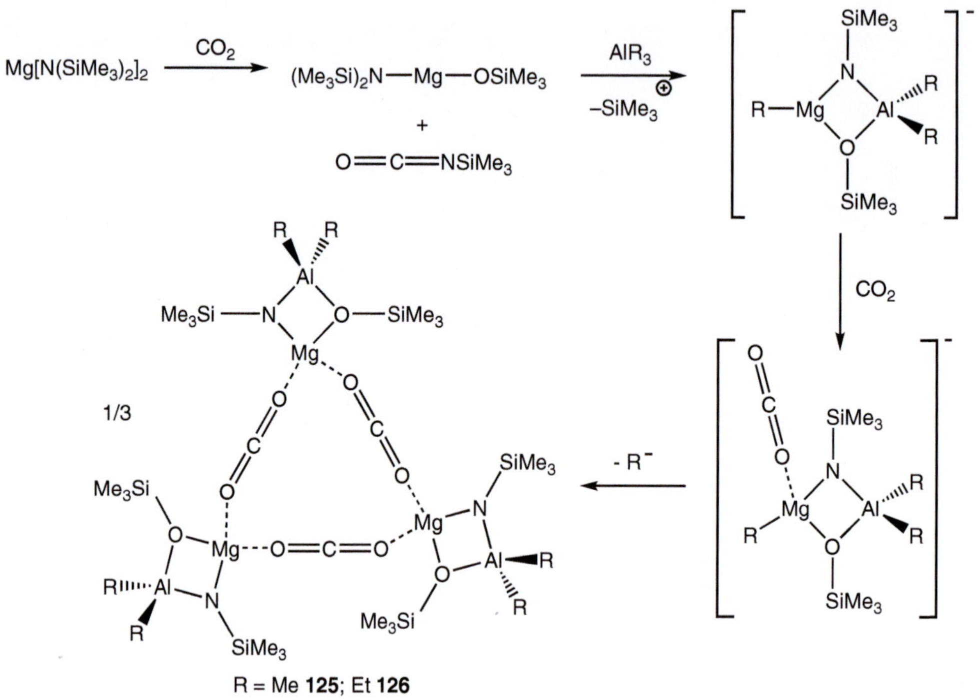

Scheme 2.36 Trapping of carbon dioxide units within a bimetallic Mg/Al complex (R = Me, Et).

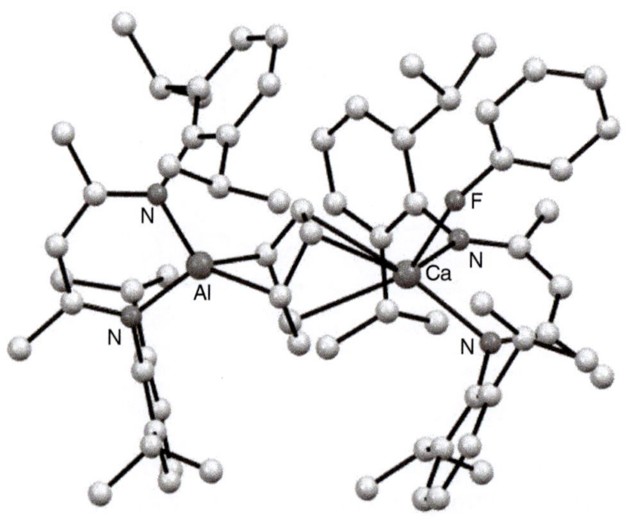

Figure 2.29 Molecular structure of $[(Nacnac)Ca^{+}(C_6H_6)^{2-}Al^{III}(Nacnac)(C_6H_5F)]^{+}$ **127**. *Source*: Adapted from Brand et al. [172].

in the C–C bond distances of this once aromatic ring. That metal cooperativity is operative here is clear since each homometallic component is stable in benzene [173] with the cationic Ca complex having previously been crystallized as a benzene solvate [174].

Low-valent metals working in concert with a second metal need not stimulate a positive synergistic effect. For example, the metal–metal bonded species $(Nacnac)Mg^{I}–Zn^{I}(Nacnac)$ **128** was

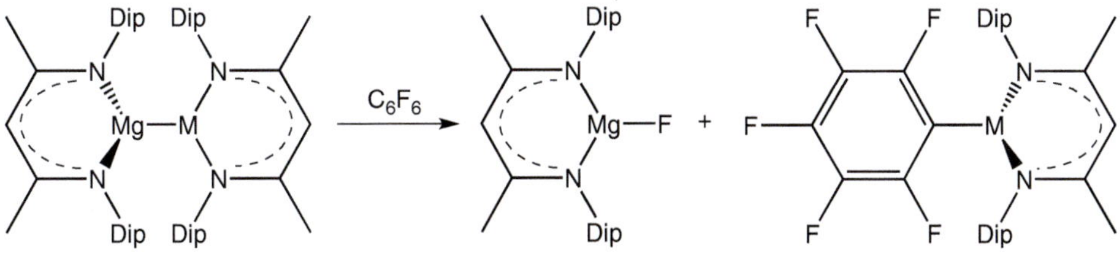

M = Zn **128**; Mg **129**

Scheme 2.37 C–F activation of perfluorobenzene with subvalent M–M bonded species **128**.

studied for its ability to effect C–F activation of fluorinated aromatics (Scheme 2.37) on the basis of the increased bond polarity with respect to the symmetrical complex (Nacnac)Mg^I–Mg^I(Nacnac) **129**. However, the study found that bond polarity was not the key factor in the C–F activation reaction and that rather it was the steric accessibility of the substrate towards the M–M unit [175]. The bimetallic complex did react with C_6F_6 but much slower than the homometallic magnesium derivative, with the magnesium complex preferentially reacting in a competition experiment.

2.4 Outlook

The insightful statement by Hill, Pécharman and Wilson in the conclusion to Chapter 2 on Turbo charging group 2 reagents for metathesis, metallation, and catalysis, 'It is evident that any evolution from pure empiricism will require the development of design principles that take into account not only the identity and roles of the individual metal centres but also the nature of the substrate species', applies equally to the larger collection of synergistic bimetallic main group compounds covered here. While many advances have been made in metallation chemistry with turbo-Grignard reagents and other selected metal–metal′ partnerships, systematic studies changing each metal in turn, as well as the substrate/s, any donor additives, bulk solvents etc. remain few in number. To gain a comprehensive understanding and put the chemistry of these intriguing bimetallic complexes on a more rational intellectual basis, future studies must include more solution characterization, more kinetic data on their reactions, and be underpinned by complementary theoretical studies. Part of the challenge is that the permutational possibilities of the components of these bases are so overwhelmingly vast that codifying chemical profiles is a mammoth task so researchers need to be selective in their choices. Though the idea of a 'Pairiodic Table', recently mooted in a review article [176], may appear fanciful, such codification may establish general patterns of reactivity that will benefit synthesis and catalysis.

References

1 Clayden, J. (**2002**). *Organolithiums: selectivity for synthesis, Vol. 23*. Amsterdam: Elsevier.
2 Luisi, R. and Capriati, V. (eds.) (**2014**). *Lithium Compounds in Organic Synthesis*. Weinheim: Wiley-VCH.
3 Benkeser, R.A., Foster, D.J., Sauve, D.M., and Nobis, J.F. (**1957**). *Chem. Rev. 57*: 867–894.
4 Schlosser, M. (**1964**). *Angew. Chem. Int. Ed. 3*: 287–306.
5 Schlosser, M. (**1964**). *Angew. Chem. Int. Ed. 3*: 362–373.
6 Schlenk, W. and Holtz, J. (**1917**). *Ber. 50*: 262–274.

7 Tidwell, T.T. (**2001**). *Angew. Chem. Int. Ed. 40*: 331–337.
8 Knochel, P. (**2013**). *Organometallics in synthesis* (ed. M. Schlosser), 223–372. John Wiley & Sons': Hoboken.
9 (**2013**). *Modern Organoaluminum Reagents*. Heidelberg: Springer.
10 Mulvey, R.E., Mongin, F., Uchiyama, M., and Kondo, Y. (**2007**). *Angew. Chem. Int. Ed. 46*: 3802–3824.
11 Setzer, W.N. and Schleyer, P.v.R. (**1985**). *Adv. Organomet. Chem. 24*: 353–451.
12 Schade, C. and Schleyer, P.v.R. (**1987**). *Adv. Organomet. Chem. 27*: 169–278.
13 Weiss, E. (**1993**). *Angew. Chem. Int. Ed. Engl. 32*: 1501–1523.
14 Seyferth, D. (**2006**). *Organometallics 25*: 2–24.
15 Reich, H.J. (**2012**). *J. Org. Chem. 77*: 5471–5491.
16 Reich, H.J. (**2013**). *Chem. Rev. 113*: 7130–7178.
17 Mulvey, R.E. and Robertson, S.D. (**2013**). *Angew. Chem. Int. Ed. 52*: 11470–11487.
18 Carl, E. and Stalke, D. (**2014**). *Lithium compounds in organic synthesis: from fundamentals to applications* (eds. R. Luisi and V. Capriati). Wiley -VCH.
19 Morton, A.A. (**1944**). *Chem. Rev. 35*: 1–49.
20 Weiss, E. and Hencken, G. (**1970**). *J. Organomet. Chem. 21*: 265–268.
21 Ohta, Y., Demura, A., Okomoto, T. et al. (**2006**). *J. Phys. Chem. B 110*: 12640–12644.
22 Dinnebier, R.E., Behrens, U., and Olbrich, F. (**1998**). *J. Am. Chem. Soc. 120*: 1430–1433.
23 Barr, D., Clegg, W., Mulvey, R.E. et al. (**1986**). *J. Chem. Soc., Chem. Commun.*: 295–297.
24 Armstrong, D.R., Barr, D., Clegg, W. et al. (**1986**). *J. Chem. Soc., Chem. Commun.*: 869–870.
25 Mulvey, R.E. (**1991**). *Chem. Soc. Rev. 20*: 167–209.
26 Gregory, K., Schleyer, P.v.R., and Snaith, R. (**1991**). *Adv. Inorg. Chem. 37*: 47–142.
27 Bond, A.D. (**2005**). *Cryst. Growth Des.*: 5, 755–771.
28 Clegg, W., Dale, S.H., Graham, D.V. et al. (**2007**). *Chem. Commun.*: 1641–1643.
29 Ogle, C.A., Huckabee, B.K., Johnson IV, H.C. et al. (**1993**). *Organometallics 12*: 1960–1963.
30 Götz, K., Gessner, V.H., Unkelbach, C. et al. (**2013**). *Z. Anorg. Allg. Chem. 639*: 2077–2085.
31 Köster, H., Thoennes, D., and Weiss, E. (**1978**). *J. Organomet. Chem. 160*: 1–5.
32 Strohmann, C. and Gessner, V.H. (**2007**). *J. Am. Chem. Soc. 129*: 8952–8953.
33 Langer, J., Köhler, M., Fischer, R. et al. (**2012**). *Organometallics 31*: 6172–6182.
34 Eckert, P.K., Schnura, B., and Strohmann, C. (**2014**). *Chem. Commun. 50*: 2532–2534.
35 Strohmann, C. and Gessner, V.H. (**2008**). *Chem. Asian J. 3*: 1929–1934.
36 Schümann, U., Kopf, J., and Weiss, E. (**1985**). *Angew. Chem. Int. Ed. 24*: 215–216.
37 Hope, H. and Power, P.P. (**1983**). *J. Am. Chem. Soc. 105*: 5320–5324.
38 Olmstead, M.M. and Power, P.P. (**1990**). *J. Am. Chem. Soc. 112*: 8008–8014.
39 Jackman, L.M. and Scarmoutzos, L.M. (**1984**). *J. Am. Chem. Soc. 106*: 4627–4628.
40 Kottke, T. and Stalke, D. (**1993**). *Angew. Chem. Int. Ed. Engl. 32*: 580–582.
41 Brown, T.L., Seitz, L.M., and Kimura, B.Y. (**1968**). *J. Am. Chem. Soc. 90*: 3245.
42 Nichols, M.A. and Williard, P.G. (**1993**). *J. Am. Chem. Soc. 115*: 1568–1572.
43 Bates, R.B., Kroposki, L.M., and Potter, D.E. (**1972**). *J. Org. Chem. 37*: 560–562.
44 Barnett, N.D.R., Mulvey, R.E., Clegg, W., and O'Neil, P.A. (**1993**). *J. Am. Chem. Soc. 115*: 1573–1574.
45 Strohmann, C. and Gessner, V.H. (**2008**). *J. Am. Chem. Soc. 130*: 11719–11725.
46 Strohmann, C. and Gessner, V.H. (**2007**). *Angew. Chem. Int. Ed. 46*: 4566–4569.
47 Eberhardt, G.G. and Butte, W.A. (**1964**). *J. Org. Chem. 29*: 2928–2932.
48 Dyker, G., Hölzer, B., and Henkel, G. (**1999**). *Tetrahedron: Asymmetry 10*: 3297–3307.
49 Ott, H., Pieper, U., Leusser, D. et al. (**2009**). *Angew. Chem. Int. Ed. 48*: 2978–2982.
50 Weiss, E., Sauermann, G., and Thirase, G. (**1983**). *Chem. Ber. 116*: 74–85.

51 Weiss, E., Corbelin, S., Cockcroft, J.K., and Fitch, A.N. (**1990**). *Angew. Chem. Int. Ed. Engl. 29*: 650–652.

52 Baillie, S.E., Clegg, W., García -Álvarez, P. et al. (**2011**). *Chem. Commun. 47*: 388–390.

53 Hitchcock, P.B., Lappert, M.F., Leung, W.P. et al. (**1993**). *J. Chem. Soc., Chem. Commun.*: 1386–1387.

54 Clegg, W., Conway, B., Kennedy, A.R. et al. (**2011**). *Eur. J. Inorg. Chem.*: 721–726.

55 von Pilgrim, M., Mondeshki, M., and Klett, J. (**2017**). *Inorganics 5*: 39.

56 Hitchcock, P.B., Khvostov, A.V., and Lappert, M.F. (**2002**). *J. Organomet. Chem. 663*: 263–268.

57 Boesveld, W.M., Hitchcock, P.B., Lappert, M.F. et al. (**2000**). *Organometallics 19*: 4030–4035.

58 Schümann, U., Behrens, U., and Weiss, E. (**1989**). *Angew. Chem. Int. Ed. Engl. 28*: 476–477.

59 Barnett, N.D.R., Mulvey, R.E., Clegg, W., and O'Neil, P.A. (**1991**). *J. Am. Chem. Soc. 113*: 8187–8188.

60 Rogers, R.D., Atwood, J.L., and Grüning, R. (**1978**). *J. Organomet. Chem. 157*: 229–237.

61 Lappert, M.F., Slade, M.J., Singh, A. et al. (**1983**). *J. Am. Chem. Soc. 105*: 302–304.

62 Hevia, E., Kennedy, A.R., Mulvey, R.E. et al. (**2013**). *Chem. Eur. J. 19*: 14069–14075.

63 Engelhardt, L.M., Jolly, B.S., Junk, P.C. et al. (**1986**). *Aust. J. Chem. 39*: 1337–1345.

64 Williard, P.G. and Salvino, J.M. (**1993**). *J. Org. Chem. 58*: 1–3.

65 Armstrong, D.R., García-Álvarez, P., Kennedy, A.R. et al. (**2011**). *Chem. Eur. J. 17*: 6725–6730.

66 Greiser, T., Kopf, J., Thoennes, D., and Weiss, E. (**1981**). *Chem. Ber. 114*: 209–213.

67 Zaragoza-Calero, S., Francos, J., Kennedy, A.R. et al. (**2015**). *Dalton Trans. 44*: 7258–7267.

68 Wittig, G., Ludwig, R., and Polster, R. (**1955**). *Chem. Ber. 88*: 294–301.

69 Schümann, U. and Weiss, E. (**1988**). *Angew. Chem. Int. Ed. Engl. 27*: 584–585.

70 Morris, J. J., Noll, B. C., Henderson, K. W. (**2007**). *Acta Cryst. E63*, m2477.

71 Brookhart, M., Green, M.L.H., and Parkin, G. (**2007**). *Proc. Natl. Acad. Sci. U.S.A. 104*: 6908–6914.

72 Williard, P.G. and Nichols, M.A. (**1991**). *J. Am. Chem. Soc. 113*: 9671–9673.

73 Nichols, M.A., Waldmüller, D., and Williard, P.G. (**1994**). *J. Am. Chem. Soc. 116*: 1153–1154.

74 Armstrong, D.R., Kennedy, A.R., Mulvey, R.E., and Robertson, S.D. (**2011**). *Chem. Eur. J. 17*: 8820–8831.

75 Lochmann, L. and Janata, M. (**2014**). *Cent. Eur. J. Chem. 12*: 537–548.

76 Morton, A.A. and Claff Jr, C.E. (**1954**). *J. Am. Chem. Soc. 76*: 4935–4938.

77 Marsch, M., Harms, K., Lochmann, L., and Boche, G. (**1990**). *Angew. Chem. Int. Ed. Engl. 29*: 308–309.

78 Clegg, W., Liddle, S.T., Drummond, A.M. et al. (**1999**). *Chem. Commun.*: 1569–1570.

79 Armstrong, D.R., Clegg, W., Drummond, A.M. et al. (**2000**). *J. Am. Chem. Soc. 122*: 11117–11124.

80 Benrath, P., Kaiser, M., Limbach, T. et al. (**2016**). *Angew. Chem. Int. Ed. 55*: 10886–10889.

81 Jennewein, B., Kimpel, S., Thalheim, D., and Klett, J. (**2018**). *Chem. Eur. J. 24*: 7605–7609.

82 Unkelbach, C., O'Shea, D.F., and Strohmann, C. (**2014**). *Angew. Chem. Int. Ed. 53*: 553–556.

83 Kennedy, A.R., MacLellan, J.G., and Mulvey, R.E. (**2001**). *Angew. Chem. Int. Ed. 40*: 3245–3247.

84 Harder, S. and Streitwieser, A. (**1993**). *Angew. Chem. Int. Ed. Engl. 32*: 1066–1068.

85 Mackenzie, F.M., Mulvey, R.E., Clegg, W., and Horsburgh, L. (**1996**). *J. Am. Chem. Soc. 118*: 4721–4722.

86 Wei, X., Dong, Q., Tong, H. et al. (**2008**). *Angew. Chem. Int. Ed. 47*: 3976–3978.

87 Whisler, M.C., MacNeil, S., Snieckus, V., and Beak, P. (**2004**). *Angew. Chem. Int. Ed. 43*: 2206–2225.

88 Fleming, P. and O'Shea, D.F. (**2011**). *J. Am. Chem. Soc. 133*: 1698–1701.

89 Das, M. and O'Shea, D.F. (**2013**). *Tetrahedron 69*: 6448–6460.

90 Manvar, A., Fleming, P., and O'Shea, D.F. (**2015**). *J. Org. Chem. 80*: 8727–8738.

91 Blangetti, M., Müller-Bunz, H., and O'Shea, D.F. (**2013**). *Tetrahedron 69*: 4285–4291.

92 Blangetti, M., Müller-Bunz, H., and O'Shea, D.F. (**2015**). *Chem. Commun. 49*: 6125–6127.
93 Haag, B., Mosrin, M., Ila, H. et al. (**2011**). *Angew. Chem. Int. Ed. 50*: 9794–9824.
94 García-Álvarez, P., Graham, D.V., Hevia, E. et al. (**2008**). *Angew. Chem. Int. Ed. 47*: 8079–8081.
95 Armstrong, D.R., García-Álvarez, P., Kennedy, A.R. et al. (**2010**). *Angew. Chem. Int. Ed. 49*: 3185–3188.
96 Wanklyn, J.A. (**1858**). *Ann. 108*: 67–79.
97 Seyferth, D. (**2001**). *Organometallics 20*: 2940–2955.
98 Hedström, A. and Lennartson, A. (**2011**). *J. Organomet. Chem. 696*: 2269–2273.
99 Mulvey, R.E. (**2006**). *Organometallics 25*: 1060–1075.
100 Wittig, G., Meyer, F.J., and Lange, G. (**1951**). *Ann. 571*: 167–201.
101 Wittig, G. (**1958**). *Angew. Chem. 70*: 65–92.
102 Tochtermann, W. (**1966**). *Angew. Chem. Int. Ed. Engl. 5*: 351–371.
103 Rathke, M. and Yu, H. (**1972**). *J. Org. Chem. 37*: 1732–1734.
104 Clegg, W., Forbes, G.C., Kennedy, A.R. et al. (**2003**). *Chem. Commun.*: 406–407.
105 Canzoneri, F. and Spica, G. (**1885**). *Gazz. Chim. Ital. 15*: 1–4.
106 Franchimont, A.P.N. and Friedmann, H. (**1905**). *Recl. Trav. Chim. 24*: 404–418.
107 Remenar, J.F., Lucht, B.L., Kruglyak, D. et al. (**1997**). *J. Org. Chem. 62*: 5748–5754.
108 Kondo, Y., Shilai, M., Uchiyama, M., and Sakamoto, T. (**1999**). *J. Am. Chem. Soc. 121*: 3539–3540.
109 Imahori, T., Uchiyama, M., Sakamoto, T., and Kondo, Y. (**2001**). *Chem. Commun.*: 2450–2451.
110 Clegg, W., Dale, S.H., Hevia, E. et al. (**2006**). *Angew. Chem. Int. Ed. 45*: 2370–2374.
111 Uchiyama, M., Matsumoto, Y., Nobuto, D. et al. (**2006**). *J. Am. Chem. Soc. 128*: 8748–8750.
112 Uchiyama, M., Matsumoto, Y., Usui, S. et al. (**2007**). *Angew. Chem. Int. Ed. 46*: 926–929.
113 Gilman, H. and Bebb, R.L. (**1938**). *J. Am. Chem. Soc. 61*: 109–112.
114 Wittig, G. and Fuhrman, G. (**1940**). *Chem. Ber. 73B*: 1197–1218.
115 Clegg, W., Dale, S.H., Drummond, A.M. et al. (**2006**). *J. Am. Chem. Soc. 128*: 7434–7435.
116 Clegg, W., Conway, B., Hevia, E. et al. (**2009**). *J. Am. Chem. Soc. 131*: 2375–2384.
117 Kondo, Y., Morey, J.V., Morgan, J.C. et al. (**2007**). *J. Am. Chem. Soc. 129*: 12734–12738.
118 Beak, P. and Brown, R.A. (**1977**). *J. Org. Chem. 42*: 1823–1824.
119 Mortier, J., Moyroud, J., Bennetau, B., and Cain, P.A. (**1994**). *J. Org. Chem. 59*: 4042–4044.
120 Metallinos, C., Nerdinger, S., and Snieckus, V. (**1999**). *Org. Lett. 1*: 1183–1186.
121 Hartung, C.G. and Snieckus, V. (**2002**). *Modern Arene Chemistry* (ed. D. Astruc), 330–367. New York: Wiley -VC H.
122 Kauch, M. and Hoppe, D. (**2006**). *Synthesis*: 1575–1577.
123 Barley, H.R.L., Clegg, W., Dale, S.H. et al. (**2005**). *Angew. Chem. Int. Ed. 44*: 6018–6021.
124 Marom, R., Amalraj, S.F., Leifer, N. et al. (**2011**). *J. Mater. Chem. 21*: 9938–9954.
125 Li, M., Lu, J., Chen, Z., and Amine, K. (**2018**). *Adv. Mater. 30*: 1800561.
126 Asako, S., Nakajima, H., and Takai, K. (**2019**). *Nat. Catal. 2*: 297–303.
127 Algera, R.F., Ma, Y., and Collum, D.B. (**2017**). *J. Am. Chem. Soc. 139*: 7921–7930.
128 Andrikopoulos, P.C., Armstrong, D.R., Barley, H.R.L. et al. (**2005**). *J. Am. Chem. Soc. 127*: 6184–6185.
129 Armstrong, D.R., Clegg, W., Dale, S.H. et al. (**2007**). *Chem. Commun.*: 598–600.
130 Clegg, W., Dale, S.H., Hevia, E. et al. (**2006**). *Angew. Chem. Int. Ed. 45*: 6548–6550.
131 Gilman, H. and Bebb, R.L. (**1939**). *J. Am. Chem. Soc. 61*: 109–112.
132 Baston, E., Maggi, R., Friedrich, K., and Schlosser, M. (**2001**). *Eur. J. Org. Chem.*: 3985–3989.
133 Armstrong, D.R., Clegg, W., Dale, S.H. et al. (**2006**). *Angew. Chem. Int. Ed. 45*: 3775–3778.
134 Clegg, W., Dale, S.H., Hevia, E. et al. (**2008**). *Angew. Chem. Int. Ed. 47*: 731–734.
135 Nobuto, D. and Uchiyama, M. (**2008**). *J. Org. Chem. 73*: 1117–1120.

136 Balloch, L., Garden, J.A., Kennedy, A.R. et al. (**2012**). *Angew. Chem. Int. Ed. 51*: 6934–6937.
137 Hevia, E., Kennedy, A.R., and McCall, M.D. (**2012**). *Dalton Trans. 41*: 98–103.
138 Hevia, E., Honeyman, G.W., Kennedy, A.R., and Mulvey, R.E. (**2005**). *J. Am. Chem. Soc. 127*: 13106–13107.
139 Mulvey, R.E. (**2013**). *Dalton Trans. 42*: 6676–6693.
140 Kennedy, A.R., Klett, J., Mulvey, R.E., and Wright, D.S. (**2009**). *Science 326*: 706–708.
141 Mosrin, M. and Knochel, P. (**2009**). *Org. Lett. 11*: 1837–1840.
142 Wunderlich, S.H. and Knochel, P. (**2007**). *Angew. Chem. Int. Ed. 46*: 7685–7688.
143 Hernán-Gómez, A., Herd, E., Hevia, E. et al. (**2014**). *Angew. Chem. Int. Ed. 53*: 2706–2710.
144 Brikci-Nigassa, N., Bentabed-Ababsa, G., Erb, W., and Mongin, F. (**2018**). *Synthesis 50*: 3615–3633.
145 Uzelac, M. and Mulvey, R.E. (**2018**). *Chem. Eur. J. 24*: 7786–7793.
146 Su, Y., Zhao, Y., Gao, J. et al. (**2012**). *Inorg. Chem. 51*: 5889–5896.
147 Frischmuth, A., Fernandez, M., Barl, N.M. et al. (**2014**). *Angew. Chem. Int. Ed. 53*: 7928–7932.
148 Akimoto, G., Otsuka, M., Takita, R. et al. (**2018**). *J. Org. Chem. 83*: 13498–13506.
149 García-Álvarez, P., Mulvey, R.E., and Parkinson, J.A. (**2011**). *Angew. Chem. Int. Ed. 50*: 9668–9671.
150 Pollard, V.A., Orr, S.A., McLellan, R. et al. (**2018**). *Chem. Commun. 54*: 1233–1236.
151 Lemmerz, L.E., McLellan, R., Judge, N.R. et al. (**2018**). *Chem. Eur. J. 24*: 9940–9948.
152 Uchiyama, M., Naka, H., Matsumoto, Y., and Ohwada, T. (**2004**). *J. Am. Chem. Soc. 126*: 10526–10527.
153 Naka, H., Uchiyama, M., Matsumoto, Y. et al. (**2007**). *J. Am. Chem. Soc. 129*: 1921–1930.
154 Crosbie, E., García-Álvarez, P., Kennedy, A.R. et al. (**2010**). *Angew. Chem. Int. Ed. 49*: 9388–9391.
155 Campbell, R., Crosbie, E., Kennedy, A.R. et al. (**2013**). *Aust. J. Chem. 66*: 1189–1201.
156 Armstrong, D.R., Crosbie, E., Hevia, E. et al. (**2014**). *Chem. Sci. 5*: 3031–3045.
157 Crosbie, E., Kennedy, A.R., Mulvey, R.E., and Robertson, S.D. (**2012**). *Dalton Trans. 41*: 1832–1839.
158 Hernan-Gomez, A., Kennedy, A.R., and Hevia, E. (**2017**). *Angew. Chem. Int. Ed. 56*: 6632–6635.
159 Wang, Y., Xie, Y., Abraham, M.Y. et al. (**2010**). *J. Am. Chem. Soc. 132*: 14370–14372.
160 Clegg, W., Crosbie, E., Dale-Black, S.H. et al. (**2015**). *Organometallics 34*: 2580–2589.
161 Armstrong, D.R., Brammer, E., Cadenbach, T. et al. (**2013**). *Organometallics 32*: 480–489.
162 Uzelac, M., Kennedy, A.R., Hevia, E., and Mulvey, R.E. (**2016**). *Angew. Chem. Int. Ed. 55*: 13147–13150.
163 McLellan, R., Uzelac, M., Kennedy, A.R. et al. (**2017**). *Angew. Chem. Int. Ed. 56*: 9566–9570.
164 Armstrong, D.R., Clegg, W., García-Álvarez, P. et al. (**2011**). *Chem. Eur. J. 17*: 4470–4479.
165 Bluemke, T.D., Clegg, W., García-Álvarez, P. et al. (**2014**). *Chem. Sci. 5*: 3552–3562.
166 Hevia, E., Chua, J.Z., García-Álvarez, P. et al. (**2014**). *Proc. Natl. Acad. Sci. U.S.A. 107*: 5294–5299.
167 Armstrong, D.R., Clegg, W., García-Álvarez, P. et al. (**2011**). *Chem. Eur. J. 17*: 8333–8341.
168 Deacy, A.C., Durr, C.B., Garden, J.A. et al. (**2018**). *Inorg. Chem. 57*: 15575–15583.
169 Garden, J.A., Saini, P.K., and Williams, C.K. (**2015**). *J. Am. Chem. Soc. 137*: 15078–15081.
170 Trott, G., Garden, J.A., and Williams, C.K. (**2019**). *Chem. Sci. 10*: 4618–4627.
171 Chang, C.-C., Liao, M.-C., Chang, T.-H. et al. (**2005**). *Angew. Chem. Int. Ed. 44*: 7418–7420.
172 Brand, S., Elsen, H., Langer, J. et al. (**2018**). *Angew. Chem. Int. Ed. 57*: 14169–14173.
173 Cui, C., Roesky, H.W., Schmidt, H.G. et al. (**2000**). *Angew. Chem. Int. Ed. 39*: 4274–4276.
174 Pahl, J., Brand, S., Elsen, H., and Harder, S. (**2018**). *Chem. Commun. 54*: 8685–8688.
175 Bakewell, C., Ward, B.J., White, A.J.P., and Crimmin, M.R. (**2018**). *Chem. Sci. 9*: 2348–2356.
176 Robertson, S.D., Uzelac, M., and Mulvey, R.E. (**2019**). *Chem. Rev. 119*: 8332–8405.

3

Turbo Charging Group 2 Reagents for Metathesis, Metalation, and Catalysis

Michael S. Hill, Anne-Frédérique Pécharman, and Andrew S. S. Wilson

Department of Chemistry, University of Bath, Claverton Down, UK

3.1 Introduction and Historical Context: Monometallic s-block Reagents and Their Utility

An abiding theme in synthetic chemistry throughout the twentieth century was the widespread use of polar organometallic and organoamide derivatives derived from the elements of groups 1 and 2 of the periodic table. As highlighted in a series of brilliant historical perspectives by Seyferth [1–3], the availability of the alkali metals, sodium and potassium by electrolysis allowed initial attempts to synthesize alkyl and aryl derivatives of these elements to date from as early as the 1850s [4]. The utility of this early chemistry, however, was limited by the pyrophoric nature and insolubility of the resultant compounds and the more widespread impact of s-block reagents was only felt with the development of Grignard's eponymous organomagnesium halide reagents at the turn of the nineteenth century [5, 6]. Grignard reagents, RMgX, are easily synthesized by the direct reaction of elemental magnesium and an organohalide and act as potent nucleophiles in reactions with both inorganic and organic electrophiles. The latter reactivity allows for ready carbon–carbon bond formation, such that RMgX reagents were so rapidly established in the arsenal of the organic chemist that Grignard was recognized by the award of the 1912 Nobel Prize for Chemistry, little over a decade after their initial discovery [7]. The outstanding synthetic utility of Grignard reagents resulted in a keen interest in their solution speciation and the classic researches of Schlenk and Schlenk revealed that RMgX the reagents undergo facile exchange in ether solution (the so-called Schlenk equilibrium, Scheme 3.1) to afford both R_2Mg and MgX_2 [8]. Subsequent studies employed an armoury of physical techniques to determine the position of the Schlenk equilibria and the aggregation states of the RMgX under various conditions of temperature, concentration, and solvent identity [9–23]. Although initially represented as monomers in the Schlenks' classic study, more commonly, halide groups in the RMgX and the MgX_2 form bridges between magnesium atoms to yield dimers and higher oligomeric structures which enable ligand exchange. A diverse variety of nuclearities and magnesium coordination modes have been verified in the solid state. Although higher coordinate structures have been observed [24], a tetrahedral geometry is most common in single-crystal X-ray structures of species such as $EtMgBr(OEt_2)_2$ **1** (Scheme 3.1) [25]. Similarly, recent computational assessment of the operant equilibria in THF solutions of CH_3MgCl has indicated that the solvated dinuclear species, $[(THF)CH_3Mg(\mu\text{-}Cl)_2MgCH_3(THF)]$ and [(THF)

Polar Organometallic Reagents: Synthesis, Structure, Properties and Applications, First Edition. Edited by Andrew E. H. Wheatley and Masanobu Uchiyama.

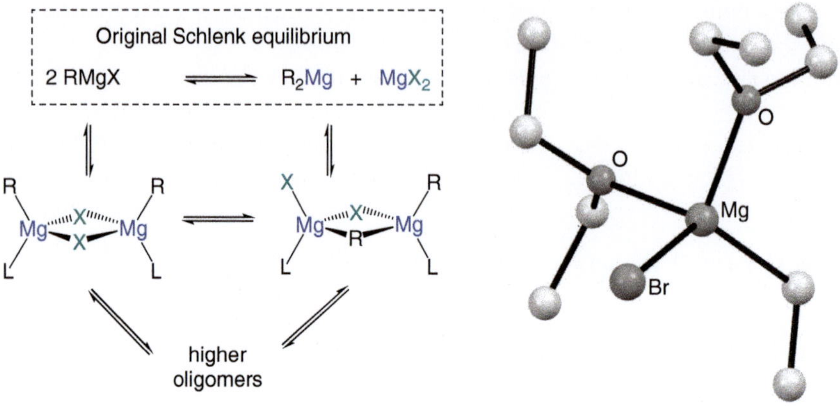

Scheme 3.1 Generalized representation of the Schlenk equilibrium and the monomeric structure of $EtMgBr(OEt_2)_2$ **1**. *Source:* Adapted from Vestergren et al. [25].

$CH_3Mg(\mu\text{-}Cl)(\mu\text{-}CH_3)MgCl(THF)]$, need to evolve to less stable asymmetrically solvated dimers, $[(THF)CH_3Mg(\mu\text{-}Cl)_2MgCH_3(THF)_2]$ and $[(THF)CH_3Mg(\mu\text{-}Cl)(\mu\text{-}CH_3)MgCl(THF)_2]$ to facilitate ligand transfer, which occurs via an axial position of a transiently pentacoordinate Mg atom [26].

While Grignard reagents react readily with numerous electrophiles such as aldehydes, ketones, epoxides, nitriles, esters, and carbon dioxide [27], and have found application in a variety of transition metal-catalyzed conjugate addition and cross-coupling reactions [28–35], their usefulness is more limited in other regards [28, 36–41]. An ability to effect both regio- and chemoselective metalation reactions, for example, either by (i) deprotonative cleavage of an acidic C–H bond or (ii) the metalation of carbon to halogen bonds is one of the most common requirements in the preparation of intermediates during the synthesis of pharmaceuticals, agrochemicals and fine chemicals. For both reaction types, the application of simple organomagnesium reagents is perturbed by the nature of the magnesium to carbon bond. Although polar, the $Mg^{\delta+}–C^{\delta-}$ interaction is significantly covalent, which limits both its basicity and the rate of halogen/magnesium exchange at usefully applicable temperatures. The means to achieve both deprotonation or metal/halogen exchange, therefore, has been historically provided by harder alkali metal compounds, in particular organolithium reagents. These commonly include the commercially available, MeLi, *n*-BuLi and *s*-BuLi and sterically congested amidolithium compounds such as lithium diisopropylamide (LDA), lithium 2,2,6,6-tetramethylpiperidide (LiTMP), and lithium hexamethyldisalazide $(Me_3Si)_2NLi$. While the reactivity and solution and solid-state structures of these latter species continues to be of keen interest, these areas have been extensively reviewed and will not detain us here. The reader is, thus, directed to selected recent advances and comprehensive reviews [42–74]. It is worth emphasizing, however, the continuing importance of organolithiums as 'first stop' strong bases in many chemical syntheses. The provenance of these species can again be traced for over 100 years, dating from the pioneering researches of Schlenk [75], who not only first described their preparation but also their deprotonative potency as bases [76, 77]. Over the subsequent decades, driven by the researches of, among others, Gilman, Ziegler, and Wittig [1, 2, 78–80], the chemistry of lithium reagents advanced to a high degree of sophistication and utility [81–90]. First introduced by Gilman and Wittig [78, 91], and then greatly extended by Beak [92], Mortier [93], Hoppe [94, 95], and Snieckus [96, 97], directed *ortho*-metalation (DoM), for example, in which substituents such as an amide, methoxy, or cyano substituent direct the site of metalation, continues to be a widely employed means to effect the regioselective functionalization of substituted aromatic and heteroaromatic compounds (Scheme 3.2) [82].

An intrinsic limitation of such strong organometallic bases, however, is their incompatibility with more delicate but synthetically important functional groups such as esters, carbonyls, nitriles, sulfoxides, and halides, which often result in competing side reactions. In comparison to lithium reagents, the greater covalency of Mg–C and Mg–N bonds within organo- and amidomagnesium derivatives results in much better functional group tolerance and enhanced chemoselectivity. As a result, the final decades of the twentieth century witnessed something of a rebirth of interest in less reactive but more selective basic and nucleophilic magnesium reagents. Although Hauser et al. had introduced amido analogues of Grignard reagents, R_2NMgCl, in the 1940s [98, 99] the synthetic potential of so-called Hauser bases only received more widespread recognition in the 1980s and 1990s [100]. Eaton et al. demonstrated that magnesiation of methyl benzoate and cyclopropyl, cyclobutyl, and even cubane carboxamides with $(TMP)_2Mg$ or TMPMgBr could be performed at 25 °C (Scheme 3.3a) [101–103]. These protocols were subsequently extended to the magnesiation of a variety of pyridine carboxamides and piperazines [104, 105], while Kondo et al. demonstrated the effectiveness of i-Pr_2NMgBr for the metalation of phenylsulfonyl-substituted

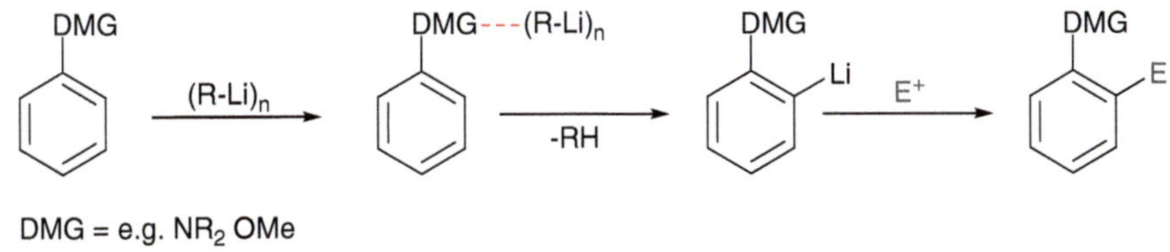

Scheme 3.2 Directed *ortho*-metalation by organolithium reagents.

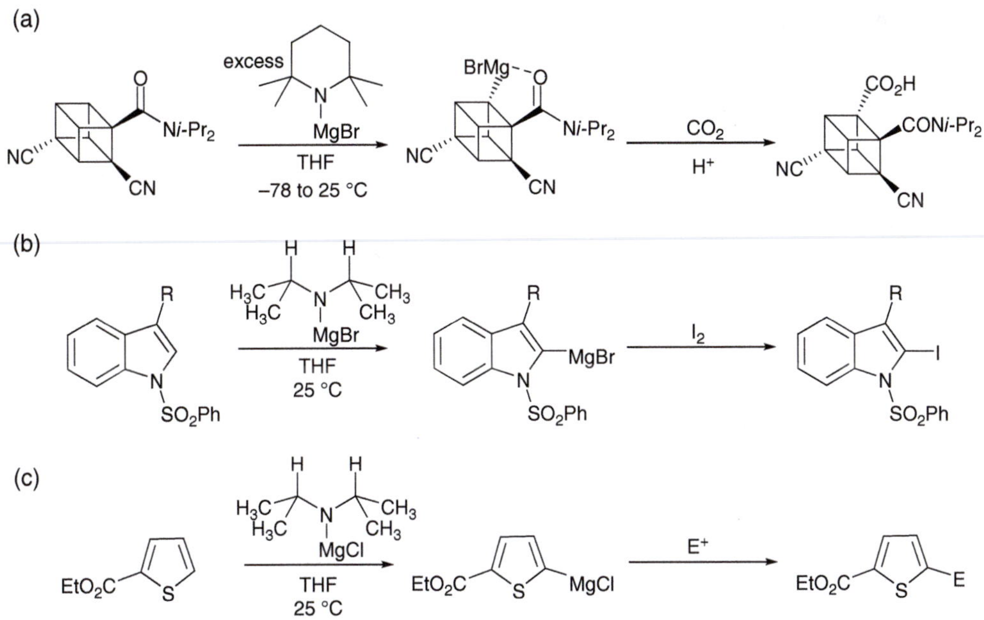

Scheme 3.3 (a) Eaton's magnesiation of a functionalized cubane and Kondo's metalation of (b) phenylsulfonyl-substituted indoles (c) thiophenes.

indoles (Scheme 3.3b, c) and the exclusive 2-metalation of thiophene [106, 107]. A significant disadvantage of Hauser bases, however, is their poor solubility in THF. As a result, metalation is kinetically slow and stoichiometries of the order of a 10-fold excess of the base are typically required. It was clear by the late 1990s, therefore, that further step changes in the activity and selectivity of s-block reagents were a necessary requirement for the continued evolution of groups 1 and 2 centred metalation chemistry. The means to address these limitations will be discussed in Section 3.2.

3.2 Heterobimetallic Reagents for Selective Metalation

3.2.1 Ate Complexes and Superbases

Wanklyn most likely prepared the first 'ate' complex, $NaZnEt_3$, during his initial studies of the reduction of alkyl halides with sodium and potassium [4]. The term 'ate complex', applied as a general descriptor to species such as Ph_3MgLi, prepared by combination of diphenylmagnesium and phenyllithium, however, appears to have been introduced by Wittig during the 1950s [108–110]. His report of the ability of such compounds to deprotonate fluorene and diphenylmethane indicated that the differential reactivity could be attained through the combination of bases derived from two dissimilar but electropositive metals. These observations did not immediately open the way to a systematic study of ate compounds as deprotonating agents, however, and a realization of the potential potency of bimetallic systems was really only uncovered with the discovery of the so-called Lochmann–Schlosser 'superbase' in 1967 [111, 112]. While neither *n*-BuLi nor *t*-BuOK in isolation can deprotonate benzene or toluene, the combination of both reagents invokes vastly enhanced basicity so as to allow the metalation of available sp^2 and sp^3 C–H bonds. Although it remains unclear whether such combinations of group 1 reagents truly qualify as ate complexes, structural assessment of superbasic systems [113, 114], such as the recently reported $[Li_4K_4Np_3(Ot\text{-}Bu)_5]$ **2** obtained from combination of neopentyllithium and *t*-BuOK (and illustrated in Chapter 1, Figure 1.7) [115] have implicated the operation of aggregated heterometallic structures. As such, it appears likely that the combination of both metals provides an element of cooperativity that drives the overall behaviour of the system.

Although these early observations provided strong hints that amplified metalation reactivity may be attained through the adoption of heterobimetallic systems, exploration of the broader utility of combinations of groups 1 and 2 metal reagents were not pursued in earnest until the early twenty-first century. Weiss et al. had earlier provided solid state structural insight into the structures of a variety of group 1 organomagnesiate complexes. These studies revealed that organomagnesiates exist in a variety of structural types and with various potential coordination numbers, either 3, 4 or 5, at the magnesium centres. The compounds $[Li(TMEDA)]_2[MgMe_4]$ **3** and $[Na(PMDETA)]_2[MgPh_4]$ **4** (Figure 3.1 and, for the solid-state structure of **3**, see Chapter 2, Figure 2.8; TMEDA = *N,N,N′,N′*-tetramethylethylenediamine, PMDETA =*N,N,N′,N″,N″*-pentamethyldiethylenetriamine), for example, are typical contact ion pair species displaying tetrahedral coordination at Mg, with the organic substituents forming asymmetric μ_2 bridges between magnesium and the solvated alkali metal atoms [46, 116]. Although the structures of compounds **3** and **4** appear analogous, it is clear that solutions of these contact ion paired species are as prone to solution equilibria as Grignard reagents themselves. The addition of TMEDA, for example, to solutions of $LiMgPh_3$ also provided $[Li(TMEDA)]_2[Ph_2MgPh_2MgPh_2]$ **5** in which two phenyl groups bridge the magnesium centres while the four remaining phenyls bridge Li and Mg [117]. In contrast, the

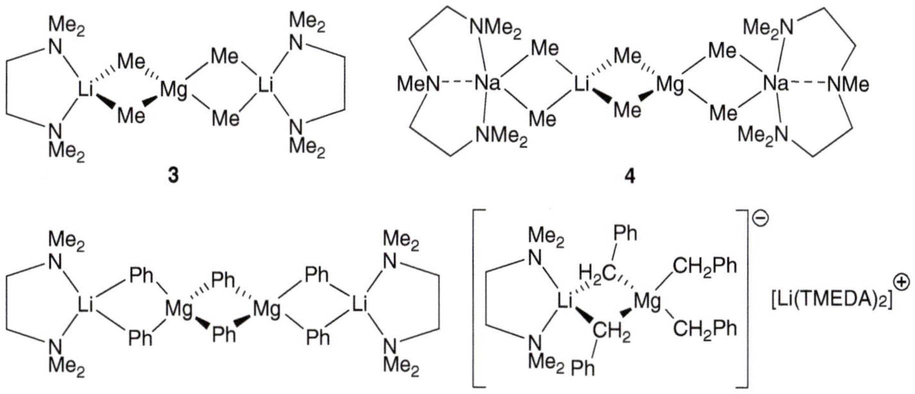

Figure 3.1 The structures of the organolithium magnesiate contact ion pairs, **3–5**, and the ion separated structure of compound **6**.

benzyl species $[Li(TMEDA)_2][(TMEDA)LiBz_2MgBz_2]$ (Bz = CH_2Ph) **6** crystallizes as an ion-separated structure in which the magnesium atom is coordinated by four benzyl groups, two of which bridge to a TMEDA-coordinated Li [118].

To contemporary eyes, the structures of compounds **3–6** may appear primed to allow cooperative engagement between the discriminated s-block centres. In retrospect, therefore, it is surprising that it was not until the latter years of the 1990s that such species attracted concerted activity as metalating reagents [119–124]. The following sections will seek to chart this progress in terms of different combinations of basic residues in conjunction with groups 1 and 2 metal centres, particularly magnesiates. The intent is to trace the development of our understanding of how combinations of groups 1 and 2 metals may modulate the reactivity of the individual components during the metalation of small molecule substrates. This overview is, thus, quite selective and for more comprehensive coverage of individual aspects of structure and reaction scope, the reader is again directed to existing more focused reviews [119–127].

3.2.2 Lithium, Sodium, Potassium Magnesiates, $MMgX_3$

Significant templating effects have been observed over the last 20 years with regard to the metalation activity and selectivity associated with the application of mixed metal reagents derived from combinations of group 1 (e.g. *n*-BuM, TMPM, *i*-Pr_2NM $(Me_3Si)_2NM$; M = Li/Na/K) and magnesium bases (e.g. *n*-Bu_2Mg and TMP_2Mg). Although not a singular focus of this chapter, it is necessary to appreciate how the solid-state aggregation behaviour of such species may impact upon and inform our understanding of their observed solution reactivity.

As an extension of their existing interest in the structural chemistry of mixed group 1/group 2 metal alkyls, alkoxides and amides [128–132], in 1998 Mulvey et al. reported that a 2:1 reaction of $(Me_3Si)_2NLi$ and $[\{(Me_3Si)_2N\}_2Mg]$ afforded the intermetallic lithium–magnesium hexamethyldisilazide $[\{(Me_3Si)_2N\}_3LiMg]$ **7** (Figure 3.2a), comprising a near-planar NMgNLi ring [133]. Although no metalation chemistry was described, the presence of trace amounts of oxygen led to the crystallization of the startling peroxide-oxide variant $[\{(Me_3Si)_2N\}_4Li_2Mg_2(O_2)_x(O)_y]$ $[x = 0.715(7); y = 0.285(7)]$ **8** (Figure 3.2b) in which the oxo and peroxide anions are coordinated by the four s-block metal centres as a component of what was later described as an 'inverse crown ether' array [134, 135].

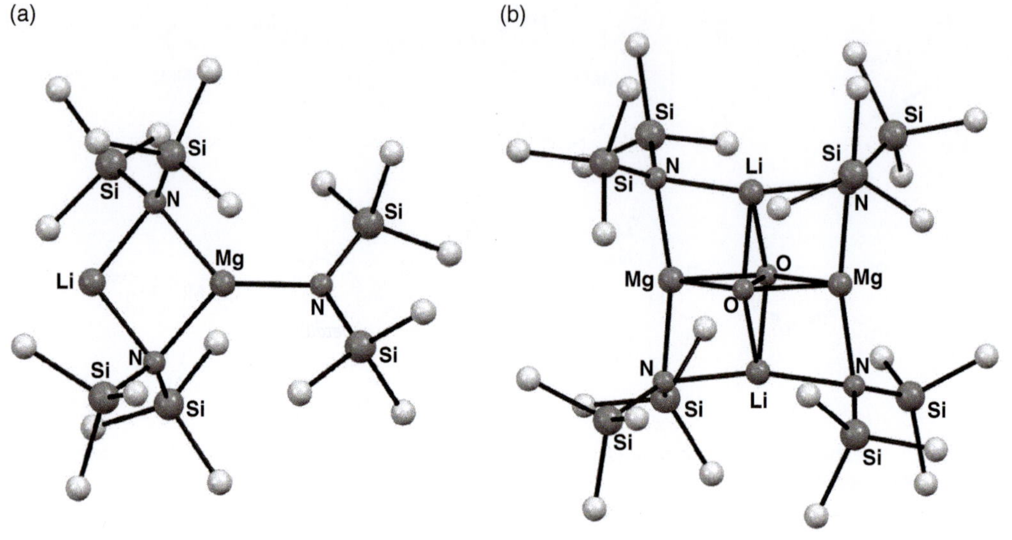

Figure 3.2 The solid-state structures of (a) Mulvey's lithium magnesiate, compound **7** (b) the peroxide component of the 'inverse crown', compound **8**. *Source:* Adapted from Kennedy et al. [133]; Kennedy et al. [134].

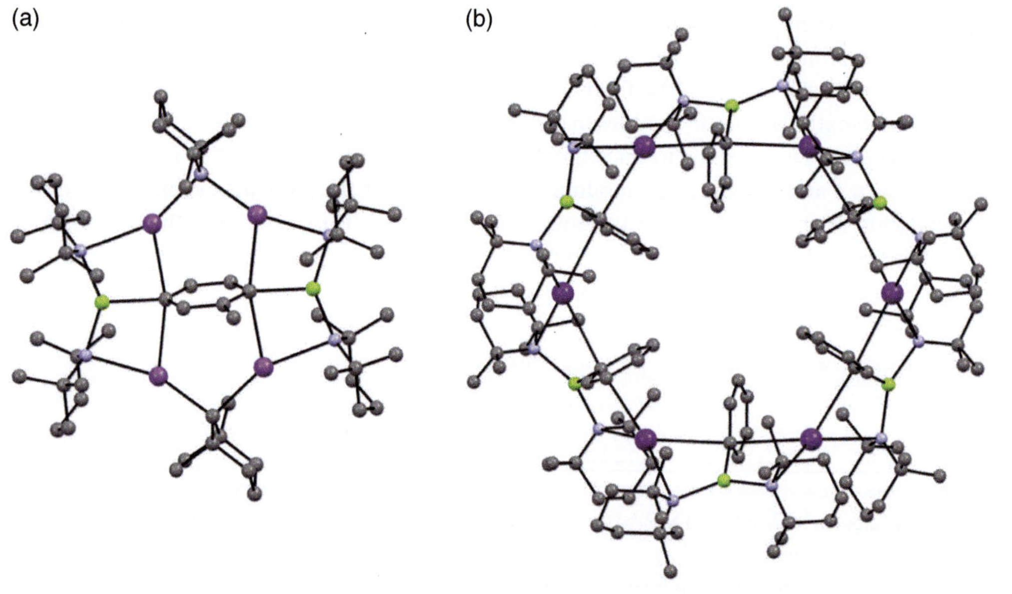

Figure 3.3 The 'inverse crown' structures of (a) Na/Mg compound **9** and (b) K/Mg compound **11**, containing doubly 2,5-deprotonated toluene and monodeprotonated benzene, respectively (K = dark purple, Mg = green, N = blue, Na = light purple). *Source*: Adapted from Armstrong et al. [136]; Andrews et al. [137].

Variation of the amide base to TMP allowed this reactivity to be rapidly expanded to encompass a heavier alkali metal congener comprising a larger twelve-membered ($N_6Na_4Mg_2$) cyclic structure. Combination of a 1:1:3 molar ratio of *n*-BuNa, Bu_2Mg, and HTMP in toluene or benzene, however, led to the formation of compounds **9** and **10** (Figure 3.3a, and for NMR analysis of these transformations see Chapter 6) resulting from double deprotonation of the toluene or benzene

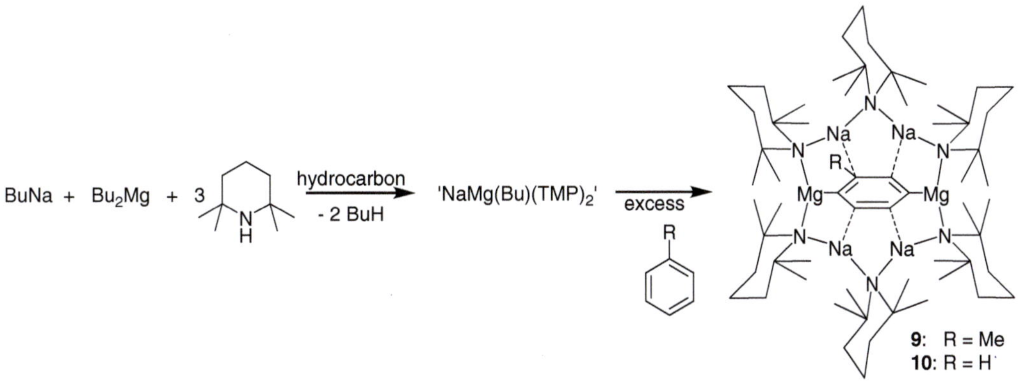

Scheme 3.4 Synthesis of the sodium magnesiate 'inverse crowns', compounds **9** and **10**, containing doubly deprotonated toluene and benzene, respectively.

solvent such that the cycles acted as hosts for the respective $[C_6H_3(CH_3)]^{2-}$ and $C_6H_4^{2-}$ dianions (Scheme 3.4) [136]. Remarkably, the most thermodynamically acidic CH_3 hydrogen atoms of toluene in the former complex were left untouched. The sites of metalation of both the toluene (2,5-, i.e. *ortho* and *meta* positions) and benzene (1,4-) dianions were, thus, interpreted as being strongly suggestive of the operation of a kinetic templating effect, which was christened as a special class of synergic activity termed alkali-metal-mediated magnesiation (AMMM).

Extension of this magnesiate inverse crown chemistry to include potassium resulted in an equally striking outcome [137]. Generation of an *in situ* mixture of $KMg(TMP)_3$ in benzene or toluene provided spectacular 24-membered hexapotassium hexamagnesium structures, compounds **11** (Figure 3.3b) and **12**, each containing six monodeprotonated arene anions. Perhaps counterintuiviely, only monodeprotonation of the arenes was achieved when switching to the more electropositive potassium. Variations of the amide base to include *i*-Pr_2N– allowed the use of the sodium magnesiate chemistry to be elaborated to the fourfold deprotonation of ferrocene (compound **13**, Figure 3.4a) [138, 139] as well as the isostructural ruthenocene (**14**) and osmocene (**15**) derivatives [140]. In a similar way, use of a *n*-BuNa/Bu_2Mg/HTMP/TMEDA system enabled the selective magnesiation of bis(benzene)-chromium (**16**; Figure 3.4b) [141, 142].

However eye-catching this chemistry, probably of more general significance was the observation that only twofold amination occurs when 3 molar equivalents of HTMP and TMEDA are reacted with a 1:1 BuNa–Bu_2Mg mixture to give the mixed alkyl amide [(TMEDA)Na(μ-Bu)(μ-TMP)Mg(TMP)] **17**. This compound affords the phenyl-bridged analogue (**18**) when reacted with benzene (Scheme 3.5) [143] and can effect the unprecedented *meta*-deprotonation of toluene (**19**) [144]. This latter transformation is especially notable as it contrasts with the normal *ortho*-, *para*-directing effect that the methyl group of toluene exerts in reactions such as electrophilic aromatic substitution.

Similar combinations of alkyl sodium and magnesium reagents were rapidly shown by the same group to effect the selective deprotonation of various *O*- and *N*-heterocyclic derivatives [145, 146], while subjecting 0.5 molar equivalents of benzene to a bisamido-monoalkyl mixture of NaTMP, *t*-BuMgTMP and TMEDA produced the 1,4-diphenylenedimagnesio complex [1,4-{(TMEDA)Na(μ-TMP)Mg(TMP)}$_2C_6H_4$] **20** [147]. The ability of bimetallic systems to allow di-deprotonation of heterocycles was observed in the reaction of furan with the heteroleptic alkyl-amido reagent [(TMEDA)Na(CH_2SiMe_3)(μ-TMP)Mg(TMP)] **21**, which produced the remarkable dodecasodium–hexamagnesium molecule, [{$(TMEDA)_3Na_6Mg_3(CH_2SiMe_3)(2,5\text{-}C_4H_2O)_3(2\text{-}C_4H_3O)_5$}$_2$] **22**, built

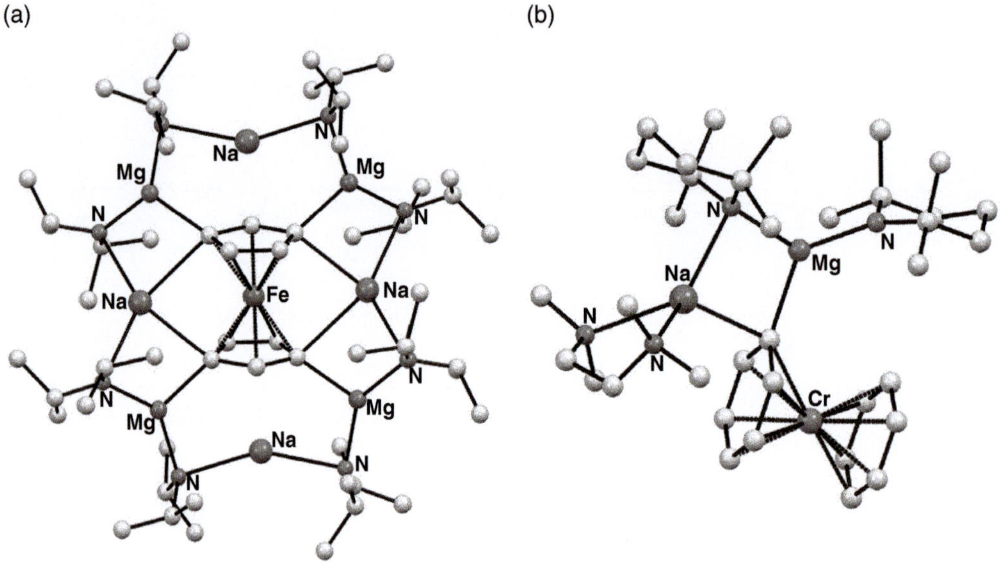

Figure 3.4 (a) Fourfold deprotonation of ferrocene and (b) monodeprotonation of bis(benzene)chromium products, **13** and **16**, resulting from the use of sodium magnesiate amide bases. *Source*: Adapted from Clegg et al. [138]; Andrikopoulos et al. [141].

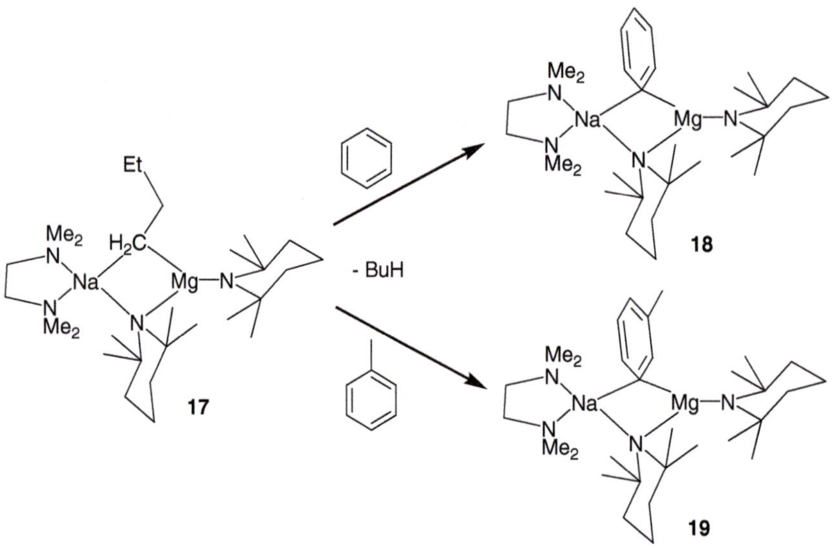

Scheme 3.5 Deprotonation of benzene and *meta*-deprotonation of toluene by the synergic butyl-TMP sodium magnesiate, compound **17**.

upon a bridge network of 10 monodeprotonated and 6 twofold-deprotonated furan ligands [148]. In contrast, the same base system (**21**) in conjunction with *S*-heterocycles was shown to effect only monodeprotonation of both thiophene and tetrahydrothiophene [149], while being capable of commiting 'chemical carnage' on THF [150]. In this latter reaction compound **21**, when treated with a stoichiometric quantity of THF, induced the cleavage of two C–O and four C–H bonds within the C_4H_8O heterocycle to yield the inverse crown, $[Na_2Mg_2(TMP)_4(O)]$ **23** and the 'cleave

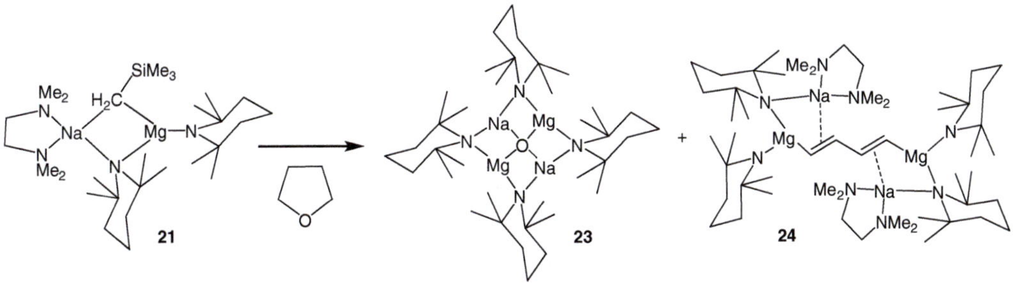

Scheme 3.6 'Cleave and capture' of THF effected by the alkyl-TMP sodium magnesiate, **21**.

and capture' dimagnesiated butadiene product, [{(TMEDA)Na(μ-TMP)}$_2$(1,4-{Mg(TMP)}$_2$C$_4$H$_4$)] **24** (Scheme 3.6). During a similar timeframe, albeit with a less structural emphasis, Mongin et al. reported that the selective magnesiation of various heterocyles and chloro- and fluoroaromatics could be achieved using the 'highly coordinated' magnesiate, 'Bu$_3$(TMP)MgLi$_2$' [151–155].

Extension of the AMM*M* protocol to potassium and monitoring of the magnesiation of anisole by [(PMDETA)K(μ-TMP)(μ-CH$_2$SiMe$_3$)Mg(TMP)] **25** revealed that the heteroleptic base reacts kinetically through its TMP component to generate first the *ortho*-magnesiated anisole product [(PMDETA)K(μ-CH$_2$SiMe$_3$)(2-C$_6$H$_4$OMe)Mg(TMP)] **26** which, in turn, reacts through its alkyl component to yield the ultimate thermodynamic *ortho*-magnesiated anisole product [(PMDETA)K(μ-TMP)(2-C$_6$H$_4$OMe)Mg(TMP)] **27** and Me$_4$Si (Scheme 3.7) [156]. Although electrophilic quenching would be expected to yield identical functionalized *ortho*-anisole products, these observations suggested that the timing of any subsequent quench of the base–substrate reaction mixture may be critical to the outcome of the reaction.

The X-ray structures of the sodium compound [NaMg(TMP)$_2$*n*-Bu] and its potassium congener [NaMg(TMP)$_2$*n*-Bu] were determined in 2015 [157]. [KMg(TMP)$_2$*n*-Bu] was isolated as three

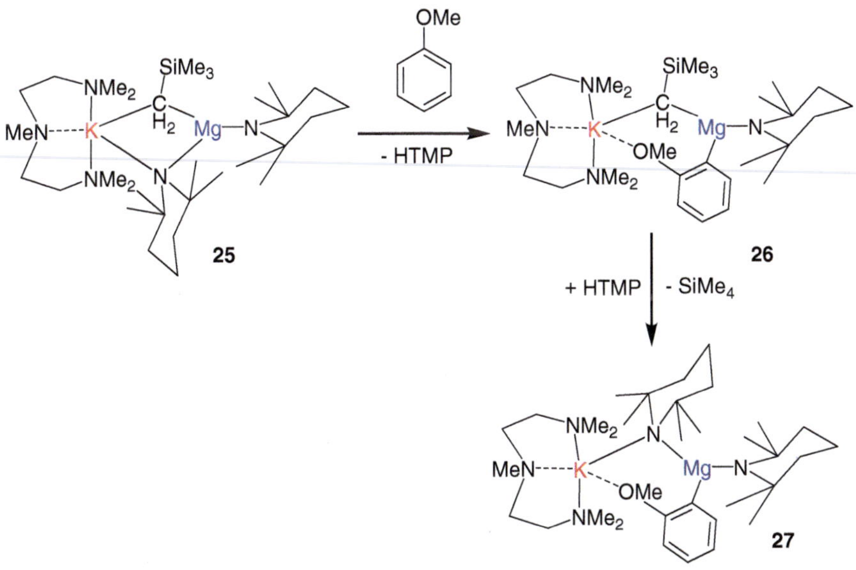

Scheme 3.7 Kinetic and thermodynamic products, compounds **26** and **27**, resulting from the *ortho*-magnesiation of anisole by the potassium-magnesium TMP-alkyl base, **25**.

polymorphic forms, a helical polymer with an infinite KNMgN chain and two cyclic structures, a tetramer with a 16-atom $[KMg(TMP)_2n\text{-}Bu]_4$ ring (**28**, Figure 3.5) and a hexamer with a 24-atom $[KMg(TMP)_2n\text{-}Bu]_6$ ring (**29**), both of which displayed *endo*-disposed alkyl substituents. The status of both magnesiate derivatives as 'pre-inverse crowns' was confirmed by their reactions with benzene and toluene to generate the known inverse crowns, compounds **9** and **10** (Scheme 3.4), providing strong mitigation of the importance of templating effects during metalation reactions performed with such alkali metal monoalkyl-bisamido magnesiates. The $[KMg(TMP)_2n\text{-}Bu]$ system was also shown to react with naphthalene to generate the further inverse crown $[KMg(TMP)_2(2\text{-}C_{10}H_7)]_6$, the molecular structure of which comprised a 24-atom $(KNMgN)_6$ cycle with six naphthalene anions regioselectively magnesiated at the 2-position. The 1,4-dimagnesiation of naphthalene was also accomplished via $[NaMg(TMP)_2n\text{-}Bu]$ **30**, albeit the resultant 12-atom $(NaNNaNMgN)_2$ inverse crown structure contained two demethylated TMP ligands as well as four that remained intact.

Exemplified by the concept of DoM, the regioselectivity of deprotonation reactions between arene substrates and basic metalating agents is usually directed to an *ortho* position adjacent to the substituent. In a compelling recent advance, O'Hara et al. utilized the further pre-inverse crown $[Na_4Mg_2(TMP)_6(n\text{-}Bu)_2]$ **31** to achieve *ortho, meta'*- and even *meta, meta'*-2,5-dimagnesiation of a range of substituted arenes [158]. Treatment of compound **31** with anisole provided the inverse crown $[Na_4Mg_2(TMP)_6(C_6H_3OMe\text{-}2,5)]$ **32** (Figure 3.6a), in which dimetalation takes place at *ortho*- and opposite-facing *meta'*-positions. Similar results were obtained with (methoxymethyl) benzene, *tert*-butoxybenzene, (trifluoromethyl)benzene, *N,N*-diisopropylbenzamide, 4,4-dimethyl -2-phenyl-4,5-dihydrooxazole, and phenyl-*N,N*-diethyl-*O*-carbamate, while quenching of the 2,5-dimetalated intermediates with I_2 provided the corresponding 2,5-diiodido trisubstituted arenes in excellent yields. In contrast, reaction of **31** with *N,N*-dimethylaniline and *N,N*-diisopropylaniline provided the corresponding inverse crowns, **33** (Figure 3.6b) and **34**, which included the 3,5-dimetalated substrates with absolute selectivity. This unprecedented *meta, meta'* selectivity was judged to be a kinetic consequence of the steric protection about the $N\text{–}C_{ipso}$ bonds, which effectively results in the methyl groups offering permanent steric protection to the *ortho*-positions. This hypothesis was afforded significant credence by the further metalation of the

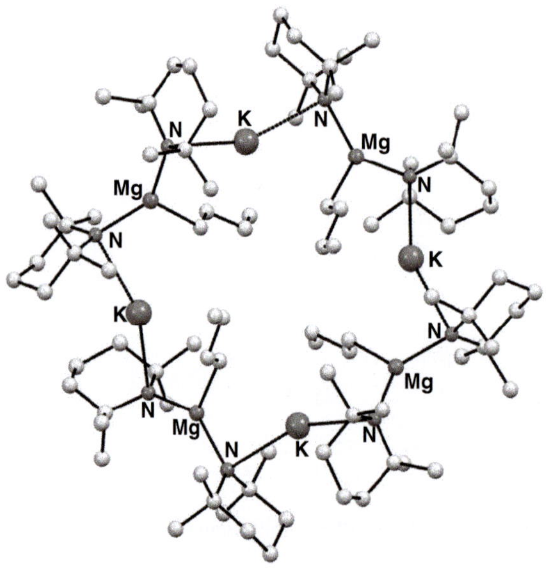

Figure 3.5 The tetrameric 'pre-inverse crown', compound **28**. *Source*: Adapted from Martinez–Martinez et al. [157].

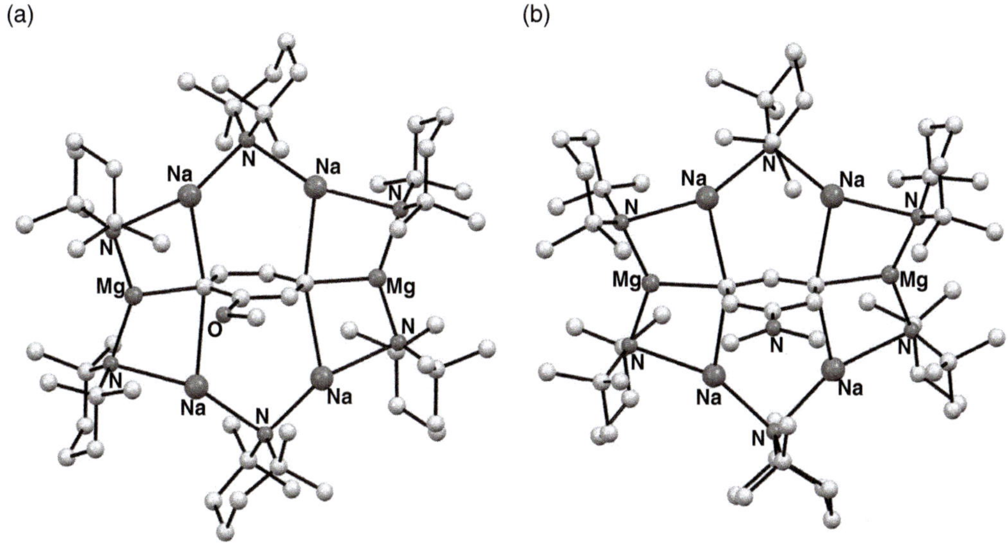

Figure 3.6 The structures of the sodium-magnesiate inverse crowns (a) compound **32** (b) compound **33**, effecting the *ortho*, *meta'* and *meta*, *meta'* dimetalation of anisole and *N*,*N*-dimethylaniline, respecively. *Source*: Adapted from Martinez–Martinez et al. [158].

nonDoM–directing *t*-butylbenzene. In this case, quenching of the reaction with iodine gave high conversion to the 1,3,5-substituted arene, 1-(*tert*-butyl)-3,5-diiodobenzene. In a subsequent advance Martinez-Martinez and O'Hara demonstrated that the use of compound **31** could be extended to the templated *meta* regioselective metalation of unactivated multiaryl substrates that lack appreciable acidity or any DoM-directing functional groups. Metalation was again reasoned to be directed by the kinetic control induced by the pre-assembly of the reactive bimetallic base into inverse crown-like arrays and, after quenching with iodine, biphenyl, *meta*- and *para*-terphenyl, 1,3,5-triphenylbenzene and biphenylene were transformed to a series of previously unobtainable multi-iodoarenes [159].

3.2.3 Salt Effects and Magnesiate Formation

It has been recognized for some time that the presence of even trace amounts of halide salts such as LiCl in organo- and amidolithium solutions can exert a profound kinetic influence over the course of metalation processes [160]. As highlighted by Hevia and Mulvey and emphasized by the painstaking studies of Collum and others, for example, LiCl loadings as low as 0.5–1.0 mol% can deliver rate enhancements of several orders of magnitude to the LDA-induced ortholithiation of arenes containing halogen-based directing groups [47, 48, 161–163]. Kinetic and spectroscopic studies have suggested that such rates of metalation are dictated by the LiCl-induced deaggregation of the lithium reagent, while diffusion ordered NMR studies on the solution speciation of MeLi–LiCl mixtures in THF solution have indicated the operation of equilibria such as that shown in Scheme 3.8 [164]. Although the precise identity (or identities) and mode of action of the lithium base activation remain uncertain, the observation that additional sources of Lewis acidic lithium and basic chloride can wield such power over the course of the metalation reactions is reminiscent of the bimetallic magnesiate base systems outlined above.

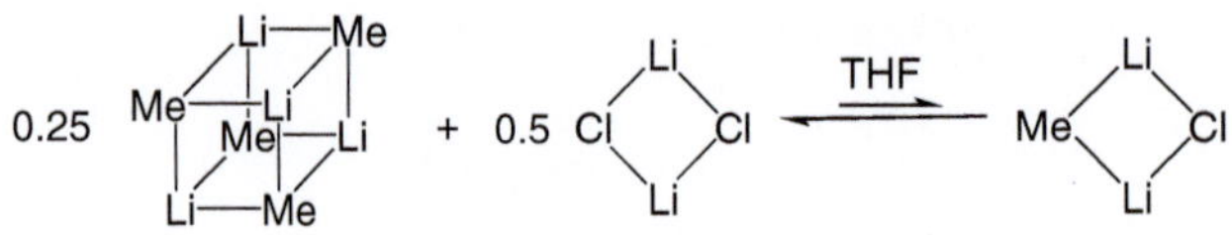

Scheme 3.8 Proposed methyllithium deaggregation equilibrium in the presence of LiCl.

3.2.3.1 'Turbo-Grignards' for Selective Metalation

In a similar manner, related 'salt effects' and positive enhancements of metalation rates are induced when stoichiometric quantities of LiCl are added to Grignard reagents or Hauser bases [165–167]. These so-called turbo-Grignard (RMgX·LiCl) and turbo-Hauser base ($R_2NMgCl{\cdot}LiCl$) reagents can deliver not only dramatic acceleration of halogen-magnesium exchange for the metalation of arenes but also significantly enhanced the tolerance of sensitive functional groups at synthetically convenient temperatures [168]. The reagents were pioneered in Knochel's 2004 report that addition of a stoichiometric equivalent of LiCl to *i*-PrMgCl provided a system which displayed a vastly improved rate for the magnesiation (84% in 68 h at 25 °C) of 4-bromoanisole in comparison to the Grignard reagent alone (18%) [169]. This initial study also demonstrated that similarly fast Mg/Br exchange could be applied to numerous other aryl and heteroaryl bromides (Scheme 3.9a). 1,2,4-Tribromobenzene for example, was shown to undergo highly selective magnesiation at the 2-position and to provide the alcohol product in 89% yield after quenching with pivaldehyde (Scheme 3.9b). The use of a stoichiometric quantity of LiCl also led the authors to postulate that salt addition disrupts the aggregation of *i*-PrMgCl, to produce a magnesiate of the general form [$i\text{-PrMgCl}_2^-\text{Li}^+$] as the actual metalating species [170].

The subsequent application of *i*-PrMgX·LiCl systems advanced rapidly and they were shown to be applicable to the magnesiation of a wide range of highly functionalized aryl or heteroaryl products [165, 171]. Knochel's group, for example, demonstrated that iodophenols were readily metalated by Mg/I exchange in an approach that could be extended to heterocyclic compounds such as pyridines and quinolones also bearing a hydroxyl substituent [172]. Such was the ease of synthesis that the protocol rapidly impacted upon the synthetic organic community as the one of the 'go to' methods for the convenient preparation of highly functionalized aryl or heteroaryl compounds. Although a complete survey of the applications of Knochel's original system is beyond the scope of this chapter, metalation by *i*-PrMgX·LiCl has been central to the successful synthesis of a diverse array of compound classes including unsaturated silylated cyanohydrins [173] tertiary aryl

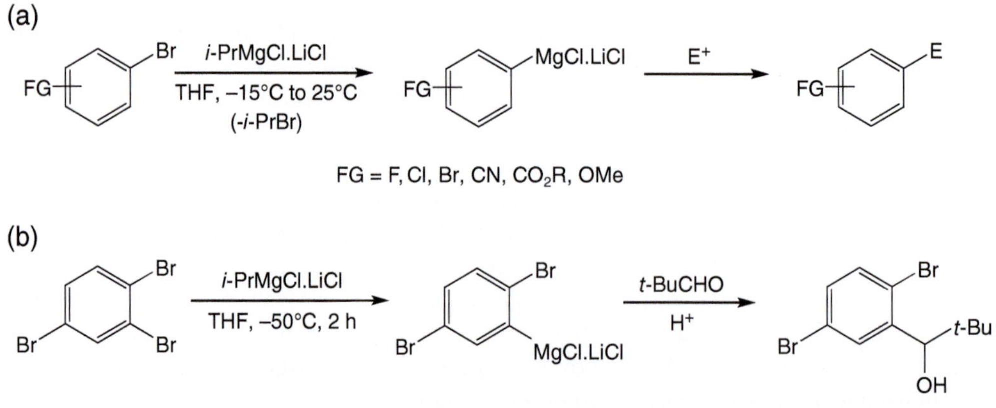

Scheme 3.9 Selective metalation of bromoarenes by the turbo-Grignard reagent, *i*-PrMgCl·LiCl.

amines [174, 175], mycophenolic acid derivatives [176], triazenes [177], indazoles [178], secondary benzylic alcohols [179–181], 4-hydroxytetrahydroquinolines [182], α-hydroxyacetophenones [183], hydroxyphenyl nucleobase derivatives [184], 2-pyridylhydroxylamines [185], 3-aryl- and alkyl-phthalides [186], aryltriazenes [187] and the functionalization of iodoporphyrins [188]. These latter molecules were shown to undergo iodine–magnesium exchange and onward functionalization without decomposition of the oxidatively sensitive porphyrin.

Turbo-Grignard systems are also applicable to metalation reactions in the presence of potentially electrophilic main group centred aromatic substituents. Whereas, for example, reactions of *para*-iodoboronic esters with *i*-PrMgCl afford only the corresponding range of iodoheteroaryl boronic esters, *i*-PrMgCl·LiCl is exquisitely selective for halogen–magnesium exchange, even in the presence of an electrophilic boron centre [189]. In these latter cases, reactions performed at −78 °C led to magnesiated boron esters, which could be reacted with a variety of electrophiles to provide highly functionalized boronic esters in high yield (Scheme 3.10a). In a further demonstration of the synthetic potential of these protocols, Suzuki cross-coupling reactions of the resulting boronic esters were shown to afford various polyfunctional aromatic and heteroaromatic compounds. Recently, a similar methodology has been extended to the synthesis of a variety of other heteroarylboronic esters [190–192].

Although replacement of CF_3 groups with SF_5 substituents has been observed to enhance the biological activity of pharmacologically active substances, this poses a problem for the synthetic chemist as this group is typically very sensitive to polar organometallic species such as organolithiums. In 2012, Knochel and coworkers demonstrated that bromine–magnesium exchange reactions starting from the commercial 1-bromo-3-pentafluorosulfanylbenzene could be readily achieved to provide the corresponding arylmagnesium halide in > 80% yield (Scheme 3.10b) [193]. This procedure was shown to be applicable in the preparation of a broad range of SF_5-substituted aromatic and heterocyclic compounds of potential interest for the pharmaceutical applications.

The use of turbo-Grignard systems is also appropriate for the magnesiation of nonaromatic and heteroaromatic sp2 C–X and sp-hybridized C–H bonds. Although, for example, stereospecific Mg/I exchange of vinyl iodides may be performed using *i*-PrMgCl alone [194], the reactions typically require high temperatures limiting the presence of functional groups such as esters and constraining substrate scope. Application of *i*-PrMgCl·LiCl, however, allows reactions to be undertaken at lower temperatures providing a greatly enhanced functional group tolerance, including esters and other halogens (Scheme 3.11) [170, 195].

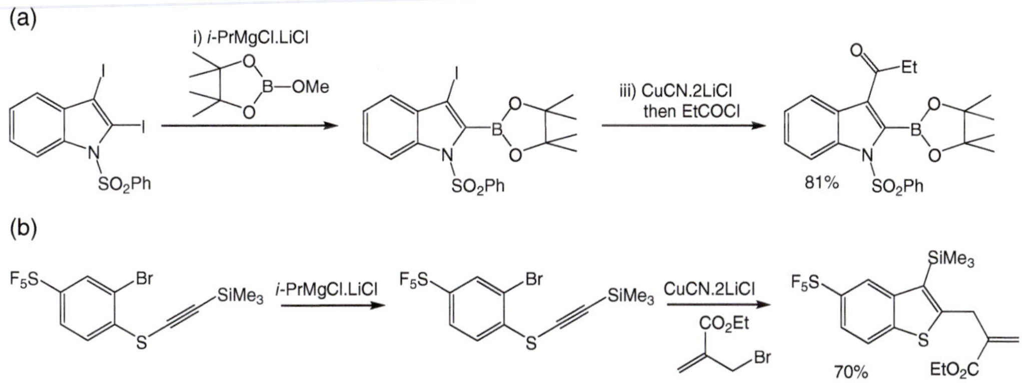

Scheme 3.10 Applications of magnesium–halogen exchange and arene metalation mediated by the turbo-Grignard reagent *i*-PrMgCl·LiCl in the presence of (a) trimethylsilyl, (b) pentafluorosufanyl substitution.

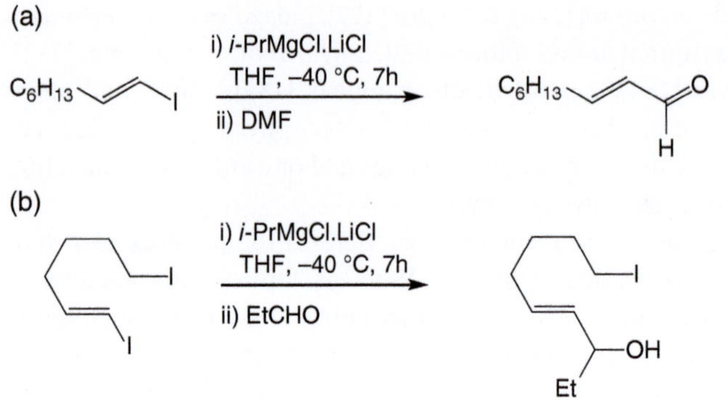

Scheme 3.11 Use of the turbo-Grignard system, *i*-PrMgCl·LiCl, for the selective metalation of vinyl sp^2-bonded halocarbons.

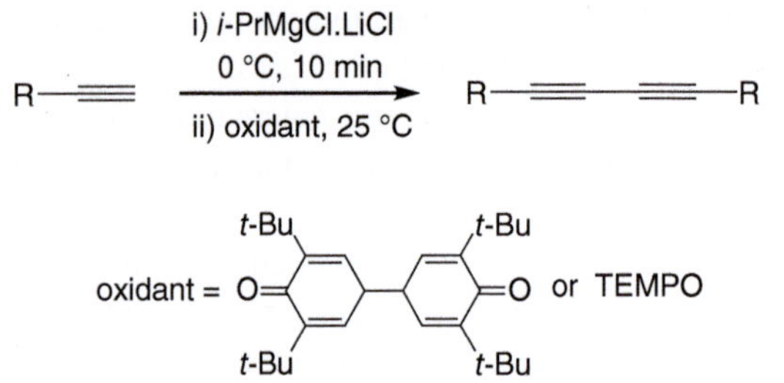

Scheme 3.12 Turbo-Grignard mediated deprotonation of terminal alkynes and oxidative C–C coupling.

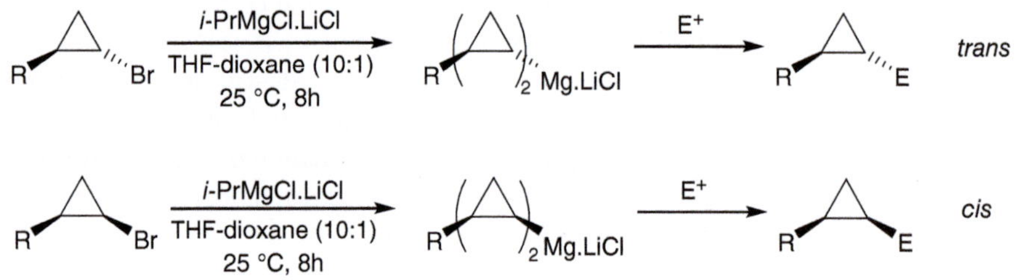

Scheme 3.13 Use of the turbo-Grignard system, *i*-Pr-MgCl·LiCl, for the metalation of the sp^3 C–Br bonds of both *trans* and *cis* cyclopropyl compounds.

While not so widely applied, acetylenic variants of turbo-Grignards are readily available by use of the parent reagent, *i*-PrMgCl·LiCl, to deprotonate terminal alkynes [196, 197]. These derivatives are useful in onward air- or TEMPO-promoted oxidative cross-coupling reactions to yield the C–C coupled bis-acetylenes (Scheme 3.12) as well as in the more conventional nucleophilic addition reactivity [198, 199].

Although turbo-Grignard-induced metalation of sp^3 carbon centres has been less commonly reported, Knochel and coworkers have described how bromine/magnesium exchange can be achieved between *i*-PrMgCl·LiCl and a range of cyclopropyl compounds (Scheme 3.13) [200–202]. The ring strain intrinsic to cyclopropyl substrates enhances the s-character of the C–Br bonds and activates them toward the exchange, albeit the reactions were quite sluggish. The resultant organomagnesiums, however, could be quenched with a range of electrophiles to provide the functionalized cyclopropanes in good yields.

In a notable recent advance, Knochel and coworkers have demonstrated that *s*-BuMgOR·LiOR (R = 2-ethylhexyl) undergoes very fast Br/Mg exchange with aryl and heteroaryl bromides, producing aryl and heteroaryl magnesium alkoxides (ArMgOR·LiOR) in hydrocarbon rather than the

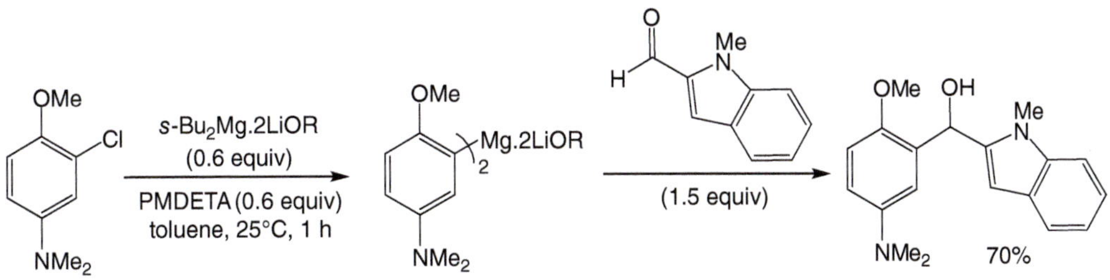

Scheme 3.14 Application of the alkoxo turbo-Grignard variant, s-$Bu_2Mg{\cdot}LiOR$ (R = 2-ethylhexyl), for the selective metalation of chloroarenes in wholly hydrocarbon (toluene) solution.

ethereal solvents required by the turbo-Grignard systems [203]. The new alkoxide reagents react with a broad range of electrophiles, including aldehydes, ketones, allyl bromides, acyl chlorides, epoxides, and aziridines in good yields. The related reagent s-$Bu_2Mg{\cdot}2LiOR$ (R = 2-ethylhexyl) undergoes Cl/Mg exchange with electron-rich aryl chlorides in toluene, producing diorganomagnesium species of the type $Ar_2Mg{\cdot}2LiOR$, which also react with aldehydes and allyl bromides (Scheme 3.14).

Although the exemplary reactions described above clearly highlight the enhanced utility that may be obtained through adoption of the turbo-Grignard formulation, only limited progress beyond Knochel's initial hypothesis has been made in terms of their mode of action and structure in both the solution and the solid state. It is clear, however, that RMgCl·LiCl reagents sit in a similar category as the synergic magnesiate base systems that have been so widely exploited by Mulvey, Mongin and others described in Section 2.2 [119–127]. Early DFT calculations performed by Straub and Knochel supported the magnesiate character of the reagents and metalation intermediates and transition states (Figure 3.7) involving the assembly of hypervalent halide species [204]. The barrier heights associated with the halogen–metal exchange reactions were predicted to decrease with increased electron density at the magnesium centre. This latter deduction was supported experimentally through the addition of Li^+ selective chelating agents to induce greater charge separation, which produced significant improvements in rate for the magnesiation of 4-bromoanisole. Particularly significant enhancements were observed after the addition of 1,4-dioxane, which was reasoned to result in the enforcement of an 'anionic Schlenk equilibrium'. The resultant precipitation of $MgCl_2$ led to the deduction that i-Pr_2Mg is, in fact, the reactive species in solution.

The dynamic nature of turbo-Grignard solutions has been underscored by several studies aimed at identifying the most persistent species both in solution and through their isolation in the solid state. Although it appears Knochel's initial hypothesis of ate complex formation is undoubtedly correct, the lability of such species toward Schlenk equilibration has so far militated against the successful crystallization of any mixed R/Li/Mg/Cl aggregates from RMgCl·LiCl solutions. Lerner et al., for example, found that the structure of $[i\text{-PrMgCl(THF)}]_2[MgCl_2(THF)_2]_2$ **35** single crystals obtained from a THF solution of i-PrMgCl·LiCl was effectively identical to that grown from the

Figure 3.7 Calculated form of the magnesiate transition state formed during the metalation of bromobenzene.

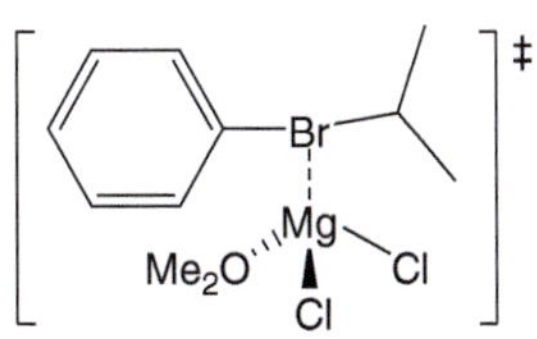

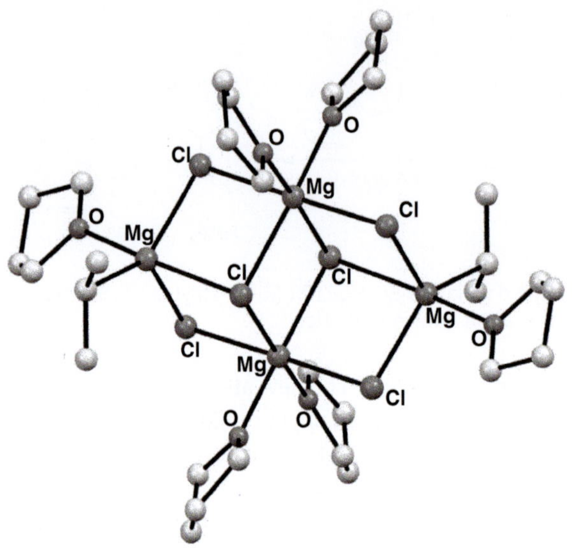

Figure 3.8 Solid state structure of [*i*-PrMgCl(THF)]$_2$[MgCl$_2$(THF)$_2$]$_2$, compound **35**, obtained from a THF solution of *i*-PrMgCl·LiCl. *Source*: Adapted from Blasberg et al. [205].

Grignard reagent without the addition of LiCl (Figure 3.8; see also Chapter 2, Scheme 2.34) [205]. More informatively, a combination of electrospray-ionization mass spectrometry, electrical conductivity measurements, NMR spectroscopy (including diffusion-ordered spectroscopy), and quantum chemical calculations performed on turbo-Grignard systems were analyzed by Koszinowski et al. [206]. The molecular constituents of RMgCl and RMgCl·LiCl were again deduced to be interrelated via Schlenk equilibria and fast intermolecular exchange processes. A small portion of the Grignard reagent was also observed to form anionic ate complexes in solution, with the abundance of such species significantly increasing upon the addition of LiCl. The action of the two reagents during the metalation process was, thus, deduced to be qualitatively similar but to differ quantitatively as a consequence of the larger proportion of more electron rich and putatively more nucleophilic ate complexes persisting in the turbo-Grignard solutions.

3.2.3.2 Turbo-Hauser Bases

In 2006, Knochel et al. developed a mixed lithium and magnesium amide bases $R^1R^2NMgCl \cdot LiCl$ through the combination of a sterically hindered and nonnucleophilic magnesium amide with LiCl. These species benefit from significantly greater solubility in THF than the Hauser bases themselves [166, 207], and are equipped with enhanced kinetic basicity for the deprotonative magnesiation of various functionalized (hetero)aromatic compounds. The most important of these so-called turbo-Hauser bases are TMPMgCl·LiCl, TMP$_2$Mg·LiCl and *i*-Pr$_2$NMgCl·LiCl, which are typically prepared by treating HTMP or *i*-Pr$_2$NH with *i*-PrMgCl·LiCl or the analogous diorganomagnesium system in THF (Scheme 3.15) [207].

The solubility of these species was suggested to reflect the existence of less kinetically labile and lower molecularity species than that in operation in the comparable turbo-Grignard systems discussed above. This supposition was later borne out by Mulvey et al. determination of the single-crystal X-ray structure of [TMPMg(μ-Cl)$_2$Li(THF)$_2$] from THF solution, compound **36**, which was shown to comprise a bimetallic motif with a terminal magnesium-bound TMP ligand and with two chlorides bridging the Mg and Li centres (and is illustrated as compound **60** in Chapter 2, Figure 2.13) [208]. In contrast, the diisopropylamide analogue, *i*-Pr$_2$NMgCl·LiCl, was later found

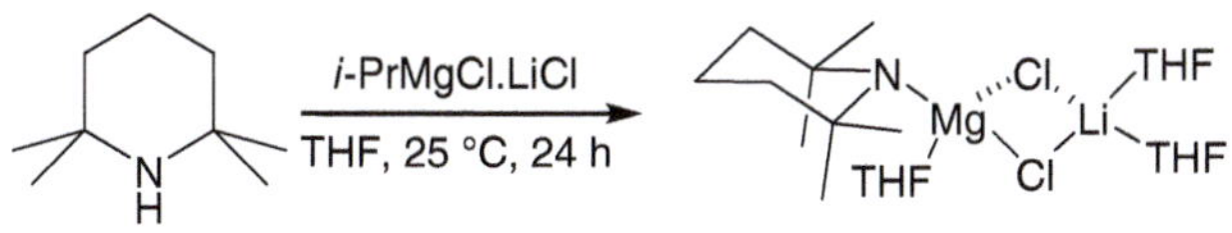

Scheme 3.15 Synthetic route to the turbo-Hauser base, TMPMgCl·LiCl.

to crystallize as the amide-bridged dimeric species, [{(THF)$_2$Li(μ-Cl)$_2$Mg(μ-N*i*-Pr$_2$)}$_2$] **37** (see again Chapter 2, Figure 2.13) [209]. These observations, in conjunction with detailed NMR studies (elaborated in Chapter 6) of both TMPMgCl·LiCl and *i*-Pr$_2$NMgCl·LiCl in THF [210, 211] have enabled a rationale to emerge for the superior kinetic competence of the former species toward deprotonation in THF solution (*vide infra*). The lithium–magnesium-containing TMP complex (**36**) has been deduced to exist as a limited number of species, predominantly in the form weakly contacted ion pairs, in solution. In contrast, *i*-Pr$_2$NMgCl·LiCl displays more complex behaviour that is largely dependent upon a dynamic equilibrium between a variety of monomeric, dimeric and even higher nuclearity species that is the reminiscent of those determined in the solid state in various lithium and sodium magnesiate systems [212]. These structural differences suggest that the superior kinetic competence of the former species toward deprotonation in the THF solution (*vide infra*) is dictated by its more pronounced tendency toward dissociation into monometallic or charge separated species in solution. Significantly, this rationale was also suggested to distinguish the typical *ortho* metalating behaviour of turbo reagents with the remarkable templated stereochemical outcomes observed for the more strongly ion paired magnesiate amido-alkyl base systems discussed in Section 3.2.2 [209].

The more pronounced tendency of *i*-Pr$_2$NMgCl·LiCl toward the assembly of higher order aggregates is also reflected in its lower solubility and consequent reduced reactivity in comparison to its TMP analogue. For example, although both reactions with isoquininoline lead to 2-quinolylmagnesium chloride in greater than 90% yield, the magnesiation requires two equivalents of *i*-Pr$_2$NMgCl·LiCl and a reaction time of 12 h at 25 °C. In contrast, the same metalation is complete within 2 h at 25 °C using only 1.1 equivalents of TMPMgCl·LiCl (Scheme 3.16) [207].

The basic properties of TMPMgCl·LiCl are also highly tolerant of more sensitive functional groups, enabling the regioselective magnesiation of a wide array of polyfunctional aromatic systems. For example, reaction with the ester diethyl bromoisophthalate at −25 °C for 30 min enables the generation of the functionalized magnesium derivative in greater than 90% yield (Scheme 3.17) [207].

Although the reader is directed elsewhere for complete coverage of the synthetic applications of these reagents in synthesis [213], such turbo-Hauser base systems have been applied to the selective metalation of an ever-growing array of highly functionalized aromatic systems including carboethoxyferrocene derivatives [214], pyridines [215], quinolines [216], pyrimidines [217, 218], protected aniline derivatives [219], various aromatic sulfones [220, 221], furans [222], thiophenes [223, 224], 2-pyridone [225], 1,4-dithiins and cyanohydrins (Scheme 3.18) [226, 227].

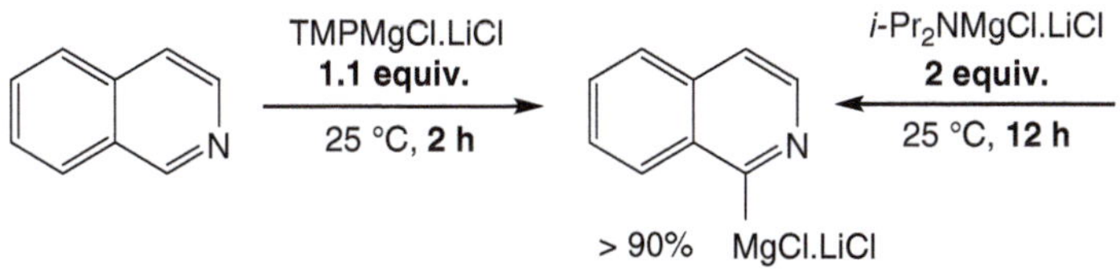

Scheme 3.16 Comparative ability of TMPMgCl·LiCl and *i*-Pr$_2$NMgCl·LiCl to effect the metalation of isoquinoline.

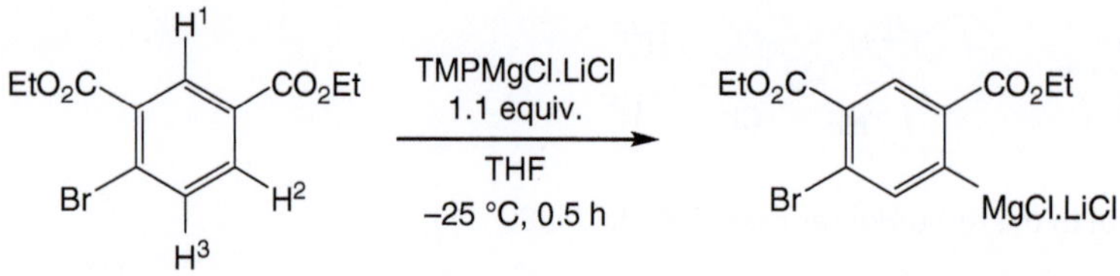

Scheme 3.17 Selective magnesiation of diethyl bromoisophthalate with TMPMgCl·LiCl.

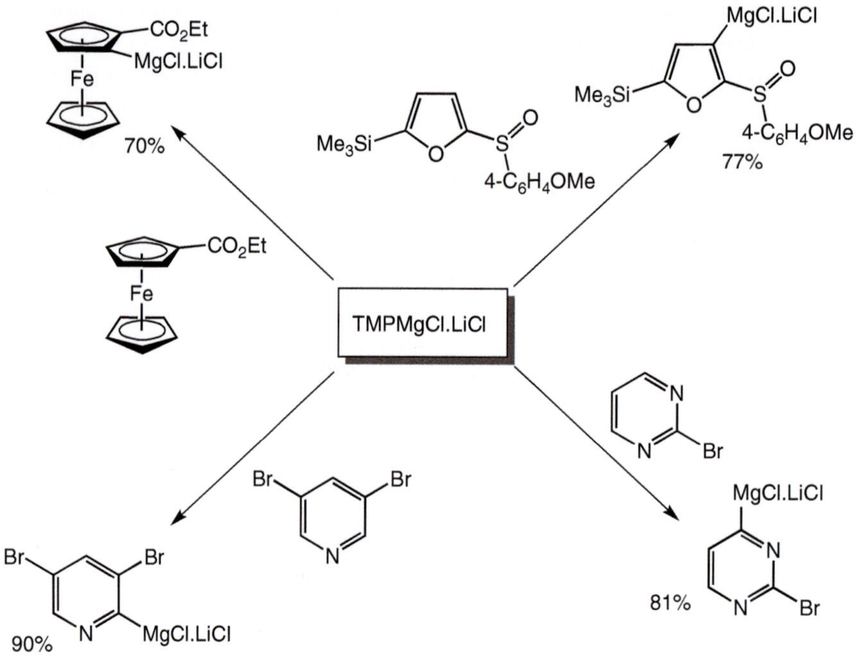

Scheme 3.18 Examples of selective deprotonation/magnesiation with the turbo-Hauser base reagent, TMPMgCl·LiCl.

3.2.4 Ate Complexes of the Heavier Alkaline Earth Elements Ca, Sr, and Ba

In contrast to the widespread study and exploitation of magnesium 'ate' species, analogous chemistry of similar systems comprising one of magnesium's heavier alkaline earth congeners, Ca, Sr, and Ba has received only scant and sporadic attention. Indeed, the organometallic chemistry of these elements in general is only at a comparably nascent stage of development [228]. While description of onward reactivity is all but non-existent, the structural features of a variety of heterobimetallic calciate, strontiate and bariate species in conjunction with an alkali metal have been described (i.e. of the form $MM'X_3$ or $M_2M'X_4$ where M and M' represent a group 1 and a heavier group 2 element, respectively, and where X is a monoionic basic residue such as an amide, alkyl, or alkoxide). Before transitioning to a more general description of recent developments in monometallic heavier alkaline earth reactivity in Section 3.3, therefore, it is germane to reflect on the current 'state-of-the-art' with regard to heavier group 2 heterobimetallic synthesis and structure to best appreciate their all but completely unexploited synthetic potential.

3.2.4.1 Alkyl Calciate, Strontiate, and Bariate Derivatives, $MM'R_3$ (M = Li, Na, K; M′ = Ca, Sr, Ba; R = alkyl)

Although the formation of the pentamethylcyclopentadienyl derivative $[LiBa(C_5Me_5)_2\{CH(SiMe_3)_2\}(THF)]$ was claimed in 1992, the insolubility of this species and consequent absence of definitive characterisation data renders its consitution somewhat doubtful [229]. The first unquestionable example of a triorgano derivative of any heavier group 2 element was thus provided by Müller's report of the sodium trialkylcalciate $[(2\text{-}MeOC_6H_4CH_2PPh'_2)_3CaNa(OEt_2)]$ **38** (Figure 3.9a), in which the sodium centre is encapsulated by the three-arrayed pendant phosphine donors of the phosphorus-stabilised carbanions [230]. The report of this compound was quickly followed by Lappert's characterisation of the trialkyl calciate, $[Ca\{CH(SiMe_3)_2\}_3K]$ **39**, which was obtained by reaction of the potassium alkyl and CaI_2 in benzene and in which the calcium centre resides in an unusual trigonal pyramidal coordination environment (Figure 3.9b) [231]. Beyond these initial examples, the sole relevant advances have been provided by Ruhlandt-Senge and Harder, who reported the structures of the tetrabenzyl dilithium species $[(TMEDA)_2Li_2Ca(CH_2Ph)_4]$ **40**, and the unique aryl calciate, $[(2,6\text{-}\{i\text{-}PrO)_2C_6H_3\}_3CaK]$ **41**, which crystallized as a molecular species and as a linear coordination polymer, respectively [232, 233]. The sole organostrontiate to be unambiguously identified appears to be Okuda's tris(allyl) potassium species $[(THF)_2KSr(C_3H_5)_3]$, which comprises a two-dimensional network of $(\mu_2\text{-}\eta^3\text{:}\eta^3\text{-}C_3H_5)$-bridged potassium and strontium centres [234].

3.2.4.2 Alkoxo and Aryloxo Calciate, Strontiate, and Bariate Derivatives, $MM'(OR/Ar)_3$ (M = Li, Na, K; M′ = Ca, Sr, Ba)

Only a handful of compounds of the form $MM'(OR/Ar)_3$ have been synthesized, primarily because of their potential as molecular precursors to mixed metal oxides by sol–gel routes. Although the resultant species does not conform to this generalized formulation, Fromm et al. have obtained the phenoxide, $[CaLi_6(\mu_3\text{-}OPh)_8(THF)_6]$ **42** from the reaction of CaI_2 with LiOPh in THF. The structure featured two heterocubane units fused via the calcium ion and provided the further aggregate $[Ca_2(DME)_2(\mu\text{-}OPh)_6\{Li(DME)\}_2]$, **43** after recrystallization from dimethoxyethane (DME) [235]. This work was subsequently extended to the synthesis of a range of phenoxides, $[MM'(OPh)_8(THF)_6]$,

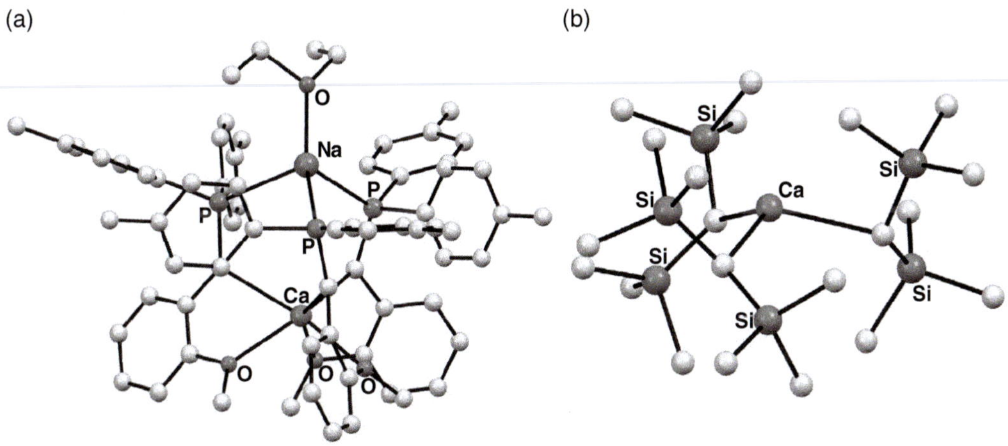

Figure 3.9 Solid state structures of (a) the sodium trialkylcalciate, compound **38**, and (b) the trigonal pyramidal anionic calciate component of compound **39**. *Source*: Adapted from Knapp and Muller [230]; Hitchcock et al. [231].

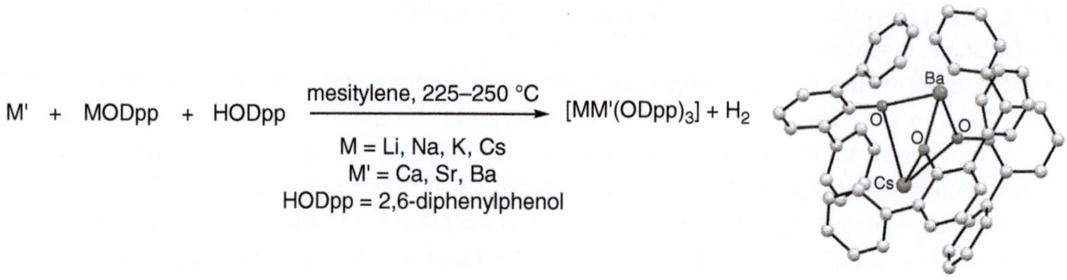

Scheme 3.19 Synthetic route to heterobimetallic diphenylphenoxides and the structure of a representative caesium bariate complex. *Source:* Adapted from Zuniga et al. [238].

and related DME adducts where M = Li, Na and M′ = Ca, Sr, Ba [236]. As a part of the same study, reaction of CaI_2 with equal amounts of PhOLi and *t*-BuOLi in the THF afforded single crystals of the mixed aryl-alkoxide derivative, $[CaLi_6(OPh)_6(Ot\text{-}Bu)_2(THF)_6]$ **44**. In a similar context, related bimetallic derivatives of perfluoro-*tert*-butoxide have been described by Ruhlandt-Senge et al. [237], while a range of compounds with the generalized formulation $[MM'(ODpp)_3]$ (M = Li, Na, K, Cs; M′ = Ca, Sr, Ba), were obtained by the same group through direct reaction of the alkaline earth metal with the more sterically demanding 2,6-diphenylphenol (HODpp) in the presence of an equivalent of the relevant alkali metal phenoxide (Scheme 3.19) [238].

3.2.4.3 Amido Calciate, Strontiate, and Bariate Derivatives, $MM'(OR/Ar)_3$ (M = Li, Na, K; M′ = Ca, Sr, Ba)

Although representative examples of amide-derived ate complexes of the heavier alkaline earths are more common, their chemistry still cannot be described as well developed and is primarily based upon use of the hexamethyldisilazide ligand. The initial examples of this class of compound were provided by Davies in 2000, who reported the synthesis and structural characterization of $[Ca\{N(SiMe_3)_2\}_3Li(THF)]$ **45**, containing two calcium-to-lithium bridging amide ligands, and the charge separated compound $[Ba\{N(SiMe_3)_2\}_2(THF)_3][Li_2\{N(SiMe_3)_2\}_2(THF)_2]$ **46** [239]. This report was rapidly followed by the synthesis of a solvent-free variant of compound **45** and compound **47**, which was observed to be based on a planar LiNCaN four-membered ring with both metals engaging in additional CH_3. . .M (where M = Li, Ca) interactions with the trimethylsilyl substituents [240]. The potassium analogue of this compound, $[KCa\{N(SiMe_3)_2\}_3(THF)]$ **48**, was reported to be composed of CaNKN four-membered rings with interdinuclear interactions between the terminal amide units and the potassium centres allowing the propagation of a polymeric chain structure. In what still appears to be a unique example of metalation reactivity, compound **48** and its stoichiometric variant, $[K_2Ca\{N(SiMe_3)_2\}_4]$ **49**, were shown to be effective for the enolization of 2,4,6-trimethylacetophenone, giving rise to $[K_2Ca_2\{OC(Mes){=}CH_2\}_6(THF)_2]$ **50** and $[K_2Ca\{OC(Mes){=}CH_2\}_4(THF)_4]$ **51**, respectively (Scheme 3.20).

It was later shown that charge separation could be induced as a more generalized phenomenon with bis(trimethylsilyl)amide derivatives through addition of a strongly σ-donating *N*-heterocyclic carbene ligand, to provide compounds of the general form $[M(NHC)_x][M'\{N(SiMe_3)_2\}_3]$, where x = 2 or 3 depending on the steric demands of the carbene ligand [241, 242].

In an advance that was eventually to lead the way into the use of group 1/group 2 heteobimetallics in catalysis, Westerhausen reported that reactions of an excess of KN(Ph)R (R = Me, *i*-Pr) with CaI_2 leads to the formation of calciates of the type $K_2Ca[N(Ph)R]_4$ [243]. The THF-free compound $[K_2Ca\{N(Ph)Me\}_4]_\infty$ **52** forms a three-dimensional net structure, whereas

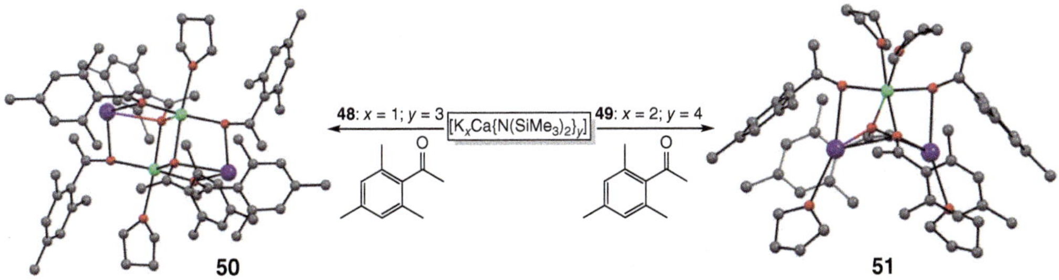

Scheme 3.20 Enolization of 2,4,6-trimethylacetophenone by the heterobimetallic bis(trimethylsilyl)amide derivatives, **48** and **49**, leading to the structurally characterized compounds **50** and **51** (Ca = green, K = dark purple, O = red). *Source:* Adapted from Kennedy et al. [240].

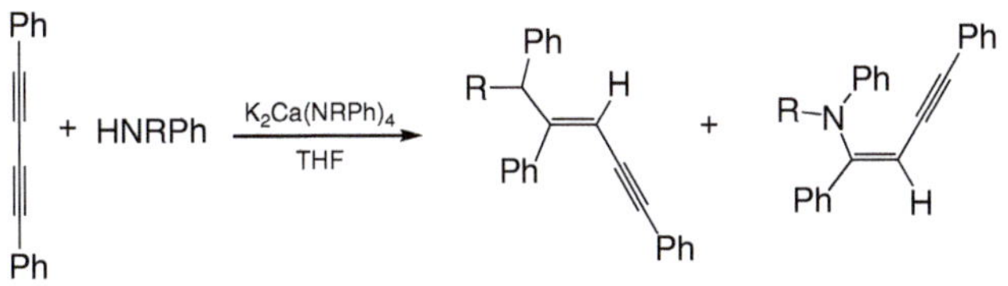

Scheme 3.21 Calciate-catalyzed (5 mol% **55**) hydroamination of diphenylbutadiyne.

the amides with the larger isopropyl group lead to the formation of the molecular compound [((THF)$_2$K{μ-N-(Ph)*i*-Pr)$_2$}$_2$Ca] **53**. Although this report was followed by a number of closely related structural studies [244], of greater potential significance is the report that compound **53** is an active catalyst for the intermolecular hydroamination of diphenylbutadiyne with secondary anilines. In reactivity reminiscent of monometallic heavier group 2 systems (*vide infra*), the formation of a mixture of the *E*- and *Z*-isomers of the corresponding 1-anilino-1,4-diphenylbut-1-ene-3-yne was observed (Scheme 3.21). Furthermore, heterobimetallic [$K_2Ca(NPh_2)_4$] **54** was shown to be an effective catalyst for the hydroamination reaction in tetrahydrofuran, albeit at elevated temperatures [245]. These observations were further advanced with the synthesis of the more sterically encumbered [$K_2Ca\{N(H)Dip\}_4]_\infty$ **55** [246]. Hydroamination of diphenylbutadiyne with 2,6-diisopropylaniline in the presence of 5 mol% of **55** provided catalytic production of the tetracyclic 2,6-di-isopropyl-9,11,14,15-tetraphenyl-8-azatetracyclo[8.5.0.0^{1,7}.0^{2,13}]pentadeca-3,5,7,9,11,14-hexaene, even at room temperature.

Although such reports of catalysis are so far very limited, it appears clear that further study is merited, particularly with regard to recent advances in the use of monometallic heavier s-block species in catalysis. To further contextualize this potential, the following section provides a brief survey of the current landscape in heavier group 2-centred homogeneous catalysis.

3.3 Homogeneous Catalysis by s-block Reagents

Some of the main drivers to the development of synthetic chemistry in the first decade of the twenty-first century were derived from considerations of sustainability and environmental benignity. Catalysis has, thus, occupied a position of prominence and the ability of catalytic vectors derived the more abundant s-block elements to attain greater atom economy at limited cost is widely recognized. As a result of magnesium and calcium's status as the 7th and 5th most abundant elements in the Earth's crust, they are significantly cheaper than many transition metals (Rh, Pd, and Ir) that are most commonly used to mediate catalytic transformations. A majority of interest thus far focused on the elements of group 2, has featured particularly magnesium and calcium [247–250]. There is also growing evidence, however, that the application of alkyl, amide, alkoxide and even hydroxide reagents derived from group 1 may display similar catalytic competence and be worthy of more concerted study [251–270]. Although a systematic survey of the relevant literature is beyond the scope of this chapter, it is worth reflecting on the main features of these elements that enable their catalytic reactivity.

Due to the absence of valence d-electrons, the chemistry of redox-inactive alkaline earth metals is dominated by two principal mechanistic steps, σ-bond metathesis and π-bond insertion, both of which proceed with the retention of the 2+ oxidation state at the metal These steps are summarized in Scheme 3.22 and, when employed sequentially, may be incorporated in catalysis.

The electronegativity of the heteroelement of the E–H and E′–H species dictates the regioselection of the species formed via a σ-bond metathesis reaction with an M–X bond. For example, when the heteroelement of a E–H (E = N, O, P, and S) bond is more electronegative than hydrogen, σ-bond metathesis yields new M–E and X–H bonds via protonolysis as illustrated in Scheme 3.22a(i). In contrast, the comparable σ-bond metathesis reaction with hydridic E′–H bonds (E′ = B, Si and Al) affords new M–H and E′–X bonds [Scheme 3.22a(ii)]. Addition of an M–X bond across an E=C (E = C, N, O and S) π-bond generates new M–E′ and C–X bonds, with retention of an E–C bond (Scheme 3.22b).

Initial investigations of alkaline earth chemistry focused on the stoichiometric steps outlined in Scheme 3.22 and the concomitant generation of catalytically relevant species. The validity of the protonolysis step was exemplified by the reactivity of homoleptic alkaline earth hexamethyldisilazides with amines, arsanes, alcohols, tellurols, pyrroles, azoles, terminal alkynes, phosphines, and acidic C–H bonds. These reactions generated the corresponding amides [271–309], arsanides [310–312], alkoxides [288, 295, 313–333], tellurolates [334], pyrrolides [281, 288, 335–339] azolides [340–347], acetylides [348–350], phosphides [351–356] and carbanions [279, 357–373] alongside the amine, hexamethyldisilazane (Scheme 3.23).

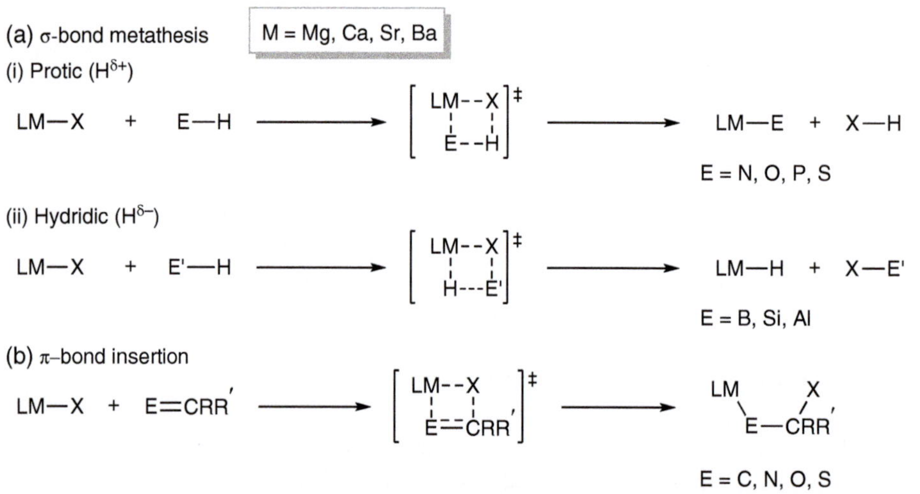

Scheme 3.22 The principal mechanistic steps invoked in alkaline earth catalysis. σ-bond metathesis (a) with a protic (i) or hydridic (ii) E(′)–H bonds and (b) insertion of an unsaturated E=C bond into an Ae–X bond.

$$[Ae\{N(SiMe_3)_2\}_2] + n\ E{-}H \xrightarrow[n\,=\,1\text{ or }2]{} E_n{-}Ae\{N(SiMe_3)_2\}_{2\text{-}n} + n\ H{-}N(SiMe_3)_2$$

Scheme 3.23 Illustrative protonolysis of alkaline earth hexamethyldisilazides.

Exploitation of hydridic σ-bond metathesis reactions, as shown in [Scheme 3.22a(ii)], of phenylsilane with hetero- or homoleptic alkaline earth alkyls (Be [374, 375] and Mg [376–389]) and hexamethyldisilazides (Mg [390], Ca [305, 391–393], Sr [389, 393], and Ba [394]) has been effective for the isolation of discrete alkaline earth hydride compounds, which occurs with the elimination of the respective silyl-alkyl or -amide by-product. Similarly, the prototypical metallophosphination of benzonitrile [395] and diphenylbutadiyne [396, 397] with the alkaline earth phosphanides $[M\{P(SiMe_3)_2\}_2(THF)_n]$ (M = Mg; n = 0 or 2, M = Ca, Sr, Ba; n = 2) provided an early indication of the propensity of group 2 complexes toward π-insertion reactions (Scheme 3.22b).

Incorporation of a combination of these σ-bond metathesis and insertion steps into independent catalytic manifolds, as shown in Scheme 3.23, has enabled the alkaline earth-mediated reduction and heterofunctionalization of a wide variety of unsaturated substrates. A plethora of reactions, including the hydrogenation [398–400] hydroamination [401–404] hydroalkoxylation [405], hydrophosphination [406] hydrosilylation [388, 407–410], hydroboration [411–414] and hydroacetylation [415, 416] of C–E multiple bonds (E = C, N, and O), have now been catalyzed by alkaline earth species. This subject has been summarized in multiple reviews and will not be discussed in depth herein [247, 249]. Significantly, however, noteworthy differences in rate and mechanism have been observed with variation of the group 2 element centre. Although no absolute reactivity scale has yet emerged, Carpentier et al., for example, have observed that, although heteroleptic anilidoimine complexes of Ca, Sr, and Ba all catalyze the regioselective intermolecular hydroamination and hydrophosphination of styrene and isoprene, the catalytic performances increased linearly with the size of the metal (Scheme 3.24) [417].

Combinations of sequential σ-bond metathesis steps enable catalytic cross-metathesis reactions, particularly dehydrocoupling processes, to be catalyzed by alkaline earth complexes. Harder et al. initially reported that the dehydrogenative Si–C coupling of Ph_3SiH with 1-hexyne was catalyzed by 5 mol% of the calcium azametallacyclopropane, $Ca(\eta^2\text{-}Ph_2CNPh)$, affording 1-hexynyltriphenylsilane in 81% conversion. This reactivity was extended to include the catalytic Si–N dehydrocoupling of Ph_3SiH and amines, RR′NH (R = alkyl, aryl; R′ = H, alkyl), affording the respective silylamines in

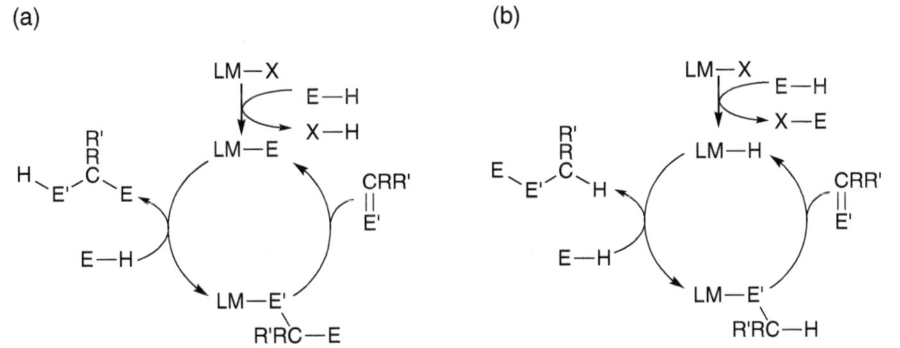

Scheme 3.24 Protic (a) and hydridic (b) alkaline earth catalytic cycles invoked in the heterofunctionalization of unsaturated substrates.

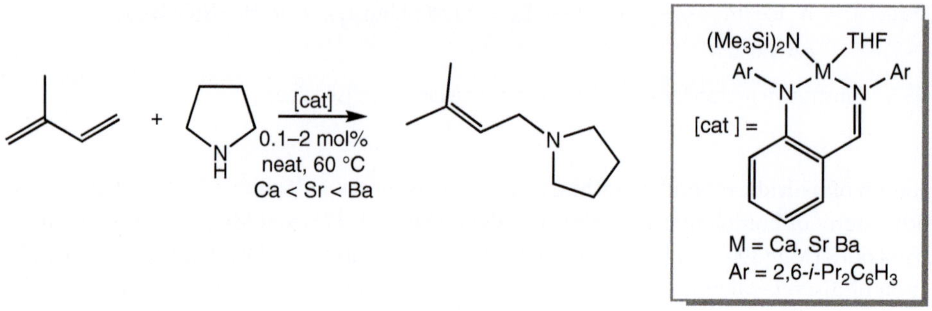

Scheme 3.25 Intermolecular hydroamination catalyzed by alkaline earth anilidoimine complexes.

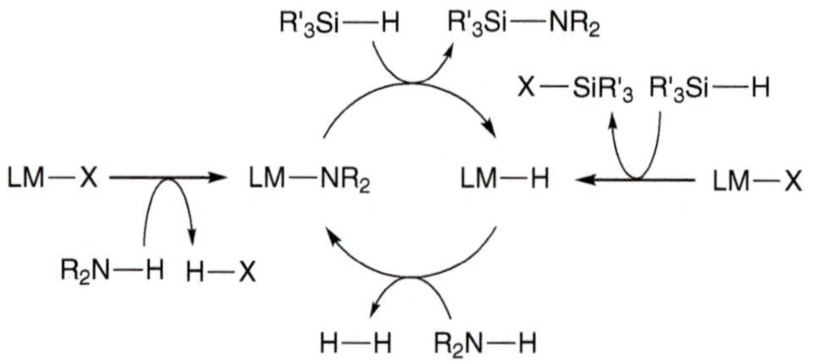

Scheme 3.26 Generic scheme for the alkaline earth-catalyzed dehydrocoupling of silanes and amines.

good conversions (55–98%) with low catalyst loadings (3–10 mol %) [418]. In subsequent advances, a variety of heteroleptic and homoleptic alkaline earth amides and alkyl pre-catalysts have been shown to mediate facile cross-dehydrocoupling of Si–H and N–H bonds (Scheme 3.25) [419–422].

Dehydrogenative B–N coupling of boranes (HBPin, 9-BBN) with amines, RR′NH (R = alkyl, aryl; R′ = H, alkyl) has also been achieved in the presence of 10 mol% of a β-diketiminato magnesium butyl pre-catalyst to afford the corresponding aminoboranes in excellent conversions (73–99%) (Scheme 3.26) [423]. Similarly, 'desilacoupling' of amines, RR′NH (R = alkyl, aryl; R′ = H, alkyl, aryl) with the polarized Si–B bond of the silaborane, pinBSiMe_2Ph, has been catalyzed by alkaline earth amides [424].

3.4 Outlook: Turbo Charging the Turbo Reagents and Prospects for Catalysis

The preceding sections demonstrate that some quite spectacular metalation reactivity may be achieved through the use of basic reagents derived from the combination of two dissimilar s-block metal centres. It is also clear from this broad overview, however, that a significant majority of these advances have been achieved with systems comprising the lighter s-block elements, with magnesium being most commonly utilized as the group 2 metal centre. From a naïve perspective, the C–H and C-halogen metalation reactions described in Sections 3.1 and 3.2 may be viewed as stoichiometric variants of the generalized σ-bond metathesis processes illustrated in Scheme 3.22. There exists vast unfulfilled potential, therefore, for the exploration of the metalation and catalytic reactivity that may be derived from bimetallic systems comprising a group 1 centre in conjunction with one of the heavier group 2 elements. While the application of heavier s-block bimetallics to further advances in both metalation and in catalysis would appear ripe for exploitation, it is also wise to introduce a note of caution. It is evident that any evolution from pure empiricism will require the development of design principles that take into account not only the identity and roles of the individual metal centres but also the nature of the substrate species. The nature of the harmonic cooperativity between the differing metal centres within even the existing systems, however, remains the subject of some conjecture. Variously described as 'black box' and 'synergic' chemistry by Mulvey, the ambiguity introduced by the facile solution equilibria, which hinder the precise interpretation of 'turbo' and even simple Grignard behaviour, are likely to be further exacerbated through the adoption of the increasingly labile heavier group 2 centres. These are major challenges. Only time will tell whether they will be met.

References

1. Seyferth, D. (**2006**). *Organometallics 25*: 2–24.
2. Seyferth, D. (**2009**). *Organometallics 28*: 2–33.
3. Seyferth, D. (**2009**). *Organometallics 28*: 1598–1605.
4. Wanklyn, J.A. (**1858**). *Justus Liebigs Ann. Chem. 108*: 67–79.
5. Grignard, V. (**1901**). *Ann. Chim. 24*: 433.
6. Grignard, V. (**1900**). *C. R. Acad. Sci. 130*: 1322–1324.
7. Grignard, V. (**1966**). *Chemistry 1921*: 234–251.
8. Schlenk, W. and Schlenk, W. (**1929**). *Chem. Ber. 62*: 920.
9. Ashby, E.C. (**1980**). *Pure Appl. Chem. 52*: 545–569.
10. Ashby, E.C., Nackashi, J., and Parris, G.E. (**1975**). *J. Am. Chem. Soc. 97*: 3162–3171.
11. Ashby, E.C. and Yu, S. (**1971**). *J. Organomet. Chem. 29*: 339–340.
12. Parris, G.E. and Ashby, E.C. (**1971**). *J. Am. Chem. Soc. 93*: 1206–1213.
13. Walker, F.W. and Ashby, E.C. (**1969**). *J. Am. Chem. Soc. 91*: 3845–3850.
14. Ashby, E.C. (**1967**). *Quart. Rev. 21*: 259.
15. Ashby, E.C. and Smith, M.B. (**1964**). *J. Am. Chem. Soc. 86*: 4363–4370.
16. Ashby, E.C. and Becker, W.E. (**1963**). *J. Am. Chem. Soc. 85*: 118–119.
17. Salinger, R.M. and Mosher, H.S. (**1964**). *J. Am. Chem. Soc. 86*: 1782–1786.
18. Smith, M.B. and Becker, W.E. (**1967**). *Tetrahedron 23*: 4215–4227.
19. Benn, R., Lehmkuhl, H., Mehler, K., and Rufinska, A. (**1984**). *Angew. Chem. Int. Ed. Engl. 23*: 534–535.
20. Ertel, T.S. and Bertagnolli, H. (**1993**). *Polyhedron 12*: 2175–2184.
21. Dessy, R.E. and Jones, R.M. (**1959**). *J. Org. Chem. 24*: 1685–1689.
22. Sobota, P. and Duda, B. (**1987**). *J. Organomet. Chem. 332*: 239–245.
23. Sakamoto, S., Imamoto, T., and Yamaguchi, K. (**2001**). *Org. Lett. 3*: 1793–1795.
24. Vallino, M. (**1969**). *J. Organomet. Chem. 20*: 1–10.
25. Vestergren, M., Eriksson, J., and Hakansson, M. (**2003**). *J. Organomet. Chem. 681*: 215–224.
26. Peltzer, R.M., Eisenstein, O., Nova, A., and Cascella, M. (**2017**). *J. Phys. Chem. B 121*: 4226–4237.
27. Kharasch, M.S. and Reinmuth, O. (**1954**). *Grignard Reactions of Nonmetallic Substances*. Editor, Prentice-Hall: New York.
28. Corriu, J.P. and Masse, J.P. (**1972**). *J. Chem. Soc. Chem. Commun.* 144–145.
29. Tamao, K., Sumitani, K., and Kumada, M. (**1972**). *J. Am. Chem. Soc. 94*: 4374–4376.
30. Furstner, A., Leitner, A., Mendez, M., and Krause, H. (**2002**). *J. Am. Chem. Soc. 124*: 13856–13863.
31. Terao, J., Kato, Y., and Kambe, N. (**2008**). *Chem. Asian J. 3*: 1472–1478.
32. Vechorkin, O., Barmaz, D., Proust, V., and Hu, X. (**2009**). *J. Am. Chem. Soc. 131*: 12078–12079.

33 Adrio, J. and Carretero, J.C. (**2010**). *Chemcatchem 2*: 1384–1386.
34 Jana, R., Pathak, T.P., and Sigman, M.S. (**2011**). *Chem. Rev. 111*: 1417–1492.
35 Cong, X., Tang, H., and Zeng, X. (**2015**). *J. Am. Chem. Soc. 137*: 14367–14372.
36 Fouquet, G. and Schlosse, M. (**1974**). *Angew. Chem. Int. Ed. Engl. 13*: 82–83.
37 Martin, D., Kehrli, S., d'Augustin, M. et al. (**2006**). *J. Am. Chem. Soc. 128*: 8416–8417.
38 Harutyunyan, S.R., den Hartog, T., Geurts, K. et al. (**2008**). *Chem. Rev. 108*: 2824–2852.
39 Tamura, M. and Kochi, J. (**1971**). *Synthesis-Int. J. Meth. Synth. Org. Chem.* 303–305.
40 Tamura, M. and Kochi, J. (**1971**). *J. Am. Chem. Soc. 93*: 1487–1488.
41 Knappke, C.E.I. and Jacobi von Wangelin, A. (**2011**). *Chem. Soc. Rev. 40*: 4948–4962.
42 Lappert, M., Power, P., Protchenko, A., and Seeber, A. (**2009**). *Metal Amide Chemistry, Editor*. Chichester: Wiley.
43 Mulvey, R.E. and Robertson, S.D. (**2013**). *Angew. Chem. Int. Ed. 52*: 11470–11487.
44 Collum, D.B. (**1993**). *Acc. Chem. Res. 26*: 227–234.
45 Gessner, V.H., Daeschlein, C., and Strohmann, C. (**2009**). *Chem. Eur. J. 15*: 3320–3334.
46 Weiss, E. (**1993**). *Angew. Chem. Int. Ed. Engl. 32*: 1501–1523.
47 Gupta, L., Hoepker, A.C., Singh, K.J., and Collum, D.B. (**2009**). *J. Org. Chem. 74*: 2231–2233.
48 Hoepker, A.C., Gupta, L., Ma, Y. et al. (**2011**). *J. Am. Chem. Soc. 133*: 7135–7151.
49 Liang, J., Hoepker, A.C., Algera, R.F. et al. (**2015**). *J. Am. Chem. Soc. 137*: 6292–6303.
50 Algera, R.F., Gupta, L., Hoepker, A.C. et al. (**2017**). *J. Org. Chem. 82*: 4513–4532.
51 Algera, R.F., Ma, Y., and Collum, D.B. (**2017**). *J. Am. Chem. Soc. 139*: 11544–11549.
52 Algera, R.F., Ma, Y., and Collum, D.B. (**2017**). *J. Am. Chem. Soc. 139*: 7921–7930.
53 Mack, K.A. and Collum, D.B. (**2018**). *J. Am. Chem. Soc. 140*: 4877–4883.
54 Williard, P.G. (**1984**). *Acta Cryst. A 40*: C289–C289.
55 Williard, P.G. and Nichols, M.A. (**1991**). *J. Am. Chem. Soc. 113*: 9671–9673.
56 Bernstein, M.P., Romesberg, F.E., Fuller, D.J. et al. (**1992**). *J. Am. Chem. Soc. 114*: 5100–5110.
57 Williard, P.G., Liu, Q.Y., and Lochmann, L. (**1992**). *J. Am. Chem. Soc. 114*: 348–350.
58 Nichols, M.A. and Williard, P.G. (**1993**). *J. Am. Chem. Soc. 115*: 1568–1572.
59 Williard, P.G. and Salvino, J.M. (**1993**). *J. Org. Chem. 58*: 1–3.
60 Williard, P.G. and Liu, Q.Y. (**1994**). *J. Org. Chem. 59*: 1596–1597.
61 Henderson, K.W., Dorigo, A.E., Liu, Q.Y. et al. (**1996**). *J. Am. Chem. Soc. 118*: 1339–1347.
62 Henderson, K.W., Dorigo, A.E., Williard, P.G., and Bernstein, P.R. (**1996**). *Angew. Chem. Int. Ed. Engl. 35*: 1322–1324.
63 Keresztes, I. and Williard, P.G. (**2000**). *J. Am. Chem. Soc. 122*: 10228–10229.
64 Li, D., Sun, C., Liu, J. et al. (**2008**). *J. Org. Chem. 73*: 2373–2381.
65 Liu, J., Li, D., Sun, C., and Williard, P.G. (**2008**). *J. Org. Chem. 73*: 4045–4052.
66 Li, D., Keresztes, I., Hopson, R., and Williard, P.G. (**2009**). *Acc. Chem. Res. 42*: 270–280.
67 Kagan, G., Li, W., Hopson, R., and Williard, P.G. (**2010**). *Org. Lett. 12*: 520–523.
68 Li, W., Kagan, G., Yang, H. et al. (**2010**). *Org. Lett. 12*: 2698–2701.
69 Socha, A.M., Kagan, G., Li, W. et al. (**2010**). *Energy Fuels 24*: 4518–4521.
70 Li, W., Kagan, G., Hopson, R., and Williard, P.G. (**2011**). *Arkivoc* 180–187.
71 Deagostino, A., Prandi, C., Tabasso, S., and Venturello, P. (**2011**). *Current Org. Chem. 15*: 2390–2412.
72 Coles, M.P. (**2008**). *Current Org. Chem. 12*: 1220–1230.
73 Coles, M.P. (**2015**). *Coord. Chem. Rev. 297*: 2–23.
74 Torvisco, A., O'Brien, A.Y., and Ruhlandt-Senge, K. (**2011**). *Coord. Chem. Rev. 255*: 1268–1292.
75 Tidwell, T.T. (**2001**). *Angew. Chem. Int. Ed. 40*: 331–337.
76 Schlenk, W. and Holtz, J. (**1917**). *Ber. Dtsch. Chem. Ges. 50*: 262–274.

77 Schlenk, W. and Bergmann, E. (**1928**). *Liebigs Ann. 463*: 98–227.
78 Gilman, H. and Bebb, R.L. (**1939**). *J. Am. Chem. Soc. 61*: 106–109.
79 Gilman, H., Langham, W., and Jacoby, A.L. (**1939**). *J. Am. Chem. Soc. 61*: 109–112.
80 Wittig, G. and Fuhrmann, G. (**1940**). *Ber. Dtsch. Chem. Ges. 73*: 1197–1218.
81 Clayden, J. (**2002**). *Organolithiums: Selectivity for Synthesis*. Elsevier, Oxford: Editor.
82 Snieckus, V. (**1990**). *Chem. Rev. 90*: 879–933.
83 Chinchilla, R., Najera, C., and Yus, M. (**2007**). *Arkivoc* 152–231.
84 Chinchilla, R., Najera, C., and Yus, M. (**2005**). *Tetrahedron 61*: 3139–3176.
85 Chinchilla, R., Najera, C., and Yus, M. (**2004**). *Chem. Rev. 104*: 2667–2722.
86 Whisler, M.C., MacNeil, S., Snieckus, V., and Beak, P. (**2004**). *Angew. Chem. Int. Ed. 43*: 2206–2225.
87 Beak, P. and Meyers, A.I. (**1986**). *Acc. Chem. Res. 19*: 356–363.
88 Turck, A., Ple, N., Mongin, F., and Queguiner, G. (**2001**). *Tetrahedron 57*: 4489–4505.
89 Mongin, F. and Queguiner, G. (**2001**). *Tetrahedron 57*: 4059–4090.
90 Chevallier, F. and Mongin, F. (**2008**). *Chem. Soc. Rev. 37*: 595–609.
91 Wittig, G., Pockels, U., and Droge, H. (**1938**). *Ber. Dtsch. Chem. Ges. 71*: 1093.
92 Beak, P. and Brown, R.A. (**1977**). *J. Org. Chem. 42*: 1823–1824.
93 Mortier, J., Moyroud, J., Bennetau, B., and Cain, P.A. (**1994**). *J. Org. Chem. 59*: 4042–4044.
94 Kauch, M. and Hoppe, D. (**2006**). *Synthesis-Stuttgart* 1575–1577.
95 Kauch, M. and Hoppe, D. (**2006**). *Synthesis-Stuttgart* 1578–1589.
96 Macklin, T.K. and Snieckus, V. (**2005**). *Org. Lett. 7*: 2519–2522.
97 Metallinos, C., Nerdinger, S., and Snieckus, V. (**1999**). *Org. Lett. 1*: 1183–1186.
98 Frostick, F.C. and Hauser, C.R. (**1949**). *J. Am. Chem. Soc. 71*: 1350–1352.
99 Hauser, C.R. and Walker, H.G. (**1947**). *J. Am. Chem. Soc. 69*: 295–297.
100 Henderson, K.W. and Kerr, W.J. (**2001**). *Chem. Eur. J. 7*: 3430–3437.
101 Eaton, P.E., Lee, C.H., and Xiong, Y.H. (**1989**). *J. Am. Chem. Soc. 111*: 8016–8018.
102 Eaton, P.E., Xiong, Y.S., and Gilardi, R. (**1993**). *J. Am. Chem. Soc. 115*: 10195–10202.
103 Eaton, P.E. and Lukin, K.A. (**1993**). *J. Am. Chem. Soc. 115*: 11370–11375.
104 Schlecker, W., Huth, A., Ottow, E., and Mulzer, J. (**1995**). *J. Org. Chem. 60*: 8414–8416.
105 Schlecker, W., Huth, A., Ottow, E., and Mulzer, J. (**1995**). *Synthesis-Stuttgart* 1225–1227.
106 Kondo, Y., Yoshida, A., and Sakamoto, T. (**1996**). *J. Chem. Soc. Perkin Trans. 1*: 2331–2332.
107 Shilai, M., Kondo, Y., and Sakamoto, T. (**2001**). *J. Chem. Soc. Perkin Trans. 1*: 442–444.
108 Wittig, G., Meyer, F.J., and Lange, G. (**1951**). *Annal. Chem. Justus Liebig 571*: 167–201.
109 Wittig, G. (**1958**). *Angew. Chem. Int. Ed. Engl. 70*: 65–71.
110 Wittig, G. (**1966**). *Quart. Rev. 20*: 191.
111 Schlosser, M. (**1967**). *J. Organomet. Chem. 8*: 9–16.
112 Lochmann, L. (**2000**). *Eur. J. Inorg. Chem.* 1115–1126.
113 Kennedy, A.R., MacLellan, J.G., and Mulvey, R.E. (**2001**). *Angew. Chem. Int. Ed. 40*: 3245–3247.
114 Unkelbach, C., O'Shea, D.F., and Strohmann, C. (**2014**). *Angew. Chem. Int. Ed. 53*: 553–556.
115 Benrath, P., Kaiser, M., Limbach, T. et al. (**2016**). *Angew. Chem. Int. Ed. 55*: 10886–10889.
116 Greiser, T., Kopf, J., Thoennes, D., and Weiss, E. (**1981**). *Chem. Ber. Rec. 114*: 209–213.
117 Thoennes, D. and Weiss, E. (**1978**). *Chem. Ber. Rec. 111*: 3726–3731.
118 Schubert, B. and Weiss, E. (**1984**). *Chem. Ber. Rec. 117*: 366–375.
119 Mulvey, R.E., Mongin, F., Uchiyama, M., and Kondo, Y. (**2007**). *Angew. Chem. Int. Ed. 46*: 3802–3824.
120 Mulvey, R.E. (**2006**). *Organometallics 25*: 1060–1075.
121 Mulvey, R.E. (**2009**). *Acc. Chem. Res. 42*: 743–755.

122 Mulvey, R.E. (**2013**). *Dalton Trans. 42*: 6676–6693.

123 Mulvey, R. E., Robertson, S. D. (**2013**). In *Alkaline-Earth Metal Compounds: Oddities and Applications*, Vol. *2*; (ed. S. Harder). 103–139 Springer.

124 Mulvey, R.E. and Robertson, S.D. (**2014**). *Organo-di-Metallic Compounds*, vol. *47* (ed. Z. Xi), 129–158.

125 Martinez-Martinez, A.J. and O'Hara, C.T. (**2016**). *Advances of Organometallic Chemistry*, vol. *65* (ed. P.J. Perez), 1–46.

126 Harrison-Marchand, A. and Mongin, F. (**2013**). *Chem. Rev. 113*: 7470–7562.

127 Mongin, F. and Harrison-Marchand, A. (**2013**). *Chem. Rev. 113*: 7563–7727.

128 Clegg, W., Henderson, K.W., Mulvey, R.E., and O'Neil, P.A. (**1993**). *J. Chem. Soc. Chem. Commun.* 969–970.

129 Clegg, W., Henderson, K.W., Mulvey, R.E., and O'Neil, P.A. (**1994**). *J. Chem. Soc. Chem. Commun.* 769–770.

130 Henderson, K.W., Mulvey, R.E., Reinhard, F.B.M. et al. (**1994**). *J. Am. Chem. Soc. 116*: 10777–10778.

131 Mackenzie, F.M., Mulvey, R.E., Clegg, W., and Horsburgh, L. (**1998**). *Polyhedron 17*: 993–998.

132 Kennedy, A.R., Mulvey, R.E., and Rowlings, R.B. (**1998**). *Angew. Chem. Int. Ed. Engl. 37*: 3180–3183.

133 Kennedy, A.R., Mulvey, R.E., and Rowlings, R.B. (**1998**). *J. Am. Chem. Soc. 120*: 7816–7824.

134 Kennedy, A.R., Mulvey, R.E., Raston, C.L. et al. (**1999**). *Chem. Commun.* 353–354.

135 Mulvey, R.E. (**2001**). *Chem. Commun.* 1049–1056.

136 Armstrong, D.R., Kennedy, A.R., Mulvey, R.E., and Rowlings, R.B. (**1999**). *Angew. Chem. Int. Ed. 38*: 131–133.

137 Andrews, P.C., Kennedy, A.R., Mulvey, R.E. et al. (**2000**). *Angew. Chem. Int. Ed. 39*: 1960–1962.

138 Clegg, W., Henderson, K.W., Kennedy, A.R. et al. (**2001**). *Angew. Chem. Int. Ed. 40*: 3902–3905.

139 Henderson, K.W., Kennedy, A.R., Mulvey, R.E. et al. (**2001**). *Chem. Commun.* 1678–1679.

140 Andrikopoulos, P.C., Armstrong, D.R., Clegg, W. et al. (**2004**). *J. Am. Chem. Soc. 126*: 11612–11620.

141 Hevia, E., Honeyman, G.W., Kennedy, A.R. et al. (**2005**). *Angew. Chem. Int. Ed. 44*: 68–72.

142 Andrikopoulos, P.C., Armstrong, D.R., Hevia, E. et al. (**2006**). *Organometallics 25*: 2415–2418.

143 Hevia, E., Gallagher, D.J., Kennedy, A.R. et al. (**2004**). *Chem. Commun.* 2422–2423.

144 Andrikopoulos, P.C., Armstrong, D.R., Graham, D.V. et al. (**2005**). *Angew. Chem. Int. Ed. 44*: 3459–3462.

145 Graham, D.V., Hevia, E., Kennedy, A.R. et al. (**2006**). *Chem. Commun.* 417–419.

146 Conway, B., Hevia, E., Kennedy, A.R., and Mulvey, R.E. (**2007**). *Chem. Commun.* 2864–2866.

147 Armstrong, D.R., Clegg, W., Dale, S.H. et al. (**2007**). *Chem. Commun.* 598–600.

148 Blair, V.L., Kennedy, A.R., Klett, J., and Mulvey, R.E. (**2008**). *Chem. Commun.* 5426–5428.

149 Blair, V.L., Kennedy, A.R., Mulvey, R.E., and O'Hara, C.T. (**2010**). *Chem. Eur. J. 16*: 8600–8604.

150 Mulvey, R.E., Blair, V.L., Clegg, W. et al. (**2010**). *Nature Chem. 2*: 588–591.

151 Awad, H., Mongin, F., Trecourt, F. et al. (**2004**). *Tetrahedron Lett. 45*: 7873–7877.

152 Awad, H., Mongin, F., Trecourt, F. et al. (**2004**). *Tetrahedron Lett. 45*: 6697–6701.

153 Bayh, O., Awad, H., Mongin, F. et al. (**2005**). *J. Org. Chem. 70*: 5190–5196.

154 Bayh, O., Awad, H., Mongin, F. et al. (**2005**). *Tetrahedron 61*: 4779–4784.

155 Mongin, F., Bucher, A., Bazureau, J.P. et al. (**2005**). *Tetrahedron Lett. 46*: 7989–7992.

156 Clegg, W., Conway, B., Garcia-Alvarez, P. et al. (**2009**). *Chem. Eur. J. 15*: 10702–10706.

157 Martinez-Martinez, A.J., Armstrong, D.R., Conway, B. et al. (**2014**). *Chem. Sci. 5*: 771–781.

158 Martinez-Martinez, A.J., Kennedy, A.R., Mulvey, R.E., and O'Hara, C.T. (**2014**). *Science 346*: 834–837.

159 Martinez-Martinez, A. J., Justice, S., Fleming, B. J., Kennedy, A. R., Oswald, I. D. H., O'Hara, C. T. (**2017**). *Science Adv. 3*, e1700832.

160 Seebach, D. (**1988**). *Angew. Chem. Int. Ed. Engl. 27*: 1624–1654.

161 Hevia, E. and Mulvey, R.E. (**2011**). *Angew. Chem. Int. Ed. 50*: 6448–6450.

162 Gupta, L., Hoepker, A.C., Ma, Y. et al. (**2013**). *J. Org. Chem. 78*: 4214–4230.

163 Ma, Y., Hoepker, A.C., Gupta, L. et al. (**2010**). *J. Am. Chem. Soc. 132*: 15610–15623.

164 Lecachey, B., Oulyadi, H., Lameiras, P. et al. (**2010**). *J. Org. Chem. 75*: 5976–5983.

165 Bao, R.L.Y., Zhao, R., and Shi, L. (**2015**). *Chem. Commun. 51*: 6884–6900.

166 Haag, B., Mosrin, M., Ila, H. et al. (**2011**). *Angew. Chem. Int. Ed. 50*: 9794–9824.

167 Barl, N.M., Werner, V., Samann, C., and Knochel, P. (**2014**). *Heterocycles 88*: 827–844.

168 Knochel, P., Dohle, W., Gommermann, N. et al. (**2003**). *Angew. Chem. Int. Ed. 42*: 4302–4320.

169 Krasovskiy, A. and Knochel, P. (**2004**). *Angew. Chem. Int. Ed. 43*: 3333–3336.

170 Ren, H.J., Krasovskiy, A., and Knochel, P. (**2004**). *Org. Lett. 6*: 4215–4217.

171 Ila, H., Baron, O., Wagner, A.J., and Knochel, P. (**2006**). *Chem. Commun.* 583–593.

172 Kopp, F., Krasovskiy, A., and Knochel, P. (**2004**). *Chem. Commun.* 2288–2289.

173 Liu, C.Y., Ren, H.J., and Knochel, P. (**2006**). *Org. Lett. 8*: 617–619.

174 Sinha, P. and Knochel, P. (**2006**). *Synlett* 3304–3308.

175 Hatakeyama, T., Yoshimoto, Y., Ghorai, S.K., and Nakamura, M. (**2010**). *Org. Lett. 12*: 1516–1519.

176 Chen, L.Q., Wilson, D.J., Labello, N.P. et al. (**2008**). *Bioorgan Med Chem. 16*: 9340–9345.

177 Peng, Z.H., Haag, B.A., and Knochel, P. (**2010**). *Org. Lett. 12*: 5398–5401.

178 Despotopoulou, C., Gignoux, C., McConnell, D., and Knochel, P. (**2009**). *Synthesis-Stuttgart* 3661–3671.

179 Salinger, D. and Bruckner, R. (**2009**). *Chem. Eur. J. 15*: 6688–6703.

180 Itakura, D. and Harada, T. (**2011**). *Synlett* 2875–2879.

181 Kadam, A., Nguyen, M., Kopach, M. et al. (**2013**). *Green Chem. 15*: 1880–1888.

182 Kobayashi, Y., Igarashi, J., Feng, C., and Toshifumi, T. (**2012**). *Tetrahedron Lett. 53*: 3742–3745.

183 McLaughlin, M., Belyk, K.M., Qian, G. et al. (**2012**). *J. Org. Chem. 77*: 5144–5148.

184 Hari, Y., Kashima, S., Inohara, H. et al. (**2013**). *Tetrahedron 69*: 6381–6391.

185 Bogart, J.A., Lee, H.B., Boreen, M.A. et al. (**2013**). *J. Org. Chem. 78*: 6344–6349.

186 Kuethe, J.T. and Maloney, K.M. (**2013**). *Tetrahedron 69*: 5248–5258.

187 Liu, C.Y. and Knochel, P. (**2005**). *Org. Lett. 7*: 2543–2546.

188 Fujimoto, K., Yorimitsu, H., and Osuka, A. (**2014**). *Eur. J. Org. Chem.* 4327–4334.

189 Baron, O. and Knochel, P. (**2005**). *Angew. Chem. Int. Ed. 44*: 3133–3135.

190 Liu, C.Y., Gavryushin, A., and Knochel, P. (**2007**). *Chem. Asian J. 2*: 1020–1030.

191 Leermann, T., Leroux, F.R., and Colobert, F. (**2011**). *Org. Lett. 13*: 4479–4481.

192 Peters, M., Trobe, M., and Breinbauer, R. (**2013**). *Chem. Eur. J. 19*: 2450–2456.

193 Frischmuth, A., Unsinn, A., Groll, K. et al. (**2012**). *Chem. Eur. J. 18*: 10234.

194 Rottlander, M., Boymond, L., Cahiez, G., and Knochel, P. (**1999**). *J. Org. Chem. 64*: 1080–1081.

195 Ren, H.J., Krasovskiy, A., and Knochel, P. (**2005**). *Chem. Commun.* 543–545.

196 Krasovskiy, A., Tishkov, A., del Amo, V. et al. (**2006**). *Angew. Chem. Int. Ed. 45*: 5010–5014.

197 Maji, M.S., Murarka, S., and Studer, A. (**2010**). *Org. Lett. 12*: 3878–3881.

198 Hirner, S. and Somfai, P. (**2009**). *J. Org. Chem. 74*: 7798–7803.

199 Melzig, L., Metzger, A., and Knochel, P. (**2011**). *Chem. Eur. J. 17*: 2948–2956.

200 Hughes, M., Boultwood, T., Zeppetelli, G., and Bull, J.A. (**2013**). *J. Org. Chem. 78*: 844–854.

201 Nishimura, R.H.V., Toledo, F.T., Lopes, J.L.C., and Clososki, G.C. (**2013**). *Tetrahedron Lett. 54*: 287–290.

202 Rauhut, C.B., Cervino, C., Krasovskiy, A., and Knochel, P. (**2009**). *Synlett* 67–70.

203 Ziegler, D.S., Karaghiosoff, K., and Knochel, P. (**2018**). *Angew. Chem. Int. Ed. 57*: 6701–6704.

204 Krasovskiy, A., Straub, B.F., and Knochel, P. (**2006**). *Angew. Chem. Int. Ed. 45*: 159–162.

205 Blasberg, F., Bolte, M., Wagner, M., and Lerner, H.W. (**2012**). *Organometallics 31*: 1001–1005.

206 Schnegelsberg, C., Bachmann, S., Kolter, M. et al. (**2016**). *Chem. Eur. J. 22*: 7752–7762.

207 Krasovskiy, A., Krasovskaya, V., and Knochel, P. (**2006**). *Angew. Chem. Int. Ed. 45*: 2958–2961.

208 Garcia-Alvarez, P., Graham, D.V., Hevia, E. et al. (**2008**). *Angew. Chem. Int. Ed. 47*: 8079–8081.

209 Armstrong, D.R., Garcia-Alvarez, P., Kennedy, A.R. et al. (**2010**). *Angew. Chem. Int. Ed. 49*: 3185–3188.

210 Neufeld, R. and Stalke, D. (**2016**). *Chem. Eur. J. 22*: 12624–12628.

211 Neufeld, R., Teuteberg, T.L., Herbst-Irmer, R. et al. (**2016**). *J. Am. Chem. Soc. 138*: 4796–4806.

212 Francos, J., Fleming, B.J., Alvarez, P.G. et al. (**2014**). *Dalton Trans. 43*: 14424–14431.

213 Samann, C., Haag, B., and Knochel, P. (**2012**). *Chem. Eur. J. 18*: 16145–16152.

214 Stoll, A.H., Mayer, P., and Knochel, P. (**2007**). *Organometallics 26*: 6694–6697.

215 Wunderlich, S.H., Rohbogner, C.J., Unsinn, A., and Knochel, P. (**2010**). *Org. Proc. Res. Dev. 14*: 339–345.

216 Boudet, N., Lachs, J.R., and Knochel, P. (**2007**). *Org. Lett. 9*: 5525–5528.

217 Mosrin, M. and Knochel, P. (**2008**). *Org. Lett. 10*: 2497–2500.

218 Mosrin, M., Petrera, M., and Knochel, P. (**2008**). *Synthesis-Stuttgart* 3697–3702.

219 Stoll, A.H. and Knochel, P. (**2008**). *Org. Lett. 10*: 113–116.

220 Rauhut, C.B., Melzig, L., and Knochel, P. (**2008**). *Org. Lett. 10*: 3891–3894.

221 Melzig, L., Rauhut, C.B., and Knochel, P. (**2009**). *Synthesis-Stuttgart* 1041–1048.

222 Piller, F.M. and Knochel, P. (**2011**). *Synthesis-Stuttgart* 1751–1758.

223 Piller, F.M. and Knochel, P. (**2009**). *Org. Lett. 11*: 445–448.

224 Kunz, T. and Knochel, P. (**2012**). *Angew. Chem. Int. Ed. 51*: 1958–1961.

225 Ziegler, D.S., Greiner, R., Lumpe, H. et al. (**2017**). *Org. Lett. 19*: 5760–5763.

226 Castello-Mico, A. and Knochel, P. (**2018**). *Synthesis-Stuttgart 50*: 155–169.

227 Castello-Mico, A., Nafe, J., Higashida, K. et al. (**2017**). *Org. Lett. 19*: 360–363.

228 Westerhausen, M. (**2001**). *Angew. Chem. Int. Ed. 40*: 2975–2977.

229 Sockwell, S.C., Hanusa, T.P., and Huffman, J.C. (**1992**). *J. Am. Chem. Soc. 114*: 3393–3399.

230 Knapp, V. and Muller, G. (**2001**). *Angew. Chem. Int. Ed. 40*: 183–186.

231 Hitchcock, P.B., Khvostov, A.V., and Lappert, M.F. (**2002**). *J. Organomet. Chem. 663*: 263–268.

232 Guino-O, M.A., Campana, C.F., and Ruhlandt-Senge, K. (**2008**). *Chem. Commun.* 1692–1694.

233 Harder, S. and Ruspic, C. (**2015**). *Eur. J. Inorg. Chem.* 5743–5750.

234 Jochmann, P., Davin, J.P., Maslek, S. et al. (**2012**). *Dalton Trans. 41*: 9176–9181.

235 Maudez, W., Haussinger, D., and Fromm, K.M. (**2006**). *Z. Anorg. Allg. Chem. 632*: 2295–2298.

236 Maudez, W., Meuwly, M., and Fromm, K.M. (**2007**). *Chem. Eur. J. 13*: 8302–8316.

237 Buchanan, W.D. and Ruhlandt-Senge, K. (**2013**). *Chem. Eur. J. 19*: 10708–10715.

238 Zuniga, M.F., Deacon, G.B., and Ruhlandt-Senge, K. (**2008**). *Inorg. Chem. 47*: 4669–4681.

239 Davies, R.P. (**2000**). *Inorg. Chem. Commun. 3*: 13–15.

240 Kennedy, A.R., Mulvey, R.E., and Rowlings, R.B. (**2002**). *J. Organomet. Chem. 648*: 288–292.

241 Hill, M.S., Kociok-Koehn, G., and MacDougall, D.J. (**2011**). *Inorg. Chem. 50*: 5234–5241.

242 Turner, Z.R. and Buffet, J.C. (**2015**). *Dalton Trans. 44*: 12985–12989.

243 Glock, C., Görls, H., and Westerhausen, M. (**2009**). *Inorg. Chem. 48*: 394–399.

244 Glock, C., Görls, H., and Westerhausen, M. (**2011**). *Dalton Trans. 40*: 8108–8113.

245 Glock, C., Görls, H., and Westerhausen, M. (**2012**). *Chem. Commun. 48*: 7094–7096.
246 Glock, C., Younis, F.M., Ziemann, S. et al. (**2013**). *Organometallics 32*: 2649–2660.
247 Hill, M.S., Liptrot, D.J., and Weetman, C. (**2016**). *Chem. Soc. Rev. 45*: 972–988.
248 Crimmin, M.R. and Hill, M.S. (**2013**). *Alkaline-Earth Metal Compounds: Oddities and Applications*, vol. *45* (ed. S. Harder), 191–241.
249 Harder, S. (**2010**). *Chem. Rev. 110*: 3852–3876.
250 Barrett, A.G.M., Crimmin, M.R., Hill, M.S., and Procopiou, P.A. (**2010**). *Proc. Roy. Soc. A-Math. Phys. Eng. Sci. 466*: 927–963.
251 Barry, C.S. and Simpkins, N.S. (**2007**). *Tetrahedron Lett. 48*: 8192–8195.
252 Quinet, C., Jourdain, P., Hermans, C. et al. (**2008**). *Tetrahedron 64*: 1077–1087.
253 Banerjee, S., Yang, Y.F., Jenkins, I.D. et al. (**2017**). *J. Am. Chem. Soc. 139*: 6880–6887.
254 Liu, W.B., Schuman, D.P., Yang, Y.F. et al. (**2017**). *J. Am. Chem. Soc. 139*: 6867–6879.
255 Smith, A.J., Young, A., Rohrbach, S. et al. (**2017**). *Angew. Chem. Int. Ed. 56*: 13747–13751.
256 Toutov, A.A., Betz, K.N., Schuman, D.P. et al. (**2017**). *J. Am. Chem. Soc. 139*: 1668–1674.
257 Toutov, A.A., Salata, M., Fedorov, A. et al. (**2017**). *Nature Energy 2*: 17008.
258 Toutov, A.A., Betz, K.N., Haibach, M.C. et al. (**2016**). *Org. Lett. 18*: 5776–5779.
259 Toutov, A.A., Liu, W.B., Betz, K.N. et al. (**2015**). *Nature Protocols 10*: 1897–1903.
260 Toutov, A.A., Liu, W.-B., Betz, K.N. et al. (**2015**). *Nature 518*: 80–84.
261 Ates, A. and Quinet, C. (**2003**). *Eur. J. Org. Chem.* 1623–1626.
262 van Otterlo, W.A.L., Pathak, R., de Koning, C.B., and Fernandes, M.A. (**2004**). *Tetrahedron Lett. 45*: 9561–9563.
263 Ong, T.G., O'Brien, J.S., Korobkov, I., and Richeson, D.S. (**2006**). *Organometallics 25*: 4728–4730.
264 Zhang, W., Werness, J.B., and Tang, W.P. (**2008**). *Org. Lett. 10*: 2023–2026.
265 Quinet, C., Sampoux, L., and Marko, I.E. (**2009**). *Eur. J. Org. Chem.* 1806–1811.
266 Deschamp, J., Olier, C., Schulz, E. et al. (**2010**). *Adv. Synth. Catal. 352*: 2171–2176.
267 Jaspers, D. and Doye, S. (**2011**). *Synlett* 1444–1448.
268 Reznichenko, A.L. and Hultzsch, K.C. (**2013**). *Hydrofunctionalization*, vol. *43* (ed. V.P. Ananikov and M. Tanaka), 51–114.
269 Rousseau, G., Lebeuf, R., Schenk, K. et al. (**2014**). *Chem. Eur. J. 20*: 14771–14782.
270 Germain, S., Lecoq, M., Schulz, E., and Hannedouche, J. (**2017**). *Chemcatchem 9*: 1749–1753.
271 Wingerter, S., Pfeiffer, M., Murso, A. et al. (**2001**). *J. Am. Chem. Soc. 123*: 1381–1388.
272 El-Kaderi, H.M., Heeg, M.J., and Winter, C.H. (**2004**). *Organometallics 23*: 4995–5002.
273 El-Kaderi Hani, M., Heeg Mary, J., and Winter Charles, H. (**2005**). *Eur. J. Inorg. Chem. 2005*: 2081–2088.
274 Datta, S., Roesky, P.W., and Blechert, S. (**2007**). *Organometallics 26*: 4392–4394.
275 Cameron, T.M., Xu, C., Dipasquale, A.G., and Rheingold, A.L. (**2008**). *Organometallics 27*: 1596–1604.
276 Datta, S., Gamer, M.T., and Roesky, P.W. (**2008**). *Dalton Trans.* 2839–2843.
277 Barrett, A.G.M., Crimmin, M.R., Hill, M.S. et al. (**2008**). *Inorg. Chem. 47*: 7366–7376.
278 Piesik, D.F.J., Range, S., and Harder, S. (**2008**). *Organometallics 27*: 6178–6187.
279 Buch, F. and Harder, S. (**2008**). *Z. Naturforsch. 63*: 169–177.
280 Sedai, B., Heeg, M.J., and Winter, C.H. (**2009**). *Organometallics 28*: 1032–1038.
281 Ho, S.-M., Hsiao, C.-S., Datta, A. et al. (**2009**). *Inorg. Chem. 48*: 8004–8011.
282 Piesik, D.F.J., Haack, P., Harder, S., and Limberg, C. (**2009**). *Inorg. Chem. 48*: 11259–11264.
283 Barrett, A.G.M., Crimmin, M.R., Hill, M.S. et al. (**2010**). *Dalton Trans. 39*: 7393–7400.
284 Yang, D., Ding, Y., Wu, H., and Zheng, W. (**2011**). *Inorg. Chem. 50*: 7698–7706.
285 Wixey, J.S. and Ward, B.D. (**2011**). *Chem. Commun. 47*: 5449–5451.

286 Wixey, J.S. and Ward, B.D. (**2011**). *Dalton Trans.* *40*: 7693–7696.
287 Buffet, J.-C., Davin, J.P., Spaniol, T.P., and Okuda, J. (**2011**). *New J. Chem.* *35*: 2253–2257.
288 Hsueh, L.F., Chuang, N.T., Lee, C.Y. et al. (**2011**). *Eur. J. Inorg. Chem.* *2011*: 5530–5537.
289 Hu, H. and Cui, C. (**2012**). *Organometallics* *31*: 1208–1211.
290 Deschner, T., Klimpel, M., Tafipolsky, M. et al. (**2012**). *Dalton Trans.* *41*: 7319–7326.
291 Huang, T.-L. and Chen, C.-T. (**2013**). *Dalton Trans.* *42*: 9255–9262.
292 Kottalanka, R.K., Anga, S., Naktode, K. et al. (**2013**). *Organometallics* *32*: 4473–4482.
293 Kottalanka, R.K., Anga, S., Jana, S.K., and Panda, T.K. (**2013**). *J. Organomet. Chem.* *740*: 104–109.
294 Deacon, G.B., Junk, P.C., and Kelly, R.P. (**2013**). *Aust. J. Chem.* *66*: 1288–1296.
295 Hsiao, M.-W., Wu, G.-S., Huang, B.-H., and Lin, C.-C. (**2013**). *Inorg. Chem. Commun.* *36*: 90–95.
296 Glock, C., Loh, C., Görls, H. et al. (**2013**). *Eur. J. Inorg. Chem.* 3261–3269.
297 Kottalanka, R.K., Adimulam, H., Bhattacharjee, J. et al. (**2014**). *Dalton Trans.* *43*: 8757–8766.
298 Loh, C., Seupel, S., Koch, A. et al. (**2014**). *Dalton Trans.* *43*: 14440–14449.
299 Naktode, K., Bhattacharjee, J., Das Gupta, S. et al. (**2013**). *Z. Anorg. Allg. Chem.* *640*: 994–999.
300 Kottalanka, R.K., Naktode, K., Anga, S., and Panda, T.K. (**2014**). *Phosphorus, Sulfur, Silicon, Rel. Elem.* *189*: 1624–1632.
301 Kalden, D., Krieck, S., Görls, H., and Westerhausen, M. (**2015**). *Dalton Trans.* *44*: 8089–8099.
302 Kottalanka, R.K., Harinath, A., and Panda, T.K. (**2015**). *RSC Adv.* *5*: 37755–37767.
303 Basalov, I.V., Yurova, O.S., Cherkasov, A.V. et al. (**2016**). *Inorg. Chem.* *55*: 1236–1244.
304 Kalden, D., Oberheide, A., Loh, C. et al. (**2016**). *Chem. Eur. J.* *22*: 10944–10959.
305 Causero, A., Ballmann, G., Pahl, J. et al. (**2016**). *Organometallics* *35*: 3350–3360.
306 Bhattacharjee, J., Das, S., Reddy, T.D.N. et al. (**2015**). *Z. Anorg. Allg. Chem.* *642*: 118–127.
307 Bhattacharjee, J., Harinath, A., Nayek Hari, P. et al. (**2017**). *Chem. Eur. J.* *23*: 9319–9331.
308 Mukherjee, D., Shirase, S., Beckerle, K. et al. (**2017**). *Dalton Trans.* *46*: 8451–8457.
309 Lapshin, I.V., Yurova, O.S., Basalov, I.V. et al. (**2018**). *Inorg. Chem.* *57*: 2942–2952.
310 Westerhausen, M., Birg, C., and Piotrowski, H. (**2000**). *Eur. J. Inorg. Chem.* *2000*: 2173–2178.
311 Westerhausen, M., Birg, C., Piotrowski, H. et al. (**2001**). *Z. Anorg. Allg. Chem.* *627*: 882–890.
312 Clobes, C., Jerabek, P., Nußbruch, I. et al. (**2015**). *Eur. J. Inorg. Chem.* *2015*: 3264–3273.
313 Becker, G., Niemeyer, M., Mundt, O. et al. (**2004**). *Z. Anorg. Allg. Chem.* *630*: 2605–2621.
314 Sarazin, Y., Howard, R.H., Hughes, D.L. et al. (**2006**). *Dalton Trans.* 340–350.
315 Boyle, T.J., Hernandez-Sanchez, B.A., Baros, C.M. et al. (**2007**). *Chem. Mater.* *19*: 2016–2026.
316 Davidson, M.G., O'Hara, C.T., Jones, M.D. et al. (**2007**). *Inorg. Chem.* *46*: 7686–7688.
317 Range, S., Piesik Dirk, F.J., and Harder, S. (**2008**). *Eur. J. Inorg. Chem.* *2008*: 3442–3451.
318 Gauvin Régis, M., Buch, F., Delevoye, L., and Harder, S. (**2009**). *Chem. Eur. J.* *15*: 4382–4393.
319 Howard, R.H., Alonso-Moreno, C., Broomfield, L.M. et al. (**2009**). *Dalton Trans.* 8667–8682.
320 Poirier, V., Roisnel, T., Carpentier, J.-F., and Sarazin, Y. (**2009**). *Dalton Trans.* 9820–9827.
321 Piesik Dirk, F.J., Stadler, R., Range, S., and Harder, S. (**2009**). *Eur. J. Inorg. Chem.* *2009*: 3569–3576.
322 Buchanan, W.D., Guino-o, M.A., and Ruhlandt-Senge, K. (**2010**). *Inorg. Chem.* *49*: 7144–7155.
323 Sarazin, Y., Poirier, V., Roisnel, T., and Carpentier, J.F. (**2010**). *Eur. J. Inorg. Chem.* *2010*: 3423–3428.
324 Sarazin, Y., Roşca, D., Poirier, V. et al. (**2010**). *Organometallics* *29*: 6569–6577.
325 Poirier, V., Roisnel, T., Carpentier, J.-F., and Sarazin, Y. (**2011**). *Dalton Trans.* *40*: 523–534.
326 Sarazin, Y., Liu, B., Roisnel, T. et al. (**2011**). *J. Am. Chem. Soc.* *133*: 9069–9087.
327 Liu, B., Roisnel, T., and Sarazin, Y. (**2012**). *Inorg. Chim. Acta* *380*: 2–13.
328 Liu, B., Roisnel, T., Carpentier, J.F., and Sarazin, Y. (**2012**). *Angew. Chem. Int. Ed.* *51*: 4943–4946.
329 Yi, W. and Ma, H. (**2013**). *Inorg. Chem.* *52*: 11821–11835.
330 Roşca, S.-C., Roisnel, T., Dorcet, V. et al. (**2014**). *Organometallics* *33*: 5630–5642.

331 Yi, W. and Ma, H. (**2014**). *Dalton Trans. 43*: 5200–5210.
332 Yang, Y., Wang, H., and Ma, H. (**2016**). *Polyhedron 117*: 569–578.
333 Basalov, I.V., Liu, B., Roisnel, T. et al. (**2016**). *Organometallics 35*: 3261–3271.
334 Gindelberger, D.E. and Arnold, J. (**1992**). *J. Am. Chem. Soc. 114*: 6242–6243.
335 Vargas, W. and Ruhlandt-Senge, K. (**2003**). *Eur. J. Inorg. Chem. 2003*: 3472–3479.
336 Panda, T.K., Yamamoto, K., Yamamoto, K. et al. (**2012**). *Organometallics 31*: 2268–2274.
337 Kottalanka, R.K., Harinath, A., Rej, S., and Panda, T.K. (**2015**). *Dalton Trans. 44*: 19865–19879.
338 Liu, N., Yao, C., Lin, F. et al. (**2015**). *Polymer 80*: 104–108.
339 Connolly, E.A., Leeland, J.W., and Love, J.B. (**2016**). *Inorg. Chem. 55*: 840–847.
340 El-Kaderi, H.M., Heeg, M.J., and Winter, C.H. (**2005**). *Polyhedron 24*: 645–653.
341 Kobrsi, I., Knox, J.E., Heeg, M.J. et al. (**2005**). *Inorg. Chem. 44*: 4894–4896.
342 O'Brien Anna, Y., Hitzbleck, J., Torvisco, A., Deacon Glen, B., Ruhlandt-Senge, K. (**2008**). *Eur. J. Inorg. Chem.*, 2007, 172–182.
343 Norman, J.A.T., Perez, M., Kim, M.S. et al. (**2011**). *Inorg. Chem. 50*: 12396–12398.
344 Nixon, T.D. and Ward, B.D. (**2012**). *Chem. Commun 48*: 11790–11792.
345 Zheng, W. and Yang, D. (**2013**). *Inorg. Chim. Acta 396*: 21–24.
346 Müller, C., Koch, A., Görls, H. et al. (**2015**). *Inorg. Chem. 54*: 635–645.
347 Schowtka, B., Görls, H., and Westerhausen, M. (**2014**). *Z. Anorg. Allg. Chem. 641*: 650–654.
348 Green David, C., Englich, U., and Ruhlandt-Senge, K. (**1999**). *Angew. Chem. Int. Ed. 38*: 354–357.
349 Guino-o Marites, A., Alexander Jacob, S., McKee Michael, L. et al. (**2009**). *Chem. Eur. J. 15*: 11842–11852.
350 Schumann, H., Steffens, A., Hummert, M., and Anorg, Z. (**2009**). *Allg. Chem. 635*: 1041–1047.
351 Westerhausen, M., Krofta, M., and Mayer, P. (**2000**). *Z. Anorg. Allg. Chem. 626*: 2307–2312.
352 Kopecky, P., von Hänisch, C., Weigend, F., and Kracke, A. (**2010**). *Eur. J. Inorg. Chem. 2009*: 258–265.
353 von Hänisch, C. and Kopecky, P. (**2010**). *Z. Anorg. Allg. Chem. 636*: 1522–1526.
354 von Hänisch, C. and Feierabend, M. (**2013**). *Z. Anorg. Allg. Chem. 639*: 788–793.
355 Lamberti, M., Botta, A., and Mazzeo, M. (**2014**). *Appl. Organometal. Chem. 28*: 140–145.
356 Reuter, K. and von Hanisch, C. (**2014**). *Chem. Commun. 50*: 7709–7711.
357 Tanner, P.S. and Hanusa, T.P. (**1994**). *Polyhedron 13*: 2417–2420.
358 Westerhausen, M., Hartmann, M., and Schwarz, W. (**1995**). *J. Organomet. Chem. 501*: 359–367.
359 Westerhausen, M., Hartmann, M., Makropoulos, N. et al. (**1998**). *Z. Naturforsch., Teil B 53*: 117–125.
360 Alexander Jacob, S. and Ruhlandt-Senge, K. (**2001**). *Angew. Chem. Int. Ed. 40*: 2658–2660.
361 Harder, S., Feil, F., and Repo, T. (**2002**). *Chem. Eur. J. 8*: 1991–1999.
362 Alexander, J.S., Ruhlandt-Senge, K., and Hope, H. (**2003**). *Organometallics 22*: 4933–4937.
363 Fichtel, K., Hofmann, K., and Behrens, U. (**2004**). *Organometallics 23*: 4166–4168.
364 Fichtel, K. and Behrens, U. (**2005**). *Z. Anorg. Allg. Chem. 631*: 2508–2516.
365 Orzechowski, L., Jansen, G., and Harder, S. (**2006**). *J. Am. Chem. Soc. 128*: 14676–14684.
366 Fichtel, K., Höxter, S., and Behrens, U. (**2006**). *Z. Anorg. Allg. Chem. 632*: 2003–2009.
367 Kling, C., Ott, H., Schwab, G., and Stalke, D. (**2008**). *Organometallics 27*: 5038–5042.
368 Garcés, A., Sánchez-Barba, L.F., Alonso-Moreno, C. et al. (**2010**). *Inorg. Chem. 49*: 2859–2871.
369 Cushion, M.G., Meyer, J., Heath, A. et al. (**2010**). *Organometallics 29*: 1174–1190.
370 Müller, C., Görls, H., Krieck, S., and Westerhausen, M. (**2013**). *Eur. J. Inorg. Chem. 2013*: 5679–5682.
371 Müller, C., Krieck, S., Görls, H., and Westerhausen, M. (**2015**). *Inorg. Chem. 54*: 2100–2102.
372 Gallegos, C., Camacho, R., Valiente, M. et al. (**2016**). *Catal. Sci. Technol. 6*: 5134–5143.

373 Koehne, I., Graw, N., Teuteberg, T. et al. (**2017**). *Inorg. Chem. 56*: 14968–14978.
374 Arrowsmith, M., Hill, M.S., Kociok-Köhn, G. et al. (**2012**). *Angew. Chem. Int. Ed. 51*: 2098–2100.
375 Arrowsmith, M., Hill, M.S., and Kociok-Köhn, G. (**2015**). *Organometallics 34*: 653–662.
376 Green, S.P., Jones, C., and Stasch, A. (**2008**). *Angew. Chem. Int. Ed. 47*: 9079–9083.
377 Bonyhady, S.J., Jones, C., Nembenna, S. et al. (**2010**). *Chem. Eur. J. 16*: 938–955.
378 Harder, S., Spielmann, J., Intemann, J., and Bandmann, H. (**2011**). *Angew. Chem. Int. Ed. 50*: 4156–4160.
379 Intemann, J., Spielmann, J., Sirsch, P., and Harder, S. (**2013**). *Chem. Eur. J. 19*: 8478–8489.
380 Arrowsmith, M., Maitland, B., Kociok-Köhn, G. et al. (**2014**). *Inorg. Chem. 53*: 10543–10552.
381 Harder, S., Spielmann, J., and Intemann, J. (**2014**). *Dalton Trans. 43*: 14284–14290.
382 Martin, D., Beckerle, K., Schnitzler, S. et al. (**2015**). *Angew. Chem. Int. Ed. 54*: 4115–4118.
383 Lalrempuia, R., Kefalidis, C.E., Bonyhady, S.J. et al. (**2015**). *J. Am. Chem. Soc. 137*: 8944–8947.
384 Xie, H., Hua, X., Liu, B. et al. (**2015**). *J. Organomet. Chem. 798*: 335–340.
385 Fohlmeister, L. and Stasch, A. (**2016**). *Chem. Eur. J. 22*: 10235–10246.
386 Langer, J., Kosygin, I., Puchta, R. et al. (**2016**). *Chem. Eur. J. 22*: 17425–17435.
387 Langer, J., Maitland, B., Grams, S. et al. (**2017**). *Angew. Chem. Int. Ed. 56*: 5021–5025.
388 Rauch, M., Ruccolo, S., and Parkin, G. (**2017**). *J. Am. Chem. Soc. 139*: 13264–13267.
389 de Bruin-Dickason, C.N., Sutcliffe, T., Alvarez Lamsfus, C. et al. (**2018**). *Chem. Commun. 54*: 786–789.
390 Arrowsmith, M., Hill, M.S., MacDougall, D.J., and Mahon, M.F. (**2009**). *Angew. Chem. Int. Ed. 48*: 4013–4016.
391 Harder, S. and Brettar, J. (**2006**). *Angew. Chem. Int. Ed. 45*: 3474–3478.
392 Causero, A., Ballmann, G., Pahl, J. et al. (**2017**). *Dalton Trans. 46*: 1822–1831.
393 Maitland, B., Wiesinger, M., Langer, J. et al. (**2017**). *Angew. Chem. Int. Ed. 56*: 11880–11884.
394 Wiesinger, M., Maitland, B., Färber, C. et al. (**2017**). *Angew. Chem. Int. Ed. 56*: 16654–16659.
395 Westerhausen, M., Digeser, M.H., and Schwarz, W. (**1997**). *Inorg. Chem. 36*: 521–527.
396 Westerhausen, M., Digeser, M.H., Nöth, H. et al. (**1998**). *J. Am. Chem. Soc. 120*: 6722–6725.
397 Westerhausen, M., Digeser, M.H., Nöth, H. et al. (**1999**). *Inorg. Chem. 38*: 3207–3214.
398 Spielmann, J., Buch, F., and Harder, S. (**2008**). *Angew. Chem. Int. Ed. 47*: 9434–9438.
399 Schuhknecht, D., Lhotzky, C., Spaniol, T.P. et al. (**2017**). *Angew. Chem. Int. Ed. 56*: 12367–12371.
400 Bauer, H., Alonso, M., Farber, C. et al. (**2018**). *Nature Catal. 1*: 40–47.
401 Crimmin, M.R., Casely, I.J., and Hill, M.S. (**2005**). *J. Am. Chem. Soc. 127*: 2042–2043.
402 Arrowsmith, M., Crimmin, M.R., Barrett, A.G.M. et al. (**2011**). *Organometallics 30*: 1493–1506.
403 Barrett, A.G.M., Brinkmann, C., Crimmin, M.R. et al. (**2009**). *J. Am. Chem. Soc. 131*: 12906–12907.
404 Brinkmann, C., Barrett, A.G.M., Hill, M.S., and Procopiou, P.A. (**2012**). *J. Am. Chem. Soc. 134*: 2193–2207.
405 Brinkmann, C., Barrett, A.G.M., Hill, M.S. et al. (**2012**). *Organometallics 31*: 7287–7297.
406 Crimmin, M.R., Barrett, A.G.M., Hill, M.S. et al. (**2007**). *Organometallics 26*: 2953–2956.
407 Buch, F., Brettar, H., and Harder, S. (**2006**). *Angew. Chem. Int. Ed. 45*: 2741–2745.
408 Leich, V., Spaniol, T.P., Maron, L., and Okuda, J. (**2014**). *Chem. Commun. 50*: 2311–2314.
409 Buch, F. and Harder, S. (**2008**). *Z. Naturforsch.B 63*: 169–177.
410 Rauch, M. and Parkin, G. (**2017**). *J. Am. Chem. Soc. 139*: 18162–18165.
411 Weetman, C., Hill, M.S., and Mahon, M.F. (**2015**). *Chem. Commun. 51*: 14477–14480.
412 Anker, M.D., Arrowsmith, M., Bellham, P. et al. (**2014**). *Chem. Sci. 5*: 2826–2830.
413 Weetman, C., Anker, M.D., Arrowsmith, M. et al. (**2016**). *Chem. Sci. 7*: 628–641.
414 Yang, Y., Anker, M.D., Fang, J. et al. (**2017**). *Chem. Sci. 8*: 3529–3537.

415 Arrowsmith, M., Shepherd, W.M.S., Hill, M.S., and Kociok-Köhn, G. (**2014**). *Chem. Commun. 50*: 12676–12679.

416 Arrowsmith, M., Hill, M.S., and Kociok-Köhn, G. (**2015**). *Chem. Eur. J. 21*: 10548–10557.

417 Liu, B., Roisnel, T., Carpentier, J.F., and Sarazin, Y. (**2012**). *Angew. Chem. Int. Ed. 51*: 4943–4946.

418 Buch, F. and Harder, S. (**2007**). *Organometallics 26*: 5132–5135.

419 Dunne, J.F., Neal, S.R., Engelkemier, J. et al. (**2011**). *J. Am. Chem. Soc. 133*: 16782–16785.

420 Bellini, C., Carpentier, J.F., Tobisch, S., and Sarazin, Y. (**2015**). *Angew. Chem. Int. Ed. 54*: 7679–7683.

421 Forosenko, N.V., Basalov, I.V., Cherkasov, A.V. et al. (**2018**). *Dalton Trans. 47*: 12570–12581.

422 Hill, M.S., Liptrot, D.J., MacDougall, D.J. et al. (**2013**). *Chem. Sci. 4*: 4212–4222.

423 Liptrot, D.J., Hill, M.S., Mahon, M.F., and Wilson, A.S.S. (**2015**). *Angew. Chem. Int. Ed. 54*: 13362–13365.

424 Liptrot, D.J., Arrowsmith, M., Colebatch, A.L. et al. (**2015**). *Angew. Chem. Int. Ed. 54*: 15280–15283.

4

Mechanisms in Heterobimetallic Reactivity: Experimental and Computational Insights for Catalyst Design in Small Molecule Activation and Polymer Synthesis

Frances N. Singer and Antoine Buchard

Department of Chemistry, University of Bath, Claverton Down, UK

4.1 Introduction and Scope of the Chapter

At the onset of this chapter, it is perhaps necessary to clarify what the term 'heterobimetallic' will encompass in the following discussions. Herein, we will employ a loose definition of heterobimetallic reagents, referring to molecules featuring two metals of a different chemical nature, but without any restrictions about the exact number of metal atoms present in the structure. Likewise, we will discuss indiscriminately examples of heterobimetallic complexes which feature metal–metal bonds as well as complexes in which metal sites have been compartmentalized by tailored ligand architecture.

Broadly speaking, heterobimetallic complexes can be divided into three categories depending on how the two different metals are brought together (Figure 4.1):

- The bimetallic complex is formed by the dimerization of two mononuclear complexes.
- The bimetallic complex consists of two mononuclear complexes that are tethered via a ligand.
- The bimetallic complex involves a di-nucleating ligand which enables coordination of two metals in close proximity.

Beyond terminology exactitude, in this chapter, we will focus instead on chemical efficiency, aiming to highlight the potential offered by heterobimetallic cooperativity to enhance catalytic activity beyond conventional ligand design.

Traditionally, advances in the preparation of homogeneous metal-based catalysts have been reliant on the design of new ligands bringing variation of stereoelectronic effects around the metal center. However, recently, potentially driven by the enhanced activity and selectivity of heterogeneous catalysts that combine multiple and diverse metal sites [1], combining two different metal centers in a single discrete molecular entity has been proposed for the mediation of complicated catalytic reactions, as well as for the combination of several catalytic reactions into a one-pot tandem process [2–5].

A logical approach for the design of one-pot tandem processes would be to combine two different metal complexes in one catalytic system, with the aim that each of the metals facilitates independent catalytic cycles. In this orthogonal approach, each metal center would sequentially

Polar Organometallic Reagents: Synthesis, Structure, Properties and Applications, First Edition. Edited by Andrew E. H. Wheatley and Masanobu Uchiyama.

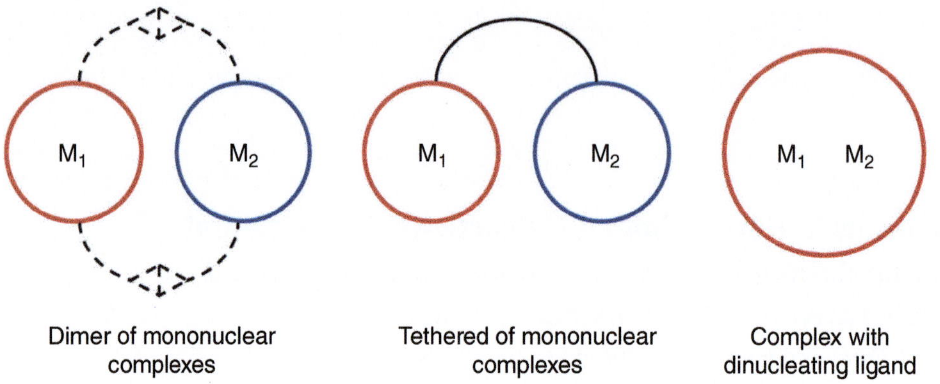

Figure 4.1 Dimeric, tethered, and di-nucleating types of heterobimetallic catalyst.

promote different catalysis, with little interaction between the two metals. However, in some cases, bringing multiple metal centers in close proximity can lead to unprecedented or improved reactivity. This synergy (or positive cooperativity) is demonstrated when a heterobimetallic system has a greater effect (activity or selectivity) than what is observed with the sum of separate corresponding homometallic components.

In nature, a number of enzyme active sites have been shown to contain two metal ions, in particular for the activation of small molecules, including Fe–Mo and V–Fe nitrogenases [6], as well as Ni–Fe hydrogenases [7]. Advances in spectroscopic techniques not only helped elucidating the structure of these active sites but also identified the nature of key intermediates in biocatalysis, highlighting that often the two metals operate cooperatively. Inspired by these natural motifs [8], in the context of developing a sustainable hydrogen economy, very efficient mimic complexes of NiFe hydrogenases have been developed [9–12], including with metal center couples not present in nature (e.g. Ru/Ni) [13].

In some existing industrial homogeneously catalysed processes, heterobimetallic reagents and cooperativity between metals have been shown, even when the pre-catalysts are not heterobimetallic. For example, in the Cativa process, developed by BP Chemicals for the production of acetic acid by carbonylation of methanol, and which involves the use of an iridium catalyst, ($[IrI_2(CO)_2]^-$) promoted by a ruthenium iodo-complex, some molecular catalytic intermediates that contain both metals have been identified by ^{13}C NMR spectroscopy [14–16].

As a result, di-nuclear complexes containing two metals in close proximity have become the subject of extensive investigations to design new heterobimetallic catalysts [17]. Beyond the aforementioned chemical processes, heterodinuclear cooperativity has been proposed as responsible for enhanced activity in homogeneous catalysis ranging from metal–halogen exchange [18], metalation [19, 20], C–H bond activation [21, 22], and C–F bond activation [23, 24]. The reader is also directed towards reviews by Buchwalter, Rosé, and Braunstein from 2014 [5], as well as by Cooper, Napoline, and Thomas from 2012 [25], which would provide a landscape on the utilisation of heterometallic complexes in catalysis.

Proving positive cooperative effects is in theory easily done through control experiments using homobimetallic or separate monometallic compounds (provided that these are accessible). However, understanding the exact nature and origin of a cooperative mechanism remains very difficult to assess. Multiple scenarios can be envisaged, including concerted or consecutive actions of the different metal centers, whether there is formal metal–metal bonding or that metal–metal interactions are facilitated by flexible bridging ligands. When a metal–metal bond is present, sharing of electrons between the metals could be at the origin of the specific reactivity observed, the metal center can also be seen as a ligand modifying the stereoelectronics of the active site. This challenge is often complicated by the fact that there is generally no structural information available for the catalytic active species *in operando*. The progress made over the past decades in computational modelling of chemical reactions have revolutionized the structure–activity understanding of catalysis and have become suitable tools to interrogate heterobimetallically catalyzed reactions.

The aim of this chapter is to provide a non-exhaustive and selected coverage of how experimental and computational investigations have been used to study mechanisms by which heterobimetallic reagents function, refining structure–reactivity understanding in organometallic catalysis, and helping develop new catalysts. With a view towards sustainability of the chemical industry and circular chemical technologies (including of plastics), and the sustainable utilisation of renewable feedstocks, heterobimetallic catalysts, with their enhanced and unique reactivity and activity, have a role to play which will greatly benefit from the insight provided by mechanistic studies.

In this chapter, a particular focus will be, therefore, placed on the transformations of small molecules such as hydrogen, nitrogen, or carbon dioxide mediated by such reagents. We will also describe how heterobimetallic catalysts have been developed for polymerization reactions, including for sustainable polymers such as the ring-opening of lactones and the ring-opening copolymerization of epoxides and carbon dioxide.

4.2 Small Molecule Activation and Catalysis

The activation of small molecules such as hydrogen, nitrogen, or carbon dioxide (CO_2) is an attractive approach to synthetic chemistry. Such reactions allow the use of abundant and renewable (N_2, CO_2) resources and may provide highly atom efficient processes. Some recent work in the area has seen the use of different metals in a single complex, in which the metals interact directly with each other. Thus, the secondary metal acts as a ligand and, hence, tunes the properties of the 'major' metal center. In cases such as these, the unusual bonding behaviour warrants further investigation, and computational chemistry can be used to great effect. Some selected examples of activation of hydrogen, nitrogen, and CO_2 by heterobimetallic complexes in which a computational element was used for investigation are presented below. In some cases, catalytic activity for certain transformations has been achieved, and these are also described.

4.2.1 Hydrogen Activation

Several prominent examples come from reports by Whittlesey et al., which describe the preparation of unsupported Ru–M bonds (M = Zn, In, Ga) [26–28]. These reactions work by reaction of a Ru–H complex with an organometallic reagent, such that liberation of a small alkane (e.g. methane or ethane) acts as a driving force, with the new alkylmetal ligand installed at the former hydride site. A key feature of the complexes formed is that they are coordinatively unsaturated at both metal centers, thus allowing for subsequent reactivity.

Complex **1** reacts with diethylzinc to produce **2** in which a formally 4 coordinate cationic Ru complex is stabilized by agostic interactions from the NHC–ligand substituents [26]. This complex can activate dihydrogen, forming **3**, which under fairly forcing conditions will release the η^2–H_2 ligand to yield **4** (Scheme 4.1). All the complexes in this study were characterized crystallographically, thus providing a good baseline comparison for calculated structures. In all cases, the agreement between crystallographic and calculated distances was good. The calculated interatomic

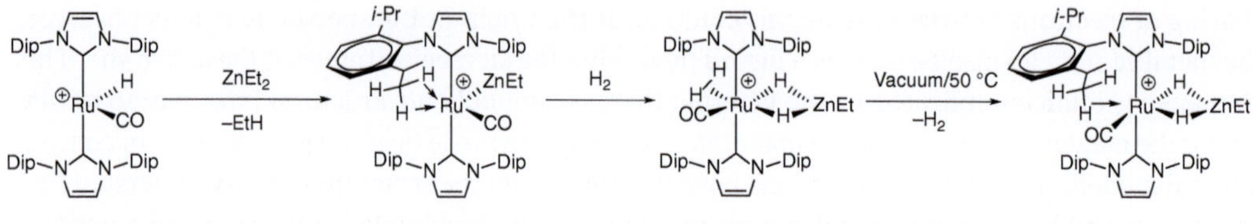

Scheme 4.1 Synthesis of RuZn complexes (**2**–**4**). Counter-ions omitted for clarity.

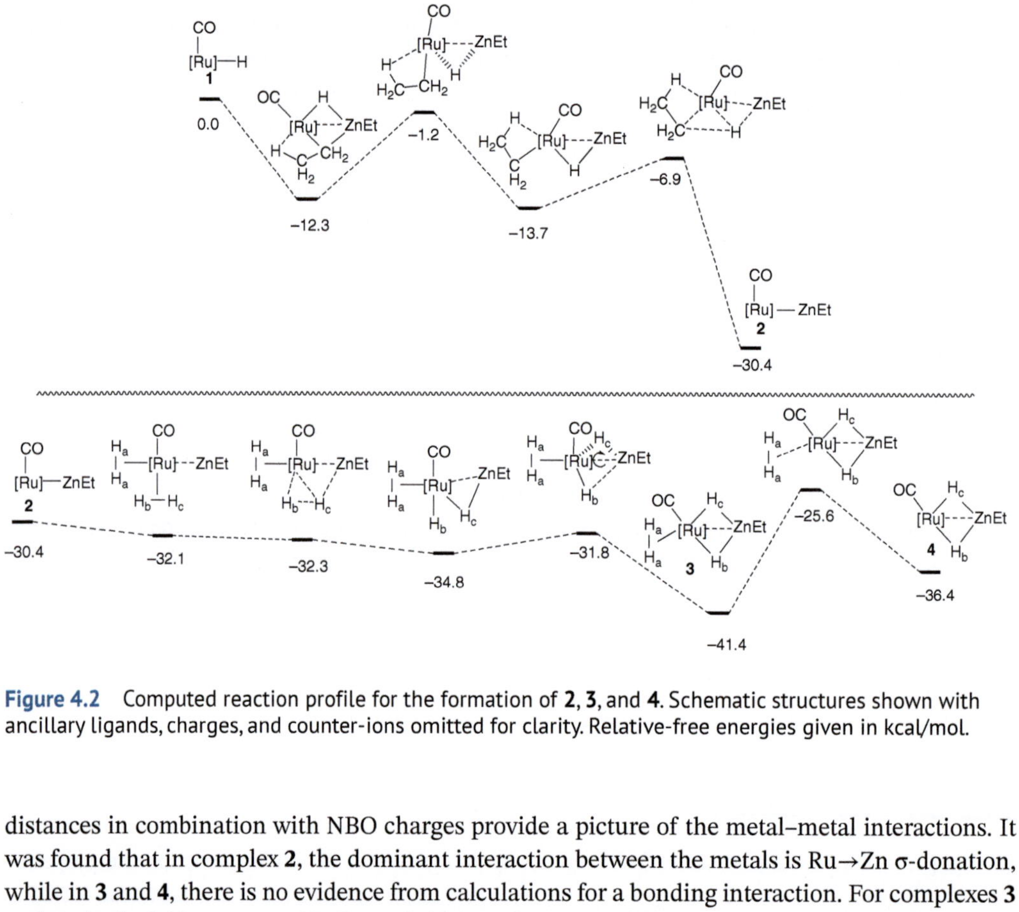

Figure 4.2 Computed reaction profile for the formation of **2**, **3**, and **4**. Schematic structures shown with ancillary ligands, charges, and counter-ions omitted for clarity. Relative-free energies given in kcal/mol.

distances in combination with NBO charges provide a picture of the metal–metal interactions. It was found that in complex **2**, the dominant interaction between the metals is Ru→Zn σ-donation, while in **3** and **4**, there is no evidence from calculations for a bonding interaction. For complexes **3** and **4**, the hydride *trans* to CO shows bridging character, while in **4** that *trans* to the vacant site behaves as a terminal Ru–H.

In addition to providing insight into the nature of the complexes, the process by which they are formed can also be usefully investigated using density functional theory (DFT) calculations. This is especially valuable in this case, given that the synthesis of a hydride bridged Ru–Zn species in this manner was, at the time, unprecedented. Calculations on the formation of **2** show a multi-step process in which an ethyl ligand initially bridges between the two metal centers, is transferred to the Ru center, and finally undergoes a β-hydride elimination step to release ethane and the product complex (Figure 4.2). Reaction of **2** with hydrogen was also investigated. Coordination of two molecules of hydrogen is shown to be favorable relative to **2**, with the cleavage of one of the H–H bonds essentially barrierless (+0.2 kcal/mol). The formation of bridged complex **3** is accomplished by rotation about the Ru–Zn axis. The barrier to formation of **4** is 15.8 kcal/mol, indicating the process is kinetically accessible; however, **4** lies 5 kcal/mol higher in energy than **3** (Figure 4.2). Hence, the formation is unfavorable, but possible using the forcing and irreversible experimental conditions described. Possibly the most interesting aspect of this computational study is the assessment of cooperativity between metals. The H–H bond cleavage is believed to be Ru centered. This is based on the large interatomic distance between the formed hydride and Zn and the lack of polarization of the H–H bond at the transition state. A calculation on the cleavage of the hydrogen ligand *trans* to Zn was also performed, which demonstrated a similar activation barrier; however, the product—in which a hydride cannot be accepted by Zn—was 6.3 kcal/mol higher in energy than that formed when the *cis*-hydrogen ligand is cleaved. This example demonstrates the use of applying DFT calculations to answer hypothetical questions, to gain a deeper understanding of a system.

The same precursor complex can be reacted with $InMe_3$ or $GaMe_3$ to produce novel bimetallic complexes [27]. In the case of $GaMe_3$, the migration of one of the NHC ligands is observed, with the aryl ring substituent acting as a π-ligand for the ruthenium center (**8**; Scheme 4.2). The reaction with $InMe_3$ is somewhat similar to that with $ZnEt_2$—yielding an unsupported Ru–In bonded complex **5**. This complex can also activate hydrogen to produce **6**. In contrast with the Ru–Zn complex, this complex was unstable in the absence of a hydrogen atmosphere. The activation of CO was also demonstrated, with 3 equiv. of CO ultimately bound to the Ru center and concomitant transfer of the methyl ligand to In (**7**; Scheme 4.2).

The mechanism of formation for complexes **5** and **7** was investigated. Of particular note is the very first step of complexation, in which a Ru–In adduct is formed. In contrast to previously reported tetrahedral In centers in Ru–In complexes, the $InMe_2$ portion of the molecule lies in the same plane as the hydride and carbonyl ligands on Ru. One possibility is that this can be attributed to the steric requirements of the NHC ligands. Further investigations, however, demonstrate that this geometry is favored even with smaller supporting ligands ($InMe_2$ or PH_3). Therefore, an alternative electronic explanation is reached based on an atoms in molecules analysis. The analysis

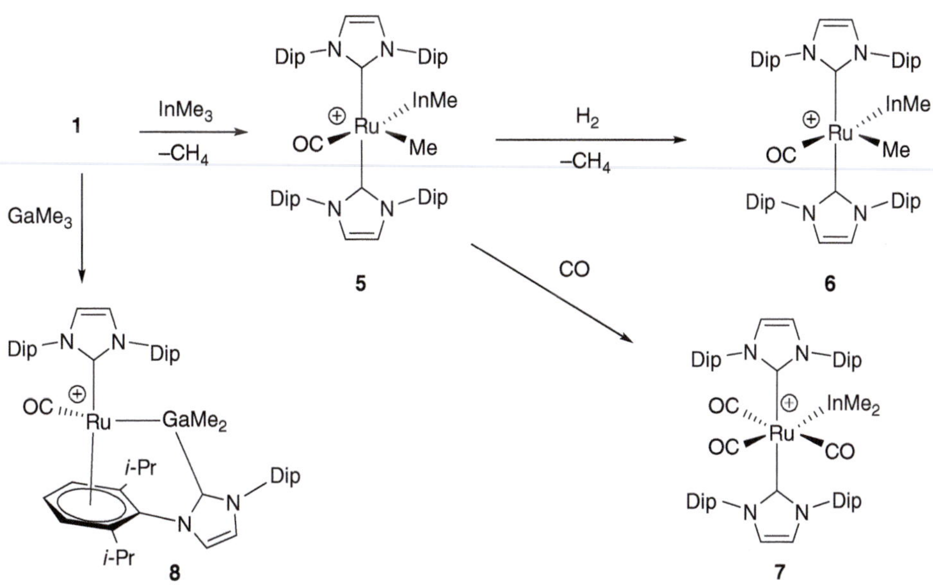

Scheme 4.2 Synthesis of Ru–In and Ru–Ga complexes (**5–8**).

shows the stabilisation of the electron-deficient Ru center is afforded by an agostic interaction with the bridging methyl group from the $InMe_3$, as well as a direct In–Ru interaction (Figure 4.3).

The combined experimental and computational approach has also been used to revisit bimetallic Ru systems which were first investigated in 1977, but could not be fully characterized at the time [28, 29]. Four complexes were prepared by reaction of $[Ru(PPh_3)_3HCl]$ with a lithium, magnesium, or zinc reagent (Scheme 4.3). The reaction causes CH activation of two of the phosphine phenyl groups, to produce a cyclometalated complex. Reaction of **9** with 12-crown-4 effectively isolates the Li ion, to form a cationic Ru species with a $[Li(12\text{-crown-}4)]^+$ counter-ion – **10**. Solutions

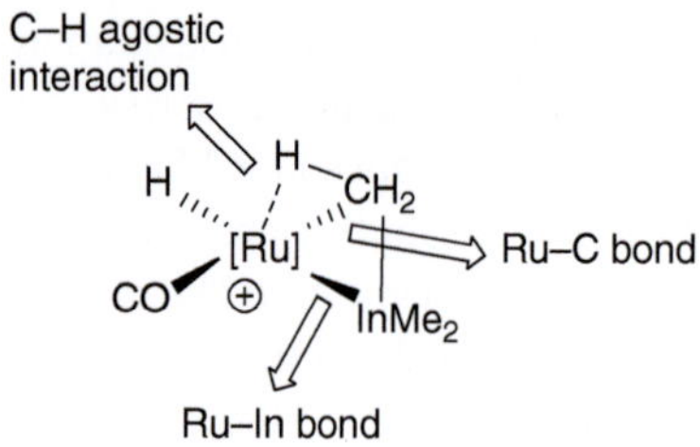

Figure 4.3 Ru–In adduct intermediate to formation of **5**, showing sources of electronic stabilisation afforded by planar arrangement of bonds to In.

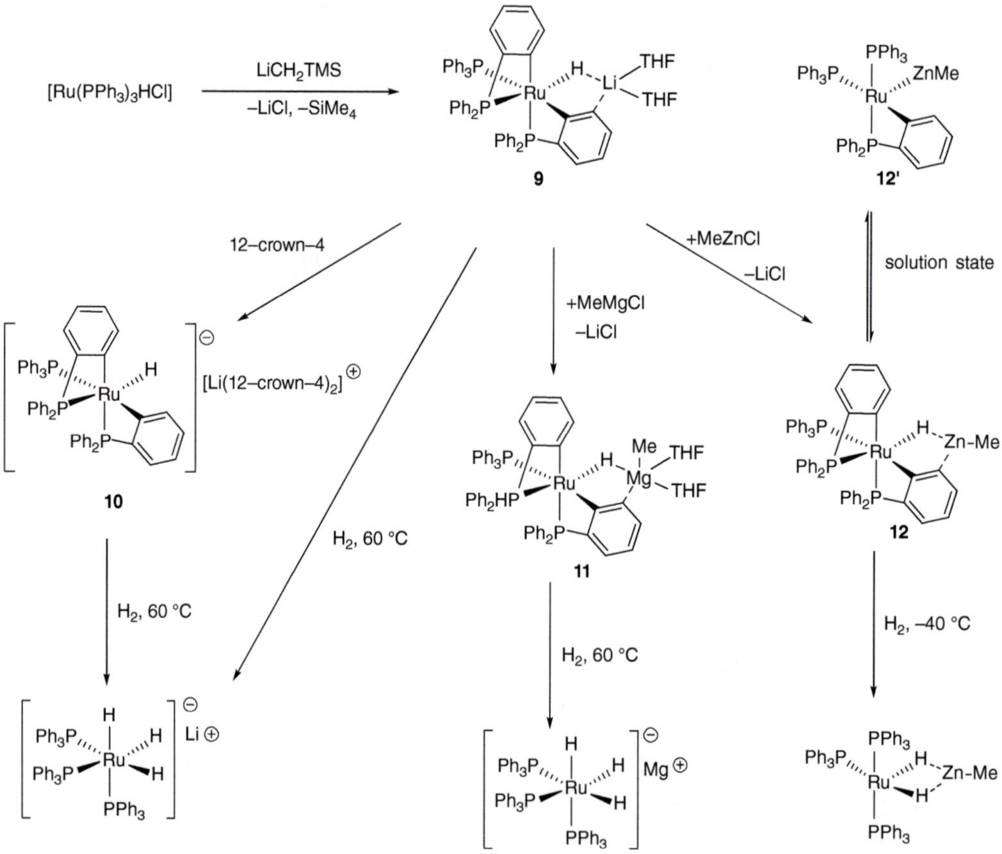

Scheme 4.3 Synthesis of Ru–M complexes (**9–12**) (M = Li, Mg, Zn) and subsequent reaction with hydrogen. Note: complexes **11** and **12** could also be prepared directly by reaction of a suitable metal salt with $[Ru(PPh_3)_3HCl]$ but purification was challenging by this method.

of complex **12** form an equilibrium made up of the *mer* and *fac* isomers of **12**, in addition to the product of a reductive elimination reaction, **12′**, the structure of which was assigned using NMR spectroscopy. Although complexes **9**, **11**, and **12** have the same general structure, the bonding interactions between Ru and M (M = Li, Mg, or Zn) were found to vary significantly. Crystallographic data show that the interatomic distance for Ru–Li is greater than the sum of the covalent radii—indicating that there is no formal bond between these centers. In the case of **11**, the Ru–Mg distance is slightly shorter than the sum of the covalent radii; in complex **12**, the Ru–Zn distance is much shorter than the sum of covalent radii (Figure 4.4a, b). It is interesting to note, however, that quantum theory of atoms in molecules (QTAIM) calculations, while indicating an increase in electron density on the bond path in the order Li < Mg < Zn, provide no evidence of a Ru–M bond, even in the case of Zn (Figure 4.4c). This discrepancy between experiment and theory could be viewed as a breakdown of theory, or as evidence that theoretical calculations can provide a more subtle analysis in some cases.

The Ru–H bonding is also shown to change with changing supporting metal, with the evidence from NMR spectroscopy in this case complementing theoretical calculations. The electron density for the bridging bond path increases in the order (**10** < **9** < **11** < **12**), which is in agreement with analysis of the $^{2}J_{HP}$ coupling constants (used to provide a measure of the terminal character of the hydride).

Upon reaction with hydrogen, complexes **9**, **10**, and **11** form anionic ruthenium trihydride complexes with the second metal becoming a counter-ion, while for **12**, the Zn remains part of bimetallic hydride-bridged complex (Scheme 4.3). It is also notable that **9**, **10**, and **11** require heating to 60 °C to react, while **12** undergoes the reaction at −40 °C. Calculations found the presence of the Zn

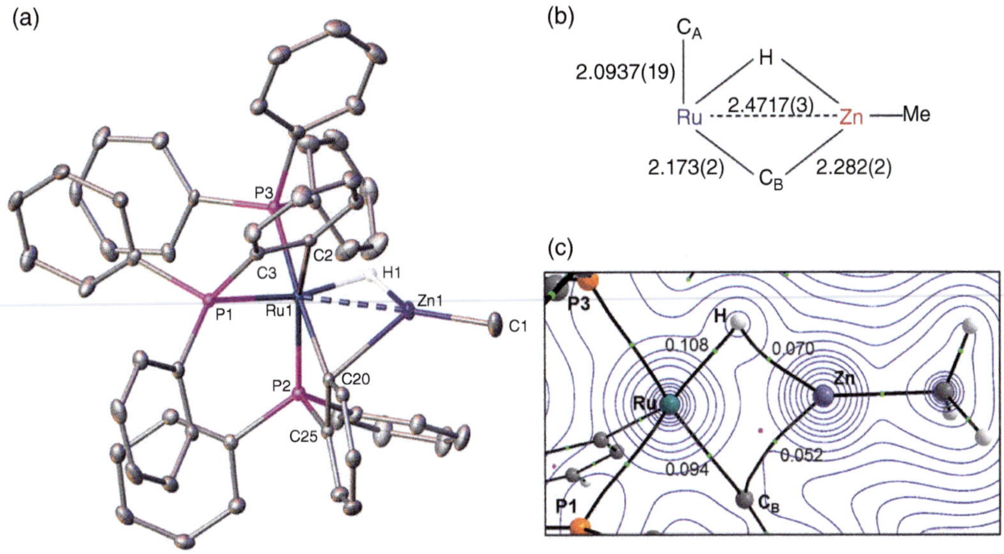

Figure 4.4 (a) Molecular structure of **12**. Ellipsoids are represented at 30% probability. Only the major component of the disordered phenyl group ligand (attached to P3) is shown. Hydrogen atoms, with the exception of H1, have also been omitted for clarity; (b) summary of key distances in **12**; (c) detail of the QTAIM molecular graph for **12** showing key bond critical points (green spheres) and ring critical points (pink spheres) with associated bond critical point electro densities, $\rho(r)$, in a.u. The experimental structure was employed with the H atoms optimized, giving Ru1–H1 and Zn1–H1 distances of 1.69 and 1.80 Å, respectively. *Source*: Miloserdov et al. [28]; © 2020 American Chemical Society.

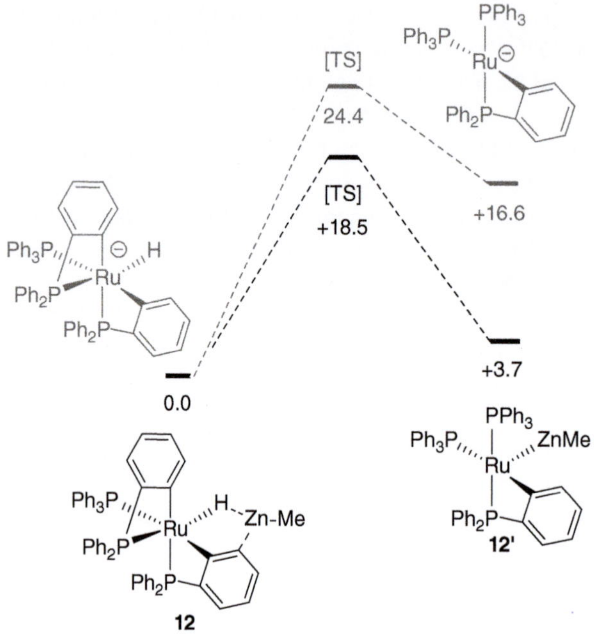

Figure 4.5 Computed energy profile for Zn-assisted and Zn-free C–H reductive elimination. Black trace = Zn-assisted pathway, formation of **12′** from **12**. Gray trace = hypothetical Zn-free pathway, with higher energy barrier. Relative free energies in kcal/mol.

ligand to lower the energy barrier of the initial reductive elimination step of the reaction from 24.4 to 18.5 kcal/mol (Figure 4.5). This demonstrates the utility of computational chemistry, allowing for the comparison of known processes with hypothetical ones.

Additionally, the bonding in hypothetical complex **12′Mg** was investigated using a range of calculations for comparison with **12′** (Figure 4.6 and Table 4.1). The results indicate that a larger interaction between the two metals results in a significantly larger bond energy in the case of Zn than Mg (129 versus 30 kcal/mol). Thus, computational chemistry has been able to provide significant insight into the reasons for reactivity of a range of bimetallic complexes. The reductive elimination step was identified as crucial to explain the reaction with hydrogen. Reductive elimination is a key step in many catalytic reactions—examining the role that a secondary metal may play in supporting this may allow for future development of better catalysts.

A related example reported by Mankad et al. applies heterobimetallic complexes of the type **13** to the catalytic semi-hydrogenation of alkynes (Scheme 4.4) [30, 31]. The inspiration for the use of these bimetallic species for hydrogenation came from a computational study of dehydrogenative borylation. One of the key steps in this process was a bimetallic reductive elimination of hydrogen which reformed the catalyst [32]. The reverse of this reaction would be hydrogen activation, and the energy barrier is low, making this pathway theoretically accessible.

Therefore, the thermodynamics of hydrogen activation by a series of model heterobimetallic complexes of the form M–Cu were calculated using DFT. In all cases, this process is endergonic, but of the eight metals screened, Fe and Ru were found to have the lowest ΔG G value. A calculation was also performed on a hypothetical Ru–Ag complex, and this was found to have a significantly lower ΔG value (14.5 kcal/mol for Ru–Ag versus 20.1 for Ru–Cu). For Ru–Cu, the transition state for this process was also modelled, showing a heterolytic H–H bond cleavage, during which

12′

12′Mg

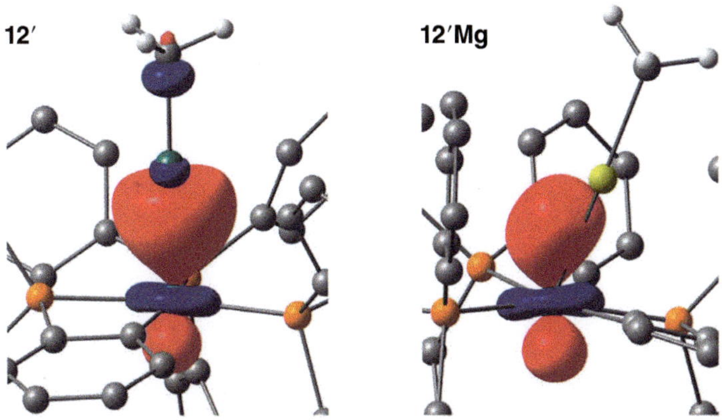

Figure 4.6 Electronic structure analysis of **12′** and its Mg cogener **12′Mg** illustrating the major natural localized molecular orbitals (NLMO) contributing to Ru–M′ bonding in each case. *Source*: Miloserdov et al. [28]; © 2020 American Chemical Society.

Table 4.1 Comparison of calculated parameters for **12′** and **12′Mg** [28].

	12′	**12′Mg**
Relative energy[a]	+3.7 kcal/mol	+12.7 kcal/mol
NLMO	78% (Ru), 13% (Zn)	88% (Ru), 5% (Mg)
$LP_{Ru} \rightarrow \sigma^*_{M'\text{-}Me}$	129 kcal/mol	30 kcal/mol
Wiberg bond index (Ru–M′)	0.42	0.22

Source: Based on Miloserdov et al. [28].
[a]Relative to parent complex, prior to C–H reductive elimination (i.e. **12′** relative to **12**).

the metal-metal bond is cleaved, but stabilisation is afforded by a bridging carbonyl ligand. The energy barrier for the process was 29 kcal/mol.

Based on the initial computational screening, Ru–Cu, Fe–Cu, and Ru–Ag complexes were tested for catalytic activity towards alkyne hydrogenation, using fairly forcing conditions. While all complexes displayed some activity, the best activity and selectivity (for E-alkenes) was achieved using the Ru–Ag complex (**13″** – Scheme 4.4), giving 96% conversion and 95% selectivity [30].

Following on from this initial report, a detailed computational study was undertaken. An intrinsic reaction coordinate calculation was performed using the transition state previously described [30, 31]. Several key parameters (interatomic distances, Wiberg bond indices, natural charges, Lewis occupation) were then compared as a function of reaction progress by calculating them for several structures. The structures chosen were the starting reactant complex, three structures on the IRC scan preceding the transition state, the transition state itself, three structures on the IRC scan succeeding the transition state, and the final product.

Information about bonding (bond distances and Wiberg bond indices; Figure 4.7) in the various structures allows for the order in which the key bond breaking and making processes take place to

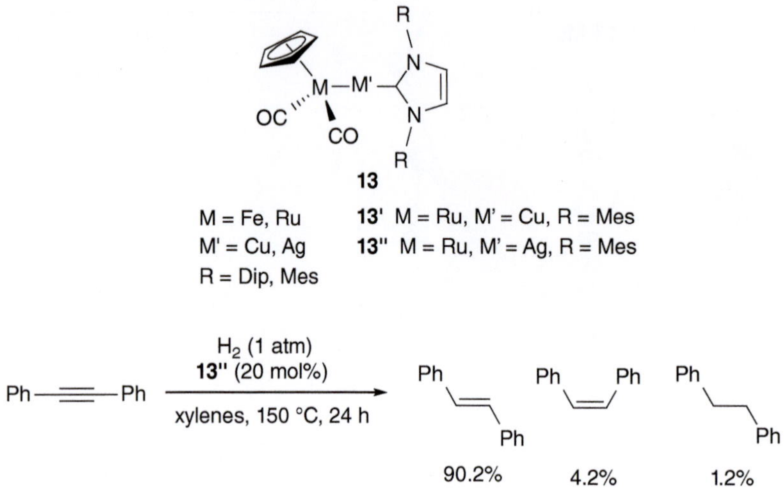

Scheme 4.4 Bimetallic complexes for catalytic hydrogenation of alkynes.

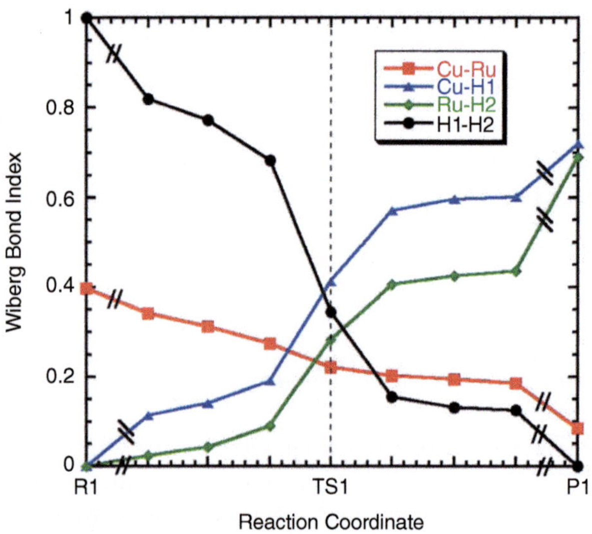

Figure 4.7 Selected Wiberg bond index values for the reaction between **13′** and H_2, plotted along the calculated intrinsic reaction coordinate scan. // = discontinuity. *Source*: Karunananda and Mankad [31]; © 2017 American Chemical Society.

be determined, with the change in orbital occupation providing supporting information. The change in natural charge throughout the reaction gives an idea of the redox processes occurring and where they are centered (Figure 4.8). Overall, a mechanistic assessment is made that the key processes are: σ-complex formation at the electrophilic metal site (comprising H–H σ-donation to Cu), followed by proton transfer to the nucleophilic metal site (donation to the H–H σ*-orbital from Ru). While this contrasts with the traditional homolytic mechanism invoked for hydrogen oxidative addition, it is in fact very similar to that described for activation of hydrogen by frustrated Lewis pairs.

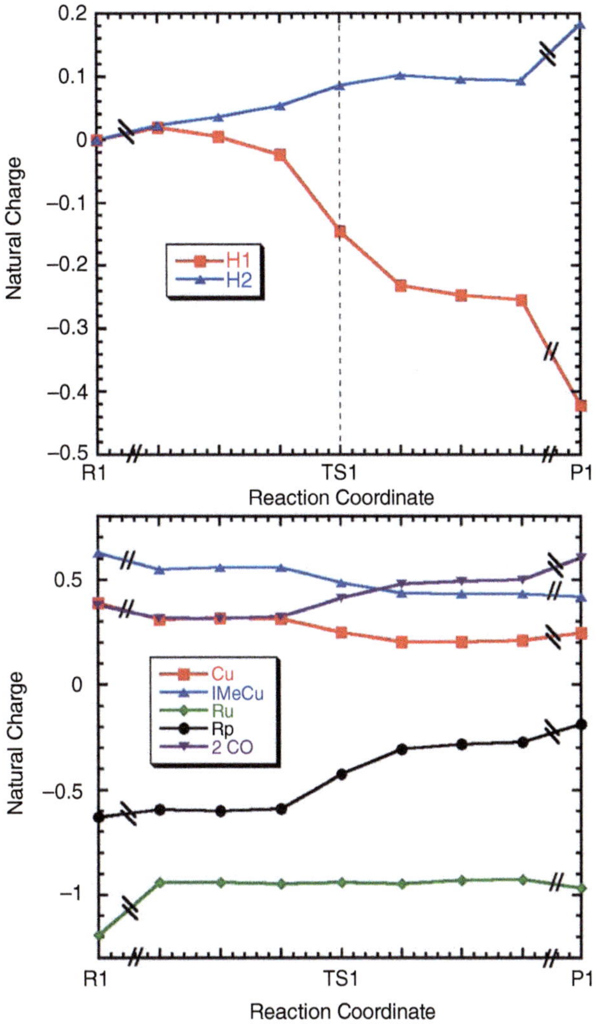

Figure 4.8 Selected natural charge values derived from natural population analysis for the reaction between **13′** and H_2 plotted along the calculated intrinsic reaction coordinate scan. // = discontinuity. *Source:* Karunananda and Mankad [31]; © 2017 American Chemical Society.

Thus, key insight into the mechanism has been obtained by using data obtained using computational methods, and it is suggested that this knowledge will inform future catalyst design.

A range of heterobimetallic complexes which activate small molecules, including hydrogen, have been reported by Lu et al. [33]. These are constructed from pincer ligands, with a phosphorus and nitrogen binding site (Figure 4.9). These complexes build on earlier work that has enhanced the catalytic activity of a metal center by installing a Lewis acidic site (typically boron) in close proximity to the metal center [34–39]. The studies typically vary either M or M′, thus, allowing for a systematic investigation into the effects of different metals [33]. A combination of experimental and computational techniques has been used, which offers a valuable insight into the unusual bonding. DFT and complete active space SCF (SCF, self-consistent field) (CASSCF) calculations have been variously applied to: geometry optimizations to examine structural parameters [40–45], calculation of spin states [40, 41], determination of MO diagrams [40–42, 46], and estimation of bond orders [40–42].

In a key study, the role of the supporting metal was investigated, by systematic preparation of a range of group 13–Ni complexes (M = Ni; M′ = Al, Ga, In). Of the three complexes that were

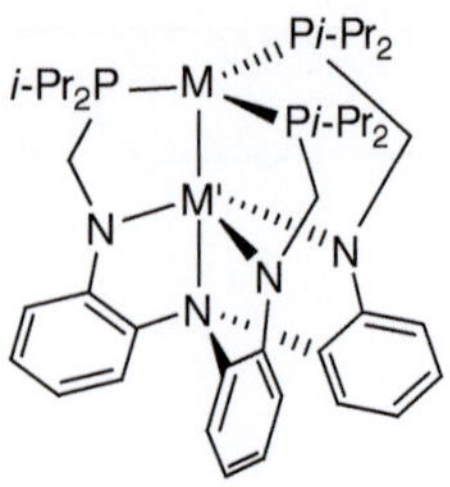

Figure 4.9 General structure of bimetallic complexes prepared by Lu et al.

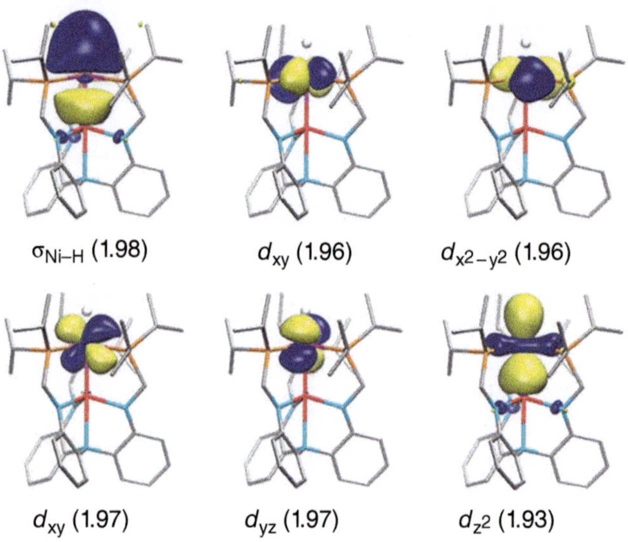

Figure 4.10 Selected natural orbitals obtained from a CASSCF calculation of $[NiGaH]^-$. Occupation numbers (shown in parentheses) indicate that these six orbitals are doubly occupied. *Source:* Cammarota et al. [48]; © 2017 American Chemical Society.

prepared, NiGa was found to be the most effective at catalysing styrene hydrogenation [47]. The same catalyst was later used for hydrogenation of CO_2, as part of a joint experimental and computational study [48]. Under high pressures of hydrogen, near quantitative conversion of CO_2 to formate was observed at a low catalyst loading (0.03 mol%), with an initial turnover frequency (TOF) of 9700/h and a turnover number (TON) of 3150. These values vastly outperform other reported homogeneous Ni catalysts. It was noted that the Ga-free analogue was inactive, and also that a mixture of the Ga-free catalyst and $GaCl_2$ was similarly ineffective. This strongly suggests that an intact Ni–Ga bond is required for turnover.

Computational chemistry was used to rationalize the unexpected bonding picture for the NiGa complex [48]. As Ni(0) has a formally full d-electron set, formation of a hydride should not be possible. The results from the bonding analysis revealed that all the d-orbitals are indeed full, and that the Ni–H bond is largely hydrogen based (51%). The donation from the hydride to Ni is $1s \rightarrow 4p_z/4s$; however, the excess electron density attributed to Ni as a result can be offset by donation from Ni to Ga ($3d_{z^2} \rightarrow 4p_z/4s$) (Figure 4.10). As further evidence of this, the stabilisation energy attributed to the Ni–Ga bond is 10 kcal/mol higher for the hydride complex than for the parent catalyst. This stabilisation is termed inverse *trans* influence; the Ga ligand provides stabilisation to a *trans* hydride ligand in a 'pull–pull' fashion.

An in-depth computational study was later completed which was able to determine the most likely mechanism for catalysis, as well as give insights into the rate determining step, and key parameters which can be investigated when considering future catalysts [49].

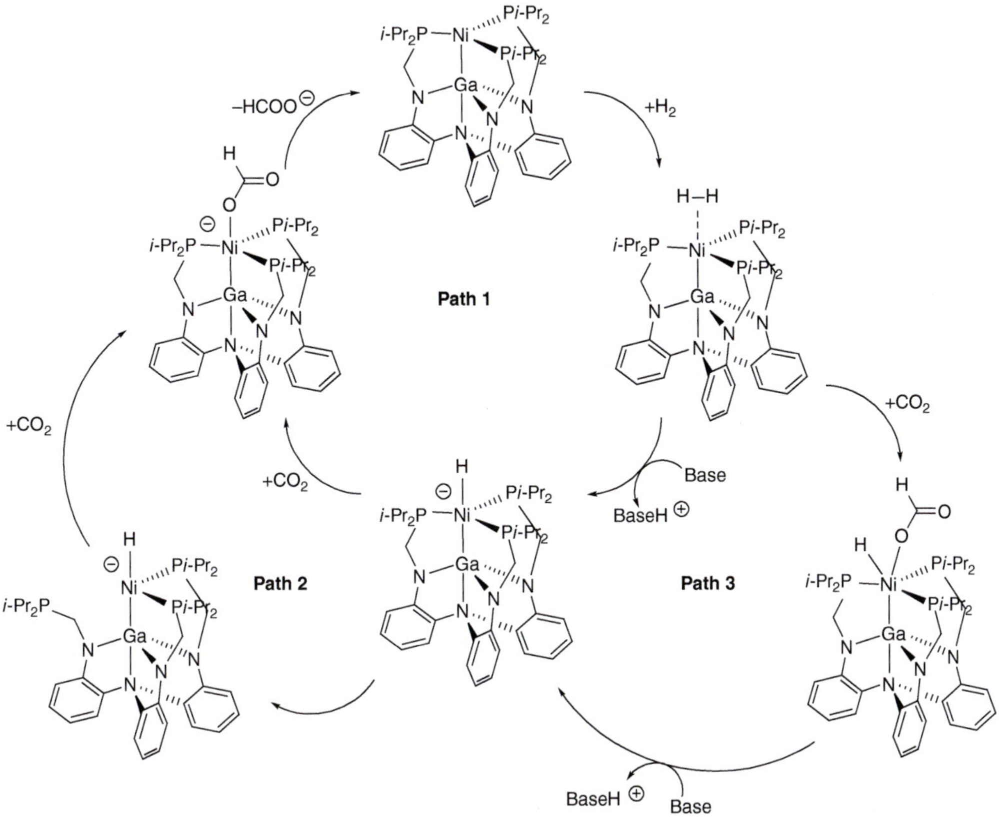

Figure 4.11 Possible pathways investigated computationally for the hydrogenation of CO_2 catalysed by NiGa complex. Base = 2,8,9-triisopropyl-2,5,8,9-tetraaza-1-phosphabicyclo[3,3,3]undecane.

Three plausible pathways for catalysis were investigated, and their calculated energy profiles compared (Figure 4.11). Pathway 1 was found to have the lowest barriers, and, hence, is considered to be the most likely to be in operation. For pathway 1, after hydrogen coordination the next step is the deprotonation of the hydrogen complex to produce the hydride complex. CO_2 then inserts directly into the Ni–H bond via a van der Waals complex. Rearrangement generates the H-bound formate which is then released to reform the initial catalyst. The rate determining step was found (based on barrier height) to be the deprotonation of the hydrogen complex; indeed the calculated barrier was inconsistent with a room temperature transformation.

Further investigation into this step revealed that the key factor affecting barrier height was the steric bulk of the base. Indeed, the distance between the Ni center and the base in the transition state correlates well with the activation energy, with higher activation energies found for larger distances (Figure 4.12). Formate had a very low energy barrier for deprotonation, and is generated during the reaction. However, catalysis ceases when the base is consumed, and so the exogenous base is still important for turnover. Hence, a mechanism was proposed in which the exogenous base is directly involved in deprotonation for a few initial cycles; however, once sufficient formate is built up, this acts as a proton shuttle. This explains the induction period observed experimentally and also the requirement for stoichiometric base. These findings demonstrate not only the enhanced understanding that can be achieved by using theoretical calculations but also their

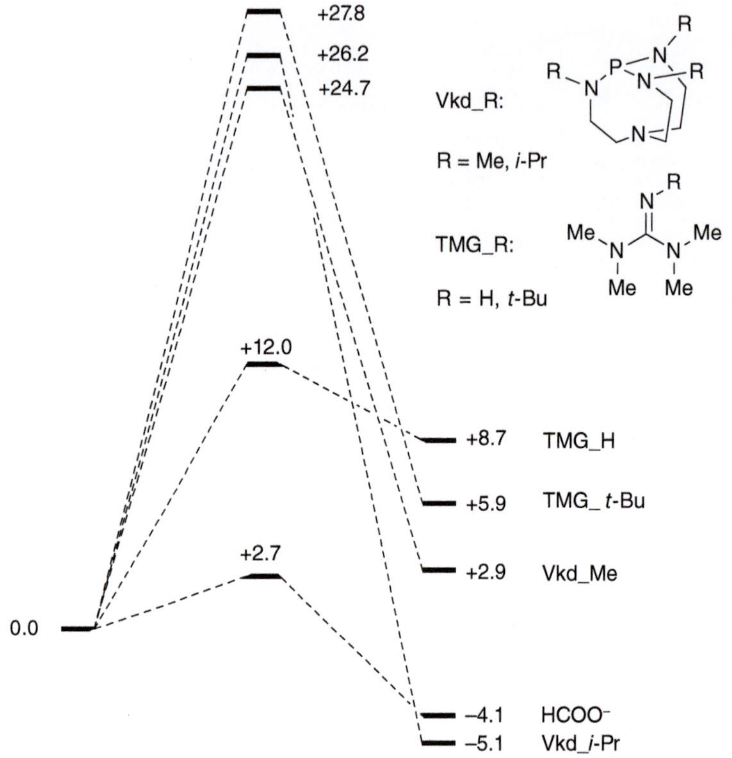

Figure 4.12 Relative free energies of the transition state for and product of the reaction between the NiGa–H_2 adduct and base ([NiGa-H_2][base] → [Hbase]$^+$[NiGa-H]$^-$). Free energies in kcal/mol.

potential predictive power. In future, potential bases could be screened simply by examining one key transition state, without the need to map the entire reaction pathway.

A number of other metrics were used to assess the feasibility of the reaction for other bimetallic complexes (M = Ni, Co, Fe, Pd, Pt; M′ = Al, Ga, In, Fe, Co). For instance, the thermodynamic hydricity (ΔG_H)—the energy change associated with release of a hydride ion from the M–H complex—of a hydride species can be a powerful measure of its reactivity. It was found that the calculated ΔG_{TS} values for the hydride transfer to CO_2 correlated well with the thermodynamic hydricity of the species, thus, providing another parameter to be investigated when considering future catalyst development. The relative binding energies of hydrogen and formate were also investigated for the same complexes. In all cases, formate binding was more favorable than hydrogen binding, which should prevent turnover. However, although the calculations in this case make an incorrect prediction about the reaction feasibility, the general trends may be used as predictors. Thus, those with the smallest increase in favorability of formate binding over hydrogen (CoAl, CoGa, NiAl, and NiGa) are predicted to perform the best. It is noteworthy that the proven NiGa catalyst is predicted to be successful by both parameters considered (thermodynamic hydricity and formate/hydrogen binding energies). Combining these assessments suggests that CoGa would also be a promising catalyst for this transformation. Subsequently, Co–Ga complex was tested for activity for CO_2 hydrogenation, and was found to be significantly more active than the NiGa complex. Under the same conditions as the NiGa catalyst the TOF for CoGa was 1100/h (versus 9700/h for NiGa) and the TON 3100 (versus 3150 for NiGa). However, under altered conditions the CoGa catalyst achieved a highest initial TOF of 27 000/h and a TON of 19 000 at an extremely low loading of just 0.004 mol%, thus demonstrating the predictive power of computational chemistry [50].

4.2.2 Dinitrogen Activation

The same class of complexes which have been used for hydrogen activation have also proved successful for nitrogen activation. An early example from Lu et al. prepared a range of the first row transition metal complexes with a Lewis acidic aluminium ligand (M = Ni, Co, Fe; M′ = Al) [51]. Extensive characterization was performed including X-ray crystallography, UV–vis, IR, and NMR spectroscopy, which revealed that the Co and Fe examples coordinated nitrogen in an end-on manner. In the case of Fe, this was revealed crystallographically to result in the formation of a dimeric complex in which the nitrogen acts as a bridging ligand. Calculations using DFT produced optimized geometries in good agreement with the crystallographic data. The calculations were also used to determine the relative d-orbital energies in the complex series, as well as the highest-occupied molecular orbital (HOMO) and lowest unoccupied molecular orbital (LUMO). Building on these results, anionic forms of the cobalt and iron complexes were prepared by reduction [52]. The new complexes also coordinated dinitrogen, but it was noted that in these cases the N–N bond was significantly weaker—believed to be due to increased electron density at the metal center enhancing back donation. The FeAl complex was investigated further by various computational techniques. Calculation of the spin density attributed one unpaired electron to the iron center, in good agreement with experimental electron paramagnetic resonance (EPR) data. The vibrational frequency of the N–N bond was also calculated. It is notable that although the calculated values are quite different to those obtained experimentally, the difference in frequency observed for the anionic and neutral complexes is very consistent. This highlights the potential utility of computational chemistry for prediction, particularly when relative measures are all that is required. The anionic FeAl complex reacts with bis(chlorodimethylsilyl)ethane at −78 °C to afford a species in which N_2 has undergone a formal four electron reduction to produce a di-silylated hydrazido group (Scheme 4.5). Although liberation of the new product from the metal center was not demonstrated, this shows the potential usefulness of heterobimetallic complexes for dinitrogen activation.

A series of Co–M complexes (M = Co; M′ = Al, Ti, V, Cr, Co) were studied to determine the effect of the secondary metal on dinitrogen binding [53]. As previously, DFT calculations were used to perform geometry optimization on all structures. Subsequently CASSCF calculations were performed to predict the MO diagrams. This allows for a comparison of the variation in d-orbital occupation across the series, as well as the polarization of the metal–metal bonding (Figure 4.13).

Of particular note is that the π^* orbitals associated with N_2 are too high in energy to be accessible for all complexes, indicating that in all cases the N–N triple bond is maintained. The stabilisation of the d_{z^2} orbital also varies a great deal, dependent on the supporting metal. In CoAl, a purely dative bond exists between the metals (Co→Al), which affords stability to the d_{z^2} orbital. CoTi,

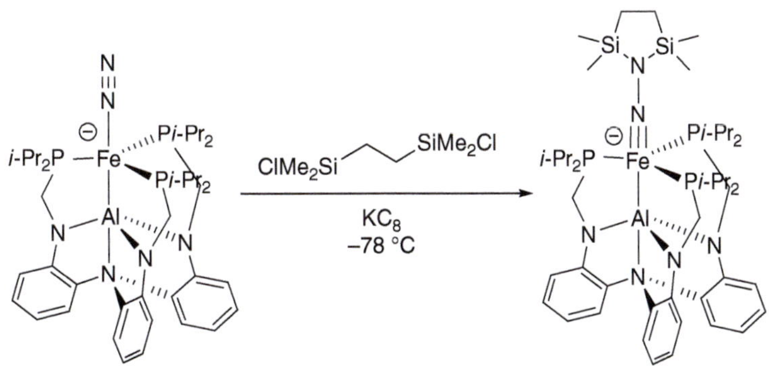

Scheme 4.5 Activation of nitrogen by $FeAl^-$ complex, and formal 4-electron reduction upon reaction with bis(chlorodimethylsilyl) ethane.

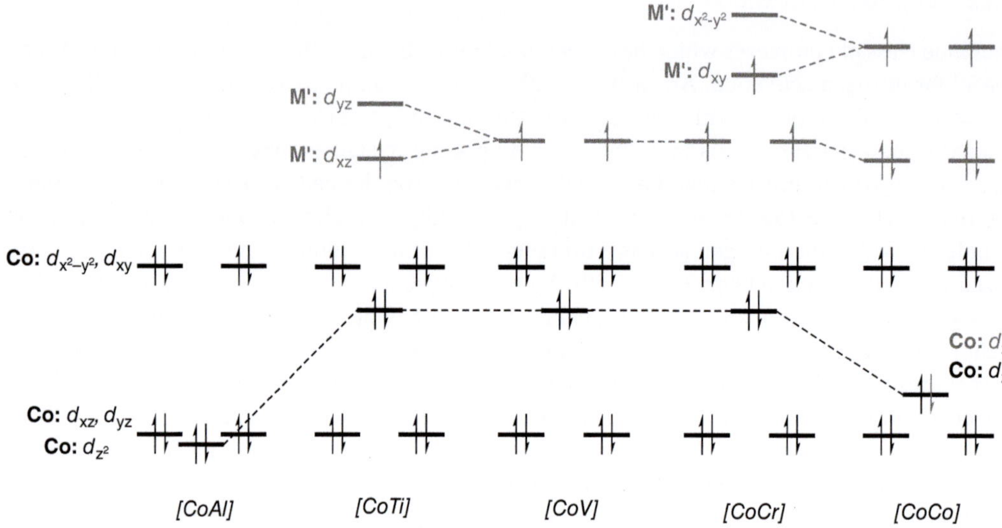

Figure 4.13 Qualitative diagram showing MO occupation for the [CoM′-N_2] – series (M′ = Al, Ti, V, Cr, Co). Black MOs = Co centered, gray MOs = M′ centered.

CoV, and CoCr have similar orbital structures, with the d_{z2} orbital lying higher in energy than in CoAl. For the CoCo complex, mixing of the two d_{z2} orbitals again allows for a lowering in energy. The oxidation states for the metals in the complexes can also be determined, suggesting Co(-I) supported by M(III) for M′ = Al, Ti, V, and Cr. In the case of the dicobalt complex, Co(0)/Co(II) is instead suggested. This difference is believed to account for structural differences, with the more reduced Co centers (M′ = Al, V, Cr) having shorter Co–P bond lengths than are observed in CoCo.

In a subsequent study, the neutral CoCo and CoAl species were tested for catalytic silylation of N_2. [54]. The dicobalt complex was found to be one of the most active catalysts reported under commonly used conditions with a TON of 195 and achieving 30% conversion (Scheme 4.6). Notably, this is significantly more active than other simple cobalt complexes screened. In addition to this, it is possible to form the complex in situ without a loss in TON. The CoAl complex by contrast achieved just 4% conversion, with a TON of 30.

DFT calculations were used to investigate a plausible mechanism for the reaction (Figure 4.14). It is believed that two silylation steps occur on the bound nitrogen with the second being the rate determining step. Subsequently, one of the phosphine ligands dissociates from the metal center, presumably to relieve steric pressure, alongside the total cleavage of the Co–Co bond. This allows for another silylation to occur on the proximal nitrogen atom. The phosphine ligand is reassociated and the Co–Co bond reformed upon substrate release. An understanding of electronic structure can be used to explain difference in activity between CoCo and CoAl. It was noted that there is a stronger interaction between Co and Al than between Co and Co, the former being an inverse dative bond (Co→Al) and the latter having an approximate bond order of a half. Additionally, more

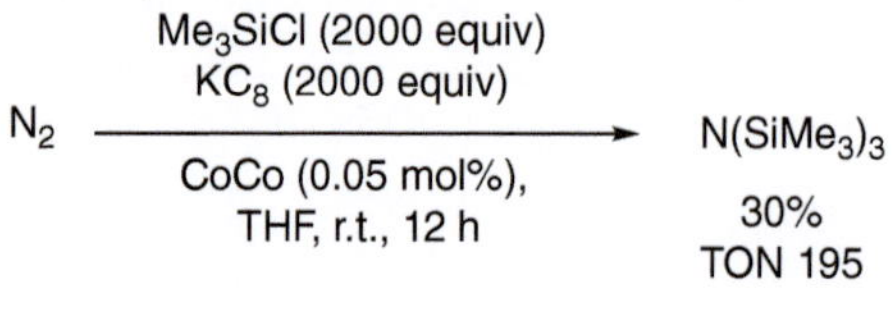

Scheme 4.6 Silylation of nitrogen catalysed by CoCo complex.

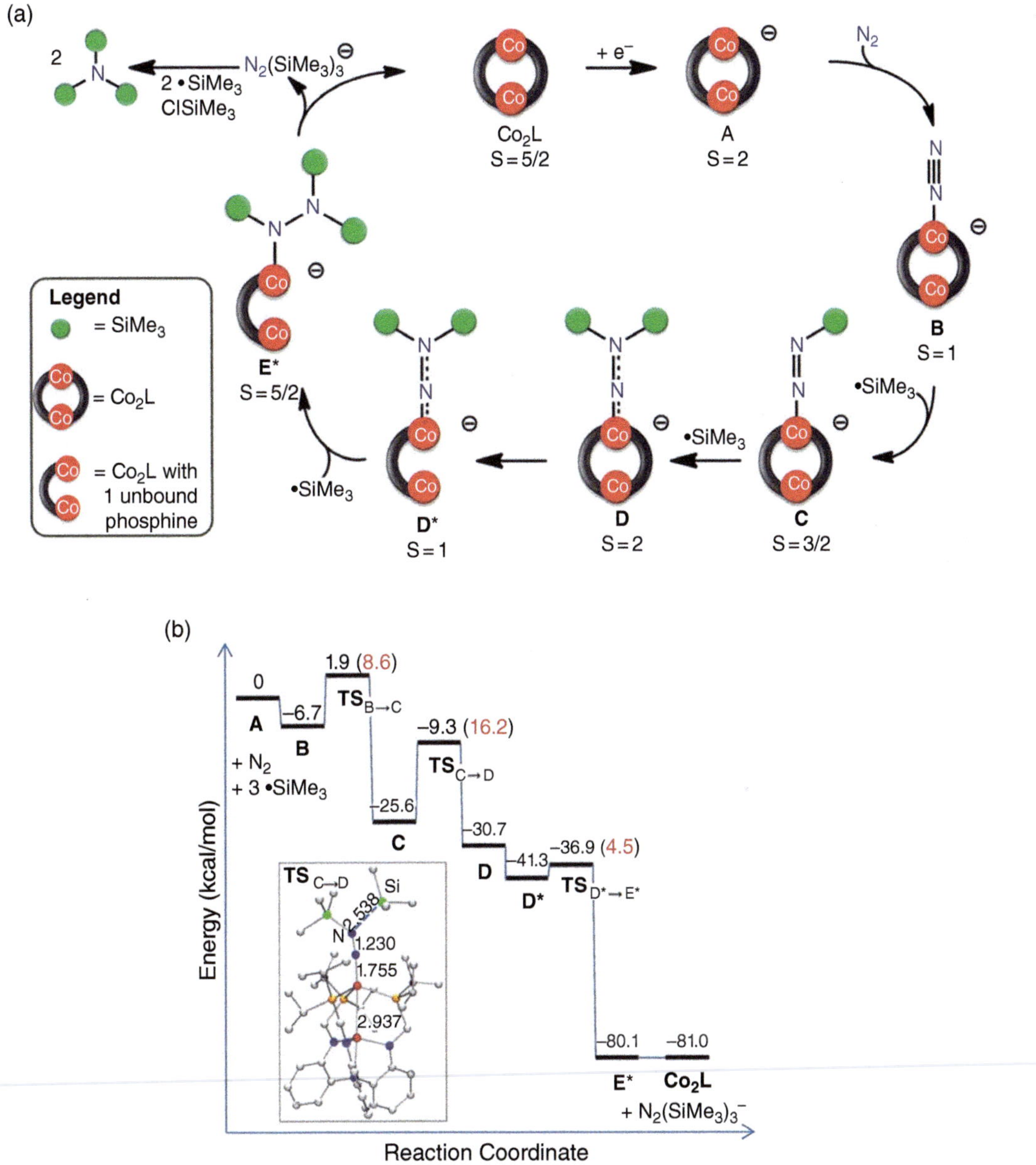

Figure 4.14 (a) DFT-calculated mechanism for the CoCo-mediated silylation of N_2. Intermediates with a dangling phosphine are labelled with an asterisk, e.g. **D***; (b) Energy profile for the proposed mechanism with activation energies in red. Inset shows $TS_{C \rightarrow D}$, the transition-state structure between **C** and **D** with interatomic distances in Å. Free energies in kcal/mol. *Source*: Siedschlag et al. [54]; © 2015 American Chemical Society.

basic phosphines were found to enhance activity in monometallic Co complexes, so it is possible that the more acidic Al support reduces activity relative to Co. That is not to say that the second Co center is irrelevant and indeed it is believed to play a crucial stabilising role, as evidenced by the gradual Co–Co bond weakening during the cycle, with the ultimate reformation of the bond on substrate release.

4.2.3 CO_2 Activation

Activation of CO_2 is also seen as a promising way to make use of an abundant and renewable C1 source. In general, such activation makes use of a reactive partner such as an epoxide, aziridine, or organometallic nucleophile [55–57]. Heterobimetallic complexes with a ligated CO_2 have been known for many years [58]. The use of bimetallic complexes for polymerization reactions involving CO_2 is well established and will be discussed later (see section 4.3.3). A few select examples of CO_2 activation using heterobimetallic complexes are presented below.

An example reported by Bourissou et al. prepares complex **14** featuring a Pt–Al interaction. Analysis of the complex using NMR spectroscopy and X-ray crystallography strongly suggests that the two metals are interacting. DFT calculations showed that the complex in which the two metals interact was 18.1 kcal/mol more stable than the 'open' form (Figure 4.15). The interaction was also confirmed using a NBO analysis, which showed donation from the Pt(5d) to the Al(3p) orbital. Complex **14** was shown to react with a range of small molecules including H_2 and CO_2; in all cases, the reaction was irreversible (Figure 4.15). A mechanistic study on the activation of hydrogen was performed and demonstrated the key role that the Al center plays in lowering the energetic span of the process [59].

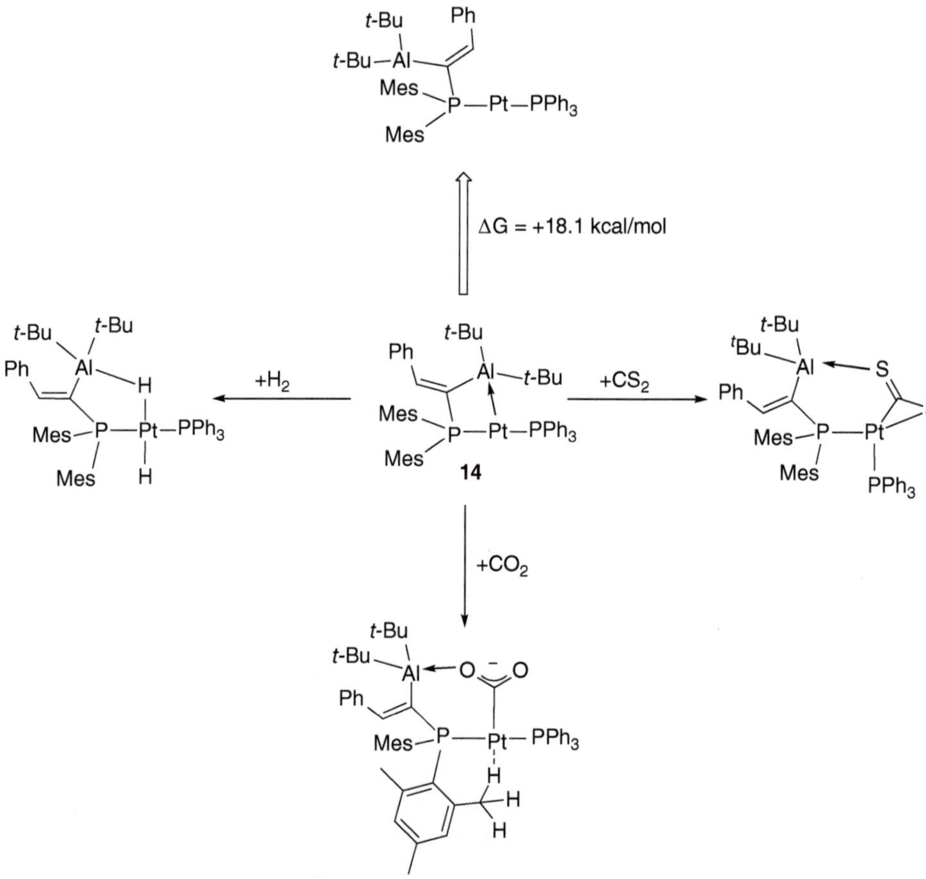

Figure 4.15 Energy difference between open and closed forms of complex **14** determined computationally, and reaction of **14** with a range of small molecules. Relative free energies in kcal/mol.

A study with Pd and varying group 13 metal (i.e. Al, Ga, or In) for hydrosilylation of CO_2 found Al to be the most beneficial partner, with the PdAl catalyst **15** the most active known for this transformation (Scheme 4.7). No computational investigations have yet rationalized the excellent activity of this catalyst. The contrast of this finding with reports from Lu et al. (see above) [47–50] highlights the need to fully understand the role of the secondary metal and also the need to tune the catalyst to the transformation [60].

A bimetallic PdZn catalyst was also reported by Tauchert et al. to be an efficient catalyst for hydrosilylation of CO_2 (Scheme 4.8) [61]. Cu and Li analogues were prepared, but these gave yields of less than 10%, compared to 100% for PdZn. By lowering the catalyst loading to 0.05 mol%, a $TOF_{1/2}$ value of 3000/h is achieved, making the catalyst one of the most active for this reaction by this measure.

NBO/NLMO analysis was used to examine the metal–metal interactions in the complexes (Figure 4.16). As for similar complexes, a donation from Pd to the second metal was observed. The Zn complex shows the greatest interaction between metals, with a contribution from the Pd $d_{x^2-y^2}$ orbital (89%) and a vacant Zn s-orbital (9%). Copper and lithium orbitals contribute just 2.6 and

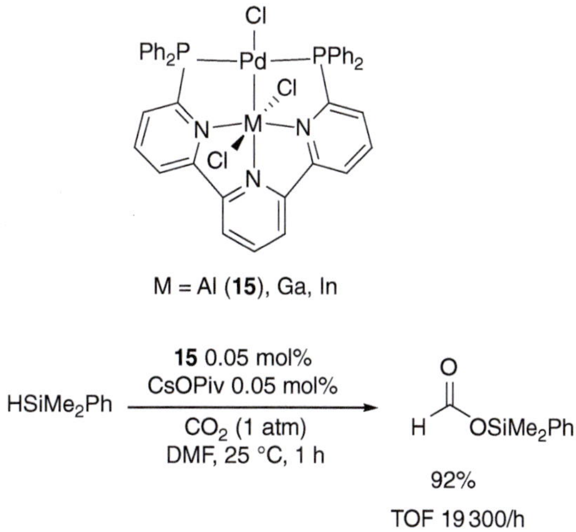

Scheme 4.7 Pd–M complexes (M = Al, Ga, In), and catalytic hydrosilylation of CO_2 promoted by PdAl complex **15**.

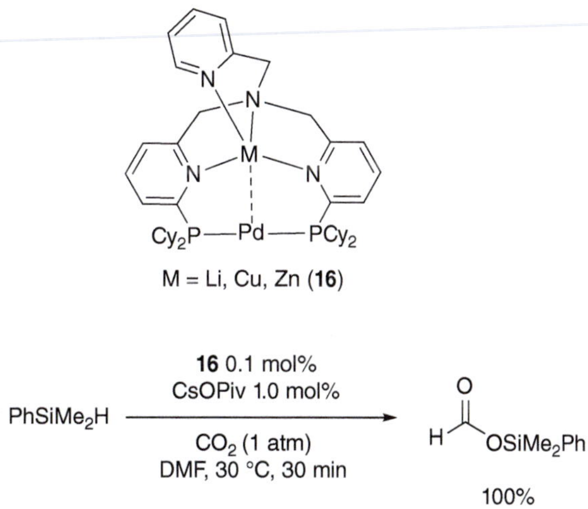

Scheme 4.8 Pd–M complexes (M = Li, Cu, Zn) and catalytic hydrosilylation of CO_2 promoted by PdZn complex **16**.

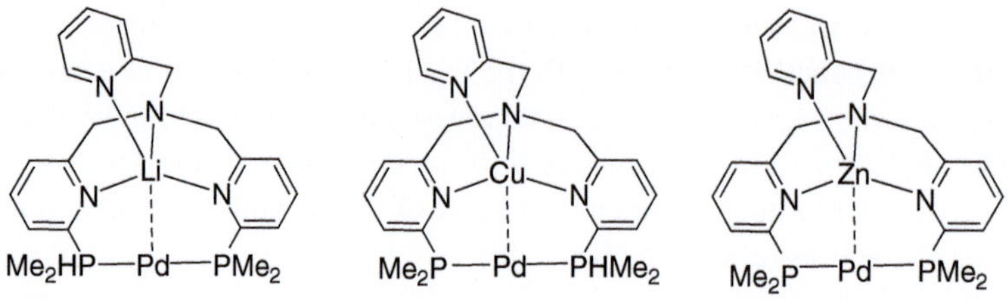

Orbital contributions:	Pd: 96.9%; Li: 1.4%	Pd: 95.3%; Cu: 2.3%	Pd: 89.0%; 9.0%
Bond order:	0.078	0.115	0.256
Stabilisation energy:	11.8 kcal/mol	27.8 kcal/mol	93.3 kcal/mol

Figure 4.16 Effect of changing supporting metal on a range of calculated parameters.

1.4% in their respective complexes. An associated decrease in calculated bond order was observed in the series, which correlated well with a decrease in NBO stabilisation energy. These results suggest that having a stronger acceptor ligand enhances the catalytic activity of the complex, although it is worth noting that the interaction of the complex with either CO_2 or silane was not explored directly [61].

4.3 Polymerization Catalysis

4.3.1 Olefin Polymerization

The annual production of plastics currently stands at almost 360 million tonnes per year, surpassing most other human-made materials. More than 60% of plastics manufactured worldwide are also made of polyolefins, and olefin polymerization continues to be a major research focus in industry [62] and academia [63], including towards monodispersed linear low-density polyethylene and polar polyolefins with high molecular weight and random functional group distribution [64, 65].

One of the most efficient methods to synthesize polyolefins is via the coordination insertion polymerization using transition metal catalysts. Such coordination insertion polymerization is initiated when an olefin inserts into a metal–alkyl/aryl bond to generate a new metal–alkyl species (Figure 4.17). Subsequent olefin binding and insertion then lead to polymer chain growth. Chain termination proceeds through chain transfer, which can involve either β-hydride elimination or β-hydride transfer.

Olefin polymerization is a field where early on heterobimetallic reagents rose to prominence. Indeed, in 1963, Ziegler and Natta were awarded the Nobel Prize in Chemistry 'for their discoveries in the field of the chemistry and technology of high polymers.' Heterogeneous Ziegler–Natta catalysts consist of combination of supported $TiCl_4$ and AlR_3 (R = alkyl, aryl, hydride) components and are today still employed to produce polyethylene and polypropylene [66]. Computational DFT studies [67, 68] showed that two different reaction paths exist for the insertion of ethylene in the titanium–carbon bond: one involving the bimetallic complex $H_2Al(\mu\text{-}Cl)_2TiCl_2CH_3$ and one involving the cationic species $Cl_2TiCH_3^+$, the two reaction paths being simultaneously available in the Ziegler–Natta polymerization reaction, and both competing towards the overall reaction rate. Other studies later showed that the most feasible polymerization mechanism occurs via an olefin-separated ion-pair.

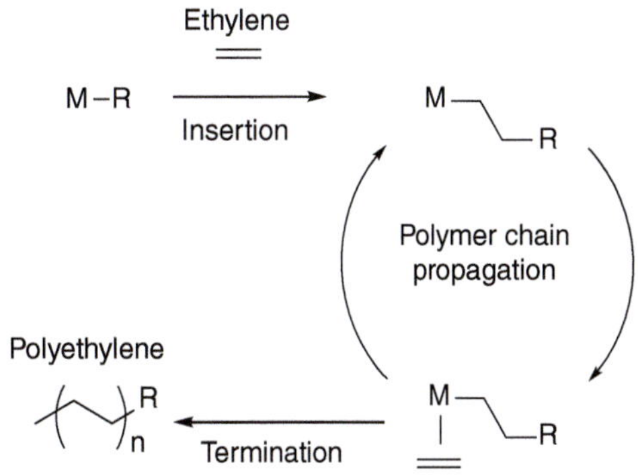

Figure 4.17 Simplified mechanism for the coordination insertion polymerization of ethylene.

Homogeneous versions of these Ziegler–Natta catalysts were also developed, based on metallocene group 4 complexes activated by alkyl aluminium reagents [69]. In their 1957 communication, Natta et al. declared: 'we have isolated a crystallizable compound containing titanium, aluminium and organometallic bonds, which causes the polymerization of ethylene.' Such catalyst species were studied later experimentally [70] as well as by computational methods [71], which revealed that the most feasible polymerization mechanism likely occurs via an olefin-separated ion pair (e.g. $Cp_2TiCH_3^+/C_2H_4/Al(CH_3)_2Cl_2^-$) rather than via a bimetallic complex. In 1976, another breakthrough came with the discovery of an efficient activator for homogeneous group 4 metallocene polymerization catalysts: a partially hydrolysed trimethyl aluminium reagent: methylaluminoxane (MAO) [72]. This breakthrough was followed by the development of isolable, structurally well-defined cocatalysts [73] and catalytically active 1:1 ion pairs [74] which are the actual catalysts. Detailed mechanistic studies of these systems, including computational [75, 76], have further emphasized the importance of the counter-anion in olefin polymerization catalysis [77], in which the strength and steric hindrance of the ion pairing can strongly influence catalytic activity, catalyst stability, chain-transfer processes, and product tacticity [78]. With these highly active new homogeneous systems, it became possible to fine-tune the product polymer polydispersity and microstructure, solely by modifying the organic ligands around the group 4 metal [79, 80].

The activity of group 4 metallocenes is, however, not suitable for some reactions such as the copolymerization of ethylene with butadiene, to produce ethylene-butadiene rubber for the tyre industry. A significant advance in this field was made in the development of lanthanide-based catalysts [81], in particular lanthanidocene complexes [82], which proved capable to efficiently homopolymerize both monomers. Like for group 4 metallocenes and derivatives, to activate the chloride or borohydride precatalysts, alkylation by an organo-magnesium, an organo-lithium, or an organo-aluminium compound is required. Dialkylmagnesium reagents are particularly attractive because they efficiently activate the catalysts but also act as a chain transfer agent (CTA), resulting in a process called chain transfer copolymerization (CCTP) of ethylene and butadiene, involving the formation of a heterobimetallic intermediate. Association/dissociation between dialkylmagnesium and neodymocene complexes is known to occur during polymerization [83, 84] and the mechanism of these reactions has also been explored computationally at the DFT level [85]. In agreement with experiments, it has been established that the active species is the neutral complex $(C_5Me_5)_2LnR$ but that chain transfer occurs between Nd and Mg via the formation of the aforementioned heterobimetallic intermediate.

Beyond such heterobimetallic intermediates generated *in situ* and catalytically active ion pairs involving different metals and main group elements, over the past two decades, the development of multimetallic catalysts has demonstrated significant cooperativity effects of positioning two polymerization-active metal centers in close proximity on catalytic activity, molecular weight, molecular weight distribution, and levels of branching. The reader is particularly directed towards some reviews that highlight the potential of homo- and heterobimetallic catalysts for olefin polymerization [86–88].

4.3.1.1 Metallocene-based Heterobimetallic Catalysts

Building on the success of metallocene chemistry for olefin polymerization, one axis of research became the development of heterobimetallic complexes based on metallocene-derived ligand architectures. A Rh center was for example linked to the *ansa*-metallocene ligand of a zirconocene unit, producing various heterobimetallic structures $(LRh(\eta^2\text{-}CH_2{=}CH)_2Si(\eta^5\text{-}C_5H_2\text{-}2,4\text{-}Me_2)_2ZrCl_2$ [L = indenyl (**17**; Figure 4.18), Cp or Cp*]) in which the late transition metal center was more than 6 Å away from the Zr atom [89, 90].

Nevertheless, this was shown to have a significant effect on catalysis: the Rh/Zr complexes exhibit higher activity for the polymerization of 1-hexene than the parent zirconocene or a mixture of the monometallic Zr and Rh compounds. Higher polymer molecular weights are also obtained with the bimetallic species. In a follow-up study, these types of Rh/Zr complexes were investigated as catalysts for the polymerization of different α-olefins and it was found that the effect of the Rh center on polymer yield and molecular weight was more pronounced in the order hexene > propylene > ethylene. The authors used cyclic voltammetry and NMR spectroscopy to conclude that Rh increases electron density at the zirconocene moiety, which is likely responsible for the improved catalytic performance of the Zr/Rh catalyst.

Another strategy has been the development of heterobimetallic dimers where one moiety is a metallocene complex. Thus, some phosphido- and arsenido-bridged Zr/M complexes $Cp_2Zr\{\mu\text{-}(Me_3Si)_2P\}_2M(CO)_n$ [M = Ni, $n = 2$; M = Mo, $n = 4$ (**18**; Figure 4.18)] and $Cp_2Zr\{\mu\text{-}(Me_3Si)_2As\}_2\text{-}Cr(CO)_4$ were studied [91, 92]. In the presence of MAO, these catalysts rival that of cationic zirconocenes but do not show any significant activity enhancement. Upon addition on MAO, the weak early transition metal–phosphide/arsenide bonds are broken and cationic methyl zirconocene complexes are formed. Although no metal–metal bond can be found in the solid-state structures of the heterobimetallic precatalysts, increased electron density at the Zr atom is seen in the ^{1}H NMR spectroscopic signals of the Cp ligands, due to the late transition metal M → Zr(d^0) donor interactions. This electronic effect imparted by the late-transition metal on the metallocene is likely to influence the formation of the cationic species: a weaker donor interaction with the oxophilic center resulting in a more reactive complex. Experimentally, the Zr/Mo catalyst is found to be substantially more active than either the Zr/Cr or Zr/Ni catalysts.

Roesky et al. exploited a similar strategy, developing the synthesis of various M–O–M′ systems bearing metallocene moieties [93, 94]. This strategy aimed at combining in a single molecular system a transition metal (which would be the active center) with a main group element (which would act as a co-catalyst), providing high catalytic activity with no co-catalyst needed or with just

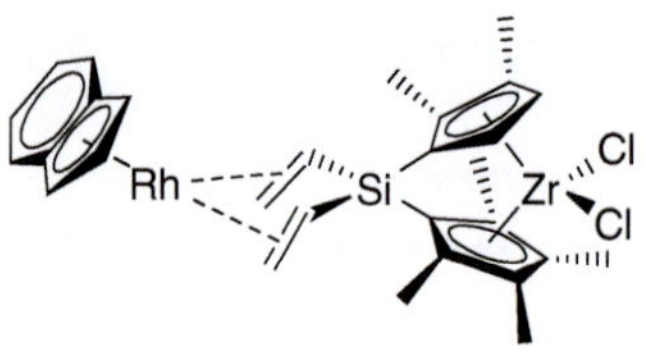

17

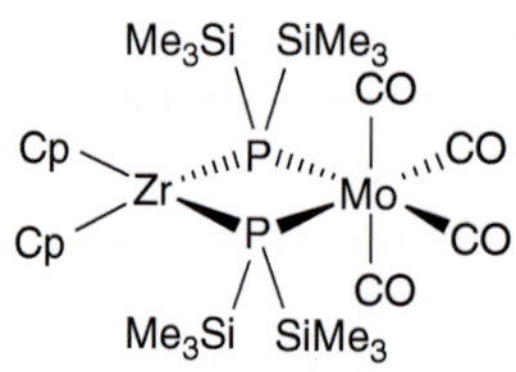

18

Figure 4.18 Selected metallocene-based heterobimetallic catalysts for olefin polymerization.

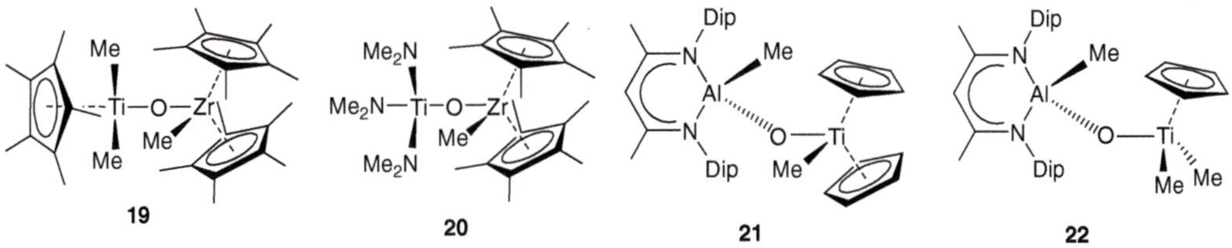

Figure 4.19 Structures of (half)-metallocene-based catalysts **19–22**.

a limited amount. In addition, the presence of a strong Lewis acid in close proximity to the transition metal would make the later more electrophilic, and activate the nucleophilic monomers more easily, resulting in higher catalytic activity.

In 2007, a covalently linked oxygen-bridged heterobimetallic complex of zirconium and titanium, **19**, was synthesized by using the precursors $Cp^*{}_2ZrMe(OH)$ and Cp^*TiMe_3 (Figure 4.19) [95].

Complex **19** still needs to be activated by MAO to polymerize ethylene, but displays high activity. NMR investigations suggested that upon reaction with MAO, a methyl group is transferred from the titanium to the aluminium, and a monocationic bridged complex of the type $[Cp^*{}_2ZrMe(\mu\text{-}O)TiMeCp^*]^+[(\mu\text{-}Me)MAO]^-$ is formed. Resonances for Zr–Me protons remain unaffected, indicating that the Zr—Me moiety does not react nor interact with MAO. In addition, this study indicated that the M—O–M′ unit remains intact, at least at low-MAO concentration. Catalysis is however usually carried out at much higher MAO concentrations so that there is no evidence that the M–O–M′ unit stays intact under true catalytic conditions. From single crystal X-ray diffraction data, it is apparent that M–O (M = Zr, Ti) bonds in **19** are shorter (2.022(4) and 1.816(4) Å, respectively) than in previously described heterobimetallic clusters, indicative of strongly polarized Zr–O and Ti–O bonds and decreased electron density at the metals, enhancing their Lewis acidic character, which may be responsible for the high catalytic activity observed (although no comparison was made with monometallic compounds, or a mixture of both). This hypothesis was corroborated by *ab initio* calculations. Natural bond orbital (NBO) analysis showed that there is a significant build up of electron density on the oxygen atom, with 90% of the electron density of the M–O–M′ unit located in a p-rich orbital of oxygen and leaving only 10% in d-orbitals of the metals. Furthermore, it was shown that that electron density on the Zr atom is more depleted compared with that of the Ti atom. This is somewhat in contradiction to what was seen by NMR spectroscopy, with the Me–Ti groups more prone to abstraction by MAO than the Me–Zr group. However, a steric argument can be made, with the steric bulk of Cp* ligands making the Zr center kinetically less active towards MAO activation compared with Ti, which has only one Cp* ligand. Once the methyl group was abstracted, the higher electron density on the titanium would also stabilize its cation more than a zirconium cation.

Another Ti/Zr heterobimetallic complex based on a zirconocene moiety was also reported by Roesky et al. The reactions of $Cp^*{}_2(Me)Zr(OH)$ with $Ti(NMe_2)_4$ lead to the formation of heterobimetallic compound $Cp^*{}_2(Me)Zr(\mu\text{-}O)Ti(NMe_2)_3$ (**20**) via concurrent elimination of Me_2NH [94]. This oxygen-bridged hybrid metallocene-(non-metallocene) complex exhibits interesting dual activity in olefin polymerization when activated with MAO. Although no enhancement of catalytic activity was found compared to mononuclear counterparts, $Cp^*{}_2ZrMe_2$ and $Ti(NMe_2)_4$ (the mixture of both was not tested), **20** is active in the polymerization of both ethylene and styrene. Control experiments carried out with $Cp^*{}_2ZrMe_2$ and $Ti(NMe_2)_4$ demonstrated that ethylene is polymerized mainly by the zirconium center, and styrene predominantly polymerized by the titanium center. DFT calculations revealed that a cation generated on titanium would be sterically more accessible for styrene monomer binding (based on LUMO shape), though it is energetically less favorable (energy needed 3.5 times higher) than a zirconium cationic species. It was concluded that at relatively lower MAO to catalyst ratio, the zirconium cation is accessible and takes control of ethylene polymerization, while at a relatively higher MAO to catalyst ratio, the titanium cation becomes accessible and enables styrene polymerization.

Roesky et al. also explored the preparation of oxygen-bridged heterobimetallic complexes based on group 4 metallocenes (Zr [96], Ti, Hf) [97] and on Al centers stabilized by β-diketiminate (BDI) ligands, towards the synthesis of highly active catalytic system that would require no or little activation by MAO. In fact, these catalysts still need to be activated by MAO, however, their benefit lies in the drastically reduced quantities of MAO needed for activity. The reaction of LAlMeOH (L = $CH\{N(Ar)(CMe)\}_2$, Ar = 2,6-i-$Pr_2C_6H_3$) with dimethylmetallocenes of Zr, Ti, and Hf results in the formation of LAl(Me)(μ-O)M(Me)Cp_2 (M = Zr; M = Ti (**21**, Figure 4.20); M = Hf). These complexes have been fully characterized and Figure 4.20 shows the solid-state structure of **21** obtained by single-crystal X-ray diffraction.

The zirconocene derivative exhibits high catalytic activity for ethylene polymerization with much lower MAO:catalyst ratios than those of the MAO:Cp_2ZrMe_2 catalytic system under similar conditions. Considering the lack of activity of parent LAlMeOH in polymerization, such a result clearly suggests the catalyst:MAO ratio can be influenced substantially by the structure of the zirconocene, and, in this case, shows cooperativity between the two metals. In the case of the titanium/aluminium heterometallic complex, which shows good catalytic activity in ethylene and styrene homo polymerization, *ab initio* calculations were carried out. HOMO and LUMO orbitals were shown to be located on different areas of the molecule (on the Ti and Al moieties, respectively), but it was the NBO analysis of the titanium cation formed during the activation by MAO which shed some light on the importance of the role of the Al center. Indeed, the bond formed between the Ti atom and the oxygen (p-rich orbital on oxygen and pure d-orbital on titanium) obtains a significant stabilization of 26 kcal/mol through a donor acceptor interaction between the Ti–O bonding orbital with an unoccupied antibonding Al–C(Me) orbital.

The same group could also prepare {Al(μ-O)M (M = Ti, Zr)}-containing half-metallocenes. In particular, the reaction of LAlMeOH with $CpTiMe_3$ yielded LAlMe(μ-O)$TiMe_2Cp$ **22**, which is stable at elevated temperatures, in contrast to its precursor $CpTiMe_3$, which readily decomposes

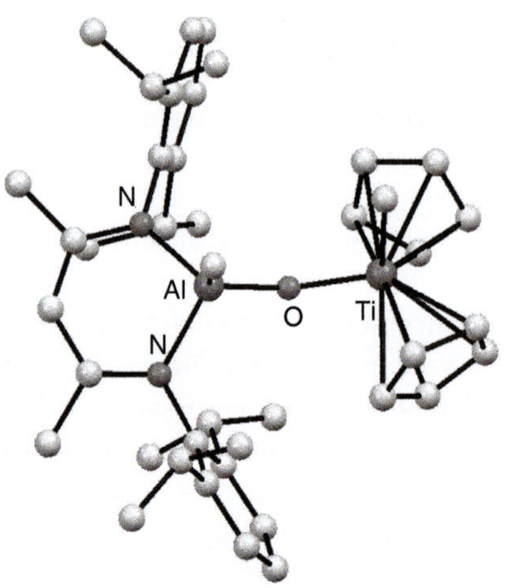

Figure 4.20 Solid-state molecular structure of **21**, obtained by X-ray diffraction. *Source:* Adapted from Gurubasavaraj et al. [97].

below room temperature [98]. However, under comparable polymerization conditions, this catalytic system shows lower activities compared to that of MAO-activated LAlMe(μ-O)Ti(Me)Cp_2.

Some mixed '^{Dip}Nacnac' heterobimetallic complexes, featuring a Ti(μ-O)AL core, were also prepared by Roesky et al., by reacting $LTiMe_3$ and LAlMe(OH) (L = HC{CMeN(2,6-*i*-$Pr_2C_6H_3$)}$_2$), which afforded intermediate $LTiMe_2$(μ-O)AlMeL in solution [99]. This heterometallic intermediate slowly degraded into titanium oxo complex LTiMe(O) and $LAlMe_2$, but could nevertheless be characterized by NMR spectroscopy, aided by NMR shifts values calculated from the DFT-optimized geometry. The Al–(μ-O) bond length (1.821 Å) is significantly longer than that in the μ-oxo-bridged titanium–aluminium complex Cp_2TiMe(μ-O)AlMeL (1.715(3) Å), while the Ti–(μ-O) distance (1.813 Å) is similar to that in Cp_2TiMe(μ-O)AlMeL (1.808(3) Å), reflecting a higher Ti–(μ-O) bond strength (1.01 valence unit in terms of bond valence theory), than that of Al–(μ-O) (0.58 valence unit). Taken collectively with the various Ti/Al complexes described previously, these results show clearly that the supporting ligands strongly influence the stability of the μ-oxo-bridged Ti(IV)–Al(III) complexes.

In comparison with the cyclopentadienyl ligand, the isoelectronic *N*-heterocyclic pyrrolyl ligand has received much less attention for olefin polymerization catalysts, including because of the tendency of electrophilic transition metals to be σ-bonded to the nitrogen atom rather than be π-bonded in a η^5-fashion. The strategy employed by Duchateau et al. involved using AlX_3 (X = halogen, alkyl), a strong Lewis acid, to lock the nitrogen atom through σ-bonding, so that π-bonding with titanium could occur, creating in the process a Ti/Al heterobimetallic complex [100]. Depending on stoichiometry, mono- and bis(aluminium-pyrrolyl) complexes can be synthesized. However, ethylene polymerization experiments of bis(aluminium-pyrrolyl) group 4 complexes with either MAO or $[CPh_3]^+[B(C_6F_5)_4]^-$ under various conditions (temperatures and ethylene pressures) were disappointing. This lack of activity has been rationalized by the fact that borane initiator can be seen by NMR spectroscopy to abstract an aluminium methyl rather than a titanium methyl group, but DFT calculations have shown that subsequently a stable coordinatively saturated species, featuring a chloride bridging between the aluminium and the titanium center, is formed after cationization and prevents ethylene coordination and, therefore, impedes catalysis (Figure 4.21).

To overcome the poor-catalytic activity of the 16-electron bis(aluminium-pyrrolyl) complexes, the more electron-deficient mono(aluminium-pyrrolyl) complexes were targeted. The reaction of a toluene solution of 2,5-dimethylpyrrole (Pyr) with $AlMe_3$, immediately followed by addition of $TiCl_4$ (Ti:Al:Pyr = 1:1:1) at room temperature, produces the desired complex (η^5-2,5-$Me_2C_4H_2NAlCl_2Me$)$TiCl_2Me$ (**23**), which has been characterized by X-ray crystallography (Figure 4.22). The analogues 2,3-dimethylindolyl (η^5-2,3-$Me_2C_8H_4AlCl_2Me$)$TiCl_2Me$ and tetrahydrocarbazolyl (η^5-3,4,5,6-$C_{12}H_{12}NAlCl_2Me$)$TiClMe_2$ complexes have also been synthesized and their solid-state structures obtained by X-ray crystallography.

The potential of these complexes for catalytic ethylene polymerization upon activation with MAO was investigated under different conditions. These catalysts, although more active than previously, were only moderately active and required high-ethylene pressures. Some insight has been obtained into the polymerization mechanism of these heterobimetallic olefin polymerization catalysts by reacting **23** with 1 equiv. of $B(C_6F_5)_3$. ^{19}F NMR spectroscopy confirms the formation of an ion pair, and the ^{1}H NMR spectrum clearly shows that the borane selectively abstracted an aluminium-methyl group rather than a titanium-bonded methyl group, resulting in the species $[(2,5\text{-}Me_2C_4H_2NAlCl_2)TiCl_2Me]^+[MeB(C_6F_5)_3]^-$. DFT calculations further showed that a constrained geometry with a chloride atom bridging the titanium and aluminium centers is the most stable configuration for the cationic species. However, to become an active ethylene

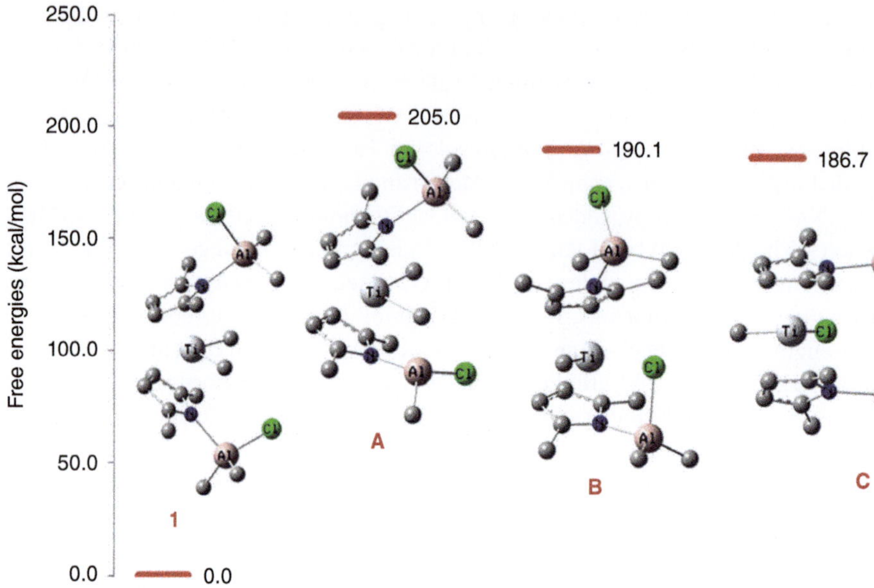

Figure 4.21 Free energies and DFT-optimized structures of (η^5-2,5-$Me_2C_4H_2NAlClMe_2)_2TiMe_2$ (1), and possible cationic species resulting from the abstraction of an aluminium methyl by MAO (A–C), showing the favorable formation of species C, featuring a chloride bridging between the aluminium and the titanium center. Hydrogen atoms in the structures were omitted for clarity. *Source:* Kulangara et al. [100]; © 2015 American Chemical Society.

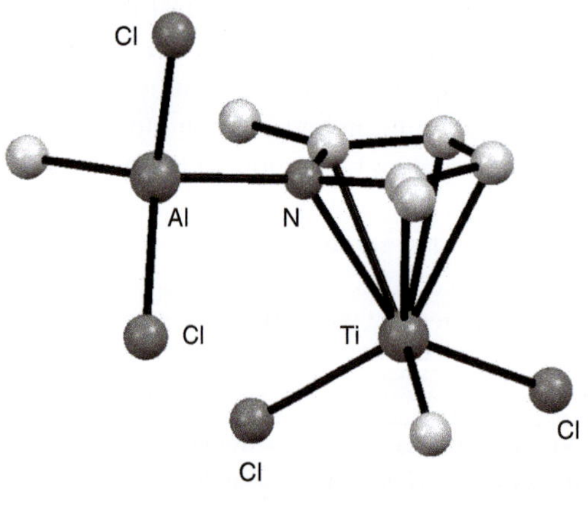

Figure 4.22 Solid-state molecular structure of **23**, obtained by X-ray diffraction analysis. *Source:* Adapted from Kulangara et al. [100].

polymerization catalyst, either release of the bridged chloride or ring slippage from η^5- to μ-bonding between aluminium and titanium of the pyrrole moiety, is thought to be needed to create a system sufficiently unsaturated to coordinate and polymerize ethylene. Calculations showed that to deco-ordinate the bridging chloride requires 12.2 kcal/mol, supporting the need for high ethylene pressure. Attempts at modelling the ring slippage from η^5- to μ-bonding of the aluminium-pyrrolyl ligand, as seen for chromium complexes [101], proved unsuccessful. It is however possible that such mechanism is at play to enable ethylene coordination to Ti. Figure 4.23 depicts the proposed mechanism based on experimental and computational work, for the activation of **23** into an active ethylene catalyst.

Some related mono(aluminium-pyrrolyl) zirconium complexes were synthesized in moderate yield by the sequential addition of $ZrCl_4$ to a freshly prepared solution of 2,5-dimethylpyrrole

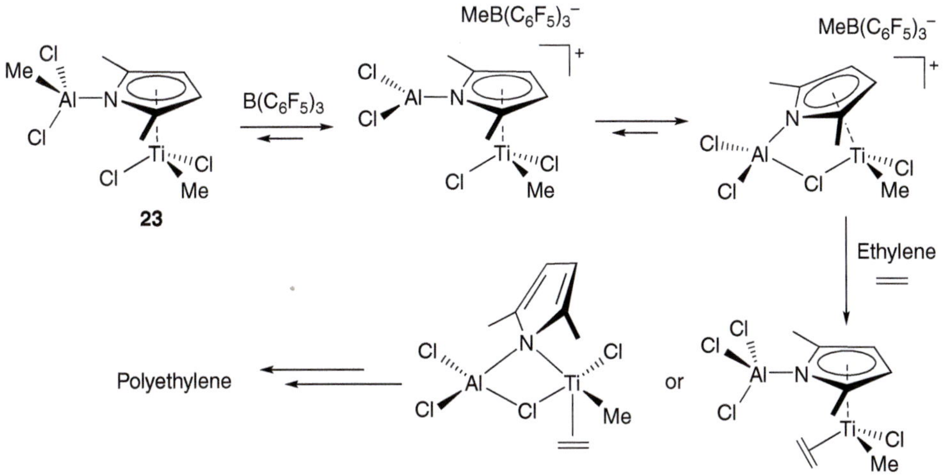

Figure 4.23 Proposed mechanism, based on experimental and computational work, for the activation of **23** into an active ethylene catalyst.

and various alkyl aluminium ($AlMe_3$ or $AlEt_2Cl$), resulting in species with bridging chlorides between aluminium and zirconium. These species are completely inactive in catalysis, even at high-ethylene pressure, and this observation supports further the negative effect that the intermolecular bridging coordination of chloride atoms have on the formation of an electron-deficient active species.

4.3.1.2 Constrained Geometries Heterobimetallic Catalysts

Marks and coworkers have developed over several decades constrained geometry catalysts (abbreviated to as CGCs in the literature) that rival metallocenes for polymerization. In particular, they can produce branched polyethylene with high productivity and selectivity. In addition, the coordinatively open nature of their active site allows for the rapid enchainment of sterically encumbered olefin co-monomers into the polyethylene backbone. These catalysts have been reviewed several times in the literature [102, 103] as well as studied computationally [104, 105].

Li and Marks highlighted early on the potential of homobimetallic group 4 constrained geometry catalysts in polymerization catalysis [86]. Significant effects (not all beneficial) of di-nuclear catalysts were found versus mononuclear controls, including on rates, and the effect can be correlated with metal–metal approach distances and ion pairing effects. In particular, one DFT study examined the metal–metal proximity effects on ethylene polymerization processes mediated by (μ-CH_2-3,3′){(η^5-indenyl)[1-H_2Si(*t*-BuN)](ZrMe_2)}$_2$ catalysts to try to understand the cooperative effect observed [106]. Although this is not related to heterobimetallic complexes pertinent to this book chapter, some conclusions drawn were informative for the development of improved heterobimetallic catalysts later on, so it is worth briefly discussing here.

Binuclear catalytic systems are distinct from mononuclear analogues in that they introduce ethyl branching, which is proposed to occur via conventional monometallic macromonomer elimination, via a chain transfer process, followed by 1,2-intermolecular reinsertion at the ethyl cation produced by chain transfer. Modelling of the propagation step of the polymerization further supports such behaviour by revealing the presence of agostic interactions between the growing chain interacting at one metal site while propagating from the second metal site, required for the proposed ethyl-branching mechanism. In this scenario, the dimensions of the oligomeric chain that would best favor the agostic interactions are directly dependent on the intermetallic distance in the pre-catalyst. Calculations also revealed that the proximity of the second metal center only slightly influences the insertion process kinetics (hence, the propagation rate), but significantly decreases the energy needed for chain transfer (hence the termination rate), in agreement with experiments with binuclear catalyst, for which increased polymer molecular weights are observed.

In 2004, Marks et al. reacted a ligand featuring CGC moieties ((μ-CH_2CH_2-3,3′)[1-(Me_2SiNt-Bu)-indenyl]$_2$; L), first with 1 equiv. of $Zr(NMe_2)_4$, then with 1 equiv. of $Ti(NMe_2)_4$, to obtain a heterobimetallic amido complex $LTiZr(NMe_2)_4$. Reaction of this amido complex with excess $AlMe_3$ afforded tetramethyl heterobimetallic Ti/Zr catalyst **24** (Scheme 4.9) [107].

24 does not display any improved catalytic activity compared to the mixture of individual Zr and Ti complexes. However, under identical conditions (64 °C, 1 min, 1 atm ethylene) and upon activation with a bisborate activator, it produces exclusively a single distribution of long-chain branched (number of carbons ≥ 6) polyethylene, in marked contrast to control experiments involving a mixture of the related Zr and Ti catalysts, which produces a bimodal distribution of high (but lower than with **24**) and low molecular weight polyolefins, with negligible branching.

In principle, mixtures of homogeneous oligomerisation and polymerization catalysts (tandem catalysts) can be used to produce such oligomeric branched polyethylene (linear low density polyethylene) using ethylene as the sole monomer, via the intermolecular coupling of intermolecular elimination (done by the oligomerization catalyst), transfer then with re-enchainment (done by the polymerization catalyst). However, this cooperation between metals is typically challenged by the low probability that the product of one catalytic center can be efficiently captured by the other

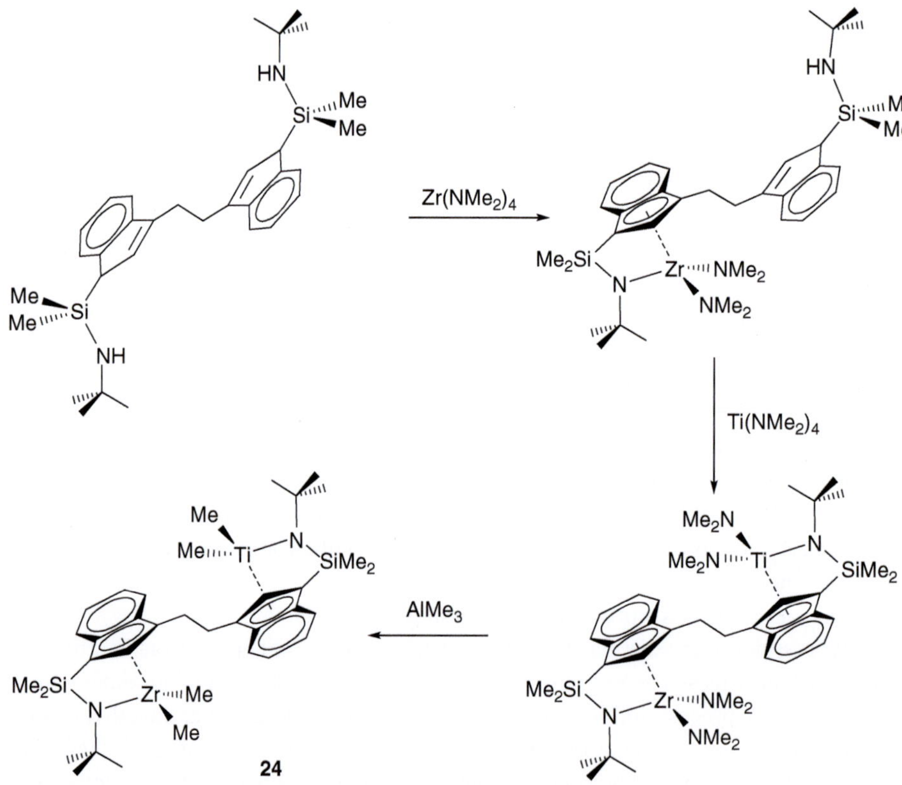

24

Scheme 4.9 Synthesis of the Ti/Zr heterobimetallic catalyst **24**.

in dilute solutions. Heterobimetallic catalysts, by reducing the distance between metals with distinct catalytic characteristics, would increase the probability of such events. The result of Marks and coworkers supports this hypothesis and demonstrates cooperativity (even if modest; there were only ~2 branches (($\geq$ C6) per 1000 C atoms) between the two metals, brought about by spatial proximity of Zr and Ti catalytic sites, towards a significant increased efficiency of intramolecular oligomer capture, transfer then enchainment, resulting in branch formation as represented in Figure 4.24. However, this has not been examined by computational modelling.

This concept was taken further by Marks et al. with the synthesis of a heterobimetallic Ti/Cr catalyst **25** based on a functionalised indene ligand, combining the polymerization characteristic of CGCTi-type catalysts, with a selective ethylene trimerisation Cr catalyst, based on an SNS tridentate ligand [108]. The catalyst is synthesized stepwise (Scheme 4.10). First, the monometallic amido complex CGCTi$(NMe_2)_2$–SNS is prepared by reaction of $Ti(NMe_2)_4$ with H_2CGC–SNS, followed by the addition of excess Me_3SiCl, which gives CGCTiCl_2–SNS, for which a solid state molecular structure was obtained by X-ray crystallography. Then, reaction with $CrCl_3(THF)_3$ afforded the heterobimetallic Ti/Cr complex, which was characterized by elemental analysis, ^{1}H NMR spectroscopy (strong paramagnetism was observed), and MALDI–ToF mass spectrometry.

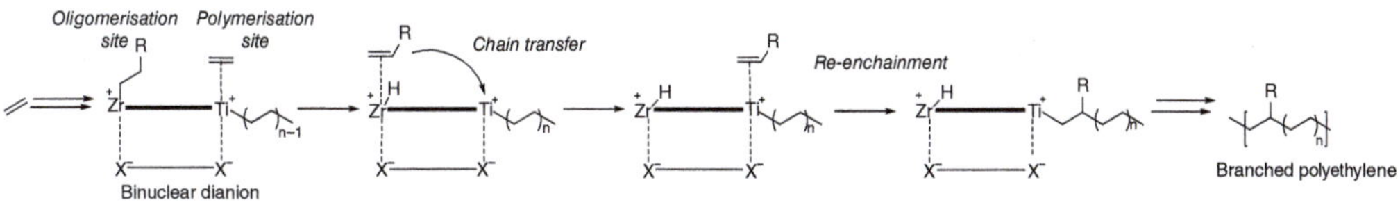

Figure 4.24 Proposed scenario for enhanced polyolefin chain branching mediated by heterobimetallic catalyst **24**.

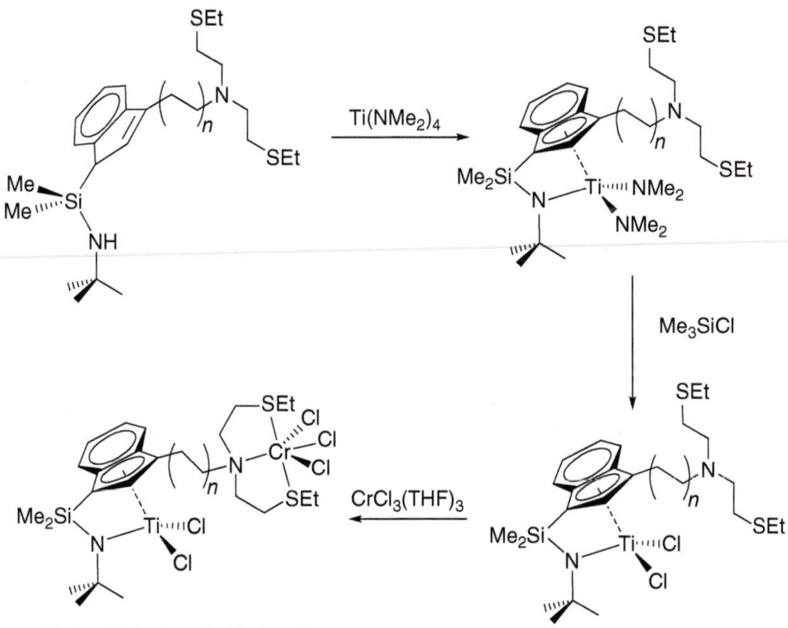

25 (n = 1), **26** (n = 0), **27** (n = 3)

Scheme 4.10 Synthesis of the heterobimetallic catalysts **25–27**.

Other related catalysts, with different lengths of the $C_{2n}H_{4n}$-bridge between the Ti and Cr component of the complex ($n = 0, 3$: **26** and **27**, respectively), have also been developed, to explore the effects of varying the metal–metal distance on the ethylene polymerization [109]. Upon activation with MAO, all these heterobimetallic complexes show lower activities than the tandem homonuclear CGC^{Et}Ti and SNSCr system for polyethylene production (480 kg(PE)/mol(M)/h/atm), and even less than with mononuclear Ti catalyst (975 kg(PE)/mol/(M)/h/atm). Under identical conditions, **26** (no aliphatic bridge) produces polyethylene with the highest activity (123.0 kg(PE)/mol(M)/h/atm), compared to 27.6 kg(PE)/mol/(M)/h/atm for **25** (C_2H_4 bridge) and 20.1 kg(PE)/mol/(M)/h/atm for **27** (C_6H_{12} bridge).

However, all produce linear low-density polyethylene as intended, with exclusively *n*-butyl branches (6.8–25.8 branches/1000 Cs), in a significant improvement, both in terms of frequency and homogeneity of branching, compared to the previous Ti/Zr complex. **26** (no aliphatic bridge) provides polymers with higher density of butyl branches (25.8 branches per 1000 carbons) compared to that by **25** (C_2H_4 bridge; 18.2 branches per 1000 carbons) and **27** (C_6H_{12} bridge; 6.8 branches per 1000 carbons), the latter being very similar to the tandem mixture of Ti and Cr complexes (6.4 branches per 1000 carbons) (Figure 4.25).

It is worth noting that the cooperation between the two metals is not total, and experimentally, some oligomers are still produced by the Cr center, in either the tandem or bimetallic configurations, and display good selectivity for ethylene trimerization. The Cr center is also responsible for the production of small amounts of (~9%) of high M_n polyethylene, but with no branching. Another interesting cooperative effect is seen in that the branching density of the polymer obtained using heterobimetallic **25** (C_2H_4 bridge) is independent of polymerization time, whereas the tandem catalysis provided polymers with greater branching density at longer reaction times.

Collectively, these results strongly suggest that close spatial proximity of Ti and Cr enables synergistic interactions between the metal centers towards *n*-butyl branched PE, and that proximity of

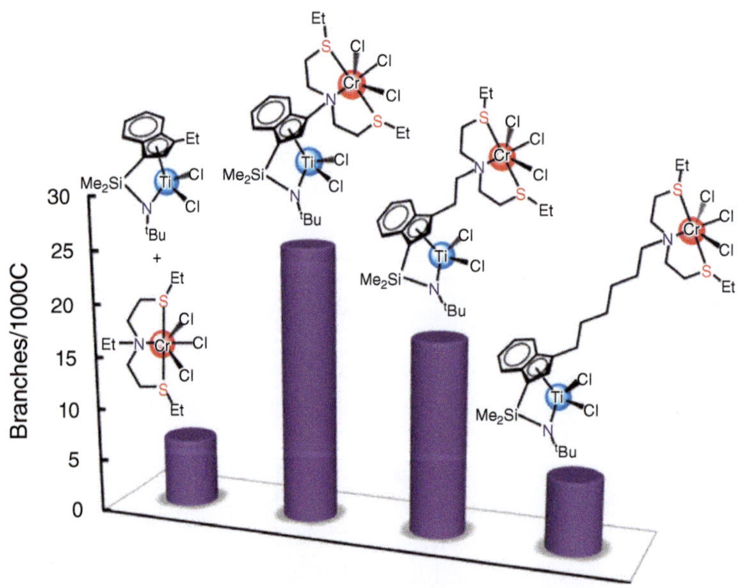

Figure 4.25 Branches/1000 C in the polyethylene produced by the mixture of mononuclear Ti and Cr complexes, **25**, **26**, and **27** catalysts, under identical reaction conditions at 80 °C. *Source:* Liu et al. [109]; © 2014 American Chemical Society.

the catalytic centers alters the propagation and chain-transfer characteristics of the heterobimetallic catalyst: having short Ti–Cr bond distances improved the shuttling of the 1-hexene monomers generated from chromium to titanium. Based on DFT calculations, the Ti...Cr distance is determined as 6.0 Å for **26**, 8.1 Å for **25**, and 13.2 Å for **27.**

DFT calculations were also performed to compare the energies of the propagation and termination pathways for mononuclear $CGC^{Et}Ti$ and binuclear **26** (no bridge) (Figure 4.26). This analysis revealed that globally, the Cr catalyst proximity in **26** stabilizes both propagation and transition compared to the mononuclear Ti complex, but does more so for the propagation transition state, so that the difference between propagation and termination is greater for the heterobimetallic complex, in agreement with the increased polyethylene M_ns achieved with the bimetallic catalysts.

To obtain better insight into the α-olefin coenchainment mechanism, which produces branching, calculations also focussed on the catalytic differences between **25** and **26**. C_6 fragments are known to be produced at the Cr center by reductive ethylene coupling and metallacyclopentane expansion to form a metallacycloheptane, followed by reductive elimination, yielding 1-hexene. Structures of intermediates in which 1-hexene molecules are coordinated to the Cr in heterobimetallic **25** and **26** have been optimized by DFT. While a π-interaction occurs between the C=C bond and the Cr, the presence of an agostic C(sp^3)–H interaction between the Ti center and the 1-hexene aliphatic chain is apparent, as evidenced by spatial conformation and the elongation of one C(5)–H bond (1.13 Å) in **26** and one C(6)–H_3 bond (1.12 Å) in **25**, compared to a standard sp^3 C–H bond distance (1.10 Å). A shorter Ti...H distance in **26** (2.11 Å) versus **25** (2.17 Å) also reflected the

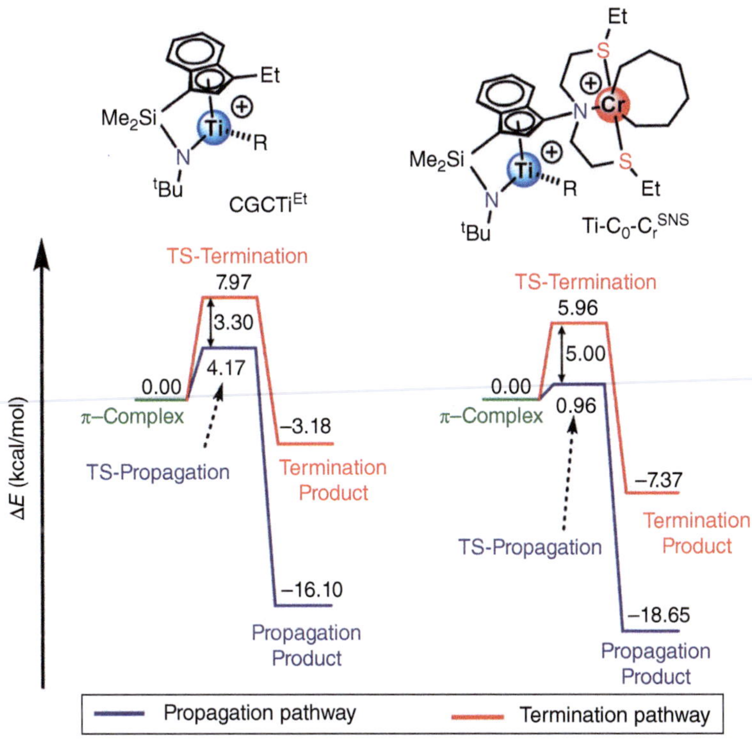

Figure 4.26 Energetic profiles (kcal/mol) for propagation (blue) and termination (chain transfer; red) pathways for ethylene homo polymerization at Ti catalysed by mononuclear and binuclear catalysts. *Source:* Liu et al. [109]; © 2014 American Chemical Society.

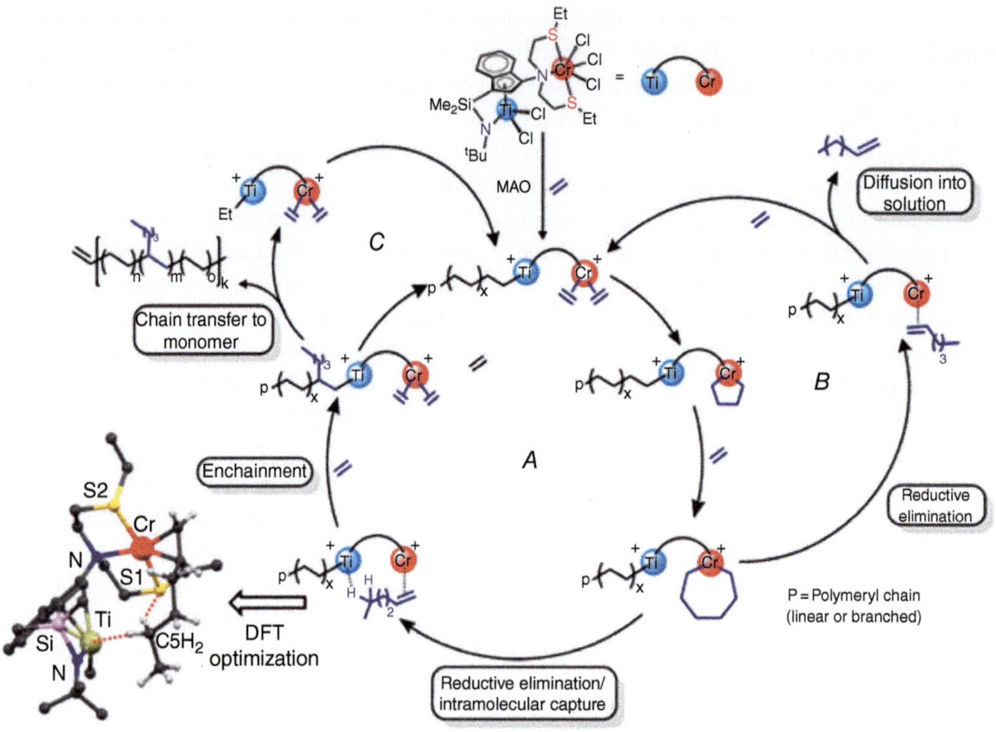

Figure 4.27 Proposed mechanism for 1-hexene generation and subsequent copolymerization with ethylene mediated by heterobimetallic catalyst **26**. *Source:* Liu et al. [109]; © 2014 American Chemical Society.

strength of these agostic interactions. The fact that the agostic interaction involves in one case the 1-hexene C5 atom and in the other case the 1-hexene C6 atom mirrors the Ti...Cr distance calculated as well as the length of the ligand bridge. Additional interactions have also been detected by NBO analysis, with the electron donation of a S lone pair to an empty Ti orbital, stabilising and constraining further the intermetallic distance between the two metal centers, and keeping 1-hexene close to the Ti center, which likely facilitates the enchainment of 1-hexene into the polyethylene chain, even more so in **26**, as seen experimentally. Based on these experimental and computational observations, a plausible pathway for the catalytic polymerization of ethylene by **26** was proposed (Figure 4.27).

4.3.1.3 Late Transition Metal Heterobimetallic Catalysts

Another breakthrough in the field of olefin polymerization came with the discovery of homogeneous late transition metal catalysts with similar activities to early transition metal catalysts [110]. These catalyst systems are capable of introducing over 50 branches/1000 carbon atoms in the polyolefin chain, yielding homopolymers with lower melting points than those produced with their group 4 counterparts, although typically at much lower molecular weights and activities. In addition, late transition metal catalysts (e.g. based on Ni and Pd) were shown to be far less susceptible to inhibition by heteroatoms, and demonstrated greater tolerance to polar monomers, solvents, and impurities [111, 112]. In particular, this ability enables the metal catalysed copolymerization of ethylene and polar vinyl monomers, even if the resulting polymers tend to have low molecular weight and achieving high-catalytic activities remains a challenge.

For this type of catalyst, a nuclearity effect was also demonstrated, and homobimetallic catalysts with enhanced catalytic activities compared to their mononuclear counterparts were developed.

The formation of heterobimetallic complexes combining late transition metals and electron-poor metals has also been explored, to adjust the activity and reactivity of late transition metal catalysts.

One example of these early-late transition heterobinuclear olefin polymerization catalysts was synthesized by cross-metathesis of an *ansa*-zirconocene having an allyl substituent with Co, Ni, and Pd (imino)pyridyine complexes having acrylate substituents [113]. Upon activation by MAO, the mononuclear *ansa*-zirconocene polymerizes ethylene into linear polyethylene of average molecular weight and high dispersity. In contrast, both the tandem mixture of zirconocene and iminopyridine Co(II) complex, and the covalently linked heterobinuclear catalyst **28** (Figure 4.28) produces polyethylene with higher molecular weight, narrower distribution, and with selective ethyl branching (4.6 and 5.6 branches/1000 carbon atoms, respectively). The mononuclear iminopyridine Co(II) complex only produces a mixture of butenes from ethylene, which demonstrates a cooperative effect between the Zr and the Co center. Under similar conditions, compared to the tandem mixture, the heterobimetallic complex displays almost twice the catalytic activity and produces polymer with slightly more branching (5.6 versus 4.6 branches per 1000 carbon atoms), higher M_n (46 900 versus 40 000), supporting the fact that the transfer and enchainment of butene in the PE chain growing from the Zr center is enhanced by the proximity of the Zr and Co active sites, similar to what is seen in the Marks Ti/Cr system. Similar results are observed for the Zr/Ni catalyst, albeit with more heterogeneity in the branching, consistent with the known formation of branched α-olefins produced by the Ni center alone. As expected, the monometallic Pd complex being inactive for oligomerization and polymerization of ethylene, the Zr/Pd catalyst only produces linear polyethylene, likely through the sole action of the Zr center.

Another example is that of Nagashima et al., who have reported azanickellacyclic complexes featuring a free di-imino moiety capable of coordinating $ZnBr_2$, $CoBr_2$, $FeBr_2$, or $NiBr_2$ [114]. These heterobimetallic complexes were tested for ethylene polymerization, and show only moderate activity enhancement compared to the mononuclear catalyst (1.7-, 1.2-, 2.7-, and 3.6-fold increase, for Ni/Zn, Ni/Co, Ni/Fe, and Ni/Ni, respectively). No significant differences could be detected in the properties of the polymer produced, albeit there was a slight reduction in branching. However, Ni/Fe and Ni/Ni catalysts yield bimodal polymer distributions, suggesting independent polymer formation and not true cooperation between the metal centers, leading to real activity enhancements for these catalysts of only 1.4 and 1.8, respectively. The activity enhancement seen is attributed to the additional rigidity of the ligand framework brought about coordination of a second

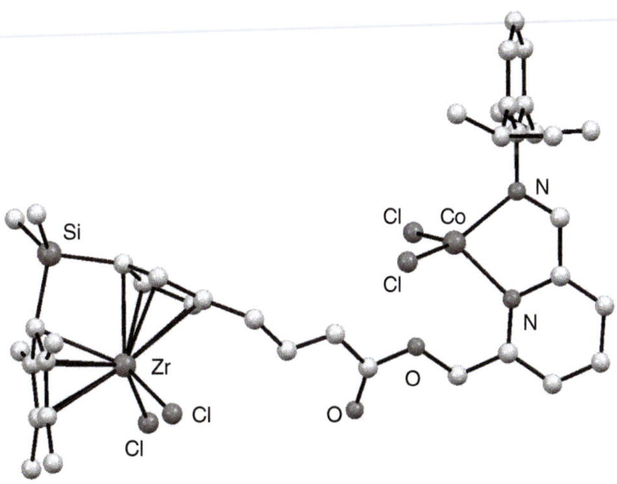

Figure 4.28 Solid-state molecular structure of **28** obtained by X-ray diffraction analysis. *Source:* Adapted from Kuwabara et al. [113].

metal. Do et al. have also demonstrated that 1,2,3-triazole-4-carboxamidate ligands could coordinate to nickel(II) centers, producing ethylene homo polymerization catalysts able to produce low molecular weight polyethylenes with about 80–130 branches per 1000 carbon atoms [115]. In addition, this ligand framework was shown to enable the binding of zinc ions at the basic β-nitrogen of the triazole ring of the ligand, even if no solid-state molecular structure of a heterobimetallic species was obtained. However, the addition of zinc salts to the nickel complexes leads mostly to catalyst inhibition, supporting a cooperative, but negative effect between the two metals. The team subsequently found that the introduction of a picolyl donor group to the ligand framework enhances the ability to coordinate zinc ions compared to the first-generation catalysts [116]. NMR spectroscopy studies of metal binding confirmed that reaction with $ZnCl_2$ leads to the formation in solution of nickel–zinc heterometallic complexes, even if the exact structure of the resulting nickel–zinc adduct could not be established. Using other zinc salts such as $ZnBr_2$, ZnI_2, or $Zn(OSO_2CF_3)_2$ has little effect on catalytic activity, despite NMR spectroscopy showing similar binding to the triazole moiety than $ZnCl_2$. This result suggests that the chloride anions in $ZnCl_2$ are functionally important in the nickel–zinc catalyst structures, although this remains to be investigated in detail.

Tonks et al. also assembled several nickel–zinc heterobimetallic catalysts for olefin polymerization, taking advantage of various ligand frameworks. First, Ni(II) phenoxyimine complexes were prepared, bearing pendant bipyridyl donors which could coordinate to $ZnCl_2$ (slow diffusion of Zn ions at low temperature was necessary rather than direct mixing), resulting in heterobimetallic Ni/Zn complexes (e.g. complex **29**, Figure 4.29) [117].

While the parent mononuclear nickel complex is completely inactive for ethylene polymerization, the isolated heterobimetallic complex **29** is active for polymerization, although slow. When generated *in situ*, similar catalytic activities, molecular weight, and melting point of the resulting polymers are observed. Based on the solid-state structure obtained for **29**, it is hypothesized that placing a bulky Zn–Cl moiety directly over the apical binding site of Ni slows down β–H transfer/termination processes and enables the synthesis of higher molecular weight polymer rather than oligomers. Polymers produced by those catalysts are mostly linear with varying amounts of methyl branching, which does not strongly support a chain transfer mechanism between the Ni and the

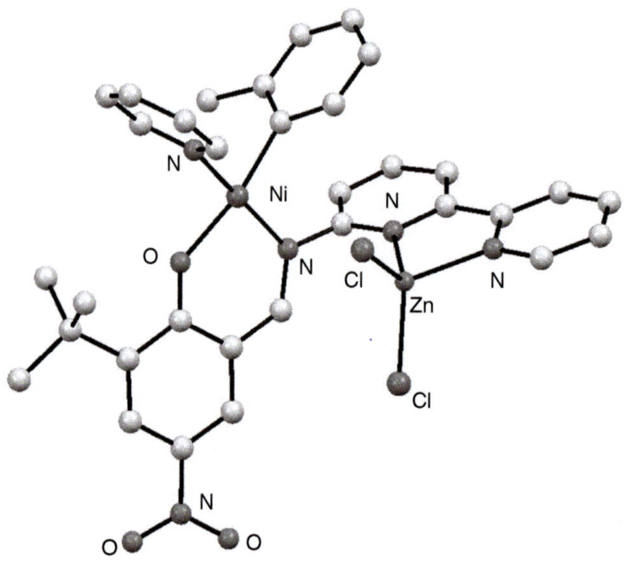

Figure 4.29 Solid-state molecular structure of **29** obtained by X-ray diffraction analysis. *Source:* Adapted from Smith et al. [117].

Zn. In addition, the heterobimetallic catalysts give polymers with broad dispersities (Đ 4–11), which supports the formation during the reaction of at least two different catalytically active nickel species, likely through dissociation of the Zn ion. It is worth noting that addition of other Lewis acids such as $CuCl_2$ or $AlCl_3$ does not result in polymerization.

The team expanded their research on the topic by developing similar ligand scaffolds, including β-oxo-δ-diimine (BODDI) ligands [118]. Depending on the method of synthesis and the functional groups on Ni, both enamine and imine tautomers of the ligand can be isolated and a large impact on ethylene polymerization catalysis has been demonstrated, despite affecting the second sphere of coordination of the nickel center. Multiple X-ray crystallographic structures show that the enamine tautomer conjugates with the Ni chelate ring and increases electron density on Ni to yield low-molecular weight polymers, while the imine tautomer arm rotates out of plane and has minimal influence on polymerization, showing catalytic characteristics similar to the single-armed β-ketoiminate analogue complex.

Upon deprotonation of the second-binding pocket with alkali bases (M(HMDS); M = Li, Na, K), Ni-alkali metal heterobimetallic complexes were made and characterized, including by X-ray crystallography. Solid-state structures reveal that for Li and Na, the second ligand arm is best described as an anionic enamide conjugated to the Ni *N*,*O*-chelate. However, for the K complex, a dimeric, bridged structure is obtained, in which the second arm is rotated away from the Ni first coordination sphere. This turns out to be a crucial observation to explain what is observed in catalysis. Indeed, heterobimetallic catalysts behave more similarly to imine-type than enamine. It is thought that upon activation and ethylene insertion into the Ni–tolyl bond, the cation–π interaction to the alkali metal is lost, resulting in the enamide arm rotating to minimize interaction with the Ni coordination sphere (as observed in the solid-state structure of the Ni/K), and ultimately resulting in catalytic behaviour similar to imine complexes.

In 2017, Tonks et al. combined these two sets of results by investigating 2,6-bis(imino)phenoxide ligands, in which the phenoxide group can bridge two metals (akin to BODDI ligands), but the rigidity of which would keep the two metal centers in close contact [119]. Protonolysis of $Ni(C_4H_7)_2$ by the NON tridentate ligand yields mononuclear complex $(NON)Ni(C_4H_7)$, which affords heterobimetallic complex, $(NON)Ni(C_4H_7)ZnBr_2$ **30**, after treatment with $ZnBr_2$. Interestingly, the metalation by $ZnBr_2$ appears to be a slow, two-step process, proceeding via an imine-bound $ZnBr_2$ species that requires extended reaction time, heating, or vacuum, for the metallacycle to close into the bridged bimetallic **30** (Scheme 4.11).

In catalysis, the mononuclear Ni(methallyl) and related heterobimetallic complex **30** are inactive for polymerization of ethylene, presumably due to slow ethylene insertion and the known stability of the methallyl unit. Only precatalysts with Ni–(*o*-tolyl) groups generate polyethylene without any activation. At room temperature, upon *in situ* reaction of mononuclear Ni–(*o*-tolyl)

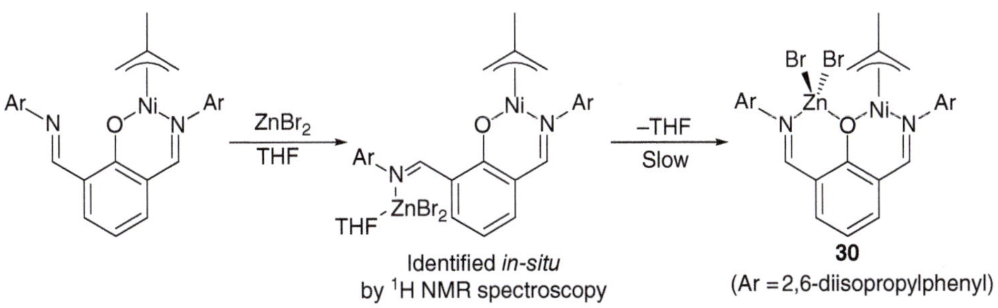

Scheme 4.11 Metalation of mononuclear complex $(NON)Ni(C_4H_7)$ by $ZnBr_2$, to give Ni/Zn heterobimetallic **30**.

complex with $ZnBr_2$, followed by the addition of ethylene, increase in catalytic activity (almost 4 times) and a polyethylene bimodal M_n distribution are observed: one mode with molecular weights similar to those from polymerizations without Zn, and one with much lower molecular weights. This indicates the presence of two active species during polymerization. However, when carried out at 50 °C, catalytic activity is approximately halved and this reaction only yields the lower M_n polymer. It is, therefore, proposed that both active species observed during room temperature polymerization are heterobimetallic: the first containing a Zn atom bound to a free imine arm of the ligand, rotated away from the Ni metallacycle, while the second species features the phenoxide bridging between Ni and Zn (Figure 4.30). DFT calculations on the bimetallic 14-electron Ni alkyl propagating species were carried out to shed some light on this hypothesis. Consistent with experimental results, the bridged bimetallic species has almost no energetic barrier for β–H elimination, which would lead to lower molecular weight polymer than the monometallic analogue, via more frequent chain termination events. This result is to contrast with the previous Ni/Zn system, where a pendent bipyridine-bound $ZnCl_2$ inhibits β–H elimination/transfer [117].

Apart from Zn ions, alkali metals have also been used to tune the reactivity of nickel catalysts for olefin polymerization. In 2015, Do and coworkers synthesized a new family of nickel phenoxy-imine–polyethylene glycol complexes that form discrete molecular species with alkali metal ions ($M^+ = Li^+, Na^+, K^+$) [120]. Alkali metals would act as an activator and binding site for polar functionalities, but without competing with the nickel site for growing the polymer chains. These complexes were fully characterized, including by X-ray crystallography, providing evidence of alkali metal ligation. Depending of the stoichiometry used, 1:1 (e.g. complex **31**, for M = Na, Figure 4.31) and 2:1 Ni/M^+ species can be formed in solution. Upon the activation by a phosphine scavenger ($Ni(COD)_2$), the mononuclear Ni complexes are active for ethylene polymerization, but the addition of alkali metals changes the properties of the polymers produced. Mononuclear nickel species yield slightly branched semicrystalline polyethylene, whereas heterobimetallic structures

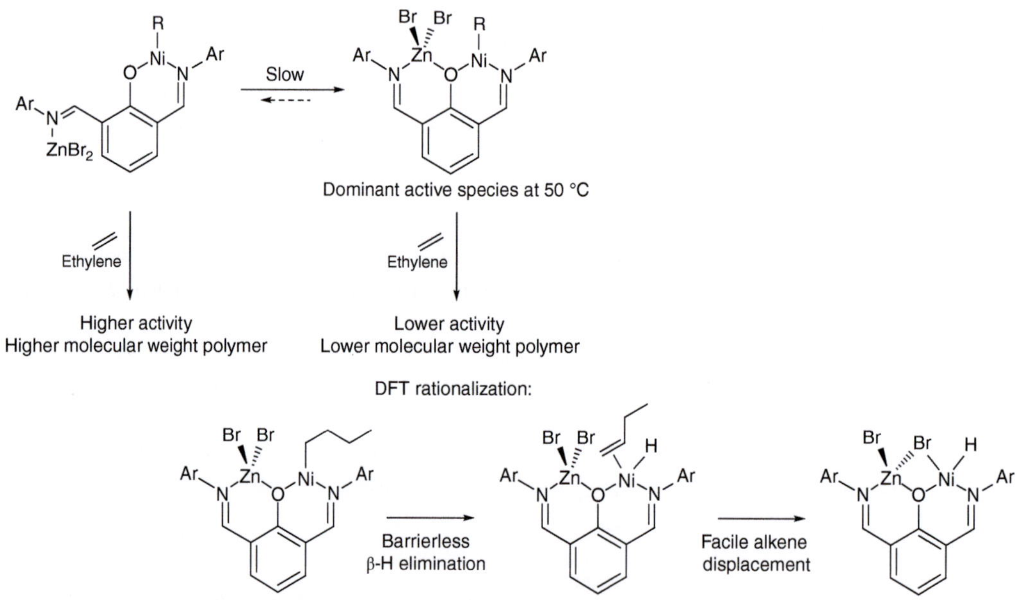

Figure 4.30 Influence of coordination pattern in the activity of heterobimetallic Ni/Zn catalyst built on 2,6-bis(imino)phenoxide ligands, as rationalized by DFT calculations.

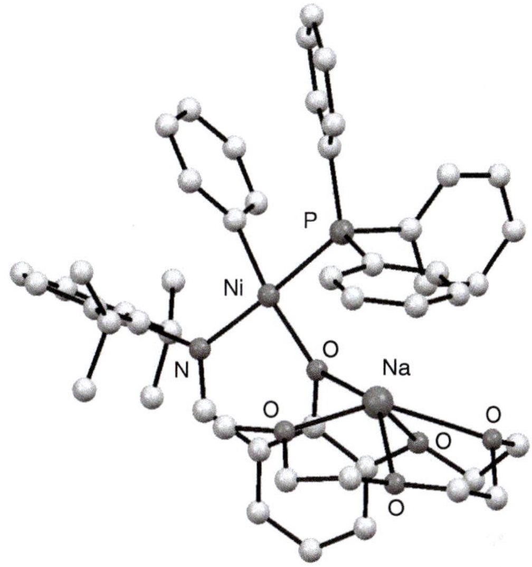

Figure 4.31 X-ray crystal structure of **31** ([NiNa(Ph)(PPh$_3$)(L3)][BAr$^F{}_4$]). L3 = *N*-(2,6-diisopropylphenyl)phenoxyimine ligand bearing a polyethylene glycol (PEG) moiety containing 3 ethylene glycol units (a range of PEG ligands were tested: L0-L4). The BAr$^F{}_4{}^-$ anion has been omitted for clarity. *Source:* Adapted from Cai et al. [120].

(generated *in situ*) yield highly branched amorphous polyethylene with enhanced molecular weights. The presence of alkali metals also results in up to a 20-fold increase in activity. When reactions are carried out using a parent mononuclear nuclear Ni complex (without PEG chains), NaBAr$^F{}_4$, and tetraethylene glycol dimethyl ether (1:1.1:2) instead of the mixture of phenoxyimine–PEG nickel complex and NaBAr$^F{}_4$, no increase in productivity is observed, indicating that the sodium–PEG group must be attached to nickel moiety to interact with the catalyst in a synergistic manner.

Data suggest that the match between the PEG chain length and the cation size determines their relative metal binding affinities, which in turn directly influence polymerization activities, suggesting that the heterobimetallic species are the catalytically active species (Figure 4.32).

Bulkier catalyst variants of the first catalyst have been later synthesized (in parallel with the replacement of the phosphine ligand by pyridine) and are generally more active, the presence of pendant strong Lewis acids still leading to dramatic increases in activities compared to the parent nickel complexes [121]. However, the changes in polymer branching and molecular weight due upon addition of Na^+ or K^+ ions are difficult to rationalize.

For proof that this strategy is also applicable to different ligand frameworks, nickel phenoxyphosphine complexes featuring PEG side arms which can chelate sodium have also been prepared and polymerize ethylene with a ~8.5-fold increase in catalytic activity compared to the parent mononuclear nickel complex [122]. Do et al. also prepared a new series of nickel phosphine phosphonate ester complexes with two polyethylene glycol (PEG) side arms, which were tested in catalysis upon activation with metal ions of different charges including alkali (Li^+, Na^+, K^+), alkaline (Mg^{2+}, Ca^{2+}), transition (Sc^{3+}, Co^{2+}, Zn^{2+}), post-transition (Ga^{3+}), and lanthanide (La^{3+}) metals [115]. Although olefin polymerization reactions are typically performed in non-polar solvents, which cannot solubilize +2 and +3 metal cations, those nickel catalysts can promote ethylene polymerization in THF. While only Ni/Na precatalysts were isolated and characterized by X-ray crystallography, several metal ions (in particular metal triflate) are capable of enhancing the catalytic activities of the Ni complexes, amongst which Co^{2+} and Zn^{2+} provide the greatest catalyst activity enhancements (10.6- and 4.8-fold, respectively). Other metals have a negative effect, with Cu^{2+} completely shutting down catalysis, and Al^{3+} and Bi^{3+} decreasing catalytic activities compared to the mononuclear Ni control. All polyethylenes produced are similar, with moderate branches

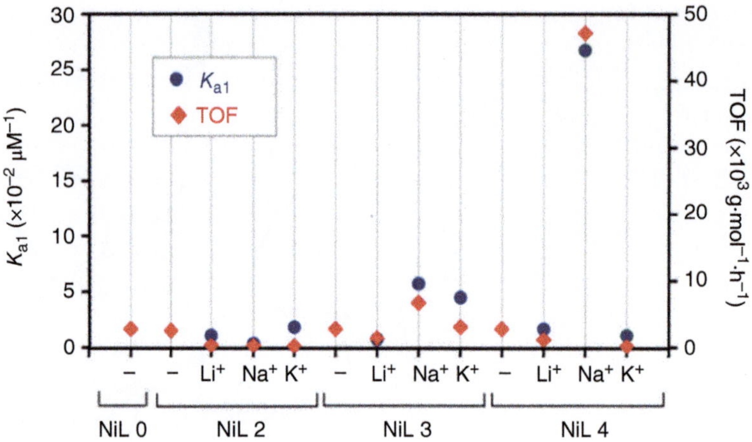

Figure 4.32 Structure–activity correlation plot showing the effect of different cations (Li^+, Na^+, and K^+) on the ethylene polymerization activity of the NiL variants. The association constants K_{a1} are shown as blue dots, whereas the TOFs are shown as red diamonds. Entries on the *x*-axis denoted with (–) indicate that no salt additives were present. *Source:* Cai et al. [120]; © 2015 American Chemical Society.

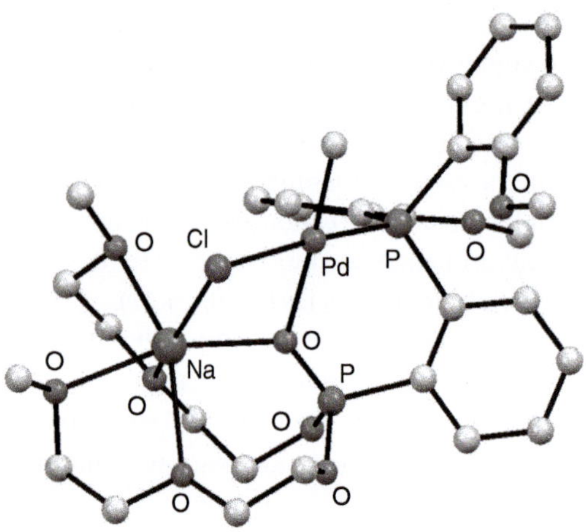

Figure 4.33 Solid-state molecular structure of **32** obtained by X-ray diffraction analysis. *Source:* Adapted from Cai and Do [124].

(18–27 per 1000 carbons), low-molecular weight ($M_n < 1 \times 10^3$), and relatively narrow dispersity ($Đ$ less than 2.1). Preliminary studies comparing the ionic radius and Lewis acidity of the cations tested do not show any obvious correlation with polymerization activity and further studies are needed to understand the secondary metal effects.

The team was also able to incorporate two PEG chains in the phosphonate group of Pd phosphine–phosphonate complexes reported by Jordan et al. [123]. As in previous studies, this method provides a second metal binding site for alkali ions (Li^+, Na^+, or K^+), resulting in Pd–M complexes which are generated *in situ*, and which favor a 1:1 stoichiometry in solution. A Pd–Na intermediate complex (**32**) has also been characterized by X-ray crystallography (Figure 4.33) [124].

As before, under similar conditions, the heterobimetallic species are more active for ethylene polymerization than the mononuclear Pd complex (between two- and threefold increase), but the influence of alkali metals on polymer branching and molecular weight is not significant. The trend observed of activity enhancement ($Na^+ > Li^+ > K^+$) is difficult to rationalize based on electrophilicity (i.e. $Li^+ > Na^+ > K^+$) and the association constant of the secondary metals (i.e. $Na^+ \approx K^+ > Li^+$) alone and likely involves a combination of several factors. It is worth noting that the addition of sodium salts to palladium phosphine diethylphosphonate complexes (without pendant PEG chains) also leads to remarkable catalyst enhancements, presumably due to similar formation of palladium–sodium species via the oxygen atoms of the phosphonate ester groups. These heterobimetallic species are also thermally robust, displaying long catalyst lifetimes at 100 °C and able to operate at temperatures as high as 140 °C.

The palladium–alkali catalysts also improve activity and molecular weights in ethylene and alkyl acrylate copolymerization reactions compared to their parent monopalladium catalysts, although they do not improve the incorporation of polar monomers (less than 2%). Although these heterobimetallic effects are relatively small, they do suggest that the addition of alkali salts is beneficial to the copolymerization reactions. The team has shown that secondary cations clearly increase the electrophilicity of the metal catalyst, but their precise role in polymerization reactions is still to be identified.

4.3.2 Ring-opening Polymerization

In 1932, Carothers reported the thermally induced ring-opening polymerization (ROP) of trimethylene carbonate [125]. Since then, many ROP processes have been developed, exploiting anionic, cationic, zwitterionic, radical, metathesis, and most relevant to this book chapter, metal coordination–insertion mechanisms (*vide infra*). ROP has now grown into one of the most powerful polymerization techniques, giving precise control over polymer molar mass, composition, tacticity and end-group functionality, enabling the synthesis of statistical, (multi-)block and topologically diverse (star-shaped, branched, cyclic) polymers. Beyond academic endeavours, many important commercial polymers are produced via ROP, including nylon-6 (from ε-caprolactam) and polycaprolactone (from ε-caprolactone). ROP has also been instrumental in the emergence of renewable polymers such as poly(lactic acid) (PLA).

Currently, industrial production of PLA is achieved in molten monomer by catalytic ROP using homogeneous tin(II) 2-ethylhexanoate, $Sn(Oct)_2$ [126]. Although this catalyst has been approved by the US FDA [127], its homogeneous nature means that complete removal of the catalyst from the final product is not feasible, and there are growing concerns over the potential toxicity of tin, which has been reported to inhibit cell growth by 50% in low doses [128]. While $Sn(Oct)_2$ is a fast catalyst and is relatively insensitive to environmental factors, such as the presence of oxygen, it requires elevated temperatures (> 120 °C) and neat conditions to give high-molecular weight product. Careful control over the polymer microstructure is also key to tuning the material properties and requires the use of an appropriate catalyst. In particular, PLA can present tacticity, which significantly influences the polymer properties (glass transition, crystallinity, degradation profiles, etc.). However, lactide ROP by $Sn(Oct)_2$ proceeds without control over polymer tacticity.

With the above in mind, there has been a significant effort to develop catalysts that are more active, less toxic, and more selective than $Sn(Oct)_2$, with a view to optimizing the properties of the polymer. For a long time, homogeneous metal catalysts have been at the forefront of this research. By extension, catalyst development efforts have also fed polymerization processes of other cyclic monomers, in particular carbonates. The field is immense, and the reader is directed towards some recent reviews on the topic [129–133]. Some of the most effective ROP catalysts are homogeneous organometallic complexes.

Many well-defined transition metal and main group metal catalysts proceed through a coordination–insertion mechanism (exemplified in Figure 4.34 in the case of lactide), for which experimental evidence was first reported by the groups of Kricheldorf [134] and of Jérôme [135]. In this

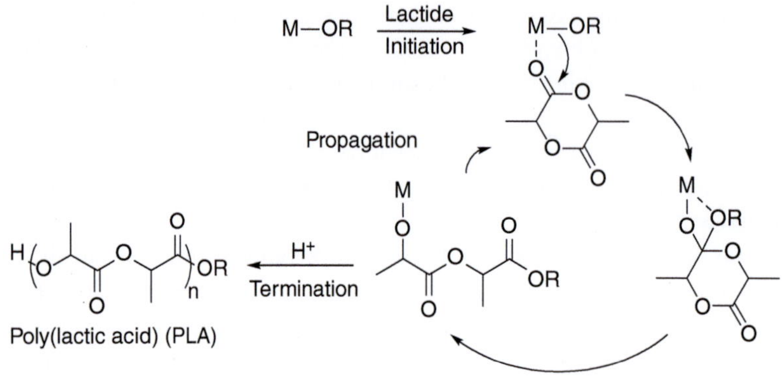

Figure 4.34 Metal-catalyzed coordination–insertion mechanism for the ROP of lactide.

mechanism, coordination of lactide by a Lewis acidic metal, typically a metal alkoxide, results in electrophilic activation of the lactide, allowing the nucleophilic attack by the alkoxide group. A tetrahedral intermediate is then formed, as is commonly observed during the interconversion of carboxylic acid derivatives. Ring-opening then occurs by collapse of this intermediate to re-form an alkoxide that now incorporates one unit of lactide. Propagation happens by subsequent lactide coordination and alkoxide insertion until the metal–alkoxide bond is cleaved by termination reactions (desired or not).

Many monometallic complexes, including (salen)Al [salen= *N,N′*-bis(salicylidene)-1,2-diaminoalkane], [amino(trisphenolato)]Zr, and (phosphasalen)Y, have accessed high-catalyst activities, broad monomer scope, and excellent control over the polymer stereochemistry. Homobimetallic catalysts have also shown exciting promise in cyclic ester ROP, and the reader is referred to a recent review on the topic. *Bis*–Zn complexes have been particularly successful [136–139]. To date, the most active *bis*–Zn catalyst reported for LA ROP is based on a macrocyclic bis(imino)diphenylamido ligand, where the activity is enhanced by the close metal–metal proximity [140]. They also significantly out-perform the mono zinc analogues, showing rates which are up to 6 times higher per active site, indicating a cooperative interaction between the two zinc ions, tuned by the ancillary ligand. For all these promising homobimetallic catalysts though, the exact nature of the cooperativity between metals remains unknown.

In parallel, recently, organocatalysis for ROP [141] has emerged as a serious alternative to metal-based catalysts, so much that some of the most active catalysts for the ROP of lactide are organocatalysts. Most organocatalysts to date tend to be good Brønsted bases, which work either by nucleophilic activation of the polymer chain end or by a general base mechanism in which the base deprotonates the alcohol initiator that then attacks a lactide monomer to initiate ring-opening. However, in some cases, in particular for triazabicyclodecene (TBD), a guanidine-type compound which features both a Lewis acid moiety (hydrogen-donor moiety; NH) and a Brønsted base moiety (hydrogen-acceptor moiety, imine N), a bifunctional mechanism is thought to occur. In such a mechanism, the catalytic system jointly exploits two distinct functional groups for more efficient polymerization: one group (usually a Lewis or Brønsted acid) activates or binds the monomer and another (Brønsted base) activates the growing polymer chain (Figure 4.35). These catalysts can be a single catalyst or a pair of catalysts with two distinct functional groups.

The strength of dual catalysis relies on the association of a Lewis acid with a Lewis base (to activate the monomer and the protic initiator, respectively) that are inactive separately but the combination of which triggers cooperation and enables ROP. In most instances, two organic molecules

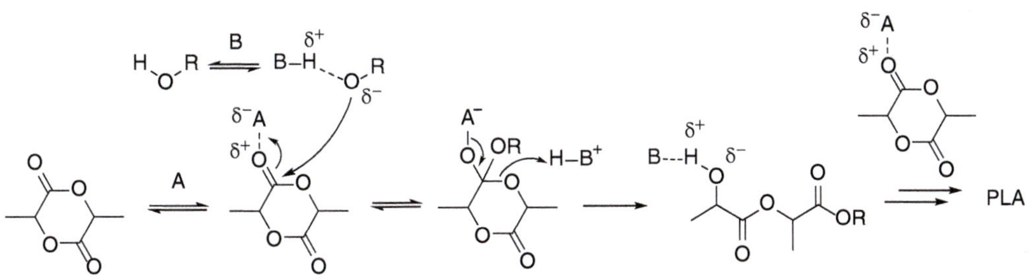

Figure 4.35 Bifunctional mechanism for the ROP of lactide (A: Lewis acid; B: Brønsted base).

are used as co-catalysts, and one of the most efficient systems to date is thiourea-based adduct and (thio)urea anions developed by Kiesewetter et al. [142–145]. However, it has also been demonstrated that dual catalysis can also operate in ROP from the cooperativity between a metallic Lewis acid and an organic Lewis base [146], including seminal reports by Arnold [147], Hillmyer and Tolman [148, 149], and more recent contributions by the groups of Bourissou [150] or Dove [151]. This field is sometimes included in what is termed ‘Lewis pair polymerization’ [152]. Most of the recent reports involve simple, readily available Lewis acids (MgX_2, ZnX_2, LiX, etc.) with either simple Lewis bases such as DMAP or triethylamine, or more sophisticated molecules such as *N*-heterocyclic carbenes, *N*-heterocyclic olefins, or cyclic isothioureas.

When considering the performance of homobimetallic complexes in ROP at the same time as the success of dual catalysis, it becomes almost natural to think of heterobimetallic catalysts as a way to achieve dual catalysis by combining two metal centers: one metal serving as Lewis acid to activate the monomer, the other one, more precisely the associated organometallic moiety, providing the Lewis base to activate a protic initiator (or to be the initiator itself, in the case of a metal alkoxide) (Figure 4.36).

However, compared to homobimetallic (and other homomultinuclear) catalysts [136], heterobimetallic complexes as catalysts for ROP remain rare. Historically, most of these species have arisen from the unexpected residual presence of alkali metals, but moving forward, these have been generated purposely. However, while promising activities have been reported, there is relatively limited mechanistic information on this type of catalyst. Often, synthetic limitations mean that no true comparison can be made with individual homonuclear complexes and with their mixture. To the best of our knowledge, no computational studies have been carried out on such systems yet. In the following section, we will discuss some selected examples of heterobimetallic complexes that have been used in the context of ROP, highlighting the existing mechanistic insight that exists (or does not). It is worth noting that we will not be discussing complexes bearing a ferrocene-based ligand, which have been used to perform redox-switchable ROP catalysis, and on that topic the reader is directed towards the work of Long [153], Diaconescu [154] (which has been examined by DFT) [155, 156], Okuda [157], and Byers [158].

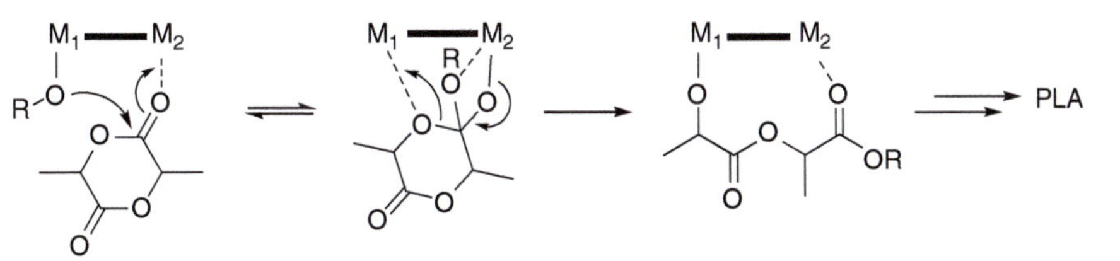

Figure 4.36 Concept of a bifunctional mechanism for the ROP of lactide involving a heterobimetallic complex.

4.3.2.1 ROP M_1–O–M_2 Heterobimetallic Catalysts

Amongst the heteromultinuclear catalysts that have been exploited for ROP, most are based on an M_1–O–M_2 framework, where the two metals can communicate electronically through the bridging oxygen and modify the properties of each other. In the context of the ROP of lactones, this may influence the Lewis acidity (i.e. monomer coordination) and metal-alkoxide nucleophilicity (i.e. propagation), thus, influencing key steps in the ROP mechanism. Several other examples in the literature have involved heterobimetallic complexes incorporating lithium in combination with another transition metal, in many instances, serendipitously and resulting from the tendency of lithium to form ate complexes (see Figure 4.37).

One example is the work by Lin et al. who reported as series of heterobimetallic titanium complexes with various other metals: Li, Na, Zn, and Mg [159]. The team used a series of bisphenol ligands (LH_2), which have been previously coordinated with Mg, Na, Li, and Zn and tested for ROP. Upon treatment of these monometallic complexes [(LH)Li$(THF)_3$, (LH)Na$(THF)_3$, $(LMg)_2$, and $(LZn)_2$] with Ti(O*i*-Pr)$_4$ in a 1:1 ratio, a series of heterobimetallic complexes have been obtained, isolated, and characterized, including by X-ray crystallography for LTi(μ-O*i*-Pr)$_3$Li$(THF)_2$. What emerges from structural and spectroscopic data obtained is that the Ti is coordinated to the bisphenolate ligand, with oxygen from isopropoxide groups involved in μ-oxo bridges between the titanium and the other metals. For comparison, the monometallic titanium complex was also synthesized. While no detailed mechanistic studies were carried out, comprehensive catalytic tests revealed that in contrast to monometallic titanium initiator, all the heterobimetallic titanium initiators (Ti–Li, Ti–Na, Ti–Zn, and Ti–Mg) show enhanced catalytic activity towards the ROP of L-lactide, while maintaining excellent control over molecular weight and narrow dispersities. The Ti–Zn (**33**) and Ti–Mg complexes are the most effective of all (with Zn providing greater enhancement than Mg), and the Ti–Li and Ti–Na complexes exhibit similar slow reactivity; slower than when the bisphenolate ligand is only coordinated to Li (which form dimers in solution) [160]. In addition to evidence that the structure of the heterobimetallic is retained in solution, these data suggest real cooperativity between metals, even if no mechanistic hypothesis has been advanced.

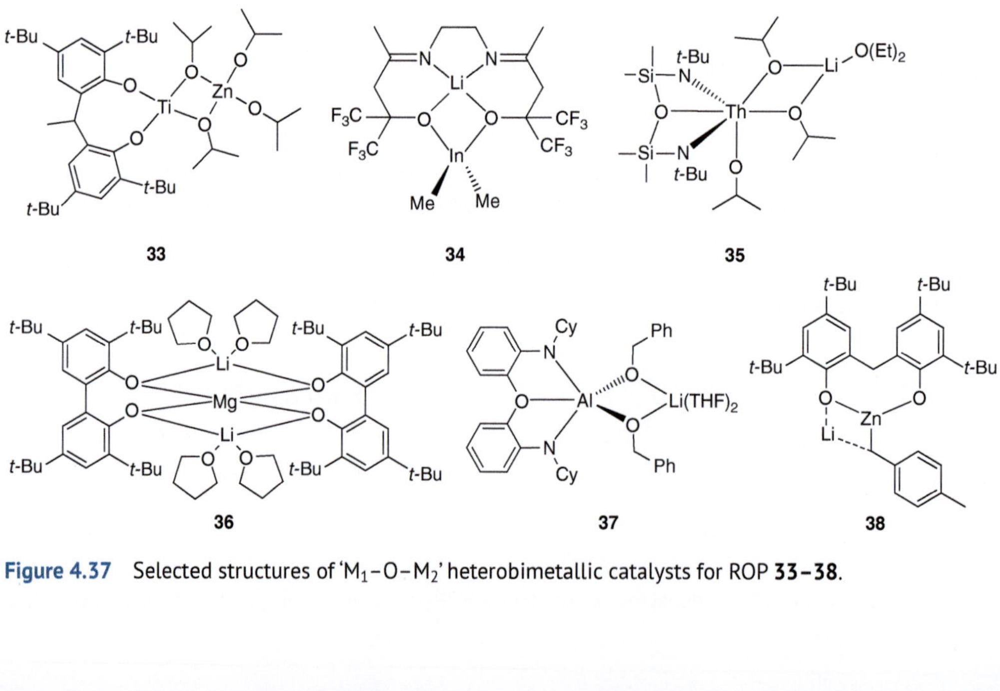

Figure 4.37 Selected structures of 'M_1–O–M_2' heterobimetallic catalysts for ROP **33–38**.

Carpentier et al. also described the reaction between $InCl_3$ and 3 equiv. of MeLi into a mixture of $InMe_3$ and Li[$InMe_4$], which upon further treatment with a fluorinated diimino-diol proligand with an ethylene bridge ({$ON^{Et}NO$}H_2), gave the ate complex [{$ON^{Et}NO$}Li]$InMe_2$ (**34**) as well as the methyl complex {$ON^{Et}NO$}InMe [161]. Both complexes could be isolated from the reaction and were characterized. However, obtaining the mononuclear methyl indium complex was not straightforward, so the team developed a more selective approach towards mononuclear alkyl indium complexes and used $In(CH_2SiMe_3)_3(THF)_{0.5}$ as the precursor in protonolysis reactions with {$ON^{Et}NO$}H_2, which yielded the desired product. Upon activation with benzyl alcohol and similar conditions (1 equiv. of alcohol) all complexes polymerize lactide, with evidence of control over molecular weight. While kinetics studies have not been carried out, the mononuclear indium complex appears slightly more active than its heterobimetallic counterpart, which also displays much broader dispersities, in many instances over 2, which could be the sign of two catalytic active sites working in parallel. Polymer end-group analysis also indicates that ring-opening of LA may occur competitively by nucleophilic attack of *i*-PrOH and Me groups. There is however no strong evidence of beneficial cooperativity between the two metals, and no catalytic tests have been done with a mixture of mononuclear alkyl indium complex and alkyl lithium.

Other heterobimetallic complexes comprising alkali metals, supported by various ligand backbones, have been reported and used in ROP. However, in a lot of instances, it is impossible to synthesize mononuclear analogues so no comparative study can be made, and the cooperation between metals can only be postulated. Such catalysts include, but are not limited to: a thorium–lithium ate complex supported by a diamido-ether ligand from Leznoff et al. (**35**) [162], heterobimetallic alkoxide clusters $Ln_2Na_8(OCH_2CF_3)_{14}(THF)_6$ (Ln = Sm, Y, Yb) from Shen et al. [163], bulky heterobimetallic Mg/Na and Mg/Li aryloxides by Wu et al. (**36**) [164, 165], as well as a *bis*-alkoxide Al/Li-ate species stabilized by a tridentate diamido-ether dianionic ligand from Dagorne et al. (**37**) [166]. For the later, while constrained tetracoordinate mononuclear Al analogues have been synthesized later on by the team [167], these were amido complexes and not alkoxide, and they were tested in the ROP of TMC and not lactide, so that no conclusion can be drawn regarding the influence of lithium.

One example worth noting is that of the magnesiate and zincate complexes reported by Thomas et al. (e.g. **38**) [168]. These are prepared by reacting a bisphenol ligand with 2 equiv. of *n*-butyllithium in THF at low temperature followed by the addition of butylmagnesium chloride or 4-methylbenzylzinc chloride. Alkoxide species are generated *in situ* and active in the ROP of *rac*-lactide. While comparison with mononuclear species cannot be done, there is some evidence that the heterobimetallic structure is retained and influences positively the polymerization. Indeed, using a THF/toluene mixture results in a substantial increase in the stereoselectivity of the polymerization in comparison with pure THF (P_r value increasing from 0.54 to 0.84). This increase in tacticity is consistent with previous observations that steric hindrance around the metal center governs tacticity in a chain-end control mechanism. It is postulated that in non-coordinating solvents (in toluene or THF/toluene mixtures) the heterobimetallic structure of the catalyst is conserved, inducing heteroselectivity, whereas in THF, these adducts likely dissociate into solvated ion pairs, resulting in loss of steric hindrance around Zn/Mg and loss of stereoselectivity.

In another study, Sarazin and coworkers reported the use of tetradendate (*R*)-Binap-based hydroxy-imine ligands (ONNO type) to assemble yttrium–lithium heterobimetallics [169]. The team performed a one-pot reaction between 2 equiv. of the protonated proligand, and 1 equiv. of $Y\{N(SiMe_3)_2\}_3$ and $LiN(SiMe_3)_2$, which afforded heterobimetallic **39** in quantitative yield. Reaction of the Binap-based ligand with $Y\{N(SiMe_3)_2\}_3$ alone unfortunately did not yield the mononuclear yttrium complex, limiting comparison. Compound **39** was characterized by NMR spectroscopy (^{1}H, ^{19}F, and ^{7}Li),

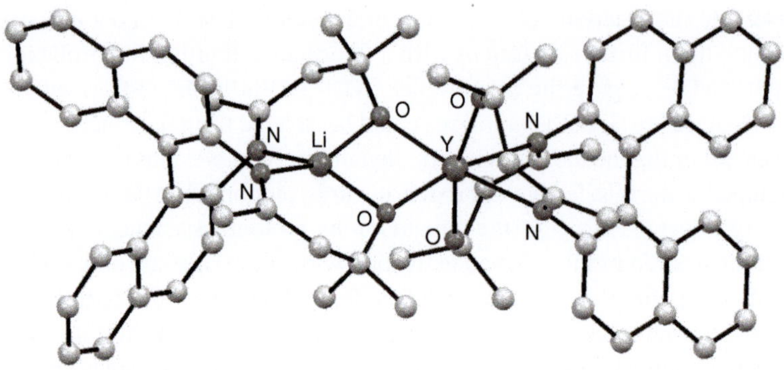

Figure 4.38 Solid-state molecular structure of **39** obtained by X-ray diffraction analysis (fluorine atoms of CF_3 groups on the ligand are omitted for clarity). *Source:* Adapted from Maudoux et al. [169].

elemental analysis, and X-ray crystallography, which reveals that in the solid state, yttrium and lithium ions are μ-oxo bridged by the alkoxide groups of the Li moiety, with the Y...Li distance being too long (3.075(6) Å) to suggest any direct metal–metal interaction (Figure 4.38). DOSY NMR spectroscopy coupled with diffusion molecular weight analysis confirmed that **40** retained its heterobimetallic structure in solution in THF. This complex is highly active in lactide ring-opening polymerization, with remarkable selectivity: 99% heterotacticity from *rac*-lactide, and 80% syndiotacticity from *meso*-lactide. However, direct reactivity comparisons with monometallic analogues were not possible.

Outside of alkali metals, lanthanides are also strong Lewis acids that have been combined with main group and late transition metals for ROP catalysis [170–172]. Roesky for example combined lanthanides with main group elements by synthesizing several oxo-bridged heterobimetallic aluminium–lanthanide alkyls complexes [173]. Reaction of Nacnac-supported LAl(OH)[C(Ph)CH(Ph)] with yttrium/samarium trialkyl Y/Sm(CH_2SiMe_3)$_3$(THF)$_2$ in *n*-hexane at 0 °C yields the oxo-bridged heterobimetallic yttrium/samarium dialkyl, which is further stabilized by coordination of the yttrium/samarium center by a bulky pyrrolylaldiminato ligand [NN]H ([2-(ArN=CH)-5-*t*-BuC_4H_2NH]). The resulting heterobimetallic complex **40** has been fully characterized, and X-ray single-crystal structural analyses indicate an almost linear Al–O–Ln geometry (Figure 4.39).

The mononuclear analogues [NN]Ln(CH_2SiMe_3)$_2$(THF)$_2$ (Ln = Y or Sm) can also be prepared by reaction of 1 equiv. of Y/Sm(CH_2SiMe_3)$_3$(THF)$_{2/3}$ with the pyrrolylaldiminato ligand. For comparison, the mononuclear aluminium and yttrium complexes were examined as initiators for the ROP of lactide. The aluminium hydroxide (LAl(OH)[C(Ph)CH(Ph)]) shows very low catalytic activity (from 25 to 70 °C) while the yttrium dialkyl complex is modestly active for the ROP reaction, with 69% conversion in 15 h at room temperature. **40** shows relatively higher reactivity, achieving 98% conversion in 5 h at room temperature. Upon activation of these alkyl complexes by alcohol, and formation of alkoxide species, the activity of the heterobimetallic complexes is improved further. Based on the low activity of the aluminium center alone, it is postulated that the aluminium center may not be directly involved in the polymerization, but that the activity enhancement provided to the lanthanide center may be due to the unique electronic properties of the Al–O ligand. However, it is also possible that the Al metal center is involved in the coordination and activation of the lactide monomer, although no mechanistic study has been carried out.

Another recent example of a heterobimetallic complex involving oxygen atoms bridging between two different metals, but this time involving the unusual combination of titanium and zinc, was developed by Williams. It was based on a hexadentate macrocycle ligand originally exploited for

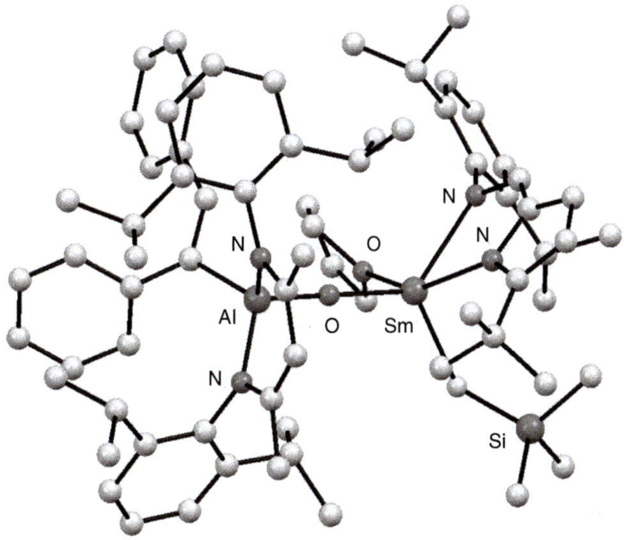

Figure 4.39 Solid-state molecular structure of **40** obtained by X-ray diffraction analysis. Adapted from Hao et al. [173]

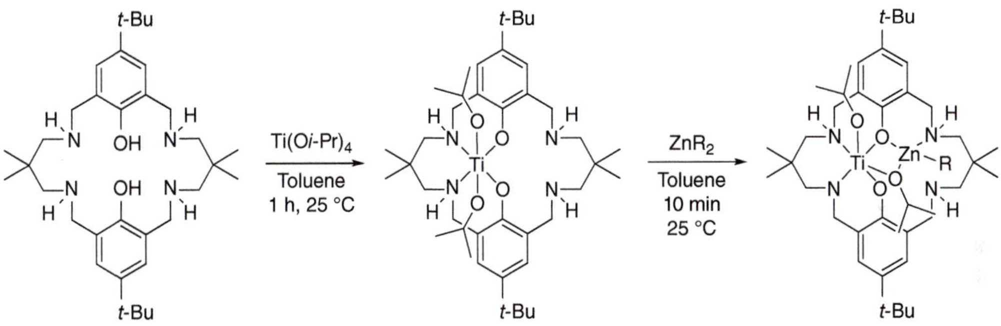

Scheme 4.12 Synthesis of Ti monometallic and Ti/Zn heterobimetallic complexes.

CO_2/epoxide co polymerization (*vide infra*) [174]. The monometallic titanium *bis*-isopropoxide complex was first synthesized, followed by reaction with zinc alkyl species, yielding a Ti/Zn species, in which the titanium retains two isoproxide groups, including one which bridges with the zinc atom, which also bears a remaining organyl group (Scheme 4.12).

While the titanium only complex is unreactive for LA ROP, the heterodinuclear complexes are moderately active. Evidence was obtained by polymer end-group analysis of ROP initiation by isopropoxide groups. The heterodinuclear complex bearing an ethyl group attached to the zinc center (**41**) gives a slightly faster rate of propagation than the one with a phenyl group, providing some evidence that the alkyl/aryl groups remain coordinated to the catalyst during the polymerization. Furthermore, the reactivity enhancement observed may have a steric origin; in the ethyl complex, the zinc center looks more exposed, which might facilitate lactide coordination (Figure 4.40). Unfortunately, no direct true comparison can be done with the monometallic zinc analogue, synthesized through the reaction of the macrocyclic ligand with 1 equiv. of either Et_2Zn or Ph_2Zn, as it does not possess an initiating co-ligand and has been shown to be inactive in related polymerization reactions [175]. The exact role of each metal during ROP is still unknown, but considering the

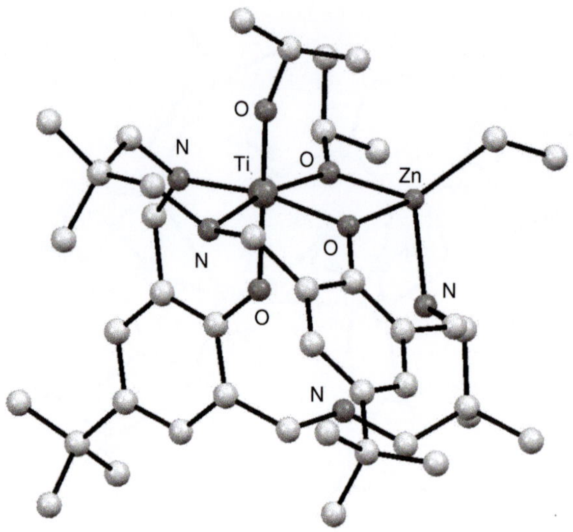

Figure 4.40 Solid-state molecular structure of **41** obtained by X-ray diffraction analysis. *Source:* Adapted from Garden et al. [174].

high activity displayed by ONN–zinc–alkyl complexes in the ROP of lactide [137, 176], the activity displayed by the Ti/Zn may very well be the result of a cooperative effect between the two metals, albeit slowing down the polymerization.

4.3.2.2 Other Heterobimetallic Catalysts for ROP

In one of the rare examples of a heterobimetallic complex which does not involve a M_1–O–M_2 motif, Mu et al. used an ethylene bridged *N*-arylanilido-imine ligand to prepare a heterobimetallic Al/Zn complex (**42**) in a stepwise fashion (Scheme 4.13) [177]. Both Al and Al/Zn complexes were characterized by ^{1}H and ^{13}C NMR spectroscopy and elemental analyses, and their molecular structures were determined by X-ray diffraction analysis. Previously, homobimetallic Al and Zn complexes had been reported [178].

Under similar conditions, in the presence of benzyl alcohol, the catalytic activity for ε-CL ROP of the heterobimetallic complex, activated by BnOH, is much higher than that of the Al-mononuclear/BnOH system, higher than the homobimetallic Al system, but lower that the similar homobimetallic Zn systems. Based on metal-dependent rate differences and on SEC analysis of the polymers, which showed molecular weight distribution with relatively narrow dispersity (1.1–1.8), it is postulated that the Zn center is the main active center, with the activity of the Al center being too negligible to be taken into account.

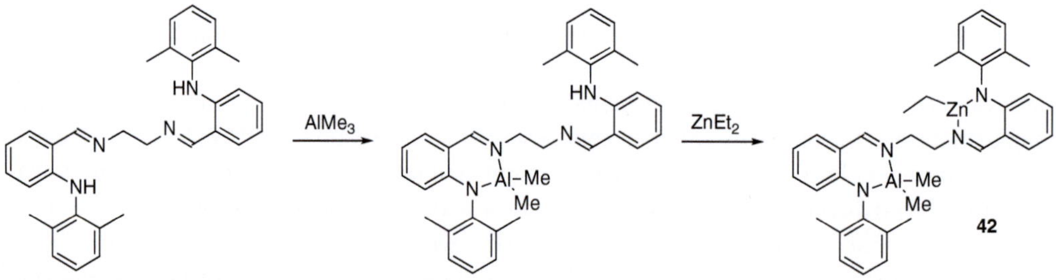

42

Scheme 4.13 Synthesis of Al/Zn heterobimetallic complex **42**.

By reacting YCl_3 or $LuCl_3$ with lithium amidocyclopentadienyl anions, Okuda et al. prepared a series of heterobimetallic Y–Li and Lu–Li complexes, which were characterized by X-ray crystallography in most cases (see example of complex **43** in Figure 4.41) [179, 180]. These air- and water-sensitive complexes are active catalysts for both ε-caprolactone and lactide polymerization, with dispersities lower than 2 but with no stereoselectivity (in the case of lactide). Attempts to abstract the Li^+ ion using crown ethers (12-crown-4) did not succeed, so direct comparison with the mononuclear analogue is impossible and conclusions cannot be drawn about the two metals' cooperativity. However, the team showed that polymerizations using the monometallic $Y\{N(SiMe_3)_2\}_3$ catalyst display less control and larger molecular weight dispersities (over 2).

Another example of heterobimetallic cooperativity outside of M_1–O–M_2 motifs was reported by Sarazin et al. [181], for which direct comparison between mononuclear complexes and the heterobimetallic catalysts is possible. Reaction of group 14 metal precursors ($M\{N(SiMe_3)_2\}_2$ (M = Ge, Sn)) with some aminophenol ligands featuring aza-15-crown-5 side-arms affords the expected mononuclear complexes [182]. These complexes are capable of further binding $Li(SO_3CF_3)$ to yield the corresponding Ge/Li (**44**) and Sn/Li (**45**) heterobimetallics. It is worth noting that starting from the lithiated aminophenol ligand, metal coordination of Ge and Sn was unsuccessful, as was the preparation of a sodium-based heterobimetallic by utilization of $Na(SO_3CF_3)$ or $[Na(OEt_2)_4][NH_2\{B(C_6F_5)_3\}_2]$ instead of $Li(SO_3CF_3)$ (Scheme 4.14). The heterobimetallic complexes were characterized in the solid state by X-ray crystallography and in solution by heteronuclear NMR spectroscopy (^{119}Sn, ^{29}Si, and/or ^{7}Li). It was shown that in solution, the lithium ion does not interact with the group 14 metals.

In the ROP of lactide ROP, under similar conditions, the Ge/Li is about twice as active and the Sn/Li is about 4 times less active than their parent monometallic complexes (without Li). Although the mechanism has not been elucidated, the team has proposed that the activity enhancement observed stems from the ability of the lithium ion (which conserves some Lewis acidity despite being trapped by the crown ether moiety) to activate the incoming monomer, a pathway very similar to the 'dual catalysis' mentioned earlier and also proposed by others [146, 183]. This hypothesis could potentially be verified experimentally by varying the strength of the anion associated with Li, which should directly influence the ability of the lithium to activate lactide.

Coates et al. have reported three generations of highly efficient Co or Cr homobimetallic catalysts for the enantioselective polymerization of epoxides [184–186], in which two Co- or Cr-salen

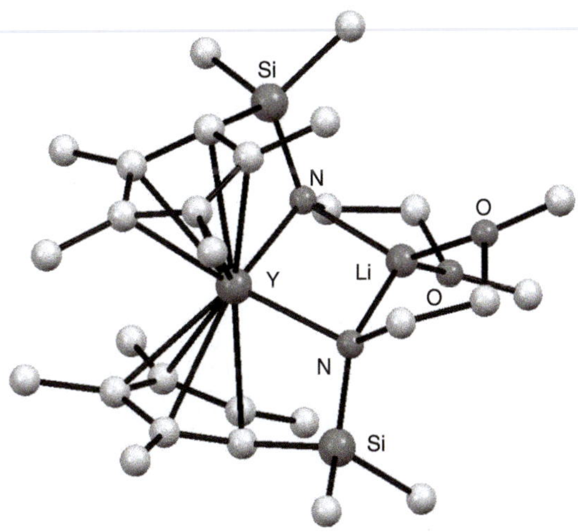

Figure 4.41 Solid-state molecular structure of **43** obtained by X-ray diffraction analysis. *Source:* Adapted from Hultzsch et al. [179].

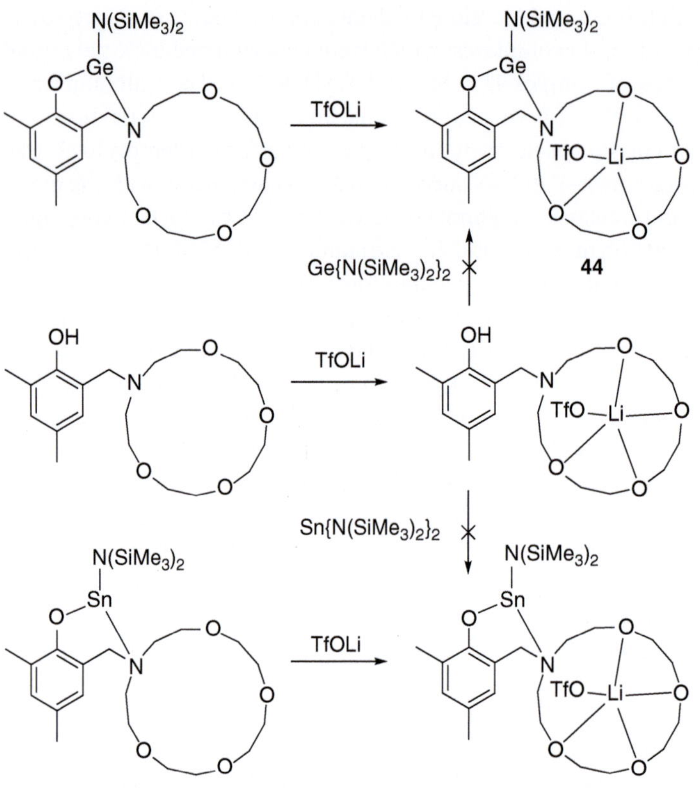

Scheme 4.14 Synthesis of Ge/Li and Sn/Li heterobimetallic complexes **44** and **45**.

complexes are held in close proximity (the optimal distance between the metals for catalysis being 6 Å) by chiral binaphthol or flexible linkers. The mechanism of these catalytic reactions was also examined by DFT simulations [187]. Building on this research, Ren et al. developed a heterodinuclear catalyst for the ring-opening copolymerization of lactones and epoxides [188]. The team designed a non-symmetric ligand featuring ONNO and ONON moieties, attached by a biphenyl linker. The synthesis of heterobimetallics is facilitated by the difference in coordination affinity of the two moieties. The mono-Co(II) complex can be obtained selectively easily, $Co(OAc)_2$ coordinating selectively the ONNO moiety rather than the ONON moiety. After oxidation of Co(II) into Co(III) by silver nitrate salts, a second metalation step using $Zn(OAc)_2$ or $EtAlCl_2$ yields the corresponding Co/Zn and Co/Al (**46**) heterobimetallic complexes. Unfortunately, no solid-state structures of these complexes have been obtained; they have only been identified by NMR spectroscopy and mass spectrometry.

These heterobimetallic complexes were developed with a view to enabling the alternating copolymerization of epoxide and lactones, based on the hypothesis that the cobalt moiety would mediate the ROP of the epoxide while the aluminium or zinc center would perform the ROP of lactones, and that spatial proximity would favor chain shuttling between the two metals (Figure 4.42). While the activity of the mixed Co/Zn complex is poor, the heterobimetallic Co(III)/Al complex in conjunction with [PPN]Cl (common cocatalyst in ROP mediated by salen-type complexes) exhibits a good activity for copolymerization of propylene oxide and ε-caprolactone, affording poly(ether-co-ester) copolymers. Even when PO and CL are used in equal feed amount, only up to 22% of ether units are incorporated in the polymer. ^{1}H NMR spectroscopy confirms the presence of

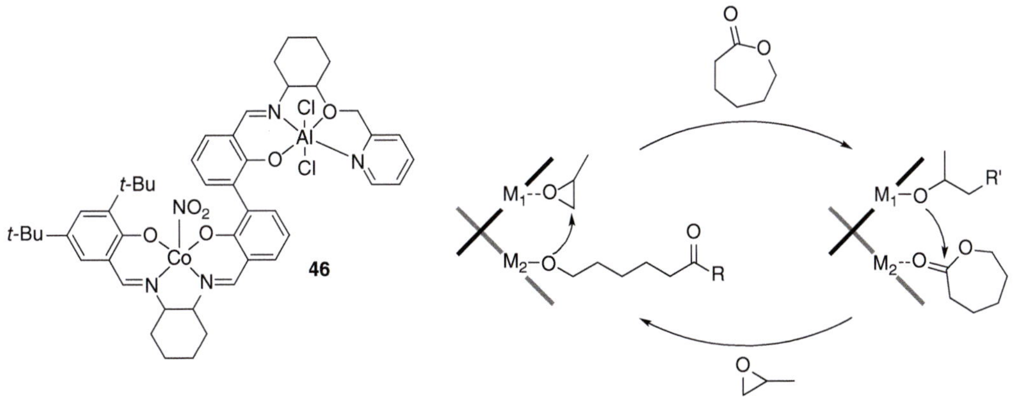

Figure 4.42 Structure of Cr/Al complex **46** and heterobimetallic catalysis strategy for the synthesis of polyether–polyester copolymers by the alternating co-polymerization of epoxides and lactones.

diagostic signals for alternating ether/ester units. Kinetics analysis and reactivity ratios confirm a tendency towards statistical copolymerization rather than alternating enchainment. The team also used DFT modelling to calculate the free enthalpies of the four transition states possible for the ring-opening of CL or PO by either Co or Al alkoxide species. However, no significant differences in activation energies are observed, in agreement with a statistical copolymerization. Crucially, by comparison, the homodinuclear Co(III) and Al complexes, the mononuclear Co(III) and Al complexes as well as their mixture result in very poor ether incorporation (up to 4%), compared to 51% under similar conditions for the heterobimetallic complex. This result alone demonstrates the cooperativity of the two metal centers in this catalytic system, enabling a different reaction outcome than the tandem mixture.

4.3.3 Ring-opening Copolymerization of Epoxides and Carbon Dioxide

The alternating copolymerization of epoxide and carbon dioxide is an attractive strategy for the preparation of polycarbonates. It proposes an alternative to polycondenzation reactions between diol and dialkyl/diaryl carbonate species (or even phosgene), which are best avoided in modern plastic production because they are energy inefficient and preclude polymerization control. It is also an alternative to the direct ROP of cyclic carbonates, which requires synthesis of a cyclic carbonate from a diol first (using previously mentioned phosgene derivative, although a new strategy using CO_2 now exists) [189–192], the polymerizability of which can be limited by ring-opening thermodynamics intrinsic to ring size and substituents. This process is also appealing because it makes use of CO_2 as an inexpensive and abundant C_1 feedstock, although it is important to state that the use of CO_2 in any polymer manufacturing process would not be able to make a large impact on overall CO_2 levels in the atmosphere. However, CO_2 utilization is worth pursuing as a means to reduce industrial emissions and as an economic driver towards carbon capture.

This reaction was discovered in 1969 when Inoue et al. reported the use of the heterogeneous $ZnEt_2/H_2O$ mixture as catalyst for the copolymerization of propylene oxide and CO_2 [193]. Since then, this reaction has attracted a lot of attention as a means to reduce pollution associated with polymer manufacture and to add value to carbon dioxide [194]. Many heterogeneous and many homogenous systems have been developed for the copolymerization of CO_2 and epoxides, including using Zn(II)(β-diiminate), Co(III)(salen) or Cr(III)(salen), Al(III) trisphenolate, and dinuclear

Zn(II) macrocycles. For a comprehensive account of all catalytic systems developed, the reader is referred to several extensive reviews on the topic [195–202]. In this section, we will discuss the mechanistic and computational insight that led to the progressive and rational development of efficient heterobimetallic catalysts for what is sometimes called, the ring-opening copolymerization (ROCOP) of CO_2 and epoxides. It is also worth mentioning from the onset that ROCOP processes also encompass the alternating copolymerization of epoxides with cyclic anhydride, which produces polyesters and which operates via a similar mechanism [203].

It is commonly accepted that the ring-opening copolymerization of CO_2/epoxide involves four steps: (i) initiation, (ii) CO_2 insertion, (iii) epoxide ring-opening, and (iv) termination (Figure 4.43).

The disadvantages of some catalysts are that they can homo polymerize epoxides, or decarboxylate to give ether polymer linkages (which alter the physical properties of the polymer), or they form cyclic carbonates through backbiting. Early ROCOP catalysis focussed on two component catalyst systems: a combination of a metal catalyst (typically chromium/cobalt porphyrin/salens) and an organic cocatalyst (typically quaternary ammonium halide salts, which could even be tethered to the metal complex for increased activity). Until recently [204], all catalysts were proposed to operate by either a bicomponent or di-nuclear mechanism. As new catalytic systems were being developed, several experimental and computational investigations hinted at the necessity of cooperation between metals, including when mononuclear metal catalysts are used, with an optimal distance between metals of around 3–4 Å during the catalysis. For a review of computational studies of carbon dioxide–epoxide copolymerization reactions, the reader is referred to the account

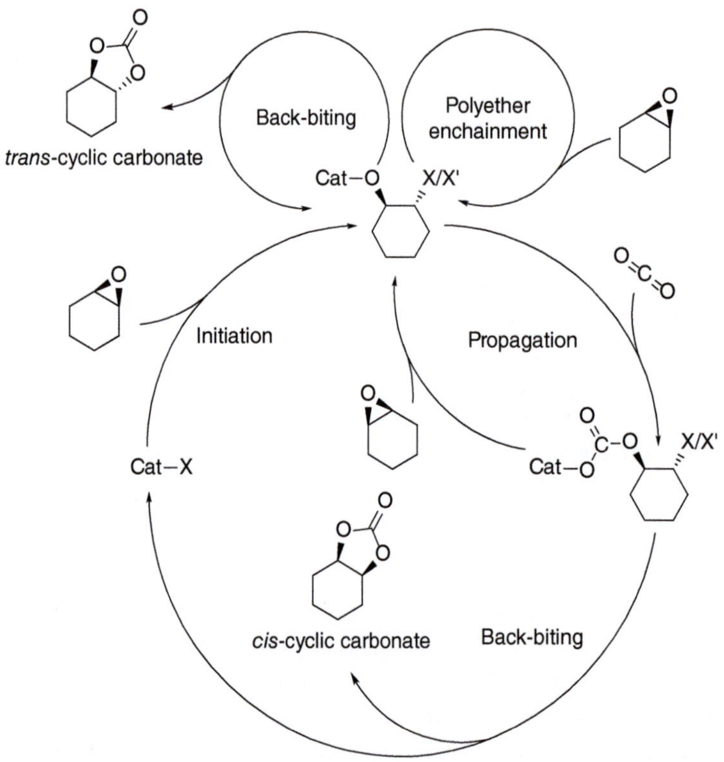

Figure 4.43 Catalytic cycle for the CO_2/epoxide copolymerization, including polyether and cyclic carbonate formation (X = initiating group, e.g. acetate, halide, alkoxide; X′=carbonate propagating chain).

written by Darensbourg and Yeung [205]. In the following section, selected significant examples, which later fed into the development of heterobimetallic catalysts, will be presented.

4.3.3.1 Mechanistic Insight into Homobimetallic Catalysts

Coates et al. pioneered highly active zinc–β-diketiminate (Zn–BDI) catalysts for CHO/CO_2 copolymerization, showing high CO_2 selectivity (99 % incorporation) and activity: TOF 729/h (50 °C, 7 bar CO_2) [206]. The team realized early on the importance of an equilibrium in solution between monomeric and dimeric complexes, influenced by temperature, concentration, and ligand substituents, and which affected greatly catalytic activities. Reduced rates are observed when ligand substituents were small, rationalized by the formation of a tightly bound dimer. Rates are also reduced when substituents were large, rationalized by the formation of a discrete monomeric species. A mechanism is proposed where the best catalysts are loosely associated dimers. Several years later, Shao et al. examined this catalytic system using density functional theory [207]. Modelling reveals that both mono- and binuclear forms of the catalyst are involved along the reaction path, with the rate determining step (ring-opening of the epoxide) being mediated by a binuclear catalyst, as suspected before (Figure 4.44). Subsequent CO_2 insertion is predicted to be kinetically facile (in agreement with the moderate CO_2 pressure required) and preferentially mediated by a mononuclear catalyst. In this study, predictions were also made towards the development of new ligands should result in increased rates and enantioselectivities for copolymerization.

In 2015, building on some work from Coates et al. who first made poly(limonene carbonate) by copolymerization of limonene oxide and CO_2 [208], Kleij et al. showed that either iron or aluminium trisphenolate catalyst systems, in combination with chloride salts (*bis*(triphenylphosphine)-iminium chloride), could also mediate effectively this reaction [209]. The aluminium catalyst is almost twice as fast as the iron catalyst, likely due to its higher Lewis acidity, and shows excellent selectivity for polycarbonate formation, with no cyclic carbonate or ether linkage formation. A DFT investigation of the reaction between CO_2 and cyclohexene oxide (CHO, a cyclic internal

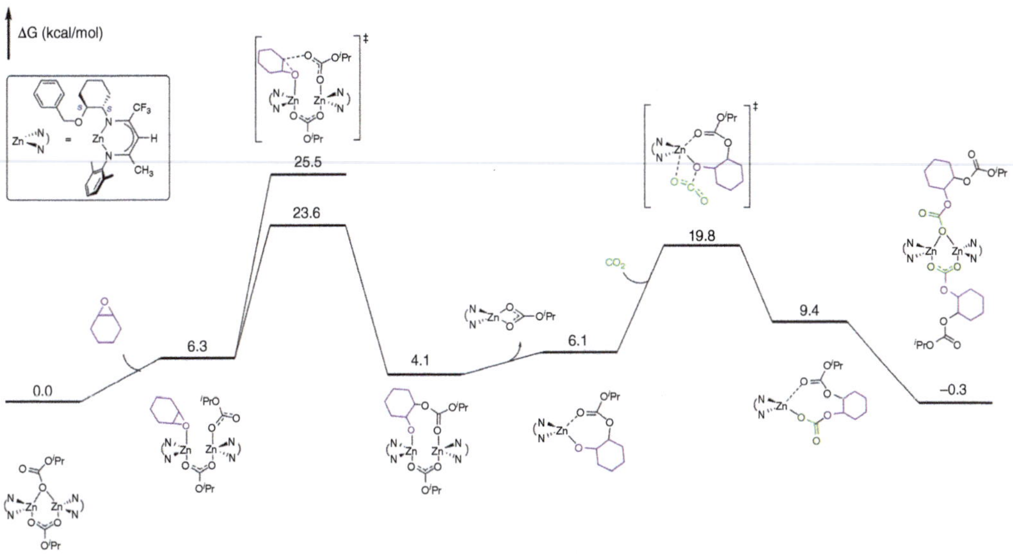

Figure 4.44 Computed potential energy surface of the first propagation cycle of the dimeric Zn-catalyzed CO_2/CHO copolymerization reaction. *Source*: Shao et al. [207]; © 2020 American Chemical Society.

epoxide often used in the field) was subsequently performed and indicates that catalyst dimerization is required to reduce the barrier to epoxide ring-opening, in agreement with experimental observations [210]. However, it is worth noting that in this study, the ammonium part of the co-catalyst was not modelled explicitly.

In addition to those mechanistic insights, kinetic analysis typically shows that for this reaction, rate laws have orders in mononuclear catalysts ranging from 1.0 to 1.8. Collectively, these findings led research towards the development of di-nuclear catalysts, to avoid the use of co-catalytic salts (expensive and corrosive to steel, which could be a problem for scale-up and commercialization). Compared to their mononuclear counterparts operating via a di-nuclear mechanism, di-nuclear catalysts would also allow for greater activity under high-dilution conditions. Lee et al. were the first to report deliberately constructed di-nuclear zinc catalysts for the ROCOP of CO_2 and CHO [211]. Other examples include systems built on Trost ProPhenol ligand [212] (a system which was also investigated by DFT later on) [213].

In 2009, Williams et al. reported the first catalyst with reasonable activity for the ROCOP of CO_2/CHO at 1 bar CO_2 pressure, a di-zinc complex featuring a macrocyclic diphenolate ancillary ligand (inset of Figure 4.45) [214]. The activity was 18/h (at 80 °C) affording perfectly alternating polycarbonate. Subsequently, a range of ligands [215], metals [216], and co-ligands [217] have been explored including the first reports of active Fe(III) [218] and Mg(II) [219] catalysts for ROCOP. Using the same symmetrical ancillary ligand, changing the metal center greatly affects the polymerization rates, with the order of activity being Co(III) > Mg(II) > Fe(III) > Zn(II).

A detailed spectroscopic and computational study shed light on the polymerization pathways and key intermediates [220, 221]. *In situ* attenuated total reflectance–infrared (ATR–IR) spectroscopy was used to characterize the reaction between the di-zinc acetate complex and CHO, and showed that *in operando* there are different environments for the acetate coligand. One of the acetate groups shows

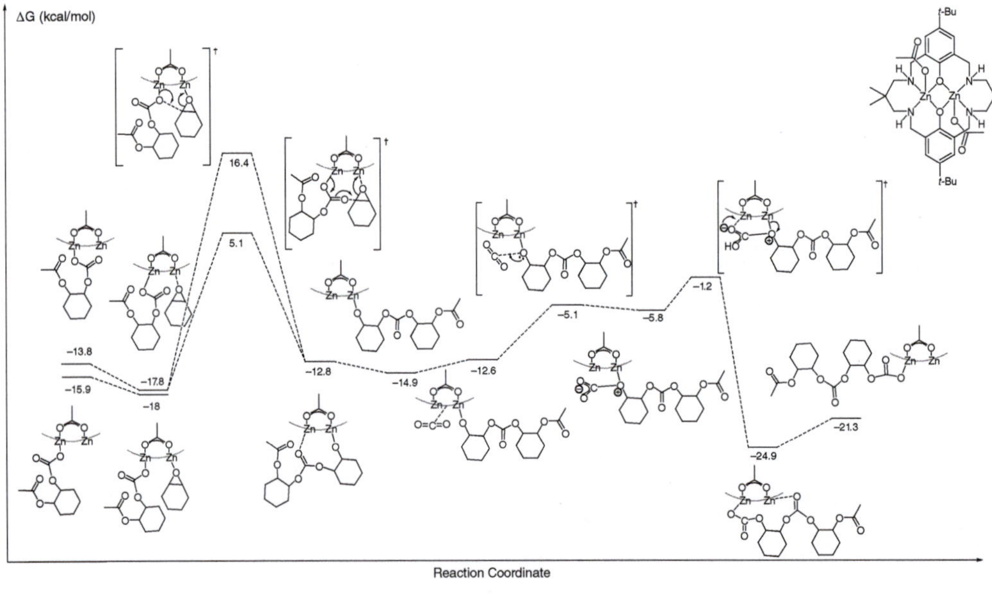

Figure 4.45 Illustration (not to scale) of the potential energy surface for the first cycle of propagation of the CO_2/CHO copolymerization reaction catalysed by Williams' di-zinc acetate complex. *Source:* Buchard et al. [221]; © 2012 American Chemical Society.

resonances consistent with attack at the CHO group, while the other has a resonance consistent with it maintaining a bridging coordination mode between the two zinc centers.

DFT substantiated this proposal and further detailed modelling of the reaction pathway has led to a proposed chain-shuttling mechanism where the growing polymer chain alternates between the two metal centers with each monomer insertion (epoxide or carbon dioxide). The chain shuttling is counter-balanced by an equal but opposite change in coordination site for the acetate co-ligand. Furthermore, the rate determining step (RDS) is identified as the ring-opening of the epoxide with a free energy barrier of 23.5 kcal/mol, in-line with the experimental barrier (25.7 kcal/mol). Sequential epoxide enchainment (to give ether linkages) is calculated to have a free energy barrier of 39.3 kcal/mol; significantly higher than for the alternating copolymer, in agreement with the selectivity observed experimentally (> 99 %). It has also been possible to use this model predictively, and by changing the acetate co-ligand, to alter the rate of polymerization as predicted, resulting in an almost twofold increase in activity.

An important implication of the chain shuttling mechanism is that there are distinct roles for the two metal centers, as sketched in Figure 4.46. One of the metal centers (M_1) coordinates the epoxide, while the other center (M_2) inserts carbon dioxide. The requirements for these processes are distinct, with epoxide coordination being accelerated by Lewis acidic/electrophilic metal centers, while the carbonate formation and attack step is favored by metals showing labile carbonate groups. As such, having distinct metals serving the two roles was postulated to be able to increase catalytic activity and thus provided an impetus to study heterobimetallic catalysts, the promise of which had been previously hinted at in older literature. It is also worth noting that amongst the heterogeneous catalysts developed for CO_2/epoxide copolymerization, mixed-metal materials (e.g., Zn/Co or Zn/Fe) are excellent heterogeneous catalysts [222–224].

In 2013, Rieger et al. reported a tethered active zinc-β-diketiminate (Zn–BDI) catalyst which showed an activity of 9130/h (40 bar, 100 °C) and high polycarbonate selectivity (> 99%) [225].

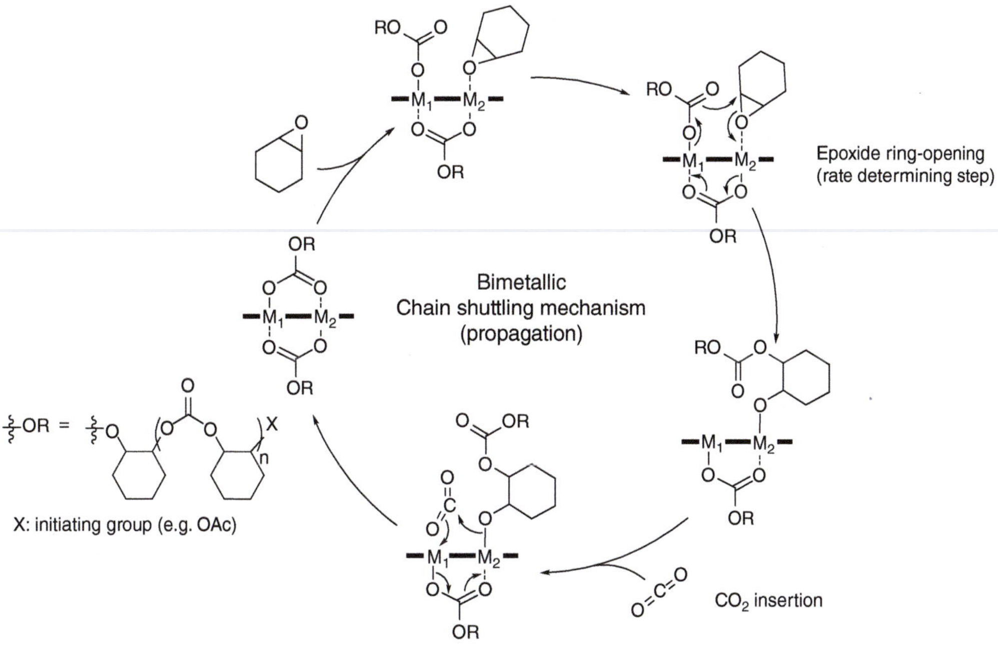

Figure 4.46 Chain shuttling mechanism.

A computational study led to a mechanistic proposal similar to the chain shuttling mechanism proposed by Williams et al. [226]. Structure–activity studies also suggested a positive correlation between ligand flexibility, electron-withdrawing substituents and activity [227].

4.3.3.2 ROCOP Heterobimetallic Catalysts

Earlier efforts go back to 2004 with the work of Bochkarev et al., who prepared a series of homogeneous *tert*-butoxide-linked bimetallic species ($(t\text{-BuO})_5Cu_2Ln$ (Ln = Y, La, Nd, Sm, Eu, Gd, Tb, Dy, Ho, Er, Tm, Yb), $(t\text{-BuO})_5MLa$ (M = Mn, Fe, Co, Ni), $(t\text{-BuO})_5ZnNd$, and $(t\text{-BuO})_4ZnFe$), that were also examined as catalysts for CO_2/epoxide copolymerization [228, 229]. While the characterization of these complexes was limited to IR spectroscopy and elemental analysis, significant variations in catalytic activity (overall limited) were observed as the late transition metal was varied. Incorporation of CO_2 can go up to with $(t\text{-BuO})_5Cu_2La$, but molecular weights are generally low (around 1000) with extremely broad dispersities (typically between 4 and 6). The team also synthesized and tested heterobimetallic complexes in which the two metals are held in close proximity by an ethylene glycol linker. The reaction of Et_2Zn with $NaOCH_2CH_2OH$ produces the bimetallic zinc complex $NaOCH_2CH_2OZnEt$, which upon reaction with Ph_3SnCl, Cp_2TiCl_2, and $Cp_2LuCl(THF)$ affords the complexes $Ph_3SnOCH_2CH_2OZnEt$, $Cp_2Ti(OCH_2CH_2OZnEt)_2$, and $Cp_2LuOCH_2CH_2OZnEt$, respectively. No solid structure was obtained for these structures, but significance differences in reactivity are observed. While the Sn/Zn complex displays no activity, and only polyether chains are generated with the Lu/Zn species, the Ti/Zn derivative is reported to catalyse the copolymerization of CO_2 with cyclohexene oxide at room temperature and atmospheric pressure, albeit with limited incorporation of CO_2 (22%), number-average molecular weight of around 1000 and large dispersities (around 4.5). No comparisons with homobimetallic systems or tandem mixtures were made in these studies.

To the best of our knowledge, similar to the field of ROP, there has been no mechanistic study involving computational modelling for any of the heterobimetallic complexes developed for ROCOP catalysis. The heterobimetallic catalytic system closest to ROCOP that has been investigated by DFT has been some rare earth (RE) metal/zinc complexes for the cycloaddition of CO_2 with epoxides to produce cyclic carbonates, a process that follows a very similar mechanism to ROCOP, except that instead of a carbonate chain attacking a new epoxide to grow the polymer further, generally back-biting happens [230].

Yao has reported a series of homonuclear RE metal complexes and heteronuclear RE/Zn complexes (e.g. Nd/complex **47**, Figure 4.47) stabilized by ethanolamine-bridged bis(phenolato) ligands [231]. Homonuclear complexes featuring a metal of small ionic radius (Y and Yb) form trinuclear structures, whereas complexes with Nd and La form dinuclear complexes. The addition of 0.5 equiv. of $Zn(OAc)_2$ to a solution of ligand and $Cp_3RE(THF)$ in THF produces heterometallic RE/Zn complexes, which have been structurally characterized. All complexes are assembled through bridging acetate and alkoxide ligands, but structures otherwise varied depending on the nature of the RE metals. Y/Zn complex is hexanuclear, with four yttrium and two zinc centers, while the Nd/Zn complex and Sm/Zn complex are trinuclear complexes, with 2 RE and 1 central zinc atoms. No ytterbium or lanthanum analogues have been isolated.

All complexes were tested for the copolymerization of CHO and CO_2, and while the mononuclear RE metal complexes show negligible activity (likely partly due to the lack of an efficient initiating group such as an acetate), heterometallic complexes are more effective, albeit with moderate activities, selectivities and molecular weights, and produce acetate-group-capped copolymers. In comparison, $Zn(OAc)_2$ shows lower activity (albeit better selectivity for carbonate linkages) than heterometallic complexes. However, no tandem mixture of $Zn(OAc)_2$ and mononuclear RE

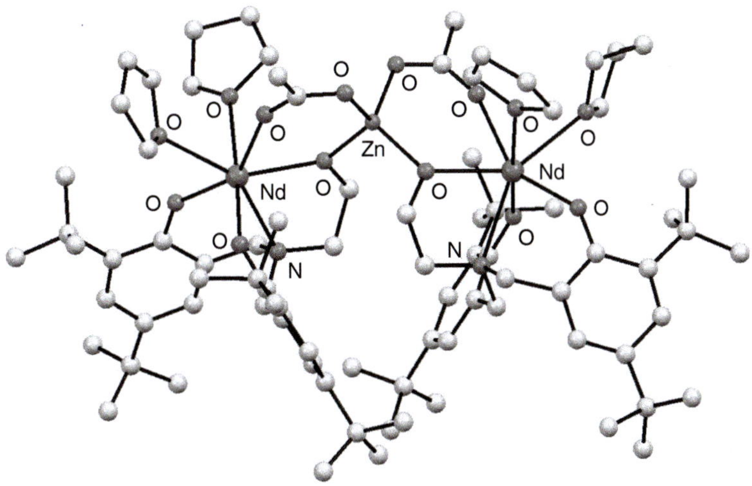

Figure 4.47 Solid-state molecular structure of **47** obtained by X-ray diffraction analysis. Adapted from Hua et al. [231].

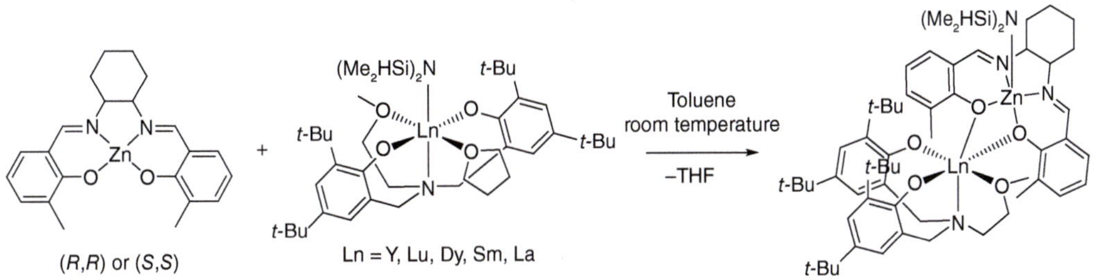

Scheme 4.15 Synthesis of RE/zinc heterometallic complexes by ligand redistribution.

complexes was tested in catalysis. The team pursued their efforts in this area by reporting in 2019 the synthesis of ten RE/zinc heterometallic complexes containing both alkoxy-amino-bis(phenolato) and chiral salen ligands, obtained from the reactions of the mononuclear complexes *via* ligand redistribution and THF disassociation (Scheme 4.15) [232]. Notably, as a result of these reactions, two oxygen atoms of the salen ligand coordinated to the oxophilic RE metal ion, the $[N(SiHMe_2)_2]^-$ group shifting from the RE metal ion to the zinc ion.

The synthesized RE/Zn heterometallic complexes show much better performances than the monometallic ones in the copolymerization of CO_2 and cyclohexene oxide. The ionic radii of the RE metals influenced the catalytic activity of the complexes: the Dy/Zn and Sm/Zn complexes, in which the RE metals have medium ionic radii, showing the highest catalytic activity. The nature of the RE also influences the selectivity towards carbonate linkages in the polymer, with again, the Dy/Zn and Sm/Zn showing the highest selectivity. On the basis of these observations, a cooperative mechanism is proposed, in which the epoxide coordinates to the RE center, before being attacked by the growing zinc–polycarbonate chain. The synthesis of the corresponding Ln/Co(Ni, Cu) heterometallic complexes in this strategy was mentioned by the authors as possible future work (see Figure 4.48).

In 2018, Okuda and coworkers reported a series of tris(salen)-trizinc-lanthanide multinuclear complexes, which were tested for the copolymerization of CO_2 with CHO [233]. Complex $LaZn_3(OAc)_3L$ (**48**), where L is a macrocyclic tris(salen)-based ligand, is prepared by the template

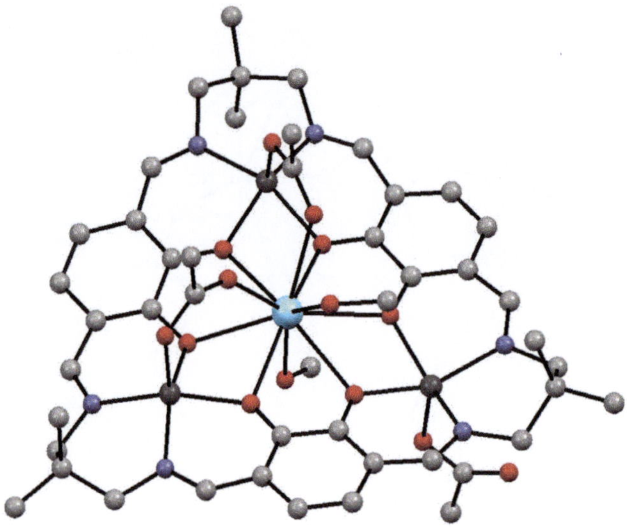

Figure 4.48 Solid-state molecular structure of La/Zn complex **48** obtained by X-ray diffraction analysis (La = light blue, N = blue, O = red, Zn = dark gray). *Source:* Adapted from Nagae et al. [233].

reaction of 1,4-diformyl-2,3-dihydroxybenzene with 2,2-dimethyl-1,3-propanediamine in the presence of 1 equiv. of $Zn(OAc)_2{\cdot}2H_2O$ and 1/3 equiv. of $La(OAc)_3{\cdot}xH_2O$ (see Figure 4.48). Other diamine sources were tested but with limited success. Nitrate and triflate analogues can also be prepared.

The acetate complex shows high activity (TOF 230/h) under 10 bar of CO_2 affording perfectly alternating PCHC (> 99 %). However, the nitrate complex shows negligible activity, and the triflate complex shows high catalytic activity (TOF 413/h) but for polyether formation with only 1% carbonate linkages. The team has synthesized other heterobimetallic complexes by varying the nature of the lanthanide precursor, and metals with a large ionic radius such as La, Ce, Pr, and Nd, offer higher catalytic activities compared with middle or late lanthanide elements, while retaining high-selectivity towards carbonate linkages and narrow molecular weight distributions. The authors propose a reaction mechanism akin to the chain shuttling mechanism proposed by Williams, with the lanthanide center being the one coordinating the epoxide, although there is no experimental or computational evidence for it.

Nozaki, Mashima and team recently built upon this work by preparing a series of heteromultimetallic complexes consisting of three Co(II) ions and one lanthanide ion (La, Ce, Pr, Nd, Eu, and Gd), coordinated by a related macrocyclic *tris*(N_2O_2) hexaoxime ligand [239]. The treatment of this ligand with 3 equiv. of $Co(OAc)_2$ and 1 equiv. of $La(OAc)_3$ affords the tetranuclear Co_3/La complex $LaCo_3(OAc)_3L(H_2O)_4(MeOH){\cdot}H_2O$ (**49**), characterized by X-ray crystallography (see Figure 4.49). The Co_3/Nd complex exhibits the highest TON of 13 000. Cooperative catalysis in **49** is supported by control experiments. Under similar conditions, mononuclear cobalt(II) salen complexes exhibit no catalytic activity and $La(OAc)_3$ only affords a trace amount of copolymer. In addition, the combination of $La(OAc)_3$ with cobalt(II) salen complexes does not show any catalytic activity either, indicating that La and Co ions being in close proximity in complex **49** is essential for the catalysis.

From 2014, based on the symmetrical macrocyclic ligand used in their previous work, Williams and coworkers have prepared several heterodinuclear complexes. First, work focussed on heterodinuclear Zn/Mg, and initially the team found that reacting the macrocyclic ligand with $ZnEt_2$, followed by treatment with $Mg(OAc)_2$, provided an intractable mixture of homobimetallic Zn and Mg_2 species, as well as a heterobimetallic Zn/Mg complex with acetate coligands (seen by mass spectrometry).

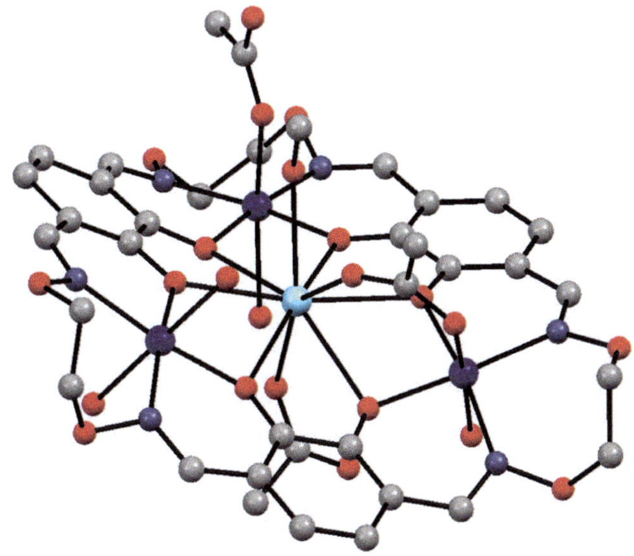

Figure 4.49 Solid-state molecular structure of Co/La complex **49** obtained by X-ray diffraction analysis (Co = dark purple, La = light blue, N = blue, O = red). *Source:* Adapted from Asaba et al. [239].

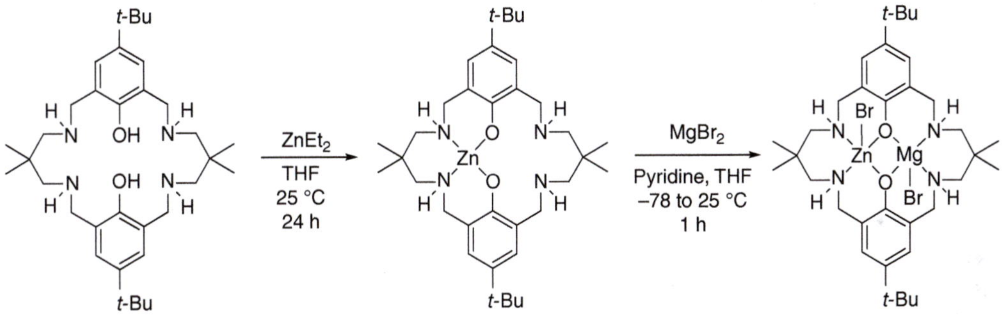

Scheme 4.16 Synthesis of heterobimetallic Zn/Mg complex **50**.

However, the mixture containing the heterodinuclear catalyst showed an activity greater than either of the homodinuclear complexes or a 1:1 combination of them [234]. In a follow-up study [175], a heterobimetallic Zn/Mg complex **50** could be isolated in pure form by using a modified two-step procedure, in which the use of coordinating solvent and low temperature was found to be crucial to the outcome of the reaction. The mono zinc species is prepared by reaction of the macrocycle with $ZnEt_2$ at room temperature, followed by addition of $MgBr_2$ in a pyridine/THF mixture at −78 °C (Scheme 4.16).

When a solution of **50** in d_8-THF is heated at 80 °C for 24 h, the ^{1}H NMR spectrum of the complex remains unchanged, suggesting that the structure and composition of the complex is retained in solution and at high temperature. In contrast, heating an equimolar mixture of di–Zn and di–Mg for 48 h at 80 °C leads to the formation of 25% of **50** (as seen in the ^{1}H NMR spectrum). These experiments suggest that the heterobimetallic species is thermodynamically favored but kinetically slow to form once some di-nuclear complexes are formed. Hence, the importance of incorporating the second metal center into the mono-Zn species. Heterobimetallic **50** significantly outperforms the homobimetallic Mg_2 and Zn_2 complexes in polymerization, and is about five times more active than using a 1:1 molar ratio of Mg_2/Zn_2 mixture. The monometallic zinc intermediate is also completely inactive, most likely because of the lack of initiating group. The polymers obtained show > 99% of carbonate linkages and display narrow molecular weight distributions, indicative of controlled polymerization. The activity enhancement is believed to stem from the distinct and complementary properties of the two metals acting on the RDS (ring-opening of the epoxide): the Lewis acid zinc ion enhancing epoxide coordination, the labile magnesium accelerating carbonate attack. Further investigation into analogous Zn/Mg catalysts has shown that by applying carboxylate ligands in place of halides, it is possible to significantly enhance turnover-frequency values [235].

Using the same sequential metalation procedure, the team further reported heterobimetallic complexes featuring both Zn(II) and a metal from group 1 (Li, Na, K), 2 (Mg, Ca) [236], and 13 (Al(III), Ga(III) or In(III)) [237] chelated by the macrocyclic diphenolate-tetraamine ligand used previously, with conversions over 90% as determined by NMR spectroscopy, and typically reasonable isolated yields (> 70%) (Figure 4.50). These complexes have been characterized in the solid state, including when possible by single crystal X-ray diffraction (see for example complex **51**; Figure 4.51), in solution using NMR spectroscopy, and by mass spectrometry. Alternatively, some heterodinuclear complexes can be accessed by reaction between the two homodinuclear

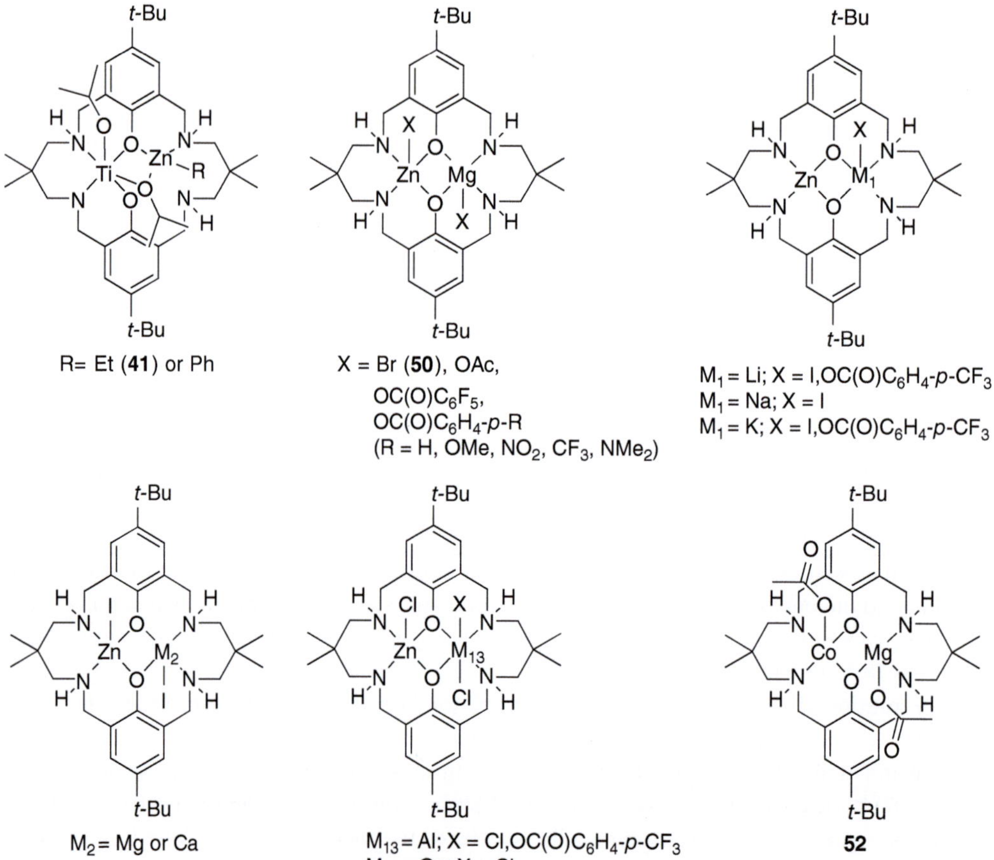

R= Et (**41**) or Ph

X = Br (**50**), OAc, $OC(O)C_6F_5$, $OC(O)C_6H_4$-*p*-R (R = H, OMe, NO_2, CF_3, NMe_2)

M_1 = Li; X = I, $OC(O)C_6H_4$-*p*-CF_3
M_1 = Na; X = I
M_1 = K; X = I, $OC(O)C_6H_4$-*p*-CF_3

M_2 = Mg or Ca

M_{13} = Al; X = Cl, $OC(O)C_6H_4$-*p*-CF_3
M_{13} = Ga; X = Cl
M_{13} = In; X = Cl, $B(C_6F_5)_4$, $OC(O)C_6H_4$-*p*-CF_3 (**51**)

52

Figure 4.50 Heterobimetallic complexes supported by a macrocyclic diphenolate ligand, synthesized by Williams.

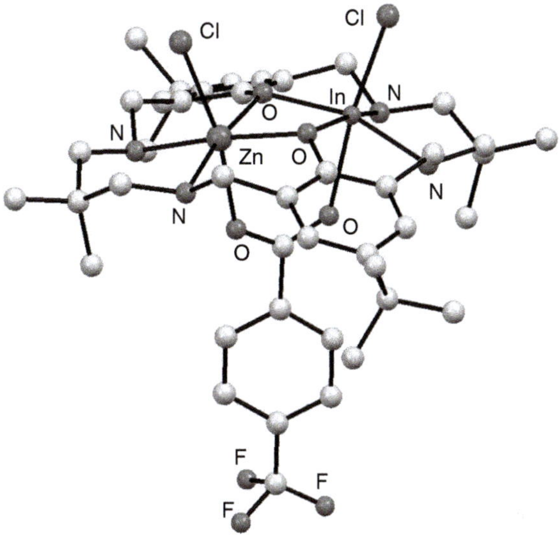

Figure 4.51 Solid-state molecular structure of Zn/In complex **51** obtained by X-ray diffraction analysis. *Source:* Adapted from Deacy et al. [237].

complexes at elevated temperatures for extended periods causing metal redistribution. This is in agreement with previous findings that some heterodinuclear complexes are the thermodynamic reaction products (with some exception such as the Na(I)/Zn(II)).

These metal combinations are not all synergic: nearly all the heterodinuclear complexes are less active than the di-Zn analogues, but the Mg(II)/Zn(II) catalyst is more active. The co-ligand influences the product selectivity, with iodide ligands resulting in cyclic carbonate formation and carboxylate ligands giving a high selectivity for polycarbonate. For group 13 complexes, CO_2 uptake and selectivity towards carbonate linkages in the polymer increases in the order Al(III) < Ga(III) < In(III), which is attributed to the lower Lewis acidity of heavier group 13 metals (high Lewis acidity promoting polyether formation). In parallel, catalytic activity also increases down the group 13 series, consistent with weaker metal–oxygen bonds, more amenable to insertion reactions.

It is worth noting that heterobimetallic Ti/Zn complex **41** mentioned previously in the ROP section is one that also uses a sequential metalation procedure, but in reverse order, starting with Ti metalation, then incorporation of the zinc center [174]. **41** was also tested in cyclohexene oxide and CO_2 copolymerization, but only exhibits moderate activity (less than the di-zinc analogue), whereas the monometallic titanium is inactive.

Capitalizing on these results, as well as on previous findings which showed that a di-Co(II) catalyst outperforms the di-zinc analogue [216], the Williams group has recently prepared a Mg/Co heterobimetallic complex **52**, via an adapted sequential metalation procedure (first $Mg\{N(SiMe_3)_2\}_2$, then $Co(OAc)_2$) [238]. The resulting complex has been characterized by elemental analysis, mass spectrometry, IR, XPS, superconducting quantum interference device (SQUID), and electrochemical measurements, which confirmed the heterobimetallic composition.

This catalyst shows the highest activity and selectivity for polymer formation of all (no cyclic carbonate was observed) at loadings as low as 1:4,000 or 0.025 mol% (catalyst/epoxide). It also shows excellent CO_2 selectivity (> 99%) across the range of temperatures tested (80–120 °C), with no ether linkages observed by ^{1}H NMR spectroscopy, as well as well-controlled, monomodal, polymer molecular weight distributions. The synergistic effect of this heterobimetallic combination is supported by catalyst benchmarking studies through the comparison of rate coefficients (k_p), other

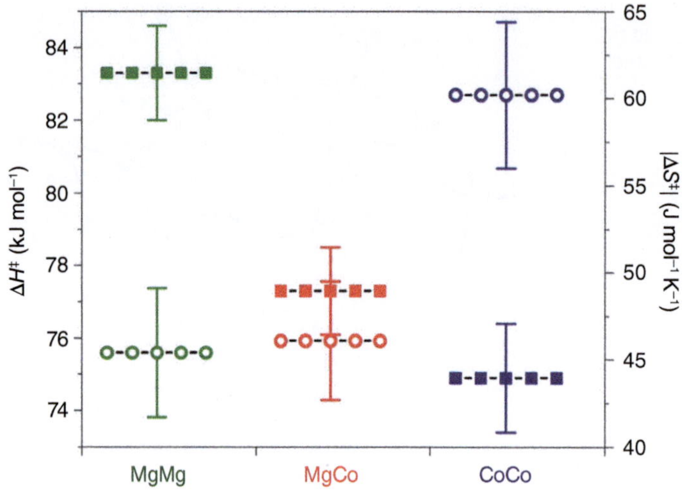

Figure 4.52 $\Delta H^{\ddagger}$ (solid squares, with errors ±1.3) and $|\Delta S\ddagger|$ (open circles, with errors ±3.7) values obtained from kinetic analysis of CHO/CO_2 ROCOP catalysed by Mg/Mg, Co/Co and Mg/Co **52**. *Source:* Deacy et al. [238]; © 2020 Springer Nature.

dinuclear complexes based on the same ligand showing the same rate law. At 80 °C, the Mg/Co complex **52** is 85 times faster than the original Zn/Zn complex, and over 1000 times faster at 120 °C. At this temperature, the Mg/Co catalyst is also twice as fast as the fastest reported Mg/Zn. It is also significantly more active than either of the homodinuclear analogues (Mg/Mg or Co/Co): with double the rate of the Co/Co complex and four times the rate of Mg/Mg.

As a control, an equimolar mixture of Mg/Mg and Co/Co was tested in catalysis. The rate of reaction was an average of the rates of the two homobimetallic complexes, and the polymerization was three times slower than for Mg/Co **52**. This experiment highlights the synergy between the Mg and Co(II) metals within the same ligand framework, and also indirectly shows that there is no appreciable conversion of the homodinuclear complexes to the Mg/Co species under the conditions of the catalysis. While computational modelling has not been performed on this system, kinetic analysis reveals that the synergy arises because the magnesium center enhances the transition state entropy through reducing the barriers to epoxide coordination, and cobalt reduces the transition state enthalpy by the enhanced nucleophilicity of the cobalt carbonate. Overall, this leads to decreased free enthalpy of transition state and high catalytic activities (see Figure 4.52).

4.4 Conclusion

In conclusion, looking at the wealth of heterobimetallic complexes that have been developed over the years and applied successfully in small molecule activation and in polymerization catalysis, there is a clear benefit in targeting metal complexes with the right metal combinations to give rise to synergic interactions. However, it is also obvious that the rational design of such catalysts still holds many challenges.

The preparation and isolation of pure heterobimetallic compounds is far from being straightforward, in particular in the case of symmetrical ligands where there is no differentiation in coordination environment for the metals, and which can lead to intractable mixtures of homobimetallic and heterobimetallic species. Another obstacle can also be metal redistribution under the conditions of catalysis, which can annihilate any previous synthetic efforts. Symmetric ligands might also not be able to accommodate specific M_1–M_2 combinations due to the different coordination preferences of the individual metals. Non-symmetric di-nucleating ligands can address this challenge by having two distinct coordination environments, targeting two different metal ions, but they typically need more synthetic effort. Equally, they would also make the synthesis of homobimetallic complexes (for control) more challenging. Evidence that the desired heterobimetallic complexes retain their structure and composition under the conditions of catalysis (in particular at high temperature or upon reaction with a substrate) is paramount to confirm any metal cooperativity and validate any mechanistic investigation that will feed into a subsequent catalyst design process. To that effect, *in operando* techniques such as DOSY NMR spectroscopy are particularly useful.

Another challenge in heterobimetallic catalysis is confirming any metal cooperative effect and understanding its origin. Compelling evidence for M_1–M_2 cooperativity usually comes from reactivity studies. If a heterobimetallic catalyst displays reactivity that is different from that of the corresponding homobimetallics or a mixture of monometallic analogues, then a case can be made for heterobimetallic interactions (beneficial or not to the reaction), enabled by a ligand framework holding both metals in close proximity. However, synthesizing suitable homobimetallic or monometallic complexes for reactivity comparisons may be impossible. In addition, even if a heterobimetallic effect is observed experimentally, investigating the specific roles of each metal is far from trivial. While a variety of modern spectroscopic tools can be employed to study further the mechanism of reactions mediated by these heterobimetallic complexes, this can be particularly challenging if key intermediates cannot be trapped or observed. In that case, but not only, modern computational modelling techniques have the potential to provide detailed insight into the origin of any metal cooperativity that might occur, synergic or not, both at the thermodynamics and kinetics levels. They can even be used predictively, saving considerable synthetic efforts.

Interest in heterobimetallic systems is growing in the catalysis community, especially as such systems are showing great promise in fields related to sustainability challenges, including the activation and utilisation of small abundant molecules such as N_2 and CO_2, energy vectors such as H_2, or the polymerization (and depolymerization) of renewable resources towards more sustainable materials. Further research in this area will undoubtedly lead to the development of new catalytic systems but also novel experimental tools and computational methods to probe the mechanisms of these reactions.

References

1 Edwards, J.K., Freakley, S.J., Carley, A.F. et al. (**2014**). *Acc. Chem. Res. 47*: 845–854.
2 Mata, J.A., Hahn, F.E., and Peris, E. (**2014**). *Chem. Sci. 5*: 1723–1732.
3 Park, J. and Hong, S. (**2012**). *Chem. Soc. Rev. 41*: 6931–6943.
4 Bratko, I. and Gómez, M. (**2013**). *Dalton Trans. 42*: 10664–10681.
5 Buchwalter, P., Rosé, J., and Braunstein, P. (**2015**). *Chem. Rev. 115*: 28–126.
6 Dance, I. (**2013**). *Chem. Commun. 49*: 10893–10907.
7 Fontecilla-Camps, J.C., Volbeda, A., Cavazza, C., and Nicolet, Y. (**2007**). *Chem. Rev. 107*: 4273–4303.
8 Steinhagen, H. and Helmchen, G. (**1996**). *Angew. Chem. Int. Ed. Engl. 35*: 2339–2342.
9 Artero, V. and Fontecave, M. (**2005**). *Coord. Chem. Rev. 249*: 1518–1535.
10 Gloaguen, F. and Rauchfuss, T.B. (**2009**). *Chem. Soc. Rev. 38*: 100–108.

11 Kaur-Ghumaan, S. and Stein, M. (**2014**). *Dalton Trans. 43*: 9392–9405.
12 Barton, B.E., Olsen, M.T., and Rauchfuss, T.B. (**2010**). *Curr. Opin. Biotechnol. 21*: 292–297.
13 Ogo, S., Kabe, R., Uehara, K. et al. (**2007**). *Science 316*: 585–587.
14 Whyman, R., Wright, A.P., Iggo, J.A., and Heaton, B.T. (**2002**). *J. Chem. Soc. Dalton Trans.*: 771–777.
15 Haynes, A., Maitlis, P.M., Morris, G.E. et al. (**2004**). *J. Am. Chem. Soc. 126*: 2847–2861.
16 Sunley, G.J. and Watson, D.J. (**2000**). *Catal. Today 58*: 293–307.
17 Chipman, J.A. and Berry, J.F. (**2020**). *Chem. Rev. 120*: 2409–2447.
18 Bluemke, T.D., Clegg, W., García-Alvarez, P. et al. (**2014**). *Chem. Sci. 5*: 3552–3562.
19 Hevia, E., Chua, J.Z., García-Álvarez, P. et al. (**2010**). *Proc. Natl. Acad. Sci. U. S. A. 107*: 5294–5299.
20 Martínez-Martínez, A.J., Kennedy, A.R., Mulvey, R.E., and O'Hara, C.T. (**2014**). *Science 346*: 834–837.
21 Andrikopoulos, P.C., Armstrong, D.R., Barley, H.R.L. et al. (**2005**). *J. Am. Chem. Soc. 127*: 6184–6185.
22 Mulvey, R.E., Mongin, F., Uchiyama, M., and Kondo, Y. (**2007**). *Angew. Chem. Int. Ed. 46*: 3802–3824.
23 Bakewell, C., Ward, B.J., White, A.J.P., and Crimmin, M.R. (**2018**). *Chem. Sci. 9*: 2348–2356.
24 Coates, G., Rekhroukh, F., and Crimmin, M.R. (**2019**). *Synlett 30*: 2233–2246.
25 Cooper, B.G., Napoline, J.W., and Thomas, C.M. (**2012**). *Catal. Rev. 54*: 1–40.
26 Riddlestone, I.M., Rajabi, N.A., Lowe, J.P. et al. (**2016**). *J. Am. Chem. Soc. 138*: 11081–11084.
27 Riddlestone, I.M., Rajabi, N.A., Macgregor, S.A. et al. (**2018**). *Chem. Eur. J. 24*: 1732–1738.
28 Miloserdov, F.M., Rajabi, N.A., Lowe, J.P. et al. (**2020**). *J. Am. Chem. Soc. 142*: 6340–6349.
29 Cole-Hamilton, D.J. and Wilkinson, G. (**1977**). *J. Chem. Soc. Dalton Trans.*: 797–804.
30 Karunananda, M.K. and Mankad, N.P. (**2015**). *J. Am. Chem. Soc. 137*: 14598–14601.
31 Karunananda, M.K. and Mankad, N.P. (**2017**). *Organometallics 36*: 220–227.
32 Parmelee, S.R., Mazzacano, T.J., Zhu, Y. et al. (**2015**). *ACS Catal. 5*: 3689–3699.
33 Cammarota, R.C., Clouston, L.J., and Lu, C.C. (**2017**). *Coord. Chem. Rev. 334*: 100–111.
34 Bouhadir, G. and Bourissou, D. (**2017**). *The Chemical Bond III* (ed. D.M.P. Mingos), 141–201. Cham: Springer International Publishing.
35 Amgoune, A. and Bourissou, D. (**2011**). *Chem. Commun. 47*: 859–871.
36 Harman, W.H. and Peters, J.C. (**2012**). *J. Am. Chem. Soc. 134*: 5080–5082.
37 Lin, T.P. and Peters, J.C. (**2013**). *J. Am. Chem. Soc. 135*: 15310–15313.
38 Fong, H., Moret, M.E., Lee, Y., and Peters, J.C. (**2013**). *Organometallics 32*: 3053–3062.
39 Harman, W.H., Lin, T.-P., and Peters, J.C. (**2014**). *Angew. Chem. Int. Ed. 53*: 1081–1086.
40 Clouston, L.J., Siedschlag, R.B., Rudd, P.A. et al. (**2013**). *J. Am. Chem. Soc. 135*: 13142–13148.
41 Rudd, P.A., Liu, S., Planas, N. et al. (**2013**). *Angew. Chem. Int. Ed. 52*: 4449–4452.
42 Clouston, L.J., Bernales, V., Cammarota, R.C. et al. (**2015**). *Inorg. Chem. 54*: 11669–11679.
43 Vollmer, M.V., Xie, J., and Lu, C.C. (**2017**). *J. Am. Chem. Soc. 139*: 6570–6573.
44 Vollmer, M.V., Xie, J., Cammarota, R.C. et al. (**2018**). *Angew. Chem. Int. Ed. 57*: 7815–7819.
45 Cammarota, R.C., Xie, J., Burgess, S.A. et al. (**2019**). *Chem. Sci. 10*: 7029–7042.
46 Vollmer, M.V., Xie, J., Cammarota, R.C. et al. (**2018**). *Angew. Chem. Int. Ed. 57*: 7815–7819.
47 Cammarota, R.C. and Lu, C.C. (**2015**). *J. Am. Chem. Soc. 137*: 12486–12489.
48 Cammarota, R.C., Vollmer, M.V., Xie, J. et al. (**2017**). *J. Am. Chem. Soc. 139*: 14244–14250.
49 Ye, J., Cammarota, R.C., Xie, J. et al. (**2018**). *ACS Catal. 8*: 4955–4968.
50 Vollmer, M.V., Ye, J., Linehan, J.C. et al. (**2020**). *ACS Catal. 10*: 2459–2470.
51 Rudd, P.A., Liu, S., Gagliardi, L. et al. (**2011**). *J. Am. Chem. Soc. 133*: 20724–20727.
52 Alex Rudd, P., Planas, N., Bill, E., Gagliardi, L., Lu, C. C. (**2013**). *Eur. J. Inorg. Chem. 2013*, 3898–3906.

53 Clouston, L.J., Bernales, V., Carlson, R.K. et al. (**2015**). *Inorg. Chem. 54*: 9263–9270.
54 Siedschlag, R.B., Bernales, V., Vogiatzis, K.D. et al. (**2015**). *J. Am. Chem. Soc. 137*: 4638–4641.
55 Liu, Q., Wu, L., Jackstell, R., and Beller, M. (**2015**). *Nat. Commun. 6*: 1–15.
56 Kleij, A.W., North, M., and Urakawa, A. (**2017**). *ChemSusChem 10*: 1036–1038.
57 Jin, X., DIng, J., Xia, Q. et al. (**2019**). *J. CO_2 Util. 34*: 115–148.
58 Gibson, D.H. (**1996**). *Chem. Rev. 96*: 2063–2095.
59 Devillard, M., Declercq, R., Nicolas, E. et al. (**2016**). *J. Am. Chem. Soc. 138*: 4917–4926.
60 Takaya, J. and Iwasawa, N. (**2017**). *J. Am. Chem. Soc. 139*: 6074–6077.
61 Steinhoff, P., Paul, M., Schroers, J.P., and Tauchert, M.E. (**2019**). *Dalton Trans. 48*: 1017–1022.
62 Chum, P.S. and Swogger, K.W. (**2008**). *Prog. Polym. Sci. 33*: 797–819.
63 Eagan, J.M., Xu, J., Di Girolamo, R. et al. (**2017**). *Science 355*: 814–816.
64 Carrow, B.P. and Nozaki, K. (**2014**). *Macromolecules 47*: 2541–2555.
65 Nakamura, A., Ito, S., and Nozaki, K. (**2009**). *Chem. Rev. 109*: 5215–5244.
66 Galli, P. and Vecellio, G. (**2001**). *Prog. Polym. Sci. 26*: 1287–1336.
67 Bernardi, F., Bottoni, A., and Pietro Miscione, G. (**1998**). *Organometallics 17*: 16–24.
68 Hooper, M.S. and Michalak, A. (**2018**). *Handbook of Transition Metal Polymerization Catalyst*, 67–130. Hoboken, NJ, USA: John Wiley & Sons, Inc.
69 Breslow, D.S. and Newburg, N.R. (**2002**). *J. Am. Chem. Soc. 79*: 5072–5073.
70 Sishta, C., Hathorn, R.M., and Marks, T.J. (**1992**). *J. Am. Chem. Soc. 114*: 1112–1114.
71 Fusco, R., Longo, L., Masi, F., and Garbassi, F. (**1997**). *Macromol. Rapid Commun. 18*: 433–441.
72 Sinn, H., Kaminsky, W., Vollmer, H.-J., and Woldt, R. (**1980**). *Angew. Chem. Int. Ed. Engl. 19*: 390–392.
73 Kaminsky, W. (**1998**). *J. Chem. Soc. Dalton Trans.*: 1413–1418.
74 Bochmann, M. (**2010**). *Organometallics 29*: 4711–4740.
75 Xu, Z., Vanka, K., Firman, T. et al. (**2002**). *Organometallics 21*: 2444–2453.
76 Kumawat, J. and Gupta, V.K. (**2020**). *ACS Catal. 10*: 1704–1715.
77 Macchioni, A. (**2005**). *Chem. Rev. 105*: 2039–2073.
78 Chen, E.Y.X. and Marks, T.J. (**2000**). *Chem. Rev. 100*: 1391–1434.
79 Möhring, P.C. and Coville, N.J. (**2006**). *Coord. Chem. Rev. 250*: 18–35.
80 Razavi, A. and Thewalt, U. (**2006**). *Coord. Chem. Rev. 250*: 155–169.
81 Hou, Z. and Wakatsuki, Y. (**2002**). *Coord. Chem. Rev. 231*: 1–22.
82 Thuilliez, J., Spitz, R., and Boisson, C. (**2006**). *Macromol. Chem. Phys. 207*: 1727–1731.
83 Bogaert, S., Chenal, T., Mortreux, A., and Carpentier, J.F. (**2002**). *J. Mol. Catal. A Chem. 190*: 207–214.
84 Pelletier, J.F., Mortreux, A., Olonde, X., and Bujadoux, K. (**1996**). *Angew. Chem. Int. Ed. Engl. 35*: 1854–1856.
85 Ribeiro, R., Ruivo, R., Nsiri, H. et al. (**2016**). *ACS Catal. 6*: 851–860.
86 Li, H. and Marks, T.J. (**2006**). *Proc. Natl. Acad. Sci. U. S. A. 103*: 15295–15302.
87 Delferro, M. and Marks, T.J. (**2011**). *Chem. Rev. 111*: 2450–2485.
88 Suo, H., Solan, G.A., Ma, Y., and Sun, W.H. (**2018**). *Coord. Chem. Rev. 372*: 101–116.
89 Yamaguchi, Y. (**1999**). *Organometallics 18*: 996–1001.
90 Takayama, C., Yamaguchi, Y., Mise, T., and Suzuki, N. (**2001**). *J. Chem. Soc. Dalton Trans.*: 948–953.
91 Lindenberg, F., Shribman, T., Sieler, J. et al. (**1996**). *J. Organomet. Chem. 515*: 19–25.
92 Shribman, T., Kurz, S., Senff, U. et al. (**1998**). *J. Mol. Catal. A Chem. 129*: 191–198.
93 Singh, S. and Roesky, H.W. (**2007**). *J. Chem. Soc. Dalton Trans.*: 1360–1370.
94 Mandal, S.K. and Roesky, H.W. (**2010**). *Acc. Chem. Res. 43*: 248–259.

95 Gurubasavaraj, P.M., Roesky, H.W., Sharma, P.M.V. et al. (**2007**). *Organometallics 26*: 3346–3351.
96 Bai, G., Singh, S., Roesky, H.W. et al. (**2005**). *J. Am. Chem. Soc. 127*: 3449–3455.
97 Gurubasavaraj, P.M., Mandal, S.K., Roesky, H.W. et al. (**2007**). *Inorg. Chem. 46*: 1056–1061.
98 Gurubasavaraj, P.M., Roesky, H.W., Nekoueishahraki, B. et al. (**2008**). *Inorg. Chem. 47*: 5324–5331.
99 Nikiforov, G.B., Roesky, H.W., Jones, P.G. et al. (**2006**). *J. Chem. Soc. Dalton Trans.*: 4149–4159.
100 Kulangara, S.V., Jabri, A., Yang, Y. et al. (**2012**). *Organometallics 31*: 6085–6094.
101 Jabri, A., Mason, C.B., Sim, Y. et al. (**2008**). *Angew. Chem. Int. Ed. 47*: 9717–9721.
102 Braunschweig, H. and Breitling, F.M. (**2006**). *Coord. Chem. Rev. 250*: 2691–2720.
103 Cano, J. and Kunz, K. (**2007**). *J. Organomet. Chem. 692*: 4411–4423.
104 Yang, S.H., Huh, J., and Jo, W.H. (**2005**). *Macromolecules 38*: 1402–1409.
105 Lanza, G., Fragalà, I.L., and Marks, T.J. (**2002**). *Organometallics 21*: 5594–5612.
106 Motta, A., Fragalà, I.L., and Marks, T.J. (**2009**). *J. Am. Chem. Soc. 131*: 3974–3984.
107 Wang, J., Li, H., Guo, N. et al. (**2004**). *Organometallics 23*: 5112–5114.
108 Liu, S., Motta, A., Delferro, M., and Marks, T.J. (**2013**). *J. Am. Chem. Soc. 135*: 8830–8833.
109 Liu, S., Motta, A., Mouat, A.R. et al. (**2014**). *J. Am. Chem. Soc. 136*: 10460–10469.
110 Ittel, S.D., Johnson, L.K., and Brookhart, M. (**2000**). *Chem. Rev. 100*: 1169–1203.
111 Younkin, T.R., Connor, E.F., Henderson, J.I. et al. (**2000**). *Science 287*: 460–462.
112 Guo, L., Liu, W., and Chen, C. (**2017**). *Mater. Chem. Front. 1*: 2487–2494.
113 Kuwabara, J., Takeuchi, D., and Osakada, K. (**2006**). *Chem. Commun.* 3815–3817.
114 Tanabiki, M., Tsuchiya, K., Motoyama, Y., and Nagashima, H. (**2005**). *Chem. Commun.* 3409–3411.
115 Xiao, D. and Do, L.H. (**2018**). *Organometallics 37*: 254–260.
116 Xiao, D. and Do, L.H. (**2018**). *Organometallics 37*: 3079–3085.
117 Smith, A.J., Kalkman, E.D., Gilbert, Z.W., and Tonks, I.A. (**2016**). *Organometallics 35*: 2429–2432.
118 Chiu, H.C., Pearce, A.J., Dunn, P.L. et al. (**2016**). *Organometallics 35*: 2076–2085.
119 Chiu, H.C., Koley, A., Dunn, P.L. et al. (**2017**). *Dalton Trans. 46*: 5513–5517.
120 Cai, Z., Xiao, D., and Do, L.H. (**2015**). *J. Am. Chem. Soc. 137*: 15501–15510.
121 Cai, Z. and Do, L.H. (**2017**). *Organometallics 36*: 4691–4698.
122 Tran, T.V., Nguyen, Y.H., and Do, L.H. (**2019**). *Polym. Chem. 10*: 3718–3721.
123 Contrella, N.D., Sampson, J.R., and Jordan, R.F. (**2014**). *Organometallics 33*: 3546–3555.
124 Cai, Z. and Do, L.H. (**2018**). *Organometallics 37*: 3874–3882.
125 Carothers, W.H., Borough, G.L., and Natta, F.J. (**1932**). *J. Am. Chem. Soc. 54*: 761–772.
126 Parwe, S.P., Warkad, S.D., Mane, M.V. et al. (**2017**). *Polym. 111*: 244–251.
127 Conn, R.E., Kolstad, J.J., Borzelleca, J.F. et al. (**1995**). *Food Chem. Toxicol. 33*: 273–283.
128 Tanzi, M.C., Verderio, P., Lampugnani, M.G. et al. (**1994**). *J. Mater. Sci. Mater. Med. 5*: 393–396.
129 Byers, J. A., Biernesser, A. B., Delle Chiaie, K. R., Kaur, A., Kehl, J. A., (**2018**). in *Synthesis, Structure and Properties of Poly(lactic acid). Advances in Polymer Science*, vol. 279 (ed. M. Di Lorenzo, R. Androsch), 67–118. Cham: Springer International Publishing.
130 Platel, R., Hodgson, L., and Williams, C. (**2008**). *Polym. Rev. 48*: 11–63.
131 Ajellal, N., Carpentier, J.F., Guillaume, C. et al. (**2010**). *Dalton Trans. 39*: 8363–8376.
132 Arbaoui, A. and Redshaw, C. (**2010**). *Polym. Chem. 1*: 801–826.
133 Guillaume, S.M. and Carpentier, J.F. (**2012**). Catal. *Sci. Technol. 2*: 898–906.
134 Kricheldorf, H.R., Berl, M., and Scharnagl, N. (**1988**). *Macromolecules 21*: 286–293.
135 Dubois, P., Jacobs, C., Jérôme, R., and Teyssté, P. (**1991**). *Macromolecules 24*: 2266–2270.
136 Kremer, A.B. and Mehrkhodavandi, P. (**2019**). *Coord. Chem. Rev. 380*: 35–57.
137 Williams, C.K., Breyfogle, L.E., Choi, S.K. et al. (**2003**). *J. Am. Chem. Soc. 125*: 11350–11359.
138 Chamberlain, B.M., Cheng, M., Moore, D.R. et al. (**2001**). *J. Am. Chem. Soc. 123*: 3229–3238.

139 Gruszka, W., Walker, L.C., Shaver, M.P., and Garden, J.A. (**2020**). *Macromolecules 53*: 4294–4302.
140 Thevenon, A., Romain, C., Bennington, M.S. et al. (**2016**). *Angew. Chem. Int. Ed. 55*: 8680–8685.
141 Dove, A.P. (**2012**). *ACS Macro Lett. 1*: 1409–1412.
142 Lin, B. and Waymouth, R.M. (**2017**). *J. Am. Chem. Soc. 139*: 1645–1652.
143 Zhang, X., Jones, G.O., Hedrick, J.L., and Waymouth, R.M. (**2016**). *Nat. Chem. 8*: 1047–1053.
144 Lin, B., Hedrick, J.L., Park, N.H., and Waymouth, R.M. (**2019**). *J. Am. Chem. Soc. 141*: 8921–8927.
145 Kazakov, O.I., Datta, P.P., Isajani, M. et al. (**2014**). *Macromolecules 47*: 7463–7468.
146 Piedra-Arroni, E., Amgoune, A., and Bourissou, D. (**2013**). *Dalton Trans. 42*: 9024–9029.
147 Patel, D., Liddle, S.T., Mungur, S.A. et al. (**2006**). *Chem. Commun.* 1124–1126.
148 Pietrangelo, A., Hillmyer, M.A., and Tolman, W.B. (**2009**). *Chem. Commun. 9*: 2736–2737.
149 Pietrangelo, A., Knight, S.C., Gupta, A.K. et al. (**2010**). *J. Am. Chem. Soc. 132*: 11649–11657.
150 Thongkham, S., Monot, J., Martin-Vaca, B., and Bourissou, D. (**2019**). *Macromolecules 52*: 8103–8113.
151 Naumann, S., Scholten, P.B.V., Wilson, J.A., and Dove, A.P. (**2015**). *J. Am. Chem. Soc. 137*: 14439–14445.
152 McGraw, M.L. and Chen, E.Y.X. (**2020**). *Macromolecules 53*: 6102–6612.
153 Gregson, C.K.A., Gibson, V.C., Long, N.J. et al. (**2006**). *J. Am. Chem. Soc. 128*: 7410–7411.
154 Broderick, E.M., Guo, N., Vogel, C.S. et al. (**2011**). *J. Am. Chem. Soc. 133*: 9278–9281.
155 Xu, X., Luo, G., Mehmood, A. et al. (**2018**). *Organometallics 37*: 4599–4607.
156 Xu, X., Luo, G., Hou, Z. et al. (**2020**). *Inorg. Chem. Front. 7*: 961–971.
157 Sauer, A., Buffet, J.-C., Spaniol, T.P. et al. (**2013**). *ChemCatChem 5*: 1088–1091.
158 Biernesser, A.B., Chiaie, K.R.D., Curley, J.B., and Byers, J.A. (**2016**). *Angew. Chem. Int. Ed. 55*: 5251–5254.
159 Chen, H.Y., Liu, M.Y., Sutar, A.K., and Lin, C.C. (**2010**). *Inorg. Chem. 49*: 665–674.
160 Hsueh, M.L., Huang, B.H., Wu, J., and Lin, C.C. (**2005**). *Macromolecules 38*: 9482–9487.
161 Normand, M., Kirillov, E., Roisnel, T., and Carpentier, J.F. (**2012**). *Organometallics 31*: 1448–1457.
162 Hayes, C.E., Sarazin, Y., Katz, M.J. et al. (**2013**). *Organometallics 32*: 1183–1192.
163 Sheng, H.T., Li, J.M., Zhang, Y. et al. (**2008**). *Polyhedron 27*: 1665–1672.
164 Wang, L., Zhang, J., Yao, L. et al. (**2011**). *Inorg. Chem. Commun. 14*: 859–862.
165 Sun, Y., Wang, L., Yu, D. et al. (**2014**). *J. Mol. Catal. A Chem. 393*: 175–181.
166 Hild, F., Haquette, P., Brelot, L., and Dagorne, S. (**2010**). *Dalton Trans. 39*: 533–540.
167 Hild, F., Brelot, L., and Dagorne, S. (**2011**). *Organometallics 30*: 5457–5462.
168 Char, J., Brulé, E., Gros, P.C. et al. (**2015**). *J. Organomet. Chem. 796*: 47–52.
169 Maudoux, N., Roisnel, T., Carpentier, J.F., and Sarazin, Y. (**2014**). *Organometallics 33*: 5740–5748.
170 Jin, W.J., Ding, L.Q., Chu, Z. et al. (**2011**). *J. Mol. Catal. A Chem. 337*: 25–32.
171 Ding, L., Jin, W., Chu, Z. et al. (**2011**). *Inorg. Chem. Commun. 14*: 1274–1278.
172 Xiao, G., Yan, B., Ma, R. et al. (**2011**). *Polym. Chem. 2*: 659–664.
173 Hao, J., Li, J., Cui, C., and Roesky, H.W. (**2011**). *Inorg. Chem. 50*: 7453–7459.
174 Garden, J.A., White, A.J.P., and Williams, C.K. (**2017**). *Dalton Trans. 46*: 2532–2541.
175 Garden, J.A., Saini, P.K., and Williams, C.K. (**2015**). *J. Am. Chem. Soc. 137*: 15078–15081.
176 Williams, C.K., Brooks, N.R., Hillmyer, M.A., and Tolman, W.B. (**2002**). *Chem. Commun. 2*: 2132–2133.
177 Gao, A.H., Yao, W., Mu, Y. et al. (**2009**). *Polyhedron 28*: 2605–2610.
178 Yao, W., Mu, Y., Gao, A. et al. (**2008**). *Dalton Trans.* 3199–3206.
179 Hultzsch, K.C., Spaniol, T.P., and Okuda, J. (**1997**). *Organometallics 16*: 4845–4856.
180 Beckerle, K., Hultzsch, K.C., and Okuda, J. (**1999**). *Macromol. Chem. Phys. 200*: 1702–1707.
181 Wang, L., Roşca, S.C., Poirier, V. et al. (**2014**). *Dalton Trans. 43*: 4268–4286.

182 Poirier, V., Roisnel, T., Sinbandhit, S. et al. (**2012**). *Chem. Eur. J. 18*: 2998–3013.
183 Brignou, P., Guillaume, S.M., Roisnel, T. et al. (**2012**). *Chem. Eur. J. 18*: 9360–9370.
184 Thomas, R.M., Widger, P.C.B., Ahmed, S.M. et al. (**2010**). *J. Am. Chem. Soc. 132*: 16520–16525.
185 Ajiro, H., Peretti, K.L., Lobkovsky, E.B., and Coates, G.W. (**2009**). *Dalton Trans.* 8828–8830.
186 Morris, L.S., Childers, M.I., and Coates, G.W. (**2018**). *Angew. Chem. Int. Ed. 57*: 5731–5734.
187 Childers, M.I., Vitek, A.K., Morris, L.S. et al. (**2017**). *J. Am. Chem. Soc. 139*: 11048–11054.
188 Ren, W.M., Wang, R.J., Ren, B.H. et al. (**2020**). *Chinese J. Polym. Sci. (English Ed.) 38*: 950–957.
189 McGuire, T.M., López-Vidal, E.M., Gregory, G.L., and Buchard, A. (**2018**). *J. CO_2 Util. 27*: 283–288.
190 Gregory, G.L., Ulmann, M., and Buchard, A. (**2015**). *RSC Adv. 5*: 39404–39408.
191 Brege, A., Méreau, R., McGehee, K. et al. (**2020**). *J. CO_2 Util. 38*: 88–98.
192 Huang, J., Worch, J.C., Dove, A.P., and Coulembier, O. (**2020**). *ChemCatChem 13*: 469–487.
193 Inoue, S., Koinuma, H., and Tsuruta, T. (**1969**). *J. Polym. Sci. Part B Polym. Lett. 7*: 287–292.
194 Hepburn, C., Adlen, E., Beddington, J. et al. (**2019**). *Nature 575*: 87–97.
195 Darensbourg, D.J. (**2007**). *Chem. Rev. 107*: 2388–2410.
196 Kember, M.R., Buchard, A., and Williams, C.K. (**2011**). *Chem. Commun. 47*: 141–163.
197 Wang, Y. and Darensbourg, D.J. (**2018**). *Coord. Chem. Rev. 372*: 85–100.
198 Childers, M.I., Longo, J.M., Van Zee, N.J. et al. (**2014**). *Chem. Rev. 114*: 8129–8152.
199 Kozak, C.M., Ambrose, K., and Anderson, T.S. (**2018**). *Coord. Chem. Rev. 376*: 565–587.
200 Romain, C., Thevenon, A., Saini, P.K., and Williams, C.K. (**2015**). *Carbon Dioxide and Organometallics, Topics in Organometallic Chemistry*, vol. *53* (ed. X.-B. Lu), 101–141. Springer International Publishing.
201 Trott, G., Saini, P.K., and Williams, C.K. (**2016**). *Phil. Trans. R. Soc. A. 374*: 20150085.
202 Klaus, S., Lehenmeier, M.W., Anderson, C.E., and Rieger, B. (**2011**). *Coord. Chem. Rev. 255*: 1460–1479.
203 Longo, J.M., Sanford, M.J., and Coates, G.W. (**2016**). *Chem. Rev. 116*: 15167–15197.
204 Thevenon, A., Cyriac, A., Myers, D. et al. (**2018**). *J. Am. Chem. Soc. 140*: 6893–6903.
205 Darensbourg, D.J. and Yeung, A.D. (**2014**). *Polym. Chem. 5*: 3949–3962.
206 Moore, D.R., Cheng, M., Lobkovsky, E.B., and Coates, G.W. (**2003**). *J. Am. Chem. Soc. 125*: 11911–11924.
207 Shao, H., Reddi, Y., and Cramer, C.J. (**2020**). *ACS Catal. 10*: 8870–8879.
208 Byrne, C.M., Allen, S.D., Lobkovsky, E.B., and Coates, G.W. (**2004**). *J. Am. Chem. Soc. 126*: 11404–11405.
209 Peña Carrodeguas, L., González-Fabra, J., Castro-Gómez, F. et al. (**2015**). *Chem. Eur. J. 21*: 6115–6122.
210 González-Fabra, J., Castro-Gómez, F., Kleij, A.W., and Bo, C. (**2017**). *ChemSusChem 10*: 1233–1240.
211 Bun, Y.L., Heon, Y.K., Su, Y.L. et al. (**2005**). *J. Am. Chem. Soc. 127*: 3031–3037.
212 Xiao, Y., Wang, Z., and Ding, K. (**2005**). *Chem. Eur. J. 11*: 3668–3678.
213 Hua, Y.Z., Yang, X.C., Liu, M.M. et al. (**2015**). *Macromolecules 48*: 1651–1657.
214 Kember, M.R., Knight, P.D., Reung, P.T.R., and Williams, C.K. (**2009**). *Angew. Chem. Int. Ed. 48*: 931–933.
215 Kember, M.R., White, A.J.P., and Williams, C.K. (**2009**). *Inorg. Chem. 48*: 9535–9542.
216 Kember, M.R., Jutz, F., Buchard, A. et al. (**2012**). *Chem. Sci. 3*: 1245–1255.
217 Kember, M.R., Copley, J., Buchard, A., and Williams, C.K. (**2012**). *Polym. Chem. 3*: 1196–1201.
218 Buchard, A., Kember, M.R., Sandeman, K.G., and Williams, C.K. (**2011**). *Chem. Commun. 47*: 212–214.
219 Kember, M.R. and Williams, C.K. (**2012**). *J. Am. Chem. Soc. 134*: 15676–15679.

220 Jutz, F., Buchard, A., Kember, M.R. et al. (**2011**). *J. Am. Chem. Soc. 133*: 17395–17405.

221 Buchard, A., Jutz, F., Kember, M.R. et al. (**2012**). *Macromolecules 45*: 6781–6795.

222 Robertson, N.J., Qin, Z., Dallinger, G.C. et al. (**2006**). *J. Chem. Soc. Dalton Trans.*: 5390–5395.

223 Darensbourg, D.J., Adams, M.J., Yarbrough, J.C., and Phelps, A.L. (**2003**). *Inorg. Chem. 42*: 7809–7818.

224 Varghese, J.K., Park, D.S., Jeon, J.Y., and Lee, B.Y. (**2013**). *J. Polym. Sci. Part A Polym. Chem. 51*: 4811–4818.

225 Lehenmeier, M.W., Kissling, S., Altenbuchner, P.T. et al. (**2013**). *Angew. Chem. Int. Ed. 52*: 9821–9826.

226 Kissling, S., Altenbuchner, P.T., Lehenmeier, M.W. et al. (**2015**). *Chem. Eur. J. 21*: 8148–8157.

227 Kissling, S., Lehenmeier, M.W., Altenbuchner, P.T. et al. (**2015**). *Chem. Commun. 51*: 4579–4582.

228 Nikitinskii, A.V., Bochkarev, L.N., Khorshev, S.Y., and Bochkarev, M.N. (**2004**). *Russ. J. Gen. Chem. 74*: 1197–1200.

229 Nikitinskii, A.V., Bochkarev, L.N., Voronin, R.V. et al. (**2004**). *Russ. J. Gen. Chem. 74*: 1194–1196.

230 Qu, L., Del Rosal, I., Li, Q. et al. (**2019**). *J. CO_2 Util. 33*: 413–418.

231 Hua, L., Li, B., Han, C. et al. (**2019**). *Inorg. Chem. 58*: 8775–8786.

232 Xu, R., Hua, L., Li, X. et al. (**2019**). *Dalton Trans. 48*: 10565–10573.

233 Nagae, H., Aoki, R., Akutagawa, S. et al. (**2018**). *Angew. Chem. Int. Ed. 57*: 2492–2496.

234 Saini, P.K., Romain, C., and Williams, C.K. (**2014**). *Chem. Commun. 50*: 4164–4167.

235 Trott, G., Garden, J.A., and Williams, C.K. (**2019**). *Chem. Sci. 10*: 4618–4627.

236 Deacy, A.C., Durr, C.B., Garden, J.A. et al. (**2018**). *Inorg. Chem. 57*: 15575–15583.

237 Deacy, A.C., Durr, C.B., and Williams, C.K. (**2019**). *Dalton Trans. 49*: 223–231.

238 Deacy, A.C., Kilpatrick, A.F.R., Regoutz, A., and Williams, C.K. (**2020**). *Nat. Chem. 12*: 372–380.

239 Asaba, H., Iwasaki, T., Hatazawa, M. et al. (**2020**). *Inorg. Chem. 59*: 7928–7933.

5

Cationic Compounds of Group 13 Elements: Entry Point to the p-block for Modern Lewis Acid Reagents

Sanjay Singh[a], Mamta Bhandari[a], Sandeep Rawat[a], and Sharanappa Nembenna[b]

[a] *Department of Chemical Sciences, Indian Institute of Science Education and Research Mohali, India*
[b] *School of Chemical Sciences, National Institute of Science Education and Research Bhubaneswar, India*

5.1 Introduction

The domain of catalysis has been heavily dominated by transition metal-based catalysts for a long time, leading to a vast variety of successful applications. In the recent past, particularly since the beginning of twenty-first century, a paradigm shift towards main group elements has been witnessed with efforts to find alternate avenues to transition elements' catalysis [1, 2]. In this endeavour, the main group elements have gained major attention where the p-block elements particularly of group 13 have received special curiosity [3]. The remarkable reactivity of group 13 compounds triggered a real explosion of interest in the catalytic field. They have a tradition of mediating various chemical transformations, the majority of which depend on Lewis acid (LA)-type activation of polar substrates. The presence of an empty orbital and unsatisfied valency in their three coordinated compounds make them potential LAs that also reflects in the tendency of these compounds to form dimers. Ligand supported well-defined compounds of trivalent group 13 elements can exhibit strong LA character particularly in three- or lower-coordinated state. The Lewis acidity and electrophilicity can be fine-tuned by varying the coordination number or converting the covalent complexes into cationic derivatives, as reflected in moderate electrophilic behaviour of four-, five- or six-coordinated complexes and high Lewis acidity of cationic complexes, respectively. The direct application of these characteristics can be realized in activity vs selectivity aspects of the reactions these compounds catalyze.

This chapter aims to provide an overview of the recent developments in the synthetic, characterization, reactivity studies, and applications of group 13 metal cationic complexes (Al, Ga, In, and Tl) that have appeared in the current decade (since 2010). These aspects of the lightest element of group 13, namely boron, have recently been covered in a few review articles, therefore, even more recent developments of boron (since 2015) will be discussed [4–7]. In spite of this, it is not possible to avoid the overlap of the contents presented in this chapter with the recent reviews for the sake of continuity and flow. An exception for inclusion of borocations in the current discussion along with cationic complexes of metals from group 13 is purposefully done for full coverage of group 13 modern cationic reagents. The rationale for including a metalloid in the discussion is because borocations generally have well-defined coordination spheres, show minimum secondary

Polar Organometallic Reagents: Synthesis, Structure, Properties and Applications, First Edition. Edited by Andrew E. H. Wheatley and Masanobu Uchiyama.

interactions with ligands and counter-anions, and yet show a similar reactivity pattern to their higher congeners. Therefore, the borocations offer simpler models to analyze the bonding and reactivity of cationic group 13 complexes that can be extended with suitable modification to their higher congeners. Our major attention will be focused on the well-defined electronically unsaturated cationic complexes (with a formal electron counting of six or less). In the selected cases, as exceptions, the examples of cationic group 13 complexes with formally complete octets i.e. tetracoordinated species will be discussed with the view that these complexes also exhibit considerable Lewis acidity and subsequent applications in catalysis. The next section of this chapter lists some of the general considerations describing the classification of the complexes, properties of the counter-ions and solvents followed by methods to quantify Lewis acidity both experimentally and computationally.

5.2 General Considerations

Based on the metallic character and electropositivity of the elements alone, the ease with which cationic complexes of group 13 elements can be formed is expected to increase down the group. However, the known number of examples of borocations and aluminium cations are far more than that of heavier elements indicating that the electropositivity of the elements cannot be the sole criterion. Instead, the size of group 13 atoms and how effectively their orbitals overlap with the ligand orbitals is a more reliable approach to judge the feasibility of formation and stability of cationic complexes with a given set of ligand donor sites. Orbitals of elements from the same period can overlap most effectively due to their matching size and energy, whereas this overlap will be less effective as the size of the group 13 element and ligand donor atom mismatch. The more effective overlap between the orbitals of lighter elements B and Al with the orbitals of ligand donor atoms like C, N, O lead to the formation of strong bonds with more directional nature and significant covalent character. Thus, the bigger atoms, Ga, In, and Tl will form more polar bonds with the ligands and their bigger size will also allow for additional interactions such as inductive or hyperconjugative, and interactions with arene–π systems, and also allow the metal atom to increase its coordination number. Therefore, the effective overlap between the central atom and the ligand orbitals is of paramount significance.

We would draw to the attention of the reader that the cationic charge on the complex in the majority of the cases is diffused and is seldom localized on the group 13 centre. Nevertheless, the Lewis acidity or electrophilic behaviour of the cationic complexes is a direct manifestation that the greater magnitude of positive charge is retained on the group 13 centre. This has also been supported by computations on natural bond orbital (NBO) analysis or the characteristics of the LUMO in a few cases. The same is also generally true for the weakly coordinating bulky counter-anions, where the negative charge is spread over the entire anion to minimize cation–anion interaction. Therefore, the reader is cautioned for the fact that positive or negative charges shown may not depict the actual charge distribution. Rather, they will represent the net or abbreviated charges on the cationic or anionic moiety.

5.2.1 Classification of Cationic Group 13 Complexes

The number of substituents connected to a cationic boron centre, that is in its coordination sphere, was used by Nöth to categorize borocations [3, 8]. Based on these considerations, three distinct classes of mononuclear borocation were clearly identified as borinium, borenium, and boronium (Figure 5.1). Highly electrophilic two coordinated borocations with two empty p-orbitals

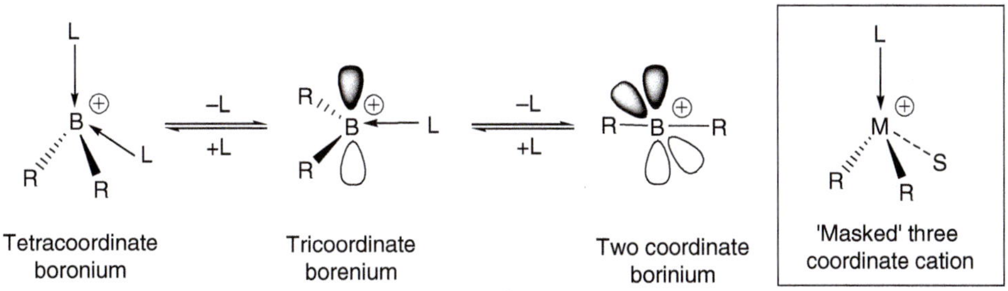

Figure 5.1 Mononuclear borocations classified on the basis of coordination number, also depicting the vacant p-orbitals at the B centre; L corresponds to a Lewis base.

are classified as borinium cations. The unprecedented reactivity and extraordinary Lewis acidity of these cations are attributed to the significant deviation from the octet rule. Due to their highly reactive nature, their condensed phase analysis is often vulnerable to significant interactions with solvent or counter-ions. The three-coordinate species in which the boron centre is bound to two substituents (R) and a neutral lone pair donor (Lewis base L) are referred as borenium cations. In spite of an incomplete electron valance shell, the donation of an electron pair from L to the boron centre reduces some of its electron deficiency. Therefore, borenium cations are notably more amenable to solution studies in comparison to their two-coordinate counterparts.

The tetracoordinated boronium cation is the most common and stable class of borocation in which the boron centre is formally electronically saturated. The stability and reduced electron deficiency compensated for by the donor ligands L is reflected in the prevalence of reports that illustrate the generation of these species or invoke them as intermediates in various reactions. Based on the electronic unsaturation, the reactivity pattern for borocations is expected to follow the order borinium > borenium > boronium, with the borenium form expected to show the best balance of stability and reactivity.

A similar classification, according to onium- (8 VE), enium- (6 VE), and inium-cations (4 VE) discussed above for boron, is not so straightforward for cationic complexes of heavier group 13 congeners, primarily due to the fact that these elements often have the tendency to expand their coordination number owing to their bigger size and are engaged in additional weak interactions either with solvent molecules, counter-ions or with the ligand substituents (e.g. arene or anagostic interactions, or σ-donation of π-density of a double bond) [3]. Most commonly encountered is the three coordinated species with additional interactions, ascribed as the 'masked' form of the cation, but this terminology has not gained much popularity (Figure 5.1).

5.2.2 General Methods for the Syntheses of Cationic Group 13 Complexes

Due to the cationic nature, the Lewis acidity on boron and aluminium centres is significantly increased, which makes for powerful electrophilic reagents. However, in some cases, the high Lewis acidity of these species makes their isolation and purification a challenging task. Due to the unoccupied p-orbital(s), the lighter group 13 elements (boron and aluminium) have a high tendency to form bridged as well as oligomeric and polymeric complexes. The unwanted side reactions and dimerization of the complexes can be controlled by: (i) increasing the steric bulk of the ligand(s) that encapsulate(s) the group 13 element centre so as to create hindrance around that element and inhibit it from forming dimers, trimers etc.; (ii) the electronic unsaturation around the element centre can be satisfied by involving strong σ- or π-donor molecules.

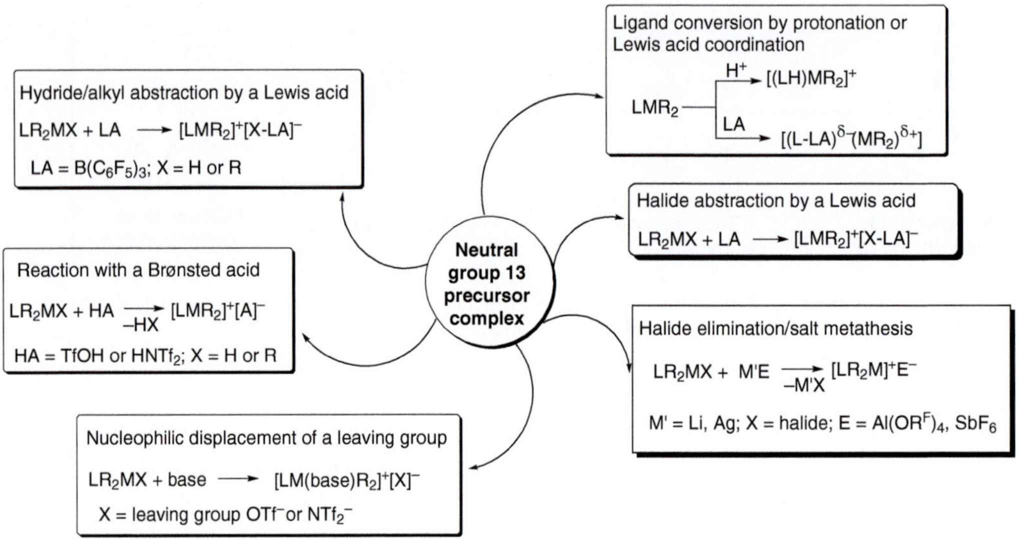

Figure 5.2 Summary of a few common synthetic strategies employed for cationic group 13 complex syntheses. L = neutral ligand or Lewis base; LA = Lewis acid; X = anionic ligand.

Different approaches have been used to synthesize the cationic complexes of group 13 by utilizing their neutral analogues as precursors. Figure 5.2 provides an indicative summary of the popular methods; a detailed discussion of each of these methods can be retrieved from the literature [6, 9, 10]. The selection and generalization of the synthetic methods are based on the availability of suitable starting material, the nature of group 13 cation to be generated, and the stability of the cation. Other methods that are not listed in Figure 5.2 are generally not observed in the group 13 precursors supported by state-of-the-art ligands, and these methods include asymmetric cleavage of the halide/hydride bridged dimers or molecules with bridges, redistribution, and auto-ionization [9].

5.2.3 Characteristics of Counter-anions and Solvents

The choice of anion is very crucial for the successful synthesis of a stable yet reactive group 13 cationic complex. In this direction, the supporting ligand and the counter-anion play the most vital roles to suppress the decomposition pathways and side reactions, for example abstraction of F^- from BF_4^- or anions containing weak C-F sp^3 bonds as in $[B\{3,5\text{-}(CF_3)_2\text{-}C_6H_3)\}_4]^-$ or abstraction of $C_6F_5^-$ from $[MeB(C_6F_5)_3]^-$ [11–13]. The use of appropriate weakly coordinating anions (WCAs) can make the synthesis more feasible [3]. The WCAs must exhibit some of the properties such as (i) they must be sufficiently bulky to prevent close contacts with the cationic centre, this is rarely achieved and sometimes the weak interactions (B···OTf, B···NTf_2 or B···Cl–$AlCl_3$) help in stabilizing the cation without quenching electrophilicity of the cation; (ii) negative charge must be delocalized over a significant portion of the anion, a symmetrical or spherical cluster anion works better, this will also minimize the Coulombic interactions; (iii) strong bonds in the anion framework and inertness towards oxidative environments and electrophiles [3, 14]. Considering the high electrophilicity of the main group cations, particularly that of group 13 elements, the anions employed are generally weakly coordinating and unreactive. These include simple ones like

$[OTf]^-$, $[BF_4]^-$, $[SbF_6]^-$ or more bulky ones like $[B(C_6F_5)_4]^-$, $[HB(C_6F_5)_3]^-$, $[Al(C_6F_5)_4]^-$, $[Sb(OTeF_5)_6]^-$, $[HCB_{11}H_5Br_6]^-$, $[HCB_{11}H_{11}]^-$, $[Al\{OC(CF_3)_3\}_4]^-$, etc.

Similarly, the properties of solvents such as chemical compatibility, coordinating ability, and polarity are very crucial factors for a successful synthesis of a cationic group 13 complex. Invariably, the majority of the cationic complexes, not only of group 13 but also of other main group cations, have a tendency to form clathrate oils which tend to settle down from arene solvents (benzene, toluene, xylene, mesitylene etc.). These clathrates contain the trapped solvent molecules when the cationic complexes are dissolved in them. The clathrate can be converted into precipitate in many cases by mixing with non-polar solvents like pentane or hexane, which tend to take away the trapped solvent molecules from the clathrate and precipitate the cationic complexes as insoluble powders. The majority of the cations can be dissolved in polar solvents like CH_2Cl_2, $CHCl_3$, etc. and polar coordinating solvents like Et_2O, THF, CH_3CN, pyridine, etc. which also coordinate if sterically not precluded. The fluorinated solvents or haloarenes with sufficient polarity, low nucleophilicity, and strong C–X bonds (chloro- or fluorobenzene) have proved to be highly useful for the synthesis, reactions, and NMR characterization of these cations.

5.2.4 Quantification of LA of Cationic Group 13 Complexes

A wide range of chemical transformations are facilitated by the Lewis acidic complexes of the main group elements. Hence, to fulfil the aim to craft a strong and tunable LA, the cationic complexes of main group elements, particularly of cationic species from group 13, have been a subject of general interest. The Lewis acidity of a chemical species can be fine-tuned (i) by modifying the substituent(s) directly connected to the group 13 centre, and (ii) by altering the steric bulk of the ligand around the active centre with an inherent tendency to exhibit the LA property [15, 16]. The strength of acidic nature can be determined by thermodynamic parameters like the bond formation enthalpy in the formation of LA–base adducts, which can also be calculated by computational methods. Similarly, the perturbation in the molecule as a consequence of Lewis adduct formation can also be studied spectroscopically using techniques such as IR and NMR (Figure 5.3).

The strength of LA–base adduct bonding is influenced by various key factors. Firstly, the nature of base used; hard bases tend to interact strongly with hard acids, which can be explained by the

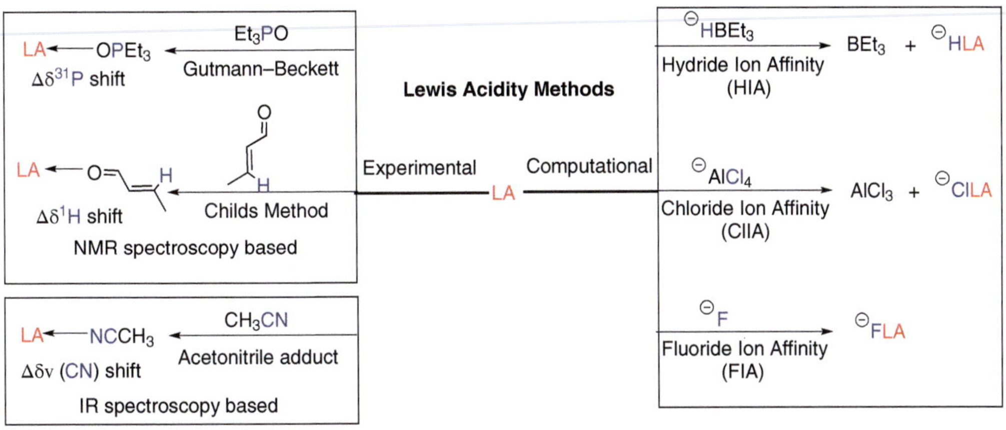

Figure 5.3 Popular experimental and computational methods for the determination of Lewis acidity.

hard and soft acid base (HSAB) principle. But, restricting the choice of base to a single molecule and varying the LA group 13 cationic species can help in generalizing the acidity scale. The second prominent factor is steric hindrance; the molecules which are encumbered by bulky ligands, experimental methods (Gutmann and Childs method) assign as weak electrophiles, but computational methods show increased affinities (fluoride and chloride ion affinity), which corresponds to stronger Lewis acidic nature. Unlike Brønsted acids, where irrespective of the substituent's steric bulk pK_a provides the measure of the acidity, it is not an easy and straightforward task to design a universal quantification method to determine the relative Lewis acidity. Nevertheless, a few methods have been designed that provide good workable models to estimate the Lewis acidity and are highly useful in the present context of cationic complexes of group 13 elements.

5.2.4.1 Experimental Methods to Quantify Lewis Acidity

5.2.4.1.1 Gutmann–Beckett Method (Adduct Formation with Triethyl Phosphine Oxide)

In this method, perturbation in the ^{31}P NMR chemical shift of Et_3PO upon complexation with a LA is recorded and compared to that of the free Et_3PO. Initially, this approach was introduced by Gutmann to quantitatively describe the electrophilic character of solvents using an acceptor number scale with two reference points relating to the ^{31}P NMR chemical shifts of Et_3PO in hexane (AN = 0, δ = 41.0 ppm) and in $SbCl_5$ (AN = 100, δ = 86.5 ppm). After that, using the formula acceptor number (AN) as {AN = 2.21×(chemical shift of sample – 41.0)} is calculated. The higher the AN, the higher the Lewis acidity [15, 16, 23–25]. The strong LAs SbF_5 and $SbCl_5$ are assigned AN of 100 and weakly interacting solvent hexane was assigned AN of zero. The accuracy of the method is a function of steric factors, as it is dependent on the interaction with the desired Lewis acidic centre. For example, AN obtained from this method for $Al[OC(C_6F_5)_3]_3$ is 72.7, which is lower than that of $B(C_6F_5)_3$ (78.1) and SbF_5 (100) but the fluoride ion affinity (FIA) for these molecules shows the opposite trend (Table 5.1). The lower AN for $Al[OC(C_6F_5)_3]_3$ can be attributed to the steric bulk provided by $-OC(C_6F_5)_3$ [28–31].

5.2.4.1.2 Childs Method

Similar to the above method, Childs method involves adduct formation between the LA in question with crotonaldehyde (CA) as the Lewis base and perturbation in the ^{1}H NMR signal of the olefinic hydrogen in the β-position (at the C3-carbon) as compared to uncomplexed CA. The difference in chemical shift value ($\Delta\delta^1H = \delta^1H_{LA \cdot CA} - \delta^1H_{CA}$) provides the corresponding Lewis acidity [17]. A positive shift difference $\Delta\delta^1H$ (downfield shift) is expected as a consequence of the electron-withdrawing (deshielding) effect of a LA.

Although Childs method has been exploited with different Lewis bases (e.g. tiglic aldehyde, pent-3-en-2-one, cyclohexenone, methyl crotonate, and crotonitrile), CA is most commonly used as the Lewis base.

The results obtained from Gutmann–Beckett and Childs methods may not corroborate each other. For example, the Lewis acidity determined by Gutmann follows the order $B(C_6F_5)_3$ < $B(C_6F_5)_2(OC_6F_5)$ < $B(C_6F_5)(OC_6F_5)_2$ < $B(OC_6F_5)_3$ but the trend is just reversed when calculated form Childs method (Table 5.1) [26, 27]. The reason for this is the hard base Et_3PO, which prefers to bind to hard acids in the Gutmann method whereas Childs method involves soft Lewis base CA and thus follows the opposite trend.

5.2.4.1.3 Adduct Formation with Acetonitrile

When any LA forms adduct with acetonitrile (the C≡N stretch of CH_3CN occurs at 2253/cm), it will affect the stretching frequency of the C≡N bond. Therefore, based on the shift in C≡N stretching frequency $\Delta\nu$(CN) the Lewis acidity can be determined. The higher the blue shift in frequency, the

Table 5.1 Perturbation in ^{31}P NMR signal for Gutmann–Beckett Lewis acidity and perturbation in ^{1}H NMR signal for Childs Method to estimate Lewis acidity.

LA	Gutmann–Beckett method		Childs method
–	^{31}P NMR δ (ppm)	(AN)	($\Delta\delta^1H$)
$B(C_6F_5)_3$	77.0 (CD_2Cl_2)	78.1	0.70
$B(C_6F_5)_3$	77.7 (C_6D_6)	81.1	–
$B(C_6H_5)_3$	65.9 (C_6D_6)	55.0	–
$B(OC_6H_5)_3$	69.4 (C_6D_6)	62.8	–
$B(C_6F_5)_2(OC_6F_5)$	80.0 (C_6D_6)	86.2	1.00
$B(C_6F_5)(OC_6F_5)_2$	80.5 (C_6D_6)	87.3	0.50
$B(OC_6F_5)_3$	80.9 (C_6D_6)	88.2	0.40
$Al(OC_6F_5)_3$	73.9 (CD_2Cl_2)	72.2	–
BCl_3	84.4 (toluene-D_8)	95.9	–
BF_3	77.0 (toluene-D_8)	79.6	–
$SbCl_5$	86.5	100.0	–
13[$B(C_6F_5)_4$]	97.6 (CD_2Cl_2)	104.5	–
26[$B(C_6F_5)_4$]	79.1 (CD_2Cl_2)	85.34	–
63[$B(C_6F_5)_4$]	75.5 (CD_2Cl_2)	76.24	–
86[$MeB(C_6F_5)_3$]	78.5 (C_6D_6)	83.1	–
87[$HB(C_6F_5)_3$]	81.3 (C_6D_6)	90.4	–
79[$B(C_6F_5)_4$]	85.1 (C_6D_6:C_6D_5Br) (1:1)	89.5	–

stronger the acid is considered to be [18, 19, 32]. The peak generated due to the Fermi coupling between ν(CN) and $\nu(CN) + \delta_s(CH_3)$ can cause difficulty in understanding of results though the problem can be eliminated by using deuterated CD_3CN [33]. This method is not very reliable because the result obtained is not consistent. Thus, the utility of this method is limited only to similar kinds of acids and is mostly applied for qualitative assessment. For example, the method correctly deduced the relative acidic strength of boron halides ($BF_3 < BCl_3 < BBr_3$) but assigned less Lewis acidity for SbF_5 [34]. Similarly, in another case, a higher blue shift was observed for $B(C_6F_5)_3$ (2367/cm) than the stronger LA $Al[OC(C_6F_5)_3]_3 \cdot NCMe$, which had a C–N stretching band at 2344/cm.

5.2.4.2 Computational Approaches to Determine Lewis Acidity

The experimental methods for Lewis acidity estimation are associated with limitations including their dependence on the steric factors. Considering these limitations, fluoride ion affinity was introduced as a reliable and benchmark method for the quantification of Lewis acidity by combining the strength of a LA in gas phase with the energy that is released upon binding a fluoride ion F^- [20, 35]. Later on, this was successfully extended to chloride ion and hydride ion affinity [21]. In this method, quantum chemical calculations are performed for the isodesmic reactions to determine the bond formation enthalpy of the molecule. Molecules with larger FIA are considered to be strong acids.

Irrespective of the steric factors, the FIA is a powerful tool to predict the fate of LAs in a reaction more accurately due to the small size and high basicity of the fluoride ion. The FIA value determined for the molecular acids by this method exceeded the FIA value of SbF_5, considered to be a Lewis superacid (LSA). In the case of hydride ion affinity, the energy released in the gas phase reaction when a hydride combines with a molecular acid generates a scale for its Lewis acidic nature and those acids whose HIA values are greater than the HIA value of $B(C_6F_5)_3$ are considered as soft LSAs (Scheme 5.1) [21]. However, a high HIA value cannot be correlated with catalytic activity as a high HIA value may prevent subsequent hydride ion delivery to the substrates.

In all these methods, quantum chemical calculations are preferred over experimental methods for isodesmic reactions to determine the bond formation enthalpy of the molecule. Molecules with larger fluoride/hydride or chloride ion affinity are generally considered to be stronger acids. Though commercially available $AlCl_3$ and $AlBr_3$ act as superacids in the gas phase only, the Lewis acidity is quenched greatly in the condensed phase due to dimerization. Some aluminium-based bulky mononuclear SLAs have been synthesized recently which shows higher FIA than SbF_5 in the gas phase [22, 28]. The gas phase values for the cationic species seem to exceed the threshold value to LSA significantly, which is attributed to charge neutralization and electrostatic attraction. But, because of the enthalpy of solvation (ΔH_{solv}) experimentally major damping in calculated FIA with cationic LAs was observed. Table 5.2 summarizes the fluoride, chloride, and hydride ion affinities for selected LAs [30, 31].

$$LA_{(g)} + [AlCl_4]^- \xrightarrow{\Delta H = -ClIA} [LA\text{-}Cl]^-_{(g)}$$

$$LA_{(g)} + F^-_{(g)} \xrightarrow{\Delta H = -FIA} [LA\text{-}F]^-_{(g)}$$

$$FIA\ (SbF_5) = 500\ kJ/mol$$

$$LA_{(g)} + [HBEt_3]^-_{(g)} \xrightarrow{\Delta H = -HIA} [LA\text{-}H]^-_{(g)}$$

$$HIA\ (B(C_6F_5)_3) = 480\ kJ/mol$$

Scheme 5.1 Isodesmic reactions to calculate chloride, fluoride, and hydride ion affinity.

Table 5.2 Fluoride ion, chloride ion, and hydride ion affinity (in kJ/mol) for a few LAs.

LA	Fluoride ion affinity (FIA)	Chloride ion affinity (ClIA)	Hydride ion affinity (HIA)
BF_3	342	146	299
BCl_3	405	183	391
AlF_3	471	306	423
$AlCl_3$	498	318	450
SbF_5	493	341	562
$B(C_6F_5)_3$	452	236	484
$Al(OC(CF_3))_3$	543	352	490
$Al(N(C_6F_5)_2)_3$	555	362	–
$Al(C_6F_5)_3$	536	348	483
43a$[AlCl_4]$	605	–	–
43b$[AlCl_4]$	603	–	–
43d$[AlCl_4]$	598	–	–
43e$[AlCl_4]$	622	–	–

5.3 Recent Developments in Cationic Group 13 Complexes

5.3.1 Advances in the Synthesis and Characterization of Borocations

In this section, we present an overview of borocations published since 2015. Further, we will focus on well-defined two-, three-, and four-coordinated mononuclear borocations excluding dinuclear and multinuclear borocations, bridged species, low-valent borocations, and ligand-centred borocations. A recent review published in 2019 provides coverage of these borocations in sufficient detail [36].

5.3.1.1 Borinium Cations: Two-coordinate Cationic Boron Complexes

As indicated previously, two-coordinated boron cations are often named as borinium [8]. These are formal 4 VE species and expected to exhibit linear geometry for a sp hybridized boron centre. These borocations are more reactive as compared to their three- and four-coordinated relatives. The first concrete attempt at the synthesis of a borinium cation dates back to 1982 when Nöth and coworkers exploited the π-donation from ligand *N*-donor centres into two vacant p-type orbitals on boron, considerably diminishing its electron deficiency [37]. However, an example very close to an ideal borinium ion was realized in 2014 when Shoji et al. in a seminal report disclosed a thermally stable dimesityl borinium cation (Mes_2B^+, Mes = 2,4,6-trimethylphenyl) in **1**[$HCB_{11}Cl_{11}$] and **1**[$B(C_6F_5)_4$], with a near-linear C–B–C (172.1 °), and where mesityl ring planes were mutually orthogonal (Scheme 5.2). The boron centre is engaged in p_π bonding though a certain degree of π donation from the mesityl groups as concluded from short B–C bonds (1.459 Å) and quinoidal character in the arene rings and also from density function theory (DFT)-based molecular orbital (MO) analysis [39, 40].

In 2017, Shoji et al. further demonstrated that when **1**[$B(C_6F_5)_4$] reacted with diphenylacetylene to undergo a two-fold 1,2-carbonboration reaction to generate a new divinylborinium cation **2**[$HCB_{11}Cl_{11}$] or **2**[$B(C_6F_5)_4$] with a slightly bent C–B–C angle (168.2 °) (Scheme 5.2) [38]. In 2018, Lei, Schaefer and coworkers presented qualitative bonding analysis of the divinylborinium cation and its hypothetical heavier congeners Al, Ga, In, and Tl [41]. Schaefer and coworkers later also made predictions, based on computational calculations, on the role of π-conjugation and hyperconjugation effects to stabilize borinium cations $[X{-}B{-}X]^+$ [42]. Recently, Stephan et al. performed reaction of H_2 and the borinium salt **1**[$B(C_6F_5)_4$] to afford an arene-stabilized mesitylborenium cation **3**[$B(C_6F_5)_4$], which isotopically scrambles HD *via* hydride delivery to B

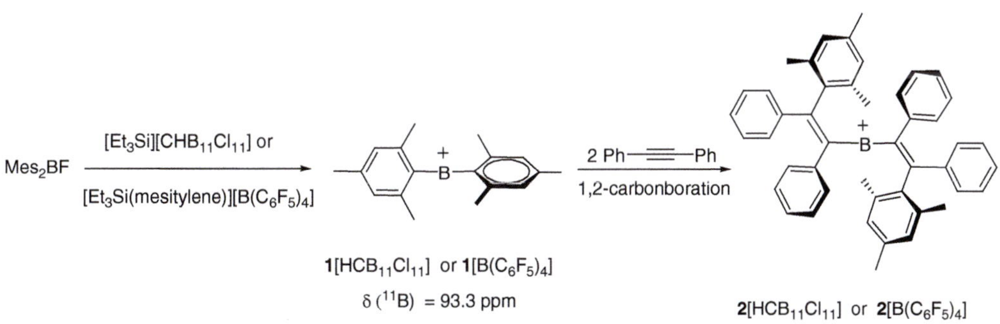

Scheme 5.2 Two coordinate C-bonded Mes_2B^+ borinium cation and its reaction with alkyne.

and protonation of a mesityl group. Similarly, the reaction of **1**[$B(C_6F_5)_4$] with $HSiEt_3$ afforded hydride delivery to boron and silylation of a mesityl group and gave the diborenium cation [MesB(μ-H)$_2$(μ-Mes)BMes]$^+$[$B(C_6F_5)_4$]$^-$ **4**[$B(C_6F_5)_4$] featuring three bridge-bonds between two boron centres (Scheme 5.3), which can also be obtained from the reaction of $(Mes_2BH)_2$ with Brønsted acid [H(MesH)]$^+$[$B(C_6F_5)_4$]$^-$. Whereas, the reaction of the salt [H(MesH)]$^+$[$B(C_6F_5)_4$]$^-$ with $(MesBH_2)_2$ yielded a triboron cation, [H_2B(μ-H)(μ-Mes)B(μ-Mes)(μ-H)BH_2]$^+$[$B(C_6F_5)_4$]$^-$ [43]. Overall, the highly Lewis acidic nature of (Mes_2B^+) and the associated B–C bond cleavage in reactions with H_2 or Et_3SiH limit the use of borinium cation in catalytic reactions. This diarylborinium ion possesses exceptional Lewis acidity, accepting a pair of electrons from CO_2 or CS_2 to cause an unusual carbon-chalcogen double bond cleavage to afford aroyl- **5**[$B(C_6F_5)_4$] and (thio)-aroyl **6**[$B(C_6F_5)_4$] cations, respectively (Scheme 5.3)[40].

Parry and coworkers reported on two coordinate borinium cation [(*i*-$Pr_2N)_2B$]$^+$[$AlCl_4$]$^-$ salts, containing linear tetraisopropyl allene-type moiety [44]. Recently, Stephan et al. extended the chemistry of this cation by preparing an analogous salt [(*i*-$Pr_2N)_2B$]$^+$[$B(C_6F_5)_4$]$^-$ **7**[$B(C_6F_5)_4$] (Scheme 5.4). Although the salt lacked a high-quality XRD characterization, but the authors stated preliminary data were good enough to show a linear geometry around B centre.

As shown in Scheme 5.4, the borinium cation **7**[$B(C_6F_5)_4$] reacts with PhCN, PhCCH, PhCCD, Ph_2CN_2, and Ph_2CO leading to hydroboration of the unsaturated moiety *via* hydride transfer from an *i*-Pr group to form their corresponding borenium complexes **8–12**[$B(C_6F_5)_4$]. In this process, an amide *i*-Pr_2N group is converted to an imine ligand. This case study showed hydroboration of unsaturated groups can take place even without the presence of a formal B–H bond in the reactant [45].

As a unique example, Chiu et al. in 2019 reported the borocation [(η^5-Cp*)B-Mes]$^+$[$B(C_6F_5)_4$]$^-$ **13**[$B(C_6F_5)_4$] as a masked potent boron LA. The thermodynamic stability of 8 VE borocation resembled the boronium cations but the two-coordinated geometry resembled that of a borinium cation. This cation readily binds to Et_3PO and the XRD confirmed a Cp* hapticity change from η^5 to η^1, making the adduct a formal 6 VE borenium cation, [(η^1-Cp*)B-Mes($OPEt_3$)]$^+$[$B(C_6F_5)_4$]$^-$ **14**[$B(C_6F_5)_4$] (Scheme 5.5). The Gutmann–Beckett AN of 104.5 also revealed its considerable LA character and the coordinatively flexible Cp* ligand promoted the catalytic activity of this cation towards the hydrosilylation and hydrodeoxygenation of aryl ketones at ambient temperature [46].

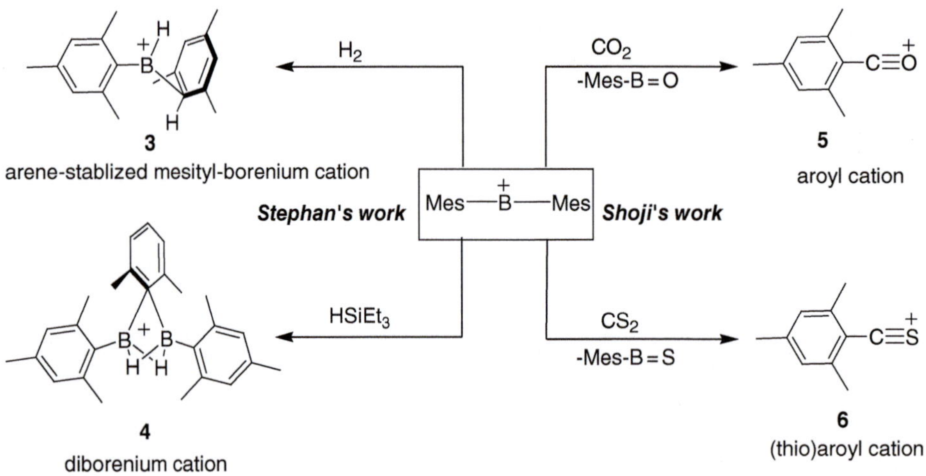

Scheme 5.3 Reactivity behaviour of Mes_2B^+ borinium cation; with H_2 and $HSiEt_3$ (at left) and with CO_2 and CS_2 (at right).

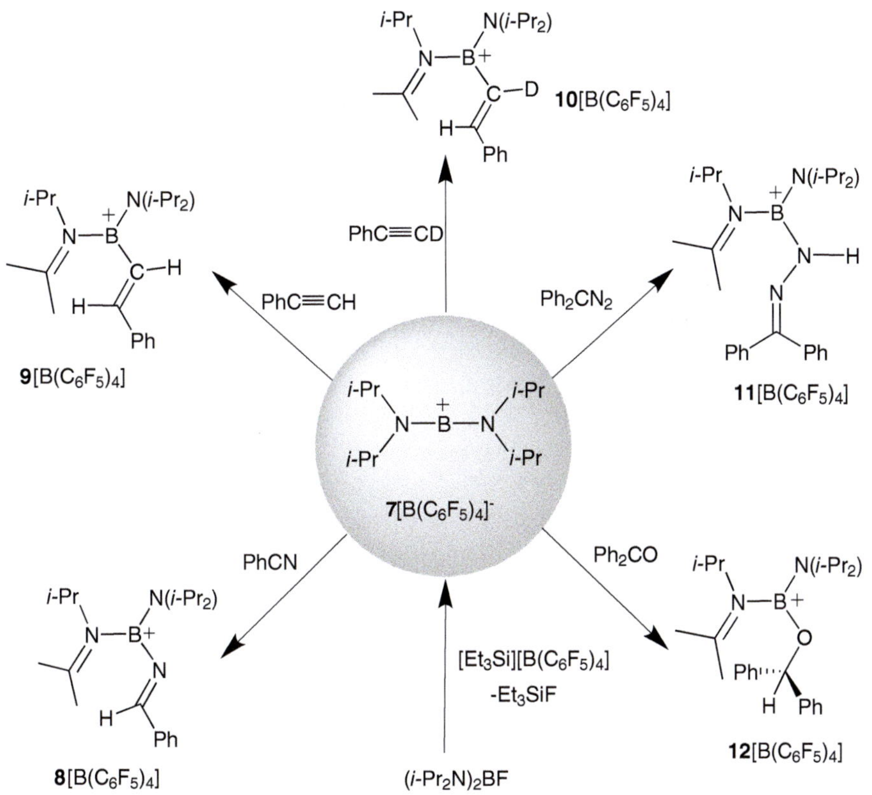

Scheme 5.4 Reactivity of $[(i\text{-}Pr_2N)_2B][B(C_6F_5)_4]$ with unsaturated molecules.

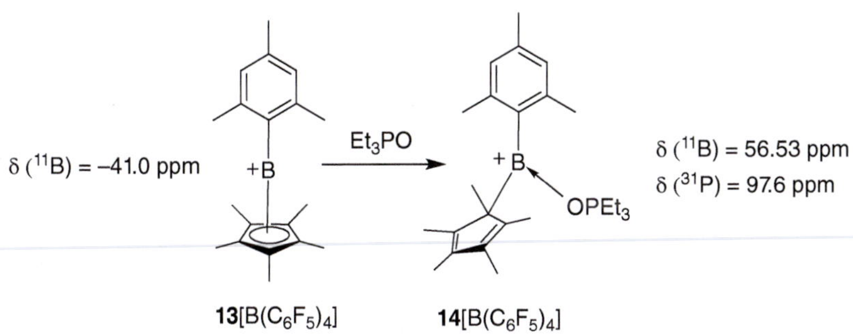

Scheme 5.5 Adduct formation between borocation **13** and Et_3PO demonstrating the high Lewis acidity of **13**$[B(C_6F_5)_4]$.

5.3.1.2 Borenium Cations: Three-coordinate Cationic Boron Complexes

Neutral three coordinated boron compounds share an electronic resemblance to carbenium ions, $RR'R''C^+$, that are also well-known electrophiles. Therefore, if the three-coordinated boron compounds also carry a positive charge, one can expect further improvements to their Lewis acidity. Three-coordinated boron cations, referred as borenium cations, are of particular interest as they provide a perfect balance of Lewis acidity and reactivity. These electronically unsaturated species have two covalent and a dative bond in three-coordinated mono-cationic form. Various sigma donor ligands such as amines, phosphines, NHCs (*N*-heterocyclic carbenes), and NHOs (*N*-heterocyclic olefins) were explored in order to determine their ability in stabilizing the borenium cations. Due to the strong σ-donor and weak π-acceptor nature of NHCs, they are strong contenders as ligands of choice to prepare electronically unsaturated borenium centres. In this regard, numerous NHC-borane adducts were synthesized and utilized in various aspects of organic transformations [47]. The neutral adduct of borane with carbene is stable and acts as a strong hydride donor. Previously, σ-donor NHC ligands were employed to stabilize the low-valent multiply bonded complexes of boron but since the last decade, they have also been widely employed for the stabilization of borenium cations.

5.3.1.3 Borenium Cations Stabilized by NHC and MIC as Neutral C-donor Ligand

In this context, since the first report by Gabbai et al. of NHC-stabilized diarylborenium cation $[Mes_2B{\cdot}NHC^{Me}][OTf]$ **15**[OTf] in 2009, (NHC^{Me} = 1,3-dimethylimidazol-2-ylidene) tremendous developments have been made. Gabbai's cation was prepared by a simple approach through the reaction of Mes_2BF with Me_3SiOTf and $[Ag(NHC^{Me})_2][Ag_2I_3]$ in chlorobenzene under refluxing conditions. The ^{11}B NMR spectrum for this product showed a signal at 66 ppm typical for three-coordinated boron. The trigonal planar geometry around the boron centre was confirmed by XRD analysis (Figure 5.4) [48]. Lindsay et al. in 2011 reported NHC-stabilized dialkylborenium cation **16**[OTf] *via* TfOH protonation of the B–H hydridic bond of 9-BBN which resulted in the asymmetric cleavage of the B–H bond. This cation showed an ^{11}B NMR signal at 81.4 ppm in CD_2Cl_2 (Figure 5.4) [49]. Tamm and coworkers reported on NHC-stabilized boryl radicals and in this endeavour, they reported related NHC–borenium cations with $[OTf]^-$ as the counter-anion, **17a–c**[OTf] (Figure 5.4) [50].

The great σ-donor potential of NHC ligands has a strong impact on the chemical behaviour of NHC–borane adducts. These adducts display higher hydridicity, which is attributed to their fast hydride transfer ability. This facilitates the synthetic protocol of using $[Ph_3C][B(C_6F_5)_4]$ or $B(C_6F_5)_3$ as hydride scavenger for the synthesis of borenium cations from the corresponding NHC–borane adducts. Using this approach, Stephan et al. in a seminal work used NHC–borane adducts **18a–h** to synthesize a series of borenium cations **19a–h**$[B(C_6F_5)_4]$ and found them to be efficient metal-free hydrogenation catalysts for imines and *N*-heterocycles. This systematic study also revealed that sterically less demanding NHCs with electron-withdrawing substituents enhance catalyst

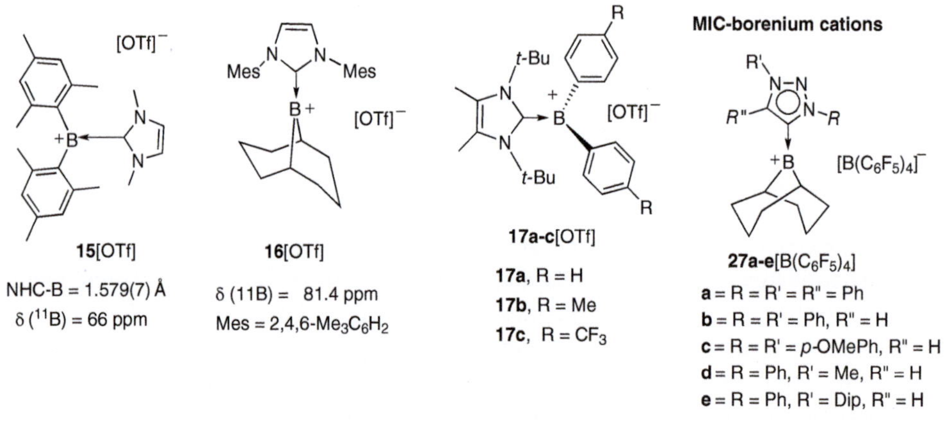

Figure 5.4 Initial examples of NHC-stabilized aryl- and alkyl-borenium cations.

activity, thus **19f**[$(BC_6F_5)_4$] was the most active catalyst of this series. Surprisingly, the hydride abstraction from $(NHC^{t\text{-}Bu})HB(C_6F_5)_2$ (**20**) led to intramolecular C–H bond activation to give **21**, highlighting the fact that the stability of the ligand itself can be a limiting factor while promoting the Lewis acidity of the borenium cations for FLP (frustrated Lewis pair) hydrogenation catalysis (Scheme 5.6, including X-ray structure of **19g** adapted from reference [51]).

Most of the three-coordinated borenium cations generally comprise an aryl, 9-BBN or pinacolato ligand scaffold. Recently, a convenient pathway for constructing NHC-stabilized iminioborenium cationic species was reported by Cui and coworkers. The sterically encumbered ligand, *N*-[2,6-bis(2,4,6-trimethylphenyl)phenyl]-*N*-(trimethylsilyl)amine was reacted with BBr_3 in Et_2O to afford the neutral precursor *N*-[2,6-bis(2,4,6-trimethylphenyl)phenyl](H)-B(OEt)Br under the elimination of Me_3SiBr. Upon dehydrobromination of this precursor with NHC^{Me_4} (1,3-dimethyl-4,5-dimethylimidazol-2-ylidene) as a Brønsted base, an iminoborane complex **22** formed which showed a resonance at 16.5 ppm in its ^{11}B NMR spectrum (Scheme 5.7). It was also characterized by single-crystal X-ray crystallography, which showed this complex to exhibit a shorter B–N bond distance [1.341(3) Å], comparable to B=N. DFT calculations revealed a high-electron density on nitrogen atom compared to Lewis base-free iminoborane, indicating a feasible site for electrophilic attack. Therefore, on subsequent reaction with Me_3SiOTf, the nitrogen atom of the B=N motif was easily silylated and generated the borenium cation **23**[OTf] that showed a broad signal at 27.6 ppm in its ^{11}B NMR consistent with a three-coordinated B centre. The solid-state structure of **23**[OTf] revealed a trigonal planar environment around boron along with elongated B–N bonds [1.426(3) Å] and a well-separated triflate counter-anion [52].

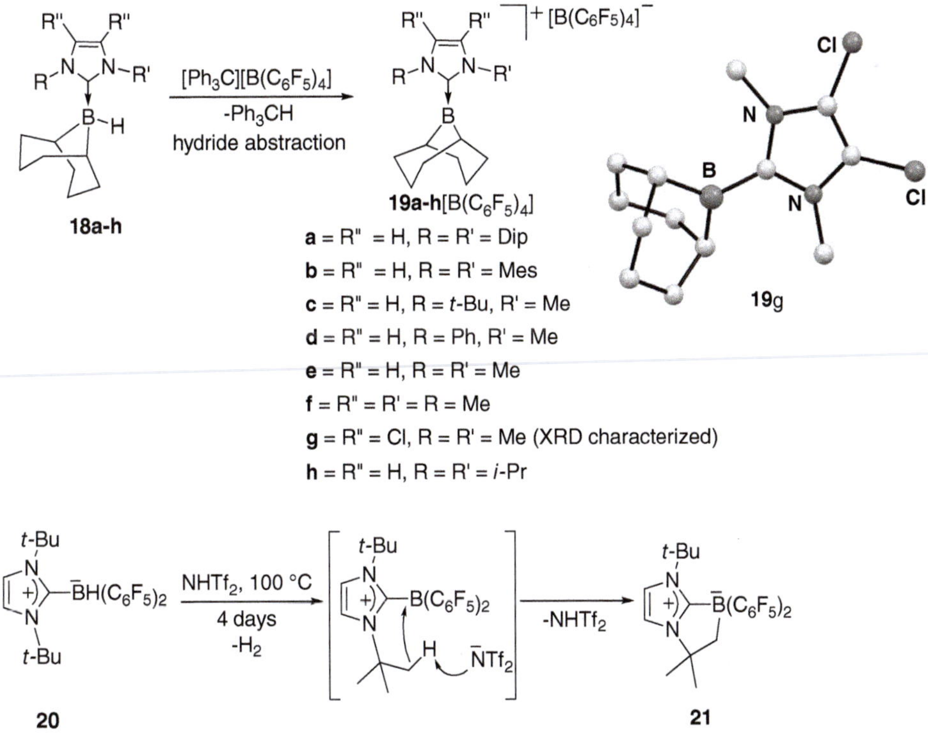

Scheme 5.6 Synthesis of NHC-stabilized borenium cations (top) and reactivity of NHC-borane adduct **20** (bottom). *Source:* Adapted from Farrell et al. [51].

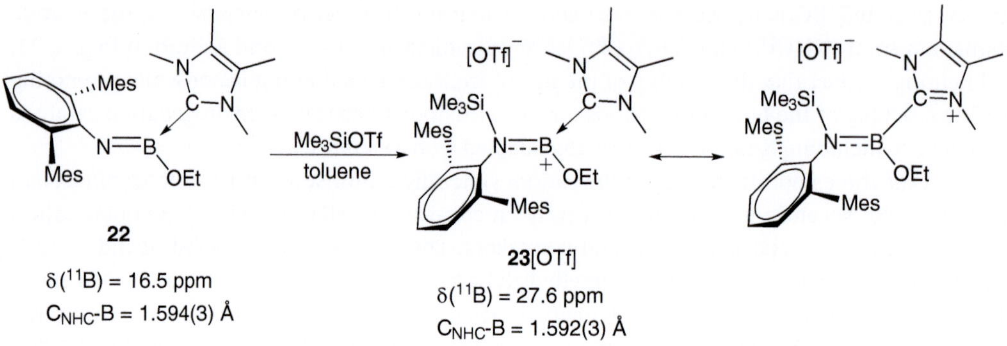

Scheme 5.7 Generation of NHC-stabilized iminioborenium cation via silylation of neutral NHC-iminioborane adduct.

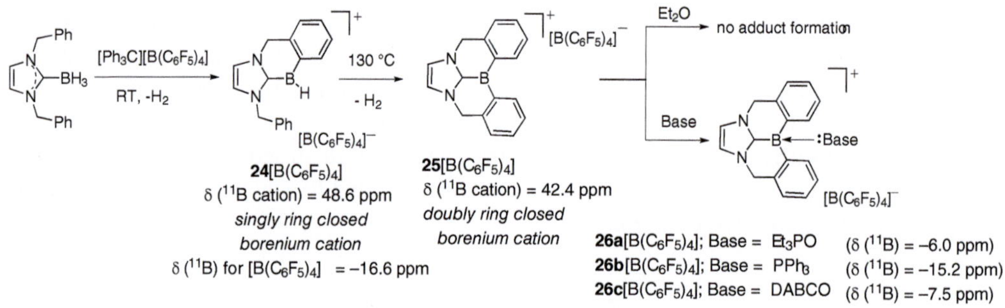

Scheme 5.8 Dehydrogenative cationic borylation of NHC-BH_3 adduct and generation of planar NHC-borenium cation and reactivity with Lewis base.

Stephan et al. reported on the reactions of the adduct $NHC^{Bn}\cdot BH_3$ (NHC^{Bn} = 1,3-dibenzylimidazol-2-ylidene) with $[Ph_3C][B(C_6F_5)_4]$ that involved sequential hydride abstraction and dehydrogenative cationic borylation to afford singly or doubly ring-closed NHC–borenium salts **24**$[B(C_6F_5)_4]$ and **25**$[B(C_6F_5)_4]$ at room temperature and 130 °C, respectively (Scheme 5.8). LA–base adducts **26a–c**$[B(C_6F_5)_4]$ were formed when **25**$[B(C_6F_5)_4]$ was treated with Et_3PO, Ph_3P, or 1,4-diazabicyclo[2.2.2]octane (DABCO), respectively. The deprotonation of the benzylic carbons occurred when **25**$[B(C_6F_5)_4]$ was reacted with *t*-Bu_3P, forming the oligomeric cationic borenium complex with one three-coordinated cationic borenium centre and two four-coordinated boron centres. The close assessment of borenium Et_3PO adduct **26a**$[B(C_6F_5)_4]$ revealed a downfield shift in the value of ^{31}P NMR at $\delta = 79.1$ ppm, suggesting a Gutmann–Beckett Lewis acidity AN of 85.34, higher than recorded for $B(C_6F_5)_3$ [53].

In the vast library of various NHCs, 1,2,3-triazolylidene-based mesoionic carbenes (MICs) possess a unique position due to even stronger σ-donor property. Thus, the boron adducts stabilized by these MICs have significantly greater hydricity compared to their NHC counterparts making the hydride abstraction from boranes even more facile. The increased σ-donor ability of the mesoionic carbene also provides greater stabilization to the electron-deficient borenium ion. Crudden et al. reported the one-pot synthesis of MIC-stabilized borane adducts and subsequent hydride abstraction with $[Ph_3C][B(C_6F_5)_4]$ or $B(C_6F_5)_3$ leading to the formation of borenium cations **27a–e**$[B(C_6F_5)_4]$ (Figure 5.4), bearing resemblance to Stephan's NHC-based borenium cations. These borenium ions also served as catalysts, *via* a FLP-type mechanism, in mild hydrogenation reactions of imines and unsaturated *N*-heterocycles at ambient pressure and temperature, and were more effective than their isosteric NHC analogues [54].

The scope of borenium cations for catalytic transformations is a very rapidly emerging area that has also been extended to materials applications. Incorporation of electron-deficient boron in extended conjugated systems with a favourable interaction between the vacant p-orbital on boron with adjacent π^*-orbitals can effectively lower the LUMO of the supporting π-conjugated moiety. This feature opens up new applications for borocations in π-conjugated networks in areas such as *n*-type semiconducting, emissive, and photovoltaic materials. Wilson, Gilliard and coworkers reported the first examples of NHC- and CAAC-stabilized [CAAC = cyclic alkyl(amino)carbene] dibenzo[b,f]borepinium and 9-borafluorenium cations **28**[SbF_6] and **29**[SbF_6] that showed thermochromic behaviour (Figure 5.5). These new benzo-fused borocations provide a method to tune the optical properties of heterocyclic borenium cations by altering the Lewis base, the substituents on the ring, and the size of the boron heterocycle. These complexes were prepared in a straightforward manner by salt metathesis of the neutral bromide precursor and $AgSbF_6$, and were also structurally characterized to reveal the planar three-coordinated boron centres in these cations [55].

In a similar study, Jäkle et al. revealed the synthesis and characterization of air-stable π-conjugated benzo[d]dithieno[b,f]borepinium cations **30a–c**[PF_6] and, upon deliberate late-stage dimerization, found the resulting dimers had very low lying LUMOs with extended conjugation and enhanced fluorescence [56]. These authors next modified the dibenzo[b,f]-borepinium cation with fused thiophene rings **31a–b**[PF_6]$_2$ by placing substituents on the backbone, which indeed provided better stabilization to the cations (Scheme 5.9).

Würthner et al. utilized (1,3-diisopropylimidazol-2-ylidene)–borane to react with $HNTf_2$, forming a transient (1,3-diisopropylimidazol-2-ylidene)–borenium species. This borenium ion in reaction with 9,10-distyrylanthracene selectively undergoes double tandem hydroboration–electrophilic C–H borylations forming four B–C bonds and leading to a crystallographically characterized polycyclic diborenium ion as its $[NTf_2]^-$ salt, **32**[NTf_2]$_2$. The XRD analysis of **32**[NTf_2]$_2$ revealed distorted trigonal geometry around the boron centre with *trans*-geometry flanking Ph rings (Figure 5.6). No interaction of the B cation with the NTf_2^- anion was observed. Interestingly, the dehydrogenation of **32** with the (2,2,6,6-tetramethylpiperidin-1-yl)oxyl (TEMPO) radical followed by acidic workup yields a 3,9-diboraperylene as boron-doped graphene and forms infinite one-dimensional π-stacks in the solid-state [57].

The NHC–borenium cation synthesis was extended to the synthesis of borenium cations with heavier carbene analogues, giving NHSi–borenium cations. Due to the poor overlap in N–Si bonds, the silylenes are considered to have higher π-acidity than simple NHCs. Attempts to synthesize NHSi–borenium cation involved the silylene and 9-BBN-OTf as starting material. Surprisingly, the reaction between NHSi and 9-BBN-OTf resulted in the formation of **33a** instead of the desired

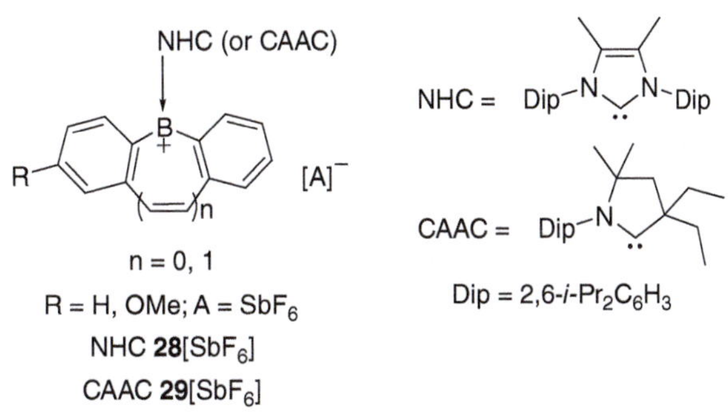

Figure 5.5 NHC- and CAAC-stabilized dibenzo[b,f]borepinium (n = 2) and 9-borafluorenium (n = 0) cations.

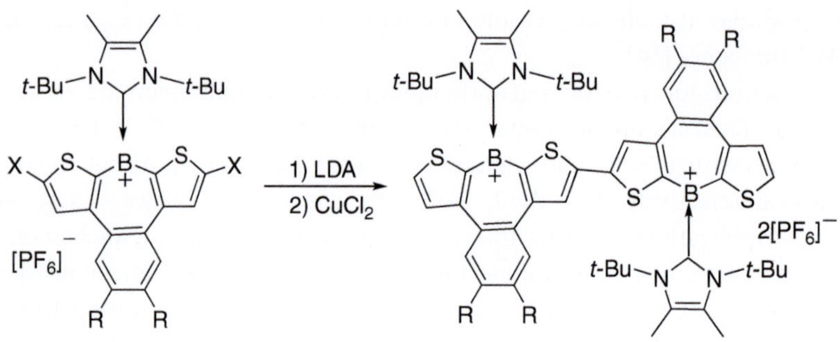

30a-c[PF$_6$]
a = R = X = H
b = R = F, X = H (XRD characterized)
c = R = F, X = SiMe$_3$

31a-b[PF$_6$]$_2$
a = R = H
b = R = F

Scheme 5.9 General scheme for the synthesis of air stable π-conjugated benzo[d]dithieno[b,f]borepinium cations.

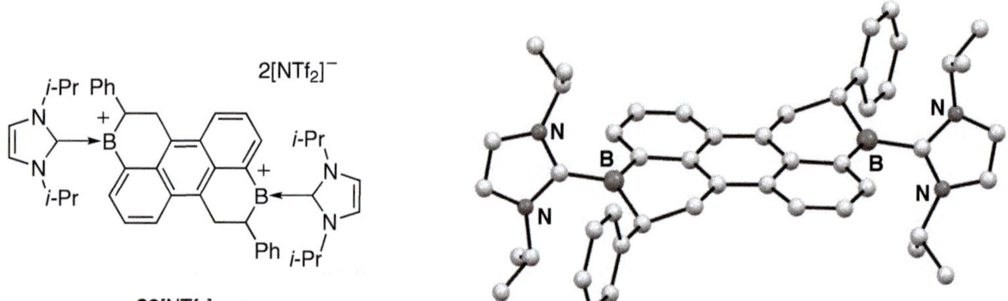

32[NTf$_2$]$_2$

Figure 5.6 Structure of polycyclic diborenium ion **32**[NTf$_2$]$_2$ and its solid-state structure. *Source:* Adapted from Farrell et al. [57].

borenium cation. The formed complex can be viewed as a contact ion pair of borenium and triflate (Figure 5.7). One can obviously infer that due to the high oxophilicity and Lewis acidic nature of silicon, it forms a direct interaction with OTf. The single-crystal X-ray of the complex determined a tetrahedral geometry around the silicon centre and a planar three-coordination at boron. An ^{11}B NMR signal at 91 ppm was recorded for the boron nucleus, which suggests a high electron deficiency at boron though, in terms of reactivity, it acts as a masked cation (Figure 5.7). Due to the poor donor ability of silicon, the silylene has formed a weak and reversible bond with boron so, under the influence of a nucleophile, the B–Si bond breaks and the nucleophile attacks the boron centre [58].

Gessner and coworkers in 2017, crafted a borenium cation complexed with two anionic carbon ylidic ligands. Compared to neutral carbodiphosphoranes or carbodicarbenes that contain two lone pairs at the neutral carbon, this ylidic ligand is formally a monoanionic version of these C(0) ligands with two lone pairs at the carbon. The strong donor ability of the ylidic ligand was crucial for the solution and solid-state stability of these borenium cations. These borenium cations, **34**[BAr$_4$], were prepared by hydride abstraction from the bis(ylidic) borane, $(Ph_3PCSO_2Tol)_2BH$, using $B(C_6F_5)_3$ or trityl salts. Further metathesis with Bu_4NPF_6 then afforded the analytically pure

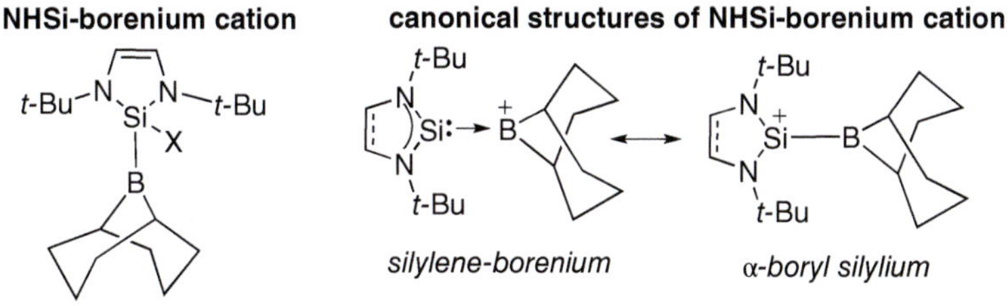

δ (^{11}B) = 91ppm
33a (X = OTf)
33b (X= NTf_2)

Figure 5.7 NHSi–borenium cation and canonical structures of silylene supported borenium cation.

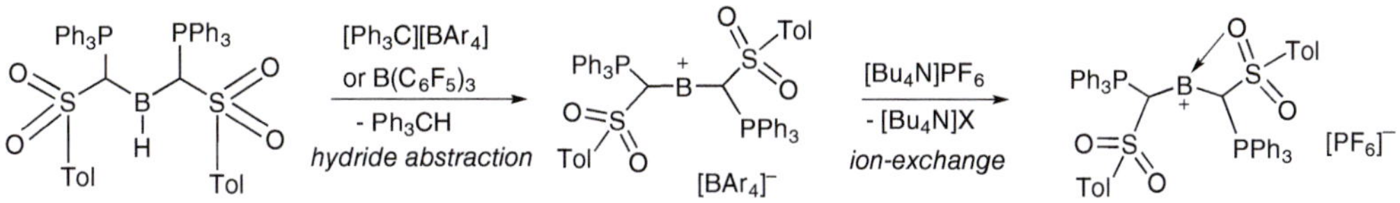

34a[BAr_4] (Ar = 3,5-$Cl_2C_6H_3$)
34b[BAr_4] (Ar = 3, 5-($CF_3)_3C_6H_3$)
34c [BAr_4] (Ar = C_6F_5)
δ (^{11}B) = 40.5 ppm for BAr_4 (broad signal)
= 39.8 ppm for $HB(C_6F_5)_3$

borenium cation
35[PF_6]

Scheme 5.10 Ylide-stabilized borenium cation.

$[(Ph_3PCSO_2Tol)_2B]^+[PF_6]^-$ **35**[PF_6] (Scheme 5.10) [59]. DFT calculations revealed that the ligand and cation interact *via* electrostatic effects as well as π-donation from the ylide ligands. Addition of external Lewis bases [dimethylaminopyridine(DMAP), DMF, and *N*-methylpyrrolidone] interfered with the $[OTf]^-$ interaction with boron and allowed the Lewis bases to afford the other three-coordinated borenium cations. However, the use of excess Lewis bases could not form the corresponding tetracoordinated boronium salts.

5.3.1.4 Phosphine-coordinated Borenium Cations

Phosphines have played an important role with neutral borane complexes in FLP chemistry by providing a hindered Lewis basic functionality. Stephan et al. reported the phosphine-stabilized borenium cation, $[C_6H_4O_2B{\cdot}P(t\text{-}Bu)_3]^+[HB(C_6F_5)_3]^-$ **36**[$HB(C_6F_5)_3$] (Figure 5.8), while trying to activate the B–H bond of catechol borane using $B(C_6F_5)_3{\cdot}P(t\text{-}Bu)_3$. The crystal structure of the species confirmed a planar three-coordinated boron centre with a longer B–P bond (1.933(5) Å), indicating significant π-donation from oxygen to boron. This complex can exist in two different canonical forms; borenium or boryl-phosphonium. Furthermore, DFT calculation revealed a significant positive charge on phosphorus and the structure in predominantly the boryl-phosphonium form [60].

In 2015, Bouhadir and coworkers prepared an intramolecularly Ph_2P-stabilized borenium cation, [Ph_2P-Naphth-BMes] **37**[A] (A = $GaBr_4$ or NTf_2) [61]. The cation was synthesized by using the bromide precursor and $GaCl_3$ or $AgNTf_2$ as halide scavenger and was characterized using NMR and HRMS methods. Further, the cation was reacted with Ph_2NH and NH_3 to give the amino-substituted borenium and NH_3 Lewis adduct as the boronium salts **38**[$GaBr_4$] and **39**[NTf_2], respectively (Scheme 5.11).

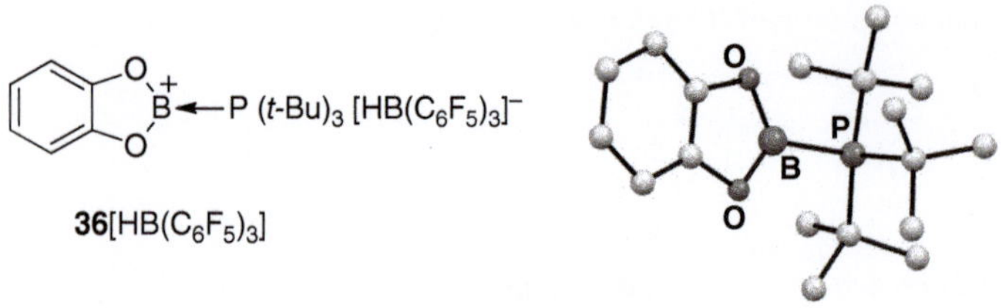

Figure 5.8 Structure of phosphine-stabilized borenium cation **36**, the counter-anion $[HB(C_6F_5)_3]^-$ has not been shown. *Source:* Adapted from Dureen et al. [60].

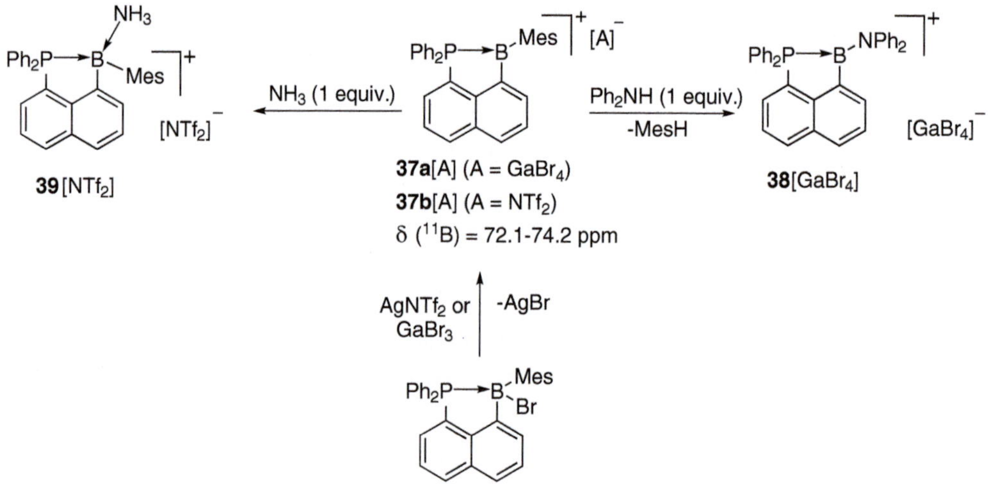

Scheme 5.11 Phosphine-stabilized borenium cation and its reactivity with ammonia and secondary amine.

The authors also explored the remote modulation of the boron Lewis acidity by changing the substituents at phosphorus and with more electron-donating *i*-Pr group and prepared the [*i*-Pr_2P-Naphth-BMes][A] (A=$GaBr_4$ or NTf_2) analogues of **37**[A]. The [*i*-Pr_2P-Naphth-BMes][NTf_2] salt reacted with H_2 to afford the neutral borane [*i*-Pr_2P-Naphth-B(H)(NTf_2)] under the elimination of MesH [62].

5.3.1.5 Borenium Cations Coordinated with *N*-donor Ligands

Singh et al. utilized sterically congested monoanionic bis(phosphinimino)amide ligands to prepare hydridoborenium cations. This ligand has some resemblance to the popular β-diketiminato ligands utilized by the groups of Cowley and Jordan for the synthesis of group 13 cations. The synthesis of a trigonal planar chloroborenium species $[\mathbf{L}BCl]^+[BCl_4]^-$ **40**[BCl_4] was demonstrated from the reaction of **L**BH_2 with three equivalents of BCl_3 [63]. Similarly, the dihydridoborane precursor on reaction with three equivalents of $BH_2Cl{\cdot}SMe_2$ or one equivalent of BCl_3 afforded the stable mono-hydridoborenium cation, $[\mathbf{L}BH]^+[HBCl_3]^-$ **41a**[$HBCl_3$]. The borenium cation **41a**[$HBCl_3$] was also obtained directly from the reaction of the ligand with three equivalents of $BH_2Cl{\cdot}SMe_2$. The solid-state structure of **41a**[$HBCl_3$] (Figure 5.9) showed a slightly puckered central N_3P_2B ring and the coordination environment around the cationic boron was distorted trigonal planar whereas the $BHCl_3^-$ anion showed a distorted tetrahedral geometry [64]. Both the examples highlight that the sterically bulky and strong donor bis(phosphinimino)amide ligand plays a crucial role in facilitating the synthesis and stabilization of these three-coordinated cationic species of boron without

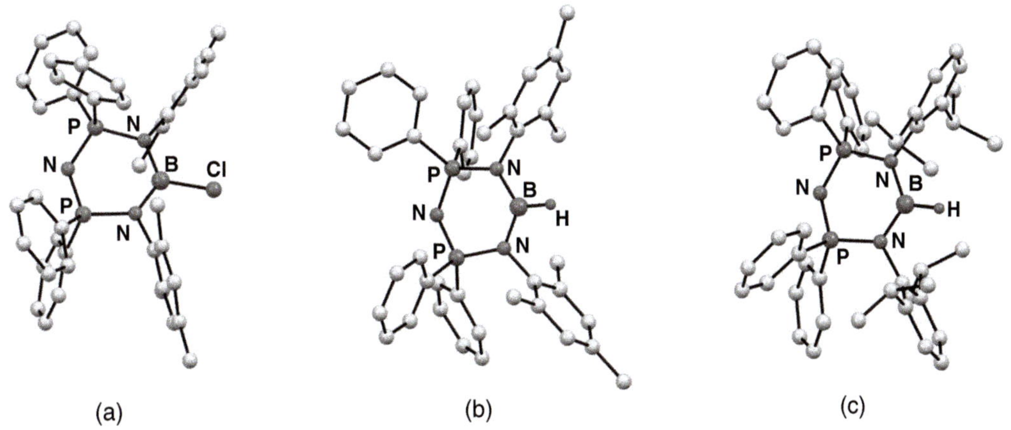

Figure 5.9 Structurally characterized bis(phosphinimino)amide-stabilized borenium cations (a) **40**[BCl_4], (b) **41a**[$HBCl_3$], and (c) **41b**[$HB(C_6F_5)_3$]. The counter-anions have been omitted. *Source:* Adapted from Jaiswal et al. [63, 64].

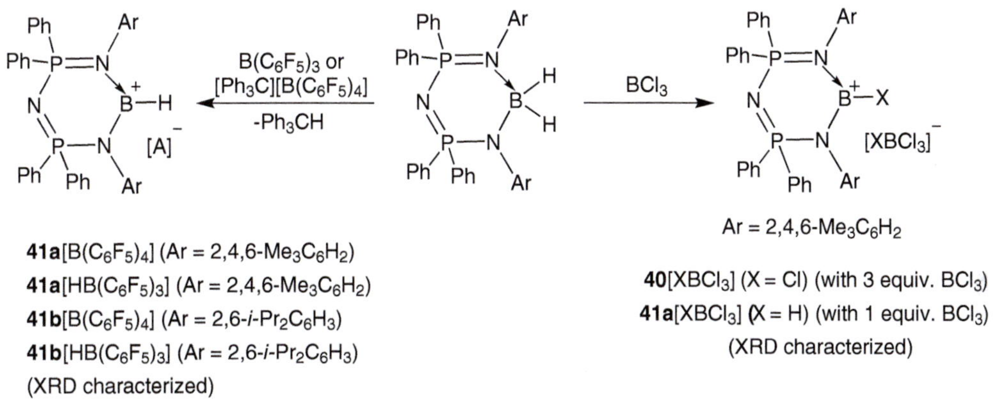

Scheme 5.12 Synthesis of bis(phosphinimino)amide supported borenium cations.

the necessity of a weakly coordinating bulky anion. Nevertheless, hydride abstraction by $B(C_6F_5)_3$ or [Ph_3C][$B(C_6F_5)_4$] from the neutral dihydroborane precursors also facilitates the formation of the borenium hydride cations **41**[$B(C_6F_5)_4$] (Scheme 5.12). The marked Lewis acidity of the borenium ion **41a**[$B(C_6F_5)_4$] was confirmed through DMAP coordination, with it resulting in the formation of its boronium adduct. Furthermore, the application of these cations was demonstrated when they acted as effective catalysts for the hydrosilylation of carbonyl compounds.

The diazadiene ligand has not been much explored towards the generation of borocation species. In 2016, Melen et al. reported a range of borocations accessible from $PhBCl_2$ or BCl_3 as boron precursors and stabilized by *N,N′*-1,4-diazabutadiene ligands. These ligands provide the control to tailor the electronic and steric properties of the formed borocationic complexes. Reaction between the diazadiene ligand with $BPhCl_2$ leads to the formation of diazaborole intermediates (**42a–f**) with Cl insertion at the carbon backbone, which on subsequent addition of $AlCl_3$ form borenium cations **43a–f**[$AlCl_4$] (Scheme 5.13). The molecular structure of the borenium complexes **43a** and **f** was determined by single-crystal X-ray diffraction, which confirmed the conversion of neutral diazadiene ligand to the monoanionic imino-amide form as also revealed by short [covalent, 1.383(3) Å] and long (dative, 1.529 Å) B–N bond lengths. However, the use of BCl_3 as the borane

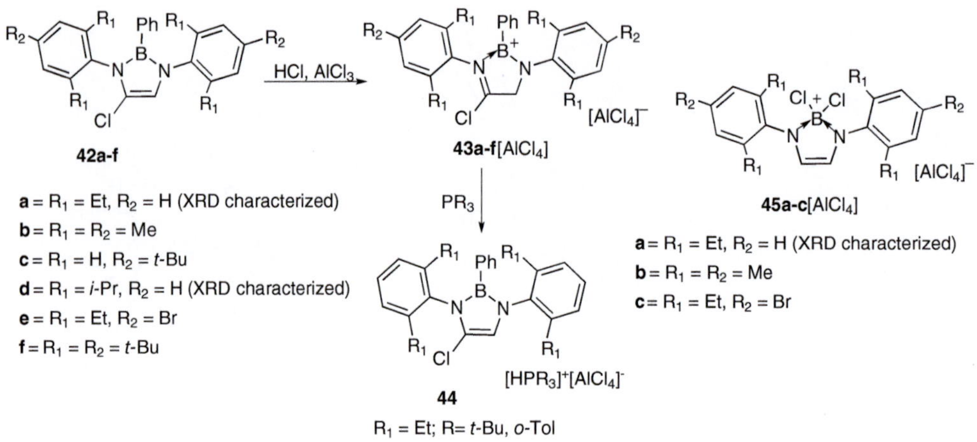

Scheme 5.13 A series of borocations stabilized by tunable diazadienes.

source generated the dichloroboronium cations (^{11}B NMR signal at 10.6 ppm) with BCl_4^- as the counter-anion. Subsequent addition of $AlCl_3$ leads to anion exchange to afford the $AlCl_4^-$ salts of the corresponding boronium ions, as was also confirmed with the aid of single-crystal X-ray diffraction of **45a**[$AlCl_4$] [65].

5.3.1.6 Boronium Cations: Four-coordinate Cationic Boron Complexes

Four-coordinated boronium cations have two σ-bonded substituents, which occupy the two coordination sites at the boron centre, with the remaining sites fulfilled by two neutral donor Lewis base ligands. The Lewis bases compensate the electronic requirement of the electron-deficient boron centre and improve its stability but it comes with a cost of quenched Lewis acidity as the positive charge often gets localized on the ligand system. Generally, these are the thermodynamically most stable among the borocations discussed above, and their Lewis acidity is most diminished due to their tetracoordinated structure and electronically saturated 8 VE configurations. Synthesis of boronium cations does not generally require very exotic state-of-the-art ligands for kinetic stabilization as evident from the earlier examples of borenium cations [4]. There have been a few reports on boronium cations since 2015 and very often the borenium cations bind with a Lewis base to convert into the boronium form. For this reason, a few of these boronium species have already appeared during the discussion on borenium cations; **26(a–c)**[$B(C_6F_5)_4$] and **45(a–c)**[$AlCl_4$].

The article published by Hill and Ward in 2017, is a good example which demonstrates the ease of synthesis and structural investigation of simple phosphine-coordinated boronium cations. These dihydrido, $[H_2B(\text{phosphine})]^+$ and monohydridoborenium $[BrHB(\text{phosphine})]^+$ cations were synthesized in high purity from two equivalents of a secondary, primary or chelating phosphine and $BrH_2B{\cdot}SMe_2$. The low reactivity of these cations can be correlated to coordinative saturation and a complete octet at the B centre. The upfield ^{11}B NMR signals for these cations were in the range of −27 to −45 ppm, correlating well with tetracoordinated boron centres. A few selected structures from this work are shown in Figure 5.10 [66].

Langer and coworkers reported on bis(phosphino)boronium salt **46**[Br] (Figure 5.11) and used it further to prepare PBP-pincer complexes by complexing phosphine end groups with Pd(II) and Ni(0) metallacycles and discussed the ligand behaviour of the boronium cation in detail [67].

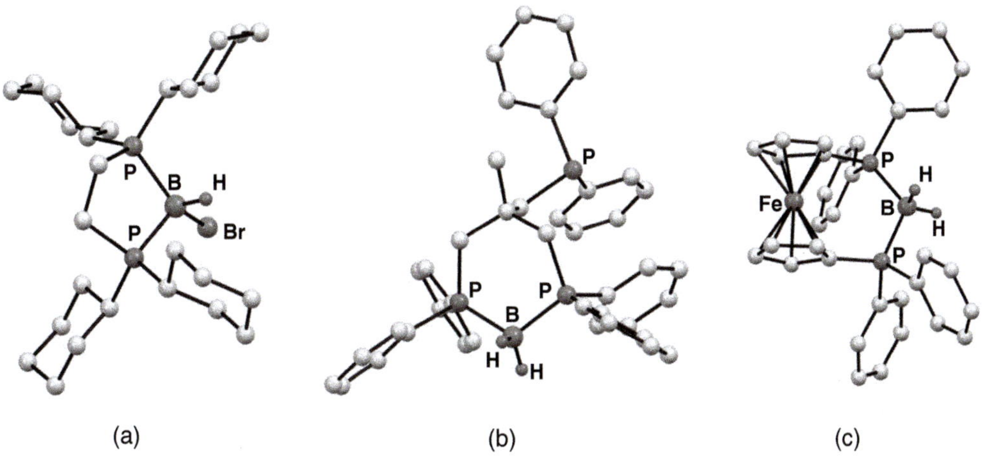

Figure 5.10 Single-crystal X-ray structure of (a) [BrHB(dcpe)]Br, (b) $[H_2B(triphos)][AsF_6]$ and (c) $[H_2B(dppf)][AsF_6]$, the anions have been omitted. dcpe = dicyclohexylphosphorylethane, triphos = triphosphine, dppf = bis(diphenylphosphino)ferrocene. *Source:* Adapted from Hill and Ward [66].

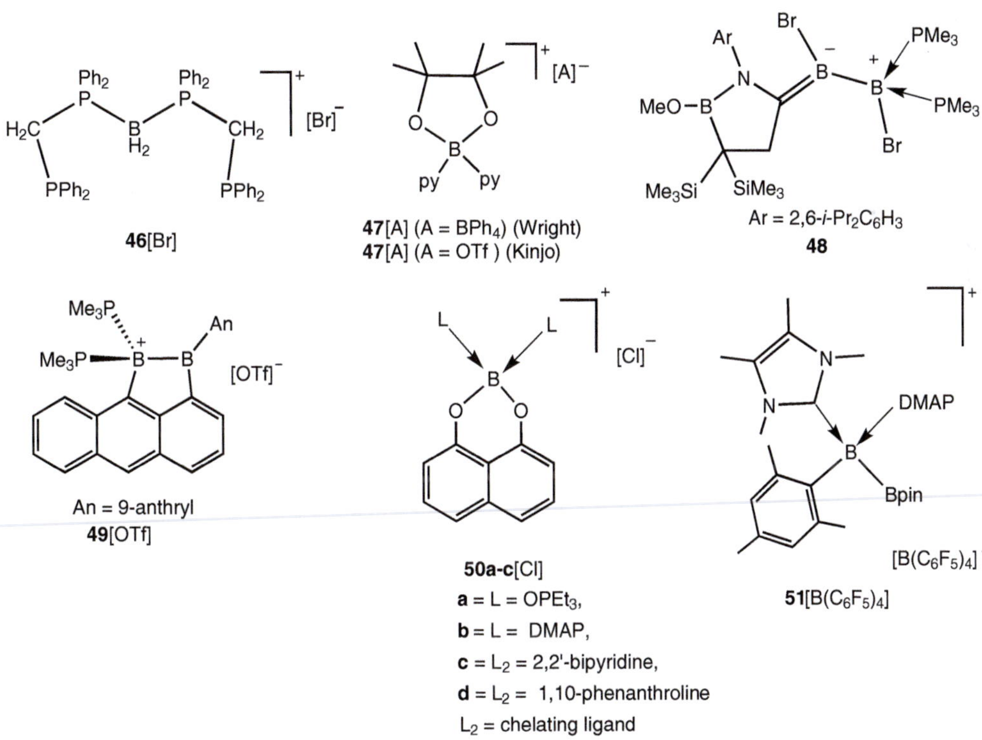

Figure 5.11 Examples of boronium cations.

Wright and coworkers showed regioselective 1,4-hydroboration of pyridines catalyzed by an acid-initiated boronium cation, $[pinB(py)_2][BPh_4]$ **47**$[BPh_4]$. This boronium cation was prepared from NH_4BPh_4, pyridine, and pinacolborane [70]. The triflate salt of the same boronium cation,

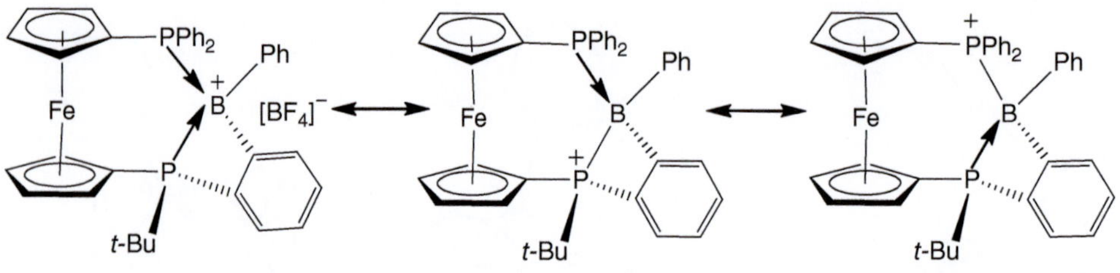

Figure 5.12 Bisphosphine-stabilized boronium cation **52**[BF_4] and its phosphonium canonical structures.

[pinB(py)$_2$][OTf] **47**[OTf] was prepared and structurally characterized by Kinjo et al. in 2018. The salt **47**[OTf] was identified as a key intermediate in 1,3,2-diazaphosphenium triflate-catalyzed regio- and chemo-selective hydroboration of pyridines [71].

Interestingly, Kinjo and coworkers prepared a zwitterionic boraalkenyl boronium **48** by KC_8 reduction of B_2Br_4 coordinated by a CAAC and Me_3P. Further reduction of this boronium cation with KC_8 afforded a neutral allenic diborane [72].

The π-complex of bis(PMe_3)-bis(9-anthryl) diborene with CuOTf slowly undergoes an intramolecular hydroarylation reaction at room temperature in several days, converting the protic C–H bond into a hydridic B–H functionality and subsequent metathesis with CuOTf has afforded a cyclic sp^2–sp^3 borylsubstituted boronium triflate salt **49**[OTf] (Figure 5.11) [68].

In a related study, Emslie and coworkers reported on 1,1′-bisphosphinoferrocene-supported intramolecularly bisphosphine-stabilized boronium cation [Fe{η^5-$C_5H_4PPh_2$}{η^5-C_5H_4P(*t*-Bu)(*o*-BPh-C_6H_4)}][BF_4] **52**[BF_4], formed from the reaction of [Fe{η^5-$C_5H_4PPh_2$}{η^5-C_5H_4P(*t*-Bu)(*o*-BPh_2-C_6H_4)}] with 2 equivalents of BF_3·OEt_2 (Figure 5.12). Although the boronium ion formulation for this complex was confirmed based on short P–B distances, nevertheless, this complex may also be described as a phosphine-coordinated borylphosphonium cation in its other canonical structures [73].

In another study, Krempner and coworkers probed the ionization of Cl-B(1,8-$O_2C_{10}H_8$) in the presence of O=PEt_3, DMAP, 1,10-phenanthroline and 2,2′-bipyridine, to afford the boronium salts, [(Et_3P=O)$_2$B(1,8-$O_2C_{10}H_8$)]Cl (**50a**), [(DMAP)$_2$B(1,8-$O_2C_{10}H_8$)]Cl (**50b**), [(2,2′-bipyridine)B(1,8-$O_2C_{10}H_8$)]Cl (**50c**), and [(1,10-phenanthroline)B(1,8-$O_2C_{10}H_8$)]Cl (**50d**), which were characterized by NMR spectroscopy and X-ray crystallography (Figure 5.11) [69].

Wang and coworkers reported on a well-characterized boronium cation **51**[B(C_6F_5)$_4$] which can be considered as a dinuclear boronium cation. The exocyclic B centre in this case is complexed with a mesityl substituent, a tetramethylimidazolidine carbene and pinacolylboryl group. The precursor borenium cation coordinated with DMAP to furnish **51**[B(C_6F_5)$_4$], and this precursor also reacted with H_2 under mild conditions accompanied by B–B bond cleavage [74].

Recently, So and coworkers in 2020 demonstrated that the amidinato-ligated silylene and its borane adducts, [(PhC{N(*t*-Bu)}$_2$){(Me_3Si)$_2$N}Si:] and [(PhC{N(*t*-Bu)}$_2$){(Me_3Si)$_2$N}Si:→BH_2(OTf)], when refluxed in toluene undergo an intramolecular C(sp^3)–H borylation and H_2 elimination to form a C–B bond in the resulting silylene–boronium ion **53a**[OTf] (Scheme 5.14). When the same reaction was performed with bulky bis(amidinato)silylene, once again C–H borylation took place to give an analogous product **53b**[OTf]. With a less bulky chlorosilylene, the boronium cation **54**[OTf] formed with silylene coordination to B and displacement of OTf. The displacement of triflate could also be affected when silylene–borane was treated with DMAP to form **55**[OTf].

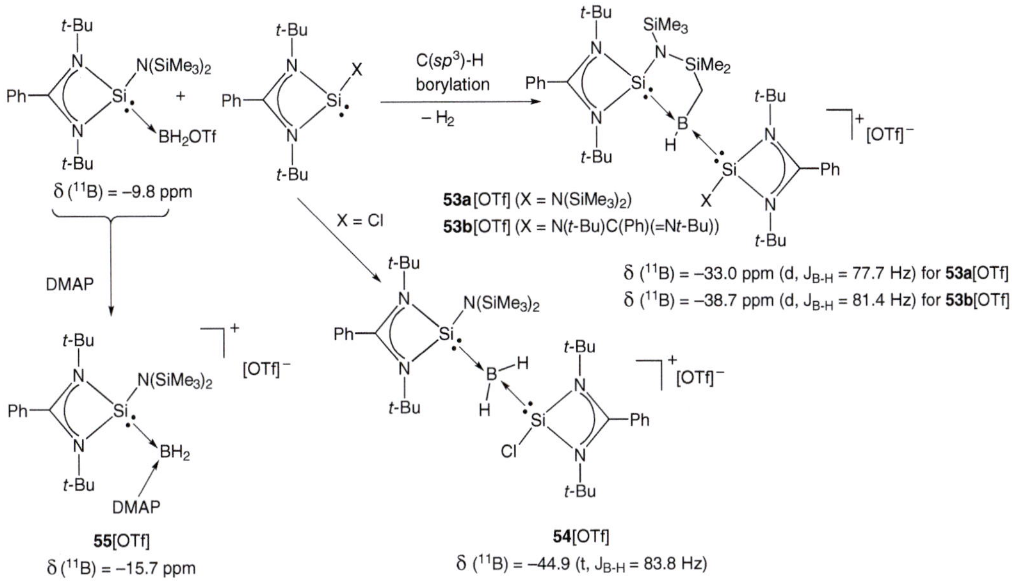

Scheme 5.14 Amidinato silylene-stabilized borenium cations.

All the boronium complexes **53–55**[OTf] were characterized with multinuclear NMR and single-crystal XRD analyses that confirmed the OTf^- as a counter-ion with no close contact with the boron centre. The ^{11}B NMR chemical shifts were upfield in the range of −9.8 to −44.9 ppm, as expected for tetracoordinated boron centres. Complex **53a**[OTf] activated H_2 gas or $NH_3{\cdot}BH_3$ at room temperature to form silylene–borane adduct [(PhC{N(*t*-Bu)}$_2$){(Me$_3$Si)$_2$N}Si:→BH$_3$] and [(PhC{N(*t*-Bu)}$_2$){(Me$_3$Si)$_2$N}SiH]OTf [75].

5.3.1.7 Miscellaneous Borocations

Some notable examples of well-defined borocations that cannot be classified by conventional routes have not been discussed here. Readers are prompted to extract the original reports of these unique borocations published since 2015. In this chapter, space limitations prevent us from a detailed discussion on the range of these borocations that include examples of dinuclear dicationic boron(III) complexes with bridging ligands and dinucleardicationic boron(II) complexes with B–B bonds [76–81]. Further, the latest miscellaneous borocation examples also include Cp*-bonded boron dication salts [82–85], high-coordinate multiboron polycations [86], monocations with B–B single bonds [79], a diboron monocation and borenium and boronium ions of the 1,2-azaborinine scaffold [74, 87].

5.3.2 Advances in the Synthesis and Characterization of Aluminium Cations

Aluminium is the most abundant metal in the earth's crust, has low toxicity levels and is inexpensive as well. These attributes of aluminium have made it a very attractive target to develop sustainable, environmentally benign catalytic systems. The developments in the fundamental study of well-defined organoaluminium compounds have seen a resurgence since the beginning of this century and the cationic Al derivatives are no exception to it. Compared to their neutral counterparts, the coordinatively and electronically unsaturated cationic Al complexes, with their enhanced Lewis acidity, are useful for catalytic applications. Well-defined new cationic aluminium complexes

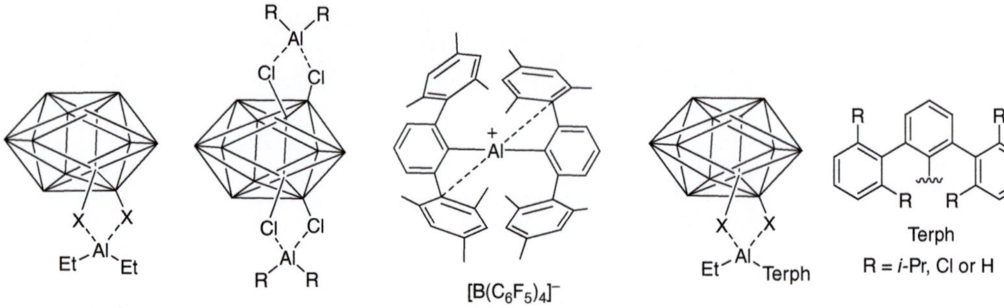

$[AlEt_2][CH_6B_{11}X_6]$
X = Cl or Br
56a (X = Cl)
56b (X = Br)
56c (X = I)

$[AlR_2]_2[B_{12}Cl_{12}]$
R = Me or Et
57a (R = Me)
57b (R = Et)

$[B(C_6F_5)_4]^-$
58$[B(C_6F_5)_4]$

$[EtAl(Terph)][CH_6B_{11}X_6]$
59a$[CH_6B_{11}Cl_6]$ (R = *i*-Pr)
59b$[CH_6B_{11}I_6]$ (R = Cl)

Figure 5.13 Examples of a few selected organoaluminium cations.

reported since 2010 will be covered in this chapter, with a major focus on three- and four-coordinated species (Figure 5.13 and the following text). Heteronuclear Al cations, aqueous Al chemistry, Al cations coordinated with multidentate ligands such as crown ethers and solid-state materials containing group 13 cations are beyond the scope of this book chapter. A few general reviews on group 13 cationic complexes have appeared (the most recent one published in 2019) covering general aspects of cationic aluminium complexes [9, 36, 92].

The class of three-coordinated cationic Al compounds provide a perfect balance of Lewis acidity and stability. Thus they are comparatively better explored than two-coordinated complexes. The three-coordinate arrangement can be obtained *via* a variety of ligand systems. The general methods of synthesis for cations have been described in Section 5.2.2. Here a brief discussion about the specific examples with key characterization data will be mentioned. Synthesis of three-coordinate Al cations is a challenging task but the choice of appropriate ligand scaffold and proper strategy can lead to success. Similar to the case of borocations, the stability of Al cations also strongly depends on the stability of the counter-anion against electrophilic attack by the same reactive cation. Unlike boron-based cations, classification of aluminium cations is not an easy or straightforward task due to the highly Lewis acidic nature of aluminium. Its cations are often found in contact with counter-anions or in coordination with the solvent, which results in a higher coordination number even if the interactions with the solvent molecules are generally weak. Due to these observations, we will not classify the cations on the basis of coordination numbers. We will classify the aluminium cations into two main classes: (i) organoaluminium cations in a mostly carbon environment and showing at least one Al–C bond; (ii) aluminium cations supported with chelating ligands with heteroatom donors (N, O, S, etc.) including examples with Al–C bonds with simple organyls.

5.3.2.1 Organoaluminium Cations

Attempts to prepare *simple* organoaluminium cations involved the reactions of $AlMe_3$ and $B(C_6F_5)_3/[Ph_3C][B(C_6F_5)_4]$. The reactions progress *via* formation of the highly electrophilic transient species $[AlMe_2][B(C_6F_5)_4]$. However, the ligand exchange process was very fast so the cation gradually degraded to a mixture of $AlMe_x(C_6F_5)_{(3-x)}$ and $BMe_{(3-x)}(C_6F_5)_x$, which ultimately converted to the neutral $Al(C_6F_5)_3$ and BMe_3 (Scheme 5.15, eq. 1) [94]. The higher aluminium alkyls, such as

$$AlMe_3 + [Ph_3C][B(C_6F_5)_4] \rightleftharpoons [AlMe_2][B(C_6F_5)_4] \rightleftharpoons AlMe_x(C_6F_5)_{3-x} + BMe_{3-x}(C_6F_5)_x \rightleftharpoons Al(C_6F_5)_3 + BMe_3 \quad ....(5.1)$$

$$AlR_3 + B(C_6F_5)_3 \xrightarrow{Et_2O} [R_2Al(OEt_2)_2][RB(C_6F_5)_3] \quad (R = Me, Et, octyl) \quad ....(5.2)$$

$$2Me_3Al + [CPh_3]_2[B_{12}Cl_{12}] \xrightarrow[80°C]{1,2\text{-difluorobenzene}} [Me_2Al]_2[B_{12}Cl_{12}] + 2MeCPh_3 \quad ....(5.3)$$

$$2Et_3Al + [CPh_3]_2[B_{12}Cl_{12}] \xrightarrow[-2\,Ph_3CH]{\text{benzene, rt}} [Et_2Al]_2[B_{12}Cl_{12}] \quad ....(5.4)$$

Scheme 5.15 A few common synthetic methods to generate cationic aluminium species.

$$RMEt_2 \xrightarrow[\text{benzene or hexane, rt}]{[Et_3Si][CHB_{11}Cl_{11}]} [RMEt][CHB_{11}Cl_{11}] + Et_4Si$$

60a$[CHB_{11}Cl_{11}]$ (M = Al; R = Et)
60b$[CHB_{11}Cl_{11}]$ (M = Al; R = 2,6-$Ph_2C_6H_3$)
61a$[CHB_{11}Cl_{11}]$ (M = Ga; R = Et)
61b$[CHB_{11}Cl_{11}]$ (M = Ga; R = 2,6-$Dip_2C_6H_3$)

Scheme 5.16 Two coordinated organoaluminium and gallium cations.

$Al(i\text{-}Bu)_3$, reacted significantly faster under β-hydride abstraction. In the presence of Et_2O, the reaction stopped after the formation of cation $[AlR_2(Et_2O)_2]^+$ (R = Me, Et, octyl), which also acts as an activator for polymerization of ethene (Scheme 5.15, eq. 2) [93].

A more controllable synthesis of aluminium cations was demonstrated by Reed et al. who used more stable and unreactive carborane as a counter-anion. The reaction was carried out between $AlEt_3$ and $[Ph_3C][CB_{11}H_6X_6]$ (X = Cl, Br), which resulted in the formation of aluminium cation $[AlEt_2][CB_{11}H_6X_6]$ *via* ethyl group abstraction and β-hydrogen elimination [88]. The single-crystal X-ray structure of $[AlEt_2][CB_{11}H_6X_6]$ (X = Cl, Br) showed that the icosahedral carbanion was weakly interacting with the $[Et_2Al]^+$ in a bidentate mode with the X groups at the 7,8-positions of the anion (**56a–b**, Figure 5.13) that was reflected in the Al–Br (2.58, 2.54 Å) and Al–Cl is (2.44, 2.43 Å) distances. Khandelwal and Wehmschulte performed deoxygenative reduction of CO_2 to methane, toluene, and diphenylmethane with the cation $[Et_2Al]^+[CH_6B_{11}I_6]^-$ as a catalyst [95]. In a continuation of this work, Wehmschulte and co-workers in 2017 used Et_3Al and terphenyl(diethyl) aluminium, (2,6-$Ph_2C_6H_3$)-$AlEt_2$, to react with *in situ* generated silylium salt $[Et_3Si][CHB_{11}Cl_{11}]$ (Scheme 5.16) [96].

These reactions afforded cationic organoaluminium compounds **60a** and **60b** with the counter-anion $[CHB_{11}Cl_{11}]^-$ which is much less basic than the previously reported complexes with $[CH_6B_{11}I_6]^-$. The aim here was that these complexes would have much weaker or no cation–anion contacts. Nevertheless, the X-ray structures of two Al compounds featured cation–anion contacts in the solid-state. The Al centre in **60a**$[CHB_{11}Cl_{11}]$ and **60b**$[CHB_{11}Cl_{11}]$ is in a distorted tetrahedral environment with two Al–C bonds (1.941 Å for **60a** and 1.939 Å for **60b**) and two Al–Cl contacts (2.453 Å for **60a** and 2.504 Å for **60b**) to the *m*- and *p*-chlorine donors from the anion (Figure 5.14). All of the cationic complexes catalyzed CO_2 reduction with Et_3SiH to form methane [96].

Similar to these, Knapp and coworkers in 2011 reacted Me_3Al and Et_3Al with the trityl salt of the weakly coordinating dianion $[Ph_3C]_2[B_{12}Cl_{12}]$ to isolate cationic dialkyl Al cations $(Me_2Al)_2(B_{12}Cl_{12})$ and $(Et_2Al)_2(B_{12}Cl_{12})$ (**57a** and **57b**) (Figure 5.13, Scheme 5.15, Eqs. 5.3 and 5.4). The crystal

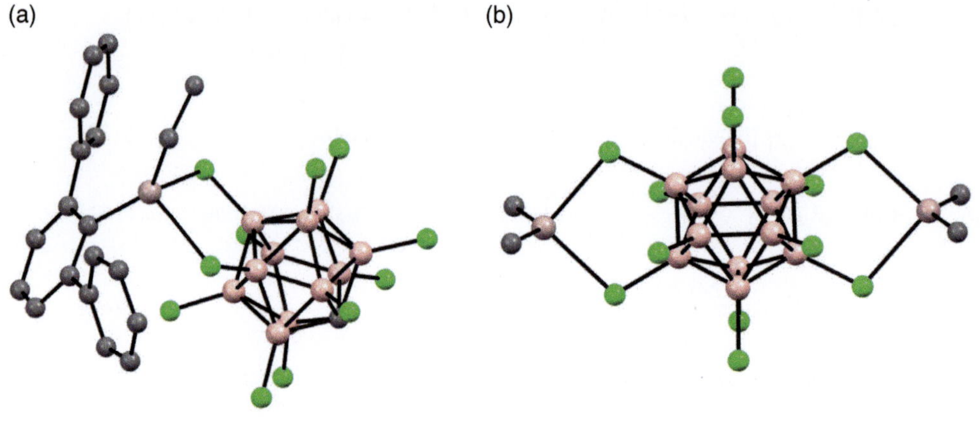

Figure 5.14 Molecular structures of (car)borane-complexed aluminium cations (a) **60b** and (b) **57a** (Al = light brown, B = light pink, Cl = green). *Sources:* Adapted from Saleh et al. [96]; Kessler et al. [89].

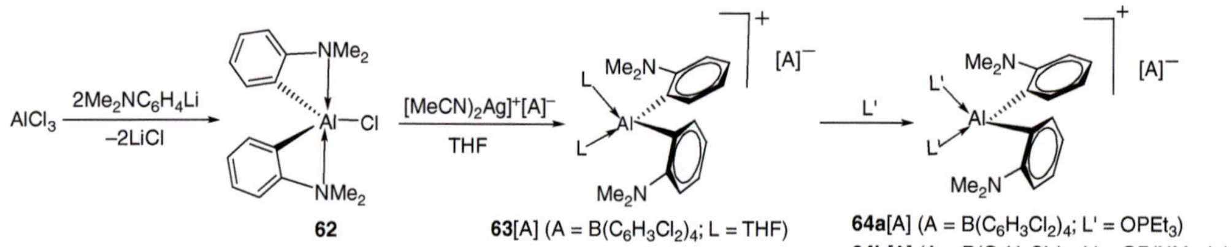

Scheme 5.17 Tetracoordinated aluminium cations and reactivity with Lewis base illustrating the labile nature of coordinated THF.

structures of **57a** and **57b** confirmed that aluminium centres are bound to two chlorine atoms, resulting in a distorted tetrahedral environment around them. The Al–Cl contacts are longer than typical Al–Cl single bonds but significantly shorter than the sum of the van der Waals radii, ascribing it as ion-like interactions [89].

Venugopal and coworkers, in their attempts to prepare tetracoordinated analogues of R_2Al^+ cations (like **56** and **58**) utilized additional lone pair donor $-NMe_2$ units in the monoanionic 2-(dimethylamino)phenyl ligand unit (Scheme 5.17) to successfully isolate the cationic Al complex $[(Me_2NC_6H_4)_2Al(THF)_2][B(C_6H_3Cl_2)_4]$ **63**$[B(C_6H_3Cl_2)_4]$ [97]. The geometry of the Al cation was described as a bicapped tetrahedron, where each of the two carbon and oxygen atoms occupies the corners of the tetrahedron and the nitrogen atoms cap the tetrahedron. THF molecules bound to the aluminium centre were labile and could be substituted with the more nucleophilic Lewis bases $OPEt_3$ and $OP(NMe_2)_3$ to give Al cations **64a**$[B(C_6H_3Cl_2)_4]$ and **64b**$[B(C_6H_3Cl_2)_4]$, respectively. The $OPEt_3$ adduct of the cation gives a sharp peak at 75.5 ppm in its ^{31}P NMR which corresponds to the acceptor number 76.24 on the Gutmann scale. The low AN of the cation can be attributed to the high coordination number around the Al centre. This cation showed excellent catalytic activity for the dimerization of aldehydes and hydrosilylation of ketones [97].

With the aim of synthesizing discrete two coordinated cations, the bulkier *m*-terphenyl ligand was employed by Wehmschulte et al. in 2004. The cation $[(2,6\text{-}Mes_2C_6H_3)_2Al]^+[B(C_6F_5)_4]^-$ **58**$[B(C_6F_5)_4]$ (Figure 5.13) was obtained in low yield *via* hydride abstraction from $(2,6\text{-}Mes_2C_6H_3)_2AlH$ with $[Ph_3C]^+[B(C_6F_5)_4]^-$. The single-crystal X-ray structure of the benzene solvate showed separate $[(2,6\text{-}Mes_2C_6H_3)_2Al]^+$ cations and $[B(C_6F_5)_4]^-$ anions, where the C–Al–C angle around Al was

159.17(5) ° and C–Al covalent bonds measured 1.9379(12) and 1.9412(12) Å. The deviation from linear structure around Al and its strong Lewis acidity was balanced by the short Al–C contacts involving *ipso* carbons of the flanking mesityl groups with the bond distance of 2.3532(12) and 2.553(1) Å [90]. The *m*-terphenyl-substituted organoaluminium precursors Dip^*AlEt_2 and $DcpAlEt_2$ underwent ethide elimination on reaction with $[Ph_3C][CH_6B_{11}X_6]$ or $[Et_3Si][CH_6B_{11}X_6]$ to afford the cations $[Dip^*AlEt][CH_6B_{11}X_6]$ [Dip^* = 2,6-(2,6-*i*-$Pr_2C_6H_3)_2C_6H_3$], $[DcpAlEt][CH_6B_{11}X_6]$ [Dcp = 2,6-(2,6-$Cl_2C_6H_3)_2C_6H_3$] where X = Cl, I (**59a** and **59b**, Figure 5.13). The crystal structures of three of these compounds revealed that the aluminium centres are four-coordinate (distorted trigonal pyramidal) with covalent bonds to the Dip^*(or Dcp), and Et substituents and weaker coordinative bonds to two of the equatorial halogen substituents of the carborane anion forming the cation–anion adducts. The compound $[DcpAlEt][CH_6B_{11}Cl_6]$ forms the bisamine adduct $[DcpAlEt\{NH_2(t\text{-}Bu)\}_2][CH_6B_{11}Cl_6]$ upon exposure to *t*-$BuNH_2$, and compounds $[Dip^*AlEt][CH_6B_{11}Cl_6]$ and $[DcpAlEt][CH_6B_{11}Cl_6]$ slowly catalyzed the alkylation of benzene with 1-hexene [91].

Zhang and coworkers described dearomatization and selective alkylation of the central pyridine ring of terpyridine at the 3′- or 5′-position to form a 1,4-functionalized product as a zwitterionic "Meisenheimer type" species [98]. These species featured pentacoordinate Al cations **65a** and **65b** (Figure 5.15) that were also structurally characterized. The Cl-substituted catalyst turned out to be extremely efficient for the selective hydroboration of C=O and C≡C functionalities with turnover numbers (TONs) as high as ~1000 [99].

Okuda and coworkers reported on bis(allyl)aluminium cations $[Al(\eta^1\text{-}C_3H_5)_2(THF)_n][BPh_4]$ (n = 2, 3) (**66**$[B(C_6H_5)_4]$) (Scheme 5.18). The single-crystal XRD of the cations showed tetrahedral and trigonal bipyramidal geometry, respectively, around Al and allyl binding in η^1-C_3H_5 fashion. Two η^1-C_3H_5 and a THF molecule reside in the equatorial plane of the pentacoordinated species. Remarkably, these aluminium cations did not exhibit very high reactivity and could be isolated as their $[BPh_4]^-$ salts without the use of the corresponding less basic perfluoro anion $[B(C_6F_5)_4]^-$ [100].

In 2016, Stephan et al. reported on reactions of carbene–alane adducts with tritylborate and demonstrated synthesis of the first dicationic aluminium hydride salts, highlighting the influence of steric congestion of the carbene substituents on the aggregation of aluminium hydrides. In their efforts, they utilized the AlH_3 adduct of Dip-substituted sterically encumbered NHC (NHC^{Dip})·AlH_3 [NHC^{Dip} = 1,3-bis(2,6-diisopropylphenyl)imidazol-2-ylidene] to react with $[Ph_3C][B(C_6F_5)_4]$. This reaction afforded the dimeric aluminium dication $[\{(NHC^{Dip})AlH(\mu\text{-}H)\}_2][B(C_6F_5)_4]_2$ **67**$[B(C_6F_5)_4]_2$, where two Al centres are connected *via* hydride bridges with an Al–Al separation of 2.584(2) Å, which closely resembles a typical Al–Al single bond length. In contrast, reaction of the alane adduct of 1,3-dibenzylimidazol-2-ylidene, (NHC^{Bn})·AlH_3 with $[Ph_3C][B(C_6F_5)_4]$ formed a planar monomeric dication $[(NHC^{Bn})_2AlH][B(C_6F_5)_4]_2$ **68**$[B(C_6F_5)_4]_2$, *via* the redistribution of carbene on the intermediate $[(NHC^{Bn})AlH_2]^+$ cationic complex (Scheme 5.19) [101]. The aluminium centre in

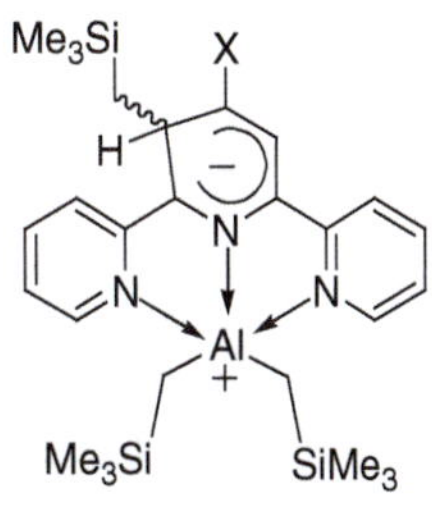

65a (X = Cl)
65b (X = Ph)

Figure 5.15 Zwitterionic "Meisenheimer type" pentacoordinated aluminium complex **65**.

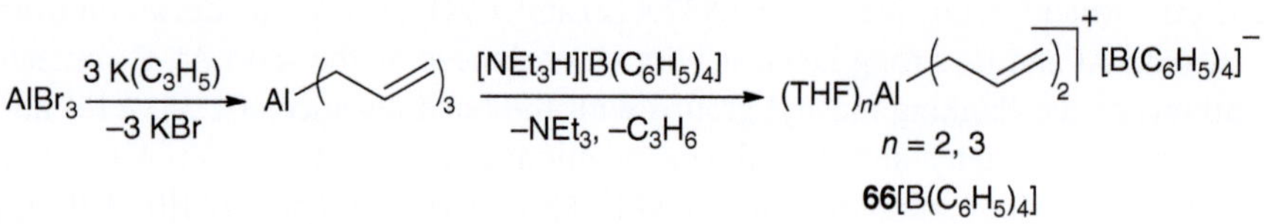

Scheme 5.18 THF adduct of bis(allyl)-substituted aluminium cations.

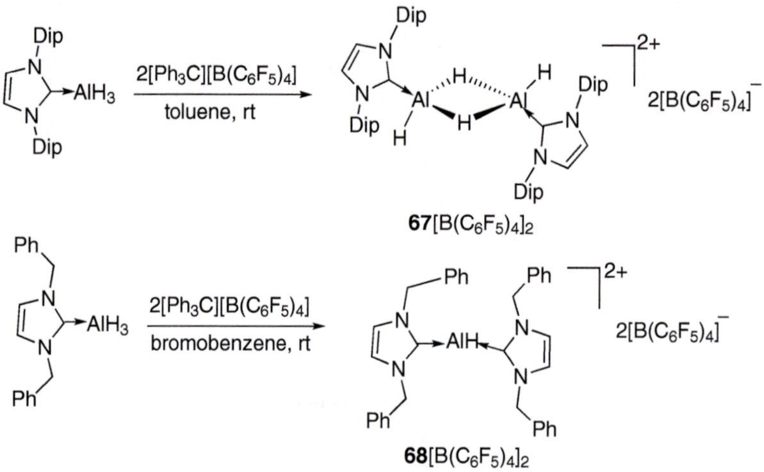

Scheme 5.19 NHC-stabilized dimeric aluminium dication **67** and monomeric aluminium dication **68**.

this complex consists of two NHC^{Bn} ligands and one hydride and adopted a trigonal planar geometry (the sum of the bond angles around the Al centre being 359 °). The stability of this complex stems from the fact that the planes of the phenyl substituents of one NHC^{Bn} ligand are arranged above and below the Al-centred trigonal plane. A weak arene interaction with the Al centre is observed and is supported by close C$_{arene}$–Al bond distances [2.572(4) and 3.124(5) Å] of one Bn with Al. The DFT calculations on this cation showed the LUMO was mostly located at the Al centre, with a marked contribution from one of the proximal arene moieties [101].

Stasch and coworkers in 2018 reported on the reactivity studies of NHC·AlX$_3$ (X = H, I) complexes with an additional NHC ligand to highlight the application of abnormal as well as normal NHC carbene ligands for the generation of cationic hydride iodide complexes of aluminium. Reaction of [(NHC^{Dip})AlHXI] (X = H, I) with another equivalent of NHC^{Dip} at 85 °C for 16 h afforded unprecedented mixed normal–abnormal NHC-coordinated ionic complexes [(NHC^{Dip})AlH$_2$(*a*NHC^{Dip})]I **69a**[I] and [(NHC^{Dip})AlHI(*a*NHC^{Dip})]I **69b**[I], where *a*NHC^{Dip} is the abnormal NHC^{Dip} carbene tautomer bonded through its four-position (Scheme 5.20). The iodide in these complexes was replaced by strong sigma donor carbene to form the cations. The molecular structures of these cationic complexes by X-ray crystallography showed slightly shorter Al–C(*a*NHC^{Dip}) distances than Al–C(NHC^{Dip}) distances [102].

In 2016, Dagorne et al. reported on (NHC^{Ar})MMe$_3$ adducts and their reactions with B(C$_6$F$_5$)$_3$ in the presence of donor solvents (THF, Et$_2$O). These adducts undergo methide abstraction to generate their corresponding cations, [(NHC^{Ar})MMe$_2$(*S*)]$^+$[MeB(C$_6$F$_5$)$_3$]$^-$ salts **70a–c**[MeB(C$_6$F$_5$)$_3$] (*S* = THF or Et$_2$O) (Scheme 5.21). Such cations, represented the first examples of stable

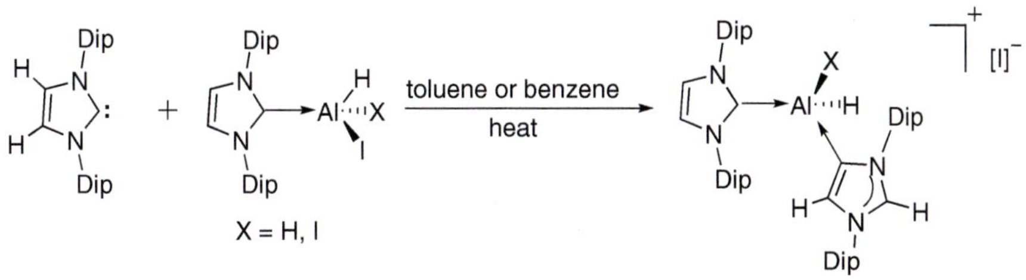

69a [I] (X = H)
69b [I] (X = I)

Scheme 5.20 Cationic aluminium hydride complexes stabilized by NHC *via* normal and abnormal mode.

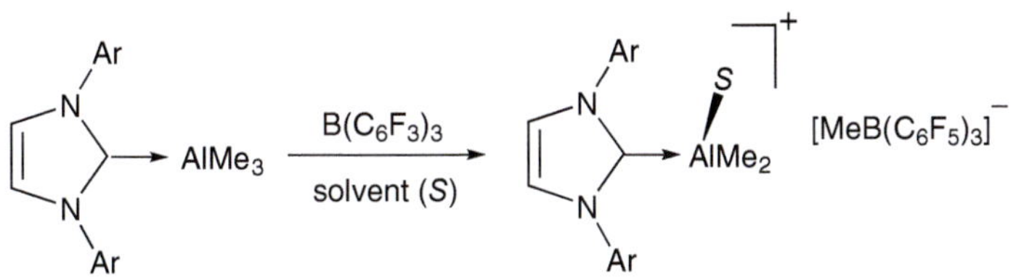

Ar = Mes = 2,4,6-$Me_3C_6H_2$
= Dip = 2,6-*i*-$Pr_2C_6H_3$

$[(NHC^{Ar})MMe_2(S)]^+[MeB(C_6F_5)_3]^-$
70a $[MeB(C_6F_5)_3]$ (M = Al, NHC^{Mes}, *S* = THF)
70b $[MeB(C_6F_5)_3]$ (M = Al, NHC^{Mes}, *S* = OEt_2) (XRD characterized)
70c $[MeB(C_6F_5)_3]$ (M = Al, NHC^{Dip}, *S* = THF)

Scheme 5.21 Selected example of NHC coordinated alkyl aluminium cations.

NHC-stabilized alkyl Al and Ga cations. The crystal structure of $[(NHC^{Mes})AlMe_2(OEt_2)]^+[MeB(C_6F_5)_3]^-$ revealed a four-coordinate Al(III) centre in a slightly distorted tetrahedral environment with dissociated $[(NHC^{Mes})AlMe_2(OEt_2)]^+$ and $[MeB(C_6F_5)_3]^-$ ions and no close anion/cation interactions. All these compounds were also tested as initiators for the ring-opening polymerization (ROP) of *rac*-lactide [103].

Recently, Fedushkin et al. reported cationic alkylaluminium complexes stabilized by the redox non-innocent BIAN (1,2-bis[(2,6-diisopropylphenyl)imino]acenaphthene) ligand. The diethylaluminium complex with radical-anion BIAN $[(BIAN)AlEt_2]$ and monoalkylaluminium complex with dianionic BIAN $[(BIAN)AlEt(Et_2O)]$ (**72**) were opted for as neutral precursor molecules. Further, $[Ph_3C][B(C_6F_5)_4]$ was used for the abstraction of ethyl with the expectation of generating cationic Al complexes (Scheme 5.22). However, oxidation of the ligand backbone was observed by the acceptance of one electron and, consequently, various cationic aluminium complexes were reported (**71, 73, 74**). The molecular geometries of **71** and **74** were also established by XRD and revealed tetrahedral geometry around aluminium centres. The reaction between the dialkylaluminium complex and $[Ph_3C][B(C_6F_5)_4]$ resulted in cationic complex **71**, where the BIAN ligand is reported as a neutral donor, which was further verified by the bond lengths obtained from the X-ray structure; for Al–N [2.006(2), 1.996(2) Å], lengths were similar to those of the neutral BIAN system. Similarly, the reaction between monoalkylaluminium complex **72** and $[Ph_3C][B(C_6F_5)_4]$ resulted in complex, **73**$[B(C_6F_5)_4]$. In contrast, employing $B(C_6F_5)_3$ for cation generation by utilizing mono alkyl

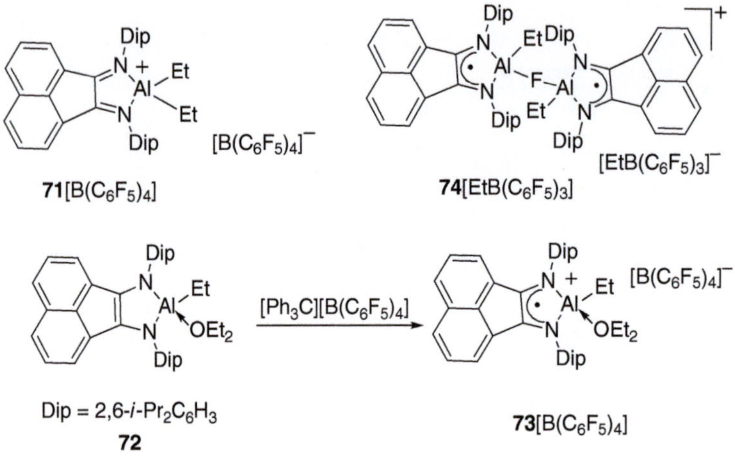

Scheme 5.22 Cationic alkylaluminium complexes stabilized by redox active non-innocent BIAN ligand.

precursor resulted in the dimeric complex **74**[EtB(C_6F_5)$_3$] as the result of fluoride abstraction. The solid-state structure of **74**[EtB(C_6F_5)$_3$] was reported as a dimer with bridging fluoride [104].

5.3.2.2 Aluminium Cations Supported by *N,N′*-donor Monoanionic Bidentate Ligands

There have not been many reports on new aluminium cations stabilized by anionic *N,N′*-donor chelating ligands ([*N,N′*]) since 2010. Nevertheless, a significant development took place in the last 10 years on the applications of these cationic Al compounds. Therefore, a glimpse of previously known compounds (Figure 5.16) is presented without elaborate discussion on them.

The early successful attempts to synthesize three-coordinated *N,N′*-chelated aluminium cations were made by the group of Jordan. In this endeavour *N,N′*-bidentate monoanionic ligands such as aminotroponiminate (ATI) (for Al cations **75**, **76**), β-diketiminate (for Al cations **77–80**) and amidinate (for Al cation **81**) ligands were utilized along with the bulky, weakly coordinating borate anions [12, 105–109]. At first, tetracoordinated covalent [*N,N′*]$AlMe_2$ and [*N,N′*]AlH_2 precursor complexes were made for further alkyl or hydride abstraction for the generation of cationic Al complexes. For example, the ATI complexes, [(*i*-Pr)$_2$ATI]AlR_2 (R = *t*-Bu, Et) were treated with [Ph_3C][B(C_6F_5)$_4$], to form the aluminium cations **75**[B(C_6F_5)$_4$] and **76**[B(C_6F_5)$_4$] under the elimination of isobutene and ethene gas [12, 108]. The ethyl derivative, **75**[B(C_6F_5)$_4$] was also characterized by single-crystal XRD, which revealed the solvent (PhCl) was in close proximity to the cationic Al centre and a weak Al–Cl interaction measuring 2.540(3) Å was found, with aluminium residing in a distorted tetrahedral geometry [108]. Alkyl bridge formation between the cation and its neutral precursor was also observed with amidinate ligands and to prevent it, alkyl groups bigger than Me were utilized. The result of this approach was evident in the formation of **77**[B(C_6F_5)$_4$], with an Et substituent on Al only obtained in the presence of an external base. Meanwhile, with a bulky *t*–Bu group on Al did not form either a bridged cation or require an external base and the cation, **80**[B(C_6F_5)$_4$], could be synthesized directly [105]. Similarly, on embedding the bulky *t*-Bu substituent on an amidinate ligand it was possible to afford three-coordinated Al–Me cation **81**[B(C_6F_5)$_4$] [106].

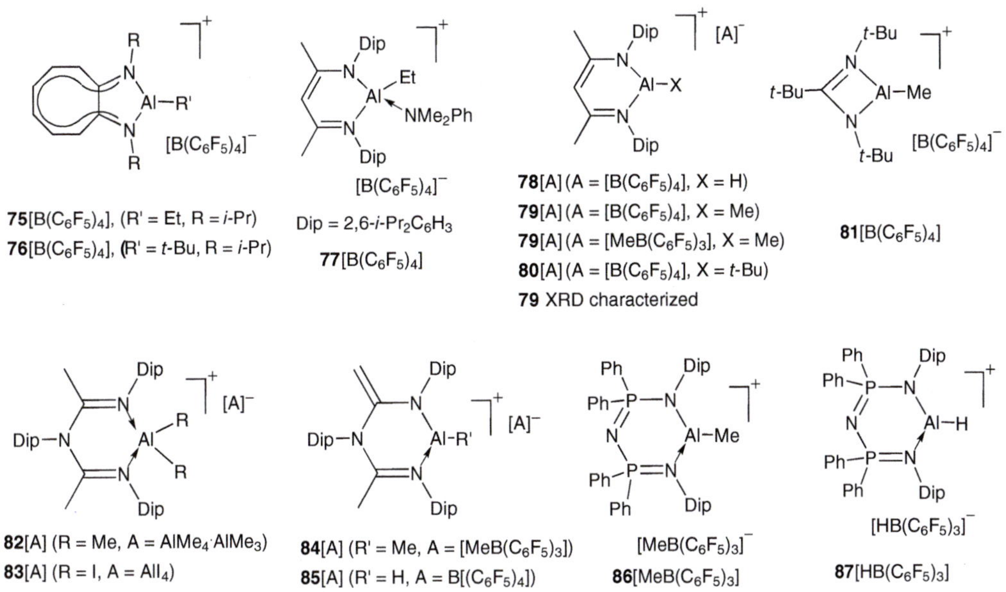

Figure 5.16 Examples of aluminium cations supported by *N,N'*-chelating ligands.

The six-membered [*N,N'*] complex of the monoanionic, bulky β-diketiminate^{Dip} ligand, (β-diketiminate^{Dip})$AlMe_2$, underwent Me abstraction by $B(C_6F_5)_3$ or $[Ph_3C][B(C_6F_5)_4]$ to afford a stable three-coordinated Al cation. The X-ray structure of the complex [(β-diketiminate^{Dip})AlMe]-$[B(C_6F_5)_4]$ **79**$[B(C_6F_5)_4]$ revealed close proximity between a *meta*-fluorine of the $[B(C_6F_5)_4]^-$ anion and the cationic Al centre [measuring the Al–F distance as 2.151(1) Å] [105–107]. LA quantification by the Guttman–Beckett method gave the acceptance number as 89.7 in 1:1 C_6H_6/C_6H_5Br solution, which is higher than for $AlCl_3$ and $B(C_6F_5)_3$ [112]. Harder et al. explored the FLP behaviour of this compound in combination with PPh_3 for the activation of small molecules such as CO_2, cyclic ethers, alkynes, and alkenes [112]. The analogous aluminium hydride cation, [(β-diketiminate^{Dip})AlH]$[B(C_6F_5)_4]$ **78**$[B(C_6F_5)_4]$ was prepared in a similar manner and utilized for hydrosilylation of olefins by the group of Nikonov. We could not retrieve any X-ray characterization on these Al cations [113].

Masuda and Stephan utilized a bulky neutral bidentate *N*-arylimidoyl(amidine) ligand to form four-coordinate cationic Al complexes, [{*N*-arylimidoyl(amidine)}$AlMe_2]^+[AlMe_4{\cdot}AlMe_3]^-$ **82**$[AlMe_4{\cdot}AlMe_3]$, and [{*N*-arylimidoyl(amidine)}$AlI_2]^+[AlI_4]^-$ (**83**$[AlI_4]$) *via* direct reaction of the ligand with two equivalents of $AlMe_3$ and AlI_3, respectively. The complex **82**$[AlMe_4{\cdot}AlMe_3]$ converts into a neutral dimethyl complex under heating conditions and further treating the complex with $B(C_6F_5)_3$ generates a new cation, **84**$[MeB(C_6F_5)_3]$. The 1:1:1 reaction of the ligand, $Me_2EtN{\cdot}AlH_3$ and $[Ph_3C][B(C_6F_5)_4]$ in toluene afforded the three-coordinated aluminium hydride cation **85**$[B(C_6F_5)_4]$, as confirmed by single-crystal XRD. In these last complexes, a remote Me group in the ligand backbone had lost a proton to convert the ligand into a formally bidentate monoanionic ligand (Figure 5.16) [110].

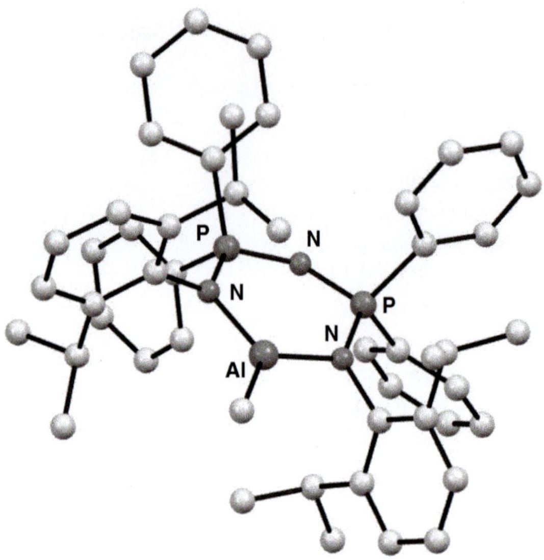

Figure 5.17 Molecular structure of organoaluminium cation **86**[MeB(C_6F_5)$_3$]. *Source:* Adapted from Prashanth et al. [111].

The P/N ligand, bis(phosphinimino)amide was utilized by Singh and coworkers to generate three-coordinated aluminium methyl and aluminium hydride cations. This ligand's steric and electronic features compare well with β-diketiminates with an additional advantage of wider bite angle at the N–Al–N centre due to longer P–N bonds (compared to narrow N–Al–N bite angle due to shorter C–N bonds of β-diketiminates), thus improving the feasibility of forming planar three-coordinate Al cations. The bis(phosphinimino)amides showed easy tunability of steric and electronic properties of the ligand to control the ability to stabilize three-coordinated aluminium methyl **86**[MeB(C_6F_5)$_3$] and hydride cations **87**[HB(C_6F_5)$_3$] [111]. The solid-state structure of aluminium methyl complex **86**[MeB(C_6F_5)$_3$] showed the sum of bond angles around Al to be 360 °, suggesting planar geometry around the metal centre (Figure 5.17). The low-lying orbitals of the cations were analysed by computation. A significant contribution of the 3p orbital from aluminium was seen and the NBO charge on Al was found to be 2.01 e. The Lewis acidity of these cations was also quantified by the Gutmann–Beckett method and the cations gave acceptance numbers of 90.4 for aluminium hydride and 83.1 for aluminium methyl cation (Table 5.1).

5.3.2.3 An Aluminium Cationic Complex Supported by a Neutral Bidentate *N,N′*-donor Ligand

In 2018, Inoue et al. isolated a four-coordinated aluminium dihydride cation in **88**[AlB_2H_8] and **88**[BH_4] by employing the neutral strong π-electron-donating ligand [1,2-({1,3-bis(mesityl)imidazoline-2-ylidene}N)$_2$C$_2$H$_4$]. The compounds were prepared by a transmetallation process after treating the diborane dication salt with an excess of $LiAlH_4$. The compounds were characterized by single-crystal X-ray and multinuclear NMR analyses (Scheme 5.23) [80].

5.3.2.4 Miscellaneous Aluminium Cations that Appeared Since 2010

Prior to 2010, multidentate ligands, e.g. O-donor phenolate-based anchor ligands, were utilized independently by the groups of Lewiński and Dagorne to investigate the synthesis of Al cations [114, 115]. The reader is directed to the reports on the well-defined aluminium cations stabilized by aza-macrocyclic, phenolate, and aminotroponate (AT) ligands that have appeared after 2010. These Al cations displayed metal coordination numbers of four or more [116–124].

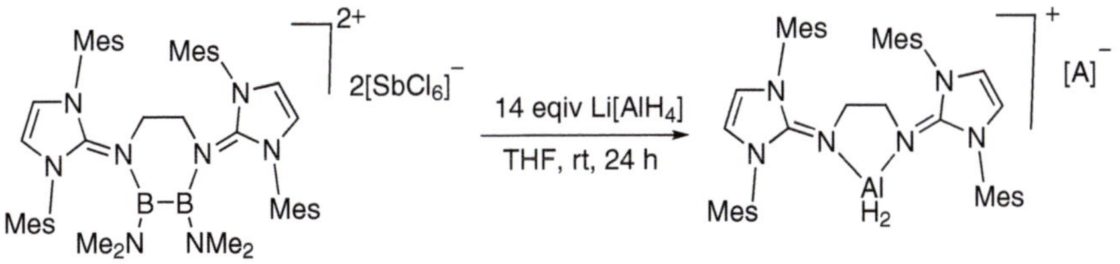

88[A] (A = AlB_2H_8)
88[A] (A = BH_4)

Scheme 5.23 Synthesis of a π-electron-donating ligand-stabilized cationic aluminium dihydride complex.

The sterically encumbered aluminium diphenolate complex $(2,6\text{-Mes}_2C_6H_3O)_2AlEt$, on reaction with the silylium salt $[Et_3Si][CHB_{11}Cl_{11}]$, underwent an ethide abstraction to afford cation complex $[(2,6\text{-Mes}_2C_6H_3O)_2Al]^+[CHB_{11}Cl_{11}]^-$ **89**$[CHB_{11}Cl_{11}]$ (Figure 5.18). The same cation can also be obtained by using $(2,6\text{-Mes}_2C_6H_3O)_2AlEt$ and $[Ph_3C]^+[CHB_{11}Cl_{11}]^-$ *via* a combination of β-hydride abstraction and concomitant ethylene elimination. X-ray structural analysis revealed well-separated ions with the Al centre in a four-coordinate distorted-tetrahedral geometry comprising two phenolate oxygens, with Al–O distances of 1.710(3) and 1.718(3) Å, and two *ortho*-carbons of the flanking mesityl groups, with Al–C distances of 2.273(4) and 2.356(3) Å [116]. The high Lewis acidity of **89**$[CHB_{11}Cl_{11}]$ was manifested in the reduction of CO_2 to d_5-toluene (that is, $C_6D_5CH_3$) and CH_4 at 82 °C in d_6-benzene (C_6D_6) solution as well as promoting the scrambling of the substituents of Et_3SiH at room temperature to give Et_4Si, Et_2SiH_2, and $EtSiH_3$ [116].

Ishii and coworkers in 2014 applied *trans*-1,2-cyclooctanediyl-bridged [OSSO]-ligated bis(phenolate)AlMe complex for methyl abstraction with $B(C_6F_5)_3$ in deuterobenzene to obtain a dicationic dinuclear aluminium species. This dimeric cation was obtained as the $[MeB(C_6F_5)_3]^-$ salt **90**$[MeB(C_6F_5)_3]_2$ (Figure 5.18). The crystal structure of the cation showed one aluminium atom in a distorted-trigonal-bipyramidal environment, in which phenoxide and μ-bridging oxygen atoms are located at two axial positions, whereas the other aluminium atom adopted a distorted-square-pyramidal geometry, which occupied the axial position with a non-bridged oxygen atom. This Al cation promoted the ROP of propylene oxide [117].

Recently, Phomphrai and coworkers utilized a potential new tetradentate aminophenolate ligand containing furfuryl groups in its side arms ([*N*,*O*] ligands), to prepare cationic Al complexes. Although these Al cations were characterized by NMR, their X-ray structures were not determined. For the tetracoordinated neutral precursors, as determined by X-ray analysis, the furfuryl groups did not interact with Al and the ligand showed only *N*,*O*-chelation. Thus, the tetracoordinated $[N,O]AlMe_2$ precursor was converted to the cationic form by methyl abstraction, using $B(C_6F_5)_3$ or $[Ph_3C][B(C_6F_5)_4]$ to form the $[[N,O]AlMe][MeB(C_6F_5)_3]$ and $[[N,O]AlMe][B(C_6F_5)_4]$ (**91**$[MeB(C_6F_5)_3]$ and **91**$[B(C_6F_5)_4]$; Figure 5.18). The cationic complex $[[N,O]AlMe][MeB(C_6F_5)_3]$ decomposed upon heating at 70 °C to afford neutral $[N,O]AlMe(C_6F_5)$ [118].

An article by Kerton et al. in 2019 reported the use of a morpholine-based dianionic pentacoordinating *N*,*N*′,*O*,*O*′,*O*′′-chelating ligand to prepare cationic aluminium complexes (e.g. **92**$[GaCl_4]$), and demonstrated their application in the ROP of ε-caprolactone (Figure 5.18). The X-ray structures of three of these $[[N,N',O,O',O'']Al]^+[A]^-$ (A = $AlEtCl_3$, $GaCl_4$, $InCl_4$) showed five-coordinate geometry around Al extremely distorted from both the ideal trigonal bipyramidal and square-based pyramidal geometries [119].

A thioacetamidate-based hetroscorpionate tridentate ligand was utilized by Otero and coworkers to prepare tetracoordinated cationic aluminium complexes. The neutral precursors were

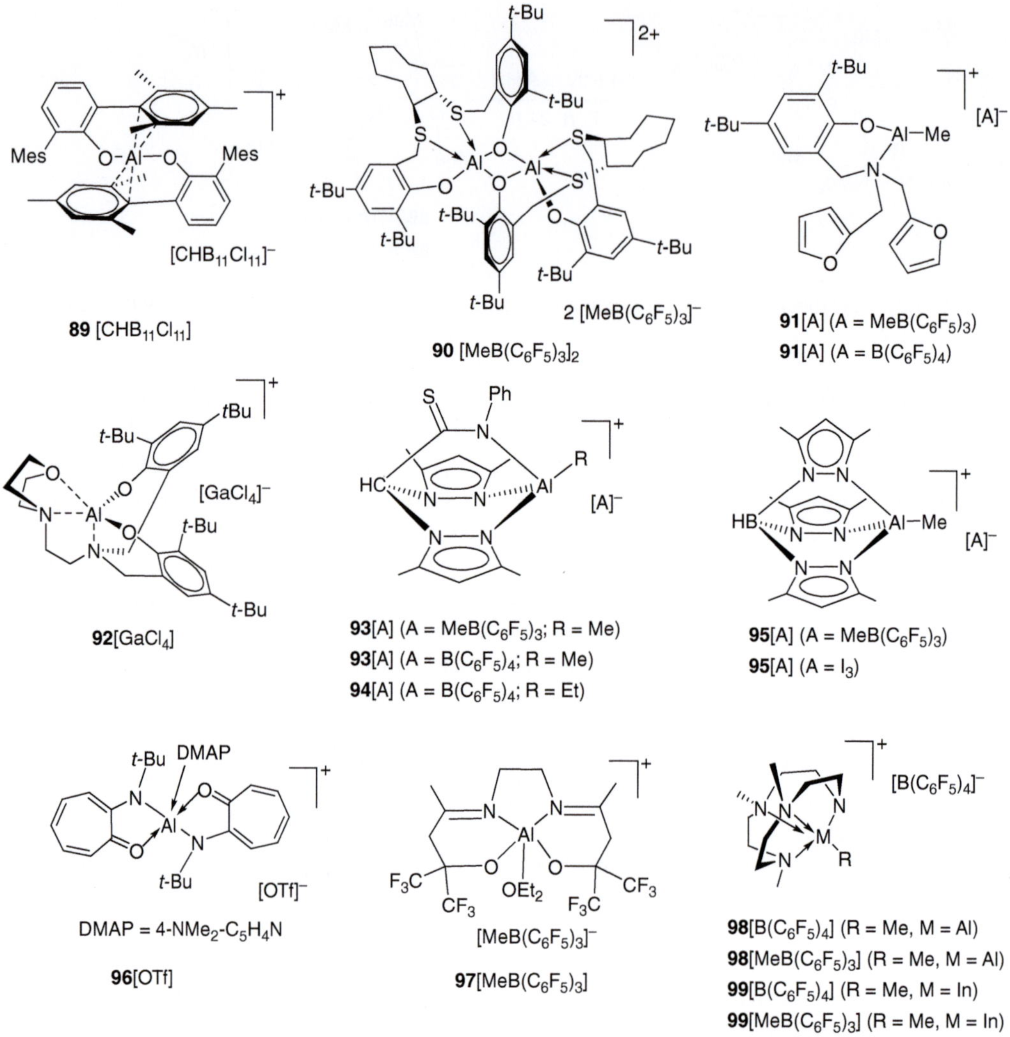

Figure 5.18 Examples of a few selected miscellaneous aluminium cations.

reacted with alkyl-abstracting reagent $B(C_6F_5)_3$ or $[Ph_3C][B(C_6F_5)_4]$ to afford complexes **93**$[MeB(C_6F_5)_3]$, **93**$[B(C_6F_5)_4]$, and **94**$[B(C_6F_5)_4]$ (Figure 5.18). NMR measurements confirmed an overall C_s-symmetric structure for these cations in solution manifested by κ^3-coordination of the heteroscorpionate ligand to the cationic aluminium centre. A few other tetracoordinate Al cations with a similar acetamidate ligand were also reported in this study. However, the solid-state structure of these complexes was not determined. Complexes **93**$[MeB(C_6F_5)_3]$ and **93**$[B(C_6F_5)_4]$ were also used for the ROP of *rac*-lactide, L-lactide, and ε-caprolactone [120]. In a related system, Koller and Bergman crafted cationic Al complexes supported with tridentate scorpionate ligand hydrotris(1,3-dimethylpyrazol-1-yl)borate (Tp*). Accordingly, metathesis between the potassium salt of the ligand (KTp*) and $AlMe_2Cl$ afforded the neutral pyrazolyl-bound dimethylaluminium precursor (Tp^*AlMe_2), which on further treatment with $B(C_6F_5)_3$ or I_2 gave the tetracoordinated cationic Al complexes $[Tp^*AlMe][MeB(C_6F_5)_3]$ and $[Tp^*AlMe][I_3]$ (**95**$[MeB(C_6F_5)_3]$ and **95**$[I_3]$; Figure 5.18). The X-ray structure of the I_3^- salt was elucidated to confirm a tetra-coordinated Al cation. Further, the activity of **95**[$MeB(C_6F_5)_3$] for the hydrosilylation of a range of carbonyls and imines was demonstrated to find alternatives for expensive late transition metal catalysts [121].

Recently, a structurally characterized pentacoordinate cationic aluminium complex $[(AT)Al(DMAP)]^+[OTf]^-$ supported by a relatively non-bulky AT ligand was reported by Nagendran et al. [DMAP = 4-(dimethylamino)pyridine] (**96**[OTf]; Figure 5.18). In spite of the five-coordination geometry, this Al cation was sufficiently Lewis acidic to catalyze the cyanosilylation of a variety of aldehydes and ketones [122].

A discrete pentacoordinate aluminium cation **97**[$MeB(C_6F_5)_3$] was reported by Dagorne and coworkers that was prepared by methyl abstraction from its neutral methylaluminium precursor using $B(C_6F_5)_3$. This cationic complex was supported by a dialkoxy-diimino salen-like ligand containing fluorinated CF_3 groups in the ligand backbone. The discrete nature of the cation with pentacoordinate Al in a slightly distorted trigonal bipyramidal geometry with Et_2O bonding was established by XRD (Figure 5.18). This complex was also used for the cationic polymerization of propylene oxide and *p*-methylstyrene at low temperatures [123].

Okuda and coworkers reported on reactions of $(^{Me3}TACD)H{\cdot}MMe_3$ adducts (M = Al, In) with Lewis [$B(C_6F_5)_3$] and Brønsted [$(PhNHMe_2)B(C_6F_5)_4$] acids to directly obtain the cations $[(^{Me3}TACD)MMe]^+$ (M = Al **98**; In **99**) with counter-anions $[MeB(C_6F_5)_3]^-$ and $[B(C_6F_5)_4]^-$. Here $(^{Me3}TACD)H$ is a macrocyclic polyamine; 1,4,7-trimethyl-1,4,7,10-tetraazacyclododecane. The molecular structure of **98**[$B(C_6F_5)_4$] showed the tetradentate bonding of the deprotonated macrocyclic ligand with AlMe to give a distorted square pyramidal geometry around Al (Figure 5.18). These compounds were also prepared by MeH elimination from the reaction of THF-solvated $[MR_2]^+[B(C_6F_5)_4]^-$ species with $(^{Me3}TACD)H$ [124].

5.3.3 Advances in the Synthesis and Characterization of Heavier Group 13 (Ga, In, and Tl) Cations

The larger group 13 elements (Ga, In, and Tl), in addition to interacting with the main donor centre of the ligand often show the tendency to do so with ligand side groups or additional donor molecules, if available. Even the use of sterically bulky ligands to restrict the extended coordination environment has been of little help. In contrast to lighter elements (B and Al) that are rather more well-behaved, exhibiting a well-defined primary coordination sphere and showing efficient orbital overlap with the ligand atoms, the interaction of heavier group 13 atoms with ligand atoms often has less pronounced covalent character. Over the past few decades heavier group 13 cations have attracted great attention due to their ability to participate in organic transformations as LA catalysts especially in the field of polymerization. Figure 5.19 and the discussion below present an overview of the well-characterized heavier group 13 cations in the last 10 years.

Wehmschulte and coworkers in 2017 reacted Et_3Ga and terphenylgallium $(2,6\text{-}Dip_2C_6H_3)GaEt_2$ with *in situ* generated silylium salt $[Et_3Si][CHB_{11}Cl_{11}]$ to prepare cationic organogallium compounds **61a** and **61b** with the $[CHB_{11}Cl_{11}]^-$ anion (Scheme 5.16). The X-ray structures of two Al and a Ga compounds were elucidated that featured cation–anion contacts in the solid-state [96]. In contrast to the structure of the Al cation in **60a**[$CHB_{11}Cl_{11}$], the analogous Ga compound **61a**[$CHB_{11}Cl_{11}$] in the solid-state consists of one-dimensional $[Et_2Ga][CHB_{11}Cl_{11}]$ chains. The coordination environment at the gallium centre is distorted octahedral with two gallium–carbon σ-bonds and four weak gallium–chlorine coordinate bonds from two anions (Figure 5.19). Notably, Wehmschulte et al. had also reported a truly linear two-coordinate gallium cation, $[(2,6\text{-}Mes_2C_6H_3)_2Ga]^+[Li(Al\{OCH(CF_3)_2\}_4)_2]^-$ **100**[$Li(Al\{OCH(CF_3)_2\}_4)_2$], in 2003 [125]. All of the cationic complexes could catalyze CO_2 reduction with Et_3SiH to form methane, the activity of

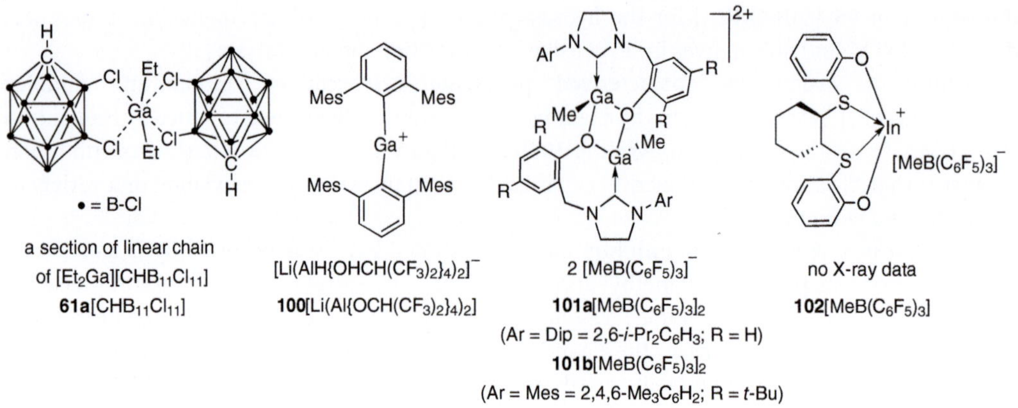

Figure 5.19 A few selected structurally characterized gallium and indium cations.

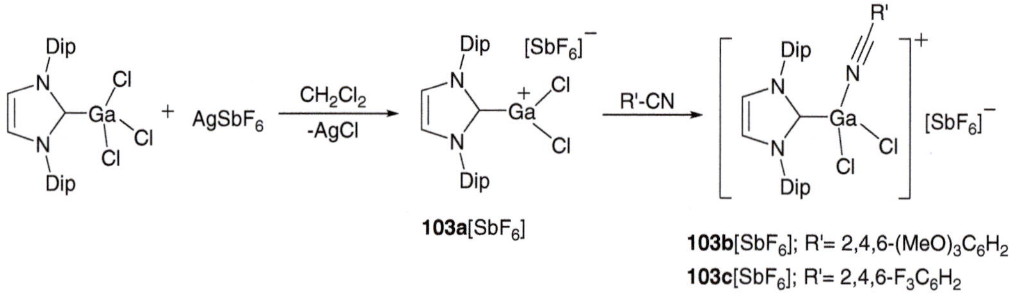

Scheme 5.24 Nitrile adduct of NHC-stabilized gallium chloride cations.

61a$[CHB_{11}Cl_{11}]$, however, was significantly lower and this was attributed to its polymeric structure and lower Lewis acidity.

The Lewis acidic nature of GaX_3 is beneficial for organic transformations such as inter- and intra-molecular nucleophilic additions to alkenes and alkynes, C=C bond formations, 1-4 addition to unsaturated ketones and allylation of imines etc. NHCs have been found to be a good ligand, and several GaX_3 adducts with NHCs proved to be easier to manipulate instead of simple GaX_3.

In 2012, Gandon and coworkers used NHC^{Dip}·$GaCl_3$ adducts to react with $AgSbF_6$ in the presence of benzonitrile derivatives as additional coordinating agents to obtain the gallium(III) dichloride cation in $[(NHC)GaCl_2(NCR')]^+[SbF_6]^-$ **103b,c**$[SbF_6]$ in quantitative yield (Scheme 5.24) [128]. The 2,4,6-trimethoxybenzonitrile and 2,4,6-trifluorobenzonitrile adducts were also structurally characterized to confirm the tetrahedral gallium cations as SbF_6^- salts. Data also emphasized the effect of electron-donating and electron-withdrawing substituents on the benzonitrile that are reflected in the Ga–N bond distance of 1.95 and 1.99 Å, respectively, indicating that electron-poor benzonitrile formed a relatively weaker adduct [129].

In 2017, Horeglad et al. reported on dinuclear dicationic alkylgalliumaryloxide cations **101a**$[MeB(C_6F_5)_3]_2$ and **101b**$[MeB(C_6F_5)_3]_2$. These complexes represent examples of NHC-stabilized alkylgalliumaryloxide cations that were prepared by methyl abstraction with $B(C_6F_5)_3$ (Figure 5.19) [126].

Following up on their earlier work on $(NHC)GaCl_2^+$ gallium cationic complexes, Gandon and coworkers in 2015 developed similar cationic indium complexes (Scheme 5.25),

Scheme 5.25 An NHC-stabilized cationic indium complex.

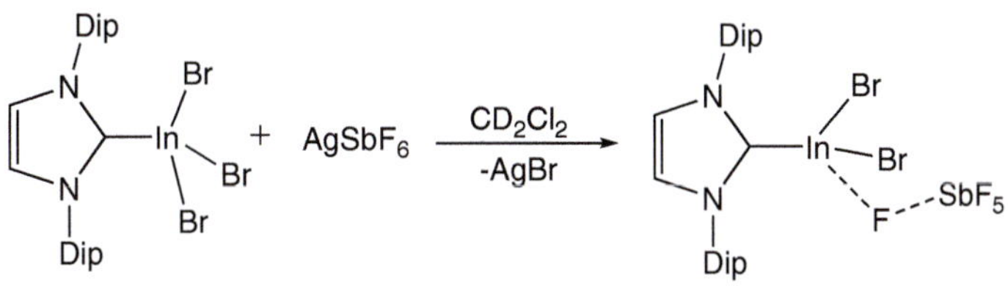

104[SbF_6]

[$(NHC^{Dip})InBr_2]^+[SbF_6]^-$ (**104**[SbF_6]), whose solid-state structure showed the fluorine bridged structure, [$(NHC^{Dip})InBr_2(\mu\text{-}F)SbF_5$], with a long In–F distance [2.189(2) Å] compared to a typical In–F covalent bond (2.05 Å) [130]. Apparently, the ^{19}F NMR spectrum of this compound showed a broad signal at 123.6 ppm, ascribed to a typical fluorine-bridging [M–F–SbF_5] moiety, where fluorine undergoes rapid exchange. This complex, when used directly or generated *in situ*, turned out to be a versatile catalyst and catalyzed reactions such as dihydroarylation of an arenyne with anisole, bimolecular dihydroarylation of arenyne with 1,2-dimethoxybenzene and 1-(phenylsulfonyl)-indole, transfer hydrogenation of cyclohexenylbenzene, hydrogenative cyclizations of arenyne and cycloisomerization of enyne.

Mehrkhodavandi et al. in 2018, utilized the anionic, tridentate ligand [2,4-dicumyl-6-({[2-(dimethylamino)cyclohexyl]imino}methyl)phenolate] ([*N,N′,O*]) with $In(CH_2SiMe_3)_3$ and obtained the dialkyl complex [*N,N′,O*]$In(CH_2SiMe_3)_2$ in good yield (78%). Protonation of this compound with one equivalent of [$HNMe_2Ph$][BAr^F] {$BAr^F = B[3,5\text{-}(CF_3)_2C_6H_3]_4$} in Et_2O or THF gave the solvated cationic complexes, [[*N,N′,O*]$In(CH_2SiMe_3)(S)$][BAr^F] **105a,b**[BAr^F] (**a**, *S* = THF; **b**, *S* = Et_2O) (Figure 5.20). The solid-state structures of Et_2O- and THF-solvated complexes were similar and confirmed the mononuclear structure that was predicted in the solution state as well. The structure of (±)-[[*N,N′,O*]$In(CH_2SiMe_3)(THF)$][BAr^F] showed a pair of cationic enantiomers inside a centrosymmetric unit cell along with non-coordinating BAr^F anions. The indium centre in either of the cations showed distorted square pyramidal geometry. The indium cation [[*N,N′,O*]$In(CH_2SiMe_3)(THF)$][BAr^F] catalyzed the quantitative conversion of a variety of

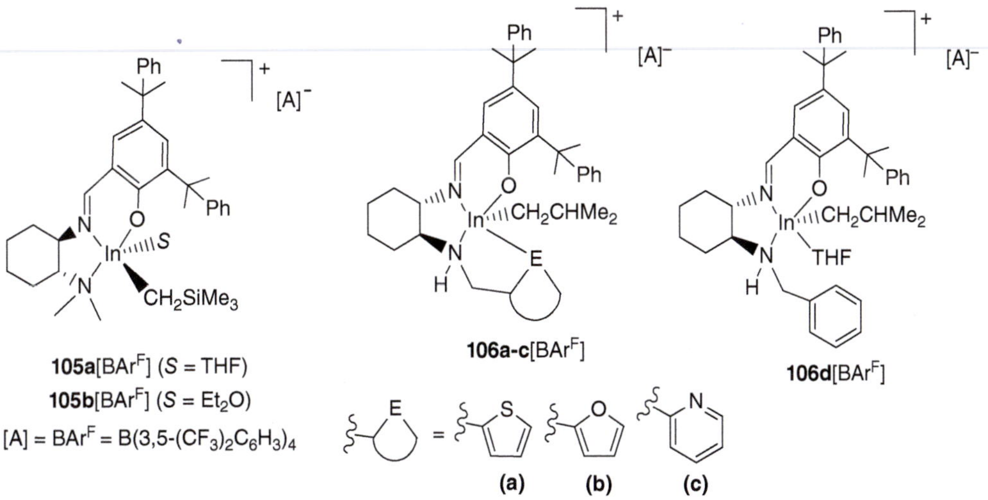

Figure 5.20 [*N,N′,O*]-coordinated indium cations.

functionalized epoxides and lactones into spiro-*ortho* esters with the absence of competing homopolymerization reactions [131].

The above indium cation work was followed up by another paper from the group of Mehrkhodavandi in 2019, that reported on cationic indium complexes supported by salen ligands. These complexes turned out to be highly active catalysts for the ROP of functionalized epoxides and their copolymerization with oxetane, THF, oxepane, and *rac*-lactide. These indium cations were the first discrete cationic group 13 complexes capable of copolymerization of epoxides and less reactive cyclic ethers, and the first indium-based system for copolymerization of *rac*-lactide and epoxides. These complexes were prepared by salt metathesis of (salen)InCl with Ag salts of the anions PF_6^-, AsF_6^-, SbF_6^-, BF_4^- (**107a–c**[A]; Scheme 5.26). The X-ray structure of three of these In cations revealed the distorted octahedral arrangement around In centres and no meaningful interaction with the counter-anions [132].

In 2020, building upon their previous discoveries, Mehrkhodavandi and coworkers reported on new cationic alkyl indium complexes **106a–d**[BAr^F] supported by hemilabile tridentate and tetradentate amino/iminophenolate ligands, and used them as catalysts for ROP reactions [133]. In their previous reports, the authors experienced the necessity of solvent molecule coordination to In centres for stabilization. However, the isolated complexes rapidly decomposed as the coordinated solvent was lost [131, 132]. In view of this, the paper emphasized the use of a hemisalen ligand architecture with a hemilabile pendant donor arm that could stabilize the cationic indium centre whenever required and would not be lost. The stability and shelf-life of the complexes followed a trend with a direct correlation to the donor ability of the pendant donor side arm containing pyridyl > furfuryl > thiophenyl > phenyl (Figure 5.20). The reactivity trend of these indium cations followed the reverse order of stability towards polymerization of epichlorohydrin and cyclohexene oxide. It was also observed that the control of polymerization followed an inverse relationship to reactivity [133].

Tetracoordinate indium complex **102**[$MeB(C_6F_5)_3$] was published by Okuda and coworkers. The metal was supported by a tetradentate (OSSO)-type bis(phenolate) ligand (Figure 5.19) [127]. Furthermore, a recent paper from Okuda and coworkers reported on the reaction of THF-solvated $[In(CH_2SiMe_3)_2]^+[B(C_6F_5)_4]^-$ with the macrocyclic polyamine (^{Me3}TACD)H to eliminate $SiMe_4$ and form the structurally characterized indium complex $[(^{Me3}TACD)In(CH_2SiMe_3)]^+[B(C_6F_5)_4]^-$ that

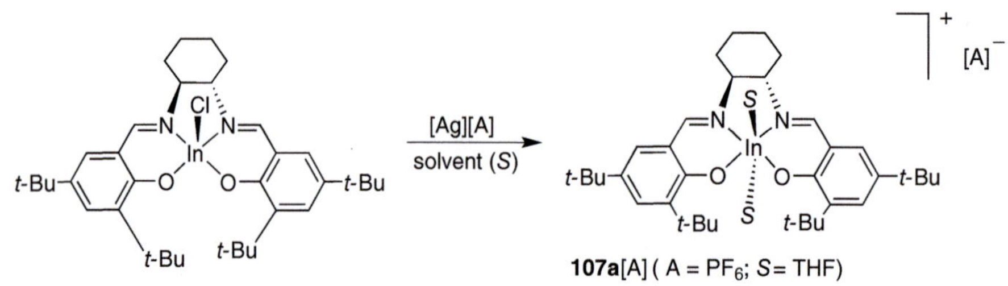

107a[A] (A = PF_6; *S* = THF)
107b[A] (A = PF_6; *S* = THF, ACN)
107a[A] (A = AsF_6; *S* = THF)
107a[A] (A = SbF_6; *S* = THF)
107a[A] (A = BF_4; *S* = THF)
107c[A] (A = SbF_6; *S* = Me-THF)
107b[PF_6], **107a**[SbF_6] and **107a**[BF_4] XRD characterized

Scheme 5.26 Cationic indium complexes supported by salen ligand.

showed a structure similar to the Al analogue (**98**[$B(C_6F_5)_4$]; Figure 5.18) with distorted square pyramidal geometry around the In centre [124].

5.3.3.1 Low Oxidation State Univalent Heavier Group 13 Cations (Ga, In, and Tl)

The propensity for the existence of low-valent states increases groups, making the formation of complexes with formal oxidation state of (I) at the group 13 metal centre increasingly favourable. In view of this, the low-valent monocations of heavier group 13 elements published since 2010 have been briefly mentioned below. A series of Ga(I) cationic species stabilized by WCA $[Al\{OC(CF_3)_3\}_4]^-$ was reported by Slattery et al. in 2010. Treatment of gallium metal in arene solvents (*o*-$C_6H_4F_2$ and toluene) under ultrasonic activation with a solution of $Ag^+[A]^-$, ([A] = $[Al\{OC(CF_3)_3\}_4]$) led to the formation of stable cations in $[Ga(C_6H_5Me)_2]^+[A]^-$, $[Ga(o\text{-}C_6H_4F_2)_2]^+[A]^-$, $[Ga(C_6H_5F)_{2.5}]^+[A]^-$ and $[Ga(o\text{-}C_6H_4F_2)_2]^+[\{(CF_3)_3CO\}_3Al\text{–}F\text{–}Al\{OC(CF_3)_3\}_3]^-$. The Ga centre in each of these cations showed a bent sandwich structure, and the arenes were slightly distorted from η^6-coordination, with longer Ga–C distances for the carbons bearing fluorine substituents. Interestingly, when $[Ga(C_6H_5Me)_2]^+[A]^-$ was treated with three equivalents of PPh_3 it afforded a Ga(I)–phosphine complex, $[Ga(PPh_3)_3]^+[A]^-\cdot 1.5\ C_6H_4F_2$ (Figure 5.21) [134].

In 2019, Wehmschulte and coworkers demonstrated that arene-solvated cationic dihydrogallium cations, $[H_2Ga(arene)_2]^+$, undergo reductive elimination of H_2 at room temperature to afford the arene-solvated gallium(I) complexes, $[Ga(PhF)_2][CHB_{11}Cl_{11}]$ and $[Ga(Ph_3CH)][B(C_6F_5)_4]$. The bent sandwich geometry with two η^6-arene ligands is common to that of Slattery's Ga(I) cations, additionally, moderate-to-weak Ga–Cl (or F) interactions were also present in the single-crystal XRD structures of these cations. These cations were also used for the oligomerization of 2,4,4-trimethyl-1-pentene and the hydrosilylation of benzophenone, 1-hexene and CO_2 [135].

In 2012, the *N*-donor 2,6-dimesitylpyridine ligand (Mes_2py), isoelectronic with the popular terphenyl system, was used to stabilize the M(I) cations [M = In and Tl]. In this endeavour, Aldridge and coworkers reacted a suspension of TlCl in fluorobenzene with less than one

Figure 5.21 Structure of the cation in $[Ga(PPh_3)_3]^+[Al\{OC(CF_3)_3\}_4]^-\cdot 1.5\ C_6H_4F_2$. *Source:* Adapted from Slattery et al. [134].

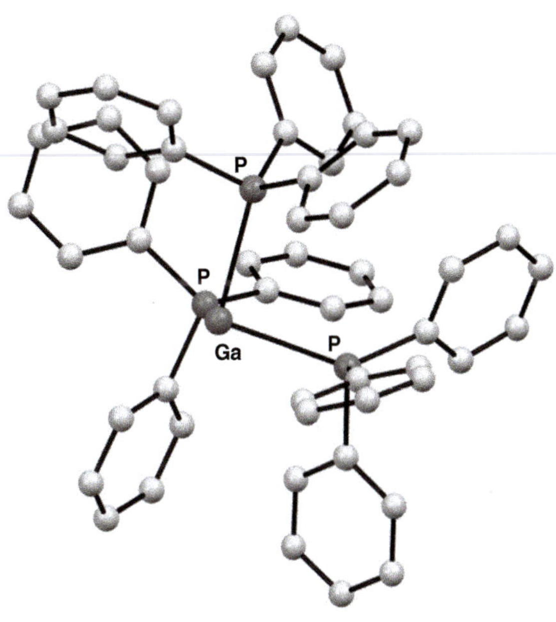

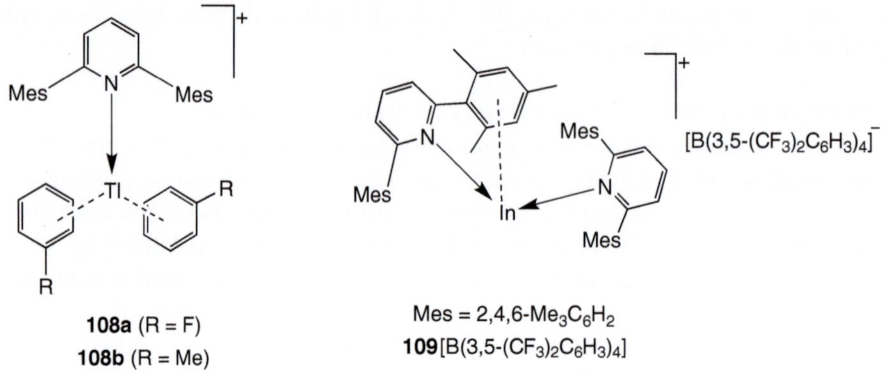

Figure 5.22 Examples of indium(I) cations.

equivalent of Na[BAr^F_4], followed by the addition of one equivalent of Mes_2py that afforded [Tl(Mes_2py)(C_6H_5F)$_2$][BAr^F_4] **108a**[BAr^F_4] in moderate yield (42%). Extraction of this cation into toluene afforded the toluene-coordinated product **108b**[BAr^F_4] (Figure 5.22) [136].

The molecular structures of these two complexes revealed weak interaction between the Tl(I) centre and nitrogen atom of the ligand with a Tl–N bond distance of 2.64 Å. Further weak interaction between cationic In(I) and nitrogen atoms (2.58 and 2.66 Å) was observed in the case of [In(Mes_2py)$_2$][BAr^F_4] **109**[B(3,5-(CF_3)$_2$$C_6H_3$)$_4$], which was obtained by reacting InBr/Na[BAr^F_4] with two equivalents of Mes_2py (yield 68%; Figure 5.22) [136]. The energy mismatch and poor overlap between orbitals of the metal centre and *N* atom caused such weak interactions which were also observed by similar work with an imino-functionalized system by Richeson and coworkers in 2009 [137].

In 2013, Richeson and coworkers reported on cationic Tl(I) complexes using pincer-type bis(imino)pyridine ligand [(ArN=CPh)$_2$(NC_5H_3)], which showed similar interactions as demonstrated by Aldridge et al. in their complexes discussed above [136]. The reaction of TlOTf with the ligand gave the cationic complexes [(ArN=CPh)$_2$(NC_5H_3)]Tl^+[OTf]$^-$ (Ar = 2,6-$Et_2C_6H_3$ (**110a**[OTf]), 2,5-(*t*-Bu)$_2C_6H_3$ (**110b**[OTf])). The weak ligand-to-metal donation allowed for additional intermolecular Tl–arene interactions in **110a**- and **110b**[OTf]. Attempts to recrystallize **110b**[OTf] from benzene (or toluene) gave inverted sandwich structures **111**[(OTf)$_2$] in which benzene (or toluene) bridged between two [({2,5-(*t*-Bu)$_2C_6H_3$}N=CPh)$_2$(NC_5H_3)]Tl^+ cations (Figure 5.23) [138, 139].

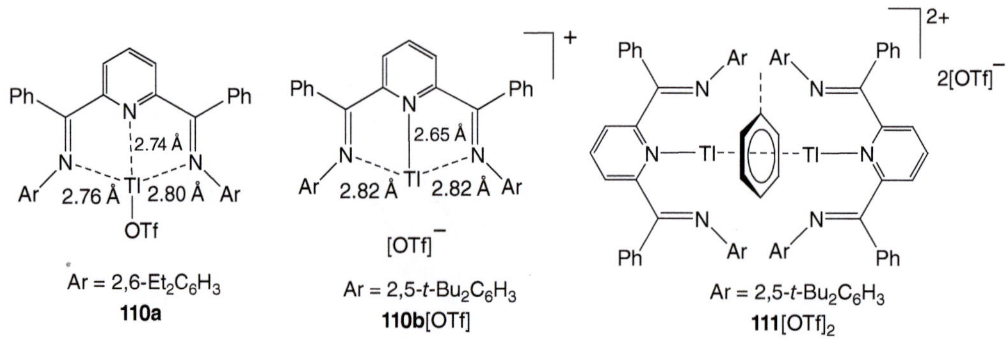

Figure 5.23 Examples of Tl(I) cations.

5.4 Recent Advancements in Catalytic Applications of Cationic Group 13 Complexes

5.4.1 Borocation in Catalysis

5.4.1.1 Cationic Boron Complexes in Catalysis

Three main areas where cationic borenium complexes have been utilized as catalysts have been identified since 2015. These are hydroboration, hydrosilylation, and hydrogenation reactions. Eisenberger and Crudden, Melen and Vedejs have recently reviewed borocation catalyzed transformations [6, 7, 140, 141].

5.4.1.2 Hydroboration Reaction

Hydroboration is one of the most prominent and fundamental reactions and has been widely explored with a range of catalysts. The prominent reports on hydroboration using borocations prior to 2015 have clearly shown that the active boranes (i.e. weak base, THF or DMS stabilized) facilitate the hydroboration reaction without a catalyst but are often associated with uncontrolled reactivity. On the contrary, NHC-stabilized boranes are difficult to activate and they require an additive initiator. In this regard, $HNTf_2$, I_2, and $B(C_6F_5)_3$ have been explored as initiators for *in situ* formation of borenium cations (Scheme 5.27) that have been recognized as active species in the course of the catalytic reactions.

The groups of Curran and Vedejs reported the hydroboration of alkene with $NHC{\cdot}BH_3$ adduct and utilized $HNTf_2$ as an activator [142]. Later on, with modification Curran et al. explored the $NHC{\cdot}BH_2I$ complex as reminiscent of masked borenium cation for the hydroboration of various olefins. The mechanistic investigation indicated olefin insertion to the B–H bond of masked borenium cation, $NHC{\cdot}BH_2I$, in a *syn* fashion as the key step [143]. In 2012, Crudden et al. reported the DABCO-coordinated borenium cation-catalyzed hydroboration of aldimines and ketimines emphasizing on the role of DABCO as a labile ligand, which stabilizes the electron-deficient borenium cation and also reinforces the hydride migration to an activated imine [144].

In 2016, *trans*-hydroboration of alkynes with NHC·9BBN adduct using $B(C_6F_5)_3$ as activator was reported by Ingleson et al. The *trans*-hydroboration products of alkynes were isolated in good yields, and a variety of alkynes including acetylene with varying electron-donating and electron-withdrawing substituents and heteroaromatic alkyne (2-ethynylthiophene) were efficiently converted to the *trans*-hydroboration product. Conjugated enynes were also subjected to hydroboration and explicitly the hydroboration occurred at the triple bond. The internal alkynes were also explored but only activated alkyne 4-(1-propynyl)-*N*,*N*-dimethylaniline showed successful

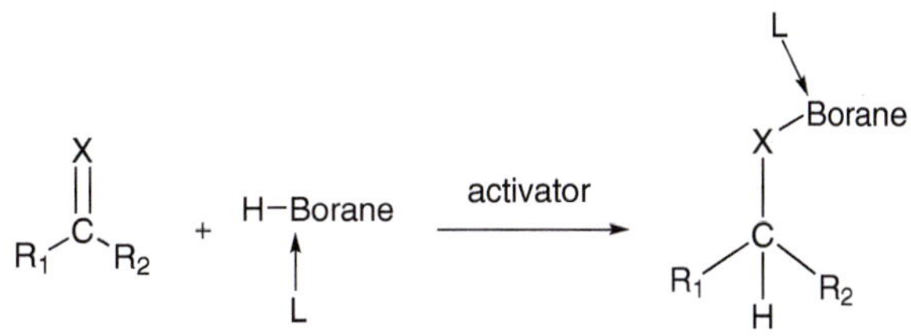

X = NR, CR_2 (R = alkyl, aryl)
activator = I_2, $HNTf_2$, L = NHC
activator = $B(C_6F_5)_3$ or $[CPh_3][B(C_6F_5)_4]$, L = DABCO
H-Borane = BH_3, 9-BBN, pinacolborane

Scheme 5.27 Generalized scheme for hydroboration reaction with active borane in the presence of activators.

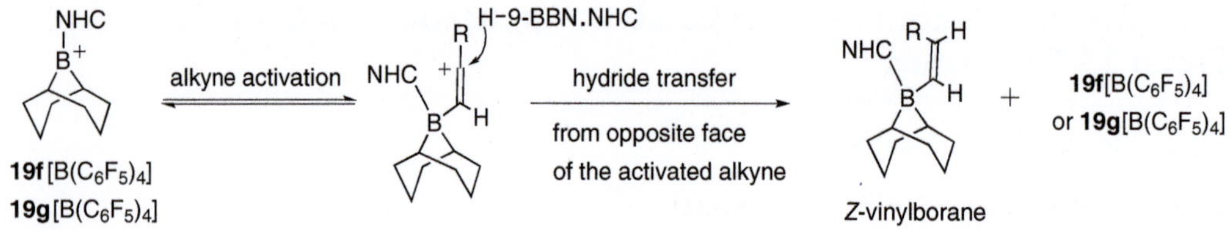

Figure 5.24 Reaction pathway for hydroboration of alkyne.

conversion at a slower rate [145]. According to this study, the choice of base (NHC) to stabilize the borenium cation and the non-basic anion is very crucial for controlling the effectiveness of this catalytic scheme. The choice of base is restricted to NHC because DABCO as a base showed reversible addition to the cation (also observed by Crudden et al. in hydroboration of imines), which interferes with the selectivity of the reaction [144]. The mechanistic insight into the reaction *via* deuterium labelling and some control experiments supported a step-wise reaction involving (i) activation of alkyne by addition of electrophilic borane (borenium ions or their functional equivalents) leading to the formation of a vinyl cation intermediate that could also be viewed as a π-complex between borenium and alkyne and, (ii) intermolecular hydride transfer from another molecule of neutral NHC–borane to the activated alkyne from the previous step (Figure 5.24).

Interestingly, Wang and coworkers reported the highly chemo- and regioselective 1,4-hydroboration of pyridines catalyzed by the bulky $Ar^F{}_2BMe$ complex, and detected the boronium $[py_2Bpin][Ar_2{}^FB(H)Me]$ as intermediate [$Ar^F = 2,4,6\text{-}(CF_3)_3C_6H_2$] [146]. In 2017, Wright and coworkers reported the regioselective 1,4-hydroboration of pyridines using pinacolborane as a reducing agent and commercially available $[NH_4][BPh_4]$ as an activator (Scheme 5.28). During the reaction, the electron-deficient HBpin readily forms adduct with pyridine and subsequently reacts with $[NH_4][BPh_4]$ to generate boroneium ion **47**$[BPh_4]$ as an active catalyst along with the evolution of H_2 and NH_3 [70].

A wide range of substituted pyridine substrates were explored for hydroboration including three-substituted pyridines, which showed a good selectivity for 1,4 addition with most of the substituents, including bromo, cyano, ester, and amide groups showing good tolerance toward the reaction. Two-substituted pyridines showed lower reactivity. Particularly, 2-phenylpyridine and 3-bromopyridine did not show any conversion due to the steric crowding of the former and poor donor ability of the latter, which disfavour adduct formation between HBpin and pyridine. The solvent also played an important role in determining the 1,4- or 1,2-selectivity. Non-polar solvents such as benzene and heptane increased the percentage of 1,2-addition product. The proposed mechanism involves the boronium cation activation of pyridine and subsequent facile hydride transfer from another molecule of the pyridine–borane adduct (Figure 5.25) [70].

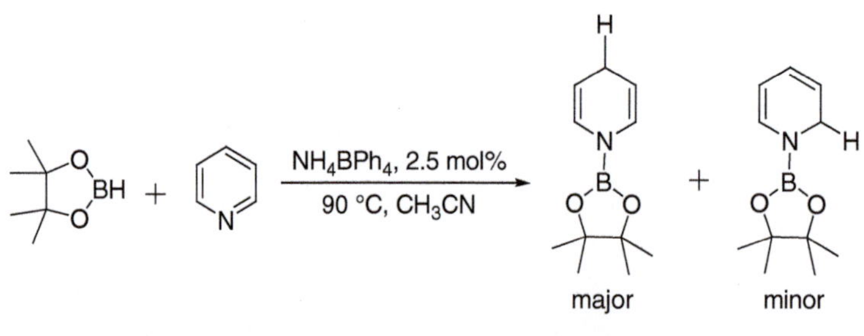

Scheme 5.28 Hydroboration of pyridines catalyzed by NH_4BPh_4.

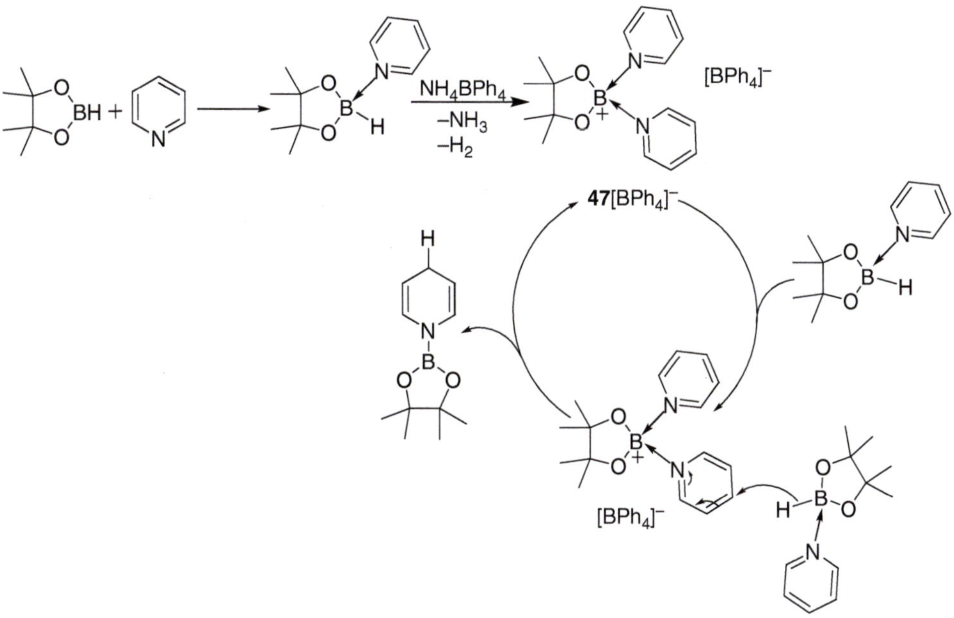

Figure 5.25 Proposed reaction pathway for regioselective 1,4-hydroboration of pyridine.

5.4.1.3 Hydrosilylation Reaction

Hydrosilylation is another reaction of great significance and in the presence of a suitable catalyst involves the addition of Si–H bonds across unsaturated bonds such as alkenes, alkynes, carbonyls, and their derivatives. According to the studies for hydroboration, activator or initiator is required for the facile conversion, and the reaction proceeds predominantly *via* the alkene/carbonyl/imine activation method. On the contrary, hydrosilylation reactions commonly require the activation of the silane moiety followed by the electrophilic attack of generated silylium cation on respective carbonyl or other suitable unsaturated functional groups.

In 2000, the pioneering work of Piers and coworkers demonstrated the hydrosilylation of carbonyl compounds catalyzed by $B(C_6F_5)_3$, the excess of silane leading to complete deoxygenation and consequently to alkane as was obtained as the final product [151]. Borenium cations are considered even more active and a strong contender for catalyzing the reaction. For the first time in 2013, Denmark and Ueki demonstrated the successful hydrosilylation of carbonyl compounds catalyzed by the 9-BBN borenium ion complex of 2,6-lutidine. The catalyst facilitated the reaction *via* classical silane activation [148]. In the same year, Jäkle and coworkers also reported the enantioselective hydrosilylation of ketones using a planar-chiral borenium LA [149].

In 2019, Tseng et al. demonstrated the capability of boron cation $[(\eta^5\text{-Cp}^*)BMes]^+[B(C_6F_5)_4]^-$ **13**$[B(C_6F_5)_4]$ for the selective deoxygenation of benzophenone and acetophenone derivatives into hydrodeoxygenation products and also converting highly deactivated aryl and aliphatic ketones into the silyl ether product by using 2.5 equivalents of $HSiEt_3$ in dry DCM as solvent at room temperature (Scheme 5.29). Although the cation was not active with amino group-containing derivatives it showed decent conversion of benzophenone to diphenylmethane (30% yield) with wet DCM (0.2% w/w water) [46]. No clear-cut trend, depending on the electronic and steric influence of substituents R_1/R_2, was observed concerning the preference for the formation of hydrodeoxygenation or silyl ether products.

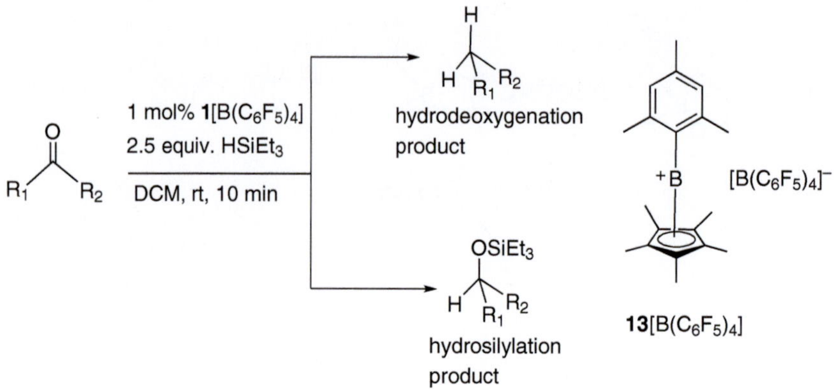

Scheme 5.29 Hydrosilylation and deoxygenation of ketones catalyzed by **13**[B(C_6F_5)$_4$].

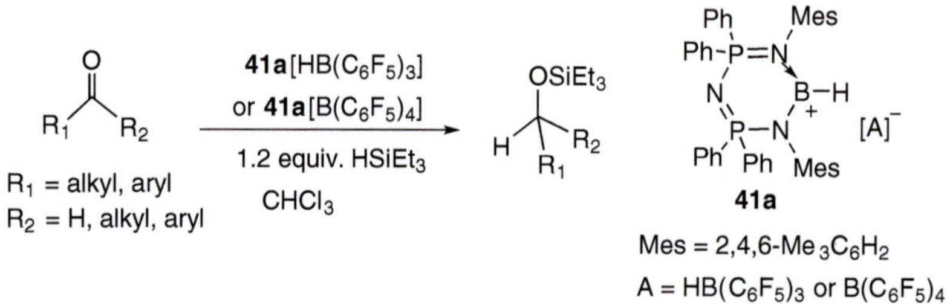

Scheme 5.30 Hydrosilylation of carbonyl with $HSiEt_3$ catalyzed by borenium cation complexes **41a**.

Recently in 2020, we performed catalytic hydrosilylation of carbonyl compounds (aldehydes and ketones) using the hydridoborenium cations **41a**[HB(C_6F_5)$_3$] and **41a**[B(C_6F_5)$_4$], and investigated the role of borenium cations and their respective counter-anion in the reaction (Scheme 5.30). A wide range of substrates were converted into their respective silyl ether derivatives with good yields (58–98%) [150]. For hydrosilylation of PhCHO as an example and using multinuclear *in situ* NMR measurements, it was established that although borenium cations played an important role in LA activation of the carbonyl group, the hydridoborate anion $[HB(C_6F_5)_3]^-$ of **41a**[HB(C_6F_5)$_3$] transferred its hydride to the activated carbonyl moiety leading to a clearly identified intermediate $[PhCH_2–O–B(C_6F_5)_3]^-$ (Figure 5.26). The addition of $HSiEt_3$ regenerated the counter-anion $[HB(C_6F_5)_3]^-$ and gave the desired silyl ether product ($PhCH_2OSiEt_3$). The proposed mechanism was also supported by DFT calculations. The involvement of $[HB(C_6F_5)_3]^-$ for hydride delivery was previously reported in a seminal work by Piers and coworkers in the $B(C_6F_5)_3$-catalyzed hydrosilylation of carbonyls [151]. However, the reaction with the borenium cation **41a**[B(C_6F_5)$_4$] where the counter-anion $[B(C_6F_5)_4]^-$ is devoid of a hydride group proceeds mainly *via* borenium ion activation of the carbonyl (typical LA activation pathway), and subsequent addition of Et_3SiH generated the product (Figure 5.26).

5.4.1.4 Hydrogenation Reaction

The field of hydrogen activation is mainly dominated by boron-based FLPs. The neutral tricoordinate *N*-heterocyclic iminoboranes, $NHC^{Dip}BR_2$ (R = Cl and/or Ph), were able to abstract H_2 from amine–boranes to provoke their dehydrocoupling [152, 153].

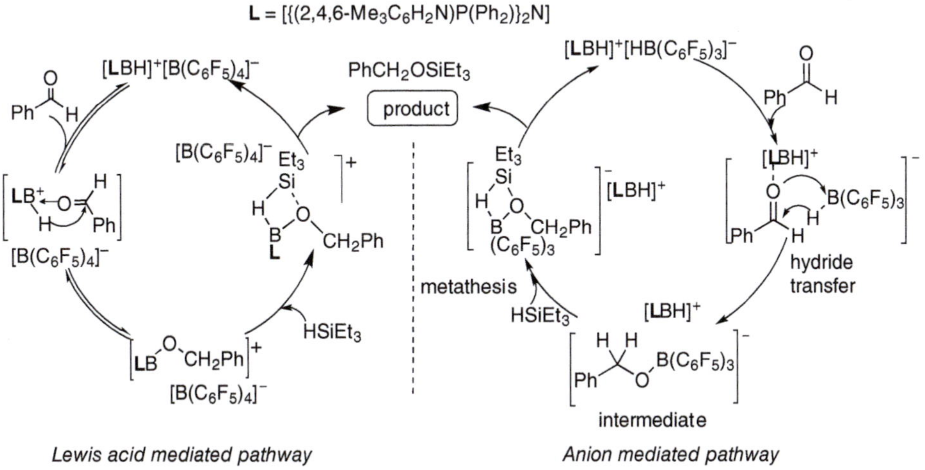

Figure 5.26 Proposed catalytic cycle for hydrosilylation of benzaldehyde catalyzed by hydridoborenium cation **41a**: choice between LA mechanism and LA mediated mechanism assisted by hydride transfer from the hydridoborate counter-anion.

Scheme 5.31 Borenium cation mediated activation of H_2 in the presence of external base.

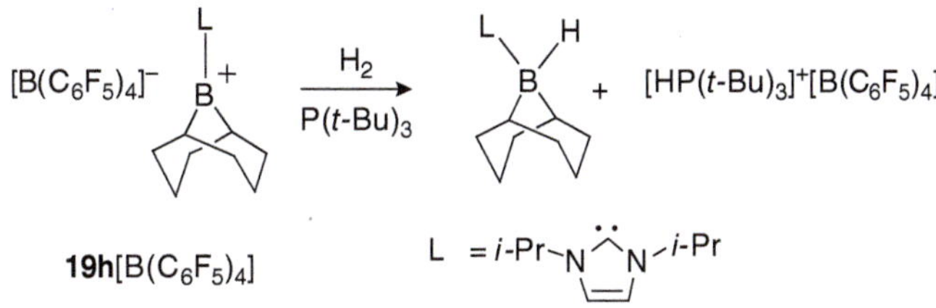

In 2012, Farrell et al. reported H_2 activation by NHC-stabilized borenium cation $[NHC^{i\text{-}Pr}\text{–}9BBN]^+[B(C_6F_5)_4]^-$ $[NHC^{i\text{-}Pr} = C_3H_2\{N(i\text{-}Pr)\}_2]$ in the presence of external base $(t\text{-Bu})_3P$. One equivalent of borenium cation and $(t\text{-Bu})_3P$, when exposed to four atm H_2 for 48 h resulted in the formation of neutral adduct $NHC^{i\text{-}Pr}H\text{–}9BBN$, and the salt $[H(t\text{-Bu})_3P][B(C_6F_5)_4]$ *via* hydrogen activation (Scheme 5.31). Subsequently, the introduction of one equivalent of imine PhCH=N(*t*-Bu), to the reaction mixture afforded its complete reduction with the regeneration of $[NHC^{i\text{-}Pr}\text{–}9BBN]^+[B(C_6F_5)_4]^-$. The observed reactivity is reminiscent of the classical FLP reactivity. The scope of the reaction was also extended toward the catalytic reduction of imines (Figure 5.27) [154].

A better understanding of the structure-reactivity relationship was revealed by systematic variation of the substituents at the N atoms of the NHC. It was found that less hindered substituents showed high-catalytic efficiency whereas bulky substituents flanking nitrogen arrested the activity (Scheme 5.6) [51].

In a related report, Eisenberger et al. introduced MIC-stabilized borenium cation for the hydrogenation of imines (Scheme 5.32) and surprisingly the strong σ-donation properties of MIC substantially enhanced the catalytic efficiency. The high efficiency of MIC borenium can be attributed to the strong sigma donating ability of MIC, which enhances the hydride donation ability of neutral MIC–borane adducts to easily form the borenium cations **27a–e** (Figure 5.4). Considering the steric demands, the results obtained in this case agreed well with the observations from NHC-stabilized borenium; less sterically encumbered MIC–borenium **27d**, with only one phenyl substituent, showed the highest reactivity. The catalytic conditions employed in this case

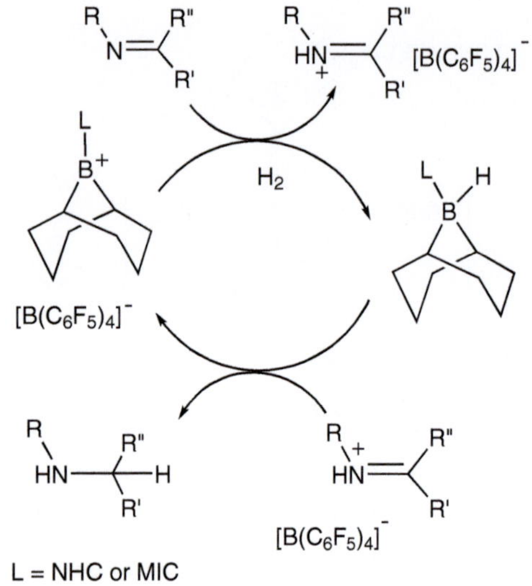

Figure 5.27 Proposed hydrogenation pathway for carbene-stabilized borenium ion.

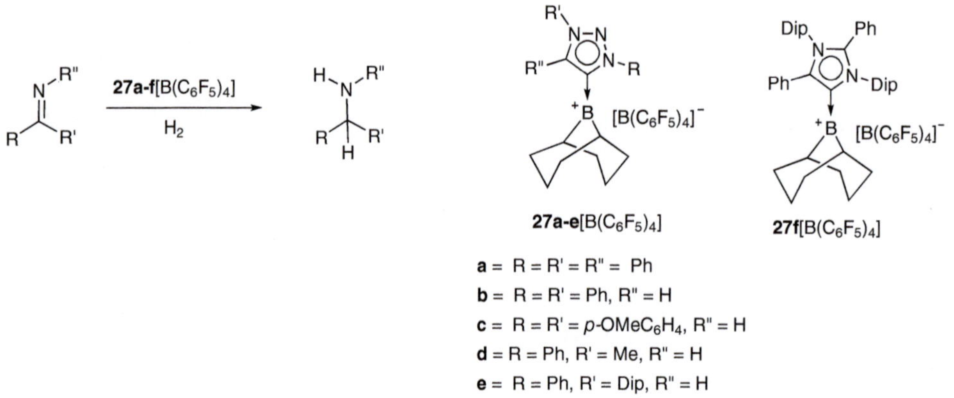

Scheme 5.32 Assorted mesoionic carbene-stabilized borenium cations as catalysts for hydrogenation reaction of imines.

were milder as compared to the previously reported results. The catalyst also proved efficient for the reduction of heterocycles [54].

The groups of Melen, Crudden, and Stephan jointly explored the synthesis and applications of chiral carbene–borane adducts to prepare chiral borenium cations and test their FLP-type activity for asymmetric hydrogenation reactions. Two approaches were adopted to prepare the active chiral catalysts: (i) use of chiral NHC to synthesize chiral catalyst and, (ii) use of chiral boranes to prepare chiral catalyst [155].

5.4.1.5 Use of Chiral NHC

In this approach, chirality was incorporated into the NHC and different chiral NHCs (**112, 113**) (Figure 5.28) were synthesized by using a variety of chiral precursors such as camphoric acid, fused oxazole rings, (*S*)-2-amino-2′-methoxy-1,1′-binaphthalene. Subsequently, chiral NHC–borane adducts and the corresponding borenium cations (**115, 116**) (Figure 5.28) were also synthesized and spectroscopically characterized. A few of these species were found to be efficient

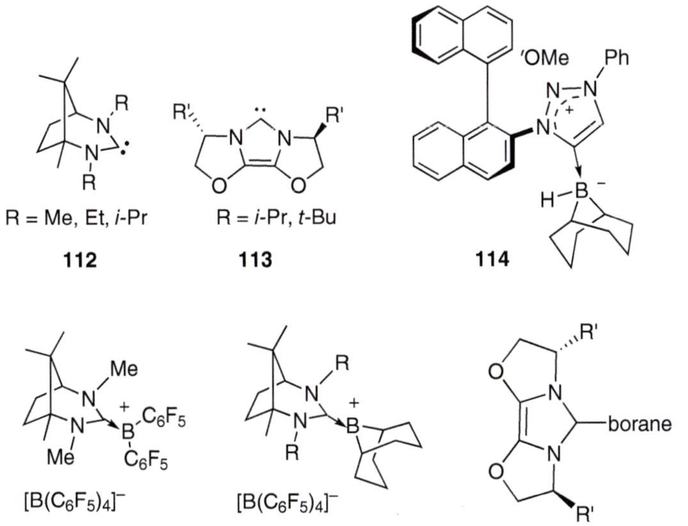

R = Me, Et, *i*-Pr

112

R = *i*-Pr, *t*-Bu

113

114

115[B(C$_6$F$_5$)$_4$]
XRD characterized

116[B(C$_6$F$_5$)$_4$]
R = Me (**116a**), Et (**116b**), *i*-Pr (**116c**)
116b is XRD characterized

R = *i*-Pr; borane = BH$_3$ (**117**), 9-BBN (**118a**), BCy$_2$Cl (**119**)
118b R = *t*-Bu; borane = 9-BBN
XRD characterized adducts

Figure 5.28 Chiral carbenes (**112** and **113**), borenium cations (**115–116**), and chiral carbene–borane adducts (**114** and **117–119**). The borenium cations were utilized as catalysts and borane adducts as catalyst precursors for asymmetric hydrogenation.

precursors for imine reduction in hydrogenation catalysis, *albeit* with poor enantioselectivity. The borenium species substituted with C_6F_5 groups on boron were not very effective for hydride delivery, whereas systems based on 9-BBN and 9-BBD derivatives suffered due to steric reasons. Bisoxazoline-derived NHC borenium cations, and triazolium–borenium derivatives with axial chirality presented examples of active hydrogenation catalysts but with limited stereoselectivity. The chiral version of MIC was also synthesized and the corresponding adduct **114** (Figure 5.28) was utilized as the catalyst but, probably due to steric congestion, the product yield was very low.

5.4.1.6 Use of Chiral Borane

In this approach, a chiral borane derived from diisopinocamphenylborane and its adducts with NHCs with a range of substituents of varying steric congestion on the *N*-centres were explored (**120a–d**) (Figure 5.29). Similarly, the adduct of (10)-phenyl-9-borabicyclo[3.3.2]decane with mesoionic triazolylidene carbene were also synthesized (**121**) (Figure 5.29). The borenium cations were synthesized using [Ph$_3$C][B(C$_6$F$_5$)$_4$] as hydride scavenger and employed for the catalysis. These chiral borenium cations also provided active hydrogenation catalysts but again with limited stereoselectivity. The low selectivities observed for these chiral borenium cation catalysts were not attributed to selectivity erosion by epimerisation of the chiral product.

Complex **121** was completely inactive for H_2 activation. On the contrary, complexes **120a–d** showed some decent conversions when the reaction was carried out at 5 mol% catalyst loading and 102 atm H_2 with Ph(Me)C=NPh as imine. The least sterically hindered complex **120a** showed the highest conversion of 55% with 12% ee, and similarly, complex **120c** gave 47% conversion and 13% ee.

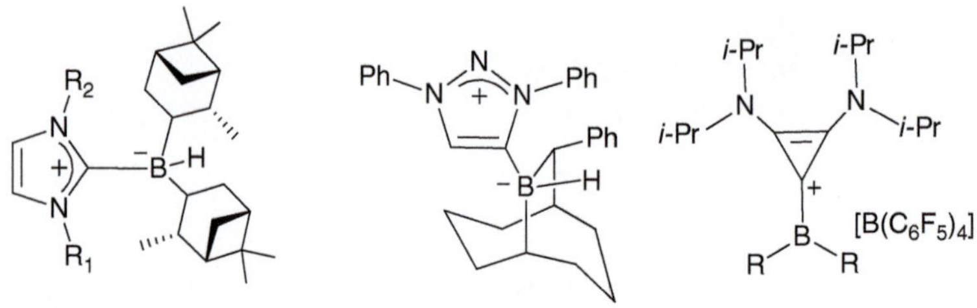

120a $R_1 = R_2 = Me$
120b $R_1 = R_2 = i\text{-}Pr$
120c $R_1 = Me, R_2 = Bn$
120d $R_1 = Ph, R_2 = t\text{-}Bu$

121

($R = C_6H_{11}$)
122a[$B(C_6F_5)_4$]

Figure 5.29 Chiral borane adducts as precursors for borenium catalysts for hydrogenation of imine.

All of the initial studies have clearly indicated that by probing the appropriate ligand and carbene an effective chiral catalyst can be synthesized [155]. Surprisingly, the complex **118a** also proved to be an excellent catalyst towards the enantioselective hydrosilylation of ketimines, yielding good enantiomeric excess. *N*-(1-phenylethylidene)pentan-1-amine substrate showed 90:10 er for both hydrosilylation and hydrogenation, reflecting the observation that an appropriate choice of substrate can also influence the enantioselectivity [156].

Furthermore, inspired by the previous studies, Speed and coworkers explored the BAC [bis(amino)cyclopropenylidene] carbene–borane adducts for stoichiometric reduction of carbonyl compounds and the respective *in situ*-generated cation was utilized for catalytic hydrogenation of imine. A decent conversion was recorded at 20 atm hydrogen pressure. The phenyl substituted analogues of **122a** (Figure 5.29) were also explored as a catalyst, but very poor conversion was obtained [157, 158].

5.4.2 Cationic Al Complexes in Catalysis

The catalytic hydro-functionalization of carbonyl compounds with HBpin/HBcat or silylhydrides is undoubtedly a facile approach for the synthesis of borate esters or silyl ethers, respectively. Both borate esters and silyl ethers are widely used as intermediates for the synthesis of alcohols. The development of cheaper and sustainable metal catalysts for both the reactions is desirable and inexpensive, earth-abundant aluminium complexes fit perfectly. Roesky and coworkers reported a pioneering work on the first aluminium hydride-catalyzed reduction of carbonyl compounds [159]. In 2016, Nembenna et al. developed aluminium monohydride catalyst to react a large number of aldehydes and ketones. Moreover, the authors reported the intra- and inter-molecular chemoselective hydroboration of aldehydes over ketones [160]. After that, new reports on organoaluminium-catalyzed hydroboration of carbonyl compounds continued to emerge [161, 162]. All the reports, as mentioned earlier, are based on neutral tetracoordinate aluminium complexes. These reports were followed by two more reports on cationic organoaluminium catalysis.

5.4.2.1 Hydroboration Reaction

For the first time in 2018, Singh and coworkers developed aluminium hydride and alkyl cations that catalyzed the hydroboration of aldehydes and ketones with catecholborane (HBcat) (Scheme 5.33). The cationic aluminium complexes **86**[$MeB(C_6F_5)_3$] and **87**[$HB(C_6F_5)_3$], possessing high Lewis acidity, proved to be efficient catalysts for the hydroboration reaction.

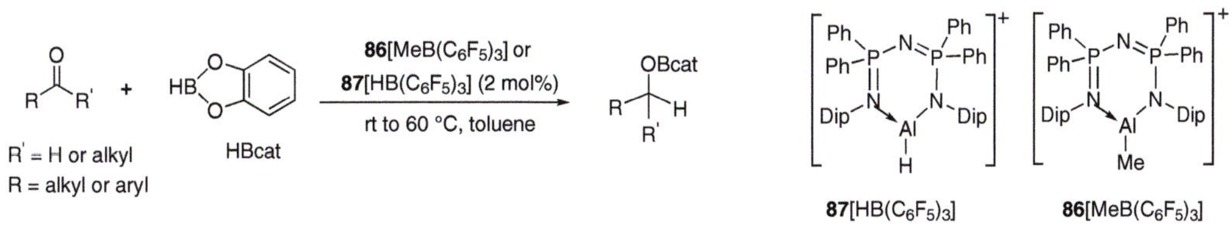

Scheme 5.33 Aluminium cation-catalyzed hydroboration of carbonyls.

Both aryl and alkyl aldehydes were efficiently hydroborated using 2 mol% catalyst at room temperature within 3–6 h. However, the aryl and cyclic alkyl ketones were hydroborated at a slightly higher temperature, i.e. 60 °C. Note that authors observed that aluminium hydride cation **87** is a superior catalyst to aluminium alkyl cation **86** for hydroboration of carbonyls.

A mechanistic investigation was done by making various *in situ* heteronuclear NMR measurements on stoichiometric reactions and adding computational calculations. Results showed adduct formation between the cationic Al centre and carbonyl oxygen is the first step (Scheme 5.34). Further, the reaction proceeds with the hydride migration and formation of cationic benzyloxide intermediate. This step is calculated as the high energy-demand step, so making it the rate-limiting step with $\Delta G = -5.84$ kcal/mol and $\Delta G^* = 20.82$ kcal/mol. The final step of the reaction involves σ-bond metathesis between benzyloxide intermediate and catecholborane along with regeneration of the catalyst [111].

Shortly after this, Zhang and coworkers reported the zwitterionic Meisenheimer aluminium complex, **65a**, as a catalyst for selective hydroboration of ketones with TONs up to 1000, which makes it a highly efficient aluminium catalyst (Scheme 5.35). However, catalytic effectiveness is lower than the other reported alkali, alkaline earth, lanthanide, and transition metal catalysts. The authors observed no interaction between **65a** and carbonyl or alkene in the absence of reducing

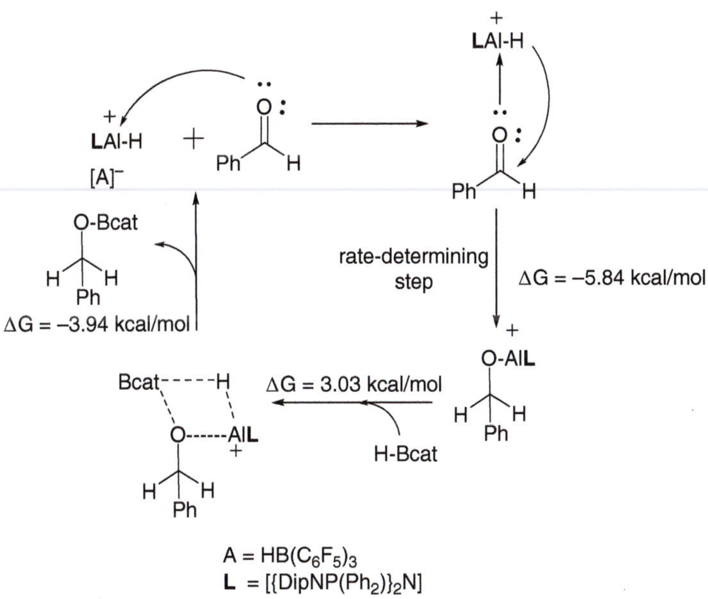

Scheme 5.34 Proposed mechanism for cationic aluminium species-catalyzed hydroboration of benzaldehyde.

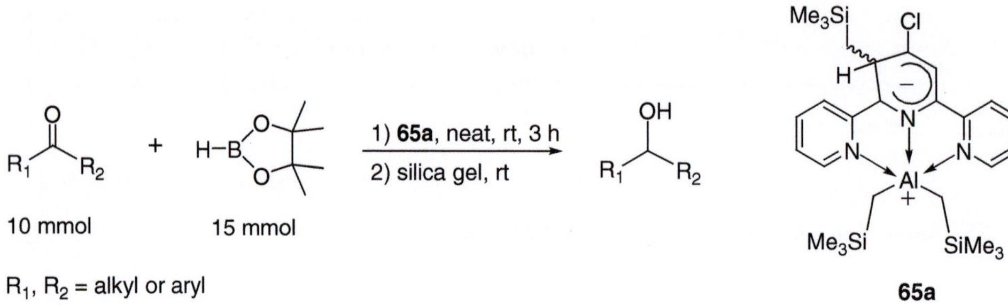

65a

Scheme 5.35 Selective hydroboration of ketones catalyzed by complex **65a.**

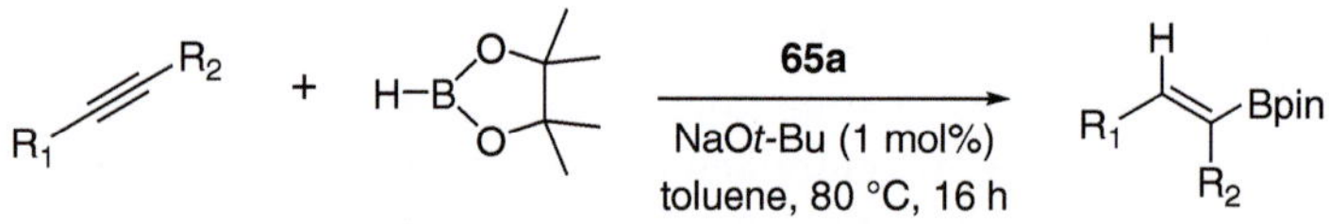

Scheme 5.36 Selective hydroboration of alkynes catalyzed by complex **65a**.

agent whereas on treating the catalyst with two equivalents of HBpin afforded the new active catalyst. The precatalyst employed was deprived of hydride functionality but the new active catalyst bears the hydride functionality at the Al centre. The proposed mechanism is similar to that described above to proceed *via* hydride migration and sigma bond metathesis with HBpin.

Further, the authors explored the catalytic performance of complex **65a** for the hydroboration of alkynes (Scheme 5.36). The complex **65a** catalyzed *cis*-selective anti-Markovnikov hydroboration of phenylacetylene *via* reaction of the proposed active aluminium monohydride species with alkyne, leading to the formation of the aluminium alkenyl complex. A high TON of 910 for the catalyst was recorded, a much higher value than TONs of ~10–33 documented for all other Al catalysts [99].

More importantly, in both hydroboration of ketones and alkyne reactions, the catalyst exhibits exceptional selectivity, leaving the C=C of alkenes and arenes, ester, nitro, and nitrile functional groups unreacted. For both hydroboration of carbonyls and alkynes, the aluminium monohydrido complex is the active catalytic species. In the case of hydroboration of carbonyls, the insertion of carbonyl functionality into the Al–H bond leads to the formation of aluminium alkoxide. Then, alkoxide reacts with HBpin to form boronate ester product and cause regeneration of the active catalytic species.

5.4.2.2 Cyanosilylation Reaction

The cyanohydrins are valuable substrates in industry and act as important intermediates for the synthesis of α-hydroxyacids, α-aminonitriles, α-hydroxyketones, β-hydroxyamines, and β-amino alcohols. Cyanosilylation of carbonyl compounds is the most common approach to prepare silyl ethers and cyanohydrin *via* C–C bond formation. To date, trimethylsilyl cyanide (Me_3SiCN) is the most widely used cyanating reagent for nucleophilic addition to carbonyl compounds to yield cyanohydrin trimethylsilyl ethers. This method avoids the usage of highly toxic hydrogen cyanide

(HCN). Therefore, a significant number of catalysts based on transition metals, main group elements, and Lewis acids and bases have been documented for the cyanosilylation of carbonyls.

Nagendran's coordinatively saturated aluminium cation **96**[OTf] was utilized to promote the cyanosilylation of a diverse range of aldehydes and ketones under mild conditions (rt, neat, 1–2 mol% catalyst) (Scheme 5.37) [122]. Notably, among the known Al catalysts, the cationic nature of coordinatively saturated pentacoordinated aluminium complexes enhances the catalytic activity in comparison to the neutral tetracoordinated Al complexes.

Recently, inspired by the enhanced catalytic efficiency of cationic catalyst **96**[OTf] and to further improve the catalytic activity in comparison to neutral Al catalysts [163], Singh and coworkers reported the use of three coordinated organoaluminium cation **86**[$MeB(C_6F_5)_3$] as a rapid and selective catalyst for the cyanosilylation of aldehydes and ketones under solvent-free conditions (Scheme 5.38) [164].

The main advantage of a cationic (electronically unsaturated) and coordinatively unsaturated (two or three coordinated) Al centre is to ensure its strong LA character and, therefore, its high reactivity. The catalyst showed stunning performance in terms of short reaction time, very low catalyst loading (0.05 mol%), solvent-free reactions and high chemoselectivity, in contrast to many important previous reports. In the mechanistic investigation, multinuclear NMR studies revealed that cyanosilylation proceeds *via* Lewis adduct formation between catalyst **86**[$MeB(C_6F_5)_3$] and Me_3SiCN, thereby activating Me_3SiCN (Si–CN bond) followed by nucleophilic attack of the carbonyl oxygen at the Si centre of the activated silane and formation of the product. The pronounced Lewis acidity of the cationic aluminium centre facilitates this reaction and opens up new avenues in the area of main group catalysis compared to expensive heavier transition elements.

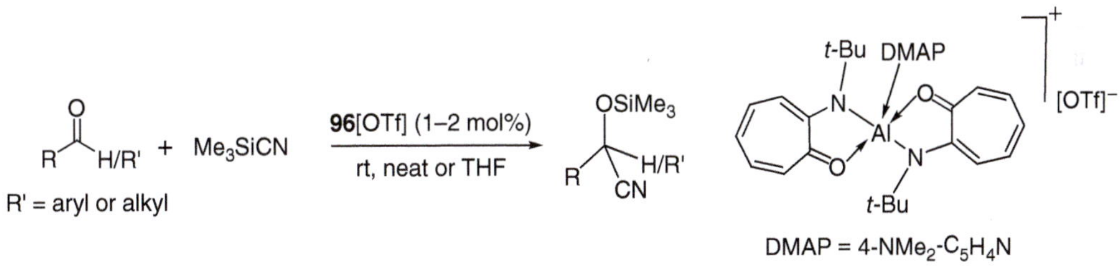

96[OTf]

Scheme 5.37 Pentacoordinated aluminium cation-catalyzed cyanosilylation of carbonyls using TMSCN.

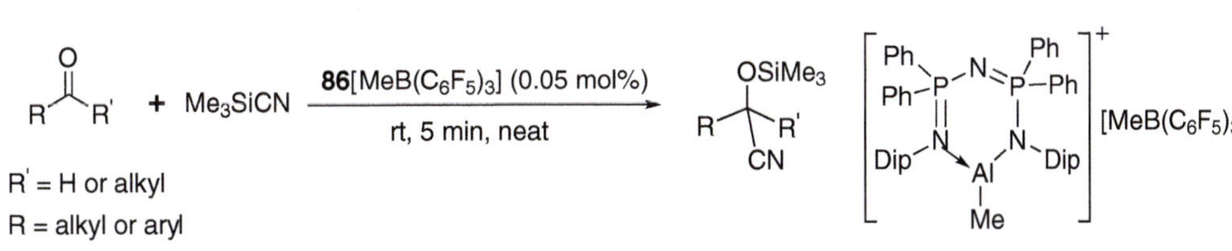

86[$MeB(C_6F_5)_3$]

Scheme 5.38 Organoaluminium cation-catalyzed cyanosilylation of carbonyls.

5.4.2.3 Hydrosilylation Reaction

In 2012, Bergmann and coworkers demonstrated the role of organoaluminium cation **95**[MeB(C_6F_5)$_3$] as the catalyst for the hydrosilylation of carbonyls and imines (Scheme 5.39). Although authors did not observe any interaction of catalyst and acetophenone during their variable temperature ^{1}H NMR studies they noticed the loss of *J* coupling in the silane ($HSiEt_3$) Si–H, which supports the interaction of catalyst with silane. The role of catalyst in activation of the Si–H bond was then demonstrated by deuterium labelling studies, which showed the H/D redistribution between $DSiEt_3$ and H_2SiPh_2 in the presence of a catalytic amount of **95**[MeB(C_6F_5)$_3$] [121].

In 2016, Nikonov and coworkers reported cationic aluminium complex [^{Dip}NacnacAlH]-[B(C_6F_5)$_4$] **78**[B(C_6F_5)$_4$] as a catalyst for the hydrosilylation of a range of alkenes and alkynes (Scheme 5.40). The authors suggested the LA activation mechanism for the hydrosilylation of olefins rather than the insertion/metathesis mechanism. It is worth noting that olefin insertion into the Al–H bond is straightforward, which is confirmed by the stoichiometric reaction between aluminium hydride cation catalyst and alkene that lead to the formation of aluminium alkyl complex.

The insertion and Si–H/Al–C metathesis mechanism was eventually excluded because the reaction between aluminium alkyl cation and $HSiEt_3$ was unsuccessful even at elevated temperatures. Stoichiometric reactions of aluminium hydride cation with alkynes showed addition to the backbone methine carbon by aluminium to form a new bridged tripodal aluminium species.

The authors proposed two different mechanisms, (i) A Piers–Oestreich type mechanism and, (ii) A Yamamato-type mechanism for the LA-catalyzed alkene hydrosilylation reaction. In the case of the Piers-Oestreich-type mechanism, first activation of the Si–H bond by the aluminium

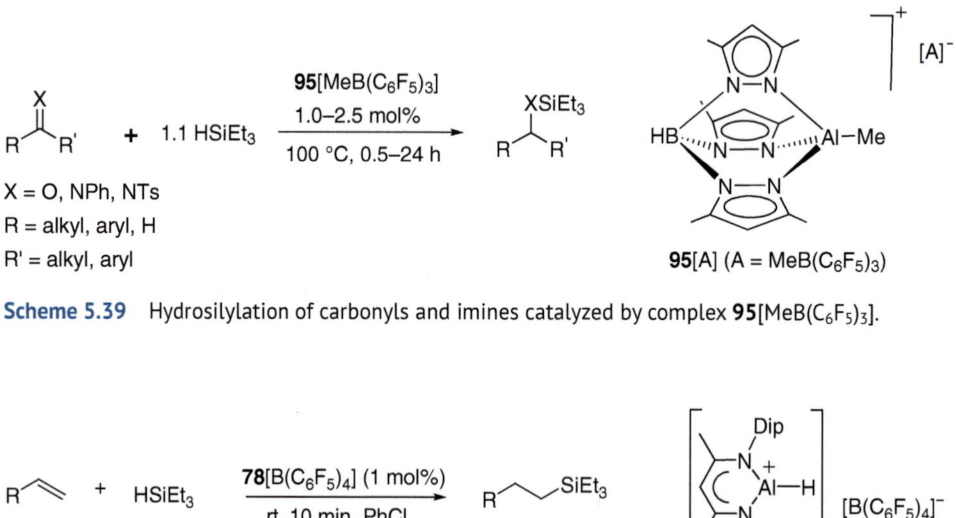

Scheme 5.39 Hydrosilylation of carbonyls and imines catalyzed by complex **95**[MeB(C_6F_5)$_3$].

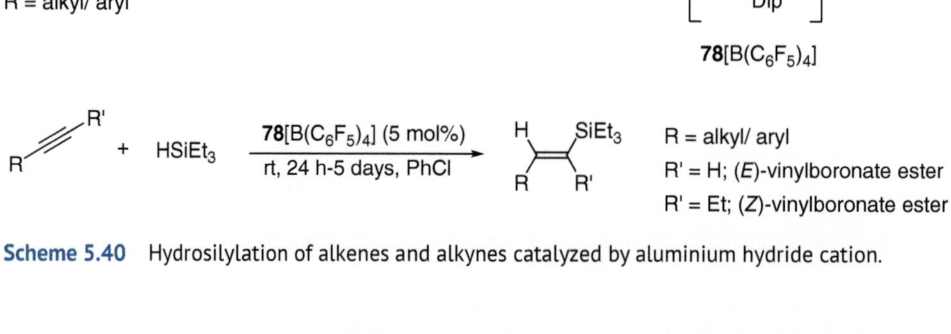

Scheme 5.40 Hydrosilylation of alkenes and alkynes catalyzed by aluminium hydride cation.

cationic centre occurs. In a second step, the olefin attacks the incipient silylium ion, generating a carbocation intermediate, which abstracts hydride from the catalyst to yield the alkylsilane and regenerates the catalyst. In the case of the Yamamato-type mechanism, first the aluminium cation coordinates with an alkene. In the next step, a transfer of hydride from the silane occurs. Finally, the Al–C bond breaks and formation of the product and regeneration of the aluminium cation takes place (Figure 5.30) [113].

Recently, Venugopal and coworkers investigated the catalytic activity of organoaluminium cation **63**[$B(C_6H_3Cl_2)_4$] for the hydrosilylation of ketones (Scheme 5.41). Besides this, the catalyst also works well for the dimerization of aldehydes (Tischenko reaction). More importantly, its catalytic efficiency in aldehyde dimerization displays turnover frequencies reaching up to 6000/h. On performing DFT calculations, the results suggest the substitution of THF molecules on the Al centre of the catalyst by PhCHO, leading to elongation and polarization of the C=O bond to produce **Int-1** with an energy change of +2.7 kcal/mol. This **Int-1** interacts with another upcoming PhCHO molecule and forms a cyclic four-membered transition state (**TS**) having an activation barrier of $\Delta E_{act} = 25.6$ kcal/mol. The rearrangement of the **TS** takes place forming ester (dimerization product) attached to the Al centre in what is a highly exothermic step ($\Delta E = -59.0$ kcal/mol, **Int-2**) The ester is then easily substituted by a new PhCHO molecule to regenerate **Int-1** along with the aldehyde dimerization product (Figure 5.31) [97].

The recent concern about global warming and shortage of energy and carbon resources demands the development of a system for the synthesis of fuels or useful organic substances from carbon

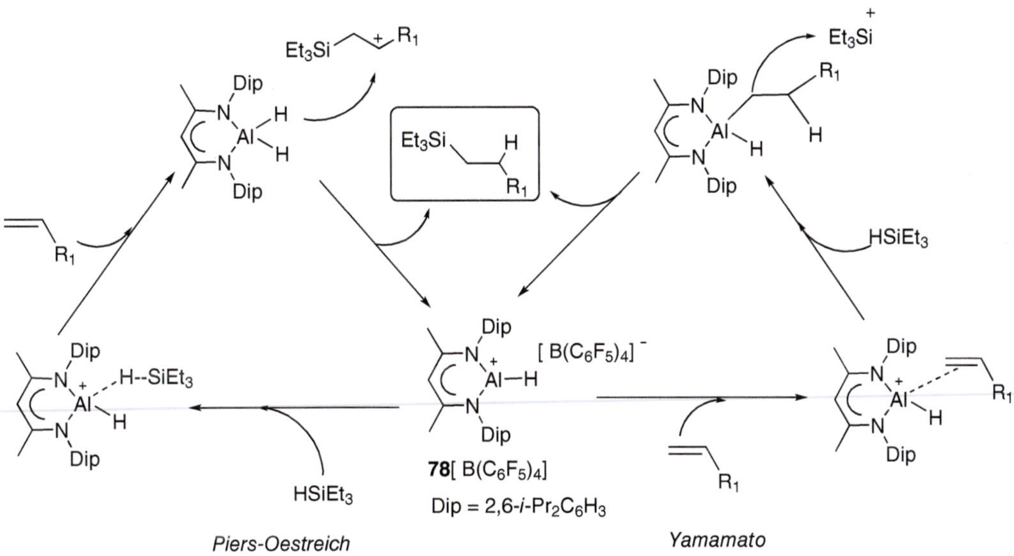

Figure 5.30 Plausible mechanisms for the hydrosilylation of alkene catalyzed by a three-coordinated aluminium cation.

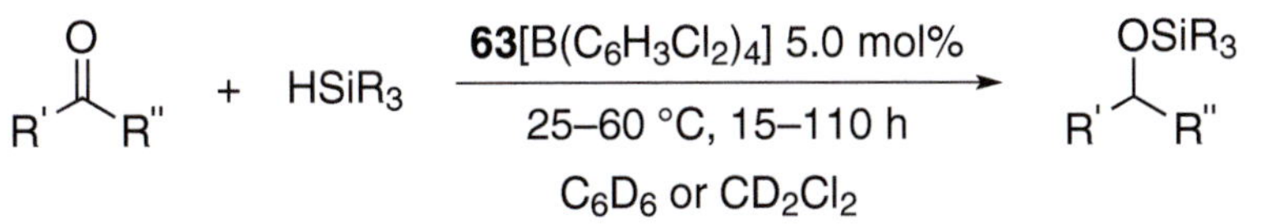

Scheme 5.41 Hydrosilylation of ketones catalyzed by a cationic aluminium complex.

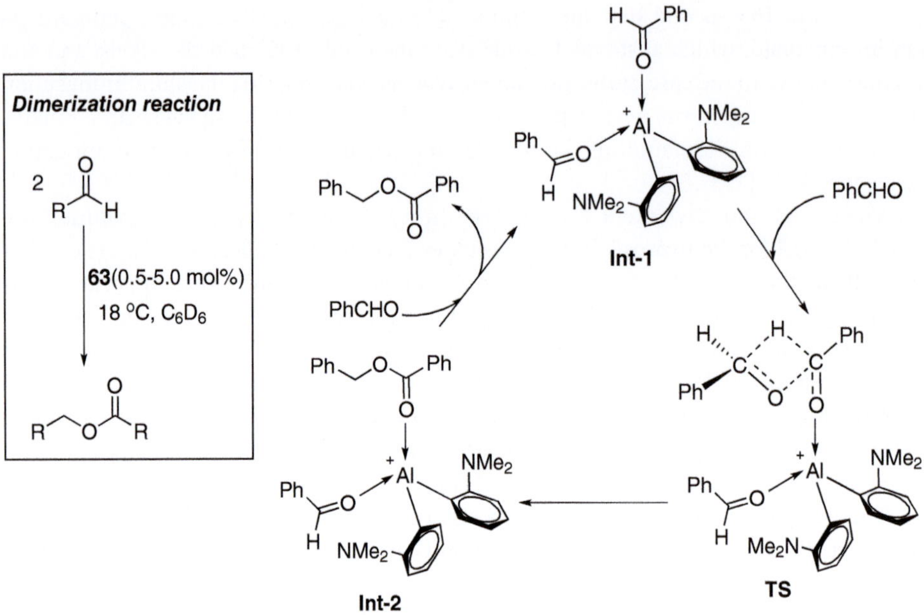

Figure 5.31 Proposed mechanism for the Tischenko reaction catalyzed by an aluminium cation.

dioxide (CO_2). In this regard, efforts have been made to develop a more sustainable catalyst design and inexpensive main group element-based catalysts are a highly appropriate choice [165].

In 2012, Wehmschulte and coworkers utilized an organoaluminium cation, $[Et_2Al]^+[CH_6B_{11}I_6]^-$, **56c**$[CH_6B_{11}I_6]$, as a catalyst for deoxygenative reduction of CO_2 with hydrosilane under mild conditions. This resulted in the formation of methane, toluene, and diphenylmethane as products [95]. Due to strong Lewis acidity of the catalyst, solvent alkylation as a side reaction was also observed, which may explain the low activity of the catalyst. To increase the catalyst reactivity, they switched to the sterically crowded diphenolate ligand-supported catalyst **89**$[CHB_{11}Cl_{11}]$, and found it to be a more efficient catalyst for the reduction of CO_2 to CH_4 and d_5-toluene ($C_6D_5CH_3$). Despite steric crowding and strong Al–O bonds, catalyst **89**$[CHB_{11}Cl_{11}]$ suffered deactivation in the course of CO_2 reduction [116]. Recently in 2017, the same group reported the diethylaluminium cation **60a**$[CHB_{11}Cl_{11}]$ with its less basic counter-anion $[CHB_{11}Cl_{11}]^-$, and found that its activity is significantly higher than those of previously reported Al cations [96].

5.4.2.4 Hydroamination Reaction

The hydroamination reaction is an atom-economical processes and a useful method to form a C–N bond. It involves the addition of primary/secondary amines across olefins. This reaction requires the use of a suitable catalyst due to the high activation energy and also to overcome the entropy loss. A number of metal catalysts have been used to promote such transformations, including main group metals (Ca and Mg), lanthanides, and transition metals. The intramolecular hydroamination of amino alkenes is synthetically useful to prepare nitrogen heterocycles. Although, is not much affected by entropy issues it still needs a catalyst to overcome high activation energy. The electronic similarity between Ca and Mg and cationic Al complexes was reasoned to offer the potential application of low-coordinated cationic Al complexes for hydroamination reactions. In 2012, Khandelwal and Wehmschulte reported cationic organoaluminium compounds as catalysts for the intramolecular hydroamination of primary and secondary amino alkenes [166]. Four

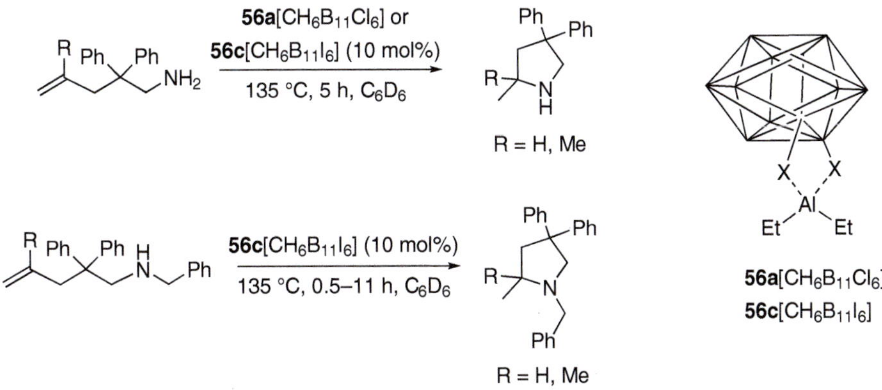

Scheme 5.42 Hydroamination of (top) primary and (bottom) secondary amino alkenes catalyzed by cationic organoaluminium cations. X = Cl (**56a**), I (**56c**).

cationic organoaluminium catalysts **56a**[$CH_6B_{11}Cl_6$], **56c**[$CH_6B_{11}I_6$] (Scheme 5.42), and **59a**[$CH_6B_{11}I_6$], and **59b**[$CH_6B_{11}Cl_6$] (Figure 5.13) were employed for the intramolecular hydroamination of aminopentenes [166]. The authors noted that among all, the bulky species **59a**[$CH_6B_{11}I_6$] was the most active catalyst. It is worth mentioning that neutral organoaluminium alkyl precursors (to the Al cations) also showed some activity for hydroamination of olefins. However, these were much less effective than their cationic counterparts. The authors envisaged the Lewis pair formation *via* amine coordination to the cationic Al centre and subsequent Al-amide complex formation (under ethane evolution) as the crucial step of the catalysis.

5.4.2.5 ROP of *rac*-Lactide, Epoxides and ε-Caprolactone

The bulky NHC-supported tetracoordinated Al cations of Dagorne, **70a–c**[$MeB(C_6F_5)_3$] (Scheme 5.21) were successfully employed as catalysts for ROP of *rac*-lactide. The cation **70a** was particularly effective at affording the polylactide at room temperature in the presence of benzyl alcohol to produce chain-length-controlled and narrow dispersed polylactide (turnover frequency = 160/h). Although the mechanistic investigations were not conclusive, nevertheless, the alcoholysis of the Al cations to form benzyloxide Al cations which then polymerize *rac*-lactide through a coordination–insertion ROP mechanism was speculated about [103].

The discrete pentacoordinate aluminium cation **97**[$MeB(C_6F_5)_3$] (Figure 5.18) was an active catalyst for the cationic polymerization of propylene oxide and *p*-methylstyrene at low temperatures. Polymerization reactions were rapidly catalyzed resulting in the complete conversion of 200 equivalents of propylene oxide (within 1 h at 20 °C, 6 h at 0 °C, and 18 h at −78 °C) with the polymer yield in the range of 80–90% as odourless, oily product. The cationic polymerization of *p*-methylstyrene by **97**[$MeB(C_6F_5)_3$] in toluene solution proceeded from −20 °C to complete the polymerization within 24 h to afford atactic poly(*p*-methylstyrene) with a narrow molecular weight distribution (M_w/M_n = 1.6) [123]. A few years later in a related study, Ishii et al. tested their dinuclear dicationic Al complex **90**[$MeB(C_6F_5)_3$]$_2$ as a catalyst for the ROP of propylene oxide. This catalyst initiated polymerization of propylene oxide under mild conditions to yield an atactic poly(propylene oxide), confirmed by $^{13}C\{^1H\}$ NMR. The gel permeation chromatography analysis of the polymer revealed the formation of a low-molecular-weight polymer (M_n = 2500) with a narrow polydispersity (PDI = 1.04) [117].

The aminophenolate-supported cationic Al complexes **91**[$MeB(C_6F_5)_3$] and **91**[$B(C_6F_5)_4$] (Figure 5.18) were found, in the presence of benzyl alcohol, BnOH, to be active in the ROP of ε-caprolactone, whereas the neutral precursors to these cationic complexes did not show any

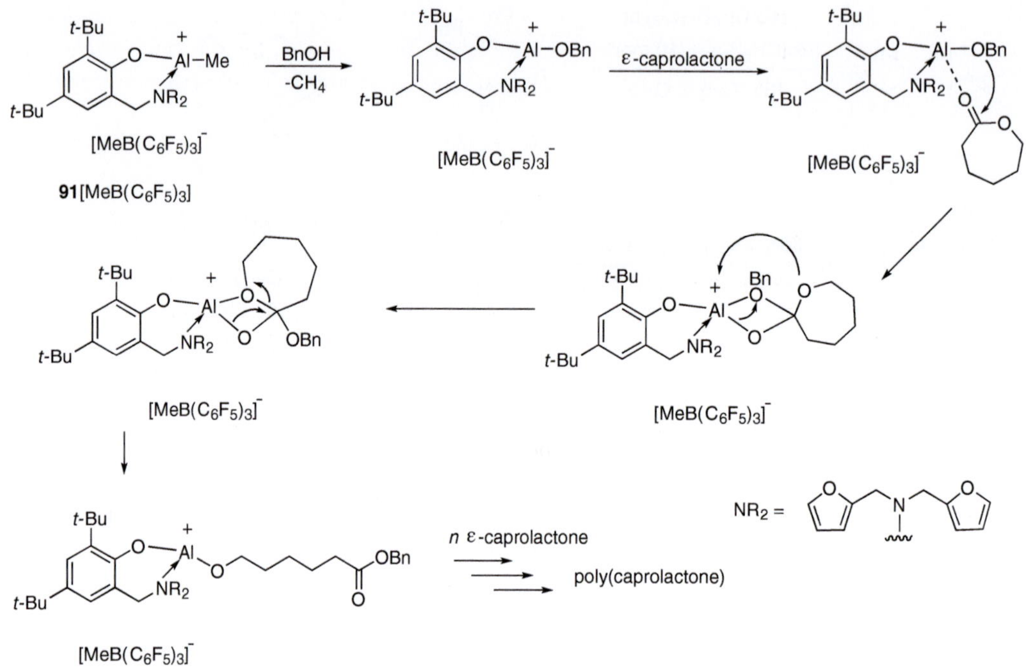

Figure 5.32 Coordination–insertion mechanism proposed for the ROP of ε-caprolactone.

activity for the ROP [118]. Polymerization of ε-caprolactone in the presence of 10 mol% of **91**[MeB(C_6F_5)$_3$] and BnOH in CH_2Cl_2 at room temperature showed 88% completion in 15 min. The polymerization mechanism (Figure 5.32) is proposed to proceed *via* the formation of a benzyloxide aluminium complex, on reaction of **91**[MeB(C_6F_5)$_3$] with BnOH. The cationic benzyloxide aluminium complex then initiates the polymerization of ε-caprolactone through a series of sequential coordination–insertion steps. The slightly lower polymerization rate of **91**[B(C_6F_5)$_4$] was attributed to its relatively bulkier anion obstructing the active site of the active species.

Kerton's Al cations, supported by morpholine-based dianionic pentacoordinating *N,N′,O ,O′,O′′*-chelation ligand, in the presence of protic co-initiators (EtOH, glycerol carbonate) were also able to catalyze ROP of ε-caprolactone to produce poly(caprolactone) with narrow dispersity [119].

The heteroscorpionate ligand-supported Al cations **93**[MeB(C_6F_5)$_3$] and **93**[B(C_6F_5)$_4$] were used as initiators for ROP of ε-caprolactone at 70 °C. Both the cations were substantially more active initiators compared to their dialkylaluminium precursors. The preference for the cationic mechanism (ring opening and propagation through a cationic species) for polymerization was consistent with the higher polydispersity of the polymer. The other possibility, *via* the transfer of nucleophilic alkyl moiety from the initiator to the monomer leading to a metal alkoxide propagating species, was less likely. Notably, both the cationic species were inactive for the ROP of lactides [120].

5.4.3 Cationic Heavier Group 13 Complexes in Catalysis

5.4.3.1 Cationic Gallium Complexes in Catalysis

Representative examples of heavier group 13 cationic complexes used in homogeneous catalysis since 2010 have been summarized below. Recently, developments that emerged on applications of group 13 metal complexes in catalysis were reviewed by Dagorne and Wehmschulte [161]. Further, the latest examples of cationic Ga complex-catalyzed reactions discussed below are activation of alcohol, olefin epoxidation in water, transfer hydrogenation of alkene, hydroarylation reaction, cycloisomerization of enyne, tandem carbonyl-olefin metathesis, and polymerization of propylene oxide and isobutylene.

5.4.3.2 Activation of Alcohols

Alcohols are used as starting material in many chemical processes, although they are quite unreactive as the hydroxyl group acts as a weak leaving group. However, metal-catalyzed activation of alcohol is a highly recommended pathway. In 2016, Gandon and coworkers reported the first cationic gallium complex [NHC^{Dip}·$GaCl_2$][SbF_6] (**103a**[SbF_6]; Scheme 5.24) to catalyze addition of nucleophiles to alcohols. With a low catalyst load of 0.7 mol%, cation **103a** efficiently deoxygenated the alcohols to alkanes with organosilanes such as Et_3SiH and Ph_2SiHCl. The reaction was also catalyzed by cationic gallium complex **103c**[SbF_6] (in 0.5 mol%), which is the 2,4,6-$F_3C_6H_2CN$ adduct of **103a**[SbF_6]. With both the cations (**103a** and **103c**), a large number of aliphatic and aromatic alcohols were reduced in high yield at room temperature, with the exception of some bulky alcohols. Overall, the cation **103a** showed better activity for the reduction of alcohols over main group salts like $InCl_3$, $Ca(NTf_2)_2$, $Bi(OTf)_3$, and $B(C_6F_5)_3$. Besides this, the gallium cation **103a** was also used to activate alcohol towards C–C bond formation. By using a two-component approach in the presence of [NHC^{Dip}·$GaCl_3$ (0.5 mol%) and $AgSbF_6$ (0.7 mol%)], **103a** reacted the propargyl alcohol with the *p*-position of anisole within 1 h at rt. The reaction was highly regioselective, unlike prior work in which both *o*- and *p*-isomers were found (Scheme 5.43) [129].

5.4.3.3 Olefin Epoxidation in Water

The epoxidation of olefin is a prime reaction in both the laboratory and chemical industry, as epoxides are used as a raw ingredient for resins, paints, and surfactants. The choice of solvents plays a significant role in the successful epoxidation of alkene derivatives. The majority of catalytic reactions are operating in a polar organic solvent such as acetonitrile. In recent years, water was developed as an alternative to organic solvents for olefin epoxidation due to economic and environmental concerns. The major drawbacks for using metal-catalyzed epoxidation of olefins in water were hydrolytic instability and the non-soluble nature of the metal catalyst, formation of oligonuclear complexes, and formation of unwanted side products (alcohols and ketones). Therefore, it was quite challenging to operate the epoxidation of alkenes in water. Goldsmith and coworkers synthesized cationic complexes **123a,b** supported by *N*-donor ligands. These complexes were used for facile epoxidation of olefin in acetonitrile. With 1 mol% catalyst load, the complexes **123a,b** methodically oxidized the electron-rich alkene substrates but performed poorly for electron-deficient olefins. The Al complex **123a** was catalytically as good as the Ga complex **123b**

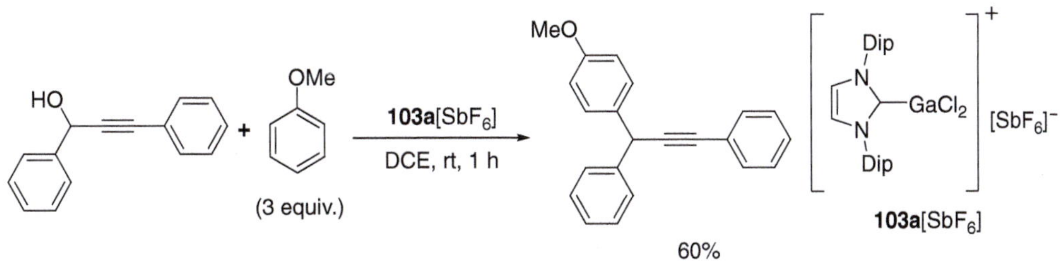

Scheme 5.43 C–C bond formation through activation of alcohols by cationic gallium catalyst.

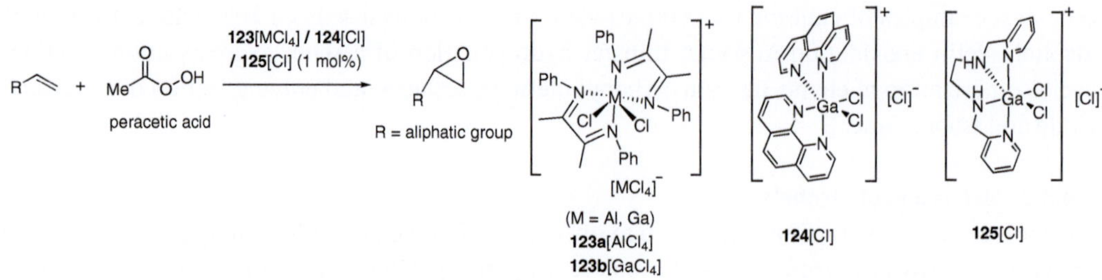

Scheme 5.44 Olefin epoxidation catalyzed by cationic gallium and aluminium catalysts.

(Scheme 5.44) [167]. Soon after this, the hydrolytically stable cationic Ga complexes **124**[Cl] and **125**[Cl] supported by *N*-donor ligands were reported and shown to be capable of olefin epoxidation by peracetic acid in water. Other common oxygen transfer agents (oxidants) such as iodosobenzene, oxone, and metachloroperbenzoic acid were found ineffective (Scheme 5.44) [168].

5.4.3.4 Transfer Hydrogenation of Alkene

The industrial applications of hydrogenation of unsaturated molecules span from fine value-added chemicals to pharmaceuticals. To avoid the commonly used hazardous H_2 gas for hydrogenation, the transfer hydrogenation process has gained much attention as it offers an alternative to the direct use of H_2 gas for reduction. In this strategy, the hydrogen-donor source is easy to handle, recyclable and easily available from either a molecule or solvent, such as isopropanol or formic acid. Numerous examples of metal-catalyzed transfer hydrogenation of polar functional groups such as imines, ketones, carbonates, etc. have been explored. However, for alkene reduction, the task was quite challenging due to its strong π-bond. In 2014, Gandon and coworkers reported the NHC-supported cationic Ga(III) complex **103a**[SbF_6] and utilized it to catalyze the reduction of alkenes. With catalyst **103a**[SbF_6], 1,4-cyclohexadiene (1,4-CHD) was found to be an excellent surrogate hydrogen-source instead of the isopropanol typically for this purpose. Both di-substituted and tri-substituted alkenes were fully hydrogenated to their corresponding alkanes. The catalyst was found to be highly efficient for the selective reduction of alkenes and showed good tolerance towards other functional groups such as esters and ketones [169]. The compound **103a**[SbF_6] was also effective at catalyzing the hydrogenation of endocyclic alkene at room temperature within 1 h. Both 3- and 4-positions in indenes were reduced with excellent yield.

To modulate the activity of the cationic gallium complex on the basis of π-Lewis acidity, in 2018 the same group reported the low-valent cationic Ga(I) complex, $[Ga(PhF)_2]^+[Al\{OC(CF_3)_3\}_4]^-$ **126**[$Al\{OC(CF_3)_3\}_4$] and examined its catalytic activity towards the transfer hydrogenation of various substituted alkenes (Scheme 5.45). In comparison to Ga(III) complex **103a** the catalytic performance of **126** was superior, and comparable to Ga_2Cl_4 [170].

5.4.3.5 Hydroarylation Reaction

The metal-catalyzed addition of nucleophiles across C–C π-bonds is a hot topic. In recent years, main group cationic complexes have revealed extraordinary performance in hydroarylation reactions due to their enhanced π-Lewis acidity.

The cationic Ga(III) **103a**[SbF_6] and In(III) **104**[SbF_6] complexes have been found to be potent homogeneous catalysts for Friedel–Crafts, intra- and/or inter-molecular hydroarylation and hydrogenative cyclization reactions.

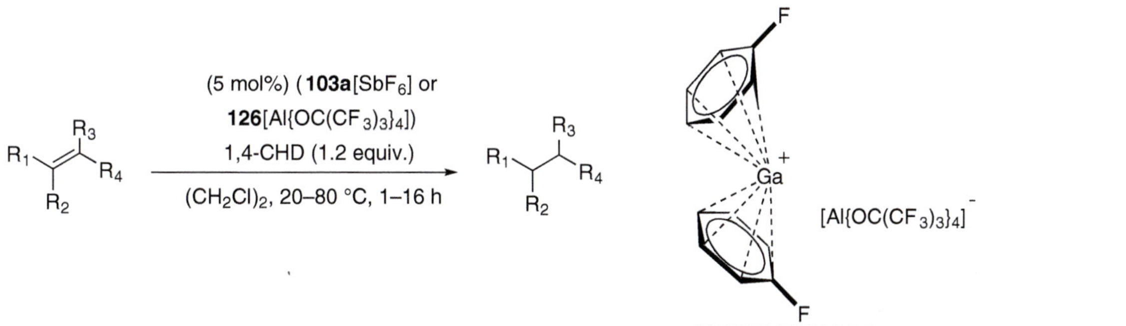

Scheme 5.45 Catalytic transfer hydrogenation of alkenes.

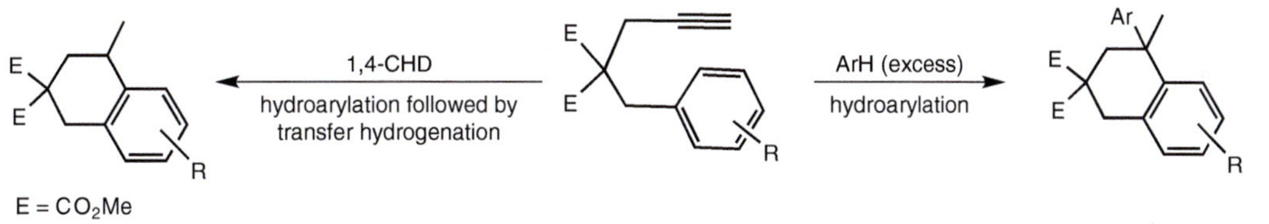

Scheme 5.46 Hydroarylation reactions catalyzed by cationic gallium and indium catalysts.

Gandon et al. compared different cationic gallium complexes for the hydroarylation reaction of alkynes. The abovementioned NHC-stabilized Ga(III) cationic complex **103a**[SbF_6] and low-valent complex **126** that were used for transfer hydrogenation of alkenes were also explored for the intramolecular hydroarylation of alkynes followed by the transfer hydrogenation of resulting styrene derivatives (Scheme 5.46). Other functionalities such as ester, ketone, and nitro were well tolerated with both the catalysts. In comparison to **103a**[SbF_6], the catalyst **126** suppressed side polymerization reactions.

Similarly, activity of the adduct $NHC^{Dip}{\cdot}GaCl_3$ and its nitrile-bonded cationic complexes **103b**[SbF_6] and **103c**[SbF_6] were compared for intramolecular hydroarylation of an alkyne (Scheme 5.46). The complex $NHC^{Dip}{\cdot}GaCl_3$ was found inactive for this catalysis reaction. Whereas, complex **103b**[SbF_6] with benzonitrile displayed less reactivity, with a 36% yield of desired arylated product made. But, interestingly with the addition of $AgSbF_6$, authors reported improved yield (81%). Therefore, they concluded that the use of additive not only abstracted the chloride to generate the dicationic centre but also removed the nitrile group. However, the adduct with more labile 2,4,6-$F_3C_6H_2CN$, **103c**[SbF_6], showed dominance and catalyzed the formation of the required product with 81% yield without the use of any additive [128].

Generally, the hydroarylation reaction is divided into two types; trimolecular and bimolecular. In the same work as their transfer hydrogenation, Gandon and team studied the ability of **103a**[SbF_6] to promote trimolecular hydroarylation. This involves the coupling of three different components alkyne, anisole, and hydrogen from 1,4-CHD and reaction afforded the desired product with good yield (Scheme 5.47)[169].

The catalytic cycle for hydroarylation followed by transfer hydrogenation as catalyzed by **103a**[SbF_6] is given in (Figure 5.33). Initially, the coordination of alkynes to the cationic Ga centre occurs. The nucleophilic attack of the arene ring leads to the formation of Wheland intermediate **B**.

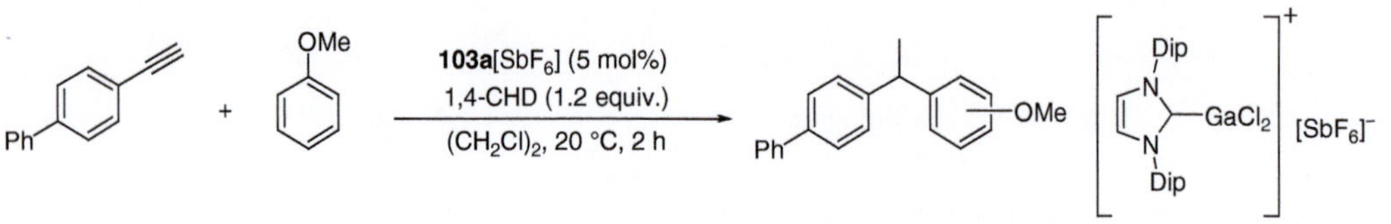

$103a[SbF_6]$

Scheme 5.47 Trimolecular hydroarylation of alkynes catalyzed by a Ga(I) complex.

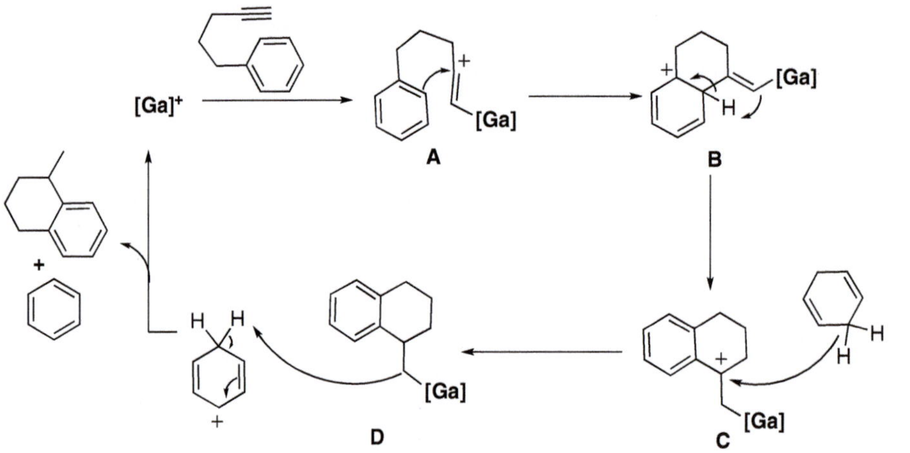

Figure 5.33 Cationic gallium-catalyzed hydroarylation of alkynes followed by transfer hydrogenation.

Then a 1,3 proton shift results in the formation of stabilized carbocation **C**. From 1,4-CHD the hydride transfers to the intermediate **C** and gives **D**, which further undergoes protodegallation and regenerates the catalyst with the elimination of final product [169].

Like Ga(III) **103a**$[SbF_6]$, the indium analogue complex **104**$[SbF_6]$ was also exploited for biomolecular dihydroarylation of alkyne with 1,2-dimethoxybenzene and 1-(phenylsulfonyl)indole. The reaction was fully catalyzed by 5 mol% **104**$[SbF_6]$ in less than 24 h time interval, and the respective product was isolated in more than 80% yield in the presence of $AgSbF_6$ [130].

5.4.3.6 Cycloisomerization of Enyne

The cycloisomerization of enyne is a complicated process due to the polymerization of substrates. In 2018, Gandon and coworkers demonstrated **126**$[Al\{OC(CF_3)_3\}_4]$ catalyzed cycloisomerization of enyne with a 45% yield. This is quite good in comparison to $GaCl_3$ and $[NHC^{Dip}GaCl_2]^+$ $[Al\{OC(CF_3)_3\}_4]^-$, where low product yield (17, 32%) was obtained (Scheme 5.48) [170].

5.4.3.7 Tandem Carbonyl–Olefin Metathesis

In 2019, Gandon and coworkers for the first time reported the tandem metathesis reaction between carbonyl and alkene by using the Ga cation complex **103a**$[SbF_6]$ in the presence of 1,4-CHD as a surrogate source for H_2. The reaction resulted in various 1,2-*cis*-disubstituted cyclopentanes and cyclohexanes in moderate yield. Substrates having dimethyl ester in the β-position of the ketone were well tolerated under the optimized reaction conditions (Scheme 5.49) [171].

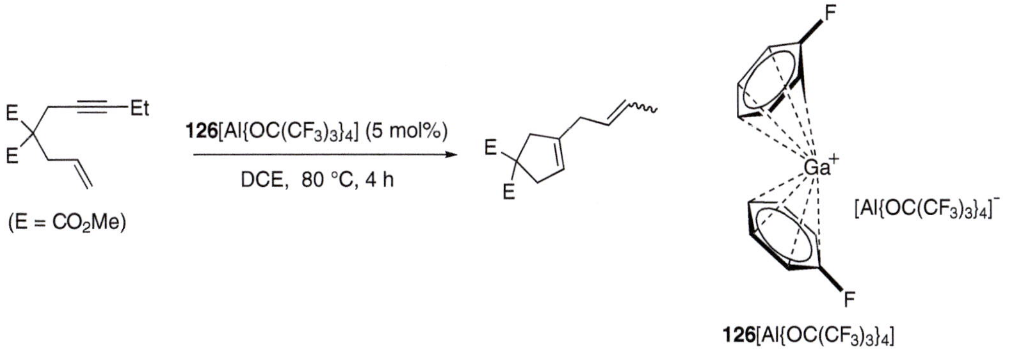

Scheme 5.48 Cycloisomerization of enyne.

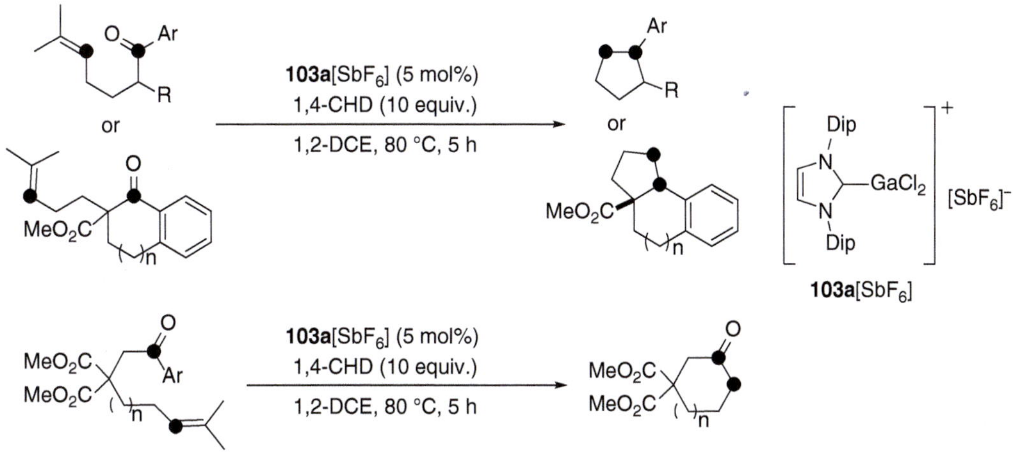

Scheme 5.49 Tandem carbonyl-olefin metathesis catalyzed by a cationic gallium complex.

5.4.3.8 Polymerization of Propylene Oxide and Isobutylene

In 2003, Wehmschulte and coworkers developed a linear unsolvated cationic gallium complex **100**[Li(Al{OCH(CF_3)$_2$}$_4$)$_2$] (Figure 5.19), which was employed for the polymerization of propylene oxide [125].

The commercial method for the preparation of highly reactive poly(isobutylene) (HR-PIB) uses the LA/base adducts between BF_3 and alcohols/ethers as an initiating component. BF_3 is a highly toxic and corrosive reagent, which makes the process more dangerous. Therefore, alternative reaction pathways were narrowed down over the years. In this regard, main group cations that behave as strong LAs can be a suitable initiator for the polymerization of isobutylene.

In 2015, Krossing and coworkers reported cationic gallium(I) arene complexes [Ga(PhC$_2$H$_4$Ph)]$^+$[Al{OC(CF_3)$_3$}$_4$]$^-$ **127** and [(C_6H_5F)Ga(μ-1,3-Ph$_2$C$_6$H$_4$)$_2$Ga(C_6H_5F)]$^{2+}$([Al{OC(CF_3)$_3$}$_4$]$^-$)$_2$ **128**, which were used as catalysts for the polymerization of isobutylene in toluene to afford HR-PIB within 2 h stirring. The synthesized HR-PIB showed high levels of terminal olefinic double bonds (84–93%), and low molecular weights (M_n = 1000–3000/gmol). [172].

5.4.3.9 Cationic Indium and Thallium Complexes in Catalysis

Cationic indium and thallium complexes have not received much attention for their application in chemical transformations including catalysis compared to their lighter congeners (B, Al, and Ga). No reports for catalytic application of lipophilic cationic thallium complexes were found since 2010; there might be some discussion on simple inorganic salts of Tl, however, that are beyond the scope of the present discussion. Nevertheless, the recent trends clearly indicate that In complexes have great potential for use in catalysis. Simple salts or ligand-supported cationic In complexes have found numerous applications. For example, the recent report by Corey et al. disclosed the applications of cationic In complexes $[InI_2]^+[SbF_6]^-$ and $[InI_2]^+[B\{3,5\text{-}(CF_3)_2C_6H_3\}_4]^-$ in selective π-activation of alkynes and synthesis of spiro ring compounds. The π-activation of alkynes by $[InI_2]^+[SbF_6]^-$ cation was also employed for the conversion of acetylenic-diols to bridged cyclic ketals [173]. As discussed previously, the cationic indium complexes were useful for hydroarylation and hydrogenation reactions (Scheme 5.25). Gandon's In analogue **104**$[SbF_6]$ of **103a**$[SbF_6]$, that was prepared by reacting neutral $[NHC^{Dip}InBr_3]$ with one equivalent of $AgSbF_6$, also showed catalytic efficiency towards transfer hydrogenation of acyclic and cyclic alkenes using 1,4-CHD as hydrogen source [130]. Additionally, the latest applications of cationic In complexes in reactions such as hydroarylation and hydrogenation reactions, ROP of epoxides, lactide, and ε-caprolactone and homopolymerization of methyl methacrylate, and copolymerization of epoxide with cyclic ethers and lactide will be discussed below.

5.4.3.10 Coupling of Epoxides and Lactones

The cationic In–alkyl complex of Mehrkhodavandi **105a**$[BAr^F]$ was exploited for the coupling of epoxides and lactones for the selective formation of functionalized spiro-*ortho* esters (SOEs) (Scheme 5.50) instead of catalyzing the ROP of epoxide. When the same reaction was performed at higher temperature, it led to the polymerization and cross-linking of SOEs, indicating that the reaction conditions play an important role in SOE formation over lactone and SOE polymerization [131].

5.4.3.11 ROP of Epoxides, Lactide, and ε-Caprolactone

The salen-supported cationic In catalyst, $[(salen)In(THF)_2]^+[SbF_6]^-$ **107a**$[SbF_6]$ reported by Mehrkhodavandi and coworkers was useful for the homopolymerization of a range of functionalized epoxides including cyclohexene oxide (Scheme 5.51). The same catalyst **107a**$[SbF_6]$ was also active for the copolymerization of larger cyclic ethers (THF, oxetane, and oxepane) with functionalized epoxides such as 1,2-epoxy-5-hexene or epichlorohydrin. This catalyst was also potent in the polymerization of epichlorohydrin and lactide, forming copolymers. Reactivity of the In cations

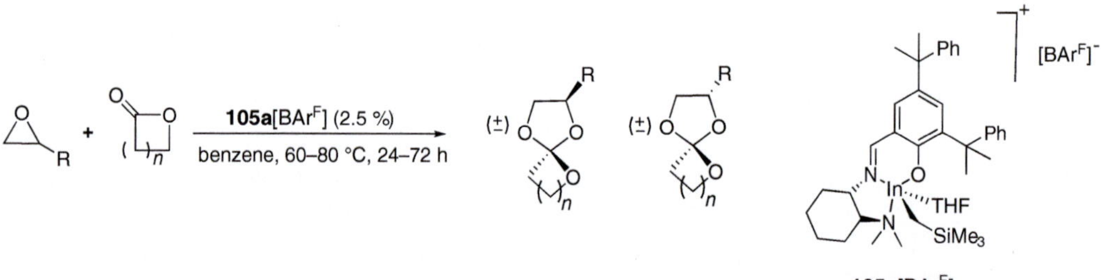

105a$[BAr^F]$

Scheme 5.50 Synthesis of spiro-*ortho* esters (SOEs).

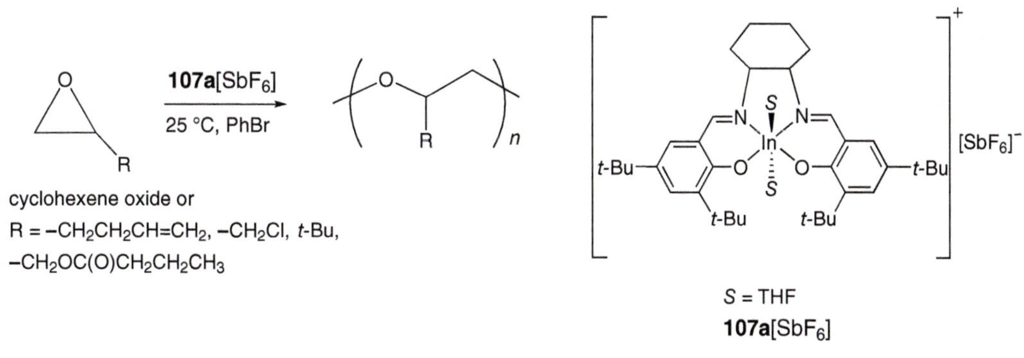

Scheme 5.51 ROP of epoxide catalyzed by a cationic indium complex.

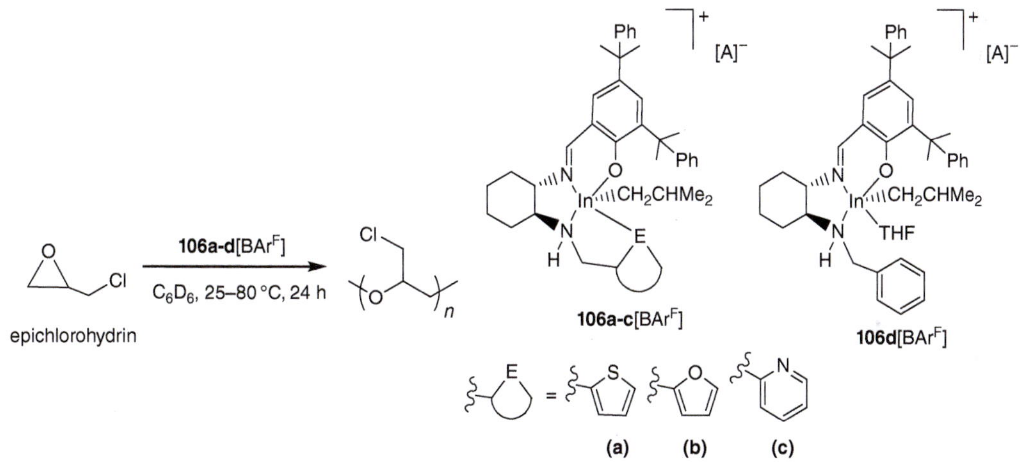

Scheme 5.52 A variety of cationic indium complexes utilized for the ROP of an epoxide.

for polymerization reactions was attributed to high Lewis acidity of the indium centres and the lability of solvent molecules. Examination of the role of the anion and coordinated solvent molecules showed that modification of the solvent/donor molecules as well as counter-ions can potentially allow for fine-tuning of catalytic activity [132].

A similar ROP to that of epoxides with In cationic complexes **106**[BAr^F] (Figure 5.20) was reported by Mehrkhodavandi and coworkers in 2020. Three of these four catalysts contained hemilabile pendant donor groups (**106a–c**) to stabilize the chiral cationic alkylindium compounds. The cationic complex **106d**[BAr^F] that lacks the pendant donor arm was able to catalyze the ROP of epichlorohydrin with an initiation efficiency (of 73% at rt) higher than that of catalysts that contain donor arms (**106a** and **106c**; initiation efficiency of less than 60%) (Scheme 5.52). Cyclohexene oxide is more reactive than epichlorohydrin, therefore, compound **106**[BAr^F] with a hemilabile arm smoothly catalyzed the ROP of cyclohexene oxide in a solventless environment with a high conversion. The catalysts **106a–d**[BAr^F] were used for ROP of *rac*-lactide and the resultant polymer showed high molecular weight (M_n = 34000) and low dispersity (M_w/M_n = 1.32) [133].

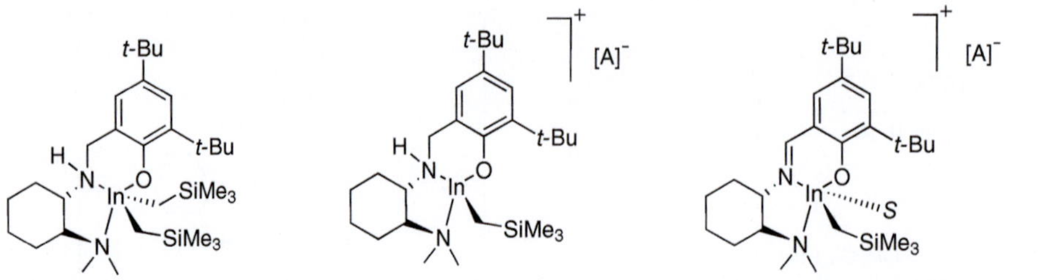

$(NN_iO_{(t\text{-Bu})})In(CH_2SiMe_3)_2$
129

$[(NN_HO_{(t\text{-Bu})})In(CH_2SiMe_3)][A]$
130[A], A = $B(C_6F_5)_4$
130[A], A = PF_6

$[(NN_iO_{(t\text{-Bu})})In(CH_2SiMe_3)][A]$
131a[A], A = $B(C_6F_5)_4$; with *S* = THF
131b[A], A = PF_6; without *S*

Figure 5.34 Cationic indium complexes for polymerization of lactide/ε-caprolactone.

Schiff base ligands, similar to that utilized earlier by Mehrkhodavandi's group, were modified to prepare new *N,N′,O*-monoanionic tridentate ligands $[NN_{H/i}O_{(t\text{-Bu})}]$, new neutral (±)-$(NN_iO_{(t\text{-Bu})})In(CH_2SiMe_3)_2$ **129** and cationic alkylindium complexes $[(NN_HO_{(t\text{-Bu})})In(CH_2SiMe_3)][A]$ **130**[A] [A = $B(C_6F_5)_4$ or PF_6] and $[(NN_iO_{(t\text{-Bu})})In(CH_2SiMe_3)][A]$ **131**[A] [A = $B(C_6F_5)_4$ (**a**-THF), PF_6 (**b**)] (Figure 5.34) [131].

The neutral indium complex (±)-$(NN_iO_{(t\text{-Bu})})In(CH_2SiMe_3)_2$ **129** was directly active as a catalyst for sequential block copolymerization of lactide/ε-caprolactone and methyl methacrylate. Meanwhile, the cationic species (**130** and **131a,b**) were used for ring-opening homo- and copolymerization of ε-caprolactone and *rac*-lactide at high temperature, but the conversion was low with respect to the neutral analogue [174]. Mechanistic investigations for the action of **129** showed its ability to activate lactide *via* both alkoxide- and alkyl-based nucleophilic attack on coordinated lactide. The methyl methacrylate polymerization was proposed to proceed *via* Michael addition by both alkyl and alkoxide nucleophiles.

5.5 Concluding Remarks

The paradigm shift of focus on group 13 metal complexes from neutral to cationic entities has made a significant impact on the prospect of how these elements have been looked at in modern time. The cationic complexes unfold new possibilities to understand structure–reactivity relationships. Until the late twentieth century, the synthesis of reactive cationic complexes was considered a difficult task. However, the design of new ligand frameworks with tuneable steric and electronic properties, which offer kinetic stabilization to the cationic fragments they encapsulate, has made a tremendous impact on this field. Since then, a plethora of cationic complexes bearing variety of ligands have been synthesized, characterized, and explored for various chemical transformations. A small variation in the structural features can fine-tune the properties of a complex and can alter the structure–reactivity relationship. The cationic complexes of elements within group 13 show tremendous differences in their reactivity patterns. Borocations, especially borenium ions, that show a perfect balance of reactivity and stability are widely explored for catalytic applications. The NHC–borenium cations have revolutionized the field of small molecule activation and showed good catalytic activity for reactions such as hydrogenation, hydrosilylation, and hydroboration of various unsaturated bonds.

Low-coordinated cationic complexes of aluminium and heavier congeners were synthesized with a bit of difficulty and the reason can always be attributed to high Lewis acidity and ineffective overlapping of metal orbitals with those of the ligand. Thus, the metals are often found in interaction with ligand side groups, solvent molecules, or anions. Well-defined, discrete aluminium cations have been effectively employed as (pre)catalysts for various organic transformations that include hydroboration of aldehydes, ketones, alkynes, and hydrosilylation of aldehydes, ketones, imines, and alkenes, hydrofunctionalization of CO_2, dimerization of aldehydes, hydroamination of amino alkenes, cyanosilylation of aldehydes and ketones, and ROP of cyclic esters. Aluminium complexes along with those of heavier metals have been extensively used for the reduction of CO_2 *via* hydroelementation reactions and can be used to make more economically efficient products such as methane or cyclic carbonates.

The recent discoveries around NHC- and anionic multidentate ligand-coordinated Ga and In cationic complexes highlight the potential of heavier group 13 elements as alternatives to Al complexes due to better resistance against hydrolytic instability. In the near future, main group chemists should focus to develop new chiral Lewis acidic cations of group 13 to facilitate medicinally important enantioselective reactions. This field is still in its infancy and holds a very strong future potential as a provider of alternatives to expensive transition metals for similar applications.

References

1 Power, P.P. (**2010**). *Nature 463*: 171–177.

2 Weetman, C. and Inoue, S. (**2018**). *ChemCatChem 10*: 4213–4228.

3 Engesser, T.A., Lichtenthaler, M.R., Schleep, M., and Krossing, I. (**2016**). *Chem. Soc. Rev. 45*: 789–899.

4 Piers, W.E., Bourke, S.C., and Conroy, K.D. (**2005**). *Angew. Chem. Int. Ed. 44*: 5016–5036.

5 Ingleson, M. J.(**2105**). *Fundamental and Applied Properties of Borocations*, Springer, (ed. E. Fernández), A. Whiting: Synthesis and Applications of Organoboron Compounds.

6 Eisenberger, P. and Crudden, C.M. (**2017**). *Dalton Trans. 46*: 4874–4887.

7 Lawson, J.R. and Melen, R.L. (**2017**). *Recent Developments And Applications Of Lewis Acidic Boron Reagents. Organomet. Chem. 41*: 1–27.

8 Kölle, P. and Nöth, H. (**1985**). *Chem. Rev. 85*: 399–418.

9 Atwood, D.A. (**1998**). *Coord. Chem. Rev. 176*: 407–430.

10 Dagorne, S. and Atwood, D.A. (**2008**). *Chem. Rev. 108*: 4037–4071.

11 Vedejs, E., Nguyen, T., Powell, D.R., and Schrimpf, M.R. (**1996**). *Chem. Commun.* 2721–2722.

12 Korolev, A.V., Ihara, E., Guzei, I.A. et al. (**2001**). *J. Am. Chem. Soc. 123*: 8291–8309.

13 Prashanth, B. and Singh, S. (**2014**). *Dalton Trans. 43*: 16880–16888.

14 Riddlestone, I.M., Kraft, A., Schaefer, J., and Krossing, I. (**2018**). *Angew. Chem. Int. Ed. 57*: 13982–14024.

15 Sivaev, B. and Bregadze, V.I. (**2014**). *Coord. Chem. Rev. 270-271*: 75–88.

16 Greb, L. (**2018**). *Chem. Eur. J. 24*: 17881–17896.

17 Childs, R.F., Mulholland, D.L., and Nixon, A. (**1982**). *Can. J. Chem. 60*: 801–808.

18 Beattie, I.R. and Gilson, T. (**1964**). *J. Chem. Soc.* 2292–2295.

19 Purcell, K.F. and Drago, R.S. (**1966**). *J. Am. Chem. Soc. 88*: 919–924.

20 Ashley, A.E., Herrington, T.J., Wildgoose, G.G. et al. (**2011**). *J. Am. Chem. Soc. 133*: 14727–14740.

21 Mallouk, T.E., Rosenthal, G.L., Müller, G. et al. (**1984**). *Inorg. Chem. 23*: 3167–3173.

22 Clark, E.R., Grosso, A.D., and Ingleson, M.J. (**2013**). *Chem. Eur. J. 19*: 2462–2466.

23 Gutmann, V. (**1976**). *Coord. Chem. Rev. 18*: 225–255.

24 Beckett, M.A., Strickland, G.C., Holland, J.R., and Varma, K.S. (**1996**). *Polym. Commun. 37*: 4629–4631.

25 Jakubczyk, M., Adamczyk-Wozniak, A., and Sporzynski, A. (**2011**). *Molecular Receptors*, 53–68. Donetsk: East Publisher House.

26 Britovsek, G.J.P., Ugolotti, J., and White, A.J.P. (**2005**). *Organometallics 24*: 1685–1691.

27 Neu, R.C., Ouyang, E.Y., Geier, S.J. et al. (**2010**). *Dalton Trans. 39*: 4285–4294.

28 Müller, L.O., Himmel, D., Stauffer, J. et al. (**2008**). *Angew. Chem. Int. Ed. 47*: 7659–7663.

29 Kögel, J.F., Timoshkin, A.Y., Schröder, A. et al. (**2018**). *Chem. Sci. 9*: 8178–8183.

30 Kögel, J.F., Sorokin, D.A., Khvorost, A. et al. (**2018**). *Chem. Sci. 9*: 245–253.

31 Böhrer, H., Trapp, N., Himmel, D. et al. (**2015**). *Dalton Trans. 44*: 7489–7499.

32 Swanson, B. and Shriver, D.F. (**1970**). *Inorg. Chem. 9*: 1406–1416.

33 Karahl, T. and Kemnitz, E. (**2006**). *J. Fluorine Chem. 127*: 663–678.

34 Shriver, D.F. and Swanson, B. (**1971**). *Inorg. Chem. 10*: 1354–1365.

35 Haartz, J.C. and McDaniel, D.H. (**1973**). *J. Am. Chem. Soc. 95*: 8562–8565.

36 Franz, D. and Inoue, S. (**2019**). *Chem Eur. J. 25*: 2898–2926.

37 Nöth, H., Staudigl, R., and Wagner, H.-U. (**1982**). *Inorg. Chem. 21*: 706–716.

38 Tanaka, N., Shoji, Y., Hashizume, D. et al. (**2017**). *Angew. Chem., Int. Ed. 56*: 5312–5316.

39 Shoji, Y., Tanaka, N., Mikami, K. et al. (**2014**). *Nat. Chem. 6*: 498–503.

40 Shoji, Y., Tanaka, N., Hashizume, D. et al. (**2015**). *Chem. Commun. 51*: 13342–13345.

41 Li, L., Lei, M., Xie, Y. et al. (**2018**). *Inorg. Chem. 57*: 7851–7859.

42 Zhang, Y., Xie, Y., Schaefer, H.F. III, and Wu, J.I.-C. (**2019**). *Inorg. Chem. 58*: 243–249.

43 Bamford, K.L., Qu, Z.-W., and Stephan, D.W. (**2019**). *J. Am. Chem. Soc. 141*: 6180–6184.

44 Higashi, J., Eastman, A.D., and Parry, R.W. (**1982**). *Inorg. Chem. 21*: 716–720.

45 Major, C.J., Bamford, K.L., Qu, Z.-W., and Stephan, D.W. (**2019**). *Chem. Commun. 55*: 5155–5158.

46 Tseng, H.-C., Shen, C.-T., Matsumoto, K. et al. (**2019**). *Organometallics 38*: 4516–4521.

47 Curran, D.P., Solovyev, A., Brahmi, M.M. et al. (**2011**). *Angew. Chem. Int. Ed. 50*: 10294–10317.

48 Matsumoto, T. and Gabbai, F.P. (**2009**). *Organometallics 28*: 4252–4253.

49 McArthur, D., Butts, C.P., and Lindsay, D.M. (**2011**). *Chem. Commun. 47*: 6650–6652.

50 Valverde, M.F.S., Schweyen, P., Gisinger, D. et al. (**2017**). *Angew. Chem. Int. Ed. 56*: 1135–1140.

51 Farrell, J.M., Posaratnanathan, R.T., and Stephan, D.W. (**2015**). *Chem. Sci. 6*: 2010–2015.

52 Cui, P., Guo, R., Kong, L., and Cui, C. (**2020**). *Inorg. Chem. 59*: 5261–5265.

53 Farrell, J.M. and Stephan, D.W. (**2015**). *Angew. Chem. Int., Ed. 54*: 5214–5217.

54 Eisenberger, P., Bestvater, B.P., Keske, E.C., and Crudden, C.M. (**2015**). *Angew. Chem. Int. Ed. 54*: 2467–2471.

55 Yang, W., Krantz, K.E., Freeman, L.A. et al. (**2019**). *Chem. Eur. J. 25*: 12512–12516.

56 Adachi, Y., Arai, F., and Jäkle, F. (**2020**). *Chem. Commun. 56*: 5119–5122.

57 Farrell, J.M., Schmidt, D., Grande, V., and Würthner, F. (**2017**). *Angew. Chem. Int. Ed. 56*: 11846–11850.

58 Tsai, H.-C., Lin, Y.-F., Liu, W.-C. et al. (**2017**). *Organometallics 36*: 3879–3882.

59 Scherpf, T., Feichtner, K.-S., and Gessner, V.H. (**2017**). *Angew. Chem. Int. Ed. 56*: 3275–3279.

60 Dureen, M.A., Lough, A., Gilbert, T.M., and Stephan, D.W. (**2008**). *Chem. Commun. 36*: 4303–4305.

61 Devillard, M., Brousses, R., Miqueu, K. et al. (**2015**). *Angew. Chem., Int. Ed. 54*: 5722–5726.

62 Devillard, M., Mallet-Ladeira, S., Bouhadir, G., and Bourissou, D. (**2016**). *Chem. Commun. 52*: 8877–8880.

63 Jaiswal, K., Prashanth, B., and Singh, S. (**2016**). *Chem. Eur. J. 22*: 11035–11041.

64 Jaiswal, K., Prashanth, B., Ravi, S. et al. (**2015**). *Dalton Trans. 44*: 15779–15785.

65 Lawson, J.R., Wilkins, L.C., André, M. et al. (**2016**). *Dalton Trans. 45*: 16177–16181.
66 Hill, A.F. and Ward, J.S. (**2017**). *Dalton Trans. 46*: 7291–7308.
67 Grätz, M., Bäcker, A., Vondung, L. et al. (**2017**). *Chem. Commun. 53*: 7230–7233.
68 Wang, S.R., Arrowsmith, M., Braunschweig, H. et al. (**2017**). *Chem. Commun. 53*: 11945–11947.
69 Manankandayalage, P., Unruh, D.K., and Krempner, C. (**2020**). *Dalton Trans. 49*: 4834–4842.
70 Keyzer, E.N., Kang, S.S., Hanf, S., and Wright, D.S. (**2017**). *Chem. Commun. 53*: 9434–9437.
71 Rao, B., Chong, C.C., and Kinjo, R. (**2018**). *J. Am. Chem. Soc. 140*: 652–656.
72 Lu, W., Li, Y., Ganguly, R., and Kinjo, R. (**2017**). *Angew. Chem. Int. Ed. 56*: 9829–9832.
73 Cowie, E. and Emslie, D.J.H. (**2018**). *Organometallics 37*: 1007–1016.
74 Zheng, J., Li, Z.H., and Wang, H. (**2018**). *Chem. Sci. 9*: 1433–1438.
75 Khoo, S., Siu, C.-K., and So, C.-W. (**2020**). *Inorg. Chem. 59*: 9551–9559.
76 Ghadwal, R.S., Schermann, C.J., Andrada, D.M., and Frenking, G. (**2015**). *Dalton Trans. 44*: 14359–14367.
77 Prokofjevs, A., Kampf, J.W., Solovyev, A. et al. (**2013**). *J. Am. Chem. Soc. 135*: 15686–15689.
78 Kong, L., Lu, W., Li, Y. et al. (**2016**). *J. Am. Chem. Soc. 138*: 8623–8629.
79 Arnold, N., Braunschweig, H., Dewhurst, R.D. et al. (**2016**). *Chem. Eur. J. 22*: 13927–13934.
80 Franz, D., Szilvsi, T., Pçthig, A. et al. (**2018**). *Chem. Eur. J. 24*: 4283–4288.
81 Widera, A., Vogler, D., Wadepohl, H. et al. (**2018**). *Angew. Chem. Int. Ed. 57*: 11456–11459.
82 Shen, C.-T., Liu, Y.-H., Peng, S.-M., and Chiu, C.-W. (**2013**). *Angew. Chem. Int. Ed. 52*: 13293–13297.
83 Huang, J.-S., Lee, W.-H., Shen, C.-T. et al. (**2016**). *Inorg. Chem. 55*: 12427–12434.
84 Lee, W.-H., Lin, Y.-F., Lee, G.-H. et al. (**2016**). *Dalton Trans. 45*: 5937–5940.
85 Lin, Y.-F., Shen, C.-T., Hsiao, Y.-T. et al. (**2016**). *Organometallics 35*: 1464–1471.
86 Litters, S., Kaifer, E., and Himmel, H.-J. (**2016**). *Angew. Chem. Int. Ed. 55*: 4345–4347.
87 Hahn, J., Biswas, S., Maichle-Mössmer, C. et al. (**2018**). *Pure Appl. Chem. 90*: 711–722.
88 Kim, K.-C., Reed, C.A., Long, G.S., and Sen, A. (**2002**). *J. Am. Chem. Soc. 124*: 7662–7663.
89 Kessler, M., Knapp, C., and Zogaj, A. (**2011**). *Organometallics 30*: 3786–3792.
90 Young, J.D., Khan, M.A., and Wehmschulte, R.J. (**2004**). *Organometallics 23*: 1965–1967.
91 Klis, T., Powell, D.R., Wojtas, L., and Wehmschulte, R.J. (**2011**). *Organometallics 30*: 2563–2570.
92 Dagorne, S. (**2005**). *Low-Coordinated Group 13 Compounds. In Encyclopedia of Inorganic Chemistry, 2nd ed* (ed. R.B. King), 2714. New York: John Wiley & Sons Ltd.
93 Klosin, J., Roof, G.R., Chen, E.Y.-X., and Abboud, K.A. (**2000**). *Organometallics 19*: 4684–4686.
94 Bochmann, M. and Sarsfield, M.J. (**1998**). *Organometallics 17*: 5908–5912.
95 Khandelwal, M. and Wehmschulte, R.J. (**2012**). *Angew. Chem., Int. Ed. 51*: 7323–7326.
96 Saleh, M., Powell, D.R., and Wehmschulte, R.J. (**2017**). *Organometallics 36*: 4810–4815.
97 Kannan, R., Chambenahalli, R., Kumar, S. et al. (**2019**). *Chem. Commun. 55*: 14629–14632.
98 Artamkina, G.A., Egorov, M.P., and Beletskaya, I.P. (**1982**). *Chem. Rev. 82*: 427–459.
99 Zhang, G., Wu, J., Zeng, H. et al. (**2019**). *ACS Catal. 9*: 874–884.
100 Lichtenberg, C., Robert, D., Spaniol, T.P., and Okuda, J. (**2010**). *Organometallics 29*: 5714–5721.
101 Cao, L.L., Daley, E., Johnstone, T.C., and Stephan, D.W. (**2016**). *Chem. Commun. 52*: 5305–5307.
102 Trose, M., Burnett, S., Bonyhady, S.J. et al. (**2018**). *Dalton Trans. 47*: 10281–10287.
103 Schnee, G., Bolley, A., Gourlaouen, C. et al. (**2016**). *J. Organomet. Chem. 820*: 8–13.
104 Dodonov, V.A., Morozov, A.G., Rumyantsev, R.V. et al. (**2019**). *Inorg. Chem. 58*: 16559–16573.
105 Radzewich, C.E., Coles, M.P., and Jordan, R.F. (**1998**). *J. Am. Chem. Soc. 120*: 9384–9385.
106 Dagorne, S., Guzei, I.A., Coles, M.P., and Jordan, R.F. (**2000**). *J. Am. Chem. Soc. 122*: 274–289.
107 Radzewich, C.E., Guzei, I.A., and Jordan, R.F. (**1999**). *J. Am. Chem. Soc. 121*: 8673–8674.
108 Ihara, E., Young, V.G. Jr., and Jordan, R.F. (**1998**). *J. Am. Chem. Soc. 120*: 8277–8278.

109 Coles, M.P. and Jordan, R.F. (**1997**). *J. Am. Chem. Soc. 119*: 8125–8126.
110 Masuda, J.D. and Stephan, D.W. (**2006**). *Dalton Trans. 17*: 2089–2097.
111 Prashanth, B., Bhandari, M., Ravi, S. et al. (**2018**). *Chem. Eur. J. 24*: 4794–4799.
112 Stennett, T.E., Pahl, J., Zijlstra, H.S. et al. (**2016**). *Organometallics 35*: 207–217.
113 Jakobsson, K., Chu, T., and Nikonov, G.I. (**2016**). *ACS Catal. 6*: 7350–7356.
114 Lewiński, J., Horeglad, P., Dranka, M., and Justyniak, I. (**2004**). *Inorg. Chem. 43*: 5789–5791.
115 Issenhuth, J.-T., Pluvinage, J., Welter, R. et al. (**2009**). *Eur. J. Inorg. Chem. 6*: 4701–4709.
116 Wehmschulte, R.J., Saleh, M., and Powell, D.R. (**2013**). *Organometallics 32*: 6812–6819.
117 Nakata, N., Saito, Y., and Ishii, A. (**2014**). *Organometallics 33*: 1840–1844.
118 Kiriratnikom, J., Chotchatchawankul, S., Haesuwannakij, S. et al. (**2018**). *New J. Chem. 42*: 8374–8383.
119 Plommer, H., Murphy, J.N., Dawe, L.N., and Kerton, F.M. (**2019**). *Inorg. Chem. 58*: 5253–5264.
120 Otero, A., Lara-Sánchez, A., Fernández-Baeza, J. et al. (**2011**). *Organometallics 30*: 1507–1522.
121 Koller, J. and Bergman, R.G. (**2012**). *Organometallics 31*: 2530–2533.
122 Sharma, M.K., Sinhababu, S., Mukherjee, G. et al. (**2017**). *Dalton Trans. 46*: 7672–7676.
123 Dagorne, S., Bouyahyi, M., Vergnaud, J., and Carpentier, J.-F. (**2010**). *Organometallics 29*: 1865–1868.
124 Mukherjee, D., Spaniol, T.P., and Okuda, J. (**2017**). *Dalton Trans. 46*: 651–655.
125 Wehmschulte, R.J., Steele, J.M., Young, J.D., and Khan, M.A. (**2003**). *J. Am. Chem. Soc. 125*: 1470–1471.
126 Dąbrowska, A.M., Hurko, A., Dranka, M. et al. (**2017**). *J. Organomet. Chem. 840*: 63–69.
127 Peckermann, I., Dols, T.S., Spaniol, T.P., and Okuda, J. (**2010**). *J. Organomet. Chem. 695*: 2325–2328.
128 Tang, S., Monot, J., El-Hellani, A. et al. (**2012**). *Chem. Eur. J. 18*: 10239–10243.
129 Michelet, B., Tang, S., Thiery, G. et al. (**2016**). *Org. Chem. Front. 3*: 1603–1613.
130 Michelet, B., Colard-Itté, J.-R., Thiery, G. et al. (**2015**). *Chem. Commun. 51*: 7401–7404.
131 Jung, H.-J., Chang, C., Yu, I. et al. (**2018**). *ChemCatChem 10*: 3219–3222.
132 Diaz, C., Ebrahimi, T., and Mehrkhodavandi, P. (**2019**). *Chem. Commun. 55*: 3347–3350.
133 Goonesinghe, C., Roshandel, H., Diaz, C. et al. (**2020**). *Chem. Sci. 11*: 6485–6491.
134 Slattery, J.M., Higelin, A., Bayer, T., and Krossing, I. (**2010**). *Angew. Chem. Int. Ed. 49*: 3228–3231.
135 Wehmschulte, R.J., Peverati, R., and Powell, D.R. (**2019**). *Inorg. Chem. 58*: 12441–12445.
136 Mansaray, H.B., Tang, C.Y., Vidovic, D. et al. (**2012**). *Inorg. Chem. 51*: 13017–13022.
137 Jurca, T., Lummiss, J., Burchell, T.J. et al. (**2009**). *J. Am. Chem. Soc. 131*: 4608–4609.
138 Jurca, T., Korobkov, I., Gorelsky, S.I., and Richeson, D.S. (**2013**). *Inorg. Chem. 52*: 5749–5756.
139 Jurca, T., Korobkov, I., Yap, G.P.A. et al. (**2010**). *Inorg. Chem. 49*: 10635–10641.
140 Stefkova, K., Gierlichs, L., Willcox, D., and Melen, R.L. (**2020**). *Borocations in Catalysis. Encyclopedia of Inorganic and Bioinorganic Chemistry* 1–37.
141 De Vries, T.S., Prokofjevs, A., and Vedejs, E. (**2012**). *Chem. Rev. 112*: 4246–4282.
142 Prokofjevs, A., Boussonnière, A., Li, L. et al. (**2012**). *J. Am. Chem. Soc 134*: 12281–12288.
143 Pan, X., Boussonnière, A., and Curran, D.P. (**2013**). *J. Am. Chem. Soc. 135*: 14433–14437.
144 Eisenberger, P., Bailey, A.M., and Crudden, C.M. (**2012**). *J. Am. Chem. Soc. 134*: 17384–17387.
145 McGough, J.S., Butler, S.M., Cade, I.A., and Ingleson, M.J. (**2016**). *Chem. Sci. 7*: 3384–3389.
146 Fan, X., Zheng, J., Hua Li, Z., and Wang, H. (**2015**). *J. Am. Chem. Soc. 137*: 4916–4919.
147 Parks, D.J. and Piers, W.E. (**1996**). *J. Am. Chem. Soc. 118*: 9440–9441.
148 Denmark, S.E. and Ueki, Y. (**2013**). *Organometallics 32*: 6631–6634.
149 Chen, J., Lalancette, R.A., and Jäkle, F. (**2013**). *Chem. Commun. 49*: 4893–4895.
150 Rawat, S., Bhandari, M., Porwal, V.K., and Singh, S. (**2020**). *Inorg. Chem. 59*: 7195–7203.

151 Parks, D.J., Blackwell, J.M., and Piers, W.E. (**2000**). *J. Org. Chem. 65*: 3090–3098.

152 Lam, J., Szkop, K.M., Mosaferi, E., and Stephan, D.W. (**2019**). *Chem. Soc. Rev. 48*: 3592–3612.

153 Lui, M.W., Paisley, N.R., McDonald, R., Ferguson, M.J., and Rivard, E. (**2016**). *Chem. Eur. J. 22*: 2134–2145.

154 Farrell, J.M., Hatnean, J.A., and Stephan, D.W. (**2012**). *J. Am. Chem. Soc. 134*: 15728–15731.

155 Lam, J., Günther, B.A.R., Farrell, J.M. et al. (**2016**). *Dalton Trans. 45*: 15303–15316.

156 Mercea, D.M., Howlett, M.G., Piascik, A.D. et al. (**2019**). *Chem. Commun. 55*: 7077–7080.

157 Huchenski, B.S.N., Adams, M.R., McDonald, R. et al. (**2016**). *Organometallics 35*: 3101–3104.

158 Huchenski, B.S.N., Christopherson, C.J., Robertson, K.N., and Speed, A.W.H. (**2019**). *Org. Biomol. Chem. 17*: 6158–6164.

159 Yang, Z., Zhong, M., Ma, X. et al. (**2015**). *Angew. Chem. Int. Ed. 54*: 10225–10229.

160 Jakhar, V.K., Kr, M., and Barman, S.N. (**2016**). *Org. Lett. 18*: 4710–4713.

161 Dagorne, S. and Wehmschulte, R. (**2018**). *ChemCatChem 10*: 2509–2520.

162 Nikonov, G.I. (**2017**). *ACS Catal. 7*: 7257–7266.

163 Yang, Z., Yi, Y., Zhong, M. et al. (**2016**). *Chem. Eur. J. 22*: 6932–6938.

164 Rawat, S., Bhandari, M., Prashant, B., and Singh, S. (**2020**). *ChemCatChem 12*: 2407–2411.

165 Chen, J., Falivene, L., Caporaso, L. et al. (**2016**). *J. Am. Chem. Soc. 138*: 5321–5333.

166 Khandelwal, M. and Wehmschulte, R.J. (**2012**). *J. Organomet. Chem. 696*: 4179–4183.

167 Koellner, C.A., Piro, N.A., Kassel, W.S. et al. (**2015**). *Inorg.Chem. 54*: 7139–7141.

168 Bronston, F., Ting, S., Zhang, Q., and Goldsmith, C.R. (**2016**). *Polyhedron 114*: 268–272.

169 Michelet, B., Bour, C., and Gandon, V. (**2014**). *Chem. Eur. J. 20*: 14488–14492.

170 Li, Z., Thiery, G., Lichtenthaler, M.R. et al. (**2018**). *Adv. Synth. Catal. 360*: 544–549.

171 Djurovic, A., Vayer, M., Li, Z. et al. (**2019**). *Org. Lett. 21*: 8132–8137.

172 Lichtenthaler, M.R., Maurer, S., Mangan, R.J. et al. (**2015**). *Chem. Eur. J. 21*: 157–165.

173 Surendra, K. and Corey, E.J. (**2014**). *J. Am. Chem. Soc. 136*: 10918–10920.

174 Jung, H.-J., Yu, I., Nyamayaro, K., and Mehrkhodavandi, P. (**2020**). *ACS Catal. 10*: 6488–6496.

6

Recent Development in the Solution Structural Chemistry of Main Group Organometallics

Alistair M. Broughton[a], Leonie J. Bole[b], Andrew E. H. Wheatley[a], and Eva Hevia[b]

[a] *Yusuf Hamied Department of Chemistry, University of Cambridge, Cambridge, UK*
[b] *Department für Chemie, Biochemie und Pharmazie, Universität Bern, Bern, Switzerland*

6.1 Introduction

Their many uses in organometallic synthesis mean that alkali metal and alkaline-earth metal reagents have been the subject of extensive study. Applications abound, with organolithium chemistry, in particular, superseding even lower (Mg) Grignard reagents in terms of number and range of transformations enabled. Since their inception in 1917 [1], through the development of the ubiquitous reagent *n*-BuLi [2], these species have become a major applied vector in fields such as fine chemical synthesis [3–6] and polymerization [7, 8]. From the perspective of correlating their reactivities with their structural chemistry, the solution identities and natures of s-block metal bases have received extensive attention. Whilst other chapters outline their solid-state analysis, major inroads have also been made in the solution study of synthetically vital organolithium- [9, 10] and Grignard- (both lower [11] and higher [12]) based reagents and reaction mechanisms.

The need to overcome the synthetic limitations of monometallic bases has also led to great interest in developing heterometallic systems capable of synergistic behaviour, in particular under mild conditions. Though mixed alkali metal systems, most famously the LICKOR superbases (LIC denotes alkyllithium, KOR denotes potassium alkoxide) [13], have been established for several decades, and a range of binary and even ternary organo (alkali metal) complexes have been developed (and are discussed in Chapter 1), a major focus of this chapter will involve the combination of alkali and non-alkali metals. In this context, although species such as Et_3ZnNa were first presented many years ago [14, 15], and alkali metal analogues have been deployed in the derivatization of α,β-unsaturated ketones [16, 17], it is only recently that the subject of non-nucleophilic, synergic main group reagents has gained traction. In 1999, the directed zincation of functionalized aromatics was developed. This work deployed the putative heterobimetallic species *t*-Bu_2Zn(TMP)Li **1** (TMP = 2,2,6,6-tetramethylpiperidide) as explained in Chapter 1 [18]. Since this report, a range of heterobimetallic systems based upon an alkali metal and a more electronegative metal have been produced. Many of these have been utilized synthetically, with the polarizing ability of the more electropositive metal underpinning synergic activity. Overall, it is indisputable that heterobimetallic compounds differ significantly from those of the parent monometallics, the latter being generally incapable of reproducing the chemistry of the former (viz. the contrast between

Polar Organometallic Reagents: Synthesis, Structure, Properties and Applications, First Edition. Edited by Andrew E. H. Wheatley and Masanobu Uchiyama.

the properties of organolithium [3—6] and diorganozinc [19] reagents and those of **1**). It stands to reason that these differences have their origins in the metal identities, and structural differences between the heterobimetallic reagents and those of their parent monometallic components. Since the original work of Kondo, heterobimetallic reagents have been the subject of a range of review articles, focusing on reactivity [20], structural [21, 22], and in some cases both [23–27] aspects of this emergent field of applied chemistry. As well as Ref. 27, Chapter 2 contains an up-to-date overview of the recent advances in solid-state structural investigations in the area. Indeed, overall, it is the case that the vast majority of structural work that has been done on these systems has involved solid-state analysis and that, therefore, discussion of reactivity has often been predicated on the retention of the often individually bimetallic complexes that have been observed by this technique. This has meant that issues of changes to geometry, aggregation state, and/or solvation (by external solvent molecules—as is evidenced by many crystal structures) in solution have not been thoroughly investigated. In addition, evidence has emerged that so-called synergic reagents can alter much more fundamentally in solution. Data to this effect started to emerge as early as 2012, with 'lithium cadmates' revealing two-step reactivity in which lithiation of a target organic was, in fact, followed by cadmium amide trapping of the initially formed carbanion [28]. Similar observations of what has become known as trans-metal-trapping (TMT) (see below) followed for lithium aluminates [29], requiring a combination of density functional theory (DFT) calculations and nuclear magnetic resonance (NMR) spectroscopic investigation to elucidate the reaction pathways on display. Most recently, similar observations have been extended to the higher elements of groups 1 and 13 [30]. It was these kinds of observations that began to highlight the potential of modern NMR techniques in not only the elucidation of homometallic species in solution but also in that of systems that were, on the face of it, individually heteropolymetallic.

This being so, recent developments in the solution analysis of organometallic reagents have played a vital role in clarifying both the structures and reactivities of a range of not only mono- but also heterobimetallic systems. While reactive intermediates in main group organometallic chemistry have been the long-time subject of study, amongst the most significant advances of recent years in this area have come from the inception of the pulse-gradient spin-echo (PGSE) [31, 33] technique. This enabled the deduction of a molecule's hydrodynamic radius from its diffusion coefficient (*D*). Subsequent to that, a 2D-refinement of the technique, diffusion-ordered spectroscopy (DOSY), enabled the discrimination of species with different diffusion rates by size [34]. As of 2000, the use of internal reference compounds in the development of internal calibration curve (ICC) DOSY [aka internally referenced (IR) DOSY] began to assume significant importance in the understanding of reactive aggregates in main group chemistry [10, 35]. The development of this work [36, 37], eventually led to the simplification of the relationship between diffusion coefficient and molecular weight by the use of external calibration curves (ECCs) [38]. This last technique has recently enabled the evaluation of several s-block organometallics in solution (see below) and, though it lies outside the remit of the current tract, has been expanded to interrogate halogenated organics [39] of a type that have hitherto resisted accurate molecular weight analysis [40]. The fundamentals that underpin diffusion methods have been the subject of review [41, 42], and the power of these techniques is suggested by their extension to fields such as paramagnetic complex chemistry [43].

The aim of this chapter is not to replicate recent reviews in the field of s-block and synergic base chemistry, but rather to offer a critical assessment of some of the most significant advances made in recent years. The last decade will be looked at in detail and, in particular, the advent of diffusion techniques will be an area of focus. Their development in the context of monometallic systems will precede a more extensive discussion of their potential in unpicking the behavioural complexities of formally heterobimetallic complexes, synthetically important alkali metal ate complexes and synergic mixtures [23–27, 44] and the LiCl-complexed 'turbo' variants on lower Grignard and Hauser reagents [45–48].

6.2 Monometallic Systems

6.2.1 Introduction

While X-ray crystallography has been instrumental in helping us to understand structure and bonding in s-block metal chemistry, NMR spectroscopy has come to the fore as a key technique in rationalizing solvent effects, aggregation, and reactivity. This section showcases a selection of examples in s-block metal chemistry, where utilizing multinuclear and multidimensional NMR techniques including ^{6}Li,^{1}H-HOESY, ^{6}Li,^{6}Li-EXSY, and DOSY has allowed researchers to shed light on the intriguing constitutions of these species in solution. Special attention is paid to DOSY NMR, outlining its development as a probe in organolithium chemistry, its extension to higher group 1 systems, and then its refinement over the last decade to enable the increasingly confident estimation of molecular weight (M_r) in several group 1 metal complexes.

6.2.2 Organo(s-block Metal) Aggregation and Reactivity

The behaviour of simple organolithium reagents in solution in the presence of either stoichiometric Lewis base additive or excess donor solvent has long attracted interest. One example of a system elucidated by a combination of 1D and 2D NMR spectroscopic techniques is PhLi-([−]-sparteine). The tetrameric ladder intercepted by diamine, $(PhLi)_4$([−]-sparteine)$_2$ **2**, observed for this complex in d_{10}-Et_2O (that is, an intercepted ladder resistant to ether coordination) was noted by a combination of ^{6}Li and ^{13}C NMR spectroscopy, ^{6}Li,^{1}H-HOESY, and ^{6}Li,^{6}Li-EXSY to convert to partially tetrahydrofuran (THF)-solvated $(PhLi)_2$([−]-sparteine)(THF) **3** and then to the dimer of PhLi(THF) **4** upon titration with the stronger Lewis base [49]. Recently, the activities of *i*-PrLi in the presence of either (−)-sparteine or an *N*-methylated surrogate of (+)-sparteine have been determined in d_{10}-OEt_2 and d_8-THF. For the former solvent, a combination of Bauer-Winchester-Schleyer analysis [50, 51] of ^{6}Li–^{13}C coupling and ^{6}Li,^{1}H-HOESY established that (−)-sparteine formed a dinuclear adduct **5** in which nOe (nuclear Overhauser effect) correlations existed between the diamine and just one of two distinct ^{6}Li signals (Figure 6.1). Meanwhile, the (+)-sparteine surrogate was evidenced by the observation of a single ^{6}Li signal and two ^{13}C quintets, with ^{6}Li–^{13}C coupling consistent with a head-to-tail dimer of **6** [52]. Titration of the more aggressive donor d_8-THF resulted in (−)-sparteine complexation of *i*-PrLi only when using > 3.0 eq. diamine (wrt Li), and at 6.0 eq. diamine, monomer **7** was characterized. In contrast, a similar monomer (**8**), revealing a characteristic 1:1:1 triplet for the carbanion centre by ^{13}C NMR spectroscopy, was seen using just 1 eq. (+)-sparteine surrogate.

An interesting series of papers have appeared in recent years that have applied the method of continuous variation (MCV) [53], an experimental mixing technique first developed by Job [54–56], in conjunction with NMR spectroscopy to elucidate the behaviour of different organolithiums in solution and understand the implications of their solution behaviour for organometallic synthesis. The MCV is well known to simply reveal the properties of a straightforward XY system through analysis of a so-called 'Job plot', a device that graphically relates the change in some physical property as a function of mol fraction of one component (e.g. X). However, the use of the MCV in

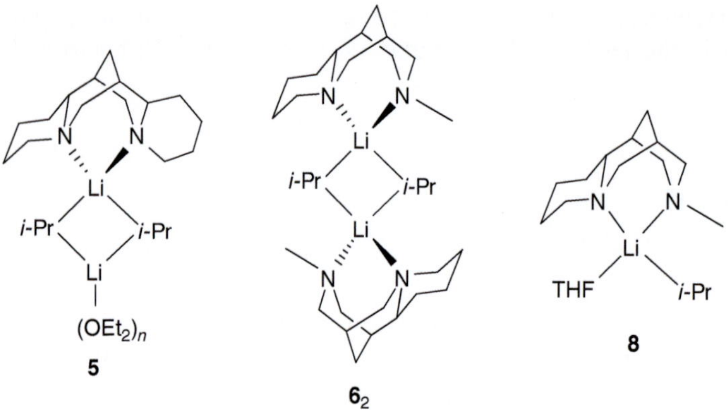

Figure 6.1 Lithium amide complexes incorporating (−)-sparteine or (+)-sparteine surrogates.

tandem with, for example, ^{1}H NMR spectroscopy has been extended to more complex systems [57], and the potential that this approach offers in elucidating processes right across organometallic chemistry that incorporate ensembles of X_aY_b upon creating a binary mixture of metalates has been both nicely outlined [58], and applied to a range of organolithium-based systems through deployment in conjunction with ^{6}Li NMR spectroscopy by Collum [59]. To focus in more detail on a representative part of this dichotomous effort, lithium enolates have long attracted the interest of chemists. Indeed, the study of pairs of these by the MCV evolved from healthcare research into aminated enolates [60–62]. In the first such probe, simple, cyclic metalates derived from 1-indanone, cyclohexanone, and cyclopentanone were looked at in the presence of mono and bidentate Lewis bases. This work established the approach to be appropriate to elucidating homo- and heteroaggregates formed from binary mixtures of enolates, leading to a range of case studies as described below. Presently, the cyclic enolates of interest were prepared from ^{6}Li,^{15}N-labelled LHMDS (HMDS = hexamethyldisilamide) [63], yielding chelated dimers in *N,N,N′,N′*-tetramethylethylenediamine (TMEDA) and pseudocubic tetramers in THF. Less expected, in dimethoxyethane (DME) pseudocubic tetramers in which the Lewis base presumably acted η^1 were detected, whereas chelated dimers were not [64]. This led to ^{6}Li experiments in solvent exchange conducted on the 1-indenone enolate **9** (Figure 6.2), which focused on whether aggregation state altered [65]. Experiments, which required distinctly different ^{6}Li environments both before and after the exchange, were based on the contrast between *fast* exchange of substitutionally labile solvent (expected to time-average the two signals) and relatively *slow* exchange of aggregation state (expected to grow one signal at the expense of the other). In the event, replacing THF with TMEDA or TMEDA with DME incurred aggregation state changes, though in the latter case the aggregate was reluctant to grow.

Moving to more complex enolates, N,O-chelation and aggregation have been monitored in salts derived from glycinimines of benzophenone and (+)-camphor in both pure toluene, and toluene containing different levels of THF, TMEDA, (*R,R*)- or (*S,S*)-*N,N,N′,N′*-tetramethylcyclohexanediamine [(*R,R*)-/(*S,S*)-TMCDA]. In the hydrocarbon only, hexamers and S_4-symmetry tetramers were noted, with hexameric benzophenone-derived enolates preferring S_6-symmetric cores and camphor-derived analogues instead of opting for D_{3d}-symmetry. The addition of Lewis bases incurred dimer and then (at elevated levels of Lewis base) monomer formation. However, of particular interest—and in contrast to solid-state structural data—solution dimers appeared to prefer

Figure 6.2 Lithium enolate **9** and the more complex head-to-head aggregated hemisolvate $\mathbf{10}_2'$(TMEDA).

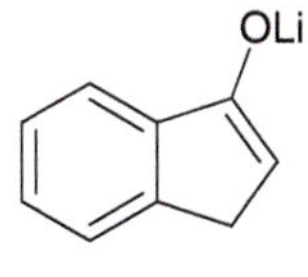

9

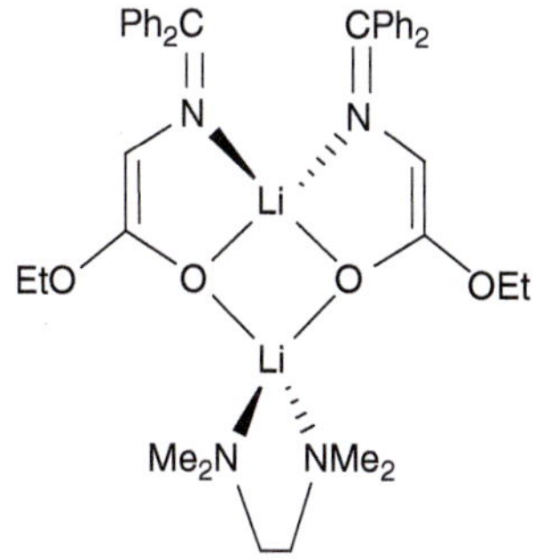

$\mathbf{10}_2'$(TMEDA)

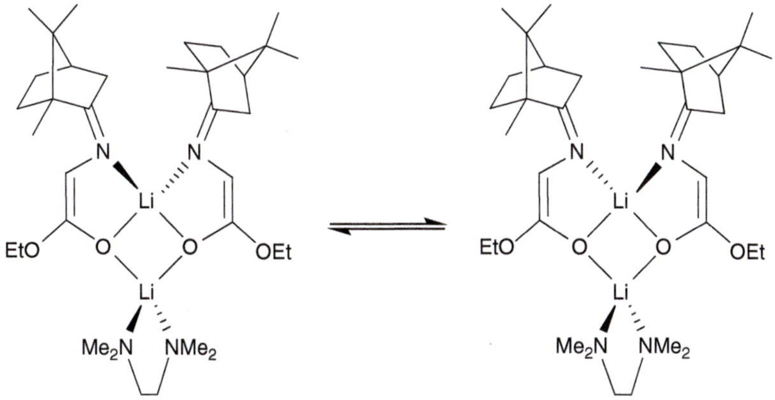

Scheme 6.1 Diasteromeric resolution in camphor-derived spirocycle $\mathbf{11}_2'$(TMEDA).

the formation of head-to-head spirocycles rather than head-to-tail dimers. This was clearly revealed by the ^{6}Li,^{15}N-labelled hemisolvate $\mathbf{10}_2'$(TMEDA) (Figure 6.2; throughout, prime denotes head-to-head aggregation), which showed two distinct ^{6}Li signals. The downfield resonance (δ 2.94 ppm) was a triplet ($^{1}J_{\text{LiN}}$ = 3.0 Hz), while the upfield peak (δ 0.72 ppm) was a singlet. Further spectral complexity arising from this spirocyclic arrangement was noted in camphor-derived systems. In this case, low TMEDA concentrations saw the stereogenic nature of the *N*,*O*-chelated metal being clearly resolvable (Scheme 6.1). Accordingly, ^{6}Li signals split into two peaks—most clearly seen for the upfield singlet, which converted to two singlets in a ca. 3:1 ratio in $\mathbf{11}_2'$(TMEDA). A similar diastereomeric resolution was possible when using (*R*,*R*)- or (*S*,*S*)-TMCDA in conjunction with benzophenone-derived enolates [66].

In addition to the study of glycinimine-derived enolates, O,O-coordinating Evans enolates [67, 68] have been probed. Priopionate enolate $\mathbf{12}^{\text{R,R}'}$(Figure 6.3) was prepared using ^{6}Li,^{15}N-labelled LDA in THF [69]. For $\mathbf{12}^{\text{Bn,Me}}$ low temperature ^{6}Li spectroscopy (−80 °C) gave a doublet ($^{1}J_{\text{LiN}}$ = 5.2 Hz) [70] that pointed to adduct $\mathbf{12}^{\text{Bn,Me}}$(LDA) (THF). The observation of just *one* doublet was attributed to fast exchange of the chelate ring. Once formed, the enolate proved capable of exhibiting both head-to-tail ($\mathbf{12}^{\text{Bn,Me}}{}_2$) and head-to-head ($\mathbf{12}^{\text{Bn,Me}}{}_2'$) dimerization in THF (Scheme 6.2), with THF concentration studies and DFT calculations suggesting trisolvation (see also below). However, depending on the choice of enolate (i.e. varying R or R′ in $\mathbf{12}^{\text{R,R}'}$), ^{6}Li NMR in THF at −80 °C showed either these dimers, a low-field signal attributable to a D_{2d}-tetramer, or else

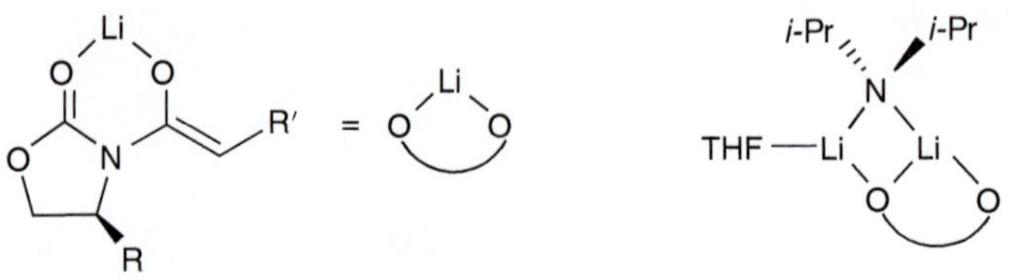

$\mathbf{12}^{Bn,Me}$(R = Bn, R′ = Me) $\mathbf{12}^{Bn,Me}$(LDA)(THF)

Figure 6.3 Recently studied Evans-type propionate enolates.

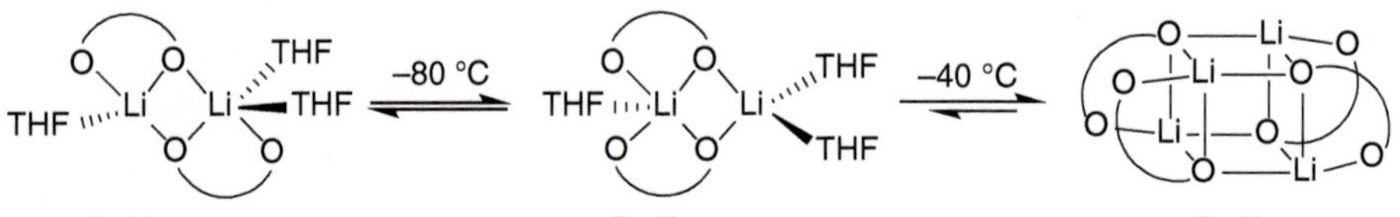

$\mathbf{12}^{Bn,Me}{}_2(THF)_3$ $\mathbf{12}^{Bn,Me}{}_2{}'(THF)_3$ $\mathbf{12}^{Bn,Me}{}_4$

Scheme 6.2 Evans enolate $\mathbf{12}^{Bn,Me}$ has demonstrated variable agglomeration behaviour in solution, favouring tetramer formation on raising temperature to −40 °C from −80 °C (shown here) or by adding toluene.

intractable oligomer signals. In the case of $\mathbf{12}^{Bn,Me}$, the tetramer increased its dominance (relative to the dimers just described) upon elevating the temperature to −40 °C or by lowering the level of THF in the system by adding toluene (Scheme 6.2). The inclusion of substitutionally labile THF in the kinetic dimers (active in aldol chemistry) and its exclusion from the tetramer (which proved correspondingly inactive) was demonstrated by titration with pyridine, which is known to leave unsolvated metal centres unchanged but to compete with THF at solvated metal centres, moving signals downfield [71, 72]. Finally, though the relative proportions of the aggregates could be varied using toluene as cosolvent, monomer formation was never seen [73].

A continuation of the Evans enolate work probed the amplification of enantioselectivity in aldol reactions offered by developing mixed enolate-alkoxide tetranuclear aggregates [74]. This focused on extending the prior art in creating mixed aggregates involving chiral salts [20]. Specifically, the combination of lithium amino alkoxides, hitherto established to form S_4-symmetric pseudocubic tetramers [75], with Evans enolates $\mathbf{12}^{R,Me}$ (R = Bn, *i*-Pr)[73] gave tetranuclear enolate-alkoxide agglomerates. These took two forms. The first was obtained from a 1:1 mixture of enolate and alkoxide in toluene. Initially forming a complex mixture, stirring at room temperature and then cooling to −80 °C resulted in only a 2:2 tetranuclear pseudocubane (illustrated representatively in Figure 6.4 for the use of enantiopure enolate (*S*)-$\mathbf{12}^{Bn,Me}$ and alkoxide **13**). With ^{15}N-labelled alkoxide, the downfield δ 1.39 ppm ^{6}Li signal for (*S*)-$\mathbf{12}^{Bn,Me}{}_2\mathbf{13}_2$ was a doublet ($^1J_{LiN}$ = 2.8 Hz), and this was matched by a 1:1:1 triplet in the ^{15}N spectrum; establishing chelation by all the alkoxide ligands. To verify tetranuclear aggregation, a 1:1 mix of **13** with $\mathbf{12}^{Bn,Me}$ [equal parts (*R*)- and (*S*)-] were combined, yielding a species identifiable as (*R*)-$\mathbf{12}^{Bn,Me}$(*S*)-$\mathbf{12}^{Bn,Me}\mathbf{13}_2$ on account of its four ^{6}Li signals in a 1:1:1:1 ratio. While preformed (*S*)-$\mathbf{12}^{Bn,Me}{}_2\mathbf{13}_2$ resisted the incorporation of THF, the formation of 3:1 mixed aggregates incorporating donor solvent [representatively illustrated in Figure 6.4 by (*S*)-$\mathbf{12}^{Bn,Me}\mathbf{13}_3$(THF)] proved possible by THF titration of a 3:1 mixture of the alkoxide and enolate, and was the only species observed if neat THF was used. The incorporation of just one enolate ligand in $\mathbf{12}^{Bn,Me}\mathbf{13}_3$(THF) was confirmed by using the same 3:1 substrate ratio but ensuring that the enolate was made up of equal amounts of (*R*)- and (*S*)-enolate; (*R*)-$\mathbf{12}^{Bn,Me}\mathbf{13}_3$(THF) and (*S*)-$\mathbf{12}^{Bn,Me}\mathbf{13}_3$(THF) were the only detectable products. Lastly, evidence that only two of the three alkoxide ligands were chelating came from the observation of two ^{6}Li doublets and one singlet matched by two ^{15}N 1:1:1 triplets and one singlet. As elsewhere in this section, pyridine titration verified the level of THF inclusion.

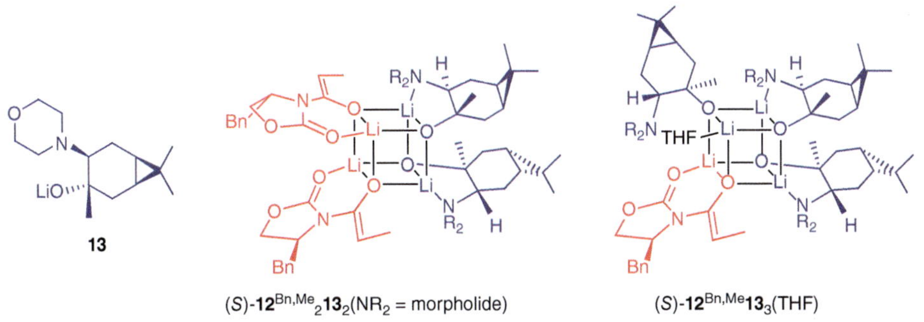

Figure 6.4 Illustrative solvent-dependent complexation of amino alkoxide **13** (blue) and Evans enolate (*S*)-$\mathbf{12}^{Bn,Me}$ (red).

Figure 6.5 Recently studied products of Weinreb amide enolization.

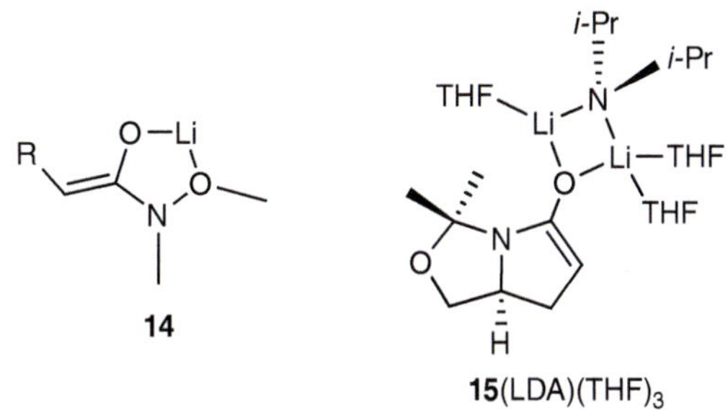

Having thoroughly looked at the structures of these mixed pseudocubanes, their potential was probed for amplification of the 12:1 syn/syn selectivity displayed by just the lithiated Evans enolates in aldol addition. In the event, reactions of mixed pseudocubanes proceeded smoothly at low temperatures even in hydrocarbon media, with the THF-incorporating 3:1 aggregates performing best. In particular, (*R*)- and (*S*)-$\mathbf{12}^{Bn,Me}\mathbf{13}_3$(THF) showed interesting stereochemical effects; though (*R*)-$\mathbf{12}^{Bn,Me}\mathbf{13}_3$(THF) eroded selectivity, its (*S*)-analogue gave yields of > 90% whilst amplifying selectivity, achieving up to 30:1 syn/syn, and avoiding even low levels of anti-product.

Moving from the six-membered chelation noted for Evans enolates, NOESY analysis of the product of enolization of Weinreb amides (O-methyl hydroxamic acids) [76] in toluene/THF disclosed the formation of only Z-isomer (generically, **14** in Figure 6.5), with binary mixtures revealing tetramers and dimers. In these systems, where chelation of the metal by a sterically undemanding methoxy group engenders five-membered ring formation, there were implications for the ease with which chelation occurred and competition between this and the incorporation of external Lewis base [77]. Of particular interest, pyridine titration established that both tetramers and dimers incorporated THF. The complexity of low-temperature spectroscopic data led to the view that external solvation of the higher aggregates was not at the expense of internal coordination and that tetramers must deviate from (expected) D_{2d} or S_4 symmetry. DFT analysis suggested THF inclusion without the loss of tetramer structure or ligand chelation to incur five-fold coordination of metal centres.

In similar work, enolates derived from potentially anti-inflammatory pyroglutaminols [78] (e.g. **15**) were studied. Initial formation from ^{6}Li,^{15}N-labelled LDA revealed a ^{6}Li doublet and a ^{15}N triplet ($^1J_{LiN}$ = 5 Hz) [70] pointing to dinuclear **15**(LDA)(THF)$_n$, with the large, concentration-dependent ^{6}Li chemical shift combining with DFT analysis to suggest n = 3 [Figure 6.5, viz.

$\mathbf{12}_2(THF)_3$ in Scheme 6.2 above]. The enolate itself was found to form tetrasolvated dimers and tetramers. The propensity of THF for inclusion in the latter could be monitored by titration of the enolate $\mathbf{16}_4(THF)_4$ in THF/hydrocarbon solution with pyridine shift reagent. Results at −95 °C suggested the gradual replacement of THF, the sequential process becoming less favourable with each substitution (Scheme 6.3).

Over a number of years, numerous authors have made enormous advances in the elucidation of the solution behaviour of N,X lithium amides—that is, amides that bear potentially chelating heteroatoms (X = O, S, P). For example, exchange of the single Lewis base molecule in the hemisolvated head-to-head dimer based on N,O lithium amide **17** [($\mathbf{17}_2'$(*Sol*)] (*Sol* = donor solvent) was followed through the observation of ^{13}C NMR shift variance in free and coordinating OEt_2 and THF [79–82]. Data obtained by Hilmersson revealed the robust nature of lithium-oxygen coordination and essentially equivalent enthalpic contributions to ether exchange for either solvent, but that only OEt_2-solvation induced a significant loss of entropy; activation parameters were $\Delta H^{\ddagger} = 11.0$ kcal/mol and $\Delta S^{\ddagger} = 12.0$ cal/K/mol (for OEt_2) and $\Delta H^{\ddagger} = 11.2$ kcal/mol and $\Delta S^{\ddagger} = 1.6$ cal/K/mol (for THF). The independence of exchange rate and concentration, and the observation of positive entropy of activation pointed to dissociative solvent exchange [83]. More recent studies recording ^{6}Li–^{31}P coupling have shown head-to-head N,P dimers comparable to those in $\mathbf{17}_2'$(*Sol*) using not pendant alkoxy groups but rather pendant phosphines [84]. Refinement of this work led to the identification of distinct steric effects associated with the choice of *N*-substituent governing the preference for either a head-to-head $\mathbf{17}_2'$-type dimer with its two distinct ^{6}Li signals appearing as a singlet and a triplet, or a head-to-tail dimer characterized by a single ^{6}Li doublet [85]. Further elucidation of structural preferences related to the retention of metal-phosphorus bonding in the presence of THF was reported; results that aligned closely with the behaviour of N,O systems [86]. This contrasted with similar synthetically useful N,S amides [87, 88], where ^{1}H,^{6}Li–HOESY showed that

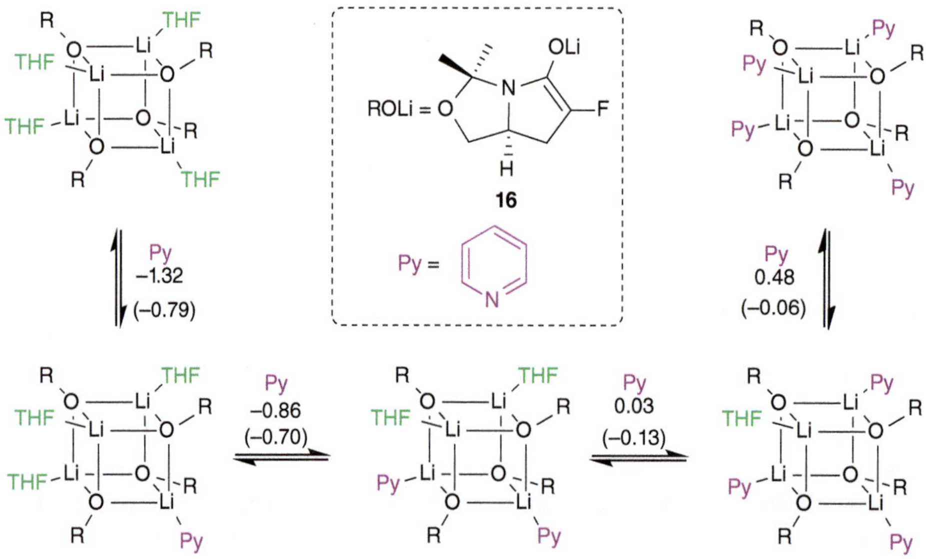

Scheme 6.3 Solvent replacement in $\mathbf{16}_4(THF)_4$ through $\mathbf{16}_4(Py)_4$ at −95 °C with free energy changes for each step extracted directly from the Job plot of relative integration vs. mol fraction Py (0.10 M **16**, 0.50 M total solvent concentration). Values in parentheses were statistically corrected for available substitution sites.

S-coordination of the metal in the salt of (*S*)-*N*-isopropyl-2-amino-2-phenyl-1-thiophenylethane was less favoured in the presence of external Lewis base, enabling head-to-tail $\mathbf{18}_2$(THF) to be produced by the titration of head-to-head $\mathbf{18}_2'(OEt_2)$ with THF [89].

The formation of adducts between N,X lithium amides and organolithium reagents was also probed by Hilmersson. A combination of ^{1}H,^{6}Li–HOESY and ^{6}Li,^{6}Li–EXSY was used to establish the behaviour of the chiral ^{6}Li-amide of (*S*)-*N*-isopropyl-*O*-methyl valinol. Its 2:1 adduct with similarly labelled *n*-BuLi, $\mathbf{19}^{Me}{}_2$(*n*-BuLi) (Figure 6.6), was found by ^{1}H,^{6}Li–HOESY to involve a long displacement between a single dicoordinated Li centre and the *n*-Bu anion in hydrocarbon media. That is, under these conditions its structure was essentially the same as that seen in the solid-state [90]. Dynamics were manifest, however, with slow metal exchange between the three positions in the adduct recorded at −33 °C, for which quantitative ^{6}Li,^{6}Li–EXSY established that $\Delta G^{\ddagger}$ = 14.7 kcal/mol. Meanwhile, upon addition of Et_2O at −90 °C clear changes were observed in both the ^{1}H and ^{13}C NMR spectra. Significantly, two α–C multiplets were seen for butyl anions. A septet at δ 10.9 ppm ($^{1}J_{LiC}$ = 5.4 Hz) suggested $(n\text{-BuLi})_4$, while a distorted quintet at δ 11.2 ppm ($^{1}J_{LiC}$ = 8.5 Hz) was attributed to there being two distinct ^{6}Li environments in a dinuclear adduct $\mathbf{19}^{Me}$(*n*-BuLi). ^{6}Li NMR spectroscopy on $\mathbf{19}^{Me}$ and *n*-BuLi in d_{10}-Et_2O revealed five signals: two for the trinuclear 2:1 adduct, two for the dinuclear adduct, and one for butyllithium. Overall, a mixture of $\mathbf{19}^{Me}{}_2$, $\mathbf{19}^{Me}$(*n*-BuLi) and $(n\text{-BuLi})_4$ was argued to form [91]. Later work linked ^{1}J coupling constant to metal coordination in lithium amide dimers and their mixed complexes with *n*-BuLi. This work involved the synthesis of a range of ^{15}N-labelled amines from correspondingly enriched (*S*)-phenylalanine and their metalation with a ^{6}Li-labelled base. Monitoring of the lithium amides upon treatment with further *n*-BuLi revealed the formation of mixed adducts through the appearance of clear ^{6}Li–^{13}C coupling. Whilst in d_8-toluene 2:1 adduct $\mathbf{20}_2$(*n*-BuLi) presented (X = O, S), in Et_2O or THF 1:1 species (**21**/**22**)(*n*-BuLi)$(Sol)_2$ resulted (Figure 6.7), as evidenced by ^{6}Li–^{15}N splitting patterns. ^{1}J coupling correlated plainly with metal coordination; dicoordinated (6.4–6.6 Hz), tricoordinated (4.4–5.2 Hz), and tetracoordinated lithium (3.9–4.3 Hz) were discernable [92].

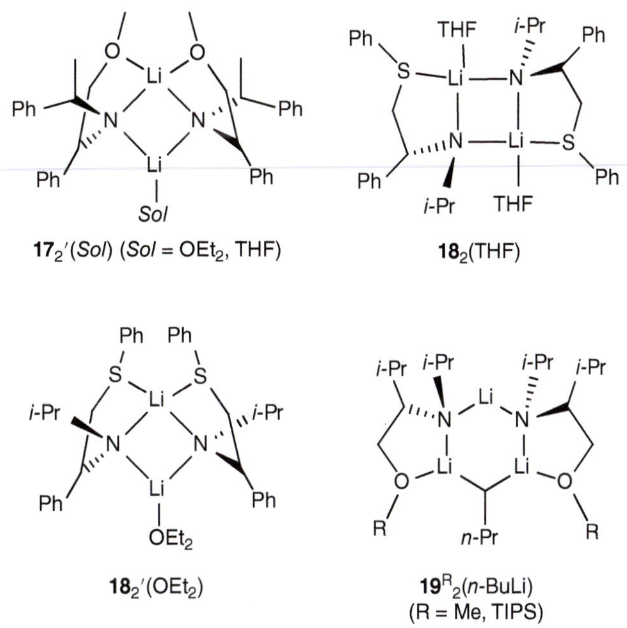

Figure 6.6 Selected N,X-lithium amides and the complex between the amide of (*S*)-*N*-isopropyl-*O*-methyl valinol and *n*-BuLi.

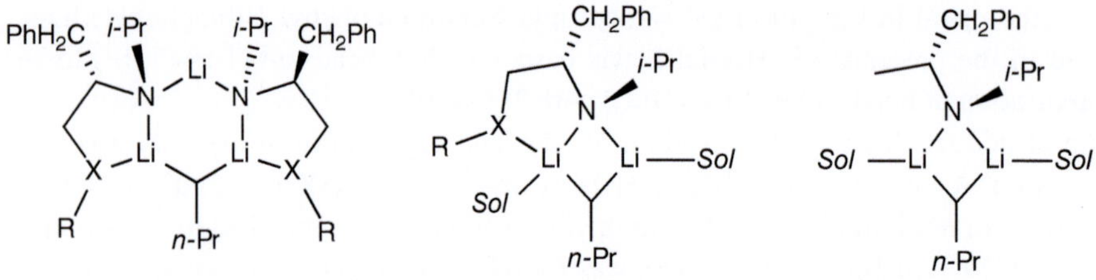

$\mathbf{20}_2$(n-BuLi) **21**(n-BuLi)(Sol)$_2$ **22**(n-BuLi)(Sol)$_2$

Figure 6.7 Selected complexes between N,X- and related lithium amides and n-BuLi.

6.2.3 DOSY on s-block Organometallics

6.2.3.1 Development and Early Applications

As recorded above, the advent of PGSE methods and their use to underpin the 2D technique DOSY that enables the identification of species with different diffusion rates by size [41, 42], have led to profound advances in our understanding of classic s-block metal reagents. The first major steps in using DOSY to interrogate monometallics of this class that find common use as reagents came courtesy of Williard through work applying ICC or IR DOSY to organolithiums. The need for internal calibrating agents, that is the addition of at least three model compounds to the sample in order to generate calibration curves, derived from there being no simple relationship between diffusion coefficient (D) and M_r. The relationship that does exist is expressed by the Stokes–Einstein equation (Eq. 6.1), where k is the Boltzmann constant, T the temperature, η the solvent viscosity and r_H the hydrodynamic radius:

$$D = \frac{kT}{6\pi\eta r_H} \tag{6.1}$$

Eq. 6.1 can be modified using the power law $r_H = cM_r^{\alpha}$ (c being a geometrical factor and α a measure of molecular compactness) [93] to relate D and M_r, with the resulting expression linearized according to Eq. 6.2:

$$\log D = \log K - (\alpha \log M_r) \tag{6.2}$$

Based on this approach, it follows that calibration curves could be prepared using molecules with different M_r but with comparable structures in a given solvent. Moreover, linearization and calibration obviated any need to determine η, c, or α.

Alluded to in Section 6.1 are overviews that Williard published in 2009 [36, 37] summarizing the work conducted since 2000 in which internal calibration agents were used to determine M_r for a range of organolithium systems [35]. This and more recent work conducted in the same vein using isotopically enriched organolithiums benefitted from the ability of ^{6}Li with its reduced quadrupole relative to that of ^{7}Li to deliver dipolar behaviour and incur sharper peaks. Work conducted on LHMDS using a range of internal reference compounds established a mass in solution within 2% of the unsolvated ^{6}Li dimer in d_8-toluene [94]. Meanwhile, the same study probed more complicated 2:1 adduct $\mathbf{19}^{TIPS}{}_2(^n$BuLi), where the lithium amide now incorporated the sterically demanding triisopropylsilyl (TIPS) group at the oxygen (see Figure 6.6 for structure representation). This extended prior art, discussed above, where ^{1}H,^{6}Li-HOESY and ^{6}Li,^{6}Li-EXSY were used to establish adduct retention in hydrocarbon media with a mixture of the less sterically congested system

Figure 6.8 Trinuclear $\mathbf{19}^{TIPS}{}_2$[n-Bu(t-Bu)CHOLi].

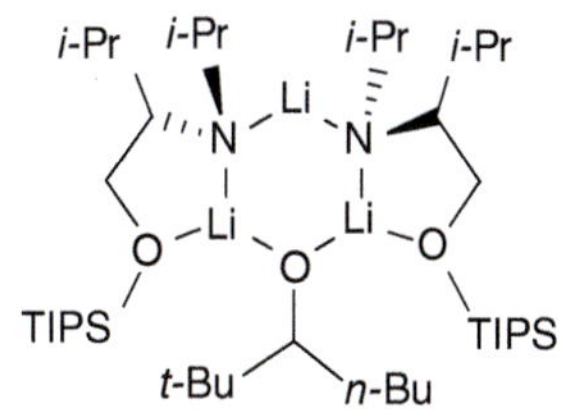

$\mathbf{19}^{TIPS}{}_2$[n-Bu(t-Bu)CHOLi]

$\mathbf{19}^{Me}{}_2$, $\mathbf{19}^{Me}$(n-BuLi), and $(n\text{-BuLi})_4$ forming upon addition of Et_2O [91]. Analysis of $\mathbf{19}^{TIPS}{}_2$(n-BuLi) by Williard revealed a single diffusion rate that pointed to structure retention—a suggestion corroborated by ^{1}H,^{6}Li–HOESY. In a similar vein, a combination of ^{1}H and ^{13}C INEPT DOSY, ^{6}Li and ^{15}N NMR, and other 2D NMR confirmed retention of this aggregate in solution, arguing this to be responsible for the asymmetric addition of *n*-BuLi to carbonyls [95]. Thus, similar diffusion experiments using $\mathbf{19}^{TIPS}{}_2$[n-Bu(t-Bu)CHOLi] (Figure 6.8) pointed to the integrity of this alkoxide complex in solution, and so to competition between reagent (organolithium) and product (lithium alkoxide) for chiral amide-complexation. These data allowed the interpretation of the enantioselectivity of asymmetric addition of *n*-BuLi to aldehydes, noting the potential for product-induced chirality inhibition [96].

Utility amide lithium 2,2,6,6-tetramethylpiperidide (LTMP) **23**, easily made and one of the most important polar organometallic reagents in synthesis (including in that of **1**), has been crystallized and structurally authenticated in two different oligomeric forms; cyclic tetrameric $\mathbf{23}_4$ [97] and trimeric $\mathbf{23}_3$ [98] (Scheme 6.4). Using isotopically labelled compounds, and a combination of ^{6}Li,^{15}N, and ^{6}Li/^{15}N HMQC NMR experiments in pentane, Collum detected a trimer:tetramer ratio of approximately 1:4 at −40 °C and a decoalescence of the tetramer resonance into several overlapping resonances at −120 °C. These data were indicative of several tetrameric conformers [99]. Mulvey has also shown that $\mathbf{23}_3$ and $\mathbf{23}_4$ were easy to distinguish from their routine ^{1}H and ^{13}C NMR spectra

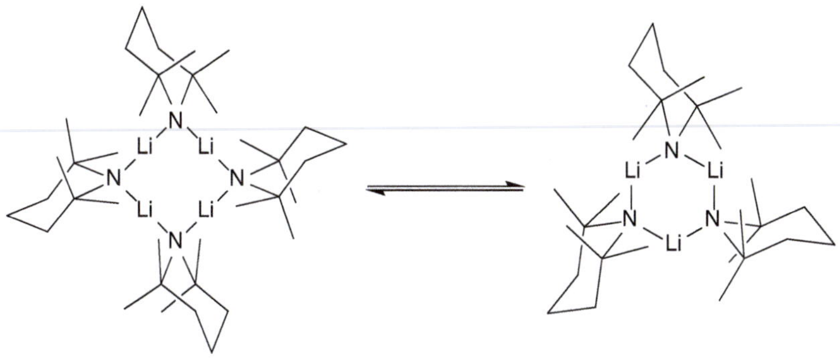

$\mathbf{23}_4$
($M_{r,calc}$ = 558 g/mol)
$M_{r,det}$(d_6-benzene) = 420 (−40%) g/mol
$M_{r,det}$(d_{12}-CyH) = 554 (−6%) g/mol

$\mathbf{23}_3$
($M_{r,calc}$ = 441 g/mol)
$M_{r,det}$ (d_6-benzene) = 348 (−27%) g/mol
$M_{r,det}$ (d_{12}-CyH) = 382 (−15%) g/mol

Scheme 6.4 LTMP oligomers $\mathbf{23}_4$ (left) and $\mathbf{23}_3$ (right) present in d_6-benzene and d_{12}-cyclohexane (d_{12}-CyH) solution and their estimated M_r values as obtained using ICC DOSY NMR.

recorded in d_6-benzene solution by virtue of the α–Me resonances enabling the differentiation of chemically distinct TMP groups [98]. Estimated M_r values for **23**$_3$ and **23**$_4$ were obtained using ^{1}H ICC DOSY NMR studies that employed tetramethylsilane (TMS), 1-phenylnaphthalene (PhN), and tetraphenylnaphthalene (TPhN; for TMS, PhN, and TPhN M_r = 88, 204, and 433 g/mol, respectively) as internal standards in d_6-benzene and in d_{12}-cyclohexane solutions. Obtained data were consistent with the expected relative size order, with those of cyclotrimer **23**$_3$ being smaller than those of cyclotetramer **23**$_4$. However, this study also reflected the limitation of this NMR method (see below). In d_6-benzene, ICC DOSY gave a determined M_r ($M_{r,det}$) of 348 g/mol for **23**$_3$ and 420 g/mol for **23**$_4$, equating to errors of −27 and −40%, respectively [cf. calculated M_r ($M_{r,calc}$) values of 441 g/mol for **23**$_3$ and 588 g/mol for **23**$_4$]. A better fit between the observed and calculated M_r values was found using d_{12}-cyclohexane as solvent (382 g/mol represents a −15% error for **23**$_3$, while 554 g/mol is a −6% error for **23**$_4$) [98]. It should also be noted that because the cyclic oligomeric structures of LTMP observed in the solid-state appeared to be retained in solution, some of the $M_{r,det}$ values fell outside the range of M_r on the calibration curve created by using the internal standards. This is a significant limitation of ICC DOSY and will be returned to later.

Moving from lithium amides to lithium phenolates and mixed amide/phenolate systems, a range of monometallic complexes of lithium or potassium based on 2-XCH$_2$-4,6-(t-Bu)$_2$-C$_6$H$_2$O$^-$, were synthesized and structurally characterized. Coordination motif and aggregation state were found to be variable (mono- or dimeric) in the solid-state. However, PGSE NMR revealed diffusion coefficients (7.62–23.11 × 10^{-10} m^2/s) and hydrodynamic radii consistent with monomeric complexes for all of **24**$^{A-D}$ (Figure 6.9) in solution (d_2-dichloromethane or d_8-THF). Meanwhile, DOSY NMR revealed a single diffusion coefficient (7.57 × 10^{-10} m^2/s), suggesting a heteroleptic dinuclear formulation **24**^{C}([HMe$_2$Si]$_2$NLi) in d_6-benzene. While this indicated solid-state structure retention in solution, variable temperature ^{7}Li NMR spectroscopy established that exchange of the metal centres did not occur. In the presence of BnOH (up to 10 eq.), **24**^{C} demonstrated the highest affinity of **24**$^{A-D}$ for controlled living and immortal ring-opening polymerisation (ROP) of L-lactide, with the authors crediting the steric hindrance and electron-richness of the ligand. The mechanistic investigation suggested that ROP reactions catalyzed by two-component **24**$^{A-D}$/BnOH catalyst systems exhibited an activated monomer mechanism rather than a coordination-insertion mechanism [100].

In the last decade, Stalke has published ICC DOSY studies on simple organolithiums, for example, verifying the partial desolvation of silylalkyllithium complexes in hydrocarbon media [101]. More complicated was the elucidation of the unusual cocrystalline [(2-Me$_2$N-C$_6$H$_4$Li)$_4$]$_4$(t-BuLi)$_4$ (**25**$_4$)$_4$**26**$_4$; that is, a crystalline lattice incorporating discrete tetramers of the aryllithium and the alkyllithium in a 4:1 ratio [102]. Submission of this material to ^{7}Li–DOSY and ^{1}H,^{7}Li–HOESY revealed that exchange retained both tetramers of 2-Me$_2$N-C$_6$H$_4$Li **25** and t-BuLi **26**, whilst also generating the three possible mixed tetramers (2-Me$_2$N-C$_6$H$_4$Li)$_{4-n}$(t-BuLi)$_n$ [n = 1 **25**$_3$**26**, 2 **25**$_2$**26**$_2$, 3 **25**(**26**)$_3$] in d_8-toluene. Meanwhile, ^{7}Li–EXSY could be used to establish rate constants

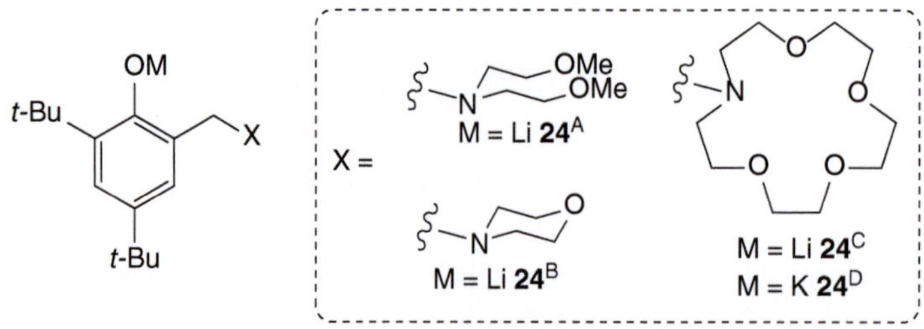

Figure 6.9 **24**$^{A-D}$ have been found by a combination of PGSE and DOSY techniques to favour monomer existence in polar solvents and, for **24**^{C}, a heteroleptic dinuclear complex with (HMe$_2$Si)$_2$NLi in hydrocarbon.

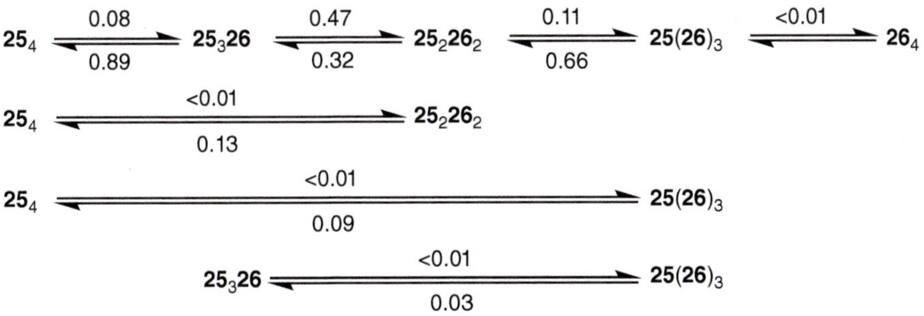

Scheme 6.5 Exchange rate constants (s^{-1}) between (2-Me_2N-$C_6H_4Li)_4$ **25_4**, (*t*-BuLi$)_4$ **26_4**, and (2-Me_2N-$C_6H_4Li)_{4-n}$(*t*-BuLi$)_n$ [n = 1-3 **25_326-$25_3(26)_3$**] in a saturated d_8-toluene solution obtained from cocrystalline (**25_4**)$_4$**26_4**.

underpinning this complex set of equilibria and to demonstrate that **25_4**, by far the dominant species in solution, could detectably exchange with mixed tetramers **25_326–$25(26)_3$** whereas **26_4** could not (Scheme 6.5).

Remaining with purely Li-containing systems, variable temperature spectroscopy of ^{13}C- and ^{6}Li-labelled methyllithium and lithium chloride has been used to elucidate the nature of a MeLi-LiCl mixture in d_8-THF. The appearance of a quintet by ^{13}C NMR spectroscopy, combined with Bauer-Winchester-Schleyer analysis of J values [50, 51], suggested a dinuclear MeLi(LiCl) aggregate **27** [103]. This, along with the simplicity of the spectral changes, contrasted with results known from comparable bromide or iodide systems [104]. Subsequent ICC DOSY analysis gave an M_r of 277, pointing to **27**(THF)$_{2.7}$. Overall, an equilibrium (for which K = ca. 0.65) between (MeLi)$_4$ **28_4**, (LiCl)$_2$ **29_2** and a *tris*-THF solvate of dinuclear adduct **27** at −100 °C was concluded.

6.2.3.2 Recent Refinements to Diffusion Techniques

The issue that there exists no simple relationship between D and M_r was noted in the previous section. So too was the fact that the use of an empirically derived power law achieved some success in overcoming this [93]. However, this solution was not applicable to diverse ranges of analyte, and attempts to deploy it widely led to estimations of M_r by DOSY with errors of up to ±30%. The approach adopted by Williard to address the issue involved internal calibrating agents and led to significant ICC DOSY studies of a range of organolithiums. Yet, the use of only a few references, often based on a limited M_r distribution, represented a limitation of this method. More fundamentally, the possibility of interference of the calibrating agents with the metal complexes being studied has proved an issue. Further, problems of peak overlap have meant that inappropriate internal standards (e.g. with geometries that deviate significantly from those of the analyte) have sometimes had to be used. All this being so, recent refinement of the DOSY technique has come from the development of the ECC method [105]. This has proved a powerful advance, yielding normalized diffusion coefficients whilst avoiding the introduction of internal referencing agents and substantially reducing errors in the analysis of small molecules with different geometries, independent of the effects of variables such as temperature, convection, viscosity, etc. on the sample. Fundamentally, it has overcome the reliance on M_r determinations underpinned by an ICC generated in the same experiment. To do this, a range of calibration curves need to be generated using model compounds independently. Linearly plotting log D vs. log M_r nevertheless revealed inaccuracies at very low or high M_r determinations. This led Stalke to evaluate D for each model compound whilst also factoring in its shape and compactness. The result was a perception that model compounds could be split into three classes: (i) compact spheres, (ii) dissipated spheres and ellipsoids (DSEs), and (iii) expanded discs [38]. Overall, by accounting for the shapes of the molecules it has proved possible to use just a single reference (which can even be the solvent) to reduce errors to less than ±10%.

Moving to the practical ramifications of the shift to ECC DOSY NMR, a diverse array of s-block systems have successfully been looked at. Studies by Stalke on the effects of simple additives such as lithium chloride (**29**) on organometallic bases and their modus operandi will be considered in detail in the final section of this chapter. Meanwhile, the same author looked at alkali metal cyclopentadienides in d_8-THF in order to establish the reliability of the DSE calibration curve, specifically in interrogating s-block organometallic compounds. Results of a DSE–ECC analysis demonstrated that while solvated monomers CpM(THF)$_n$ (M = Li **30**, Rb **31**, n = 2; M = Na **32**, K **33**, n = 3) formed mostly, caesium gave cyclic aggregates (CpM)$_n$(THF)$_{2n}$ (n = 5, 6) **34** [106]. The same diffusion approach also helped to elucidate the predominance of trimer and tetramer structures for unsolvated LDA over a temperature range of −75 to 100 °C [107]. Recently, the solution structure and stability of a bulky organometallic potassium complex was studied. The chemistry was based on the bis(4,6-(t-Bu)$_2$-benzoxazol-2-yl)methane ligand, which was used to generate the highly unusual crystalline material [{4,6-(t-Bu)$_2$-C_6H_2OCN}$_2$CH]K(18-crown-6)(OH_2) **35** (Scheme 6.6) [108]. The inclusion of water (with a site occupation factor of ca. 35% and contributing to a hydrogen-bonded network) was, of course, highly unusual—particularly given the presence of an organic residue bearing an acidic CH group [109]. Though rare, structures such as these have initiated interest in the reactivity of water in such systems [110] and this has evolved into an interest in deep eutectic solvents as versatile 'green' media for sustainable organometallic chemistry. This in turn has led to major new avenues in s-block chemistry using biorenewable protic media [111–113].

A ^{1}H ECC DOSY analysis, using the DSE calibration curve technique, was undertaken to establish how **35** behaved in d_8-THF solution. Evidence for a CIP (contact-ion pair) structure in solution immediately came from the observation of a single diffusion coefficient for all **35**-related components. Thereafter, M_r values obtained from the DSE and the merge calibration curves (that is, ECCs obtained with either just DSE calibrating agents or all calibrating agents employed, respectively) were 870 g/mol (DSE) and 967 g/mol (merge), respectively. These values both eliminated the retention of polymeric structures. Moreover, the $M_{r,DSC}$ deviated by just 2 and 6%, respectively, from expectation if monomeric **35** (less H_2O) were to be solvated by either one or two THF molecules. Further spectroscopic investigation of this system was undertaken by water titration, revealing astonishing resistance to hydrolysis. Aromatic signals attributable to the free parent ligand emerged only slowly at downfield shifts relative to those of the metalated ligand, the latter moving gradually downfield upon addition of water, yet remaining clearly separated from signals from the protonated form. Full hydrolytic conversion of the bis(benzoxazolyl)methane ligand was only achieved after introducing >114 eq. water over 7 days.

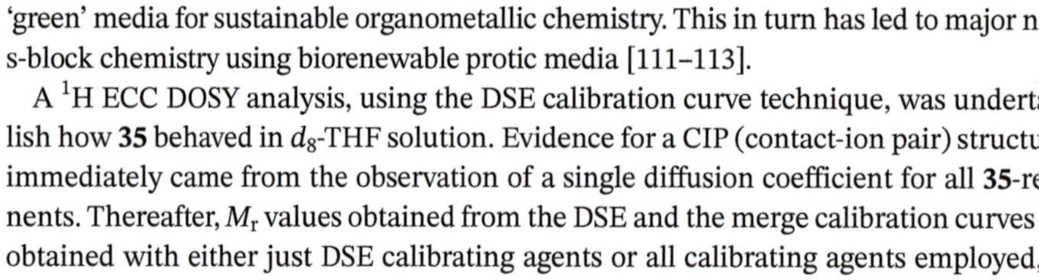

35

Scheme 6.6 **35** (Less water) was found to be a THF-solvated CIP in that solvent, with analysis of aromatic chemical shifts pointing to sustained stability on adding a large excess of water.

This demonstrable power of ECC DOSY in eradicating errors has led to its uptake by other groups. For example, it has been applied by Hevia to estimate the aggregation of lithium arylamides Ph_2NLi **36**, PhN(H)Li **37**, and PhN(Me)Li **38** in THF and to correlate their structures with those of these amides in biomass-derived 2-methyl-THF (2-MeTHF) [114]. They have shown remarkable solubility in this ethereal solvent, and in the case of **38**, it was demonstrated that effective reaction with a wide range of esters, R′C(=O)OEt, could furnish various synthetically relevant carboxamides, R′C(=O)N(Me)Ph. Reactions in 2-MeTHF took place without the need for external additives in just 20 seconds (Scheme 6.7i) [114]. This approach proved also to be compatible with activated organic amides such as $PhC(=O)NBoc_2$ (Boc = *tert*-butoxycarbonyl), which contain a good leaving group and are capable of undergoing transamidation with some of the lithium anilides aforementioned (Scheme 6.7i). Remarkably, these reactions could be successfully carried out under air, at room temperature, and in the presence of moisture; a trio of conditions typically disallowed in polar organometallic chemistry [115]. The choice of 2-MeTHF as reaction medium appeared significant. In particular, this solvent looked to favour the formation of small, kinetically activated aggregates capable of reacting rapidly with the organic substrate before decomposition reactions in air or moisture could compete.

To shed light on the potential mechanism of these transformations and on whether they occurred via precoordination of the lithium amide to the organic substrate, the constitution of equimolar mixtures of **36** with *N*,*N*-dimethylbenzamide **39** were investigated by ^{1}H DOSY NMR in d_8-toluene and d_8-THF. Since the reactions being probed were found to be very fast, **39**

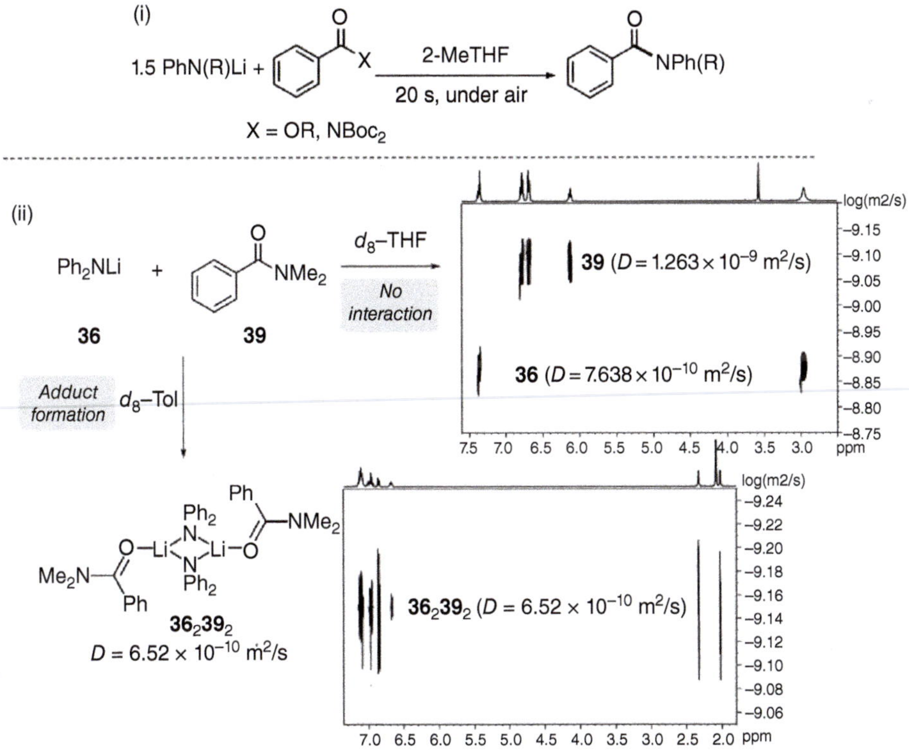

Scheme 6.7 (i) Lithium amide-mediated amidation and transamidation processes in 2-MeTHF and under air; (ii) ^{1}H ECC DOSY NMR spectra of equimolar mixtures of Ph_2NLi **36** and model amide $PhC(=O)NMe_2$ **39** in d_8-THF and d_8-toluene.

was chosen as a substrate because its lack of a good leaving group was felt likely to maximize opportunities to spectroscopically detect any possible intermediate(s). Results showed that in the nondonor solvent toluene, a coordination adduct between the lithium amide and the carboxamide substrate arose. This was evidenced by codiffusion of their respective resonances ($D = 6.52 \times 10^{-10}$ m^2/s), suggesting them to be part of the same molecular entity in solution (Scheme 6.7ii). This observation was emphasized by the isolation and X-ray crystallographic analysis of a dimer of complex **36(39)**; intermediates of this type having been previously proposed on the basis of DFT studies [115]. Conversely, in d_8-THF, compounds **36** and **39** were found to diffuse independently of each other, demonstrating distinct diffusion coefficients of 7.638×10^{-10} and 1.263×10^{-9} m^2/s, respectively. ECC analysis against TMS allowed for the determination of molecular weights, with results suggesting uncomplexed benzamide **39** (M_r = 153 g/mol) and monomeric lithiate **36** (solvated by three molecules of THF, M_r = 363 g/mol) to coexist in donor solvent (Scheme 6.7ii).

As the reactions under investigation were conducted in 2-MeTHF, which bears similar donor properties to THF, it was proposed that a coordination adduct of the type observed in toluene (**36**$_2$**39**$_2$) would be unlikely to form in the Lewis basic environment. This has led to the suggestion that such an adduct might be inert in amidation chemistry. However, an alternative explanation has been posited that the formation of small, monomeric (presumed highly active) lithium amide species (such as **36**[d_8-THF]$_3$, as suggested by DOSY) could facilitate the rapid reactivity observed under air in 2-MeTHF.

To further advance the understanding of the coordinating ability of 2-MeTHF, correlations between the solid-state structures of the 2-MeTHF solvates of **36–38** established by X-ray crystallographic studies and their constitution in 2-MeTHF solutions were drawn using ^{1}H DOSY NMR experiments. In the solid-state, the three lithium amides all adopted dimeric structures with three or four molecules of solvating 2-MeTHF to satisfy the coordination sphere of the lithium centres (Figure 6.10) [114, 116].

Given the solid-state data related to Figure 6.8, ^{1}H ECC DOSY was used to estimate the aggregation and to examine the solvation effects when d_8-THF was applied to **36–38** as a solvent. In contrast to the crystallographic studies, complete deaggregation to monomeric units AMLi(*Sol*)$_3$ (AM = amide; *Sol* = d_8-THF) was found to be the most likely solution scenario for each system [$M_{r,det}$ for **36** = 393 g/mol (error < 1%), for **37** = 331 g/mol (−5%), and for **38** = 335 g/mol (−2%)]. A tandem study was also conducted that used nondeuterated 2-MeTHF to examine the solution behaviours of the same lithium amides. As 2-MeTHF is not a common solvent for the ECC method, the ICC DOSY technique was now employed with TMS, PhN, and TPhN reference compounds. Interestingly, it was seen that, as with the case of non-substituted THF, lithium amides **36–38** all appeared to become monomeric in 2-MeTHF solution, each hosting three molecules of solvent [AMLi(*Sol*)$_3$] despite 2-MeTHF being a slightly more sterically demanding Lewis base ($M_{r,det}$ for **36** = 451 g/mol [−4%], for **37** = 393 g/mol [−9%], for **38** = 363 g/mol [+2%]).

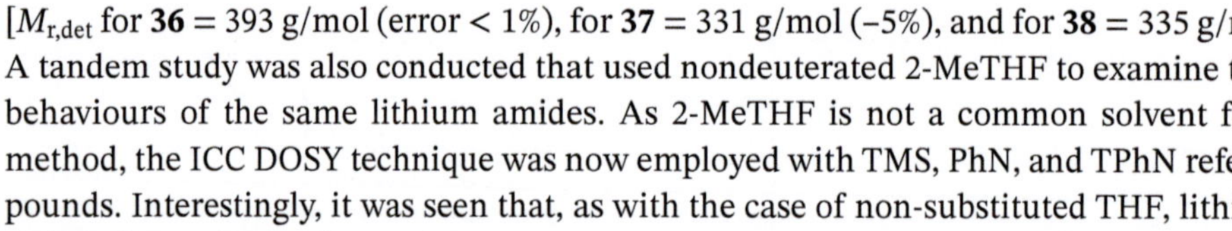

36$_2$(*Sol*)$_3$ **37**$_2$(*Sol*)$_4$ **38**$_2$(*Sol*)$_4$

Figure 6.10 Dimeric structures for 2-MeTHF complexes of **36–38** as established by X-ray crystallography.

With consistent evidence for monomeric complexes of **36–38** in both d_8-THF and 2-MeTHF, it was proposed that the formation of these small solvates could be key to the near-instant amidation reaction observed under the unusual (for organometallic chemistry) conditions that can be employed (under air, room temperature, stoichiometric conditions, without the aid of transition metal catalysis, or other additives) [114]. Monomer formation would logically enhance the nucleophilic power of the lithium amides under consideration, nucelophilicity that can be viewed as being structurally 'locked in' in solvents like toluene. The view is that this enhanced reactivity is then available to promote addition across the unsaturated C=O bond of the carboxylic amide substrate before the lithium amide succumbs to the expected limitations of degradation or hydrolysis.

6.3 Heteropolymetallic Systems

6.3.1 Introduction

The solid-state analysis of a range of heterobimetallic or ate complexes is presented in the earlier chapters of this book. These data have allowed the development of an excellent understanding of associative behaviour (either aggregation or adduct formation). Studies of the extent to which these are retained in solution have benefitted from advances made in multidimensional NMR spectroscopy in recent years, and this section discusses some model cases in which this has been so. The solution dichotomy between individually heterobimetallic complexes and cooperative monometallic mixtures, as well as (de)aggregation and conformational flexibility have been the subject of study.

6.3.2 s/s-block Systems

6.3.2.1 Alkali Metal/Magnesium

The analysis of lithium phenolate **24**^{C} in the ROP of L-lactide is reported in Section 2.3.1 [100]. However, synergistic effects have also been probed in the ROP of L- and *rac*-lactide with BnOH coinitiator by comparing the activity of PhN=N($C_{10}H_6$)OLi **40**, [PhN=N($C_{10}H_6$)O]$_2$Mg **41**, and [PhN=N($C_{10}H_6$)O]$_6$$Mg_2Li_2$ **42**. Whilst **40** and **41** proved difficult to analyse, they did add Lewis bases to give, for example, crystalline [**40**(THF)]$_2$ and **41**(THF)$_2$. Meanwhile, **42** was based on two truncated cubes sharing a Mg_2O_2 ring as one of the cube sides (Scheme 6.8). ^{1}H DOSY against three internal references established the retention of this last motif in hydrocarbon media; $D = 3.80 \times 10^{-10}$ m^2/s gave an M_r value with just a 1% error with respect to crystallography. In the same way, as was seen in reference 100, ^{1}H NMR spectroscopy established that ROP by **42** occurred via an activated monomer mechanism.

Scheme 6.8 [PhN=N($C_{10}H_6$)O]$_6$$Mg_2Li_2$ **42**, based on two truncated cubes sharing a Mg_2O_2 ring, retains its structure in hydrocarbon solution.

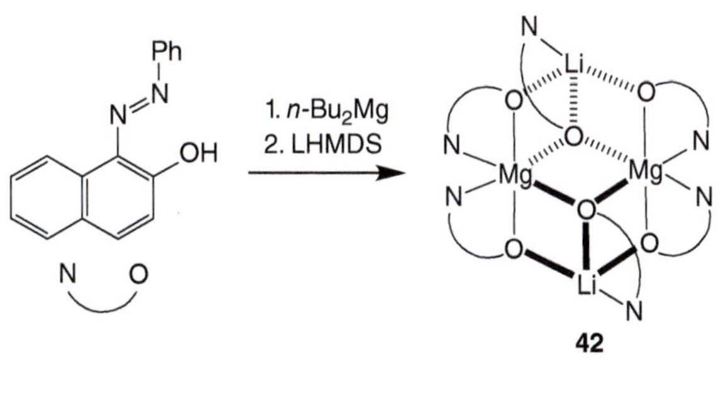

42

Moving down group 1, studies have been conducted on the solution constitution of heteroleptic sodium and potassium magnesiates such as n-Bu(TMP)$_2$MgNa **43**, which contain alkyl and amide ligands. Data indicated that in the absence of a Lewis base such bimetallic systems exhibited highly oligomeric cyclic structures capable of templating the mono-, di-, and even tetramagnesiation of a range of aromatics. Over a number of years, the deprotonation of the simple aromatics benzene and toluene [117, 118], as well as more complex ferrocene [119], anisole [120], polyaryls [121], and naphthalene [122] have been reported using both Mg/Na and Mg/K combinations, with reaction offering unique regioselectivities. In the example shown representatively in Scheme 6.9, structural studies on metalated intermediates of the type (arene-2H)(TMP)$_6$Mg$_2$Na$_4$ **44** have uncovered novel supramolecular arrangements, comprising a multiatom cationic skeleton of alternating metal and nitrogen atoms, centred around a doubly deprotonated arene core (arene-2H).

Alkali metal magnesiates such as **43** have been labelled as 'preinverse crowns' [122]. Combining equimolar amounts of NaTMP **45** and n-BuMgTMP **46** in methylcyclohexane at room temperature allowed for isolation of this in a spectroscopically pure form, with the MCH$_2$ environment giving a single resonance at δ −0.79 ppm by ^{1}H NMR spectroscopy. Although a good starting point, it was acknowledged that this compound contained an incorrect Na:Mg ratio insofar as the final inverse crown products contain a 2:1 Na:Mg ratio (Scheme 6.9). That notwithstanding, **43** was combined with an equimolar amount of **45** in d_{12}-cyclohexane, resulting in complete dissolution of the latter (normally insoluble in hydrocarbon solvents) to give n-Bu(TMP)$_3$MgNa$_2$ **47** (Scheme 6.10). Subsequent analysis by ^{1}H NMR spectroscopy showed the presence of peaks characteristic of **43** as well as a new resonance at δ −0.43 ppm. This was assigned to the higher-Na:Mg ratio co-complex **47**. The authors rationalized this as indicative of an equilibrium being manifested between **43** and **47**, facilitated by solubilized **45**.

In work that tested the deprotonative capabilities of the mixture outlined above with respect to more extensive aromatic frameworks than those in Scheme 6.9, its exposure to one equivalent of naphthalene gave a 76% yield of the dimetalation product in an unprecedented 1,4-regioselectivity.

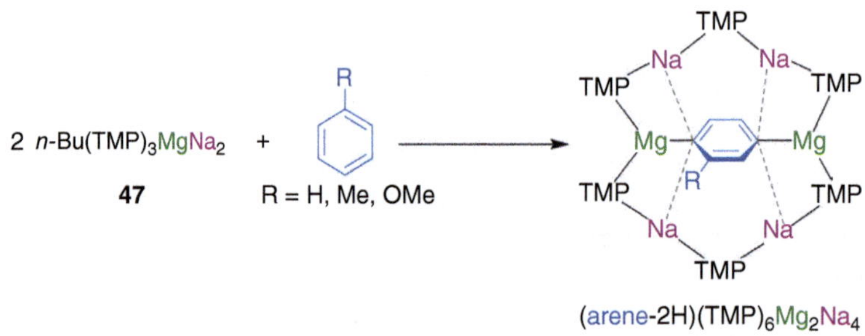

Scheme 6.9 The selective dimagnesiation of un- and monosubstituted arenes such as benzene, toluene, and anisole by sodium magnesiate **47** to give inverse crown complex **44**.

$$\underset{\mathbf{43}}{n\text{-Bu(TMP)}_2\text{MgNa}} + \underset{\mathbf{45}}{\text{NaTMP}} \rightleftharpoons \underset{\mathbf{47}}{n\text{-Bu(TMP)}_3\text{MgNa}_2}$$

Scheme 6.10 Proposed equilibrium in d_{12}-cyclohexane solution between sodium magnesiates **43** and **47**.

This result has allowed **47** to be labelled as a potential precursor to the di-metalated-naphthalene intermediate, the view being that it dimerises prior to metalation to yield a 12-membered pre-inverse crown *template*, n-$Bu_2(TMP)_6Mg_2Na_4$ $\mathbf{47}_2$ [122].

As already noted, magnesiates of the type exemplified by **43** have also proved achievable using Mg/K combinations. Accordingly, O'Hara and coworkers investigated an equimolar mixture of KTMP **48** and **46,** and were able to isolate and structurally characterize different oligomers to those described above for Mg/Na systems. Crystallisation from methylcyclohexane at room temperature gave rise to the tetrameric polymorph $[n\text{-Bu(TMP)}_2\text{MgK}]_4$ $\mathbf{49}_4$, where the n-Bu groups were arranged in an *anti*-conformation, alternating above and below the central ring plane. Cooling the same mixture down to 8 °C resulted in the crystallisation of a higher degree, hexameric form of the same monomeric unit – $\mathbf{49}_6$ again having the butyl groups pointing above and below the central ring system. With these data in hand, spectroscopic analysis by ^{1}H ICC DOSY NMR of a 1:1 mixture of **46** and **48** in d_{12}-cyclohexane revealed the presence of a single oligomer. At $M_{\mathrm{r,det}} = 2478.7$ g/mol, this was deemed to be an excellent fit for the hexameric polymorph, $\mathbf{49}_6$ ($M_{\mathrm{r,calc}} = 2406.1$ g/mol, error = −3%) [122].

6.3.2.2 Turbo-Hauser Chemistry

A specific class of mixed s/s-block systems that have recently been subjected to extensive solution interrogation is Turbo-Hauser bases—these reagents also have an established, strong track record in organic synthesis. To achieve them, Hauser reagents (where a Mg halide fragment bears an amide; R_2NMgCl) [123] were subjected to a lithium salt (typically LiCl **29**) to notionally give $R_2NMgCl(LiCl)$. This showed enhanced solubility [124, 125] and reaction rates, enabling also controllable activity at room temperature and kinetic basicity [126]. The results have seen significant strides in a range of traditionally challenging deprotonations and the moniker 'turbo-Hauser' chemistry. To extract the influence of LiCl in the reactivity of these bases, they have been compared to their Hauser congeners using ECC DOSY. Concerning the turbo variant, Knochel introduced both putative TMPMgCl(LiCl) **50** and DAMgCl(LiCl) **51** in 2006 [126]. Though they revealed solid-state CIP structures [127, 128], limitations to the ICC DOSY methodology (discussed above) meant that it proved impossible to confidently assert whether or not LiCl coordinated to the Mg component in solution [128]. However, the more recent development of ECC DOSY prompted a return to this issue. In the case of **50**, dissolution in THF was proposed to potentially maintain the solid-state structure [i.e. **50**$(THF)_3$] or else result in desolvation [**50**$(THF)_2$], deaggregation (**52-54**) or aggregation (**55, 56**) as shown in Scheme 6.11(i).

^{1}H,^{7}Li–HOESY analysis of **50** in d_8-THF immediately pointed to coordination between lithium and magnesium amide in solution, while variable temperature ^{7}Li NMR spectroscopy eliminated the (observable) presence of **23** and $\mathbf{53}^+$ (i.e. of **53**). Next, ^{1}H and ^{7}Li ECC DOSY analysis of **50** gave $M_{\mathrm{r,det}} = 403$ g/mol (^{1}H) and 387 g/mol (^{7}Li). This made *in situ* desolvation of trisolvate **50**$(THF)_3$ (cf. the crystalline structure) to disolvate **50**$(THF)_2$ ($M_{\mathrm{r,calc}} = 387$ g/mol, error = 4% [^{1}H], 0% [^{7}Li]) by far the most compelling scenario (Scheme 6.11ii), the lability of THF being explained by the steric bulk of the TMP ligand [129].

The attribution of steric bulk in the amide as the driver for THF elimination when **50**$(THF)_3$ was dissolved in THF, also explained the lack of dimerization in either the solution or solid-state in that system. This represented a point of contrast with DAMgCl(LiCl) **51**, a base that had previously revealed contrasting reactivity [126] and regioselectivity [128], and which had been documented as dimeric in the solid state. Recently, ECC DOSY has been applied to the study not only of **51**, but also of LiCl-free DAMgCl **57**, in etherate solution over a wide temperature range. The interest in elucidating the last of these stemmed from there being an extensive body of work on the behaviour of alkyl and aryl Grignard reagents in common ethers [130–132], but no comparable studies on

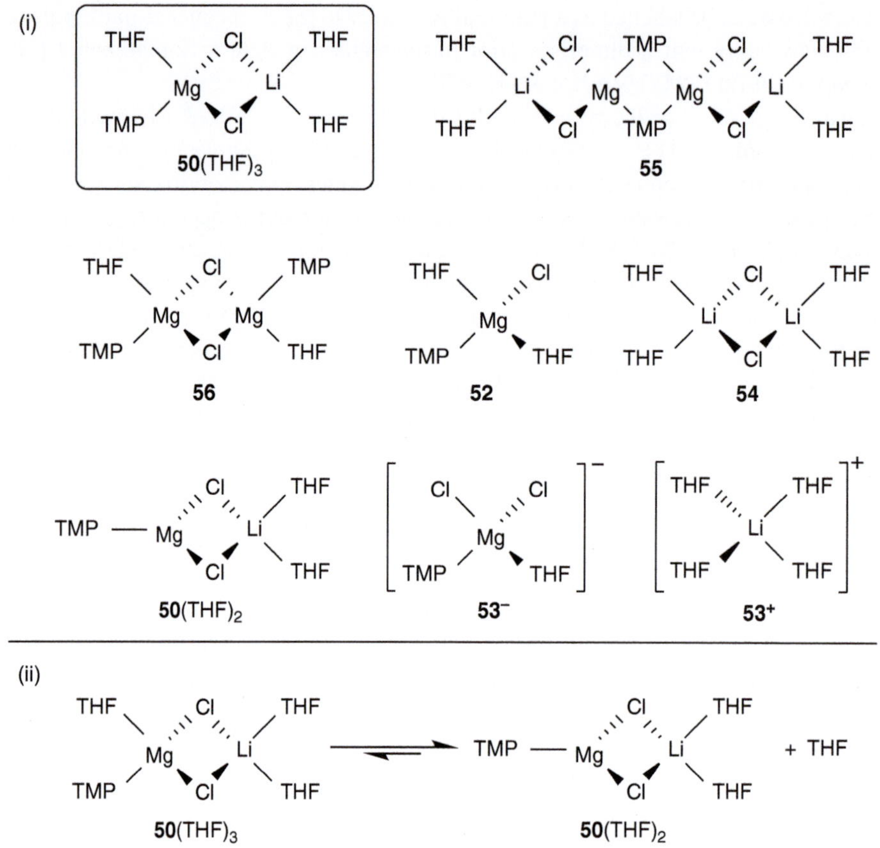

Scheme 6.11 (i) Potential solution structures of **50** in THF (solid-state structure of the THF-solvate boxed), and (ii) the process of desolvation predicted based on ECC DOSY analysis.

Hauser bases. As noted above, the one prior attempt at probing Hauser-related chemistry suffered from its dependence on ICC DOSY [128]. However, having advanced our understanding of LDA in solution by using a more elaborate diffusion ordered technique [107], in 2016 Stalke sought to compare **51** and **57** and probe the effects of LiCl **29** on Schlenk equilibria. The solution behaviour of **57** was established to be based on a temperature-dependent equilibrium between monomers **58** and **59** (Scheme 6.12i), with **58** being dominant above −20 °C. However, now using DA in place of TMP (see above) as the amide component, ^{1}H ECC DOSY also detected species with $M_{r,det}$ = 435 and 450 g/mol, introducing dimers **60** and **61** ($M_{r,calc}$ = 464 g/mol, error = 6 and 3%, respectively) as well as $(DAMgCl)_x(MgCl_2)_y(THF)_z$ ($M_{r,det}$ = 408–578 g/mol) and DAMg(μ-Cl)$_3$Mg(THF)$_5$ **62** ($M_{r,det} = M_{r,calc}$ = 616 g/mol, error = 0%). The introduction of **29** shifted equilibrium towards the heteroleptic side. In spite of the crystallographic evidence for the heterobimetallic nature of **51** in the solid-state [127], its spectroscopic interrogation in d_8-THF below −50 °C revealed an equilibrium dominated by ([THF]DAMg)$_2$(μ-Cl)$_2$ **63** and the simple dimer of LiCl(THF)$_2$ **29**(THF)$_2$, with ^{1}H,^{7}Li-HOESY instrumental in clarifying the existence of Li-free species. Meanwhile, room temperature analysis showed **58(29)**(THF) **64** ($M_{r,det}$ = 425 g/mol, $M_{r,calc}$ = 419 g/mol, error = −2%) and **60(29)**$_2$(THF)$_2$ **65** ($M_{r,det}$ = 525 g/mol, $M_{r,calc}$ = 623 g/mol, error = 24%) to dominate (Scheme 6.12ii). The latter species was particularly favoured at elevated concentrations (ca. 0.5 M) of **51** in solution and quite unlike any species seen in the analogous TMP system [133].

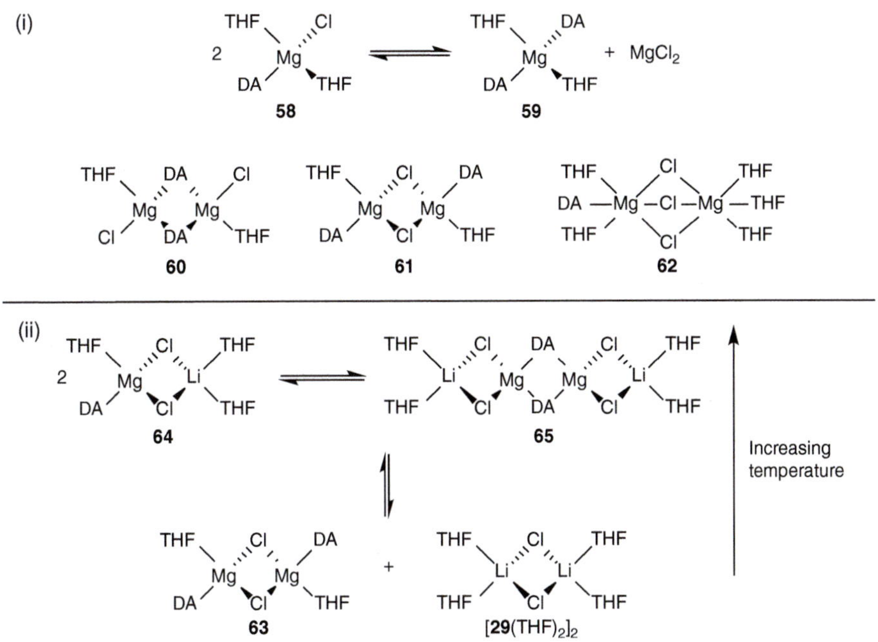

Scheme 6.12 (i) Solution behaviour of LiCl-free DAMgCl **57** in THF, and (ii) that of its 'turbo' congener as a function of temperature based on ECC DOSY.

6.3.3 s/p-block Systems

6.3.3.1 Lithium/Aluminium Chemistry and Trans-metal-trapping

As described previously in this book, *i*-$Bu_3Al(TMP)Li$ **66** and *i*-$Bu_2Al(TMP)_2Li$ **67** have been straightforwardly produced from **23** with R_3Al [134–136]. Whilst both species have been nominally shown to be responsible for the 'simple' alumination of functionalized aromatics, in neither case was this process shown to occur directly [137]. Indeed, whereas logic might dictate (and X-ray diffraction might imply) direct *ortho* alumination of the aromatic substrate, the truth was somewhat more complex. Ultimately, re-examination of the behaviour of mixed Li–Al systems in solution yielded surprising results about the stepwise nature of aromatic alumination by these reagents, and demonstrated the power of DOSY in mechanistic elucidation.

Dissolution of crystalline solvate *i*-$Bu_3Al(TMP)Li(THF)$ **66**(THF) in d_8-THF enabled DOSY analysis to establish a single diffusion coefficient ($6.95 \times 10^{-10} \pm 0.09 \times 10^{-10}$ m^2/s) in relation to both TMP and *i*-Bu ligands, indicating a single solution structure, with the spectroscopically-seen lability of THF suggesting this to be ion-separated [*i*-$Bu_3Al(TMP)$][$Li(THF)_4$] **66**$(THF)_4$ [29]. Reactivity studies probing the stoichiometry of reaction between **66**$(THF)_4$ and model reagent anisole further hinted at the aluminate undergoing a significant level of dismutation in bulk THF to give **67**$(THF)_4$ and [*i*-Bu_4Al][$Li(THF)_4$] **68**$(THF)_4$. The first of these products was shown to be active whilst the latter (the presence of which was verifiable by ^{27}Al NMR spectroscopy) was inert in deprotonation reactions (Scheme 6.13) [127] Interestingly, these attempts at aluminating anisole also produced evidence for the stoichiometry-dependent observation of both lithiated and aluminated organic.

Of particular interest, spectroscopic studies rapidly established that the activity of **67**$(THF)_n$ in deprotonation reactions was not a result of it being TMP-rich, but rather an artefact of its deconvolution as THF-solvated LTMP **23** and *i*-$Bu_2Al(TMP)$ **69**. ^1H NMR spectroscopy pointed to this notion, showing that a d_8-THF sample containing these two components represented merely the

$$2[i\text{-}Bu_3Al(TMP)][Li(THF_4)] \xrightleftharpoons{THF} [i\text{-}Bu_2Al(TMP)_2][Li(THF)_4] + [i\text{-}Bu_4Al][Li(THF)_4]$$

66(THF)$_4$ → **67**(THF)$_4$ + **68**(THF)$_4$

Scheme 6.13 **66**(THF)$_4$ has been shown to undergo dismutation in THF to give **67**(THF)$_4$ and [i-Bu$_4$Al][Li(THF)$_4$] **68**(THF)$_4$, of which the first could be used in deprotometalation.

$$\mathbf{66}(THF)_4 \xrightleftharpoons{THF} LTMP\ (\mathbf{23}) + i\text{-}Bu_3Al$$

$$\mathbf{66}(THF_4) + i\text{-}Bu_3Al \xrightleftharpoons{THF} i\text{-}Bu_2Al(TMP)\ (\mathbf{69}) + \mathbf{68}(THF)_4$$

Scheme 6.14 The convoluted equilibrium resulting from the rearrangement of **66**(THF)$_4$ in solution. Of the products formed, only **23** demonstrated activity in deprotometalation, leading to the notion of *trans-metal-trapping*.

superposition of the individual spectra of each of the two components. This viewpoint was confirmed by DOSY, which revealed diffusion coefficients for **23** and **69** of 1.22×10^{-9} m^2/s and 8.37×10^{-10} m^2/s respectively; these data were in line with DFT studies, and spoke against an ability to create a heterobimetallic complex between the two species. Now revisiting **66**, it proved possible to assert that once added to the reaction medium (typically bulk THF) lithium aluminate rearranged to produce a convoluted set of equilibria (Scheme 6.14), the individual components of which were identifiable by ^{1}H NMR spectroscopy. Of these, only **23** demonstrated activity in deprotonation, meaning that the observation of alumination resulted from an indirect process, termed TMT, of the conjugate base of the organic substrate.

Stemming from the report of the deprotonation of 1-aryl-1*H*-benzotriazoles by Li–Zn mixtures [138], the strategic use of **69** as a TMT agent has been developed in the derivatization of 1-phenyl-1*H*-benzotriazole. Utility lithium amide **23** could be used as the sole deprotometalation agent, in which case a low yield of 4-R-1-(2-lithiophenyl)-1*H*-benzotriazole-tris(THF) **70** [R = 2-C_6H_4(Ph)NLi] resulted after an unstable lithio-intermediate underwent destructive triazole ring-opening and N_2 extrusion processes. In contrast, by conducting the reaction in the presence of **69** rapid TMT of the initial lithiate enabled stabilization of the metalated intermediate, with reaction stoichiometry dictating which of two crystallographically distinct monometalates (**71** or **72** in Scheme 6.15) was isolated [139].

Whilst **71** represented a minor product and proved spectroscopically straightforward, major species **72** proved more complex, but ultimately revealed itself to remain intact and exhibit a

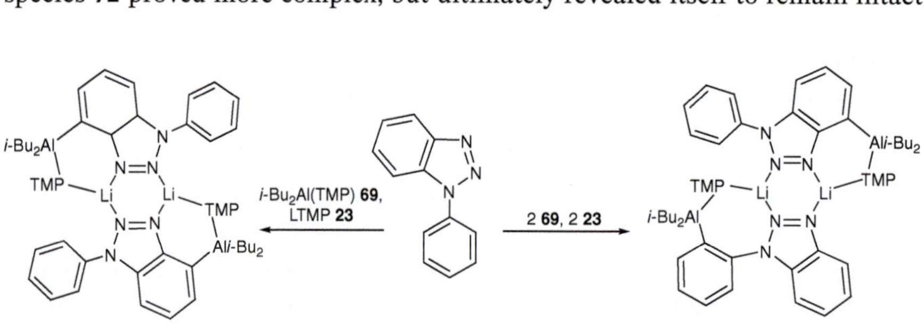

71 **72**

Scheme 6.15 Using lithium amide **23** along with **69** as a TMT agent in the derivatization of 1-phenyl-1*H*-benzotriazole enabled the isolation of **71** and **72**.

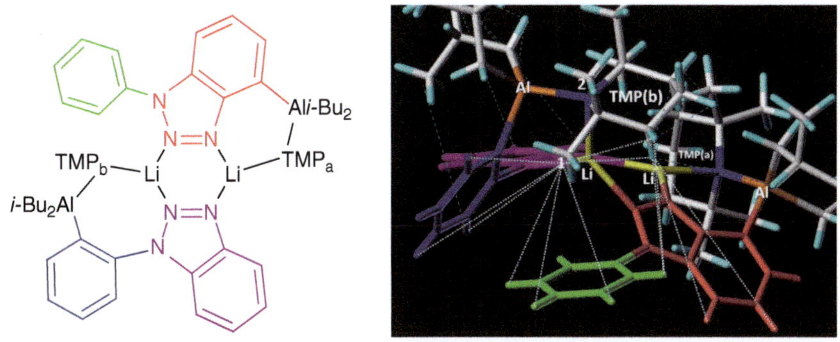

Figure 6.11 Colour-coded structure of **72** (left) and selected nOes associated with one of the four Me groups on TMP_b (dashed white lines, right) mapped onto the colour-coded 3D structure. *Source:* Adapted from Fuentes et al. [139].

slow chemical exchange process in solution. Variable temperature $^1H,^1H$–EXSY showed only aliphatic correlations and enabled the calculation of $\Delta H^{\ddagger}$, $\Delta G^{\ddagger}{}_{298}$ and E_a values. These data suggested the process to be a conformational rearrangement, and this view was probed using NOESY in d_8-toluene at low temperature to arrest dynamics. The observation of nOes between one of the four Me groups (labelled '1' in Figure 6.11) of TMP_b, and the green aromatic ring in that figure plainly suggested retention of aggregation. Meanwhile, the observation of two chemically distinct ^1H NMR resonances for this Me group at low temperature pointed to conformational flexibility absent in the six-membered rings formed in **71** and **72**. This was deemed to occur when amine interacted with an alkali metal in the seven-membered ring in **72** [140, 141].

6.3.3.2 Alkali Metal/Gallium Systems

The extension of the TMT concept to other alkali metal amides and alternative Group 13 trapping agents has been extensively investigated by Mulvey and Hevia [30, 142–144]. Thus, tris(trimethylsilylmethyl)gallium (GaR_3, R = CH_2SiMe_3), has proved to be a powerful trapping and stabilising agent. Initial spectroscopic experiments revealed that **23** and GaR_3 did not form a cocomplex in d_6-benzene solution. ^1H NMR spectroscopic analysis of the mixture revealed that the signal characteristic of uncomplexed GaR_3 was visible at δ 0.13 ppm (both CH_2 and Me groups coincidentally overlap in d_6-benzene) [144]. Additionally, **23** was present in its tetra [97] and trimeric [98] forms, which were both unaltered by the presence of the Lewis acid [144]. The mixture was reported to not evolve over time. This resistance to co-complexation was backed up by DFT calculations, which revealed that an energetic penalty of +9.0 kcal/mol would be required for adduct formation.

Comparative experiments with the sodium amide **45** enabled interesting observations. The first was that GaR_3 promoted the complete dissolution of **45**. As noted previously, this amide is typically insoluble in non-donor solvents. Additionally, ^1H NMR analysis revealed that the normally distinctive singlet at δ 0.13 ppm for pure GaR_3 was no longer present. After 3 days, clear cocomplexation between the metal amide and the Lewis acid was detected by ^1H NMR; analysis showing two distinct new resonances for GaR_3 at δ 0.35 ppm (Me) and δ −0.6 ppm (CH_2), alongside well-defined signals for the **45**-component of the adduct. Attempted metalation of anisole with co-complex

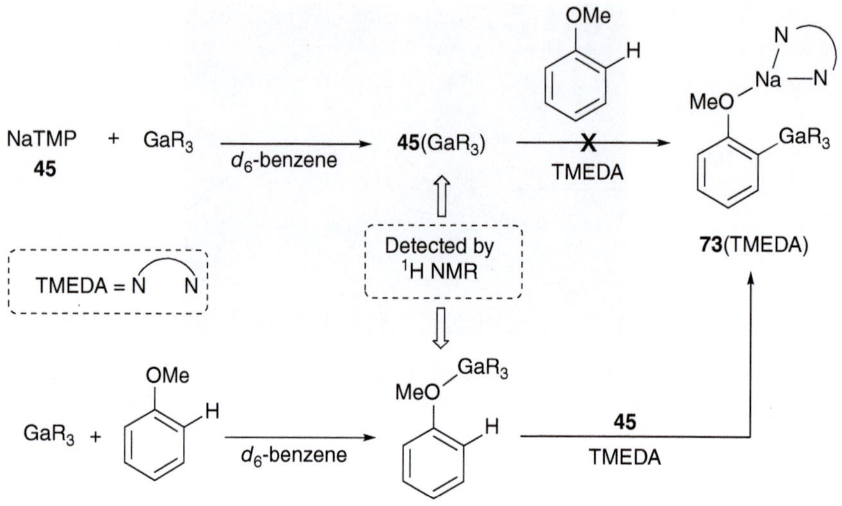

Scheme 6.16 The effects of order on the addition of the single-metal components (R = CH_2SiMe_3) for the indirect gallation of anisole via TMT approaches.

45(GaR_3) was unsuccessful suggesting that, once the adduct was afforded, it was inactive towards further reactivity.

Building on the information gathered during these spectroscopic studies, it was found that controlling the order of addition of the reagents represented a major factor in accomplishing the successful metalation of anisole (Scheme 6.16) [142]. Thus, the combination of equimolar amounts of GaR_3 and anisole in d_6-benzene was followed by ^{1}H NMR spectroscopy. Experiments revealed the apparent coordination of Lewis acidic GaR_3 to the aromatic substrate, as indicated by a slight upfield shift of the signals attributed to the latter. Subsequent addition of one equivalent of **45** resulted in a 74% yield of metalated anisole (against C_6Me_6 internal standard) with concomitant release of 2,2,6,6-tetramethylpiperidine (HTMP). The ability of GaR_3 to offer stability to the metalated aryl was evidenced by the crystallographic analysis of *ortho* gallate **73**, with the coordination sphere of Na^+ found to be satisfied by the bidentate donor TMEDA [that is, **73**(TMEDA) was formed]. On the other hand, using *lithium* amide **23** as the basic component did not require such restrictions to the order of addition as it failed to form a cocomplex with GaR_3 [cf. **45**(GaR_3)] [144]. Thus combining **23** with GaR_3 and anisole in equimolar quantities allowed for the isolation of R_3Ga(2-anisyl)Li(PMDETA) **74**(PMDETA) (PMDETA = *N,N,N′,N′′,N′′*-pentamethyldiethylenetriamine) in a 55% crystalline yield.

6.3.4 s/d-block Systems

6.3.4.1 Lithium/Cadmium

Though TMT is discussed in the section above, the question of whether the deprotonation of an organic by a nominally heterobimetallic ate complex is necessarily by the more electronegative metal was actually first addressed in 2012. This involved developing studies by Uchiyama and Mongin that had introduced a base described as putative lithium cadmate '$(TMP)_3CdLi$' **75** [145–151]. Though a combination of ^{13}C NMR spectroscopy and DFT calculations led to the inference that the ate formulation was preferred to a co-operative mixture of metal amides, the details of the species active in deprotonation remained speculative. In seeking to clarify these points, Mulvey and coworkers noted that the interrogation of a 1:3 mixture of $CdCl_2$ and **23** in THF displayed ^{1}H NMR resonances consistent with both **23** and $(TMP)_2Cd$ **76** [28]. To reinforce the message of this data, the ICC DOSY was attempted at room temperature in a THF solution of 1:3 $CdCl_2$:**23** that included three calibration standards. The lack of any TMP-containing species with an M_r greater than that of standard TPhN quickly became apparent, arguing against the presence of **75** in the solution. Corroboration of the M_r values obtained was aided by the calculation of the DOSY-determined formula weight of serendipitously formed amine HTMP generated by the reaction of the metalated species with trace moisture. Two unknown entities were seen, with data taken to indicate the presence of **76** and $LTMP(LiCl)_2$ **23(29)**$_2$ (Scheme 6.17, top). Whilst data for the latter species agreed well with expectation ($M_{r,det} = 392$ g/mol vs. $M_{r,calc} = 401$ g/mol), at 327 g/mol the observed formula weight of **76** was only 83% as massive as expected. This was attributed to the mass of the central atom, but is indicative of the nature and extent of the accuracy of ICC DOSY. The suggestion that an organic compound could be lithiated by **23** in the presence of independently coexisting **76** led the authors to investigate the possibility of trans-metalating *ortho*-lithiated anisole **77** to give cadmate **78** (Scheme 6.17, bottom). The observation of satellites on either side of the ^{1}H NMR resonance at δ 7.49 ppm that suggested $^3J_{CdH} = 39$ Hz was certainly consistent with this view. Finally, ^{1}H–^{7}Li HOESY lent further weight, demonstrating a correlation between metal and TMP-Me but *not* aromatic or methoxy.

6.3.4.2 Lithium/Copper

Advances in the solid-state structural chemistry of lithium cuprates and recent interest in the use of these species in halogen–copper exchange and the deprotometalation of aromatic rings are covered in detail in Chapter 1. Examples of crystallographically elucidated lower-order cuprates include CIP $(DMBA)_2Cu(CN)Li_2(THF)_4$ (DMBA = 2-$Me_2NCH_2C_6H_4$) and SIP (separated ion-pair) $[t\text{-}Bu_2Cu][CN\{Li(THF)(PMDETA)\}_2]$ (see Chapter 1, complexes **114** and **115** and also Figure 1.16) [152, 153]. Recorded displacements in these structures pointed to the capacity to discriminate between structure-types in solution based on the observation or not of nOes between lithium and the methyl groups of the cuprate. This being so, evidence for these structure-types as well as aggregation state [154], salt effects [155], and CIP–SIP equilibria in ^{6}Li-labelled model dialkylcuprates have emerged from ^{1}H,^{6}Li–HOESY experiments. These data corroborated existing views on the behaviour of cuprates such as Me_2CuLi **79**, long held to be a cyclic dimer in Et_2O solution [156]. Hence, in Et_2O (in this work donor solvents were 80% deuterated) ^{6}Li-labelled **79** now showed correlations between the alkali metal, the solvent and (strongly to) the cuprate methyl groups. Taken together, these data indicated a CIP structure [157]. In contrast, the spectra of ^{6}Li-labelled t-$Bu_2Cu(CN)Li_2$ **80** in more strongly coordinating THF revealed nOes only with the solvent, pointing to SIP behaviour. More subtly, weak dipolar interactions between lithium and methyl groups were revealed by **79** and $Me_2Cu(CN)Li_2$ **81** in THF. This was interpreted as being due to direct dipolar interactions, and established an equilibrium between CIP (e.g. **79**$_2$) and SIP

$$\underset{\mathbf{23}}{CdCl_2 + 3\ LTMP} \xrightarrow{THF} \underset{\mathbf{76}}{(TMP)_2Cd} + \underset{\mathbf{23(29)_2}}{LTMP(LiCl)_2}$$

$$\mathbf{76} + n\ \underset{\mathbf{77}}{(2\text{-}C_6H_4OMe)Li} \xrightarrow{THF} \underset{\mathbf{78}}{(2\text{-}C_6H_4OMe)_n(TMP)_{2-n}Cd} + n\ \mathbf{23}$$

Scheme 6.17 Proposal for the reaction of $CdCl_2$ with **23** and a demonstration that ortho-cadmation can result from a TMT process.

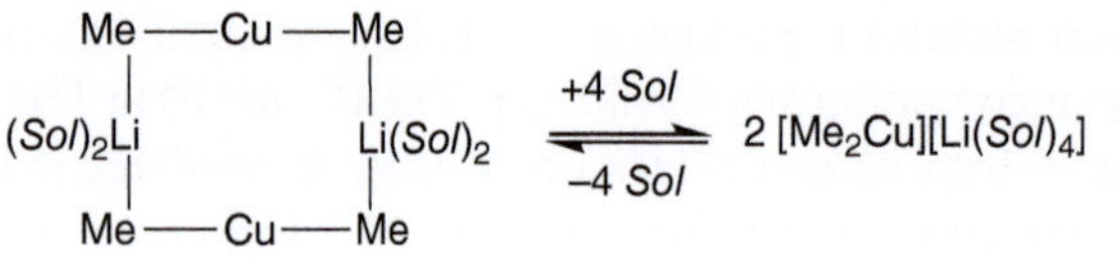

79_2 79*

Scheme 6.18 Proposed solution equilibrium between CIP 79_2 and its SIP form **79*** (*Sol* = solvent).

(e.g. **79***) behaviour at −60 °C (Scheme 6.18) [154, 157]. Depressing the temperature was expected to reduce the CIP contribution [158], and this proved the case, with $\mathbf{79}_2$ correlations unobservable at −110 °C [157]. The prevalence of CIP and SIP behaviour in Et_2O and THF, respectively, was correlated with the logarithmic reactivity profiles established for lithium cuprates with α,β-unsaturated ketones [159, 160].

^{13}C NMR spectroscopic analysis of $Me_2Cu(I)Li_2$ **82** led to the proposition of an equilibrium between two structure-types in solution; a six-membered metalacycle $\mathbf{79}_2$ and a dinuclear adduct structure **79**(LiI) based on a six-membered ring. Of these, the former was suggested to be the major contributor [161]. The ^{1}H,^{6}Li–HOESY substantiation of the CIP structure $\mathbf{79}_2$ led to interest in applying nOe techniques to the issue of competitive structure formation for salt-containing lithium cuprate systems in solution. This resulted in an elegant study that considered competition between retention of the $\mathbf{79}_2$ motif and the formation of a seven-membered cyanide-containing metalacycle in **81**. Monitoring the initial build-up rates of ^{1}H,^{6}Li hOes (heteronuclear Overhauser effects) and motional correlation times in Et_2O solutions of ^{6}Li-labelled **79** and **81** suggested similar dimeric core structures in either case, i.e. $\mathbf{79}_2$. Confirmation came from the use of a new pulse sequence designed to enable the detection of long-range nOes and rOes (rotating frame Overhauser effects) between equivalent protons. The observation of similar ^{1}H–^{1}H dipolar interactions reinforced the view that these systems predominantly displayed a common core structure [155].

More recent work in a similar vein compared **81** with **82** in Et_2O, and the effect of THF introduction on their 1,4-addition to 4,4-dimethylcyclohex-2-enone [162]. In contrast to the cyanide, whose reactivity continuously decreased with the addition of THF, kinetic studies on the iodide system revealed an acceleration effect that depended both on the level of donor and reagent concentration. Pulsed field gradient measurements and nOe analysis revealed that the iodide system incorporated small salt units solvent-separated from cuprate aggregates. This view emerged from the observation that though reactivity was highly dependent upon THF dose, D and r_H of the cuprate oligomers were not. Meanwhile, measurement of the ^{1}H,^{7}Li–HOESY cross-peak volume integrals for **82**, Et_2O and THF suggested that adding THF increased the number of coordination sites but did not displace Et_2O, while a drop in relative ^{1}H,^{1}H–NOESY cross-peak volume integrals pointed to a reduced fraction of complexing THF at increased dosage. Contrastingly, cyanide system **81** appeared to incorporate oligomers coordinated by the salt. Reactivity now declined since adding THF incurred cuprate deaggregation—evidenced by an increase in D. ^{1}H,^{7}Li–HOESY cross-peak volume integrals now suggested the replacement of Et_2O by THF, with retention of a similar number of coordination sites. Meanwhile, ^{1}H,^{1}H–NOESY cross-peak volume integrals were not dependent on THF dosage. The view that cyanide was a coordinating salt was verified by ^{1}H,^{13}C–HOESY using **81** prepared with $Cu^{13}CN$. A Et_2O solution of labelled **81** and THF (3 eq.) gave $D = 0.83 \times 10^{-9}$ m^2/s at −34 °C, indicating a dimer of the type concluded previously [155] with one LiCN attached. Under these conditions, splitting of the THF signals allowed unambiguous observation of donor molecules complexed to the dimer, enabling the identification of a strong correlation between these and the ^{13}C centre of the salt.

Chapter 1 features discussion of solid-state structural analysis of a range of Lipshutz(-type) cuprates; that is structurally related cuprates incorporating cyanide or other salts. The discussion there alludes to the conflicting explanations for an apparent contradiction in cuprate reactivity; the apparently higher reactivity of Lipshutz (i.e. cyano) cuprates in only certain circumstances. An interesting new approach to this issue has deployed diffusion techniques to interrogate how the differing solubility of organic groups may contribute to observed variations in reactivity. Alkylcopper by-products were argued to be incorporated into iodide-containing cuprates but not into their cyanide-containing congeners during synthetic applications. It was established that during reactions and in the presence of iodide, cuprates bearing relatively soluble substituents yielded a range of 'extended' copper-rich complexes. These last species incorporated less polarized (i.e. less reactive) alkyl groups, yet in forming also consumed iodocuprate; both effects logically leading to depleted yields. This thesis was interrogated by preparing a variety of highly soluble copper-rich complexes of the type $R_{1\text{-}4}Cu_{1\text{-}3}Li_{0,1}X_{0,1}$ (R = Me_3SiCH_2, X = I, ^{13}CN) by combining RLi and CuI. At RLi:CuI > 2:1, exclusively R_2CuLi **83** was recorded. Reducing the RLi content moved equilibrium towards more Cu-rich complexes, as shown by the increasingly downfield shifts of their signals in the corresponding ^1H NMR spectrum. These data indicated stepwise-reduced polarization (and thus reactivity) as follows: **83**, R_3Cu_2Li **84**, R_4Cu_3Li **85**, RCu **86**, and R_2Cu_3I **87** (whilst formally LiI units should be incorporated into these formulae, they are excluded based on prior art [163] and DOSY data, see below). In contrast, a similar approach to preparing cyanocuprate complexes gave just $R_2Cu(^{13}CN)Li_2$ **88** or (using less RLi) $RCu(^{13}CN)Li$ **89**, but no detectable Cu-rich complexes (Scheme 6.19).

A combination of $^1H,^{13}C$–HMBC and DOSY measurements at low temperature (between −100 and −90 °C to ensure the satisfactory resolution of Cu-rich complexes) was used to relate the molecular weights, the position of equilibria, and the ability of the various Cu-rich complexes to exist as monomers or to aggregate in Et_2O solution. Evidence for the nature of 'extended' copper-rich cuprates in the iodide system emerged as exemplified in Scheme 6.19, and it was further shown that these extended cuprates acted to reduce yields when cross-coupling reactions were attempted with MeI. Overall, therefore, it became possible to assert that in the case of small substituents the produced alkylcopper precipitates, ensuring similar reactivities for both iodide and cyanide containing systems. Meanwhile, in the presence of iodide, cuprates with large substituents incorporate alkylcopper units, giving copper-rich species with reduced reactivities [164].

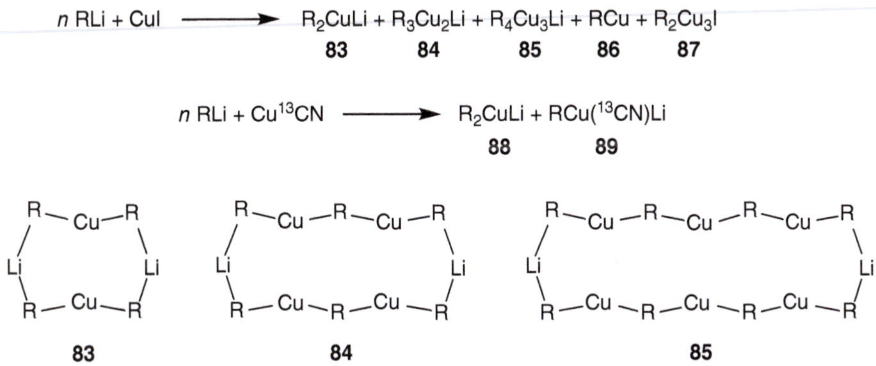

Scheme 6.19 Top, reactions in Et_2O at −100 °C of RLi (n = 2 or less, R = Me_3SiCH_2) to give Cu-rich complexes only when using iodide (upper reaction) or to avoid them when using cyanide (lower reaction). Bottom, exemplar extended or Cu-rich cuprates as identified by $^1H,^{13}C$–HMBC and DOSY analysis.

The intermediacy of Cu(III) in a range of carbon–carbon bond-forming reactions has long been suggested, and the use of RI (rapid injection) NMR spectroscopy has enabled elucidation of thus oxidized copper in conjugate addition [165]. This was done by Bertz and coworkers by combining 2-cyclohexanone with **81** or **82** in d_8-THF at −100 °C (Scheme 6.20) and monitoring changes to the ^{1}H and ^{13}C NMR chemical shifts of Cu-bonded Me groups as reaction progressed, via observable intermediates **90** and **91**, towards the tetracoordinate square planar (TCSP) complex lithium cyanobis(methyl)(3-trimethylsiloxycyclohex-2-en-1-yl)cuprate(III) **92**, which could be isolated or thermally decomposed to the expected enolate. Subsequently, the RI NMR approach was extended to probe the reactions of a range of $Me_2Cu(X)Li_2$ (X = CN **81**, I **82**, SCN **93**, SPh **94**) with the model alkyl halide, EtI. By this route, it proved possible to observe a series of TCSP intermediates with varying stabilities [166]. At the same time, Gschwind and coworkers achieved a Cu(III) intermediate by substituting a Gilman cuprate using an alkyl halide. The same work established the necessary conditions for stabilizing Cu(III) intermediates sufficiently for extended 2D NMR spectroscopic characterization [167]. Originally, an unexpected and low yield product formed when treating 4,4-dimethylcyclohexen-2-one with ^{13}CN-labelled **81**. However, the preparation of $Me_3Cu(^{13}CN)Li$ **95** was subsequently refined by using heteroleptic $MeCu(^{13}CN)Li$ **96** (Scheme 6.21). The TCSP geometry of **95** was clearly demonstrated by ^{1}H,^{13}C–HMBC investigation. $Me_3Cu(X)Li$ (X = Me **97**, CN **95**) has served as a source of a range of Cu(III) complexes, including $Me_3Cu(PPh_2)Li$ **98**, the structural integrity of which was verified by ^{1}H,^{31}P–HMBC analysis [168]. Meanwhile, RI techniques have also been expanded to generate a range of neutral Cu(III) complexes of the type $EtMe_2CuX$ **99**, by combining EtI with **82** in the presence of X (= phosphines, pyridines, imidazoles) [169].

The formation of a π-complex between **79** and chalcone was instrumental in the slow synthesis and isolation of a prototypical Posner (phenylthiocuprate) reagent that existed predominantly as a CIP monomer in THF solution [170]. Meanwhile, π-intermediates formed during cuprate addition reactions to more extensive and chiral α,β-unsaturated ketones have been observed by interrogating the exposure of one such substrate to **73, 81,** or **82** in Et_2O. Using **79**, it proved possible to record the formation of intermediate π-complexes on *both* sides of the double bond in 4,4a,5,6,7,8-hexahydro-4a-methyl-naphthalen-2(3*H*)-one. The two identified complexes had very different populations (Figure 6.12). The major species was found by ^{1}H,^{1}H–NOESY to be the β-face π-complex, with the conformation adopted by the organic explaining the exclusively *syn* addition known to follow *en route* to the β-methyloctalone product. This work reported also the development of INEPT-INADEQUATE spectroscopy as a means of observing ^{13}C,^{13}C scalar coupling

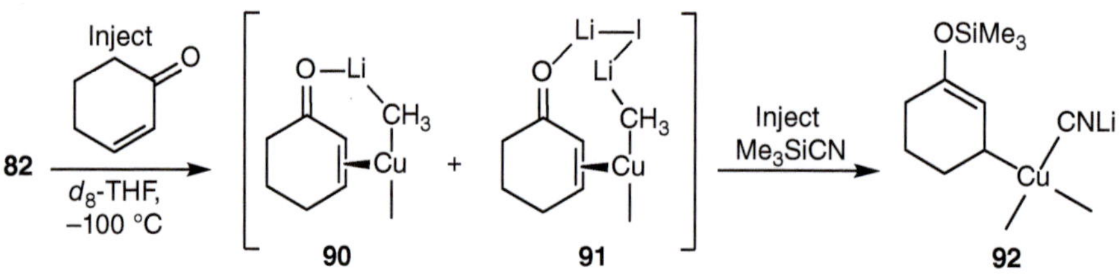

Scheme 6.20 Rapid injection (RI) synthesis of Cu(III) intermediate **92** using $Me_2Cu(I)Li_2$ **82**.

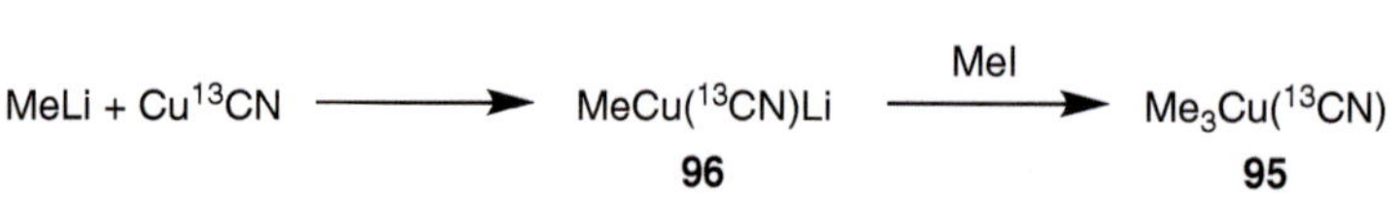

Scheme 6.21 Synthesis of Cu(III) intermediate **95** using $MeCu(^{13}CN)Li$ **96**.

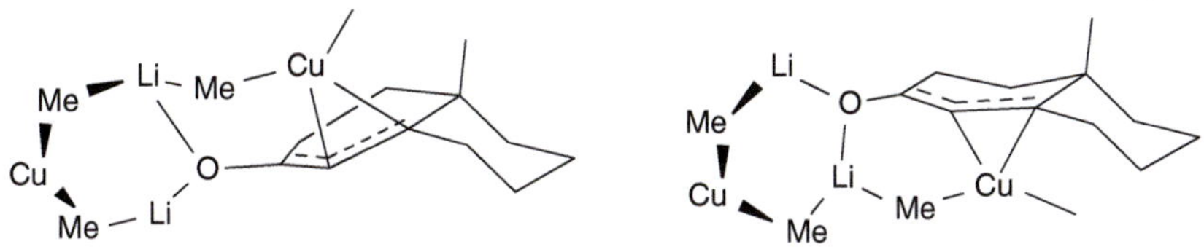

Figure 6.12 Proposed major (left) and minor (right) π-complexes achieved by introducing **79** to an extensive chiral α,β-unsaturated ketone.

between the cuprate and organic without the need to label the latter. Experiments with **81** and **82** established that though the salt was included in the π-complexed intermediate, neither relative populations nor the basic structures of the complexes were altered by salt inclusion [171].

Heteroleptic lithium amidocuprates have seen major interest in the last decade since they combine excellent organyl transfer ability with that of the amido group to act as a dummy ligand [172] or a chiral auxiliary [172–174] affording applications in C–C, C–O, and C–N bond formation [26, 175]. Inroads into our understanding of the solution behaviour of these systems were first made by Davidsson, who probed the cuprate formed by reacting lithium (*S*)-(^{15}N)methyl-1-phenyl-2-(1-pyrrolidinyl)ethanamide with BuCu in Et_2O [176]. ^{6}Li,^{15}N–COSY established coupling between two ^{6}Li doublets, at δ 2.01 and 1.91 ppm, and one ^{15}N centre, pointing to **100**. This view was in line with previous predictions [177]. Upon titration with excess THF, a new triplet emerged at δ 1.61 ppm, which was attributed to **101** (Figure 6.13).

Davies has worked extensively on modified Ullmann amination, and in so doing has probed both copper amides and lithium amidocuprates in solution [178]. His work with species of this type has also led to a series of papers on aggregated and/or solvated mesitylamidocuprates that revealed a new structural motif. MesCu(NBn$_2$)Li **102** (Mes = mesityl, Bn = benzyl) exhibited a solid-state head-to-tail dimer **102**$_2$ (see **160**$_2$ in Chapter 1). However, ^{7}Li NMR spectroscopy pointed to *five* lithium environments in d_8-toluene solution. While this suggested disintegration of the dimer, cryoscopic data argued against this having happened substantively. These data were explained by using inverse-detected ^{1}H,^{7}Li–HOESY to show the dominant solution species to be the dimer seen crystallographically, with the ^{7}Li NMR shift (δ −5.79 ppm) pointing to η^6-complexation by an aromatic ring [179] and strong nOes seen to both methyl and methylene groups [180]. Using similar logic, ^{7}Li signals at δ −9.93 and −11.88 ppm were assigned to the known dimer of Mes$_2$CuLi **103** with its η^1,η^6-coordination of the alkali metal [181] and to the η^6,η^6-coordinated head-to-*head* dimer of **102**, **102**$_2$′ (Scheme 6.22). A broad, complex signal at δ 1.11 ppm was accounted for by the amine-coordinated Li centre in the last of these species as well as the necessary presence of cuprate

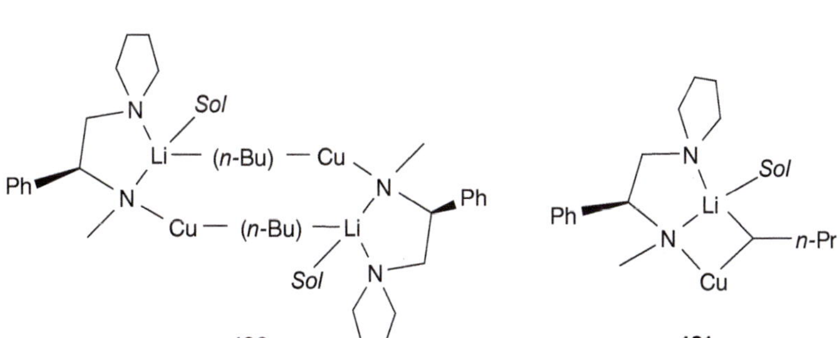

100 **101**

Figure 6.13 In **100**, *Sol* = d_{10}-OEt$_2$, which gradually changes to d_8-THF as monomer **101** is produced by titration.

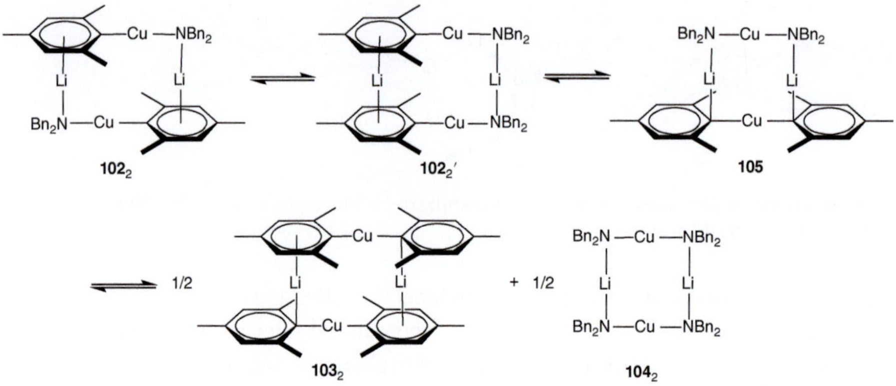

Scheme 6.22 The structural possibilities elucidated for cuprate **102** in hydrocarbon solvent.

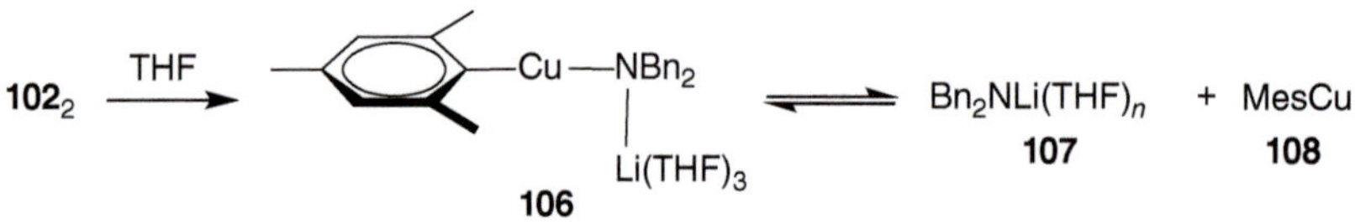

Scheme 6.23 The effect of gradual THF addition to a hydrocarbon solution of **102**.

104. Finally, a signal at δ −4.06 ppm that showed correlations to benzylic and mesityl ^{1}H signals was suggested to be due to a mixed complex, **105**.

Subsequent to the study of **102**$_2$ in toluene, titration with THF was probed, with only head-to-head aggregate **102**$_2$′proving at all durable. Finally, inverse-detected ^{1}H,^{7}Li–HOESY enabled assignment of the only two peaks that remained in the ^{7}Li NMR spectrum (δ −0.53 and 1.41 ppm) after adding a full 10 eq. THF [182]. The major (highfield) signal correlated strongly with methylene groups in Bn and THF but only weakly with Mes and Ph groups, pointing to CIP structure **106** (Scheme 6.23), while the minor signal corresponded to the parent lithium amide **107** [176], suggesting the formation also of MesCu **108**.

Since 2007 [183], work on lithium amidocuprates has developed the notion of directed *ortho*-cupration, as described in Chapter 1. Recently, with the recognition that Gilman amidocuprates are active in this process and that they are accessed from Lipshutz cuprates [184], attempts have been ongoing to replace the cyano-component in the latter with a less toxic moiety. As part of a study into the potential of cyanate in this respect, the reaction of **23** with CuOCN yielded **109**; a cocrystalline mixture of (TMP)$_2$Cu(OCN)Li$_2$(THF) **109a** and (TMP)$_2$(OCN)Li$_3$(THF) **109b** (Scheme 6.24) [185]. ^{7}Li NMR spectroscopy showed that in d_6-benzene **109** formed a complex mixture containing at least five lithiated species. Meanwhile, ^{1}H NMR analysis revealed the dominance of three TMP^{Me} singlets at δ 1.76, 1.57, and 1.39 ppm in a 1:2:1 integral ratio. It was concluded that the mixture retained **109b** but also incorporated additionally solvated **109b**(THF) as well as known dimer of Gilman cuprate (TMP)$_2$CuLi **110** (see **156** in Chapter 1) [184]. However, based in particular on the ^{1}H NMR spectroscopy, which suggested three types of TMP ligand, the major solution species formed by dissolving **109** was proposed to be an isomer of **110**$_2$; adduct (CuTMP)$_2$(LTMP)$_2$ **111**.

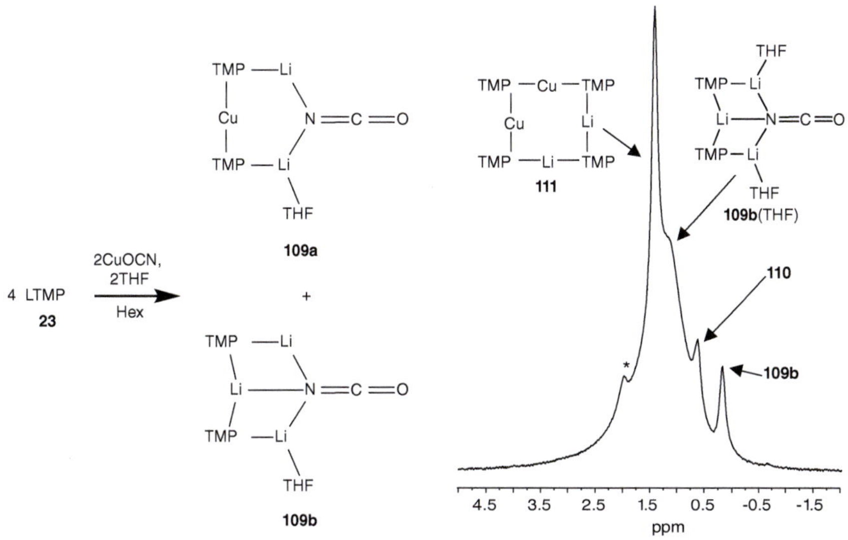

Scheme 6.24 Synthesis of **109** and the ^{7}Li NMR spectrum obtained upon dissolving this co-complex in d_6-benzene, ***23**.

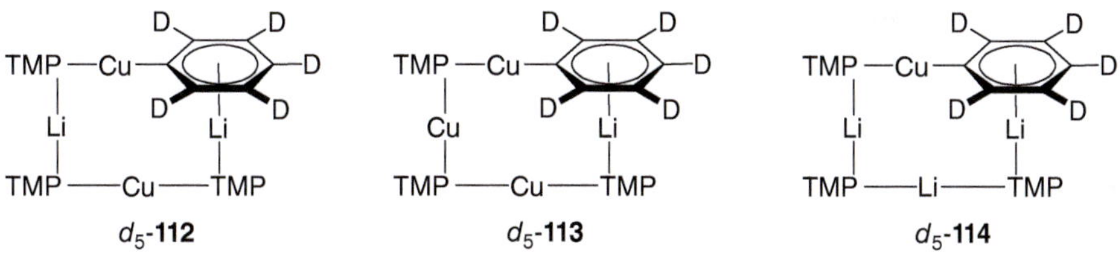

Figure 6.14 Depending on their relative populations, heating LTMP **23** and CuTMP **115** in d_6-benzene for ca. 24 h has been found to selectively generate d_5-**112–114**.

The identity of **111** was later confirmed by the preparation, crystal structure determination, and room temperature spectroscopic analysis of an authentic sample (see **174** in Chapter 1) [186]. Moreover, ^{7}Li NMR spectroscopy in d_6-benzene showed the replacement of a signal at δ 1.64 ppm for **111** with one at δ 0.90 ppm for **110** at 50 °C, establishing the former to be a *kinetic* variant on the latter. However, by the end of the experiment, **110** was also being *consumed*, with dominant signals in a 1:1 integral ratio emerging at δ 1.41 and −2.93 ppm. The concomitant observation of d_1-HTMP led to the suggestion that reaction of **111** with the deuterated solvent was yielding d_5-Ph(TMP)$_3$Cu$_2$Li$_2$ d_5-**112** [186].

The selective *in situ* syntheses of d_5-**112** and related Cu- and Li-rich species d_5-**113** and d_5-**114** (Figure 6.14) were subsequently shown by heating LTMP **23** and CuTMP **115** in a range of molar ratios in d_6-benzene to 50 °C for ca. 24 h. ^{7}Li NMR spectroscopy revealed a gradual transition from d_5-**113**, through d_5-**112**, to d_5-**114** as Cu was replaced by Li in the mixture (see **175–177** in Chapter 1). The deliberate synthesis of **112–114** also proved achievable; that is **112**, formally an aggregate of Gilman cuprate **110**, and PhCu(TMP)Li **116**, could be made through an explicit combination of these two preisolated species, with **113** and **114** resulting if **116** was combined in 1:2 ratios with **115** or **23**, respectively. These observations, combined with the failure of ^{7}Li NMR spectroscopy to observe significant **116** in the *in situ* reaction mixtures provided evidence for the formation of

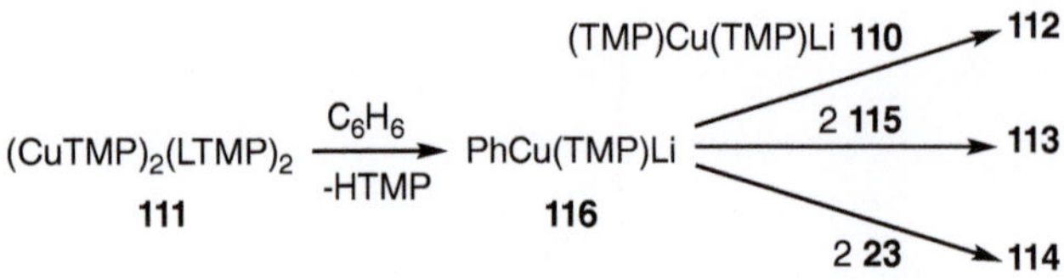

Scheme 6.25 Synthesis of **112–114** via intermediate **116**.

112–114 through multiple steps, suggesting reaction of **111** or its components with benzene followed by re-equilibration with remaining **110**, **115**, or **23**, respectively (Scheme 6.25).

A variable temperature ^{1}H,^{1}H-NOESY and ^{1}H,^{7}Li-HOESY study of **112** and **113** demonstrated that they retained their solid-state structures, notwithstanding that conformational changes were noted for amide ligands. For **112**, correlations observed in d_8-toluene at −10 °C were exactly in line with those expected from the crystal structure. In contrast, at 25 °C exchange correlations were seen between the axial and equatorial hydrogens in both the Me groups and the ring of the TMP ligand *trans* to the Ph group. This evidenced ring inversion of the TMP ligand at the higher temperature. A similar pattern emerged for Cu-rich **113**, with the Ph-*trans* TMP showing exchange correlations at relatively high temperature (though this now meant 80 °C, with no exchange seen at 25 °C). Study of Li-rich **114** revealed a very different picture. ^{7}Li,^{7}Li-NOESY at 25 °C pointed to the exchange of Li^{A-C} in solution (as labelled in Scheme 6.26). This fluxional picture was reinforced by ^{1}H,^{1}H-NOESY evidence for substantial exchange between TMP-Me groups, which established conformational fluxionality for $TMP^{A/C}$ (i.e. ring inversion, as seen in **112**) and the chemical exchange of $TMP^{B/C}$. These data were explained by a dissociative pathway, with path **I** occurring more slowly than path **II**. Substantiating data came from lower temperature EXSY work. This revealed neither $TMP^{B/C}$ or Li exchange. Overall, data suggested the exchange of inequivalent TMP and Li sites in **114** was mediated by a common intermediate.

6.3.4.3 Alkali Metal/Zinc

Lithium zincate t-Bu$_2$Zn(TMP)Li **1**, first developed by Uchiyama and Kondo [18] and discussed in detail in Chapter 1, has proved to be a versatile base capable of inducing the deprotozincation of a huge range of substituted aromatic and heteroaromatic substrates. In addition to introducing excellent functional group tolerance and chemoselectivity, **1** has often proved to be useable at room temperature, making its reactions more efficient and cost-effective than those of many more traditional organometallic bases. Structural studies on **1** [187, 188] and of some key zincated intermediates that follow from its reaction with arene substrates unequivocally established these metalations to be genuine direct zincations [189–191], where the position on the aromatic substrate previously occupied by a hydrogen atom became occupied by zinc. This is representatively depicted for the zincation of anisole to afford *ortho*-zincate **117** in Scheme 6.27 [192]. These types of deprotonations, where zinc affects the metalation but the presence of an alkali-metal, in this case Li, is nevertheless essential for the reaction to take place, have been coined in the literature as examples of alkali metal-mediated zincation (AMM*Zn*) [23, 193]. It was early recognized that the heteroleptic constitution of **1** made it possible that it could perform as either an alkyl or as an amido base. Definitive structural elucidations of metalates obtained from the treatment of aromatics with lithium zincate **1** have suggested reaction overall as an alkyl base, where ultimately the bridging t-Bu group in **1** (for solid-state structural evidence see Chapter 2) was replaced by, in one representative example using **1**(THF) and shown in Scheme 6.27, an *ortho*-metalated anisole fragment in the final product. Related to this work, insightful computational studies by Uchiyama, Wheatley, and

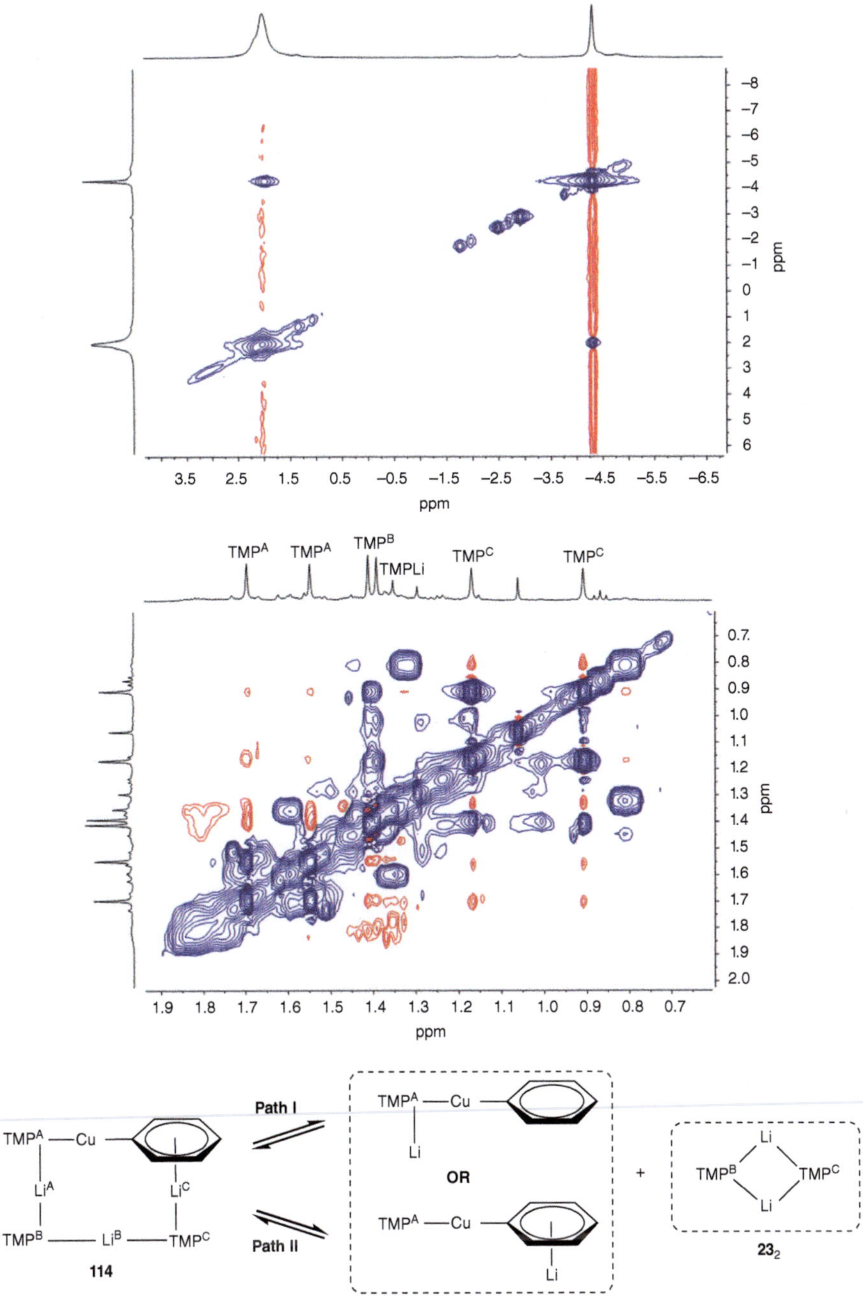

Scheme 6.26 Dissociative interchange of Li and TMP in **114**, mediated by symmetrically dimeric lithium amide as evidenced by room temperature ^{7}Li,^{7}Li-NOESY of **114** (upper spectrum; d_8-toluene, exchange correlations in blue, τ = 0.05 s, TMPLi shoulder at δ 2.23 ppm) and ^{1}H,^{1}H-NOESY of **114** (lower spectrum; d_6-benzene, exchange correlations in blue, τ = 0.6 s).

Morokuma, showed that kinetically, amide basicity was preferred to the experimentally observed overall alkyl basicity [192], and that this was due to the process of cleaving a Zn–N bond having a significantly lower activation energy than that of cleaving a Zn–C bond [194, 195]. Thus a two-step

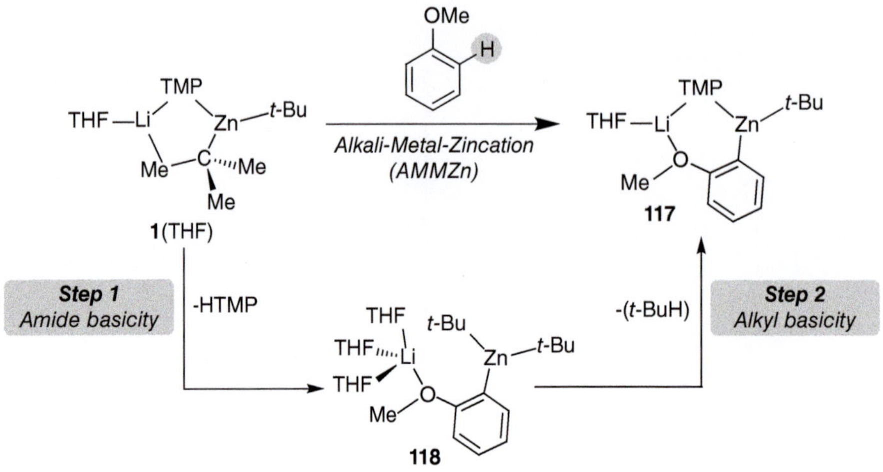

Scheme 6.27 Proposed two-step mechanism for AMM*Zn*, in this case of anisole, by lithium zincate **1**.

mechanism was proposed, where the aromatic substrate was initially deprotonated by the TMP ligand, affording a reaction intermediate such as *t*-Bu_2Zn(2-anisyl)Li(THF) **118**, which could then react with concomitantly generated HTMP in a second step. Such a sequence would furnish **117** and *t*-BuH.

While the formation of **118** during the reaction of **1** and anisole could not be detected when reactions of this type were monitored by ^{1}H NMR spectroscopy, Hevia prepared **118** via the cocomplexation of lithaited anisole **77** and *t*-Bu_2Zn in THF and assessed its reactivity towards HTMP in order to mimic the second step proposed in the computational studies [196]. ^{1}H NMR monitoring of the reaction of **118** with HTMP in d_6-benzene revealed the almost immediate deprotonation of the amine, as evidenced by the inequivalence of α-Me resonances developed at δ 1.21 and 1.42 ppm (*vs*. a single resonance at δ 1.06 ppm in native HTMP). In addition, the appearance of a distinct doublet at δ 0.87 ppm and a multiplet at δ 1.56 ppm was observed, confirming the formation of isobutane (*t*-BuH). These data established that one of the *tert*-butyl groups of compound **118** had reacted with HTMP to afford the metalation product **117** and isobutane as coproduct (Scheme 6.28i). These coincided with the products detected upon reaction of anisole with **1**. The spectroscopic findings not only provided solid experimental support to the two-step mechanism proposed by computational studies but also shed light on the key role of the *t*-Bu groups in **1** in facilitating the zincation of anisole. Thus, while ^{1}H NMR spectroscopic monitoring of the reaction of the methyl analogue of **118**, Me_2Zn(2-anisyl)Li$(THF)_2$ **119**, with one equivalent of HTMP in d_6-benzene also showed metalation of the amine, as evidenced by the α-Me resonances of the TMP fragment appearing at δ 1.15 and 1.45 ppm, free anisole was also released (multiplets at δ 7.12, 6.84, 6.81 ppm). These spectroscopic findings indicated an alternative reaction pathway for **119**, furnishing Me_2Zn(TMP)Li(THF) **120** and anisole instead of MeZn(2-anisyl)(TMP)Li(THF) **121** and methane (Scheme 6.28ii). This idea provided a rationale for the lack of reactivity observed for **120** towards anisole metalation, since the zincate would react first as an amide base, generating **119** and HTMP that in a second step would react as shown in Scheme 6.28ii to regenerate **120** and anisole. Overall, no net reaction would be observed.

Moving from lithium to sodium, performing the metalation of trifluoromethylbenzene with bimetallic base *t*-Bu_2Zn(TMP)Na(TMEDA) **122** [197] at 0 °C has enabled isolation of the kinetic

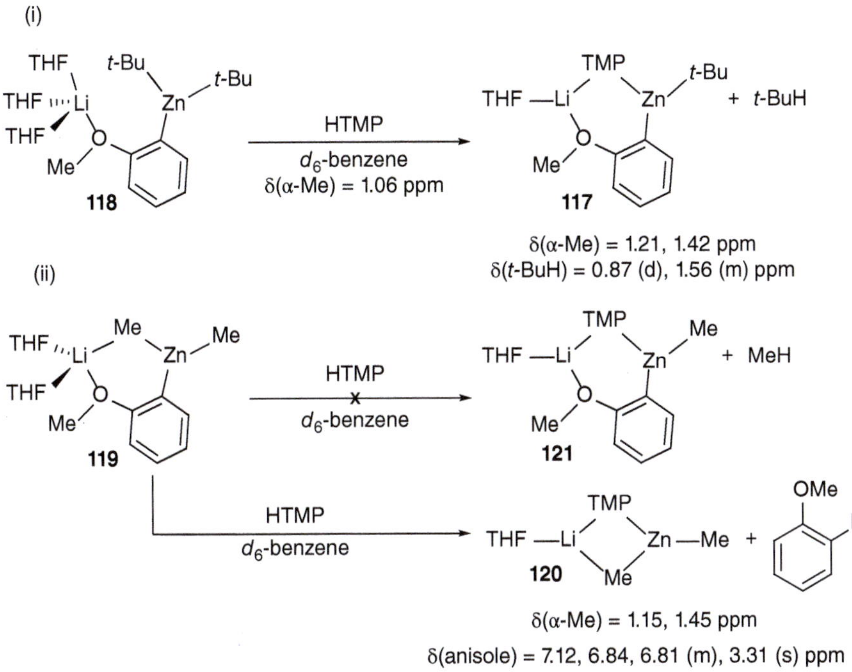

Scheme 6.28 The contrasting reactivities of **118** and **119** towards HTMP in d_6-benzene, as monitored by ^{1}H NMR spectroscopy. The chemicals shift of the most diagnostic signals are given.

zincate [*t*-$Bu_2Zn(C_6H_4$-CF_3-2)][Na(TMEDA)$_2$] **123**. This resulted from **122** acting as an amide base via its TMP group (Scheme 6.29) [198, 199]. Interestingly, if the reaction was carried out at room temperature, ^{1}H, ^{19}F, and ^{1}H,^{1}H–COSY NMR data showed the formation of a complex mixture of products, indicating the presence of *all three* possible regioisomers for the monometalation of $PhCF_3$. Thus, along with some kinetic *ortho* product **123**, the work reported four multiplets at δ 7.85, 7.54, 7.19, and 6.98 ppm for the major thermodynamically *ortho*-metalated product *t*-BuZn(C_6H_4-CF_3-2)(TMP)Na(TMEDA) $\mathbf{124}_{ortho}$, three multiplets at δ 7.74, 7.34, and 7.06 ppm and a singlet at δ 8.14 ppm for the second most abundant thermodynamic (*meta*) product *t*-BuZn(C_6H_4-CF_3-3)(TMP)Na(TMEDA) $\mathbf{124}_{meta}$, and two doublets at δ 7.71 and 7.42 ppm for the minor *para* product *t*-BuZn(C_6H_4-CF_3-4)(TMP)Na(TMEDA) $\mathbf{124}_{para}$. These were recorded in a 20:11:1 ratio (see Scheme 6.29 and Table 6.1). Supporting the presence of four different organometallic compounds in solution, four resonances, at δ −59.9, −61.5, −61.7, and −61.8 ppm, were observed in the ^{19}F NMR spectrum of the reaction mixture [198, 199].

Since kinetically obtained **123** could be isolated as a crystalline solid, its reactivity towards amine HTMP could be investigated by ^{1}H NMR spectroscopy, in order to replicate the second step of the AMM*Zn* of $PhCF_3$ (Scheme 6.29). At room temperature, the almost instantaneous metalation of HTMP was observed alongside the formation of isobutene. At the same time, the aromatic region of the spectrum was dramatically changed, revealing the presence of a complex mixture of $\mathbf{124}_{ortho}$, $\mathbf{124}_{meta}$, and $\mathbf{124}_{para}$ whilst retaining evidence for some unreacted **123** and trifluoromethylbenzene. The ratio of species revealed by this region of the spectrum was the same as when the reaction of $PhCF_3$ with **122** was carried out at room temperature (Scheme 6.29). These spectroscopic studies, therefore, revealed that for this combination of base and aromatic substrate, the second step of the AMM*Zn* process, far from being a simple substitution of a *t*-Bu ligand by a TMP ligand (see

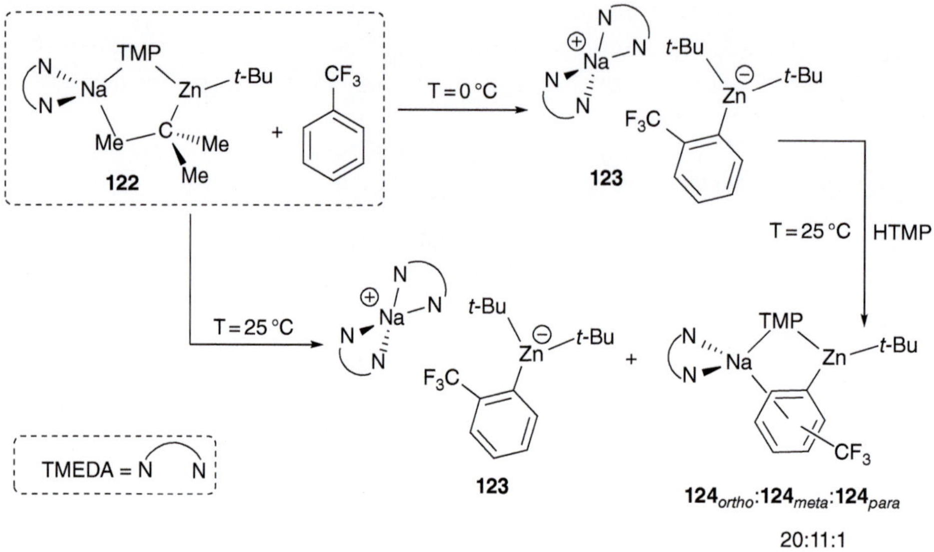

Scheme 6.29 Contrasting outcomes of the AMM*Zn* of fluoromethylbenzene by sodium zincate *t*-Bu_2Zn(TMP)Na(TMEDA) **122** at *T* = 0 and 25 °C.

Table 6.1 Aromatic chemical shifts in the ^1H NMR spectra of the products of metalation of $PhCF_3$ by sodium zincate **122** at 0 °C.

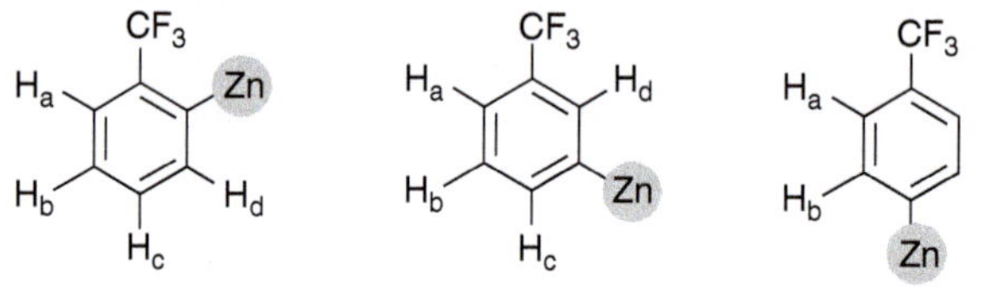

Compound	δ(H_a)	δ(H_b)	δ(H_c)	δ(H_d)
[*t*-$Bu_2Zn(C_6H_4$-CF_3-2)][Na$(TMEDA)_2$] **123**	7.51 (d)	7.20 (t)	6.96 (t)	8.06 (d)
t-BuZn(C_6H_4-CF_3-2)(TMP)Na(TMEDA) $\mathbf{124}_{ortho}$	7.54 (d)	7.10 (t)	6.98 (t)	7.85 (d)
t-BuZn(C_6H_4-CF_3-3)(TMP)Na(TMEDA) $\mathbf{124}_{meta}$	7.34 (d)	7.06 (t)	7.74 (d)	8.14 (s)
t-BuZn(C_6H_4-CF_3-4)(TMP)Na(TMEDA) $\mathbf{124}_{para}$	7.42 (d)	7.71 (d)	–	–

Source: Adapted from García-Álvarez et al. [207].

Scheme 6.27), also involved isomerization of the deprotonated aryl group. This had profound consequences for the final regioselectivity of the reaction.

Moving further down group 1, Mulvey has also investigated the reactivity of the potassium zincate Et_2Zn(TMP)K(PMDETA) **125** towards 4-substituted pyridines. Interestingly, in this case, as indicated by NMR analysis of the reaction crudes, the zincation stopped after the first step of the AMM*Zn* process, with there being no observable incorporation of TMP in the final products [200].

As noted above, the introduction of LiCl **29** has allowed the tractability and effectiveness of lower-Grignard and Hauser reagents to be improved—leading to the expression 'turbo reagent' [45–48]. Whereas the introduction of **29** is often explicit, examples have emerged of heterobimetallic systems in which salt-metathesis has generated it (and therefore the turbo-formulation) *in situ*. Hence, returning to lithium chemistry, a Li/Zn reagent for which NMR spectroscopic studies have been key to understanding its constitution and rationalizing its reactivity has been made up by the combination of three molar equivalents of LTMP **23** with $ZnCl_2$(TMEDA). This system was developed by Mongin, who showed that it efficiently deprotonated a range of aromatic substrates, offering greater yields and improved chemoselectivities than related monometallic systems. These data overall hinted at synergistic behaviour on the part of the metals [201–205]. Since the active base was prepared *in situ* upon combining the single-metal components and TMEDA in THF solvent, initially very little was known about its constitution. At first, an individually heterobimetallic CIP ate '$(TMP)_3$ZnLi' **126** structure was proposed [208]. However, later, and with the aid of DFT calculations, the base was reinterpreted as a mixture of **23** and *in situ* generated $(TMP)_2$Zn **127** cooperating in sequence. This view posited that the aromatic substrate first be lithiated by **23** and subsequently trapped and stabilized by the bis(amido)zinc (Scheme 6.30) [201–203, 206].

DOSY NMR experiments by García-Álvarez and Mulvey on the 3:1 mixture of **23** and $ZnCl_2$(TMEDA) in d_8-THF solution have been key to shedding light on the real constitution of '**126**' and understanding its ability to act as a zincating reagent/combination [207]. Combining ^{1}H and ^{7}Li ICC DOSY NMR data obtained using TPhN, PhN, and TMS at 27 °C in d_8-THF, it was found that at no point did TMEDA, $ZnCl_2$, and **23** belong to the same species. Rather, each exhibited a distinct diffusion coefficient. Interestingly, ^{7}Li NMR analysis of the reaction mixture displayed a broad signal at δ 1.09 ppm. This potentially corresponded to two species in solution; **23** but also LiCl **29**; the presence of the latter would verify that salt-metathesis had occurred. Correspondingly, low-temperature NMR studies revealed that at −80 °C this broad signal split into *three* resonances, suggesting a fast equilibrium between adduct **23(29)**, and possibly free **23** and **29** [207]. Further insights into the intriguing composition of this mixture were gained by comparing the estimated M_r values of the different species present in solution with those of **23**, **29**, **23(29)**$_2$, **23**(TMEDA), **23(29)**$_2$(TMEDA), **(29)**$_2$(TMEDA), **127**, and TMEDA, which were all sourced independently. The M_r values for these compounds and adducts were obtained by DOSY NMR in the presence of the same internal standards and under the same conditions of temperature and concentration as used for the parent 3:1:1 mixture of **23**, $ZnCl_2$, and TMEDA and results are summarized in Table 6.2 [207].

On the basis of the spectroscopic study outlined here, the active bimetallic reagent was described as a combination of **127** (which in solution did not interact with either **23** or **29**) and **23** (which formed a co-complex with two equivalents of **29**), with a degree of solvation by THF that was in equilibrium with TMEDA inclusion. These findings supported the mechanism proposed based on DFT calculations (Scheme 6.30) whereby lithiation preceded *trans*-metalation by *in situ*-formed

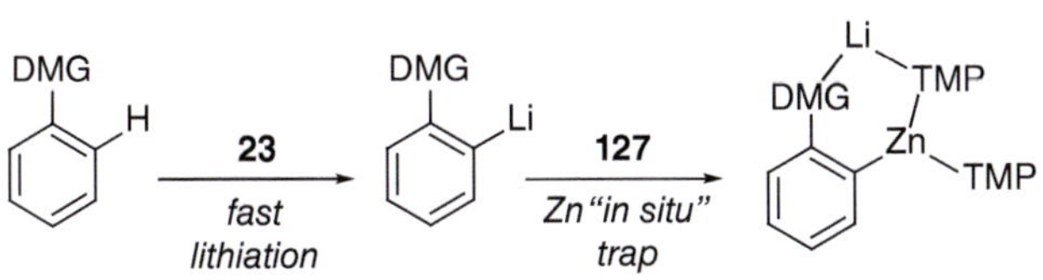

Scheme 6.30 Proposed mechanism (from DFT calculations) for the directed metalation of a monosubstituted arene using **23** and $ZnCl_2$(TMEDA) (3:1) to provide both base and source of carbanion trap **127** (TMEDA omitted from scheme for clarity).

Table 6.2 Estimated M_r from ICC DOSY NMR (expressed in g/mol) for all reasonable combinations of each component of a 3:1:1 mixture of LTMP **23**, $ZnCl_2$ and TMEDA[a].

Entry	Combination	"TMEDA"	"$(TMP)_2Zn$"	"LTMP"	"Li"
1	3:1:1 LTMP **23**:$ZnCl_2$:TMEDA	141	277	425	407
2	TMEDA, 116[b]	110	–	–	–
3	$Zn(TMP)_2$ **127**, 346[b]	–	280	–	–
4	**23**, 147[b]	–	–	336	–
5	LiCl **29**, 42[b]	–	–	–	342
6	**23(29)**$_2$	–	–	401	400
7	**23(29)**$_2$(TMEDA)	148	–	406	393
8	**23**(TMEDA)	116	–	361	–
9	**29**$_2$(TMEDA)	140	–	–	326

[a] The species in quotation marks are the components of these entities in solution and "Li" refers to the unique lithium-containing species extracted from the ^{7}Li DOSY data. [b] M_r value of the pure compound. *Source:* Adapted from García-Álvarez et al. [207].

127. However, they went further; establishing the active base composition to be **23(29)**$_2$ ± TMEDA, meaning that it is in fact best envisaged as a turbo-lithium amide reagent.

Lastly in this section, Knochel has developed a family of heteroleptic lithium zincates, $R_2Zn(OR')_2Li_2$, made by combining diorganozinc reagents R_2Zn (R = Et, *s*-Bu, *t*-Bu, 4-MeC_6H_4) with two equivalents of the lithium alkoxide R′OLi **128** [R′ = $Me_2N(CH_2)_2N(Me)(CH_2)_2$]. These have shown remarkable reactivities in promoting direct Zn–halogen exchange reactions in wide ranges of both aryl and heteroaryl iodides that bear sensitive functional groups, as well as in aryl bromides [208]. From a structural perspective, the formation of each of the ate complexes $Et_2Zn(OR')Li$ **129** and $Et_2Zn(OR')_2Li_2$ **130** was established by multinuclear ^{1}H, ^{13}C, and ^{7}Li NMR as well as DOSY NMR studies by assessing the cocomplexation reactions of Et_2Zn (i.e. Et = R) with variable amounts of R′OLi in d_8-toluene (Scheme 6.31). The most informative resonances observed were those attributable to the $ZnCH_2$ group. These were seen at δ 0.11 and 4.3 ppm in the ^{1}H and ^{13}C NMR spectra, respectively, in triorganozincate **129**, whereas in the tetraorganozincate **130** these signals were more shielded, appearing at δ 0.08 and 2.8 ppm (Scheme 6.31). ^{1}H NMR spectroscopic monitoring of the reaction of each of zincates **129** and **130** with two molar equivalents of two-iodoanisole revealed that while for **129** only 23% conversion to **131** was observed, for **130** virtually full conversion of the Zn-attached Et groups to give EtI occurred. Thus, $Ar_2Zn(OR')_2Li_2$ **131**(R′OLi) (Ar = 2-OMe-C_6H_4) was now obtained in 95% yield. These data demonstrated that the reaction took place with optimal atom economy, that is, both Et groups in **130** were active towards zinc–halogen exchange. The putative formulation of **131**(R′OLi) was next verified by ^{1}H DOSY NMR in d_8-THF, with results showing the codiffusion of both the aromatic and the alkoxide components and a mean D of 7.364×10^{-10} m^2/s observed. Both sets of resonances were clearly attributable to the same molecular entity, which evidently retained its integrity in this solvent system [208].

6.3.4.4 Magnesium/Zinc

Previous work by Ishihara, amongst others, has shown that the addition of substoichiometric amounts of $ZnCl_2$ to lower Grignard reagents can greatly enhance the chemoselectivity of these organometallics towards ketones [209]. These reactions have been proposed to take place via an active magnesium

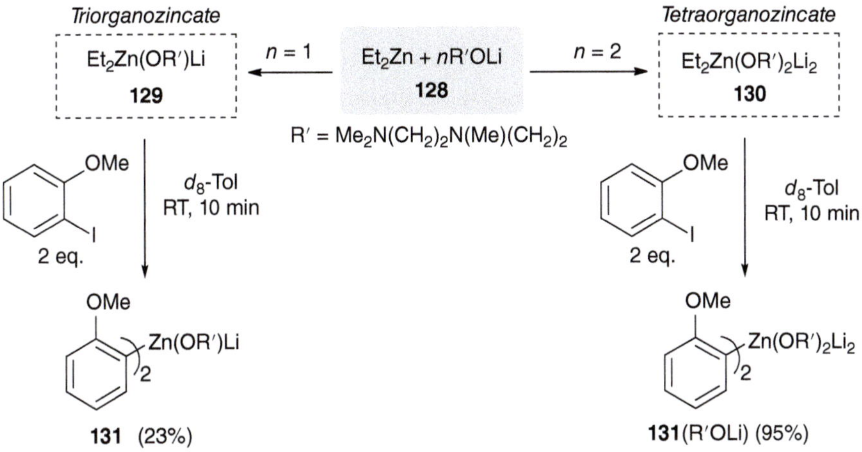

Scheme 6.31 Synthesis of heteroleptic, alkoxide-containing lithium zincates **129** and **130**, and the effect of ligand stoichiometry on direct Zn–I exchange reactions.

zincate species of the type $R_3ZnMgCl$. Structural and spectroscopic studies by Hevia have subsequently shown that the treatment of 3 eq. *t*-BuMgCl **132** with $ZnCl_2$ in THF allows the isolation of SIP [*t*-Bu_3Zn][$Cl_3Mg_2(THF)_6$] **133** [210]. Interestingly, in d_8-THF **133** was found by ^{1}H NMR to exist in equilibrium with its single metal components; **132** (δ 0.80 ppm) and *in situ*-created *t*-Bu_2Zn (δ 0.88 ppm). The coexistence of these three organometallic species was also observed *in situ* for mixtures prepared in THF, using non-deuterium NMR to avoid the removal of any potentially volatile components, such as *t*-Bu_2Zn, which could influence the composition of the mixture of species present (Scheme 6.32) [211]. These studies further indicated the equilibrium observed between **133** and its homometallic components to be concentration-dependent, being significantly driven towards the presence of the monometal species under dilute conditions. Interestingly, NMR spectroscopic monitoring of the addition of three equivalents of 4-iodotoluene to a **132**/$ZnCl_2$ mixture showed the almost instantaneous formation of [$(4\text{-MeC}_6H_4)_3Zn$][$Cl_3Mg_2(THF)_6$] **134**, with the disappearance of all resonances attributed to metalated *tert*-butyl groups explained by their transformation into components of *t*-BuI, the appearance of which was identified in the ^{1}H NMR spectrum by a distinct singlet at δ 1.91 ppm. The most notable spectroscopic feature of **134** was the presence of an *ipso* carbon signal at δ 162.8 ppm in the ^{13}C NMR spectrum. Now bonded directly to zinc, this carbon centre produced a signal remarkably more downfield than its analogue in the aryl iodide (δ 91.1 ppm) [210]. The work also noted the synthetic importance of introducing $ZnCl_2$; in contrast with the smooth reactivity of **134** towards zinc-iodine exchange, neither *t*-Bu_2Zn nor **132** underwent exchange with 4-iodotoluene on their own.

Further investigations into the constitution of magnesium zincates made by the salt metathesis of $ZnCl_2$ with Grignard reagents revealed that reaction of 3 eq. EtMgCl with $ZnCl_2$ allows the isolation and characterisation (X-ray crystallography and ^{1}H and ^{13}C NMR spectroscopy) of novel magnesium 'zinc-rich' zincate [Et_5Zn_2][$Cl_3Mg_2(THF)_6$] **135**. The complicated constitution of this species in THF solution was assessed by variable-temperature ^{1}H DOSY NMR studies [212]. At room temperature, the ^{1}H NMR spectrum in d_8-THF showed a single set of signals for the Et groups, with a well-resolved quartet at δ −0.21 ppm attributable to $ZnCH_2$. However, variable temperature studies revealed a dramatic change in the shape and the resolution of this resonance, which upon cooling to −43 °C had become broad. In addition, ^{1}H ICC DOSY NMR studies using TPhN, PhN, and TMS at 25 and −40 °C showed significantly different values for coefficient *D* for

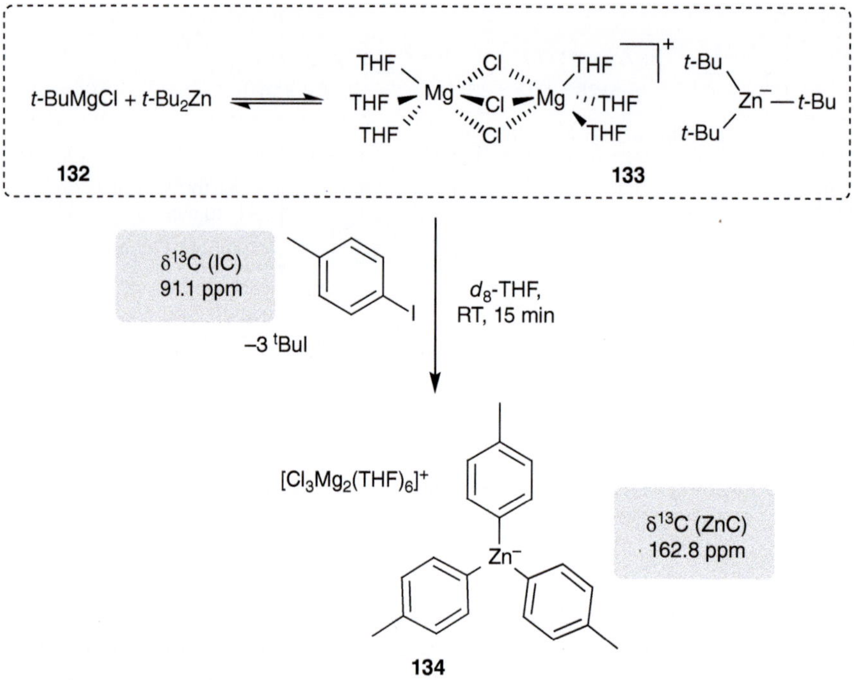

Scheme 6.32 Concentration-dependent, dynamic equilibrium of **133** with its monometallic components and its reactivity towards 4-iodotoluene to give magnesium zincate **134**.

the Et groups: $D(25\ °C) = 1.24(3) \times 10^{-9}\ m^2/s$, $D(-40\ °C) = 2.69(1) \times 10^{-10}\ m^2/s$. These data suggested that in solution, the constitution of the anion present in **135** changed dramatically as the temperature decreased. Highlighting the complexity of these bimetallic systems in THF solution, these findings were interpreted as indicative of the presence of a temperature-dependent equilibrium between **135** and a combination of triorganozincate $[Et_3Zn][Cl_3Mg_2(THF)_6]$ **136** and free Et_2Zn, with exchange being fast at room temperature.

Further support for the exchange thesis outlined above was found when introducing ferrocene as an internal standard in the ^1H NMR spectroscopic analysis of **135**. At room temperature, the spectrum now initially displayed a quartet at $\delta = -0.25$ ppm for the $ZnCH_2$ group, the integration of which accounted for 10H, pointing to five ethyl groups. As Et_2Zn is volatile, this solution was put under vacuum for one hour and the solid residue obtained was then redissolved in d_8-THF. Interestingly, the ^1H NMR spectrum of this sample showed a significantly upfield quartet at δ −0.75 ppm attributable to $ZnCH_2$. Its relative integration against the internal standard of 6.3H (ca. three ethyl groups) supported the presence of the $(Et_3Zn)^-$ ion in **136** and also the creation of Et_2Zn or a volatile THF-solvate thereof from **135** (Scheme 6.33) [212].

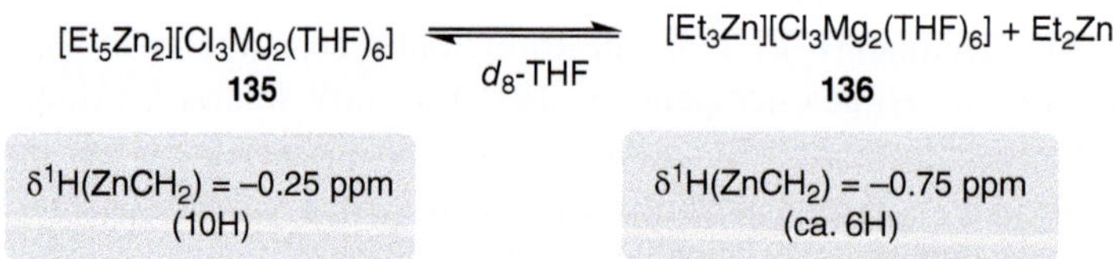

Scheme 6.33 A temperature-dependent equilibrium process affecting **135** in THF solution was evaluated by integrating $ZnCH_2$ resonances against an internal standard.

6.4 Concluding Remarks

The development of new applications of s-block metals and heterobimetallic complexes in the functionalization of organic molecules has benefitted considerably from recent advances made in our understanding of the solution constitutions of these systems. This knowledge has been largely accrued through a variety of multidimensional and multinuclear NMR techniques that probe not just bonding connectivity but also through space and exchange interactions, and which have deciphered the key roles played by aggregation and solvation in defining and controlling reaction mechanisms. Rising to a position of significance amongst these techniques in recent years, DOSY NMR spectroscopy has enabled the correlation of solid-state crystal structures determined by X-ray diffraction with solution structures, uncovering new aggregates and, in some cases, cooperative combinations of reagents not reflected by the solid-state data yet which may be active in solution. DOSY NMR has been successfully used to separate NMR signals according to the diffusion coefficient (D) of individual species in a complex system, though the lack of a simple relationship between D and molecular weight (M_r) has had to be addressed. Whereas initial studies required the use of multiple internal references in order to generate ICCs, a major development was seen with the introduction of the alternative ECC method. Requiring the use of just a single reference (even potentially the solvent), this method has provided increasingly accurate and reliable M_r data for diverse organometallics. Analysis has been compatible with the use of a wide range of temperatures, enabling monitoring under 'reaction conditions' and the more confident identification of intermediates at both relatively low and high temperatures; the former often needed with traditional reagents such as organolithiums and use of the latter synonymous with the advent of new heterobimetallic complexes. By way of an example, ECC DOSY NMR analysis has been used to shed light on the superficially simple Turbo-Hauser bases, establishing that these exist in THF solution as several hetero- and homometallic species in equilibrium with each other.

The use of multinuclear NMR in the reaction monitoring of mixed-metal ligand combinations for the deprotonation of aromatic molecules has also been instrumental in aiding understanding of how different metals can operate together synergistically and how such cooperative effects can be maximized for specific applications. Two main types of mechanistic pathway have emerged from this work; namely (i) synchronized cooperation and (ii) stepwise cooperation. In the former type, both metals must be integrated within the same molecular scaffold, giving an ate complex formulation which, having enhanced basicity through resulting cooperativity, can react with a substrate. Contrastingly, in the latter type, two single-metal entities fail to or barely undergo cocomplexation. The group 1 reagent now acts first as a Brønsted base, deprotonating a substrate to give a conjugate base that can then undergo *in situ* transmetallation to afford a crossover complex that incorporates the lower polarity metal. This approach has been described in the literature as TMT.

Looking at the broader picture, the use of NMR techniques to assess constitution and monitor reactions of s-block homo- and heterometallic complexes has gone a long way in recent years towards filling a major knowledge gap. This has historically existed between X-ray crystallographic determinations that fail to necessarily provide an accurate representation of an organometallic's active formulation in solution, and synthetic studies where organometallic species have often gone unidentified or their formulations have at best been extrapolated, subsequent to reactions being quenched and final product(s) fully characterized. Building on the recent advances showcased in this chapter, it seems certain that the use of advanced NMR techniques in delineating structural constitutions in solution will stimulate many more synthetic applications of main group organometallics in the future.

References

1 Schlenk, W. and Holtz, J. (**1917**). *Ber. Dtsch. Chem. Ges. 50*: 262–274.
2 Ziegler, K. and Colonius, H. (**1930**). *Liebigs Ann. Chem. 479*: 135–149.
3 Clayden, J. (**2002**). *Organolithiums: Selectivity for Synthesis, Pergamon*. Oxford.
4 Schlosser, M. (**2002**). *Organometallics in Synthesis*, 2nd ed. (Ed.: M. Schlosser), Chapter 1. New York: Wiley.
5 Schlosser, M. (**2005**). *Angew. Chem. Int. Ed. 44*: 376–393.
6 Wu, G. and Huang, M. (**2006**). *Chem. Rev. 106*: 2596–2616.
7 Smid, J., Van Beylen, M., and Hogen-Esch, T.E. (**2006**). *Prog. Polym. Sci. 31*: 1041–1067.
8 Carlotti, S., Desbois, P., Billouard, C., and Deffieux, A. (**2006**). *Polym. Int. 55*: 1126–1131.
9 Reich, H.J. (**2013**). *Chem. Rev. 113*: 7130–7178.
10 Oulyadi, H. (**2018**). *Synthesis 50*: 3603–3614.
11 Heard, P. J. (**2008**). In *The Chemistry of Organomagnesium Compounds (Patai Series: The Chemistry of Functional Groups)*, (Eds.: Z. Rappoport, I. Marek), pp. 131–154. Hoboken: Wiley.
12 Westerhausen, M., Koch, A., Görls, H., and Krieck, S. (**2017**). *Chem. Eur. J. 23*: 1456–1483.
13 Lochmann, L. and Janata, M. (**2013**). *Chem. Listy. 107*: 777–782.
14 Wanklyn, J.A. (**1858**). *Liebigs Ann. 108*: 67–79.
15 Wanklyn, J.A. (**1859**). *Proc. R. Soc. London 9*: 341–345.
16 Tuckmantel, W., Oshima, K., and Nozaki, H. (**1986**). *Chem. Ber. 119*: 1581–1593.
17 Isobe, M., Kondo, S., Nagasawa, N., and Goto, T. (**1977**). *Chem. Lett.*: 679–682.
18 Kondo, Y., Shilai, M., Uchiyama, M., and Sakamoto, T. (**1999**). *J. Am. Chem. Soc. 121*: 3539–3540.
19 Knochel, P. and Jones, P. (**1998**). *Organozinc Reagents*. Oxford University Press.
20 Harrison-Marchand, A. and Mongin, F. (**2013**). *Chem. Rev. 113*: 7470–7562.
21 Harrison-Marchand, A. and Mongin, F. (**2013**). *Chem. Rev. 113*: 7563–7727.
22 Davies, R.P. (**2011**). *Coord. Chem. Rev. 255*: 1226–1251.
23 Mulvey, R.E., Mongin, F., Uchiyama, M., and Kondo, Y. (**2007**). *Angew. Chem. Int. Ed. 46*: 3802–3824.
24 Mulvey, R.E. (**2009**). *Acc. Chem. Res. 42*: 743–755.
25 Mulvey, R.E. (**2013**). *Dalton Trans. 42*: 6676–6693.
26 Harford, P.J., Peel, A.J., Chevallier, F. et al. (**2014**). *Dalton Trans. 43*: 14181–14203.
27 Robertson, S.D., Uzelac, M., and Mulvey, R.E. (**2019**). *Chem. Rev. 119*: 8332–8405.
28 Armstrong, D.R., Kennedy, A.R., Mulvey, R.E. et al. (**2012**). *Chem. Sci. 3*: 2700–2707.
29 Armstrong, D.R., Crosbie, E., Hevia, E. et al. (**2014**). *Chem. Sci. 5*: 3031–3045.
30 Uzelac, M., Kennedy, A.R., Hevia, E., and Mulvey, R.E. (**2016**). *Angew. Chem. Int. Ed. 55*: 13147–13150.
31 Stejskal, E.O. and Tanner, J.E. (**1965**). *J. Chem. Phys. 42*: 288–292.
32 Tanner, J.E. (**1965**). *Rev. Sci. Instrum. 36*: 1086–1087.
33 Stejskal, E.O. (**1965**). *J. Chem. Phys. 43*: 3597–3603.
34 Morris, K.F. and Johnson, C.S. (**1992**). *J. Am. Chem. Soc. 114*: 3139–3141.
35 Keresztes, I. and Williard, P.G. (**2000**). *J. Am. Chem. Soc. 122*: 10228–10229.
36 Li, D., Kagan, G., Hopson, R., and Williard, P.G. (**2009**). *J. Am. Chem. Soc. 131*: 5627–5634.
37 Li, D., Keresztes, I., Hopson, R., and Williard, P.G. (**2009**). *Acc. Chem. Res. 42*: 270–280.
38 Neufeld, R. and Stalke, D. (**2015**). *Chem. Sci. 6*: 3354–3364.
39 Kreyenschmidt, A.-K., Bachmann, S., Niklas, T., and Stalke, D. (**2017**). *ChemistrySelect 2*: 6957–6960.
40 Evans, R., Deng, Z., Rogerson, A.K. et al. (**2013**). *Angew. Chem. Int. Ed. 52*: 3199–3202.

41 Macchioni, A., Ciancaleoni, G., Zuccaccia, C., and Zuccaccia, D. (**2008**). *Chem. Soc. Rev.* *37*: 479–489.
42 Raya-Barón, Á., Oña-Burgos, P., and Fernández, I. (**2019**). *Ann. Rep. NMR Spect.* *98*: 125–191.
43 Crockett, M.P., Zhang, H., Thomas, C.M., and Byers, J.A. (**2019**). *Chem. Commun.* *55*: 14426–14429.
44 Balkenhohl, M. and Knochel, P. (**2020**). *Chem. Eur. J.* *26*: 3688–3697.
45 Barl, N.M., Werner, V., Sämann, C., and Knochel, P. (**2014**). *Heterocycles* *88*: 827–844.
46 Benischke, A.D., Ellwart, M., Becker, M.R., and Knochel, P. (**2016**). *Synthesis* *48*: 1101–1107.
47 Zhu, M., Liu, L., Yu, H.-T. et al. (**2018**). *Chem. Eur. J.* *24*: 19122–19135.
48 Ziegler, D.S., Wei, B., and Knochel, P. (**2019**). *Chem. Eur. J.* *25*: 2695–2703.
49 Sott, R., Håkansson, M., and Hilmersson, G. (**2006**). *Organometallics* *25*: 6047–6053.
50 Bauer, W., Winchester, W.R., and Schleyer, P.v.R. (**1987**). *Organometallics* *6*: 2371–2379.
51 Bauer, W. and Schleyer, P.v.R. (**1992**). *Adv. Carbanion Chem.* *1*: 89–175.
52 Carbone, G., O'Brien, P., and Hilmersson, G. (**2010**). *J. Am. Chem. Soc.* *132*: 15445–15450.
53 Gil, V.M.S. and Oliveira, N.C. (**1990**). *J. Chem. Educ.* *67*: 473–478.
54 Job, P. (**1925**). *Compt. Rend.* *180*: 928.
55 Job, P. (**1928**). *Ann. Chim.* *9*: 113–203.
56 Job, P. (**1936**). *Ann. Chim.* *6*: 97–144.
57 Kissling, R.M. and Gagne, M.R. (**2001**). *J. Org. Chem.* *66*: 9005–9010.
58 Renny, J.S., Tomasevich, L.L., Tallmadge, E.H., and Collum, D.B. (**2013**). *Angew. Chem., Int. Ed.* *52*: 11998–12013.
59 De Vries, T.S., Goswami, A., Liou, L.R. et al. (**2009**). *J. Am. Chem. Soc.* *131*: 13142–13154.
60 McNeil, A.J. and Collum, D.B. (**2005**). *J. Am. Chem. Soc.* *127*: 5655–5661.
61 McNeil, A.J., Toombes, G.E.S., Chandramouli, S.V. et al. (**2004**). *J. Am. Chem. Soc.* *126*: 5938–5939.
62 McNeil, A.J., Toombes, G.E.S., Gruner, S.M. et al. (**2004**). *J. Am. Chem. Soc.* *126*: 16559–16568.
63 Romesberg, F.E., Bernstein, M.P., Gilchrist, J.H. et al. (**1993**). *J. Am. Chem. Soc.* *115*: 3475–3483.
64 Liou, L.R., McNeil, A.J., Ramirez, A. et al. (**2008**). *J. Am. Chem. Soc.* *130*: 4859–4868.
65 Qu, B. and Collum, D.B. (**2006**). *J. Am. Chem. Soc.* *128*: 9355–9360.
66 Jin, K.J. and Collum, D.B. (**2015**). *J. Am. Chem. Soc.* *137*: 14446–14455.
67 Evans, D.A., Bartroli, J., and Shih, T.L. (**1981**). *J. Am. Chem. Soc.* *103*: 2127–2129.
68 Evans, D.A., Ennis, M.D., and Mathre, D.J. (**1982**). *J. Am. Chem. Soc.* *104*: 1737–1739.
69 Ma, Y., Hoepker, A.C., Gupta, L. et al. (**2010**). *J. Am. Chem. Soc.* *132*: 15610–15623.
70 Collum, D.B. (**1993**). *Acc. Chem. Res.* *26*: 227–234.
71 Dean, R.K., Recklinga, A.M., Chena, H. et al. (**2013**). *Dalton Trans.* *42*: 3504–3520.
72 Tomasevich, L.L. and Collum, D.B. (**2014**). *J. Am. Chem. Soc.* *136*: 9710–9718.
73 Tallmadge, E.H. and Collum, D.B. (**2015**). *J. Am. Chem. Soc.* *137*: 13087–13095.
74 Jermaks, J., Tallmadge, E.H., Keresztes, I., and Collum, D.B. (**2018**). *J. Am. Chem. Soc.* *140*: 3077–3090.
75 Bruneau, A.M., Liou, L., and Collum, D.B. (**2014**). *J. Am. Chem. Soc.* *136*: 2885–2891.
76 Nahm, S. and Weinreb, S.M. (**1981**). *Tetrahedron Lett.* *22*: 3815–3818.
77 Houghton, M.J. and Collum, D.B. (**2016**). *J. Org. Chem.* *81*: 11057–11064.
78 Wright, S.W., Choi, C., Chung, S. et al. (**2015**). *Org. Lett.* *17*: 5204–5207.
79 Lucht, B.L. and Collum, D.B. (**1994**). *J. Am. Chem. Soc.* *116*: 6009–6010.
80 Collum, D.B. and Lucht, B.L. (**1995**). *J. Am. Chem. Soc.* *117*: 9863–9874.
81 Hilmersson, G. and Davidsson, Ö. (**1995**). *J. Org. Chem.* *60*: 7660–7669.
82 Hilmersson, G., Ahlberg, P., and Davidsson, Ö. (**1996**). *J. Am. Chem. Soc.* *116*: 3539–3540.
83 Hilmersson, G. (**2000**). *Chem. Eur. J.* *6*: 3069–3075.
84 Rönnholm, P. and Hilmersson, G. (**2011**). *ARKIVOC*: 200–210.

85 Rönnholm, P., Nilsson Lill, S.O., Gräfenstein, J. et al. (**2012**). *ChemPlusChem 77*: 799–806.

86 Granander, J., Sott, R., and Hilmersson, G. (**2002**). *Tetrahedron 58*: 4717–4725.

87 Granander, J., Sott, R., and Hilmersson, G. (**2003**). *Tetrahedron: Asymmetry 14*: 439–447.

88 Granander, J., Eriksson, J., and Hilmersson, G. (**2006**). *Tetrahedron: Asymmetry 17*: 2021–2027.

89 Sott, R., Granander, J., Dinér, P., and Hilmersson, G. (**2004**). *Tetrahedron: Asymmetry 15*: 267–274.

90 Williard, P.G. and Sun, C. (**1997**). *J. Am. Chem. Soc. 119*: 11693–11694.

91 Hilmersson, G. and Malmros, B. (**2001**). *Chem. Eur. J. 7*: 337–341.

92 Granander, J., Sott, R., and Hilmersson, G. (**2006**). *Chem. Eur. J. 12*: 4191–4197.

93 Augé, S., Schmit, P.-O., Crutchfield, C.A. et al. (**2009**). *J. Phys. Chem. B. 113*: 1914–1918.

94 Kagan, G., Li, W., Hopson, R., and Williard, P.G. (**2010**). *Org. Lett. 12*: 520–523.

95 Li, D., Sun, C., Liu, J. et al. (**2008**). *J. Org. Chem. 73*: 2373–2381.

96 Liu, J., Li, D., Sun, C., and Williard, P.G. (**2008**). *J. Org. Chem. 73*: 4045–4052.

97 Lappert, M.F., Slade, M.J., Singh, A. et al. (**1983**). *J. Am. Chem. Soc. 105*: 302–304.

98 Hevia, E., Kennedy, A.R., Mulvey, R.E. et al. (**2013**). *Chem. Eur. J. 19*: 14069–14075.

99 Remenar, J.F., Lucht, B.L., Kruglyak, D. et al. (**1997**). *J. Org. Chem. 62*: 5748–5754.

100 Roşca, S.C., Roşca, D.A., Dorcet, V. et al. (**2013**). *Dalton Trans. 42*: 9361–9375.

101 Tatic, T., Meindl, K., Henn, J. et al. (**2010**). *Chem. Commun. 46*: 4562–4564.

102 Pöppler, A.C., Meinholz, M.M., Faßhuber, H. et al. (**2012**). *Organometallics 31*: 42–45.

103 Lecachey, B., Oulyadi, H., Lameiras, P. et al. (**2010**). *J. Org. Chem. 75*: 5976–5983.

104 Fox, T., Hausmann, H., and Günther, H. (**2004**). *Magn. Reson. Chem. 42*: 788–794.

105 Bachmann, S., Neufeld, R., Dzemski, M., and Stalke, D. (**2016**). *Chem. Eur. J. 22*: 8462–8465.

106 Bachmann, S., Gernert, B., and Stalke, D. (**2016**). *Chem. Commun. 52*: 12861–12864.

107 Neufeld, R., John, M., and Stalke, D. (**2015**). *Angew. Chem. Int. Ed. 54*: 6994–6998.

108 Koehne, I., Bachmann, S., Herbst-Irmer, R., and Stalke, D. (**2017**). *Angew. Chem. Int. Ed. 56*: 15141–15145.

109 Lambert, C., Schleyer, P.v.R., Pieper, U., and Stalke, D. (**1992**). *Angew. Chem. Int. Ed. Engl. 31*: 77–79.

110 Capriati, V., Pernaa, F.M., and Salomone, A. (**2014**). *Dalton Trans. 43*: 14204–14210.

111 García-Álvarez, J., Hevia, E., and Capriati, V. (**2015**). *Eur. J. Org. Chem.*: 6779–6799.

112 Vidal, C., García-Álvarez, J., Hernán-Gómez, A. et al. (**2016**). *Angew. Chem. Int. Ed. 55*: 16145–16148.

113 Cicco, L., Rodriguez-Alvarez, M.J., Perna, F.M. et al. (**2017**). *Green Chem. 19*: 3069–3077.

114 Fairley, M., Bole, L.J., Mulks, F.F. et al. (**2020**). *Chem. Sci. 11*: 6500–6509.

115 García-Álvarez, J., Hevia, E., and Capriati, V. (**2018**). *Chem. Eur. J. 24*: 14854–14863.

116 Su, C., Guang, J., and Williard, P.G. (**2014**). *J. Org. Chem. 79*: 1032–1039.

117 Armstrong, D.R., Kennedy, A.R., Mulvey, R.E., and Rowlings, R.B. (**1999**). *Angew. Chem. Int. Ed. 38*: 131–133.

118 Andrews, P.C., Kennedy, A.R., Mulvey, R.E. et al. (**2000**). *Angew. Chem. Int. Ed. 39*: 1960–1962.

119 Clegg, W., Henderson, K.W., Kennedy, A.R. et al. (**2001**). *Angew. Chem. Int. Ed. 40*: 3902–3905.

120 Martínez-Martínez, A.J., Kennedy, A.R., Mulvey, R.E., and O'Hara, C.T. (**2014**). *Science 346*: 834–837.

121 Martínez-Martínez, A.J., Justice, S., Fleming, B.J. et al. (**2017**). *Sci. Adv. 3*: 1–9.

122 Martínez-Martínez, A.J., Armstrong, D.R., Conway, B. et al. (**2014**). *Chem. Sci. 5*: 771–781.

123 Hauser, C.R. and Walker, H.G. (**1947**). *J. Am. Chem. Soc. 69*: 295–297.

124 Piller, F.M., Appukkuttan, P., Gavryushin, A. et al. (**2008**). *Angew. Chem. Int. Ed. 47*: 6802–6806.

125 Stern, D., Finkelmeier, N., and Stalke, D. (**2011**). *Chem. Commun. 47*: 2113–2115.

126 Krasovskiy, A., Krasovskaya, V., and Knochel, P. (**2006**). *Angew. Chem. Int. Ed 45*: 2958–2961.

127 García-Álvarez, P., Graham, D.V., Hevia, E. et al. (**2008**). *Angew. Chem. Int. Ed. 47*: 8079–8081.

128 Armstrong, D.R., García-Álvarez, P., Kennedy, A.R. et al. (**2010**). *Angew. Chem. Int. Ed. 49*: 3185–3188.

129 Neufeld, R. and Stalke, D. (**2016**). *Chem. Eur. J. 22*: 12624–12628.

130 Ashby, E.C. and Smith, M.B. (**1964**). *J. Am. Chem. Soc. 86*: 4363–4370.

131 Smith, M.B. and Becker, W.E. (**1967**). *Tetrahedron 23*: 4215–4227.

132 Walker, F.W. and Ashby, E.C. (**1969**). *J. Am. Chem. Soc. 91*: 3845–3850.

133 Neufeld, R., Teuteberg, T.L., Herbst-Irmer, R. et al. (**2016**). *J. Am. Chem. Soc. 138*: 4796–4806.

134 García-Álvarez, J., Graham, D.V., Kennedy, A.R. et al. (**2006**). *Chem. Commun.*: 3208–3210.

135 Naka, H., Uchiyama, M., Matsumoto, Y. et al. (**2007**). *J. Am. Chem. Soc. 129*: 1921–1930.

136 Mulvey, R.E., Armstrong, D.R., Conway, B. et al. (**2011**). *Inorg. Chem. 50*: 12241–12251.

137 Klatt, T., Groll, K., and Knochel, P. (**2013**). *Chem. Commun. 49*: 6953–6955.

138 Nagaradja, E., Chevallier, F., Roisnel, T. et al. (**2014**). *Org. Biomol. Chem. 12*: 1475–1487.

139 Fuentes, M.Á., Kennedy, A.R., Mulvey, R.E. et al. (**2015**). *Chem. Eur. J. 21*: 14815–14822.

140 Clegg, W., Liddle, S.T., Snaith, R., and Wheatley, A.E.H. (**1998**). *New J. Chem. 22*: 1323–1326.

141 Armstrong, D.R., Davies, J.E., Davies, R.P. et al. (**1999**). *New J. Chem. 23*: 35–41.

142 McLellan, R., Uzelac, M., Bole, L. et al. (**2019**). *Synthesis 51*: 1207–1215.

143 McLellan, R., Uzelac, M., Kennedy, A.R. et al. (**2017**). *Angew. Chem. Int. Ed. 56*: 9566–9570.

144 Uzelac, M., Kennedy, A.R., and Hevia, E. (**2017**). *Inorg. Chem. 56*: 8615–8626.

145 L'Helgoual'ch, J.-M., Bentabed-Ababsa, G., Chevallier, F. et al. (**2008**). *Chem. Commun.*: 5375–5377.

146 Snégaroff, K., L'Helgoual'ch, J.-M., Bentabed-Ababsa, G. et al. (**2009**). *Chem. Eur. J., 15*: 10280–10290.

147 Snégaroff, K., Komagawa, S., Yonehara, M. et al. (**2010**). *J. Org. Chem. 75*: 3117–3120.

148 Bentabed-Ababsa, G., Ely, S.C.S., Hesse, S. et al. (**2010**). *J. Org. Chem. 75*: 839–847.

149 Dayaker, G., Sreeshailam, A., Chevallier, F. et al. (**2010**). *Chem. Commun. 46*: 2862–2864.

150 Sreeshailam, A., Dayaker, G., Chevallier, F. et al. (**2011**). *Eur. J. Org. Chem.*: 3715–3718.

151 Snégaroff, K., Nguyen, T.T., Marquise, N. et al. (**2011**). *Chem. Eur. J. 17*: 13284–13297.

152 Kronenburg, C.M.P., Jastrzebski, J.T.H., Spek, A.L., and van Koten, G. (**1998**). *J. Am. Chem. Soc. 120*: 9968–9688.

153 Boche, G., Bosold, F., Marsch, M., and Harms, K. (**1998**). *Angew. Chem. Int. Ed. 37*: 1684–1686.

154 John, M., Auel, C., Behrens, C. et al. (**2000**). *Chem. Eur. J. 6*: 3060–3068.

155 Gschwind, R.M., Xie, X., Rajamohanan, P.R. et al. (**2001**). *J. Am. Chem. Soc. 123*: 7299–7304.

156 Pearson, R.G. and Gregory, C.D. (**1976**). *J. Am. Chem. Soc. 98*: 4098–4104.

157 Gschwind, R.M., Rajamohanan, P.R., John, M., and Boche, G. (**2000**). *Organometallics 19*: 2868–2873.

158 Hoffman, D., Bauer, W., and Schleyer, P.v.R. (**1990**). *J. Chem. Soc, Chem. Commun.*: 208–211.

159 Bertz, S.H., Eriksson, M., Miao, G., and Snyder, J.P. (**1996**). *J. Am. Chem. Soc. 118*: 10906–10907.

160 Bertz, S.H., Chopra, A., Eriksson, M. et al. (**1999**). *Chem. Eur. J. 5*: 2680–2691.

161 Bertz, S.H., Vellekoop, A.S., Smith, R.A.J., and Snyder, J.P. (**1995**). *Organometallics 14*: 1213–1220.

162 Henze, W., Vyater, A., Krause, N., and Gschwind, R.M. (**2005**). *J. Am. Chem. Soc. 127*: 17335–17342.

163 Xie, X., Auel, C., Henze, W., and Gschwind, R.M. (**2003**). *J. Am. Chem. Soc. 125*: 1595–1601.

164 Neumeier, N. and Gschwind, R.M. (**2014**). *J. Am. Chem. Soc. 136*: 5765–5772.

165 Bertz, S.H., Cope, S., Murphy, M. et al. (**2007**). *J. Am. Chem. Soc. 129*: 7208–7209.

166 Bertz, S.H., Cope, S., Dorton, D. et al. (**2007**). *Angew. Chem. Int. Ed. 46*: 7082–7085.

167 Gärtner, T., Henze, W., and Gschwind, R.M. (**2007**). *J. Am. Chem. Soc. 129*: 11362–11363.

168 Bertz, S.H., Murphy, M.D., Ogle, C.A., and Thomas, A.A. (**2010**). *Chem. Commun. 46*: 1255–1256.

169 Bartholomew, E.R., Bertz, S.H., Cope, S. et al. (**2008**). *Chem. Commun.*: 1176–1177.

170 Bertz, S.H., Hardin, R.A., Heavey, T.J. et al. (**2013**). *Chem. Eur. J. 19*: 10138–10141.

171 Henze, W., Gärtner, T., and Gschwind, R.M. (**2008**). *J. Am. Chem. Soc. 130*: 13718–13726.

172 Yamanaka, M. and Nakamura, E. (**2005**). *J. Am. Chem. Soc. 127*: 4697–4706.

173 Rossiter, B.E. and Swingle, N.M. (**1992**). *Chem. Rev. 92*: 771–806.

174 Dieter, R. K. (**2002**). Modern Organocopper Chemistry, (Ed.: N. Krause), Wiley-VCH, Weinheim, pp. 79–144.

175 Tezuka, N., Shimojo, K., Hirano, K. et al. (**2016**). *J. Am. Chem. Soc. 138*: 9166–9171.

176 Eriksson, J., Arvidsson, P.I., and Davidsson, O. (**2000**). *J. Am. Chem. Soc. 122*: 9310–9311.

177 Rossiter, B.E., Eguchi, M., Miao, G. et al. (**1993**). *Tetrahedron 49*: 965–986.

178 Sung, S., Braddock, D.C., Armstrong, A. et al. (**2015**). *Chem. Eur. J. 21*: 7179–7192.

179 H. Günther in Encyclopedia of Nuclear Magnetic Resonance, (**1996**). Vol. *5* (Eds.: D. M. Grant R. K. Harris), pp. 2807–2828. New York: Wiley.

180 Davies, R.P., Hornauer, S., and Hitchcock, P.B. (**2007**). *Angew. Chem. Int. Ed. 46*: 5191–5194.

181 Davies, R.P. and Hornauer, S. (**2007**). *Chem. Commun.*: 304–306.

182 Bomparola, R., Davies, R.P., Hornauer, S., and White, A.J.P. (**2009**). *Dalton Trans.*: 1104–1106.

183 Usui, S., Hashimoto, Y., Morey, J.V. et al. (**2007**). *J. Am. Chem. Soc. 129*: 15102–15103.

184 Komagawa, S., Usui, S., Haywood, J. et al. (**2012**). *Angew. Chem. Int. Ed. 51*: 12081–12085.

185 Peel, A.J., Ackroyd, R., and Wheatley, A.E.H. (**2017**). *Chem. Sci. 8*: 4904–4916.

186 Peel, A.J., Tezuka, N., D'Rozario, J.M. et al. (**2019**). *Chem. Sci. 10*: 3385–3400.

187 Uchiyama, M., Matsumoto, Y., Nobuto, D. et al. (**2006**). *J. Am. Chem. Soc. 128*: 8748–8750.

188 Clegg, W., Dale, S.H., Hevia, E. et al. (**2006**). *Angew. Chem. Int. Ed. 45*: 2370–2374.

189 Clegg, W., Dale, S.H., Harrington, R.W. et al. (**2006**). *Angew. Chem. Int. Ed. 45*: 2374–2377.

190 García, F., McPartlin, M., Morey, J.V. et al. (**2008**). *Eur. J. Org. Chem.*: 644–647.

191 Baillie, S.E., Blair, V.L., Blakemore, D.C. et al. (**2012**). *Chem. Commun. 48*: 1985–1987.

192 Clegg, W., Dale, S.H., Drummond, A.M. et al. (**2006**). *J. Am. Chem. Soc. 128*: 7434–7435.

193 Mulvey, R.E. (**2006**). *Organometallics 25*: 1060–1075.

194 Uchiyama, M., Matsumoto, Y., Usui, S. et al. (**2007**). *Angew. Chem. Int. Ed. 46*: 926–929.

195 Kondo, Y., Morey, J.V., Morgan, J.C. et al. (**2007**). *J. Am. Chem. Soc. 129*: 12734–12738.

196 Clegg, W., Conway, B., Hevia, E. et al. (**2009**). *J. Am. Chem. Soc. 131*: 2375–2384.

197 Andrikopoulos, P.C., Armstrong, D.R., Barley, H.R.L. et al. (**2005**). *J. Am. Chem. Soc. 127*: 6184–6185.

198 Armstrong, D.R., Blair, V.L., Clegg, W. et al. (**2010**). *J. Am. Chem. Soc. 132*: 9480–9487.

199 Conway, B., Graham, D.V., Hevia, E. et al. (**2008**). *Chem. Commun.*: 2638–2640.

200 Clegg, W., Conway, B., Graham, D.V. et al. (**2009**). *Chem. Eur. J. 15*: 7074–7082.

201 Chevallier, F., Halauko, Y.S., Pecceu, C. et al. (**2011**). *Org. Biomol. Chem. 9*: 4671–4684.

202 Snégaroff, K., Komagawa, S., Chevallier, F. et al. (**2010**). *Chem. Eur. J. 16*: 8191–8201.

203 L'Helgoual'c, J.-M., Seggio, A., Chevallier, F. et al. (**2008**). *J. Org. Chem. 73*: 177–183.

204 Seggio, A., Chevallier, F., Vaultier, M., and Mongin, F. (**2007**). *J. Org.Chem. 72*: 6602–6605.

205 Seggio, A., Lannou, M.-I., Chevallier, F. et al. (**2007**). *Chem. Eur. J. 13*: 9982–9989.

206 Akimoto, G., Otsuka, M., Takita, R. et al. (**2018**). *J. Org. Chem. 83*: 13498–13506.

207 García-Álvarez, P., Mulvey, R.E., and Parkinson, J.A. (**2011**). *Angew. Chem. Int. Ed. 50*: 9668–9671.

208 Balkenhohl, M., Ziegler, D.S., Desaintjean, A. et al. (**2019**). *Angew. Chem. Int. Ed. 58*: 12898–12902.

209 Hatano, M., Suzuki, S., and Ishihara, K. (**2006**). *J. Am. Chem. Soc. 128*: 9998–9999.

210 Hevia, E., Chua, J.Z., García-Álvarez, P. et al. (**2010**). *Proc. Natl. Acad. Sci. U.S.A. 107*: 5294–5299.

211 Bluemke, T.D., Clegg, W., Garc, P. et al. (**2014**). *Chem. Sci. 5*: 3552–3562.

212 Armstrong, D.R., Clegg, W., García-Álvarez, P. et al. (**2011**). *Chem. Eur. J. 17*: 4470–4479.

7

Chemistry of Boryl Anions: Recent Developments

Makoto Yamashita

Department of Molecular and Macromolecular Chemistry, Graduate School of Engineering, Nagoya University, Nagoya, Japan

7.1 Introduction

Neutral three-coordinate boron-containing molecules generally exhibit a Lewis acidity on the boron center due to a vacant p-orbital on the boron atom being capable to accept lone pair electrons from a Lewis basic compound. In contrast, the recent development of boron-centered anions having a nucleophilicity on the boron atom in the past 15 years have led to a new pathway for the synthesis of boron-containing compounds. As a result, many new boryl-substituted organic, main group, and organometallic compounds have been synthesized and characterized. This chapter focuses on the preparative methods of boryl anion derivatives having boron-metal bonds and nucleophilicity on the boron atom. However, most of their applications that were recently reviewed in other review articles [1–5] are omitted here because of the limitation of space. Related chemistry of sp^2–sp^3 diborane(4) as a boryl anion equivalent has also been reviewed in the recent literature [4, 6–9].

7.2 Boryl Anions as a Salt of Alkali Metals

7.2.1 Early Examples of Base-stabilized Boryl Anions and Borylcopper Species

The parent boryl anion, BH_2^-, has a singlet ground state as theoretical calculations indicated [10], in contrast to carbene, CH_2, which has a triplet ground state [11]. This unstable boryl anion was calculated to be stabilized by complexation with Li^+ cation and introduction of electronegative atoms such as nitrogen to the boron center. Although the detailed history of boryl anions, before we isolated the first example, was summarized in another review article [3], this paragraph describes its brief history as follows. Schmidbaur and Imamoto independently reported the generation of Lewis base-stabilized boryl anion **1** having a nucleophilicity on the boron center (Scheme 7.1) [12, 13]. Later, Ito and Hosomi reported a catalytic β-borylation of enones in which the generation of borylcopper species **2**, having nucleophilicity on the boron center, was proposed [14, 15]. Subsequently, Sadighi reported an isolation of a similar borylcopper species **3** and applied it to catalytic reduction of carbon dioxide with B_2pin_2 and catalytic borylation reaction of aldehyde [16, 17]. The first structurally characterized B–Cu bond length, in **3**, was 2.002(3)

Polar Organometallic Reagents: Synthesis, Structure, Properties and Applications, First Edition. Edited by Andrew E. H. Wheatley and Masanobu Uchiyama.

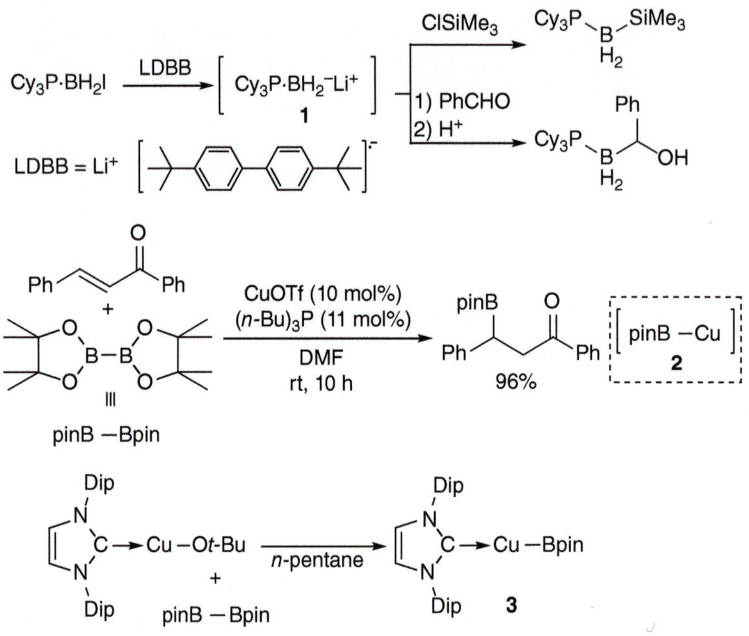

Scheme 7.1 Reports of nucleophilic boron compounds prior to boryllithium [Dip = 2,6-(i-Pr)$_2$C$_6$H$_3$].

Å. Although a nucleophilic attack of the boryl group in the borylcopper species to the carbon atom of aldehyde C=O bond was not experimentally observed, Density Functional Theory (DFT) calculations by Lin and Marder provided information about nucleophilicity of the boron center in borylcopper **3** toward the carbonyl functionality [18, 19].

7.2.2 Diaminoboryl Anions as a Lithium Salt

Considering the stability of monomeric *N*-heterocyclic carbene [20, 21], having six-valence electrons around the carbon atom, we tried to synthesize a boryl anion as a lithium salt with a five-membered ring structure and bulky substituents on the nitrogen atoms. Reduction of bromoborane precursor possessing a nitrogen-containing five-membered ring and two bulky Dip [2,6-(i-Pr)$_2$C$_6$H$_3$] groups afforded boryllithium **4**, which reacted with water to give hydroborane **5** (Scheme 7.2) [22–24]. The ^{11}B nuclear magnetic resonance (NMR) spectrum of **4** in THF-d_8 exhibited a broad signal at 45 ppm. The structures of DME- or THF-solvated boryllithium **4** were unambiguously characterized by single-crystal X-ray diffraction analysis (Figure 7.1). The B–Li bonds in **4**·(DME)$_2$ (DME = 1,2-dimethoxyethane) and **4**·(THF)$_2$ (THF = tetrahydrofuran) are 2.291(6) and 2.276(5) Å, which is slightly longer than the sum of covalent radii of boron and lithium atoms (2.11 Å) [25]. The N–B–N bond angle in (**4**·DME)$_2$ [99.2(2)°] and **4**·(THF)$_2$ [98.7(2)°] are smaller than that of **5**,

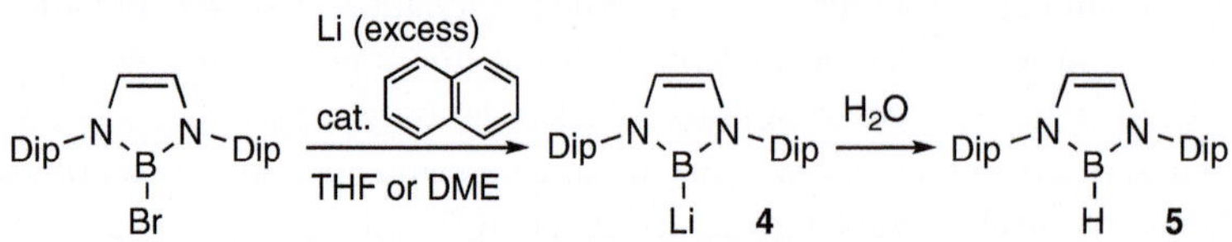

Scheme 7.2 Synthesis of boryllithium **4** and its reaction with water.

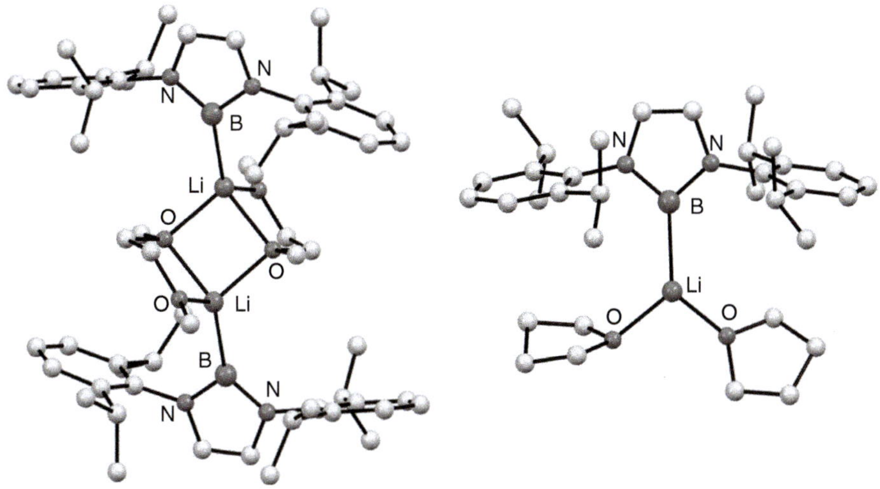

Figure 7.1 Crystal structures of $(\mathbf{4}\cdot DME)_2$ and $\mathbf{4}\cdot(THF)_2$. *Source:* Adapted from Segawa et al. [23].

indicating that the B–N bonds in **4** have higher p-character than those in **5** to stabilize the lone pair on the boron atom with a high s-character. DFT calculations revealed that the the HOMO (highest occupied molecular orbital) of $\mathbf{4}\cdot(THF)_2$ reflects lone-pair character at the boron center. AIM (atoms in molecules) analysis [26–28] of $\mathbf{4}\cdot(THF)_2$ disclosed an ionic nature of the B–Li bond.

Boryllithium **4** reacted with a variety of electrophiles to produce substituted products and adducts. The reaction of **4** with MeOTf ($Tf = SO_2CF_3$) and *n*-BuCl gave the corresponding alkylborane derivatives through S_N2 reaction. In the case of the reaction with benzyl chloride, 1-bromobutane, and benzyl bromide, haloborane products were obtained with/without alkylborane products. The formation of haloborane products could be explained by halophilic attack of the boryl anion as DFT calculations by Marder and Lin indicated. [29]. Reaction of **4** with benzaldehyde followed by quenching with protons resulted in the formation of α-borylbenzyl alcohol **6**. It should be noted that α-borylbenzyl alcohol **6** underwent Brook-type rearrangement involving a migration of the boryl group from carbon to oxygen in the presence of a catalytic amount of *n*-BuLi to furnish benzyloxyborane product **7** [30]. Reaction of **4** with benzoyl chloride gave benzoylborane formed through a nucleophilic acyl substitution. Boryllithium **4** also reacted with *t*-Bu- or phenyl-carbonate to result in the formation of boryl-substituted ester derivatives. A nucleophilic aromatic substitution reaction of **4** with C_6F_6 took place to afford pentafluorophenylborane. Treatment of **4** with 1 atm CO_2 followed by protonation furnished a mixture of borylcarboxylic acid and hydroxyborane. The latter would form through a Brook-type rearrangement liberating carbon monoxide. Furthermore, **4** reacted with toluene to deprotonate at the benzylic position, giving benzylborane as a product through deprotnation, nucleophilic attack by the resulting benzyllithium intermediate to the hydroborane, and loss of metal hydrides as precipitates [31]. The strong basicity of **4** was also high enough to deprotonate H_2 to produce hydroborane **5** and lithium hydride as judged by X-ray photoelectron spectroscopy (XPS). Thus, boryllithium **4** could be used as an equivalent of a boryl anion (Scheme 7.3).

Boryllithium could also be prepared in alkane solvent (Scheme 7.4a) [32]. Stirring Li powder in alkane solvent followed by stirring with bromoborane precursor and Na/K alloy afforded potassium diboryllithate **8**. X-ray crystallographic analysis revealed **8** has a slightly bent B–Li–B angle of 153.2(3)° to form an ate complex of lithium with two boryl anions, while a potassium cation was coordinated by two Dip rings (Figure 7.2). All of the four Dip signals of **8** appeared to be a single

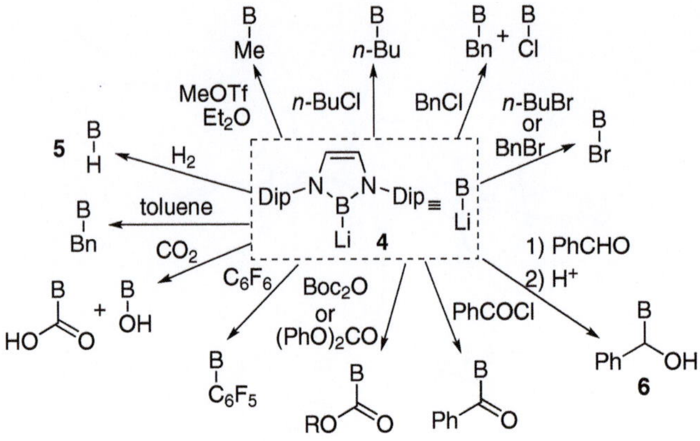

Scheme 7.3 Reactivity of boryllithium **4** as a boron nucleophile.

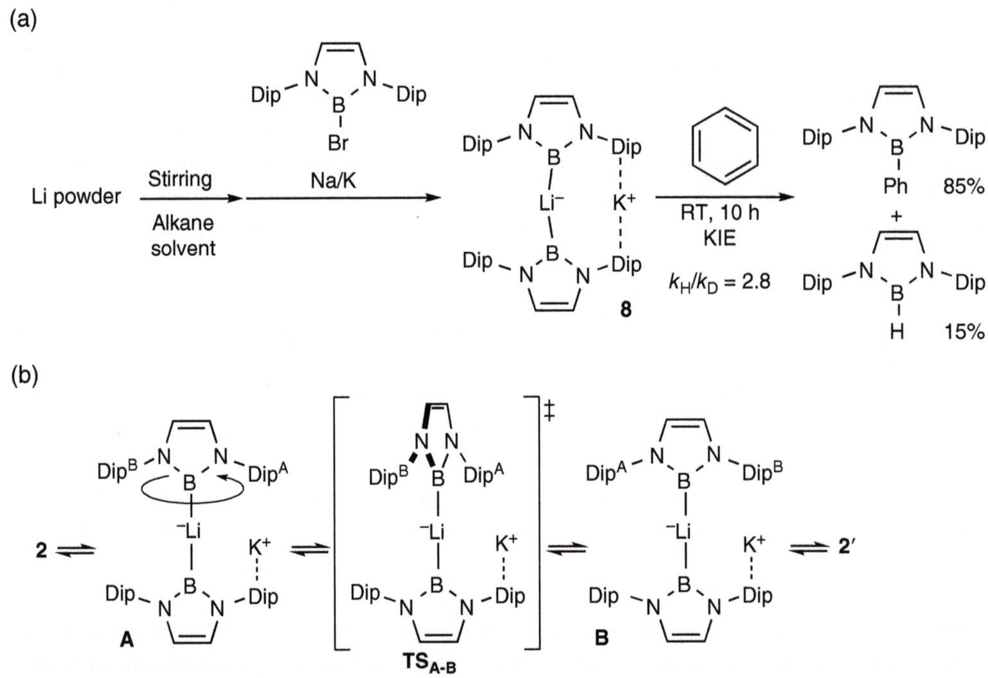

Scheme 7.4 (a) Generation of boryl anion **8** as an ate complex of lithium in alkane solvent (KIE = kinetic isotope effect) [32]; (b) possible dynamic exchange of the Dip rings coordinating to the K$^+$ cation in **8**.

resonance in the ^{1}H NMR spectrum, while the ^{7}Li NMR spectra in solution and solid states exhibited almost the same chemical shift [δ_B 4.7 ppm in cyclohexane-d_{12} solution, δ_B 5.5 ppm under Cross-polarization/Magic angle spinning (CPMAS) conditions]. These results indicate that the Dip groups spontaneously dissociate from K$^+$, and boryl groups rotate along the B–Li bonds to exchange Dip groups on the timescale of NMR spectroscopy, but two boryl anion units are bonding to the Li$^+$ cation in solution (Scheme 7.4b). DFT calculations revealed the electronic structure of **8** to have a

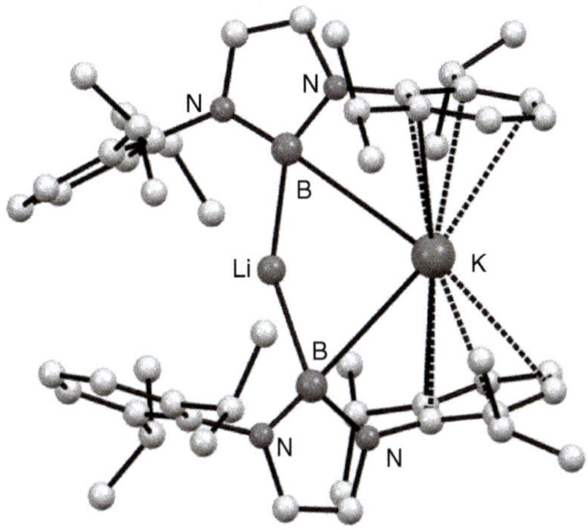

Figure 7.2 Crystal structure of **8**. *Source:* Adapted from Ohsato et al. [32].

significant contribution of the lone pair electrons to the HOMO. AIM analysis provided information on the polar B–Li bond, the existence of ionic interaction between B and Li^+/K^+ cations, and the absence of interaction between Li^+ and K^+ cations. Dissolution of **8** in benzene and leaving it at room temperature for 10 hours led to formation of phenylborane (major) and hydroborane (minor). DFT calculations proposed a mechanism initiating with deprotonation of benzene to form hydroborane followed by a nucleophilic attack of the resulting phenyl anion to the hydroborane. NBO (natural bond orbital) analysis for the structure of the transition state disclosed a coordination of benzene to the unoccupied orbital on K^+ after dissociation of the Dip group to promote deprotonation of benzene by the boryl anion moiety.

Recently, synthesis of 1,2,4,3-triazaborol-3-yllithium species **9** by a reduction of the corresponding bromoborane was reported by Kinjo et al. (Scheme 7.5a) [33]. DFT calculation for a parent compound of **9** without solvent and substituents disclosed the enhancement of ionic character in the B–Li bond in comparison with that of **4**. They also reported the reactivity of **9** toward electrophiles. Treatment of **9** with *n*-Bu-substituted isocyanide furnished *n*-butylborane. Although the structure of this product seems to form through S_N2 reaction liberating cyanide as a leaving group, the authors proposed another reaction mechanism consisting of formation of imidoyllithium, releasing *n*-BuLi as a leaving group and giving cyanoborane, the second nucleophilic attack of the resulting *n*-BuLi on cyanoborane then making *n*-butylborane. Similarly, CO and other isocyanides also reacted with **9** to form diborylalkene and boryl-substituted indole derivatives (Scheme 7.5b). The authors also proposed the reaction mechanism of these reactions initiated by a formation of acyl- or imidoyl-lithium, followed by isomerization to oxy- and amino-substituted carbene derivatives, and proton or hydride migration. The related transmetalation chemistry of **9** with other metals will be described in the later part of this chapter.

7.2.3 Base-stabilized Boryl Anion with π-delocalization

7.2.3.1 Lewis Base-stabilized Borole Anion

As described in Scheme 7.2, coordination of Lewis base should have a stabilizing effect on six-electron boryl anion species. If one of the three substituents on the boron atom or the Lewis base coordinating to the boron atom would have a π-acceptor character, the base-stabilized boryl anion should be further stabilized through a delocalization of the lone pair electrons on the boron center.

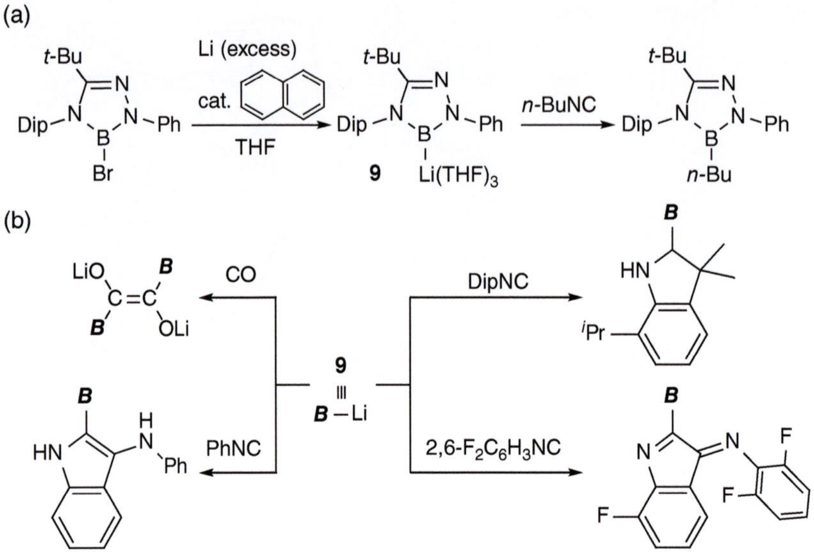

Scheme 7.5 Synthesis and reactivity of 1,2,4,3-triazaborol-3-yllithium **9**.

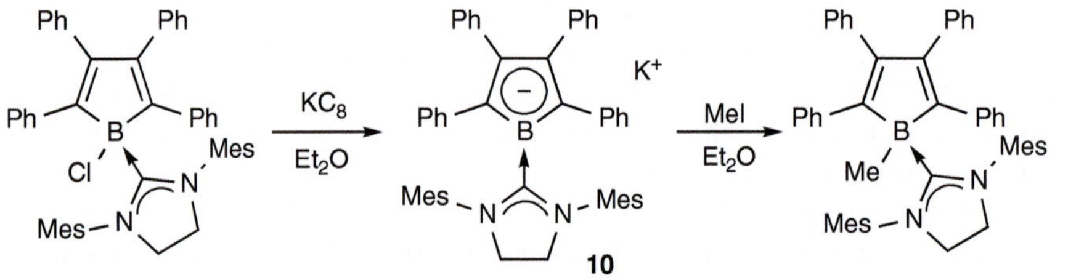

Scheme 7.6 Synthesis and reactivity of NHC-stabilized borole anion **10** (Mes = 2,4,6-$Me_3C_6H_2$).

This idea was first proven by Braunschweig et al., who synthesized *N*-heterocyclic carbene (NHC)-stabilized borole anion **10** by a reduction of NHC-coordinated chloroborole (Scheme 7.6) [34]. X-ray crystallographic analysis and DFT calculations for **10** revealed that the lone pair electrons on the boron atom were delocalized over the five-membered borole ring and the carbon atoms of NHC. Quenching this reactive species with MeI gave the corresponding *B*-methylated borole stabilized with NHC, implying a nucleophilicity of the boron center of **10**. In contrast, reaction of **10** with Ph_3SnCl or Ph_3PbCl, one-electron transfer from **10** to Ph_3ECl (E = Sn, Pb) proceeded to form Ph_3E–EPh_3 and borole-centered radical species [35]. Thus, delocalization of the lone pair electrons on the boron atom of the boryl anion led to stabilization. It should be emphasized that **10** is the first example of a carbon-substituted boryl anion.

7.2.3.2 Carbene-stabilized Boryl Anion

Addition of carbene to boryl anion was also applied to synthesize an NHC-coordinated parent boryl anion [36]. Curran and Lacôte reported a reduction of NHC-coordinated iodoborane with lithium 4,4′-di-*tert*-butylbiphenylide to generate NHC-coordinated parent boryl anion **11** (Scheme 7.7). Although they did not report an X-ray crystal structure of **11**, DFT calculations disclosed localization of the lone pair electrons on the BH_2 moiety toward the carbon atom in NHC. The resulting **11** reacted with a variety of electrophiles, such as carbonate, ester, aldehyde,

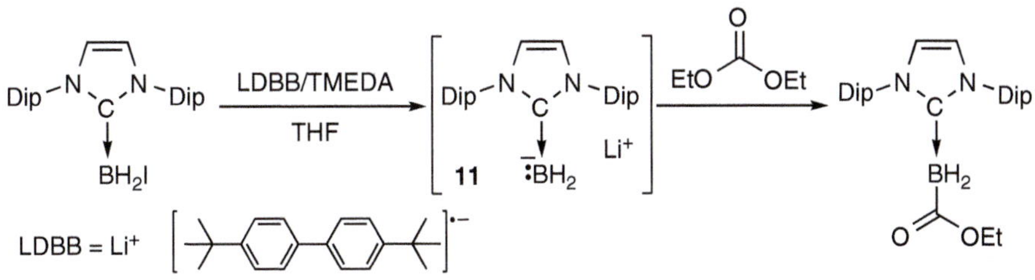

Scheme 7.7 Synthesis and reactivity of NHC-stabilized parent boryl anion **11**.

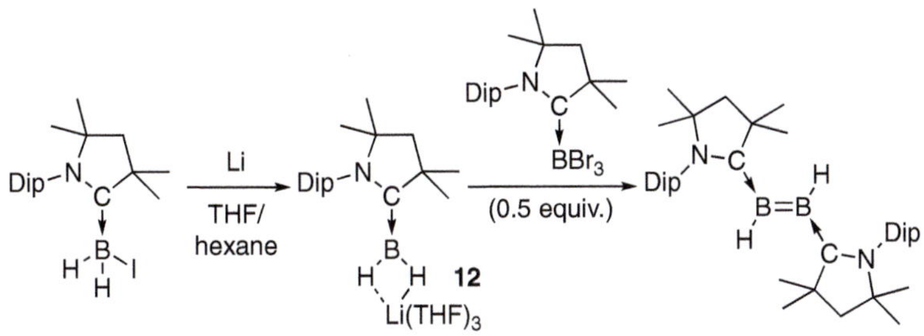

Scheme 7.8 Synthesis and reactivity of cAAC-stabilized parent boryl anion **12**.

epoxide, nitrile, alkyl halide, allyl halide, and C_6F_6, to furnish the corresponding adducts and substituted products through a formal nucleophilic borylation. The reaction mechanism for these reactions has not been estimated, but it would presumably be expected that the direct nucleophilic attack of the boron center on the electrophile would initiate the reaction. Later, similar cyclic alkyl(amino)carbene (cAAC)-stabilized parent boryl anion **12** was synthesized by a reaction of cAAC-coordinated iodoborane precursor (Scheme 7.8) [37]. The resulting **12** reacted with cAAC-coordinated bromoborane to produce cAAC-coordinated dihydrodiborene.

7.2.3.3 Stabilization with Cyanide

A cyanide ion is also known to stabilize the boryl anion as a π-accepting ligand [38]. Reduction of tetracyanoborate salt with metallic potassium or sodium furnished alkali metal salts of λ^3-tricyanoborandiide **13a** or **13b** via reductive cleavage of the B–CN bond (Scheme 7.9). A lithium salt **13c** with the same dianion was prepared by reaction of a similar precursor with *n*-BuLi, with concomitant formation of Li[(*n*-Bu)$_2$CN]. X-ray crystallographic analysis showed that **13a** has multiple interionic contacts between the nitrogen atom of cyanide in $[B(CN)_3]^{2-}$ dianion and the potassium cation. The B–C bond distances (av. 1.513 Å) in **13** are much shorter than that (av. 1.595 Å) in $K[B(CN)_4]$ due to the delocalization of the lone pair electrons on the boron center in **13** to the cyanide moiety. An improved synthesis of **13a** on a larger scale was achieved by a reduction of $K[FB(CN)_3]$ with potassium metal or deprotonation of $K[HB(CN)_3]$ with $K[N(SiMe_3)_2]$ [39, 40]. Reaction of **13a** with ethyl iodide, allyl halide, C_6F_6 and other multiply fluorinated arenes, and CO_2 resulted in the formation of the corresponding nucleophilic substitution products [39–41]. It should be noted that these reactions produced anionic products after nucleophilic substitution because the starting **13a** is dianionic.

The combination of cAAC with cyanide also produced a base-stabilized boryl anion (Scheme 7.10) [42]. Deprotonation of cAAC-coordinated dicyanoborane by using a strong base in the presence of crown ether gave cAAC-stabilized dicyanoboryl anion **14** (Scheme 7.11). This

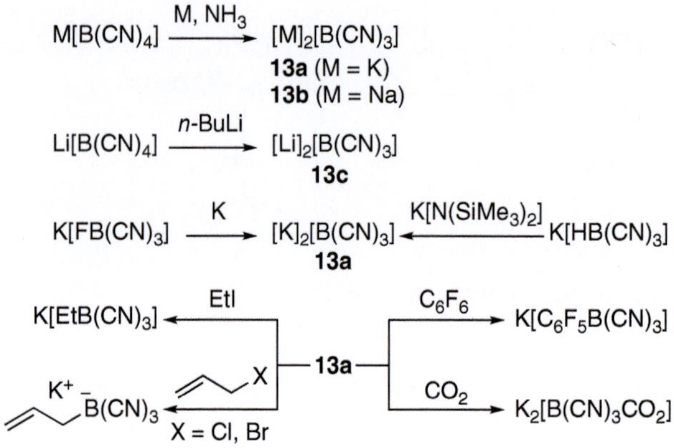

Scheme 7.9 Synthesis and reactivity of cyanide-stabilized λ^3-tricyanoborandiide dianion **13a–c**.

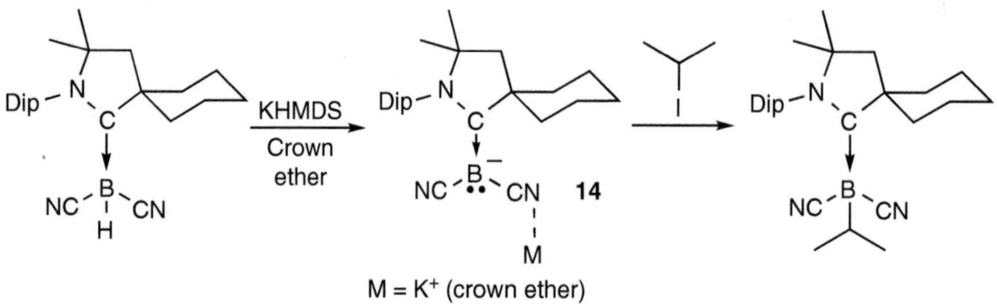

Scheme 7.10 Synthesis and reactivity of cAAC-stabilized dicyanoboryl anion **14**.

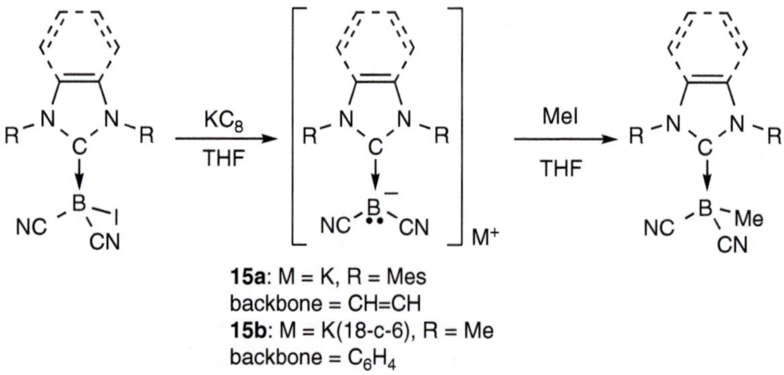

Scheme 7.11 Synthesis and reactivity of NHC-stabilized dicyanoboryl anion **15**.

reaction is the first example of deprotonation of hydroborane. X-ray crystallographic analysis revealed a contact ion pair structure of **14** through coordination of the nitrogen atom of the cyano group to the K^+ cation. Among three B–C bonds in **14**, the B–C(cAAC) bond [1.473(2) Å] is shorter than the B–CN bonds [1.605(6), 1.595(6) Å], reflecting the strong π-acceptor character of cAAC and strong delocalization of the lone pair electrons on the boron center toward the carbon atom in the cAAC unit. This anion **14** could react with 2-iodopropane to generate an S_N2 substituted product via nucleophilic borylation. The boron center in **14** was also capable of forming a B–Au complex through a reaction with the gold halide complex. Similar to the case of **14**, a combination of

NHC and cyanide was also effective in producing NHC-stabilized dicyanoboryl anion **15**, generated by a reduction of NHC-coordinated dicyanoiodoborane precursor with KC_8 (Scheme 7.10) [43]. Single-crystal X-ray diffraction analysis revealed that the B–CN bond lengths of **15a** and **15b** [1.588(3) and 1.584(2) Å] were longer than the B–C(carbene) bonds [1.495(2) and 1.487(2) Å] reflecting that the backdonation of lone pair electrons on the boron center went to the unoccupied orbital on the carbene carbon atom. These anions reacted with MeI to furnish the corresponding substituted methylborane. Reactions with other electrophiles, such as $(PMe_2Ph)AuCl$, $ClSiMe_3$, $ClSnMe_3$, and $ClP(t\text{-}Bu)_2$, also afforded borylated products.

7.2.3.4 Metal-substituted Boryl Anion

Metal substituents on the boron atom were also effective to stabilize boryl anions. Reduction of chloroborylene-bridged dimanganese complex with metallic lithium gave an anionic dimetalloborylene complex **16**, which can be considered as a stabilized boryl anion with two manganese substituents (Scheme 7.12) [44]. The Mn–B–Mn bond angle was almost linear (176.1°) and two Mn–B bond lengths were short [1.8812(14) and 1.8809(14) Å] to reflect the delocalization of the lone pair electrons on the boron center over two manganese metals. Treatment of **16** with MeI led to formation of a B–Me bond, implying the nucleophilicity of the boron atom in **16**.

7.3 Boryl Anions as a Salt of Magnesium, Zinc, and Copper as Relatives of Carbanions

Section 7.2 described a preparation of boryl anions as a salt with alkali metal cations. This section summarizes preparative methods for a series of boryl anions with other metals such as magnesium, copper, and zinc, as the family of carbanions has these metals.

7.3.1 Transmetalation of Boryllithium to Magnesium, Copper, and Zinc to Form Borylmetals

Treatment of boryllithium **4** with 1 equivalent of magnesium bromide-diethyl ether complex gave the corresponding borylmagnesium bromide as a mixture of **17** and its LiBr adduct **17·LiBr** (Scheme 7.13, top), which were physically separated as single crystals [45]. In THF-d_8 solvent, **17** and **17·LiBr** showed the same ^1H NMR spectra, and the ^7Li NMR spectrum of **17·LiBr** was identical to that of LiBr in THF-d_8. These results indicate that LiBr dissociates from **17·LiBr** in solution. The unprecedented 2-center 2-electron B–Mg bonds [**17**: 2.281(6) Å, **17·LiBr**: 2.282(6) Å] were slightly longer than the sum of covalent radii of boron and magnesium atoms (2.24 Å) [25]. On the other hand, decreasing stoichiometry of magnesium bromide-diethyl ether complex to

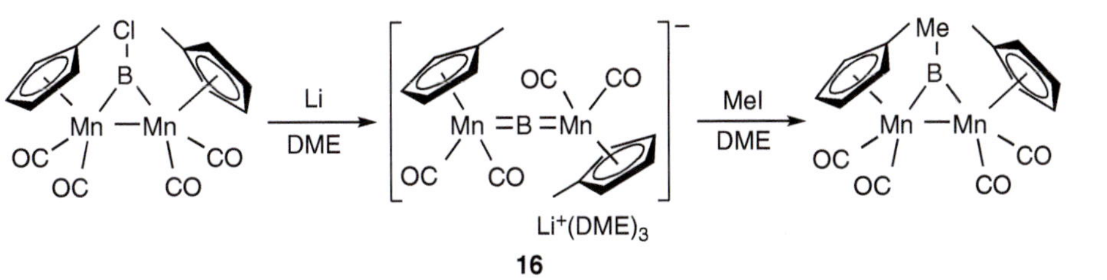

16

Scheme 7.12 Linear dimetalloborylene complex **16** having a nucleophilicity on the boron atom.

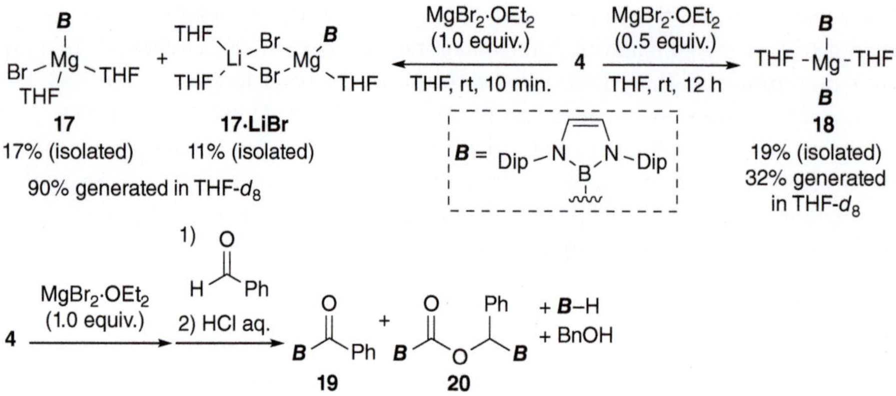

Scheme 7.13 Synthesis and reactivity of borylmagnesium.

0.5 equivalents in the reaction of **4** led to formation of the corresponding diborylmagnesium **18**. In this case, the B–Mg bond in **18** [2.377(4) Å] was further longer than those of **17** and **17·LiBr**, probably due to steric repulsion between two bulky boryl groups on the Mg atom. Borylmagnesium bromide **17** exhibited a characteristic reactivity toward benzaldehyde (Scheme 7.13, bottom). After the generation of borylmagnesium bromide **17**, the reaction mixture was treated with 1 equivalent of benzaldehyde followed by quenching with a proton source to give a mixture of benzoylborane **19** and boron-substituted ester **20**. After the nucleophilic addition of a boryl group to the aldehyde functionality, the resulting boryl-substituted magnesium benzyloxide intermediate would undergo an intramolecular hydride transfer to form boryl-substituted ester **20**, as reported for the magnesium–Oppenauer oxidation [46]. It should be noted that **19** and **20** are the first examples of boryl-carbonyl compounds.

Transmetalation of boryllithium is also effective to prepare the corresponding borylcopper and borylzinc species. Treatment of boryllithium **21** bearing a saturated backbone with 1 equivalent of CuBr afforded the corresponding borylcopper **22** as an adduct of the concomitantly formed LiBr (Scheme 7.14) [47] that can also be considered as lithium boryl(bromo)cuprate. The resulting **22** is the second example of an isolated borylcopper species. Increasing stoichiometry of CuBr to 2 equivalents in the same reaction furnished boryl- and bromo-bridged tetranuclear copper complex **23**. The B–Cu bond distances in **22** (av. 1.983 Å) were shorter than those in **23** [2.093(4) and 2.073(5) Å], reflecting the difference in the coordination mode of the boryl ligand in **22** (2-center 2-electon bond) and **23** (3-center 2-electron bond as a bridging ligand). Using 1 equivalent of zinc bromide in the reaction of **21** resulted in the formation of borylzinc bromide **24** as an adduct of LiBr, where the product can be named lithium boryl(dibromo)zincate. **24** is the first example of a borylzinc species. Decreasing zinc bromide to 0.5 equivalents in the same reaction gave diborylzinc **25** as the first example of the homoleptic borylmetal complex. Based on the textbook chemistry of organocopper and organozinc species, reactivity of the resulting borylcopper and borylzinc species toward α,β-unsaturated ketones was examined. The reaction of borylcopper **23** or borylzinc **24** with 2-cyclohexen-1-one afforded the corresponding β-borylated product **26** after aqueous workup. During the reaction, the lithium cation would behave as a Lewis acid to promote the conjugate addition of boron nucleophile to the β-position of 2-cyclohexen-1-one. This result is in contrast to boryllithium **21** that underwent deprotonation with 2-cyclohexen-1-one as a strong base, rather than exhibiting nucleophilic addition. It should be noted that 1,2,4,3-triazaborol-3-yllithium

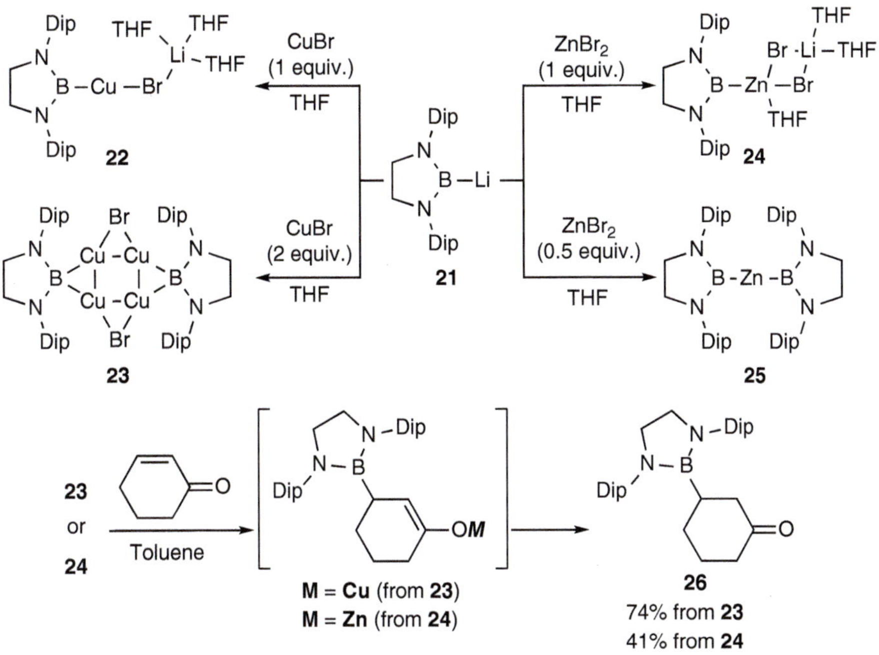

Scheme 7.14 Generation of borylcopper and borylzinc species from boryllithiums and their reactivity.

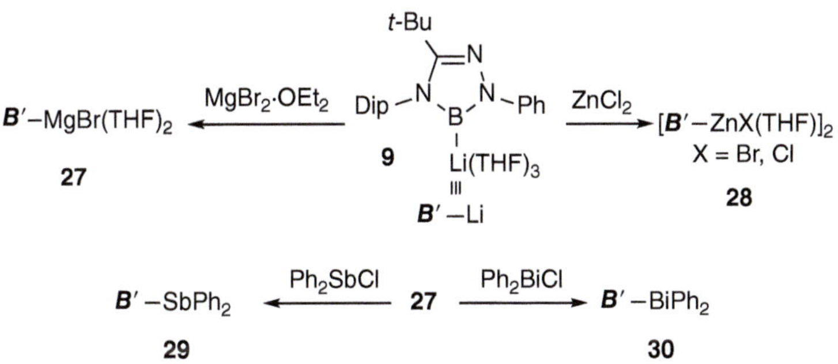

Scheme 7.15 Generation of borylmagnesium and borylzinc species **9** from 1,2,4,3-triazaborol-3-yllithium and subsequent reactions.

species **9** can also be used for the transmetalation of a boryl anion from Li^+ to Mg^{2+} or Zn^{2+} to provide the corresponding borylmagnesium **27** and borylzinc **28** (Scheme 7.15) [33]. The resulting borylmagnesium **27** reacted with Ph_2SbCl and Ph_2BiCl to furnish the corresponding substituted products, borylstibine **29** and borylbismuthine **30**, through a nucleophilic borylation of the main group element. Similarly, using borylzinc **31**, derived from boryllithium **21**, was also effective to construct a B–P bond in boryl(dichloro)phosphine **32** (Scheme 7.16) [48–50]. This compound was converted to the corresponding boryl-substituted diphosphene **33**. Because of the low-lying π^*-orbital of the P=P bond, diphosphene **33** could react with nucleophilic *n*-BuLi to afford stable *n*-BuLi adduct **34**, in which the (Li)P–B bond has a double bond character as judged by single-crystal X-ray diffraction analysis. The electron deficiency of **33** also allowed one- or two-electron

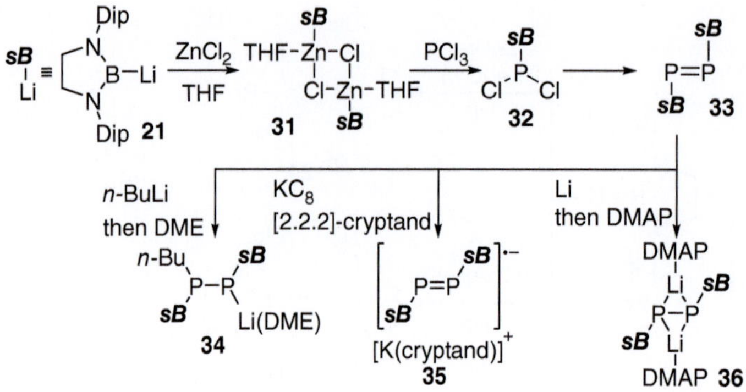

Scheme 7.16 Generation of borylzinc species from boryllithium **21** and subsequent reactions.

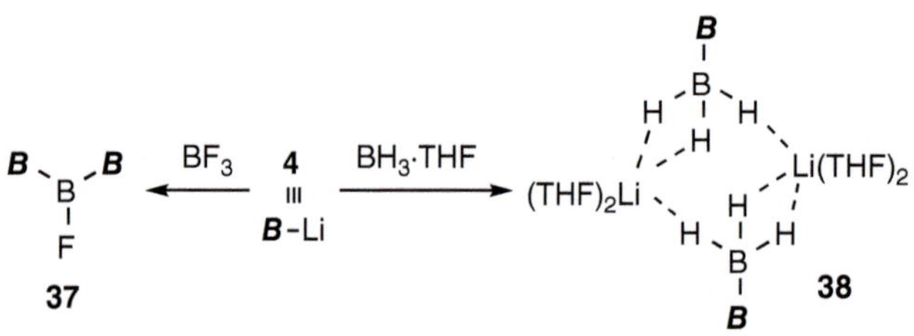

Scheme 7.17 Nucleophilic borylation of boron compounds by using boryllithium **4**.

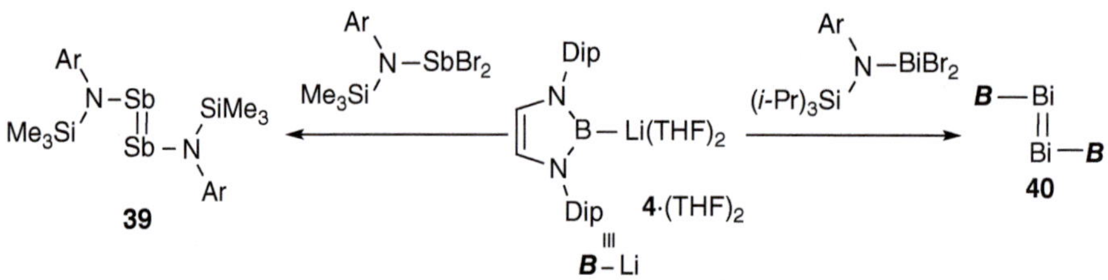

Scheme 7.18 Reaction of boryllithium **4** with amino(dibromo)pnictogen [Ar = 2,6-$(CHPh_2)_2$-4-(*i*-Pr)C_6H_2].

reduction with KC_8/[2.2.2]-cryptand or Li/(4-*N,N*-dimethylamino)pyridine (DMAP) to furnish radical anion **35** and dianion **36**, which have multiple-bond character between boron and phosphorus atoms, reflecting the π-accepting character of the boryl substituent. In the case of relatively hard main group electrophiles, such as BF_3 and BH_3, boryllithium **4** can be directly used in the nucleophilic borylation of main group element compounds (Scheme 7.17) [51, 52]. Treatment of **4** with BF_3 or BH_3·THF afforded fluorotriborane(5) **37** or boryltrihydroborate **38** via nucleophilic borylation. These results in Schemes 7.14–7.16 about construction of B–(main group element) bonds are in stark contrast to the reaction of boryllithium **4**·$(THF)_2$ with aminodibromo-stibine and -bismuthine, which gave distibene **39** and dibismuthene **40** (Scheme 7.18) [53] probably through a halophilic attack of the boryl anion to reduce the group 15 element center from the oxidation state of three to one.

7.3.2 Transmetalation of Diborane(4) to Magnesium and Zinc to Form Borylmetals

Recently, transmetalation of diborane(4) reagent was proven to be effective in preparing borylmagnesium [54]. Reaction of diketiminato-stabilized *n*-butylmagnesium with bis(pinacolato)diborane(4), B_2pin_2, gave a new species **41** consisting of the sp^2–sp^3 pinB–Bpin(*n*-Bu) anion and diketiminato-Mg cation (Scheme 7.19). Addition of DMAP to **41** induced an elimination of *n*-Bu(pinacolato)borane to liberate diketiminato- and DMAP-stabilized borylmagnesium **42**. A single-crystal X-ray diffraction analysis of **42** showed the tetracoordinate Mg^{2+} cation environment and the Mg–B bond distance of 2.324(2) Å, which is slightly longer than those of **17**, **17·LiBr**, and **18**.

Although one would expect that **42** would react as a boron nucleophile due to the presence of a B–Mg bond, **41** was also active as a boryl anion equivalent (Scheme 7.20) [54–58]. In fact, addition of **41** to 9-BBN or BPh_3 led to formation of a new B–B bond to furnish pinB–(H)BBN **43** or pinB–BPh_3 anion **44** coordinating to (diketiminato)Mg cation. This **41** also reacted with 1 equivalent of benzophenone to give a B–C bonded species **45**, and the resulting **45** further reacted with 1 equivalent of benzophenone to furnish spirocyclic borate **46** through C–C bond formation between two benzophenone units. Treatment of **41** with Dip- or *t*-Bu-substituted isocyanates resulted in the formation of the corresponding adducts **47** and **48** through the creation of a B–C bond. It was also demonstrated that **42** reacted with MeI to produce methylpinacolborane as a nucleophilic substitution product, in which the reaction mechanism was supported by control experiments with radical trapping agent and by DFT-based energy profile with reasonable activation barriers. The pinB-Mg species **42** also reacted with benzophenone to furnish the corresponding (boryl)(diphenyl)methoxide **49** caused by B–C bond formation through nucleophilic addition of the boryl group to the carbonyl. Interestingly, the addition of DMAP to **45** also afforded **49** through migration of the boryl group, indicating that migration between carbon and boron would be reversible, as we have reported in the bora-Brook rearrangement (Scheme 7.3) [30]. Similarly, treatment of **42** with cyclohexyl-substituted carbodiimide resulted in the formation of borylamidinato-Mg complex **50**. Thus, conversion of the B–B single bond in B_2pin_2 to the B–Mg bond provides a new method to construct boryl anion equivalents.

Similar to the preparative method for borylmagnesium species by transmetalation of diborane(4) (Scheme 7.19), it was reported that borylzincate could be generated by transmetalation from diborane(4) and organozinc reagent. Aryl halides were converted to the corresponding arylboronate

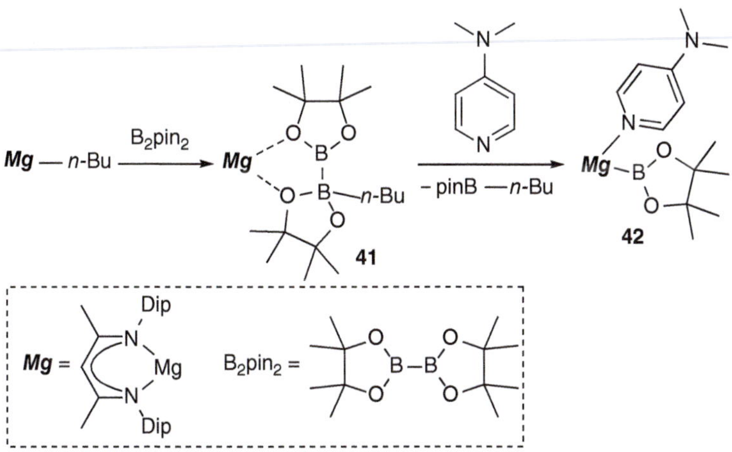

Scheme 7.19 Generation of borylmagnesium species by transmetalation of B_2pin_2.

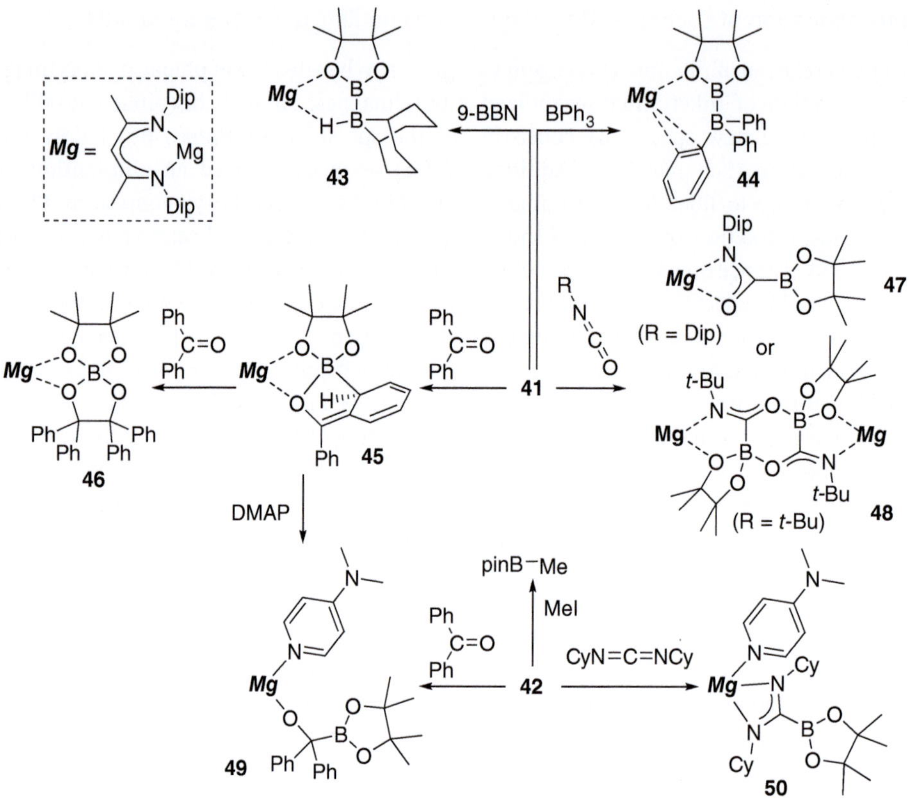

Scheme 7.20 Reactivity of **41** and **42**.

derivatives in the presence of a base, zinc catalyst, and B_2pin_2 (Scheme 7.21) [59]. In this report, DFT calculations were performed to predict the generation of boryl(dialkyl)zincate **51** from dialkylzinc, B_2pin_2, and NaO*t*-Bu as an additive, although it was not observed and isolated. On the basis of the calculations, a plausible mechanism was proposed to consist of transmetalation of diborane(4) to zinc, aryl-boryl exchange on zinc to liberate haloborane, and nucleophilic substitution at boron by the resulting aryl(dialkyl)zincate. The same paper also reports the reaction of the present boryl(dialkyl)zincate with benzyne intermediate followed by trapping with an electrophile to introduce two substituents into the benzene ring.

7.4 Application of Borylcopper and Borylzinc Species for Synthetic Organic Chemistry

Borylcopper and borylzinc species were also applied to synthetic organic chemistry to prepare boron-containing compounds. We report a reaction of boryllithium **4** with copper(I) cyanide to give boryl(cyano)cuprate **52** as a crystalline solid (Scheme 7.22) [60]. The B–Cu bond length of **52** [1.973(6) Å] was relatively shorter than those of other borylcopper species in this review, and that would come from the π-electron-accepting character of cyanide. Addition of diethyl acetylenedicarboxylate to **52** followed by treatment with various electrophiles afforded a regioisomeric

Scheme 7.21 Zinc-catalyzed borylation of aryl halide involving borylzincate **51** as a potential intermediate.

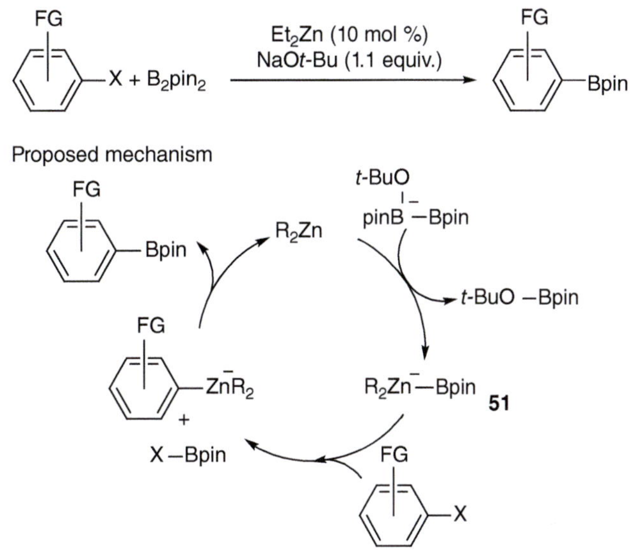

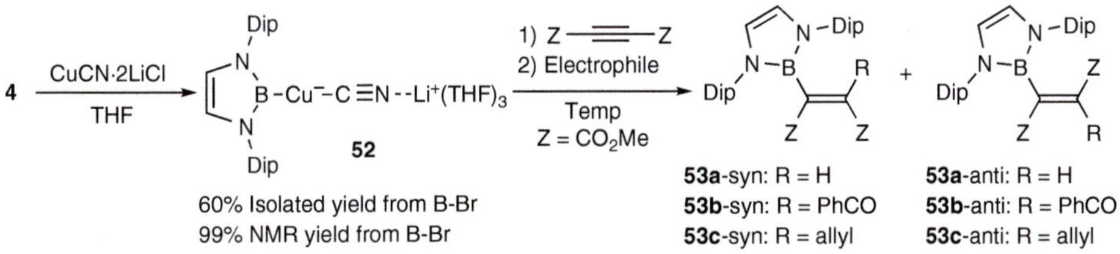

Scheme 7.22 Generation of boryl(cyano)cuprate **52** and subsequent reaction with alkyne and electrophiles.

mixture of tetrasubstituted borylalkenes **53** having substituents coming from the electrophiles. The regioisomeric ratio of the product depends on electrophiles and reaction temperature. This result suggests that the kinetically favored *syn*-isomer of borylalkenylcopper intermediate would isomerize to the corresponding *anti*-isomer through a boryl-substituted allenolate intermediate. Alkynes without ester substituents are also effective for borylation by using borylcuprate species (Scheme 7.23) [61]. The reaction of boryl(cyano)cuprate **54**, which is generated from boryllithium **21** with its saturated backbone, with 1-phenylpropyne derivatives followed by a reaction with allyl bromide furnished tetrasubstituted alkenylboranes. A large *t*-Bu substituent on the alkyne moiety shut the reaction down, while substituents on the benzene ring of 1-phenylpropyne did not alter the yield of the carboborated product. Diamino substituents of the resulting alkenylborane **55** were converted to pinacolato substituents to give alkenylboronate **56** that can be used as a substrate for the subsequent Suzuki–Miyaura cross-coupling reaction to result in the formation of a tetrasubstituted alkene **57**. Reactions of borylzinc bromide dimer **58**, isolated by a treatment of boryllithium **4** with $ZnBr_2{\cdot}(THF)_2$, with bromoarene or acid chloride in the presence of Pd(II) catalyst, gave the corresponding diaminoborylbenzenes and acylboranes (Scheme 7.24) [62]. The latter product was treated with $(n\text{-Bu})_4N^+{\cdot}F^-{\cdot}4H_2O$ to result in the formation of acyltrifluoroborate derivative in good yield.

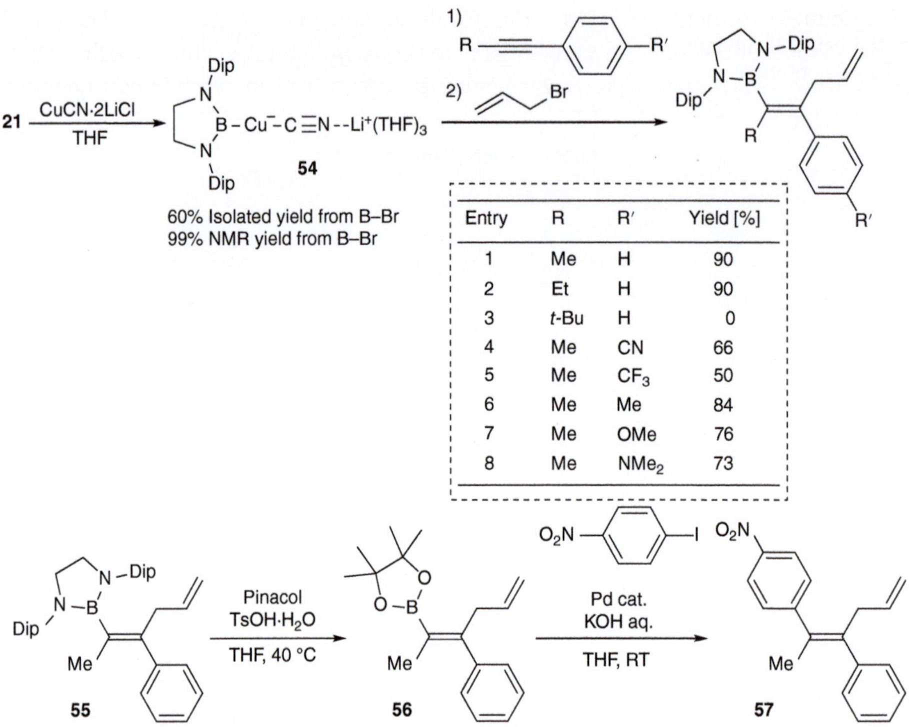

Entry	R	R′	Yield [%]
1	Me	H	90
2	Et	H	90
3	*t*-Bu	H	0
4	Me	CN	66
5	Me	CF_3	50
6	Me	Me	84
7	Me	OMe	76
8	Me	NMe_2	73

Scheme 7.23 Direct carboboration of 1-phenylpropyne derivatives by using boryl(cyano)cuprate **54** and subsequent Suzuki-Miyaura cross-coupling.

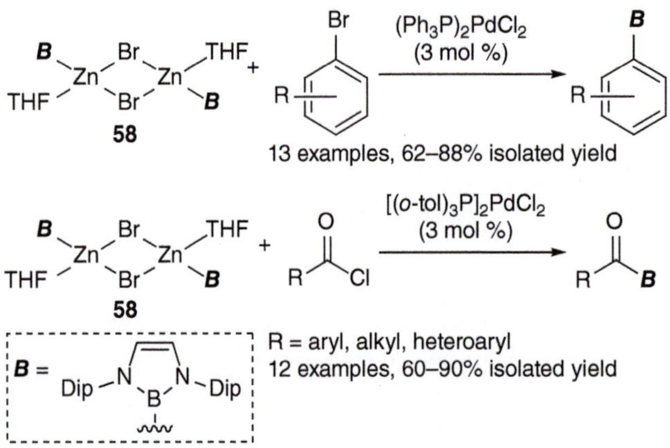

Scheme 7.24 Pd-catalyzed coupling of borylzinc **58** with bromoarenes and acid chlorides.

7.5 Summary

Nowadays, many types of boryl anion possessing boron–metal bonds are available, and these can be used as a source of boryl substituents toward organic, main group, and organometallic molecules. In this chapter, all the available methods to prepare boryl anions were classified into several types depending on the substituents and counter cations. In addition to boryl anions, the chemistry of group 13 element anions, such as of aluminium [63–76], gallium [77–91], and indium [92], is also expanding in recent years. Mixing these group 13-element anions with electrophiles would provide a wide range of molecules containing group 13 elements with new bonding, structure, and functions.

References

1 Weber, L. (**2017**). *Eur. J. Inorg. Chem. 2017*: 3461–3488.
2 Stephan, D.W. (**2017**). *Angew. Chem. Int. Ed. 56*: 5984–5992.
3 Yamashita, M. and Nozaki, K. (**2015**). *Boryl Anions in Synthesis and Application of Organoboron Compounds* (eds. E. Fernández and A. Whiting), 1–37. Springer International Publishing.
4 Cid, J., Gulyás, H., Carbó, J.J., and Fernández, E. (**2012**). *Chem. Soc. Rev. 41*: 3558–3570.
5 Asay, M., Jones, C., and Driess, M. (**2010**). *Chem. Rev. 111*: 354–396.
6 Pietsch, S., Neeve, E.C., Apperley, D.C. et al. (**2015**). *Chem. Eur. J. 21*: 7082–7098.
7 Dewhurst, R.D., Neeve, E.C., Braunschweig, H., and Marder, T.B. (**2015**). *Chem. Commun. 51*: 9594–9607.
8 Cuenca, A.B., Shishido, R., Ito, H., and Fernandez, E. (**2017**). *Chem. Soc. Rev. 46*: 415–430.
9 Gulyas, H., Bonet, A., Pubill-Ulldemolins, C. et al. (**2012**). *Pure Appl. Chem. 84*: 2219–2231.
10 Wagner, M., van Eikema Hommes, N.J.R., Noeth, H., and Schleyer, P.v.R. (**1995**). *Inorg. Chem. 34*: 607–614.
11 Schaefer, H.F. III (**1986**). *Science 231*: 1100–1107.
12 Blumenthal, A., Bissinger, P., and Schmidbaur, H. (**1993**). *J. Organomet. Chem. 462*: 107–110.
13 Imamoto, T. and Hikosaka, T. (**1994**). *J. Org. Chem. 59*: 6753–6759.
14 Ito, H., Yamanaka, H., Tateiwa, J., and Hosomi, A. (**2000**). *Tetrahedron Lett. 41*: 6821–6825.
15 Ishiyama, T. and Miyaura, N. (**2000**). *J. Organomet. Chem. 611*: 392–402.
16 Laitar, D.S., Mueller, P., and Sadighi, J.P. (**2005**). *J. Am. Chem. Soc. 127*: 17196–17197.
17 Laitar, D.S., Tsui, E.Y., and Sadighi, J.P. (**2006**). *J. Am. Chem. Soc. 128*: 11036–11037.
18 Zhao, H., Lin, Z., and Marder, T.B. (**2006**). *J. Am. Chem. Soc. 128*: 15637–15643.
19 Zhao, H., Dang, L., Marder, T.B., and Lin, Z. (**2008**). *J. Am. Chem. Soc. 130*: 5586–5594.
20 Arduengo, A.J. III, Harlow, R.L., and Kline, M. (**1991**). *J. Am. Chem. Soc. 113*: 361–363.
21 Arduengo, A.J. III, Krafczyk, R., and Schmutzler, R. (**1999**). *Tetrahedron 55*: 14523–14534.
22 Segawa, Y., Yamashita, M., and Nozaki, K. (**2006**). *Science 314*: 113–115.
23 Segawa, Y., Suzuki, Y., Yamashita, M., and Nozaki, K. (**2008**). *J. Am. Chem. Soc. 130*: 16069–16079.
24 Yamashita, M., Suzuki, Y., Segawa, Y., and Nozaki, K. (**2008**). *Chem. Lett. 37*: 802–803.
25 Emsley, J. (**1998**). *The Elements*, 3rde. New York: Oxford University Press.
26 Bader, R.F.W. (**1990**). *Atoms In Molecules - A Quantum Theory*. New York: Oxford University Press.
27 Biegler-Konig, F., Schonbohm, J., and Bayles, D. (**2001**). *J. Comput. Chem. 22*: 545–559.
28 Keith, T. A. (**2014**). (version 14.04.17) ed., TK Gristmill Software, Overland Park KS, USA.
29 Cheung, M.S., Marder, T.B., and Lin, Z. (**2011**). *Organometallics 30*: 3018–3028.
30 Kisu, H., Sakaino, H., Ito, F. et al. (**2016**). *J. Am. Chem. Soc. 138*: 3548–3552.
31 Dettenrieder, N., Aramaki, Y., Wolf, B.M. et al. (**2014**). *Angew. Chem. Int. Ed. 53*: 6259–6262.
32 Ohsato, T., Okuno, Y., Ishida, S. et al. (**2016**). *Angew. Chem. Int. Ed. 55*: 11426–11430.
33 Lu, W., Hu, H., Li, Y. et al. (**2016**). *J. Am. Chem. Soc. 138*: 6650–6661.
34 Braunschweig, H., Chiu, C.-W., Radacki, K., and Kupfer, T. (**2010**). *Angew. Chem. Int. Ed. 49*: 2041–2044.
35 Bertermann, R., Braunschweig, H., Dewhurst, R.D. et al. (**2014**). *Angew. Chem. Int. Ed. 53*: 5453–5457.

36 Monot, J., Solovyev, A., Bonin-Dubarle, H. et al. (**2010**). *Angew. Chem. Int. Ed. 49*: 9166–9169.

37 Arrowsmith, M., Mattock, J.D., Hagspiel, S. et al. (**2018**). *Angew. Chem. Int. Ed. 57*: 15272–15275.

38 Bernhardt, E., Bernhardt-Pitchougina, V., Willner, H., and Ignatiev, N. (**2011**). *Angew. Chem. Int. Ed. 50*: 12085–12088.

39 Landmann, J., Keppner, F., Hofmann, D.B. et al. (**2017**). *Angew. Chem. Int. Ed. 56*: 2795–2799.

40 Landmann, J., Sprenger, J. A. P., Bertermann, R., Ignat'ev, N., Bernhardt-Pitchougina, V., Bernhardt, E., Willner, H., Finze, M. (**2015**). *Chem. Commun.*, *51*, 4989–4992.

41 Landmann, J., Hennig, P.T., Ignat'ev, N.V., and Finze, M. (**2017**). *Chem. Sci. 8*: 5962–5968.

42 Ruiz, D.A., Ung, G., Melaimi, M., and Bertrand, G. (**2013**). *Angew. Chem. Int. Ed. 52*: 7590–7592.

43 Böser, R., Haufe, L.C., Freytag, M. et al. (**2017**). *Chem. Sci. 8*: 6274–6280.

44 Braunschweig, H., Burzler, M., Dewhurst, R.D., and Radacki, K. (**2008**). *Angew. Chem. Int. Ed. 47*: 5650–5653.

45 Yamashita, M., Suzuki, Y., Segawa, Y., and Nozaki, K. (**2007**). *J. Am. Chem. Soc. 129*: 9570–9571.

46 Meerwein, v.H. and Schmidt, R. (**1925**). *Justus Liebigs Ann. Chem. 444*: 221–238.

47 Kajiwara, T., Terabayashi, T., Yamashita, M., and Nozaki, K. (**2008**). *Angew. Chem. Int. Ed. 47*: 6606–6610.

48 Asami, S.-s., Okamoto, M., Suzuki, K., and Yamashita, M. (**2016**). *Angew. Chem. Int. Ed. 55*: 12827–12831.

49 Asami, S.-s., Ishida, S., Iwamoto, T. et al. (**2017**). *Angew. Chem. Int. Ed. 56*: 1658–1662.

50 Asami, S.-s., Suzuki, K., and Yamashita, M. (**2017**). *Chem. Lett. 46*: 686–689.

51 Nozaki, K., Aramaki, Y., Yamashita, M. et al. (**2010**). *J. Am. Chem. Soc. 132*: 11449–11451.

52 Hayashi, Y., Segawa, Y., Yamashita, M., and Nozaki, K. (**2011**). *Chem. Commun. 47*: 5888–5890.

53 Dange, D., Davey, A., Abdalla, J.A.B. et al. (**2015**). *Chem. Commun. 51*: 7128–7131.

54 Pécharman, A.-F., Colebatch, A.L., Hill, M.S. et al. (**2017**). *Nat. Commun. 8*: 15022.

55 Pécharman, A.F., Hill, M.S., McMullin, C.L., and Mahon, M.F. (**2017**). *Angew. Chem. Int. Ed. 56*: 16363–16366.

56 Pecharman, A.-F., Hill, M.S., and Mahon, M.F. (**2018**). *Dalton Trans. 47*: 7300–7305.

57 Pécharman, A.-F., Hill, M.S., McMullin, C.L., and Mahon, M.F. (**2018**). *Organometallics 37*: 4457–4464.

58 Pécharman, A.F., Hill, M.S., and Mahon, M.F. (**2018**). *Angew. Chem. Int. Ed. 57*: 10688–10691.

59 Nagashima, Y., Takita, R., Yoshida, K. et al. (**2013**). *J. Am. Chem. Soc. 135*: 18730–18733.

60 Okuno, Y., Yamashita, M., and Nozaki, K. (**2011**). *Angew. Chem. Int. Ed. 50*: 920–923.

61 Okuno, Y., Yamashita, M., and Nozaki, K. (**2011**). *Eur. J. Org. Chem. 2011*: 3951–3958.

62 Campos, J. and Aldridge, S. (**2015**). *Angew. Chem. Int. Ed. 54*: 14159–14163.

63 Hicks, J., Vasko, P., Goicoechea, J.M., and Aldridge, S. (**2018**). *Nature 557*: 92–95.

64 Hicks, J., Mansikkamäki, A., Vasko, P. et al. (**2019**). *Nat. Chem. 11*: 237–241.

65 Schwamm, R.J., Anker, M.D., Lein, M., and Coles, M.P. (**2019**). *Angew. Chem. Int. Ed. 58*: 1489–1493.

66 Schmidt, E.S., Jockisch, A., and Schmidbaur, H. (**1999**). *J. Am. Chem. Soc. 121*: 9758–9759.

67 Schmidt, E.S., Schier, A., and Schmidbaur, H. (**2001**). *J. Chem. Soc., Dalton Trans.*: 505–507.

68 Baker, R.J., Farley, R.D., Jones, C. et al. (**2002**). *J. Chem. Soc., Dalton Trans.*: 3844–3850.

69 Baker, R.J., Jones, C., and Platts, J.A. (**2003**). *Dalton Trans.*: 3673–3674.

70 Baker, R.J., Jones, C., Kloth, M., and Platts, J.A. (**2004**). *Organometallics 23*: 4811–4813.

71 Baker, R.J. and Jones, C. (**2005**). *Coord. Chem. Rev. 249*: 1857–1869.

72 Baker, R.J., Jones, C., and Kloth, M. (**2005**). *Dalton Trans.*: 2106–2110.

73 Green, S.P., Jones, C., Lippert, K.A. et al. (**2006**). *Inorg. Chem. 45*: 7242–7251.

74 Arnold, P.L., Liddle, S.T., McMaster, J. et al. (**2007**). *J. Am. Chem. Soc. 129*: 5360–5361.

75 Green, S.P., Jones, C., Mills, D.P., and Stasch, A. (**2007**). *Organometallics 26*: 3424–3430.

76 Jones, C., Rose, R.P., and Stasch, A. (**2007**). *Dalton Trans.*: 2997–2999.

77 Jones, C., Stasch, A., and Woodul, W.D. (**2009**). *Chem. Commun.*: 113–115.

78 Liddle, S.T., Mills, D.P., Gardner, B.M. et al. (**2009**). *Inorg. Chem. 48*: 3520–3522.

79 Bonello, O., Jones, C., Stasch, A., and Woodul, W.D. (**2010**). *Organometallics. 29*: 4914–4922.

80 Protchenko, A.V., Saleh, L.M.A., Vidovic, D. et al. (**2010**). *Chem. Commun. 46*: 8546–8548.

81 Schwamm, R.J., Anker, M.D., Lein, M. et al. (**2018**). *Angew. Chem. Int. Ed. 57*: 5885–5887.

8

Novel Chemical Transformations in Organic Synthesis with Ate Complexes

Keiichi Hirano[a] *and Masanobu Uchiyama*[a,b]

[a]*Graduate School of Pharmaceutical Sciences, The University of Tokyo, Tokyo, Japan*
[b]*Research Initiative for Supra-Materials (RISM), Shinshu University, Nagano, Japan*

8.1 Introduction

Ate complexes are anionic organometallic complexes, whose central metals have increased valence due to acceptance of Lewis basic ligands at vacant orbitals. They have been attracting much attention not only in research on inorganic coordination chemistry but also in synthetic organic chemistry, since they offer tunable reactivity due to the scope for choosing central metals, counter-cations, and coordination environments as required. In this chapter, we review the development of novel synthetic methodologies utilizing various ate complexes, especially zincates, aluminates, cuprates, and borates, in our laboratory.

8.2 Ate Complexes

The term "ate complex" was coined by Wittig (the original German description is At-komplex) in 1958 [1], but the first mono-anion-type zincate, $Na[ZnEt_3]$, was discovered as early as 1859 by Wanklyn [2], and even a di-anionic zincate, $Li_2[ZnMe_4]$, had already been synthesized by Hurd in 1948 [3]. After these seminal reports, zincates attracted little interest for a long time; however, they are currently used extensively in chemical synthesis owing to their unique reactivity and chemoselectivity. Neutral organozinc reagents (ZnR_2 and RZnX) alone show low reactivity (in other words, they exhibit low ligand (R) transferability), but the reactivity of these reagents is often significantly boosted by coordination of external ligands, i.e. ate complex formation.

The difference in reactivity profiles between neutral organozincs and their ate congeners can be understood in terms of the electronic structures in the outer shell of the central zinc atom (Figure 8.1). In neutral organozinc compounds, the outer shell of zinc has 14 electrons. This coordinatively unsaturated zinc of RZnX or ZnR_2 tends to act more as a Lewis acid, accepting another ligand, than as a nucleophile transferring the R anion. Coordination of an additional anionic ligand at the vacant orbital of zinc leads to the formation of a more stable zincate species with a 16-electron state on zinc, resulting in enhanced thermodynamic stability and reduced Lewis acidity. From the point of view of nucleophilicity, incoming anionic ligands make the whole

Polar Organometallic Reagents: Synthesis, Structure, Properties and Applications, First Edition. Edited by Andrew E. H. Wheatley and Masanobu Uchiyama.

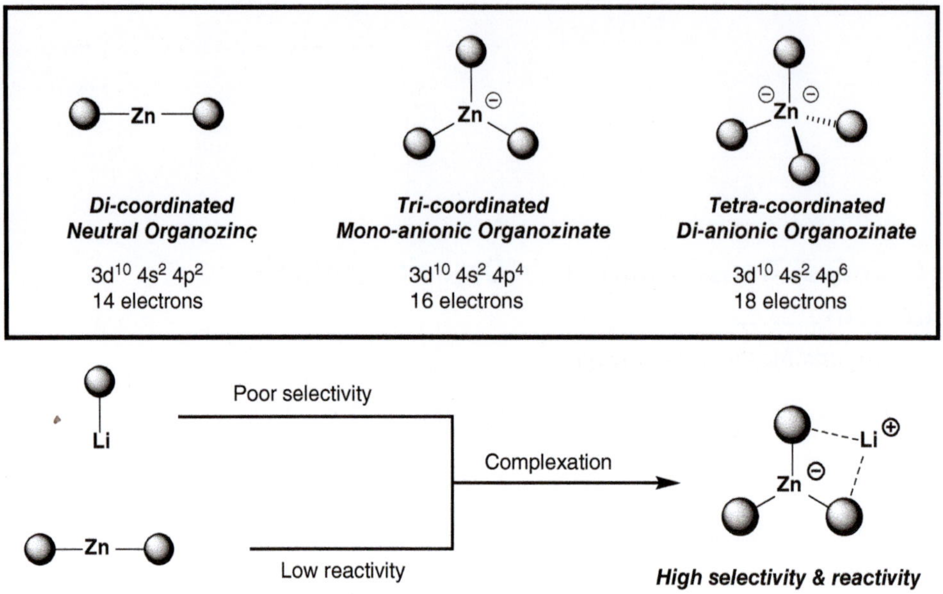

Figure 8.1 Zinc ate complexes.

species more electron-rich, facilitating ligand transfer (enhanced kinetic reactivity). Although neutral organozinc species are highly chemoselective reagents and are widely employed in organic synthesis [4–8], they sometimes suffer from lack of reactivity. In contrast, organolithiums are extremely reactive nucleophiles/bases to the extent that chemoselectivity is compromised. Zincates offer the best of both worlds, having a good balance of properties [9, 10].

8.3 Di-anion-type Zincate

8.3.1 Mono-anion-type Zincates and Di-anion-type Zincates

In 1994, Kondo et al. reported that the halogen–zinc exchange reaction of aryl iodides *3-1*-**1** with $Li[ZnMe_3]$ proceeds smoothly at −78 °C and the generated aryl zincates can be trapped with electrophiles (Scheme 8.1) [11]. On the other hand, synthetically more desirable aryl bromides remain intact under these reaction conditions.

In 1996, we discovered that di-anion-type zincates, $Li_2[Zn(X)Me_3]$ (X = Me, CN, or SCN), smoothly react with aryl bromides *3-1*-**6** at an ambient temperature or 0 °C, and the resultant aryl zincates *3-1*-**7** react with benzaldehyde in high yields (Table 8.1, entries 1–6) [12]. Base-susceptible methyl ester is not compatible, and these results illustrate the high reactivity of the di-anionic zincates (entries 7–9). In addition, tellurium–zinc exchange is also achievable with di-anionic zincates.

Di-anion-type aryl zincates, $Li_2[ArZnMe_3]$ *3-1*-**7**, exhibit much higher reactivity than mono-anion-type $Li[ArZnMe_2]$ *3-1*-**2** in their reactions with carbon–carbon double bonds (Scheme 8.2). In contrast with the reaction between $Li[ZnMe_3]$ and aryl iodide *3-1*-**10**, which ends only with iodine–zinc exchange, the treatment of *3-1*-**10** with $Li_2[ZnMe_4]$ provides the dihydroindole product *3-1*-**12** via iodine–zinc exchange followed by intramolecular conjugate addition to α,β-unsaturated ester. Moreover, zincation of 2-allyloxyiodobenzene (*3-1*-**13**) with $Li_2[ZnMe_4]$ provides the dihydrobenzofuran *3-1*-**15** via carbozincation reaction of the unactivated carbon–carbon double bond [13].

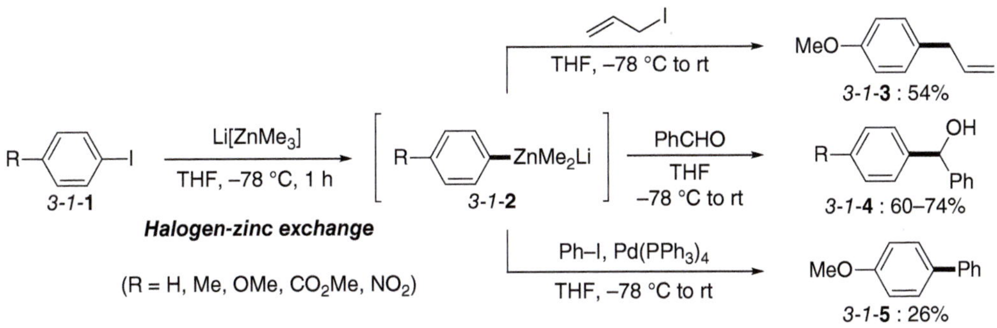

Scheme 8.1 Halogen–zinc exchange reaction of aryl iodides with $Li[ZnMe_3]$.

Table 8.1 Metalation using di-anion-type zincate: $Li_2[Zn(X)Me_3]$.

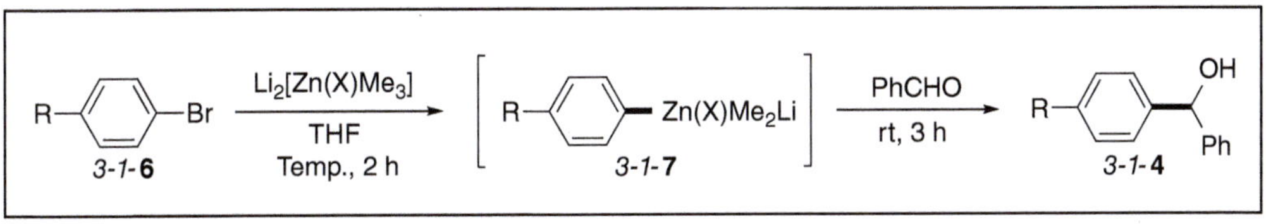

Entry	R	X	Temp. [°C]	Yield [%]	Entry	R	X	Temp. [°C]	Yield [%]
1	H	Me	0	90	6	MeO	SCN	rt	92
2	H	CN	rt	90	7	CO_2Me	Me	0	trace
3	H	SCN	rt	89	8	CO_2Me	CN	0	trace
4	MeO	Me	0	84	9	CO_2Me	SCN	0	trace
5	MeO	CN	rt	90					

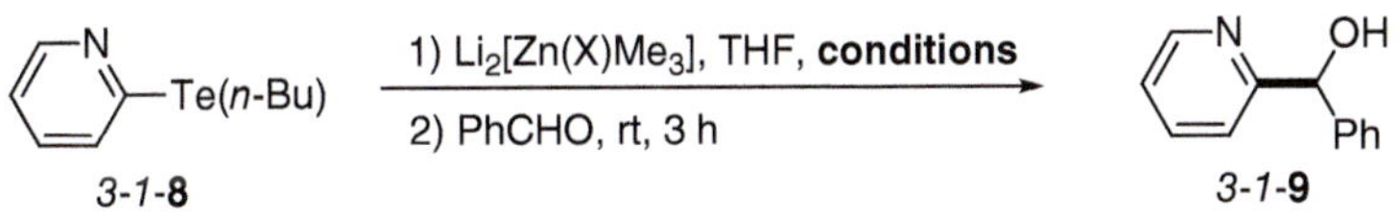

conditions:
X = CN: rt, 2 h, 63%　X = SCN: rt, 2 h, 58%　X = Me: 0 °C, 2 h, 81%

8.3.2 Highly Bulky Di-anion-type Zincate: $Li_2[Znt\text{-}Bu_4]$

8.3.2.1 Halogen–Zinc Exchange in the Presence of Proton Sources

In 2006, we developed a highly chemoselective metalation methodology for aryl bromides and iodides *3-2-1*-**1**, utilizing a newly designed, extremely sterically encumbered di-anion-type zincate, $Li_2[Znt\text{-}Bu_4]$ (Scheme 8.3) [14]. The most characteristic feature of this zincate is its unprecedented compatibility with protic functional groups such as alcohols, as well as the even more acidic phenols (*3-2-1*-**3a**, -**3b**) and the N–H groups in carboxyamides (*3-2-1*-**3c**). Rapid halogen–zinc exchange and this unique proton tolerance enable aromatic functionalizations without protection of those groups.

The regioselectivity of the nucleophilic substitution reaction of propargyl bromide is dictated by *t*-Bu ligands. In sharp contrast to the Grignard reagent and the cuprate (*3-2-1*-**5**), $Li_2[ArZnt\text{-}Bu_3]$ *3-2-1*-**2** was found to deliver its aryl ligand highly selectively in an S_N2' manner to afford the corresponding allene product *3-2-1*-**4** (Table 8.2) [15].

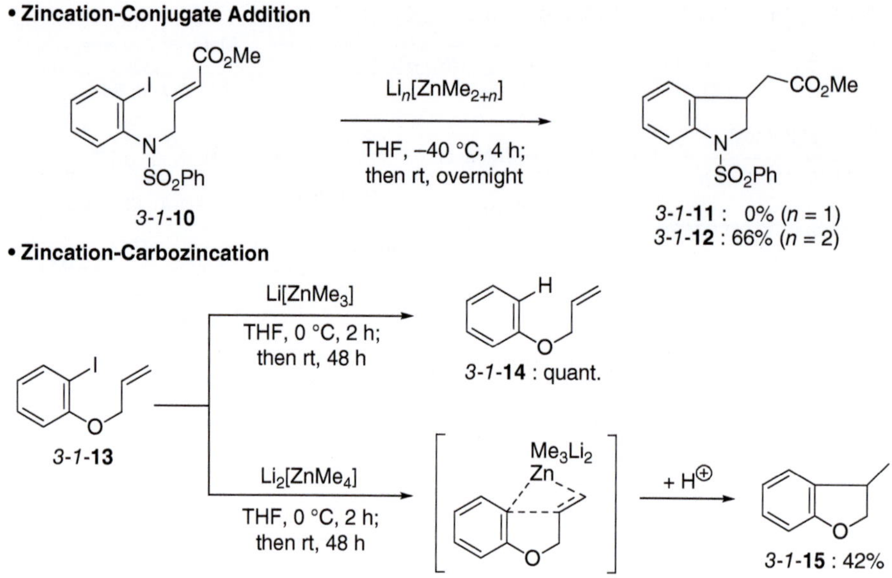

Scheme 8.2 Enhanced reactivity of di-anion-type zincate.

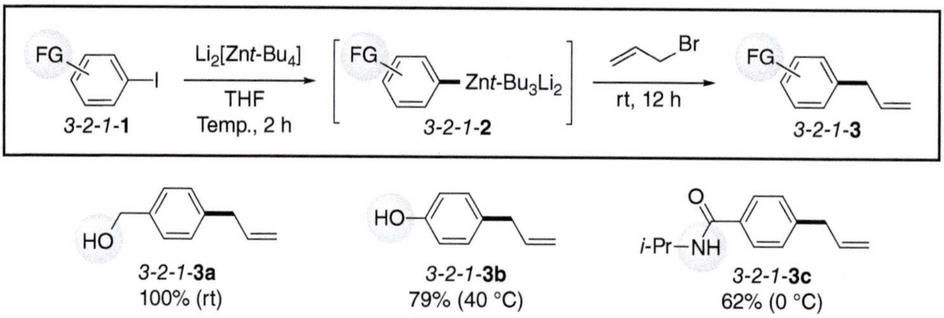

Scheme 8.3 $Li_2[Znt\text{-}Bu_4]$: proton-proof metalating agent.

8.3.2.2 Anionic Polymerization in Water

The di-anion-type zincate $Li_2[Znt\text{-}Bu_4]$ acts as an initiator of anionic polymerization [16]. Polymerization of *N*-isopropylacrylamide (NIPAm, *3-2-1-***6**) proceeds rapidly, especially in protic solvents such as MeOH and H_2O, to give poly(NIPAm) *3-2-1-***7** in high yields (Table 8.3). It is remarkable that the reaction is completed within 15 minutes in H_2O. Other acrylic acid analogues such as *N,N*-dimethylacrylamide, acrylamide, and 2-hydroxyethyl methacrylate are also applicable to this zincate-initiated polymerization.

8.3.3 Cross-coupling Reaction via C–O Bond Cleavage

We reported the Negishi-type coupling reaction using aryl ethers as electrophiles in 2012 (Scheme 8.4) [17]. The di-anionic zincates $Li_2[ArZnMe_3]$ *3-3-***2** generated from various aryl halides and $Li_2[ZnMe_4]$ couple with aryl ethers to give biaryl products *3-3-***3** in good yields at an ambient temperature in the presence of a nickel catalyst. This cross-coupling reaction is uniquely specific to di-anionic zincates *3-3-***2**, and other classes of arylzincs such as neutral- (ArZnX or $ZnAr_2$) and

Table 8.2 $Li_2[Znt\text{-}Bu_4]$ for nucleophilic substitution of propargyl bromide.

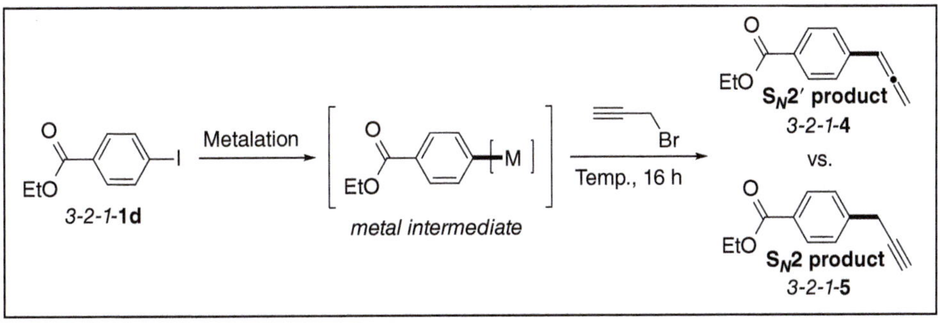

Entry	Metalation Reagent and Conditions	Temp. [°C]	Yield [%]	Ratio [S_N2'/S_N2]
1	$Li_2[Znt\text{-}Bu_4]$ (1.1 eq.), THF, 0 °C, 2 h	25	100	98:2
2	*i*-PrMgBr (1.1 eq.), THF, –40 °C, 1 h, then CuCN (10 mol%)	–40	76	79:21
3	$Li_2[Cu(CN)Me_2]$ (1.1 eq.), THF, 0 °C, 2 h	25	54	63:37

Source: Based on Furuyama et al. [15].

Table 8.3 Anionic polymerization of *N*-isopropylacrylamide using $Li_2[Znt\text{-}Bu_4]$.

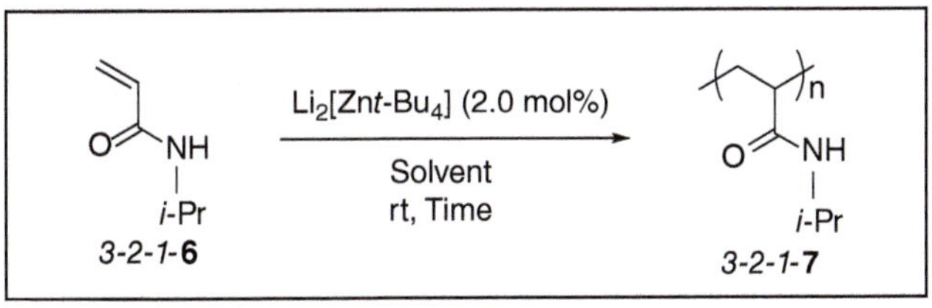

Entry	Solvent	Time [h]	Yield [%]	M_n	PDI[a]
1	THF	24	8	7000	1.50
2	THF	168	33	7000	2.71
3	MeOH	3	76	18 000	1.65
4	H_2O	< 1	92	27 000	2.72

[a] Polydispersion index.

mono-anionic zinc species ($Li[ArZnMe_2]$) do not support biaryl formation. The high chemoselectivity of the zincate, as well as the mild reaction conditions of this methodology, allows the substitution of the methoxy group of (+)-naproxene *3-3-***1f** with a phenyl group without loss of optical purity, affording the engineered naproxene *3-3-***3fa**. Since aryl ethers are ubiquitous among natural products and biologically active substances, this cross-coupling reaction is a powerful tool with which to fine-tune the activity of those compounds through facile modification of the parent molecular architectures.

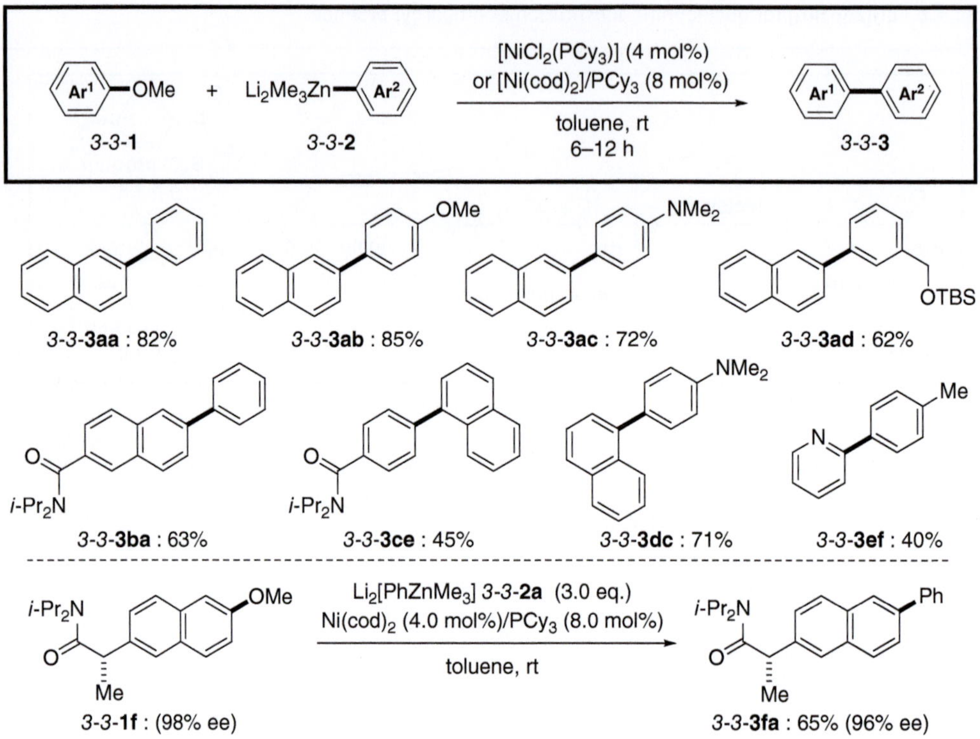

Scheme 8.4 Negishi-type cross-coupling reaction via C–O bond cleavage.

8.4 Heteroleptic Zinc Ate Complexes

8.4.1 Deprotonative Metalation of Aromatic C–H Bonds

Deprotonative metalation is one of the most important and effective ways to functionalize aromatic rings. This chemical process, however, usually requires bases such as organolithiums and lithium amides [18–20] and therefore often suffers from low functional group tolerance. Ate complex formation is a good solution to address this issue since the reactivity and selectivity of ate complexes depend on the central metal and ligands and can be flexibly tuned, since huge numbers of combinations are available [21]. "Heteroleptic" ate complexes are especially fine-tunable by using ligands with different functions (Figure 8.2) [22].

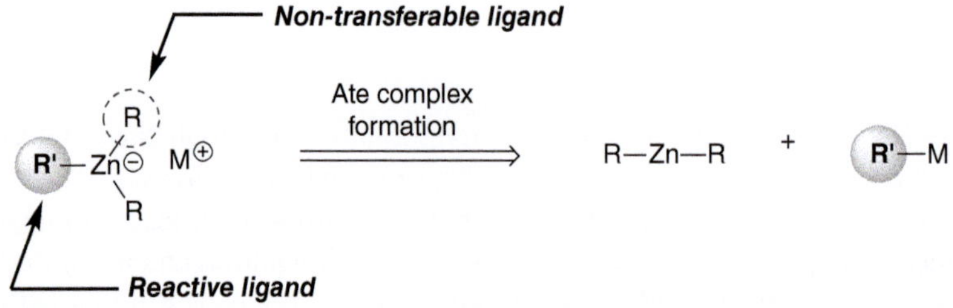

Figure 8.2 Heteroleptic zincates: an enormous range of possibilities.

8.4.1.1 Amidozincate Base: Li[(TMP)ZnR$_2$]

In 1999, we developed the deprotonative zincation reaction of arenes *4-1-1-***1** using a newly designed amidozincate base, Li[(TMP)Zn*t*-Bu$_2$] (Scheme 8.5) [23]. This amido complex is easily prepared from Zn*t*-Bu$_2$ and LiTMP and enables the directed *ortho*-metalation of aromatic rings with ester (*4-1-1-***4a-4d**, **-4g**, **-4f**, **-5a**, **-5b**), amide (*4-1-1-***4e**), and cyano (*4-1-1-***3**) groups as the directing groups; the generated aryl zincates *4-1-1-***2** can engage in various chemical transformations. The polar groups are not damaged during the reaction processes. Moreover, π-electron-rich heteroaromatics such as thiophene (*4-1-1-***4f**) and furan (*4-1-1-***4g**), as well as electron-deficient ones such as pyridine (*4-1-1-***4h**), quinoline (*4-1-1-***4i**), and isoquinoline (*4-1-1-***4j**), are smoothly metalated in good yields.

In general, *ortho*-lithiated bromobenzenes instantly generate the corresponding benzynes via facile elimination of lithium bromide, even at rigorously controlled cryogenic temperatures. Notably, deprotonation of the C–H bond *ortho* to the bromine atom of bromobenzenes *4-1-1-***6** by Li[(TMP)Zn*t*-Bu$_2$] smoothly proceeds without benzyne formation, and the resultant aryl zincate *4-1-1-***7** can be trapped with electrophiles in high yields (Scheme 8.6) [24, 25].

On the other hand, *ortho*-metalation of *4-1-1-***6** with Li[(TMP)Zn**Me**$_2$], which has much smaller alkyl ligands, smoothly generates benzynes *4-1-1-***9**. This provides efficient access to functionalized benzynes, taking advantage of the highly chemoselective nature of the amidozincate base. The Diels–Alder reaction with 1,3-diphenylisobenzofuran (*4-1-1-***10**) has been adopted to efficiently access benzynes from aromatics (Table 8.4).

8.4.1.2 Amidoaluminate Base: Li[(TMP)Al*i*-Bu$_3$]

Organoaluminum reagents have long been utilized in organic synthesis, but their usage has been limited mainly to aliphatic chemistry [26–28]. In general, arylaluminum compounds are prepared via transmetalation from aryllithiums and Grignard reagents using aluminum halides. This protocol is facile and well accepted to date, albeit limited in the scope of functional groups on the

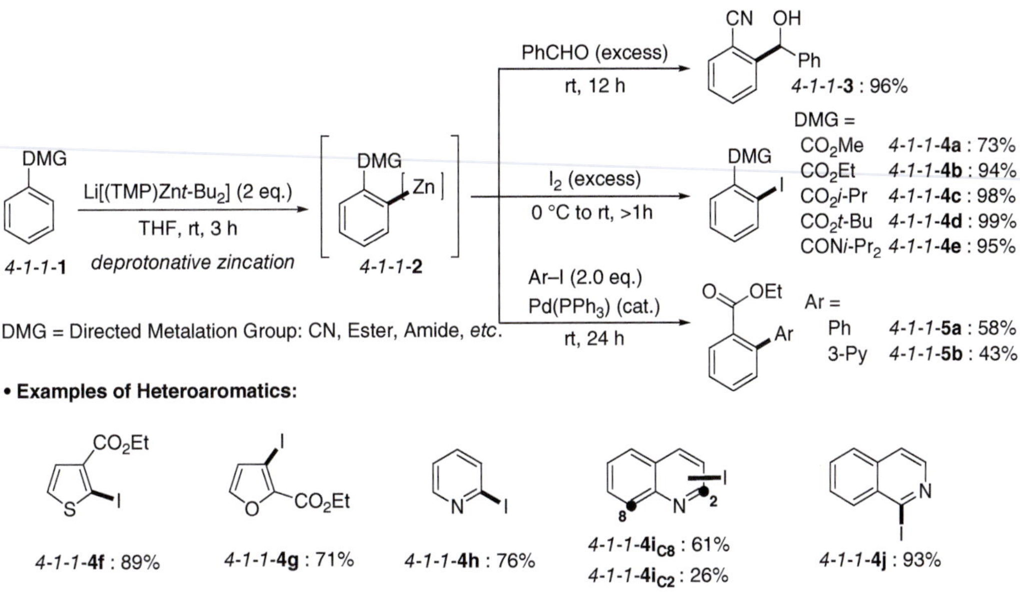

Scheme 8.5 Highly regio- and chemoselective zincation of (hetero)aromatics.

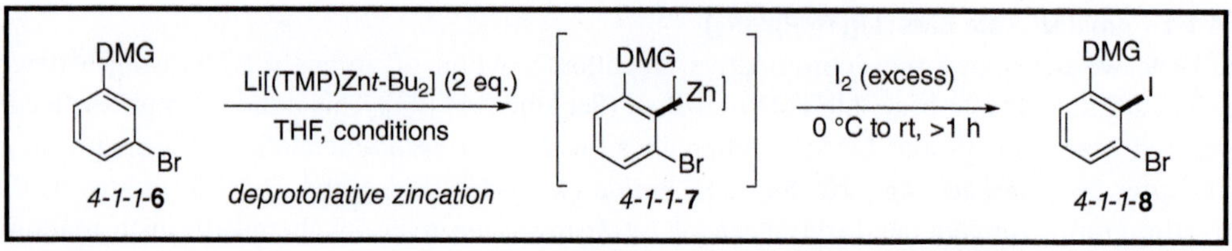

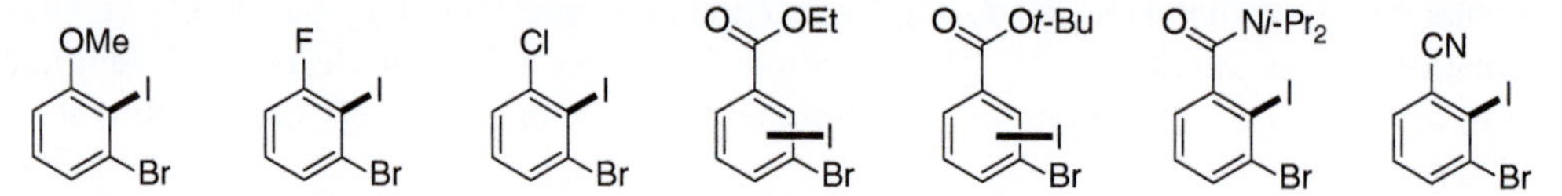

*4-1-1-*8a : 95% (−30 °C, 12 h) *4-1-1-*8b : 93% (−30 °C, 2 h) *4-1-1-*8c : 77% (−20 °C, 2 h) *4-1-1-*8d : 92% (rt, 3 h) *4-1-1-*8e : 98% (rt, 3 h) *4-1-1-*8f : 84% (0 °C, 3 h) *4-1-1-*8g : 96% (0 °C, 3 h)

Scheme 8.6 *Ortho*-iodination of bromobenzenes without benzyne formation.

Table 8.4 Regioselective generation of functionalized benzynes with Li[(TMP)ZnMe$_2$].

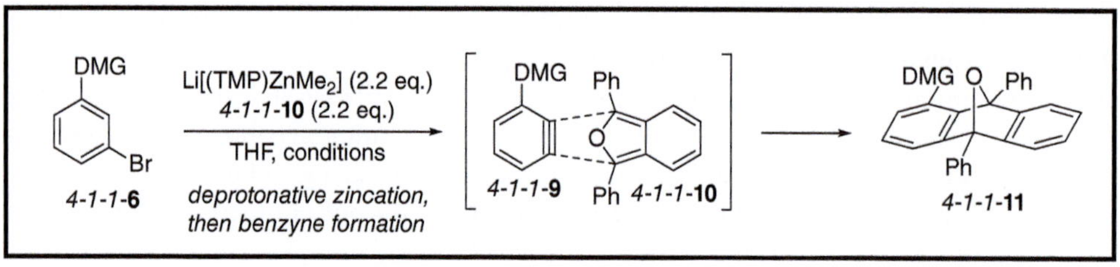

Entry	Substrate	Conditions	Benzyne	Yield [%]	Entry	Substrate	Conditions	Benzyne	Yield [%]
1	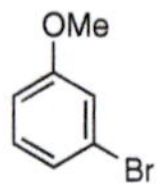 *4-1-1-*6a	rt, 12 h	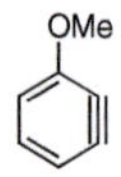 *4-1-1-*9a	100	5	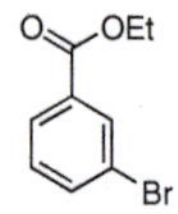 *4-1-1-*6d	reflux, 3 h	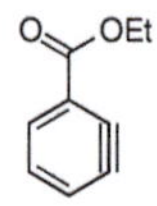 *4-1-1-*9d	55
2	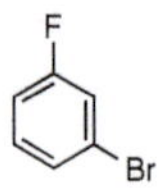 *4-1-1-*6b	60 °C, 15 h	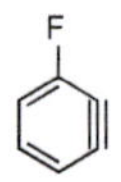 *4-1-1-*9b	71	6	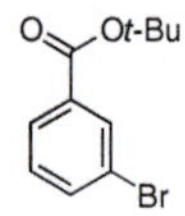 *4-1-1-*6e	reflux, 6 h	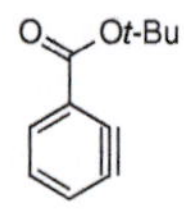 *4-1-1-*9e	88
3	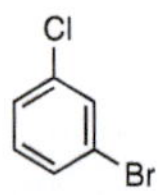 *4-1-1-*6c	rt, 48 h	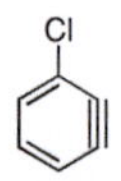 *4-1-1-*9c	72	7	 *4-1-1-*6f	reflux, 12 h	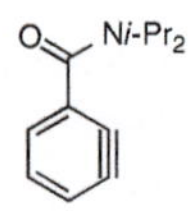 *4-1-1-*9f	100
4	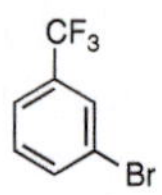 *4-1-1-*6h	rt, 12 h	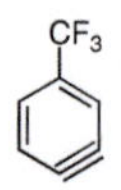 *4-1-1-*9h	100	8	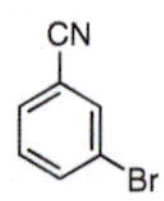 *4-1-1-*6g	reflux, 12 h	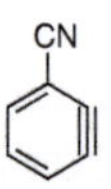 *4-1-1-*9g	90

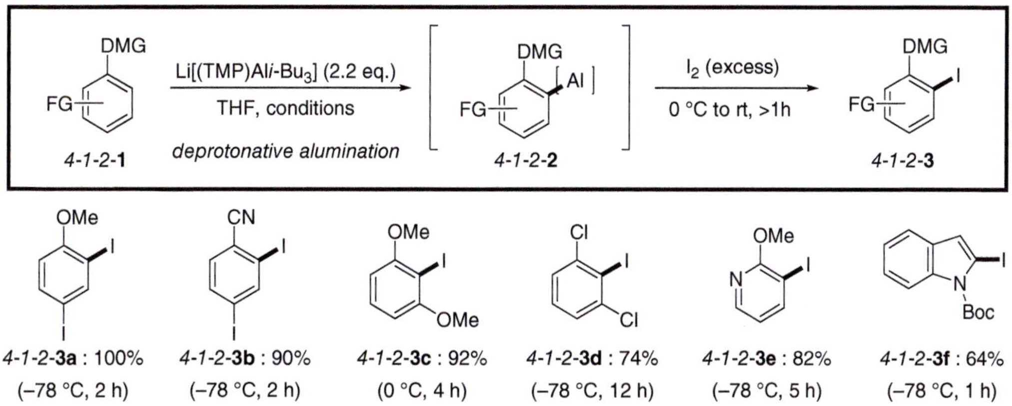

Scheme 8.7 Chemoselective deprotonative *ortho*-alumination with Li[(TMP)Al*i*-Bu$_3$].

aromatic rings. We developed a preparative method for functionalized arylaluminums via deprotonative metalation adding our amidoaluminate, Li[(TMP)Al*i*-Bu$_3$]. Many functional groups including heteroaromatic rings (*4-1-2-*-**3e**, -**3f**) are tolerated in this metalation reaction (Scheme 8.7) [29, 30]. A notable feature of this aluminate is the low halogen–metal exchange ability of the isobutyl ligands. Halogen-containing arenes especially enjoy this unique reactivity of the aluminate, and deprotonative *ortho*-metalation of 4-iodoanisole (*4-1-2-*-**1a**) and 4-iodobenzonitrile (*4-1-2-*-**1b**) proceeds chemo- and regioselectively in high yields, while the C–I bond at the 4-position remains intact. Details on mechanistic insights into aluminate chemistry are discussed in Chapters 2 and 6 [31].

8.4.1.3 Amidocuprate Base: Li$_2$[(TMP)Cu(CN)R]

Organocuprates are frequently used for various types of C–C bond-forming reactions based on the redox activity of copper [32–34]. Amido ligands behave as dummy ligands in cuprate chemistry and have not often been utilized as reactive ligands [35]. Lipshutz amidocuprate bases, Li$_2$[(TMP)Cu(CN)R] [36, 37], developed by our group undergo chemoselective deprotonative metalation of a series of aromatics with polar functional groups, such as methoxy (*4-1-3-*-**3a**), cyano (*4-1-3-*-**3b**), and amide (*4-1-3-*-**3c**) groups. Heteroaromatics are also regioselectively deprotonated with Li$_2$[(TMP)Cu(CN)Me] (*4-1-3-*-**1d**, -**1e**, -**1f**) (Scheme 8.8).

The arylcuprate reacts with various electrophiles in excellent yields owing to its high nucleophilicity (Scheme 8.9).

This direct cupration of aromatics *4-1-3-*-**1** can be extended to the synthesis of densely functionalized phenols *4-1-3-*-**9** (Scheme 8.10). The reaction of arylcuprates prepared by deprotonation of aromatic compounds with Li$_2$[(TMP)$_2$Cu(CN)] and *t*-BuOOH (TBHP) affords the corresponding phenol products *4-1-3-*-**9** in high yields [38]. Cumene hydroperoxide (CHP) can be used as an oxygen source as well (*4-1-3-*-**9k**). Iodine and vinyl groups survive under these hydroxylation conditions (*4-1-3-*-**9a**, -**9c**), and the high functional group compatibility and regioselectivity have been employed in a total synthesis of (+)-pancratistatin *4-1-3-*-**11** by Sarlah et al. [39].

Density functional theory (DFT) calculation suggests that this oxygenation reaction occurs via the redox reactions of copper, oxidative addition of peroxide to copper (**CP2** to **CP3**), and reductive elimination (**CP3** to **CP4**), rather than direct nucleophilic attack of an aryl anion on peroxide (Scheme 8.11).

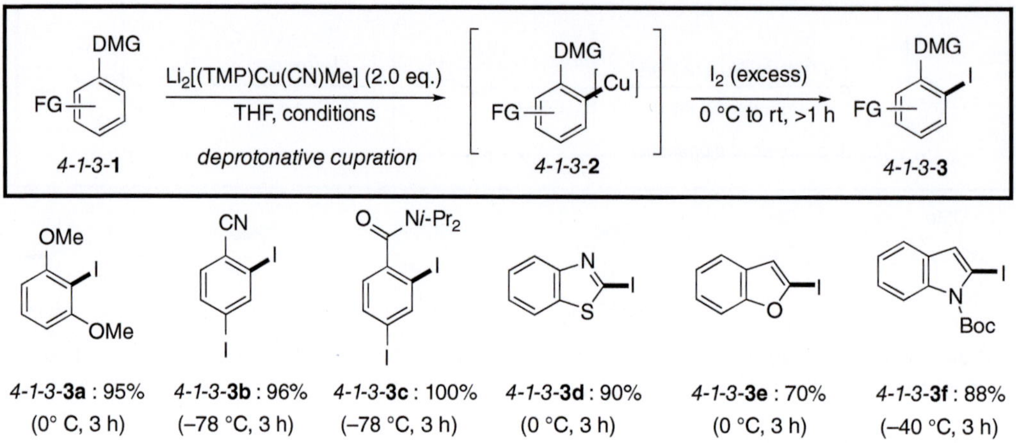

Scheme 8.8 Directed *ortho*-cupration with Li_2[(TMP)Cu(CN)Me].

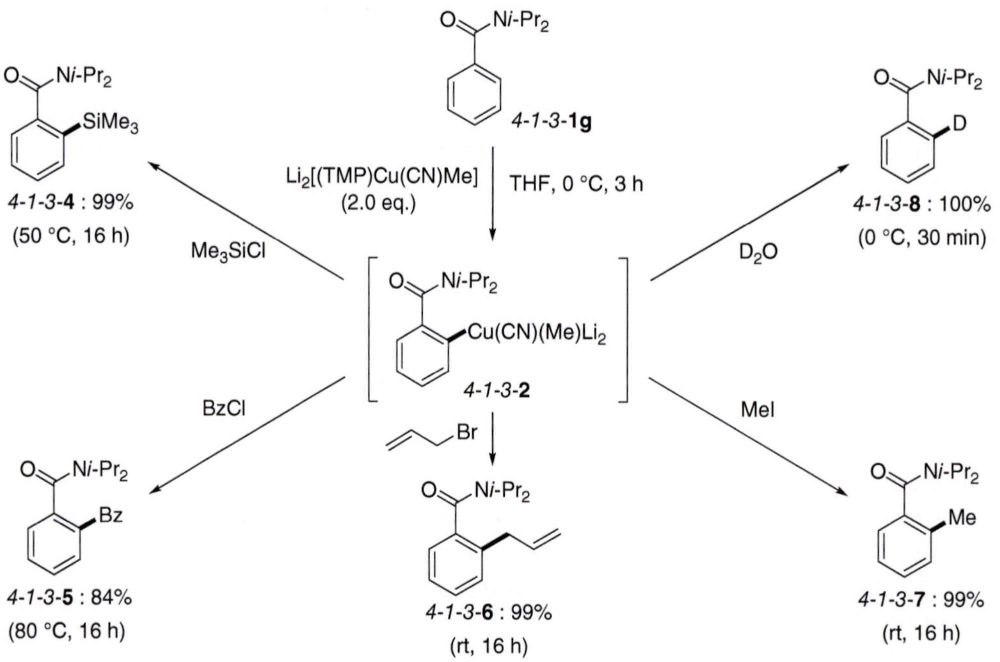

Scheme 8.9 Arylcuprate reactivity.

The direct amination reaction of aromatics was also achieved with a broad substrate scope by using $BnONH_2$ instead of ROOH (Scheme 8.12). This amination is considered to proceed in the same manner as the hydroxylation.

8.4.2 Hydridozincate: M[$HZnMe_2$]

Although trihydrido-type zincates are known in the fields of inorganic chemistry and coordination chemistry, the reactivities of heteroleptic hydridozincates have been little explored. We discovered that the hydridozincates M[$HZnMe_2$] (M = Li *4-2-***1a** or Na *4-2-***1b**) generated from $ZnMe_2$ and

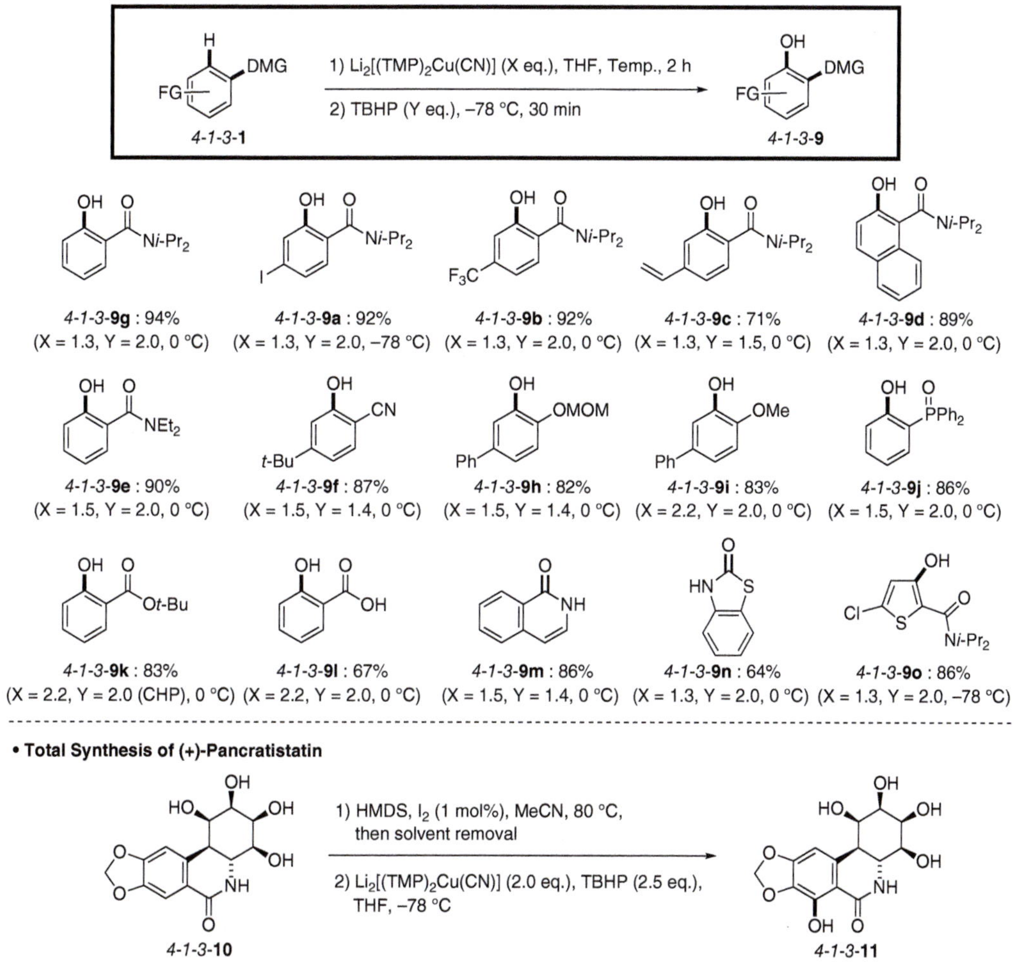

Scheme 8.10 Amidocuprate hydroxylation of aromatics.

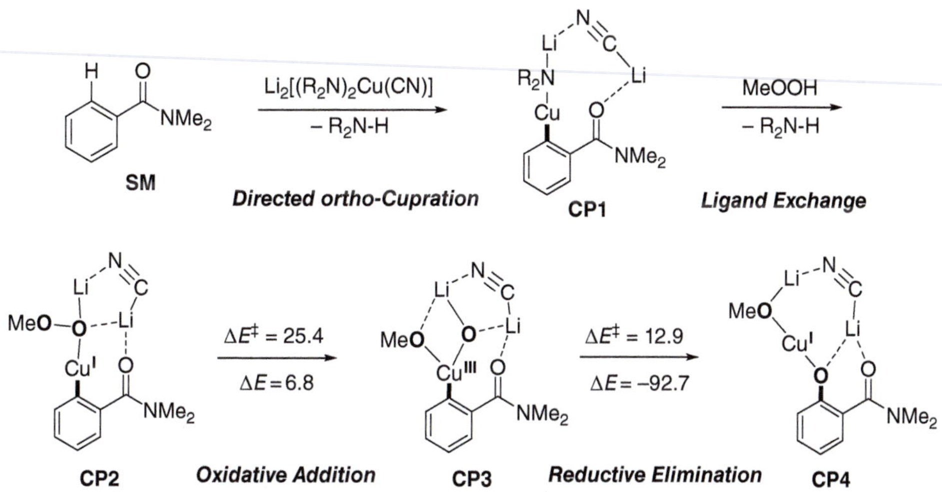

Scheme 8.11 DFT calculations (kcal/mol) to assess the reaction mechanism for amidocuprate deprotonation of an aromatic ring; M06/SVP(Cu) and 6-31+G*(other atoms).

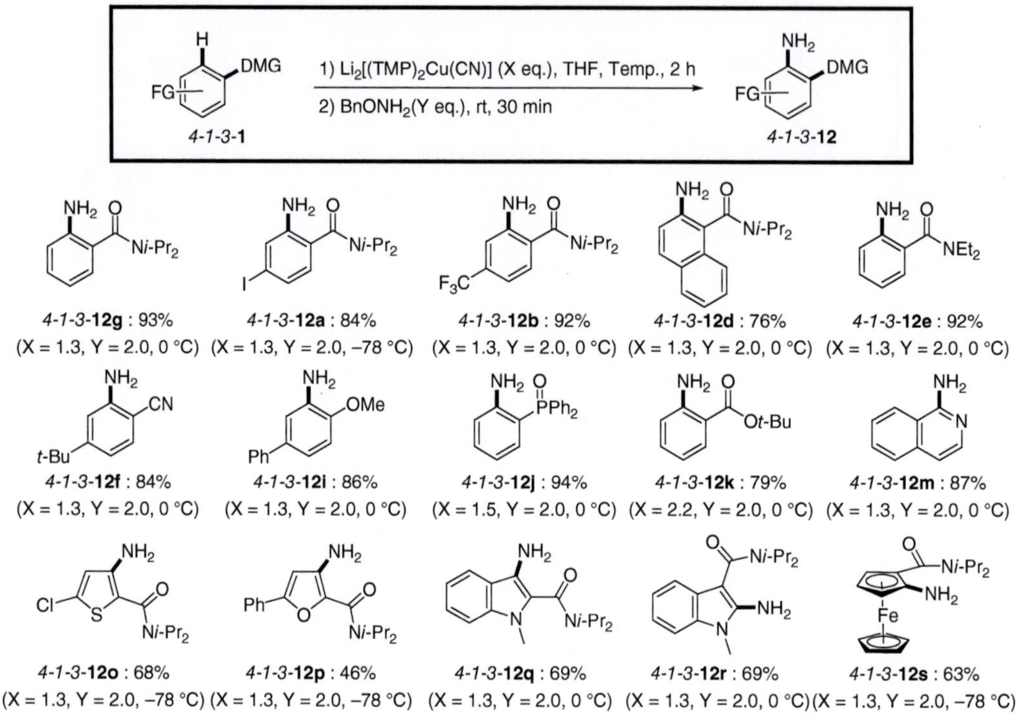

Scheme 8.12 Amidocuprate amination of aromatics.

metal hydrides deliver the hydride ligand preferentially to Me ligands to carbonyl compounds to give the corresponding alcohols *4-2-***2** (Scheme 8.13) [40, 41]. Notably, no methylation product is obtained at all. In other words, $ZnMe_2$ regenerates after hydride transfer and can be used as a catalyst. Indeed, the reduction of carbonyl compounds with NaH/LiH in the presence of catalytic $ZnMe_2$ proceeds smoothly to give the corresponding alcohols in high yields.

Additionally, $Li[HZnMe_2]$ *4-2-***1a** uniquely semi-reduces carboxylic acids *4-2-***3** to aldehydes *4-2-***5** without the formation of any trace of over-reduced alcohols (Scheme 8.14). This can be explained by the formation of the zincioacetal intermediate *4-2-***4** after hydride transfer to the carbonyl group, with this tetrahedral species *4-2-***4** being stable in the reaction mixture. Aldehydes *4-2-***5** are released upon quenching.

This mode of semi-reduction can be applied to the direct transformation of carboxylic acids *4-2-***3** to ketones *4-2-***7**. Di-anion-type zincate $Li_2[R^2ZnMe_3]$ is considered to form the stable zincioketal intermediate *4-2-***6** upon reaction with carboxylic acid *4-2-***3**, with the tetrahedral structure *4-2-***6** presumably avoiding over-reaction to give the undesired tertiary alcohol. Mono-anion-type zincates are also available for this ketonization reaction by the pre-formation of a lithium carboxylate (Scheme 8.15) [42].

8.4.3 Silylzincates

Carbon–carbon multiple bonds are versatile chemical platforms for myriad transformations and reactions with organometallic reagents, and these are fundamental reactions for increasing molecular complexity. Chemo- and regioselective silylzincation reactions of alkynes and alkenes are efficient tools with which to obtain synthetically versatile vinyl and alkylsilanes [43, 44].

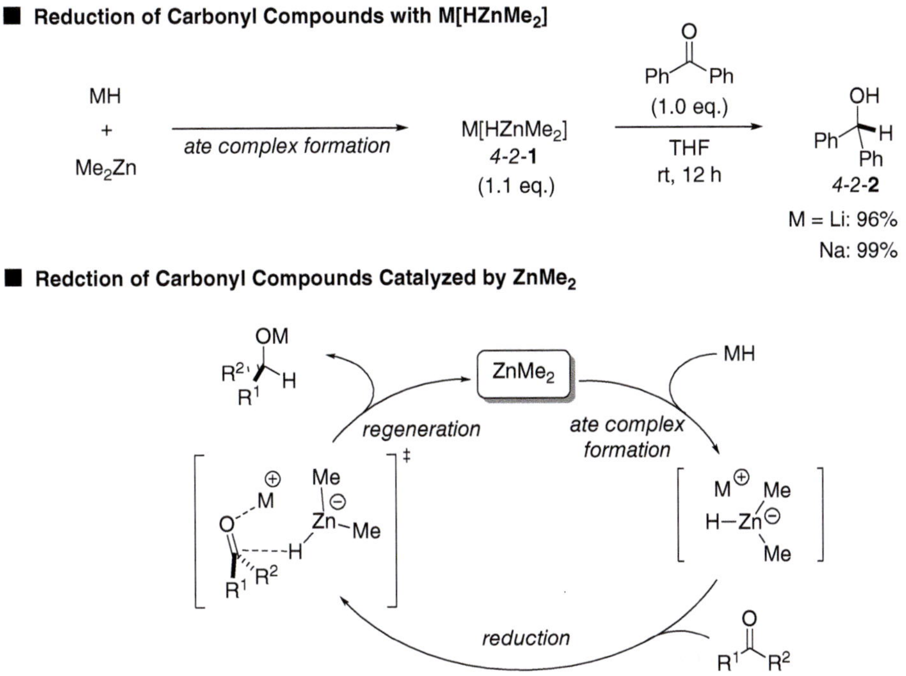

Scheme 8.13 Reduction of carbonyl compounds with M[HZnMe₂].

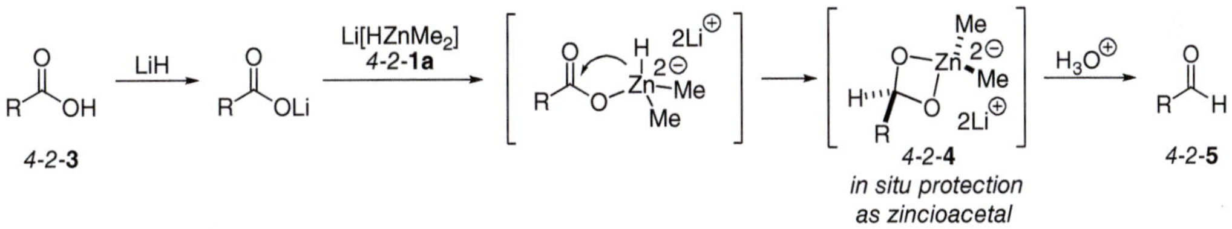

Scheme 8.14 Semi-reduction of carboxylic acids to aldehydes.

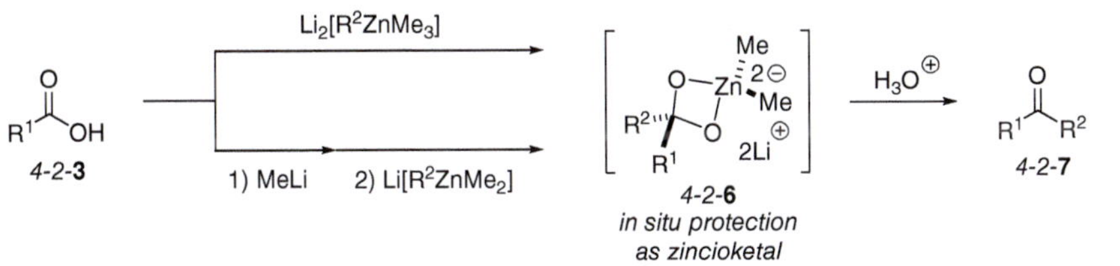

Scheme 8.15 Direct conversion of carboxylic acids to ketones by zincates.

8.4.3.1 Silylzincation of Alkynes

The di-anion-type silyl zincate, SiBNOL-Zn-ate *4-3-1-***3**, adds chemoselectively across various terminal alkynes *4-3-1-***1** without the aid of transition-metal catalysts (Scheme 8.16) [45]. Functionalized vinylsilanes *4-3-1-***2** are obtained in high yields with high regioselectivity for branched products.

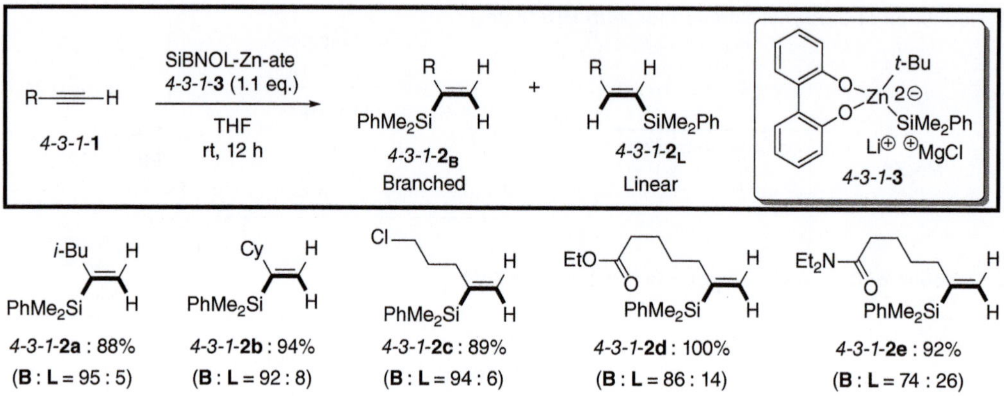

Scheme 8.16 Silylzincation of alkynes with SiBNOL-Zn-ate.

8.4.3.2 Silylzincation of Alkynes via Si–B Activation

Activation of silylborane, $PhMe_2Si–B(pin)$ *4-3-2-***2**, by $ZnMe_2$ in combination with *n*-Bu_3P generates silylzinc species *in situ*, and this adds regioselectively across various terminal alkynes *4-3-2-***1** (Scheme 8.17) [46]. The resultant vinylzincs *4-3-2-***3** can be trapped by electrophiles (*4-3-2-***4**, **-5**, **-6**). Additionally, $ZnMe_2$-catalyzed regioselective silaboration (*4-3-2-***7**) is achieved by using 2.0 eq. of silylborane *4-3-2-***2**. Triphenylphosphine is the best activator for the catalytic transformations.

8.4.3.3 Silylzincation of Alkenes (1): Synthesis of Allylsilanes

A catalytic amount of Cp_2TiCl_2 promotes the silylzincation reaction across terminal alkenes *4-3-3-***1** of SiSiNOL-Zn-ate *4-3-3-***3** (Scheme 8.18) [47]. In contrast to the reaction with alkynes, silylzincates by themselves are not reactive with alkenes *4-3-3-***1**. Regioselective addition of the SiSiNOL-Zn-ate *4-3-3-***3** to alkenes *4-3-3-***1** followed by β-hydride elimination affords *Z*-configured allylic silanes *4-3-3-***2** in preference to *E*-isomers. Neither alkylsilane formation nor double bond migration is observed in this chemical process.

8.4.3.4 Silylzincation of Alkenes (2): Synthesis of Alkylsilanes

SiBNOL-Zn-ate *4-3-4-***3** participates in the silylzincation of alkenes *4-3-4-***1** in the presence of CuCN as a catalyst (Scheme 8.19) [48]. Linear alkylsilanes *4-3-4-*$\mathbf{2_L}$ are obtained as major products without β-hydride elimination. On the other hand, highly coordinating functional groups, such as cyano group and phosphine oxide, switch the regioselectivity, generating the branched isomers *4-3-4-*$\mathbf{2_B}$ in essentially pure form.

8.4.4 Perfluoroalkylzincates Li[R_FZnMeCl] and R_FZnR

Perfluoroalkyl (R_F) organometallics *4-4-***2** are thermally unstable in general and readily decompose via α- (*4-4-***3**) or β-fluoride elimination (*4-4-***4**). As a result, the generation and use of these species require cryogenic reaction conditions and strict temperature control (Figure 8.3) [49–56].

The C–Zn bond is relatively stable, and perfluoroalkylzinc reagents are thus generally much more convenient to handle [57–61]. Results of the metalation of C_4F_9–I *4-4-***1a** with various zinc reagents *4-4-***5** at 0 °C and trapping experiments on the generated R_F–zincs *4-4-***6** with benzaldehyde *4-4-***7a** at *ambient temperature* are shown in Scheme 8.20 [62]. Ate complexes Li[$ZnMe_3$]

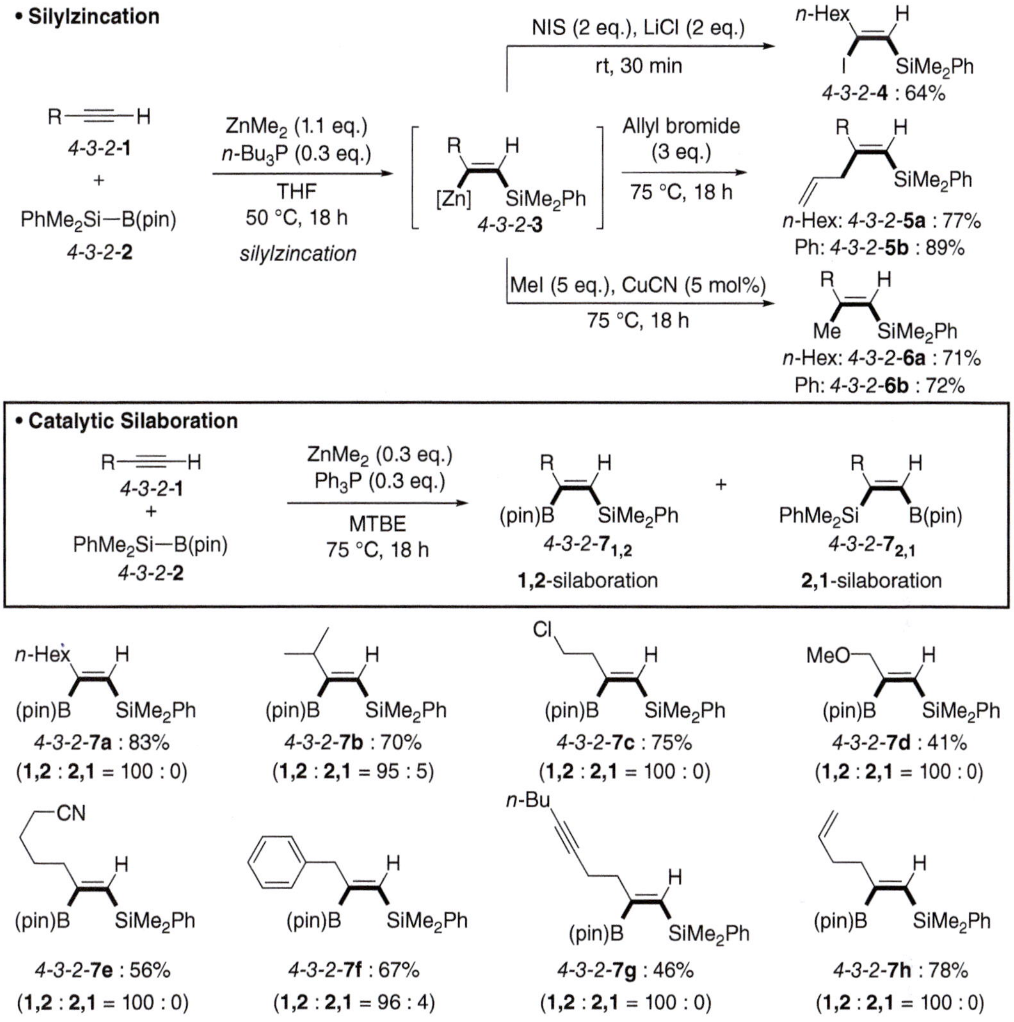

Scheme 8.17 Silylzincation of alkynes via Si–B bond activation.

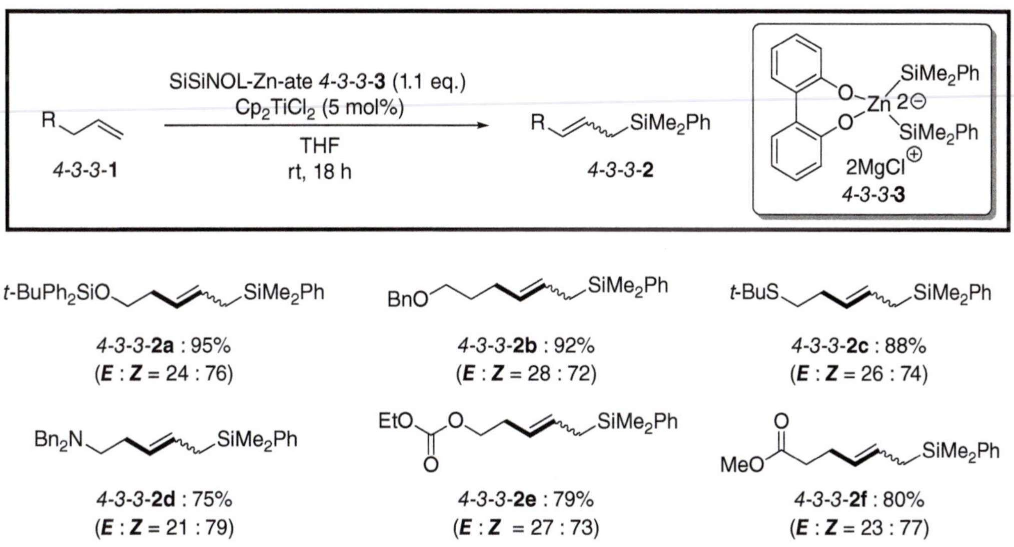

Scheme 8.18 Silylzincation of alkenes with SiSiNOL-Zn-ate catalyzed by Cp_2TiCl_2.

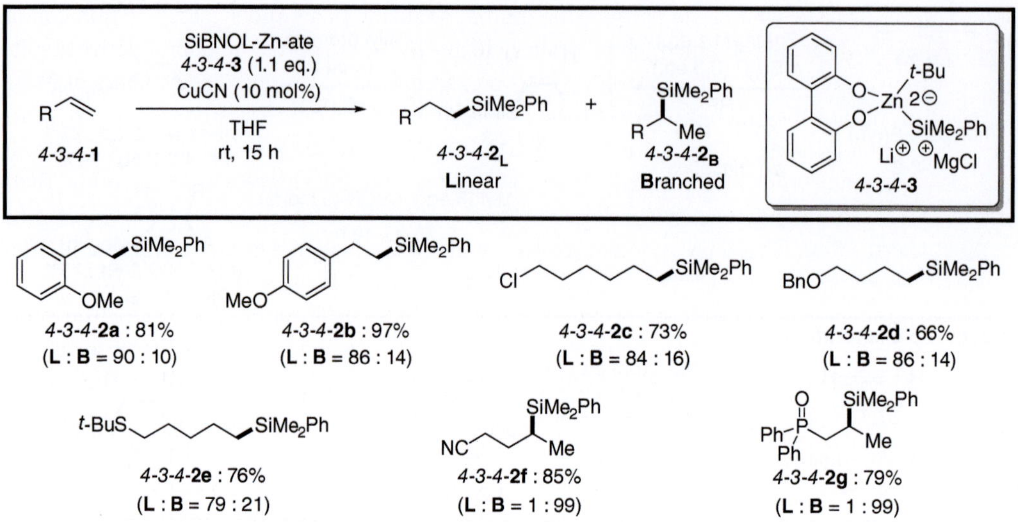

Scheme 8.19 CuCN-catalyzed silylzincation of alkenes.

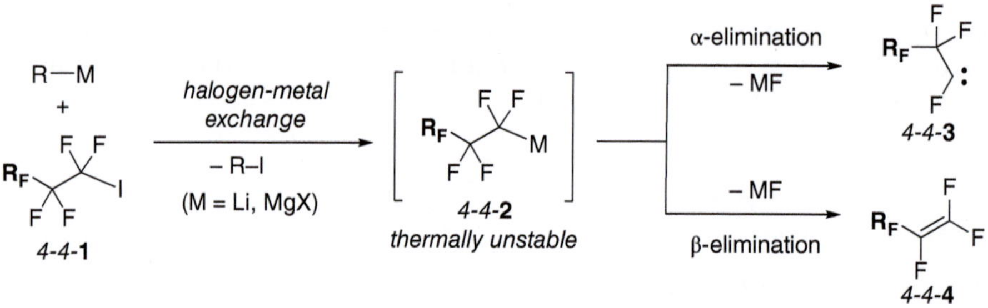

Figure 8.3 Decomposition of R_F-organometallics.

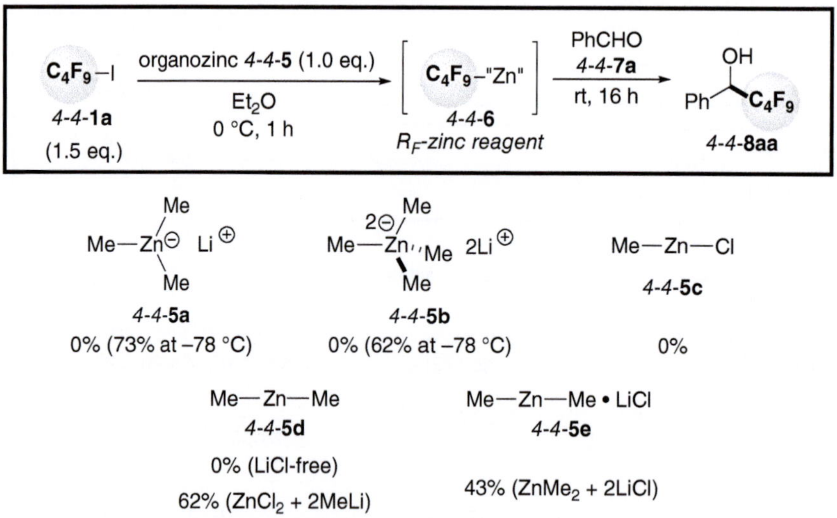

Scheme 8.20 Zincation of perfluoroalkyl iodide.

4-4-**5a** and $Li_2[ZnMe_4]$ *4-4*-**5b** do not give the desired adduct *4-4*-**8aa** under these reaction conditions, whereas *4-4*-**8aa** is obtained in moderate yield at −78 °C. In contrast to unproductive results using MeZnCl *4-5*-**5c** and $ZnMe_2$ *4-5*-**5d**, *4-5*-**5d** prepared from $ZnCl_2$ and 2 equivalents of MeLi successfully affords *4-4*-**8aa** in a 62% yield. *4-4*-**8aa** is also obtained by the addition of 2 equivalents of LiCl to $ZnMe_2$ (*4-4*-**5e**). These results imply both that the perfluoroalkylzincate $Li[R_FZnMeCl]$ *4-4*-**6e** is generated from $Li[Me_2ZnCl]$ *4-4*-**5e** [63, 64] and that ate complex *4-4*-**6e** is both "thermally stable" and "highly reactive" at the same time.

R_F–zincates add to aldehydes and ketones *4-4*-**7** with good functional group tolerance and offer facile access to Csp^3–R_F and Csp^3–Ar_F bond formation (*4-4*-**8**) (Scheme 8.21).

Cross-coupling using perfluoroalkylzinc with aryl halide *4-4*-**9** represents one of the most efficient ways to install a perfluoroalkyl chain on a benzene ring with controlled regioselectivity [65]. Ate complexes used for carbonyl addition are not eligible for copper-catalyzed cross-coupling at an elevated temperature due to their instability [62]. Neutral Lewis basic solvents are suitable for activation and stabilization of zinc species [66–70], and *N,N′*-dimethylpropyleneurea (DMPU) promotes the reaction most effectively, giving *4-4*-**10aa** in an 89% yield [71–73]. On the contrary, non-coordinating solvents such as toluene and dichloromethane hardly promote the desired reaction at all.

The perfluoroalkylation reaction described here displays a wide substrate scope, and various R_F groups can be introduced onto aromatic rings (Scheme 8.22) [71]. Scaling-up of this reaction is facile, and the coupling product *4-4*-**10aa** can be obtained in a 93% yield on a gram-scale. A perfluoroaryl group can also be installed (*4-4*-**10da**). This method is applicable to multiple perfluoroalkylations of substrates possessing more than one reactive site (*4-4*-**10at**, -**10au**, -**10av**, -**10aw**, **10dt**).

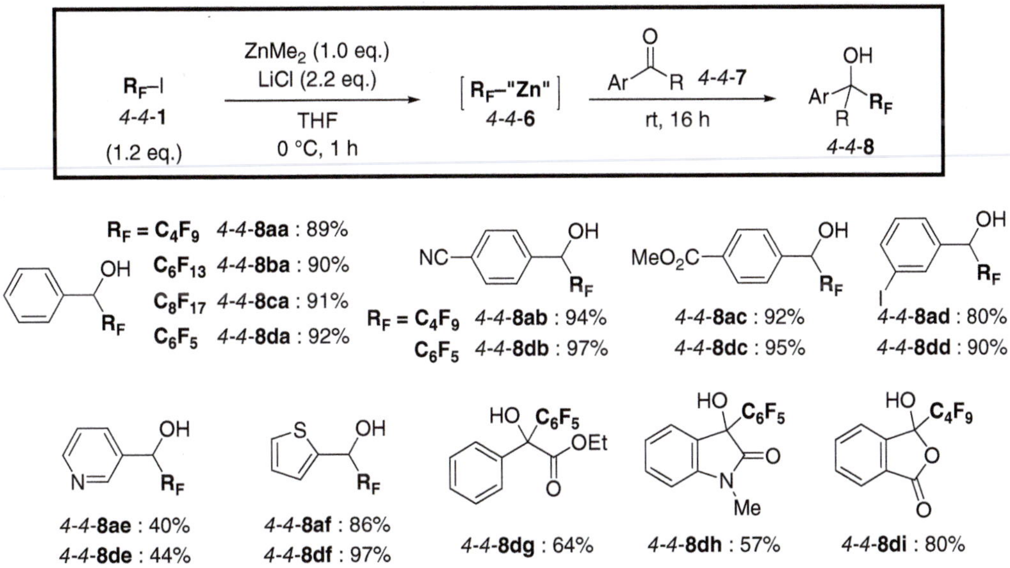

Scheme 8.21 Perfluoroalkylation and -arylation of carbonyl compounds.

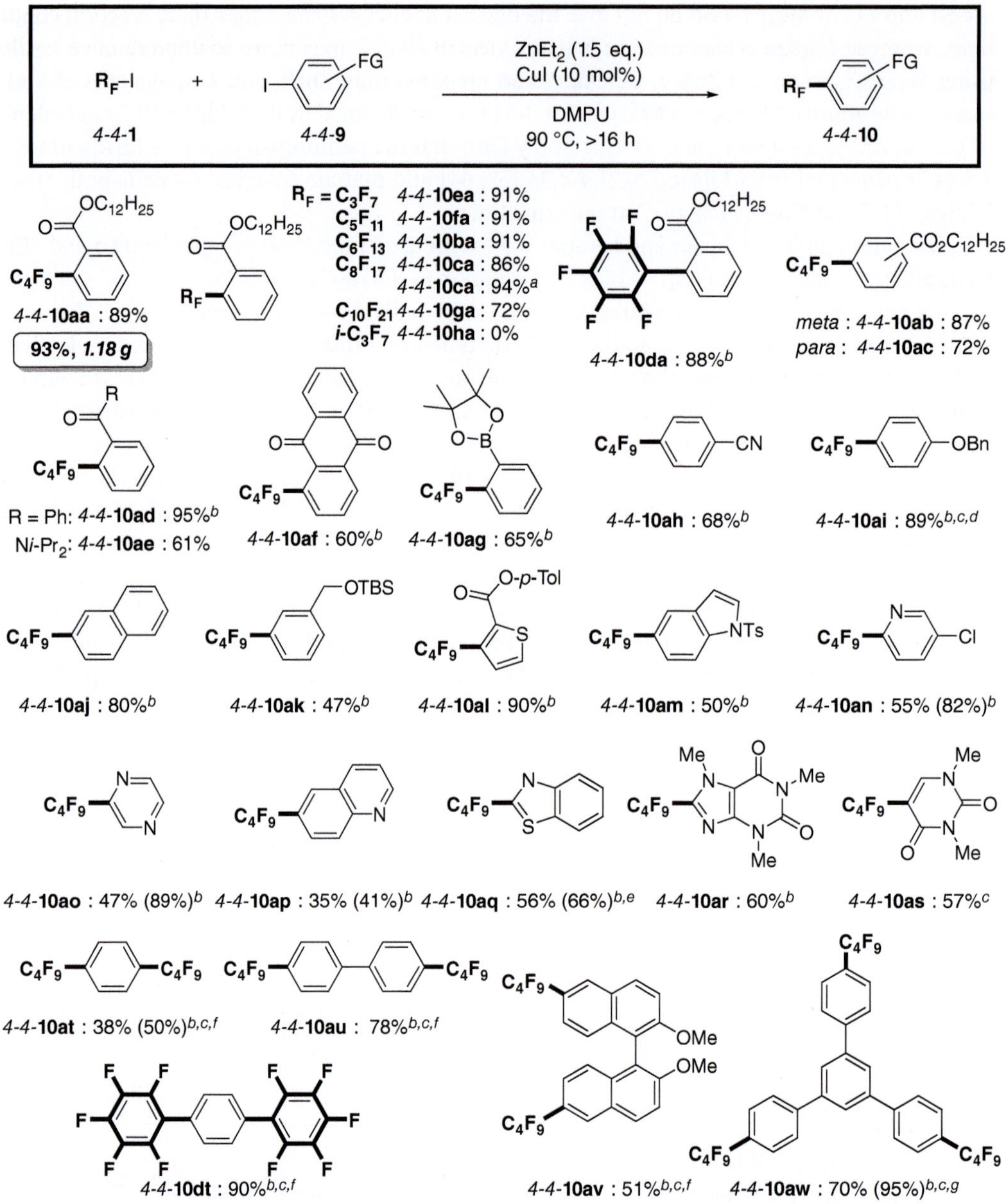

Isolated yields. Nuclear magnetic resonance (NMR) yields are given in parentheses. [a]C_8F_{17}–Br is used. [b]1,10-Phenanthroline (20 mol%) is added. [c]120 °C. [d]CuI (20 mol%) is added. [e]$ZnMe_2$ is used instead of $ZnEt_2$. [f]Amounts of the reagents are doubled. [g]Amounts of the reagents are tripled.

Scheme 8.22 Aromatic perfluoroalkylation.

8.4.5 Design of Boryl Anion Equivalents and Applications in Synthetic Chemistry

Boryl anions are highly nucleophilic species that potentially represent attractive tools with which to synthesize boron-containing compounds. However, boryl anions have been underutilized to date [74–78], and only a few applications have been described, such as a finely stabilized boryl anion reported by Yamashita and Nozaki [79] and a borylcopper complex reported by Sadighi [80]. Controlling the reactivity of boryl anions through ate complex formation should provide an avenue for novel nucleophilic boration methodologies. In this chapter, we focus on our contributions in this research field.

8.4.5.1 Borylzincate: M[(pinB)ZnEt$_2$]

Heteroleptic ate complexes bearing boryl anions as a ligand of zinc have not been investigated [81, 82] despite their potential utility in organic synthesis. We hypothesized that borylzincate could be generated via transmetalation of the boryl group from an alkoxide-activated diboron to a dialkylzinc (Figure 8.4). DFT calculations have implied that the formation of borylzincate is kinetically feasible, with an activation barrier of 15.8 kcal/mol, although the small stabilization energy for the formation of the borylzincate appears to be a hurdle to the utilization of this species for chemical synthesis.

Based on this computational analysis, we designed a catalytic boration of aryl halides via a halogen–zinc exchange reaction with borylzincate (Figure 8.5). In this sequence, energetically disfavored borylzincate formation is compensated for by the stepwise formation of stable C–Zn and C–B bonds.

Various aryl iodides *4-5-1*-**1** are converted to the corresponding aryl boronates *4-5-1*-**3** by reaction with ZnEt$_2$ (10 mol%), bis(pinacolato)diboron *4-5-1*-**2a** (B$_2$(pin)$_2$; 1.1 eq.), and NaO*t*-Bu (1.1 eq.) at 75 °C in THF (Scheme 8.23) [83]. Notably, sterically hindered mesityl substrate (*4-5-1*-**1e**), aryl iodides containing the transition-metal-susceptible allyloxy group (*4-5-1*-**1f**), base-susceptible ester group (*4-5-1*-**1l**, -**1m**) and cyano group (*4-5-1*-**1n**), and various heteroaromatic substrates (*4-5-1*-**1q**, -**1r**, -**1s**, -**1t**) can be employed. Moreover, aryl bromides are also available for this boration reaction, if the reaction temperature is elevated to 120 °C (*4-5-1*-**3aa**, -**3ca**, -**3ga**). Diborons are not limited to bis(pinacolato)diboron (*4-5-1*-**3ab**, -**3ic**).

Another strategy by which to attain a large stabilization energy in the borylzincate system involves reaction between borylzincate and benzyne. Benzynes are highly unstable reaction

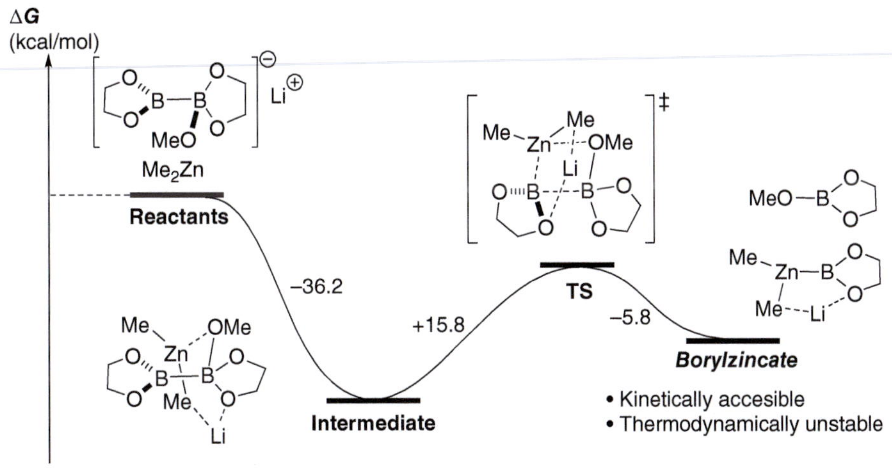

Figure 8.4 Model DFT calculation on borylzincate formation M06/SVP (Zn) and 6-31+G* (other atoms).

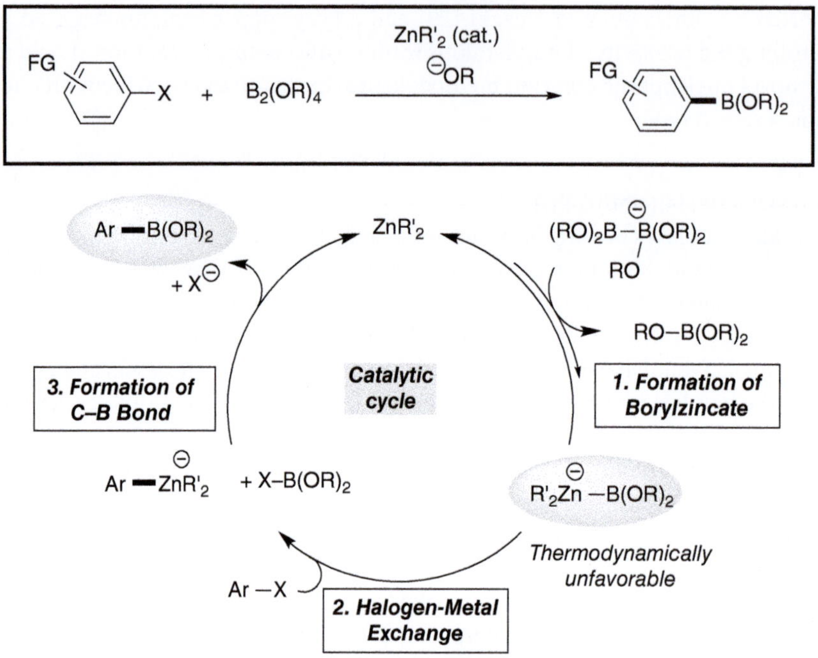

Figure 8.5 Design of a catalytic boration cycle for aryl halides.

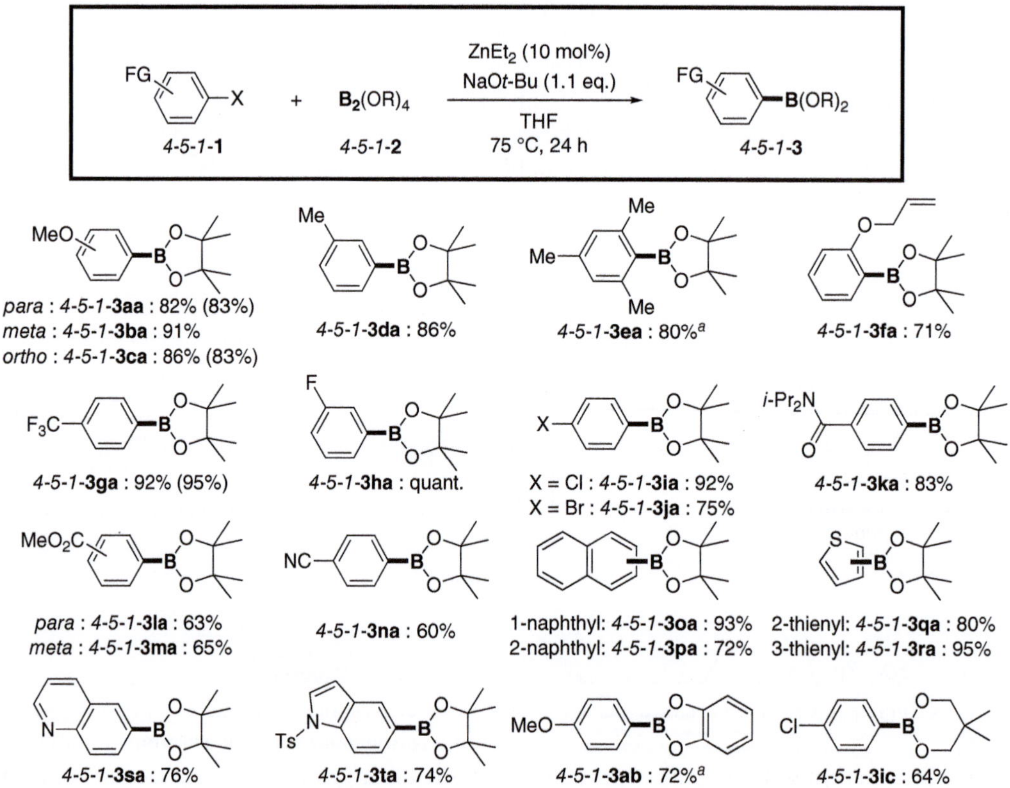

Isolated yields. Yields in parentheses are of the reaction with aryl bromides at 120 °C. [a]Reaction at 120 °C.

Scheme 8.23 Substrate scope of aromatic boration.

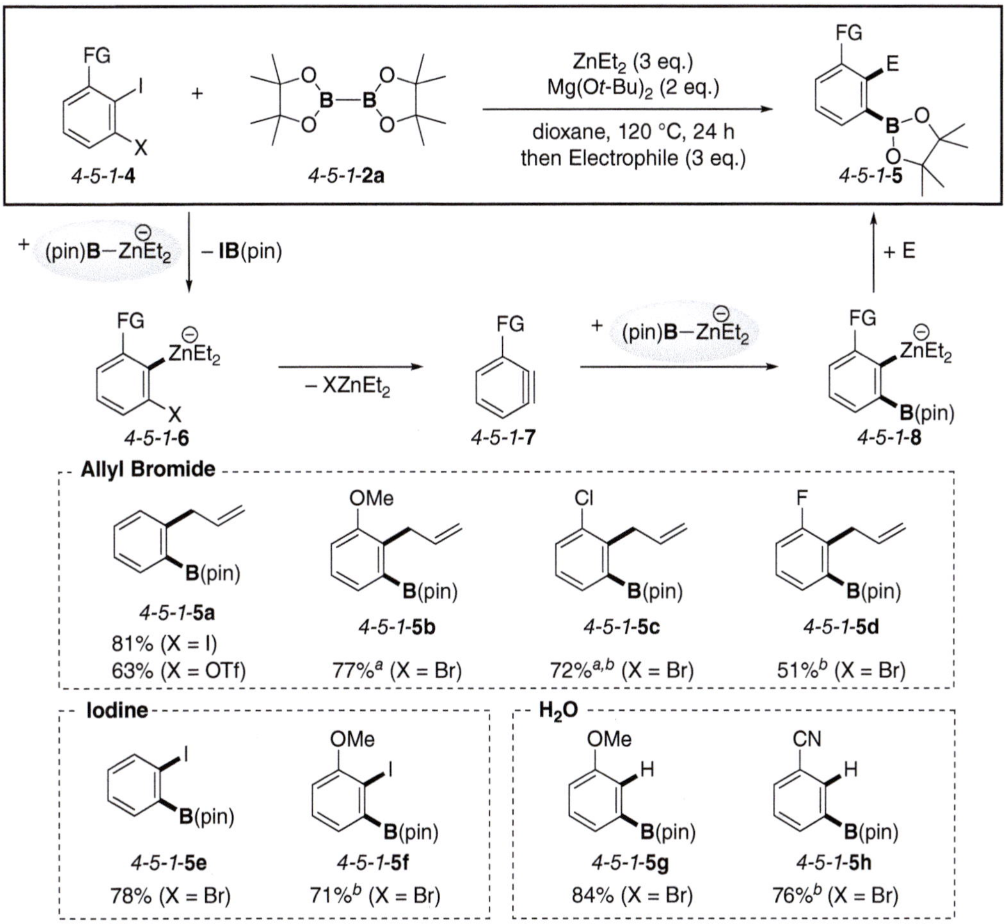

Isolated yields. [a]NMR yields. [b]2 eq. of $B_2(pin)_2$ is used.

Scheme 8.24 Borylzincation reaction of benzynes.

intermediates, and stable benzene ring formation upon reaction with nucleophiles should be highly energetically favorable. Thus, highly nucleophilic borylzincate undergoes the iodine–zinc exchange reaction of aryl halide *4-5-1-***4** followed by elimination of an *ortho* leaving group of *4-5-1-***6** to give a benzyne *4-5-1-***7** (Scheme 8.24). Subsequent addition of another borylzincate across the benzyne generates functionalized arylzincate *4-5-1-***8**, which can be trapped by various electrophiles to give multiply substituted benzenes *4-5-1-***5** with high regioselectivities [83].

The reactions with electrophiles were carried out at room temperature with H_2O for 5 min or with I_2 for 3 h, or at 75 °C with allyl bromide for 12 h.

8.4.5.2 *Trans*-Diboration of Alkynes via *pseudo*-Intramolecular Activation

Vinylboronates are important and useful synthetic intermediates in organic synthesis. These compounds are prepared mainly through the hydroboration [84, 85], haloboration [86, 87], or diboration of alkynes [88–94]. These reactions are normally triggered by the interaction between π-electrons of the triple bond and the vacant p-orbital of boron, giving C–B bond formation such

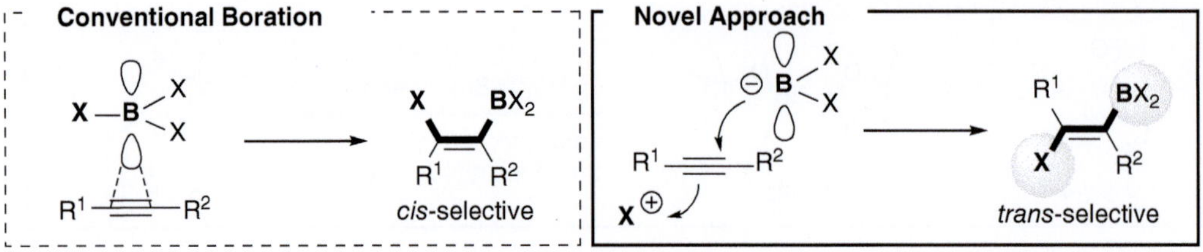

Figure 8.6 Concept for the *trans*-selective boration of triple bonds.

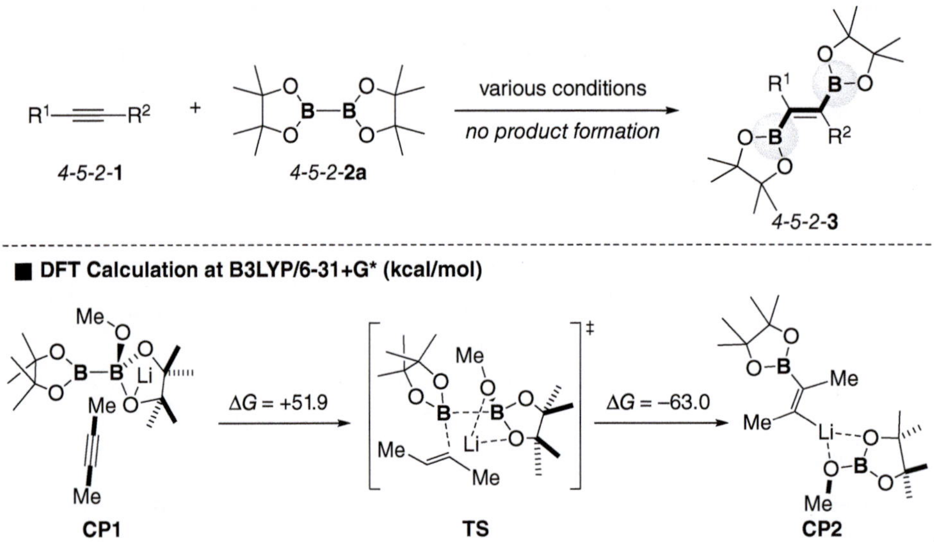

Scheme 8.25 Fruitless intermolecular diboration of alkynes.

that *cis*-configured alkene products are generally obtained (Figure 8.6). In transition-metal-catalyzed processes, a d-orbital on the metal interacts with the triple bond to afford *cis*-product formation as well. In order to realize unprecedented *trans*-selectivity in a diboration reaction [95], the nucleophilic addition of the boryl anion (equivalent) to a triple bond would be a rational strategy, following the sense of stereoselection of carbolithiation and stannyllithiation of alkynes [96, 97].

Intermolecular diboration between internal alkynes *4-5-2-***1** and bis(pinacolato)diboron *4-5-2-***2a** activated by Lewis bases does not lead to the formation of the desired diborated products *4-5-2-***3** (Scheme 8.25) [98]. In accordance with the experimental results, model DFT calculations have revealed that the activation barrier for the initial C–B bond formation step is more than 50 kcal/mol.

"*Pseudo*-intramolecular reaction" using alkynes possessing a Lewis basic activator within the molecule can address the above issue [99]. Reaction of *4-5-2-***2a** with propargylic alkoxide now affords the desired diborated product *4-5-2-***5aa** in a 77% yield (Scheme 8.26) [100]. This diboration reaction can be easily operated on a gram scale and has a broad substrate scope. Triple bonds proximal to the alkoxide group react preferentially over the distal multiple bonds in the cases of enyne and diyne substrates (*4-5-2-***4p**, **-4q**). Secondary alcohols are also converted to oxaborole products

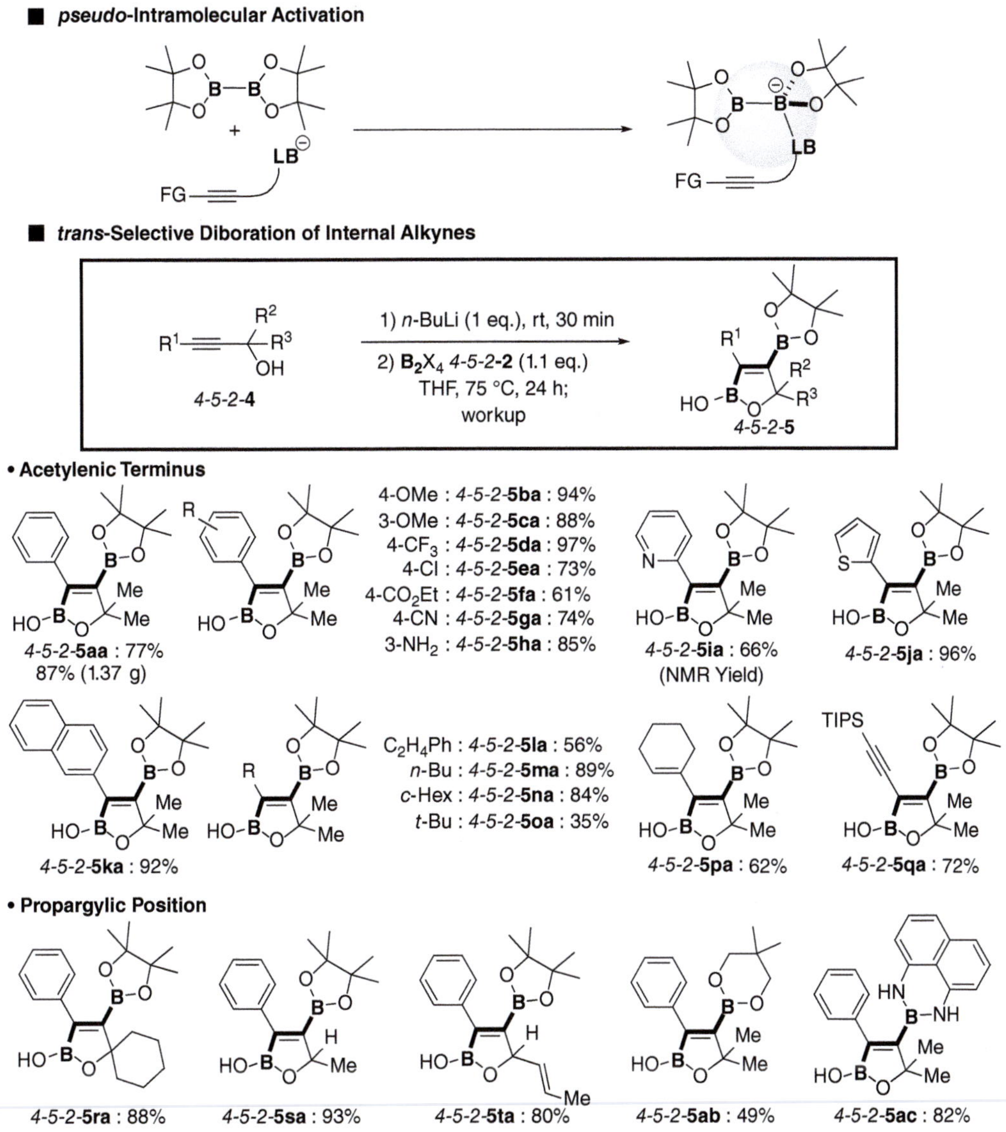

Scheme 8.26 *Trans*-selective diboration of alkynes.

in high yields (*4-5-2-***4s**, -**4t**). On the other hand, primary alcohols and homopropargylic alcohols are not available for this process.

The diboration can be performed in one-pot, starting from terminal alkynes *4-5-2-***6** (Scheme 8.27). Lithium alkoxides *4-5-2-***7**, generated *in situ* by the reaction of a corresponding lithium acetylide with acetone, react with diboron *4-5-2-***2a** smoothly to give *trans*-diborated oxaboroles *4-5-2-***5ba**, -**5ja**, and -**5ka** with high efficiency.

A fully substituted olefin *4-5-2-***10** is synthesized through a sequential diboration/Suzuki–Miyaura cross-coupling reaction with *p*-iodotoluene *4-5-2-***9** in one-pot without isolation of the borate intermediate *4-5-2-***8** or oxaborole (Scheme 8.28). This tandem process offers rapid and versatile access to densely functionalized olefins in a perfectly regioselective manner [100].

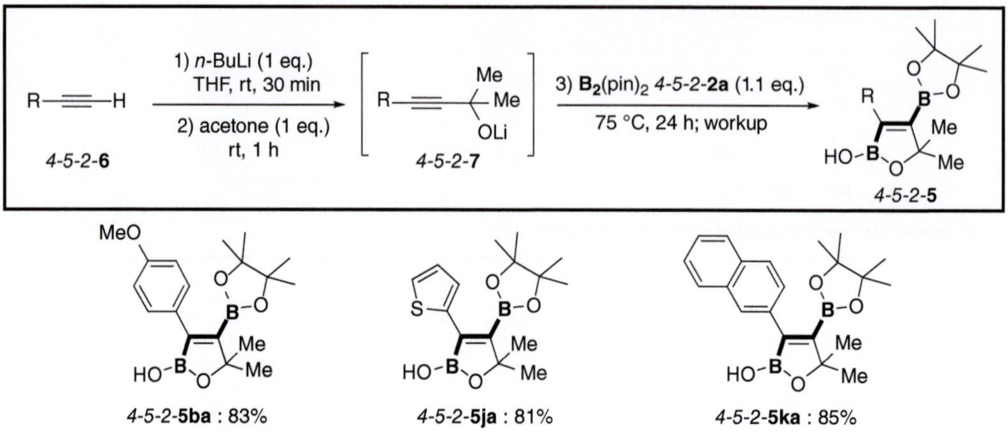

Scheme 8.27 One-pot diboration reactions.

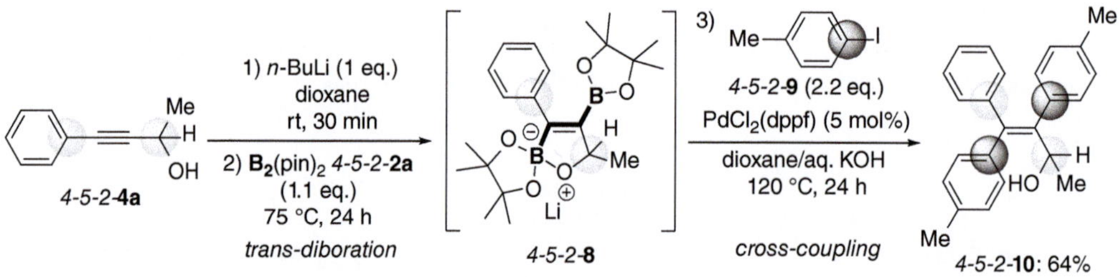

Scheme 8.28 Sequential diboration/Suzuki–Miyaura cross-coupling, leading to a multisubstituted olefin.

8.4.5.3 *Trans*-Alkynylboration of Alkynes

The concept of *pseudo*-intramolecular activation can be extended to the alkynylboration of alkynes. Alkynylboronate *4-5-3-***2** transfers the carbon–carbon triple bond and boron to the internal alkynes *4-5-3-***1** to give a variety of alkynylated oxaboroles *4-5-3-***3** (Scheme 8.29) [101].

Oxaboroles *4-5-3-***3** can be converted to a wide range of useful compounds through simple transformations (Scheme 8.30). The Suzuki–Miyaura cross-coupling gives tetrasubstituted olefin *4-5-3-***4** in high yield, and trisubstituted olefin *4-5-3-***5** can be obtained by silver-catalyzed protodeboration in quantitative yield. These transformations are powerful for the preparation of perfectly stereo- and regiocontrolled multiply-substituted olefins. By modifying ligands on the boron atom, trifluoroborate formation (*4-5-3-***6**) and mesitylation (*4-5-3-***7**) are easily conducted. Oxidation of the C–B bond yields the α-alkynylated aldol product *4-5-3-***8**.

8.5 Conclusion

In this chapter, we have outlined the design and reactivity of novel functional ate complexes. Unprecedented chemo-, regio-, and stereoselectivities in various chemical transformations can be realized by combining "element chemistry" (understanding and utilizing the

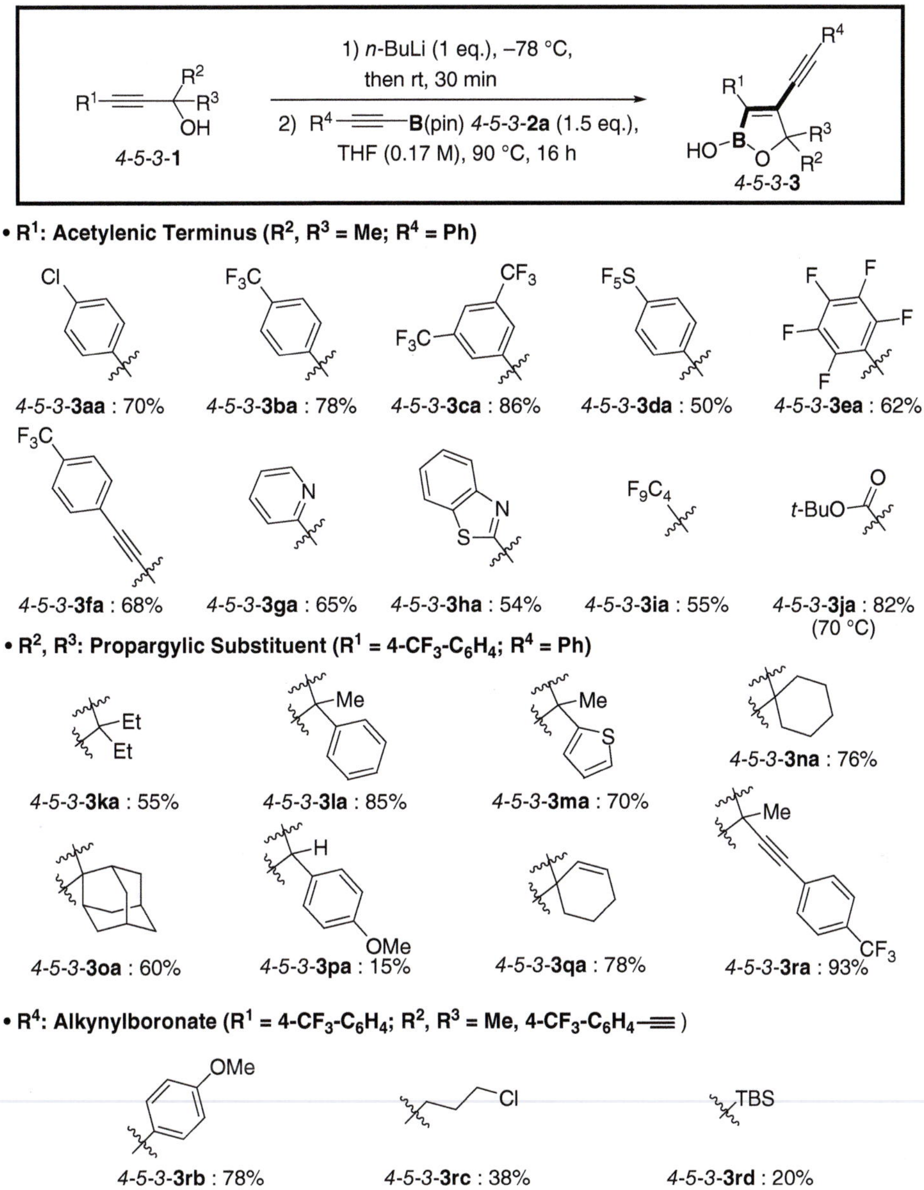

Scheme 8.29 *Trans*-alkynylboration of alkynes.

characteristics of elements) with "ate complex formation" (bringing out the potential reactivities of these elements).

Ate complexes will surely continue to fascinATE and motivATE chemists, and we believe further developments in this field will keep ate chemistry at the frontiers of synthetic organic chemistry, to the benefit of medicinal chemistry, materials science, and more.

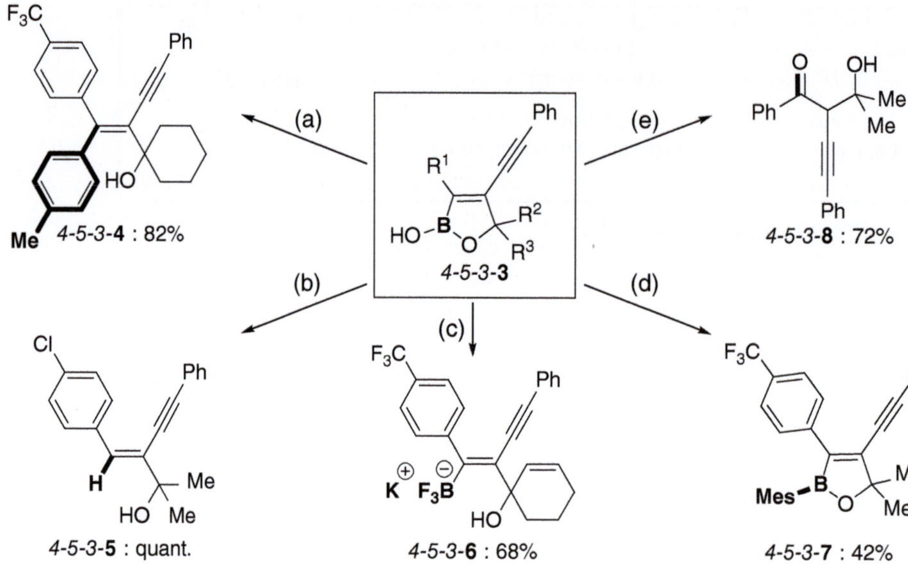

(a) 4-Iodotoluene (1.1 eq.), Pd(Pt-Bu_3)$_2$ (5 mol%), 5 M NaOH aq. (2 eq.), dioxane, 90 °C, 48 h. (b) $AgNO_3$ (6 mol%), EtOH/H_2O, 80 °C, 20 min. (c) sat. KHF_2 aq. (ca. 5.4 eq.), MeOH, rt, 4.5 h. (d) MesMgBr (3 eq.), THF, 75 °C, 24 h. (e) $NaBO_3$•$4H_2O$ (5 eq.), THF/H_2O, rt, 90 min.

Scheme 8.30 Transformation of oxaboroles.

References

1 Wittig, G. (**1958**). *Angew. Chem. 70*: 65–71.
2 Wanklyn, J.A. (**1858**). *Liebigs Ann. 108*: 67–79.
3 Hurd, D.T. (**1948**). *J. Org. Chem. 13*: 711–713.
4 Erdik, E. (**1992**). *Tetrahedron 48*: 9577–9648.
5 Knochel, P. and Singer, R.D. (**1993**). *Chem. Rev. 93*: 2117–2188.
6 Knochel, P. (**1995**). *Synlett*: 393–403.
7 Knochel, P., Almena Perea, J.J., and Jones, P. (**1998**). *Tetrahedron 54*: 8275–8319.
8 Knochel, P., Millot, N., Rodriguez, A.L., and Tucker, C.E. (**2001**). *Org. React. 58*: 417–731.
9 Uchiyama, M. (**2008**). *J. Syn. Org. Chem. 67*: 4–16.
10 Uchiyama, M. and Wang, C. (**2014**). *Top. Organomet. Chem. 47*: 159–202.
11 Kondo, Y., Takazawa, N., Yamazaki, C., and Sakamoto, T. (**1994**). *J. Org. Chem. 59*: 4717–4718.
12 Uchiyama, M., Koike, M., Kameda, M. et al. (**1996**). *J. Am. Chem. Soc. 118*: 8733–8734.
13 Uchiyama, M., Kameda, M., Mishima, O. et al. (**1998**). *J. Am. Chem. Soc. 120*: 4934–4946.
14 Uchiyama, M., Furuyama, T., Kobayashi, M. et al. (**2006**). *J. Am. Chem. Soc. 128*: 8404–8405.
15 Furuyama, T., Yonehara, M., Arimoto, S. et al. (**2008**). *Chem. Eur. J. 14*: 10348–10356.
16 Kobayashi, M., Matsumoto, Y., Uchiyama, M., and Ohwada, T. (**2004**). *Macromolecules 37*: 4339–4341.
17 Wang, C., Ozaki, T., Takita, R., and Uchiyama, M. (**2012**). *Chem. Eur. J. 18*: 3482–3485.
18 Mallan, J.M. and Bebb, R.L. (**1969**). *Acc. Chem. Res. 69*: 693–755.
19 Beak, P. and Snieckus, V. (**1982**). *Chem. Rev. 15*: 306–312.
20 Snieckus, V. (**1990**). *Chem. Rev. 90*: 879–933.

21 Mulvey, R.E., Mongin, F., Uchiyama, M., and Kondo, Y. (**2007**). *Angew. Chem. Int. Ed. 46*: 3802–3824.
22 Tückmantel, W., Oshima, K., and Nozaki, H. (**1986**). *Chem. Ber. 119*: 1581–1593.
23 Kondo, Y., Shilai, M., Uchiyama, M., and Sakamoto, T. (**1999**). *J. Am. Chem. Soc. 121*: 3539–3540.
24 Uchiyama, M., Miyoshi, T., Kajihara, Y. et al. (**2002**). *J. Am. Chem. Soc. 124*: 8514–8515.
25 Uchiyama, M., Kobayashi, Y., Furuyama, T. et al. (**2008**). *J. Am. Chem. Soc. 130*: 472–480.
26 Maruoka, K. and Yamamoto, H. (**1988**). *Tetrahedron 44*: 5001–5032.
27 Negishi, E. and Kondakov, D.Y. (**1996**). *Chem. Soc. Rev. 25*: 417–426.
28 von Zezschwitz, P. (**2008**). *Synthesis*: 1809–1831.
29 Uchiyama, M., Naka, H., Matsumoto, Y., and Ohwada, T. (**2004**). *J. Am. Chem. Soc. 126*: 10526–10527.
30 Naka, H., Uchiyama, M., Matsumoto, Y. et al. (**2007**). *J. Am. Chem. Soc. 129*: 1921–1930.
31 Armstrong, D.R., Crosbie, E., Hevia, E. et al. (**2014**). *Chem. Sci. 5*: 3031–3045.
32 House, H.O. (**1976**). *Acc. Chem. Res. 9*: 59–67.
33 Normant, J.F. (**1978**). *Pure Appl. Chem. 50*: 709–715.
34 Lipshutz, B.H., Wilhelm, R.S., and Kozlowski, J.A. (**1984**). *Tetrahedron 40*: 5005–5038.
35 Yamamoto, Y., Asao, N., and Uyehara, T. (**1992**). *J. Am. Chem. Soc. 114*: 5427–5429.
36 Usui, S., Hashimoto, Y., Morey, J.V. et al. (**2007**). *J. Am. Chem. Soc. 129*: 15102–15103.
37 Komagawa, S., Usui, S., Haywood, J. et al. (**2012**). *Angew. Chem. Int. Ed. 51*: 12081–12085.
38 Tezuka, N., Shimojo, K., Hirano, K. et al. (**2016**). *J. Am. Chem. Soc. 138*: 9166–9171.
39 Hernandez, L.W., Pospech, J., Klöckner, U. et al. (**2017**). *J. Am. Chem. Soc. 139*: 15656–15659.
40 Uchiyama, M., Furumoto, S., Saito, M. et al. (**1997**). *J. Am. Chem. Soc. 119*: 11425–11433.
41 Uchiyama, M., Nakamura, S., Ohwada, T. et al. (**2004**). *J. Am. Chem. Soc. 126*: 10897–10903.
42 Murata, R., Hirano, K., and Uchiyama, M. (**2015**). *Chem. Asian J. 10*: 1286–1290.
43 Okuda, Y., Wakamatsu, K., Tückmantel, W. et al. (**1985**). *Tetrahedron Lett. 26*: 4629–4632.
44 Wakamatsu, K., Nonaka, T., Okuda, Y. et al. (**1986**). *Tetrahedron. 42*: 4427–4436.
45 Nakamura, S., Uchiyama, M., and Ohwada, T. (**2004**). *J. Am. Chem. Soc. 126*: 11146–11147.
46 Nagashima, Y., Yukimori, D., Wang, C., and Uchiyama, M. (**2018**). *Angew. Chem. Int. Ed. 57*: 8053–8057.
47 Nakamura, S., Uchiyama, M., and Ohwada, T. (**2005**). *J. Am. Chem. Soc. 127*: 13116–13117.
48 Nakamura, S. and Uchiyama, M. (**2007**). *J. Am. Chem. Soc. 129*: 28–29.
49 Pierce, O.R., McBee, E.T., and Judd, G.F. (**1954**). *J. Am. Chem. Soc. 76*: 474–478.
50 Johncock, P. (**1969**). *J. Organomet. Chem. 19*: 257–265.
51 Gassman, P.G. and O'Reilly, N.J. (**1985**). *Tetrahedron Lett. 26*: 5243–5246.
52 Pierce, O.R., Meiners, A.F., and McBee, E.T. (**1953**). *J. Am. Chem. Soc. 75*: 2516.
53 Denson, D.D., Smith, C.F., and Tamborski, C. (**1974**). *J. Fluorine Chem. 3*: 247–258.
54 Dua, S.S., Howells, R.D., and Gilman, H. (**1974**). *J. Fluorine Chem. 4*: 409–413.
55 Haszeldine, R.N. (**1952**). *J. Chem. Soc.*: 3423–3428.
56 Burton, D.J. and Yang, Z.-Y. (**1992**). *Tetrahedron 48*: 189–275.
57 Haszeldine, R.N. and Walaschewski, E.G. (**1953**). *J. Chem. Soc.*: 3607–3610.
58 Miller, W.T. Jr., Bergman, E., and Fainberg, A.H. (**1957**). *J. Am. Chem. Soc. 79*: 4159–4164.
59 Chambers, R.D., Musgrave, W.K.R., and Savory, J. (**1962**). *J. Chem. Soc.*: 1993–1999.
60 Keller, T.M. and Tarrant, P. (**1975**). *J. Fluorine Chem. 6*: 297–310.
61 Sekiya, A. and Ishikawa, N. (**1977**). *Chem. Lett.*: 81–84.
62 Wang, X., Hirano, K., Kurauchi, D. et al. (**2015**). *Chem. Eur. J. 21*: 10993–10996.
63 Fleckenstein, J.E. and Koszinowski, K. (**2011**). *Organometallics 30*: 5018–5026.
64 Koszinowski, K. and Böhrer, P. (**2009**). *Organometallics 28*: 100–110.

65 Kitazume, T. and Ishikawa, N. (**1985**). *J. Am. Chem. Soc. 107*: 5186–5191.
66 Lange, H. and Naumann, D. (**1984**). *J. Fluorine Chem. 26*: 435–444.
67 Naumann, D., Schorn, C., Tyrra, W. (**1999**). *Z. Anorg. Allg. Chem.* 625: 827–830.
68 Schorn, C., Naumann, D., Scherer, H., and Hahn, J. (**2001**). *J. Fluorine Chem. 107*: 159–169.
69 Kaplan, P.T., Xu, L., Chen, B. et al. (**2013**). *Organometallics 32*: 7552–7558.
70 Kaplan, P.T., Chen, B., and Vicic, D.A. (**2014**). *J. Fluorine Chem. 168*: 158–162.
71 Kato, H., Hirano, K., Kurauchi, D. et al. (**2015**). *Chem. Eur. J. 21*: 3895–3900.
72 Popov, I., Lindeman, S., and Daugulis, O. (**2011**). *J. Am. Chem. Soc. 133*: 9286–9289.
73 Aikawa, K., Nakamura, Y., Yokota, Y. et al. (**2015**). *Chem. Eur. J. 21*: 96–100.
74 Auten, R.W. and Kraus, C.A. (**1952**). *J. Am. Chem. Soc. 74*: 3398–3401.
75 Köster, R. and Benedikt, G. (**1963**). *Angew. Chem. Int. Ed. Engl. 2*: 219.
76 Parsons, T.D., Self, J.M., and Schaad, L.H. (**1967**). *J. Am. Chem. Soc. 89*: 3446–3448.
77 Smith, K. and Swaminathan, K. (**1976**). *J. Chem. Soc. Dalton Trans.*: 2297–2300.
78 Weber, L., Schnieder, M., and Lönnecke, P. (**2001**). *J. Chem. Soc., Dalton Trans.*: 3459–3464.
79 Segawa, Y., Yamashita, M., and Nozaki, K. (**2006**). *Science 314*: 113–115.
80 Laitar, D.S., Müller, P., and Sadighi, J.P. (**2005**). *J. Am. Chem. Soc. 127*: 17196–17197.
81 Kajiwara, T., Terabayashi, T., Yamashita, M., and Nozaki, K. (**2008**). *Angew. Chem. Int. Ed. 47*: 6606–6610.
82 Kumar Bose, S. and Marder, T.B. (**2014**). *Org. Lett. 16*: 4562–4565.
83 Nagashima, Y., Takita, R., Yoshida, K. et al. (**2013**). *J. Am. Chem. Soc. 135*: 18730–18733.
84 Brown, H.C. and Campbell, J.B. Jr. (**1981**). *Aldrichmica Acta 14*: 3–11.
85 Beletskaya, I. and Pelter, A. (**1997**). *Tetrahedron 53*: 4957–5026.
86 Wang, C., Tobrman, T., Xu, Z., and Negishi, E.-I. (**2009**). *Org. Lett. 11*: 4092–4095.
87 Wang, C. and Uchiyama, M. (**2012**). *Eur. J. Org. Chem.*: 6548–6554.
88 Takaya, J. and Iwasawa, N. (**2012**). *ACS Catal. 2*: 1993–2006.
89 Ishiyama, T., Matsuda, N., Miyaura, N., and Suzuki, A. (**1993**). *J. Am. Chem. Soc. 115*: 11018–11019.
90 Lesley, G., Nguyen, P., Taylor, N.J. et al. (**1996**). *Organometallics 15*: 5137–5154.
91 Marder, T.B. and Norman, N.C. (**1998**). *Top. Catal. 5*: 63–73.
92 Iwadate, N. and Suginome, M. (**2010**). *J. Am. Chem. Soc. 132*: 2548–2549.
93 Lillo, V., Fructos, M.R., Ramírez, J. et al. (**2007**). *Chem. Eur. J. 13*: 2614–2621.
94 Yoshida, H., Kawashima, S., Takemoto, Y. et al. (**2012**). *Angew. Chem. Int. Ed. 51*: 235–238.
95 Adams, C.J., Baber, R.A., Batsanov, A.S. et al. (**2006**). *Dalton Trans.*: 1370–1373.
96 Fressigné, C., Girard, A.-L., Durandetti, M., and Maddaluno, J. (**2008**). *Angew. Chem. Int. Ed. 47*: 891–893.
97 Tsuji, H., Ueda, Y., Ilies, L., and Nakamura, E. (**2010**). *J. Am. Chem. Soc. 132*: 11854–11855.
98 Bonet, A., Pubill-Ulldemolins, C., Bo, C. et al. (**2011**). *Angew. Chem. Int. Ed. 50*: 7158–7161.
99 Yamamoto, Y., Ishii, J.-I., Nishiyama, H., and Itoh, K. (**2005**). *J. Am. Chem. Soc. 127*: 9625–9631.
100 Nagashima, Y., Hirano, K., Takita, R., and Uchiyama, M. (**2014**). *J. Am. Chem. Soc. 136*: 8532–8535.
101 Nogami, M., Hirano, K., Kanai, M. et al. (**2017**). *J. Am. Chem. Soc. 139*: 12358–12361.

9

Isolable Alkenylcopper Compounds: Synthesis, Structure, and Reaction Chemistry

Liang Liu[a], Chao Wang[b], and Zhenfeng Xi[a]

[a] *College of Chemistry, Peking University, Beijing, China*
[b] *Graduate School of Pharmaceutical Sciences, The University of Tokyo, Tokyo, Japan*

9.1 Introduction

Copper-catalyzed and -mediated reactions, including a range of coupling reactions, azide–alkyne cycloaddition reactions, and the classic nucleophilic substitution or addition of organocuprates, are widely applied to construct C–C or C–heteroatom bonds [1–11]. In these reactions, organocopper species are often proposed as the key intermediates [12–17]. Hence, the isolation and structural characterization of organocopper compounds are highly significant for understanding the mechanism of copper-based reactions and will be insightful for designing new reactions. For many years, the synthesis and structural analysis of organocopper complexes have attracted much attention. Compared to alkynyl and aryl organocopper complexes, the alkenyl analogues remain less explored [12, 13] due to their air- and moisture-sensitivity, thermal instability towards decomposition into copper mirrors, and low solubility in common organic solvents. So, it is rather difficult to obtain their pure crystals, which results in the lack of well-defined single-crystal structures. The efficient strategy to facilitate the successful isolation of alkenylcopper compounds is to introduce stabilization factors into the structures. Up to date, *N*-heterocyclic carbenes (NHCs), intramolecular coordinating heteroatoms, alkynyl groups, aryl substituents, and *cis*-butadienyl skeletons have been reported to work well in this regard.

This chapter mainly covers the reports on the synthesis, solid structures, and preliminary reactivity of well-defined alkenylcopper compounds containing Cu–C σ-bonds. The work regarding copper olefin π-complexes and cyclopentadienyl copper π-complexes has already been summarized [13, 18].

9.2 Well-defined Alkenylcopper Compounds

In organic synthesis, alkenylcopper compounds are generally prepared *in situ* by a range of methods, such as transmetalation reaction from more polar organometallic reagents [13], the addition of Cu–X (X = H, boryl, aryl, alkyl, *etc.*) to alkynes [19–21], copper-halide exchange

Polar Organometallic Reagents: Synthesis, Structure, Properties and Applications, First Edition. Edited by Andrew E. H. Wheatley and Masanobu Uchiyama.

reaction [22–29], and oxidative addition of zero-valent activated Cu* to R–X [30, 31]. The majority of successfully isolated alkenylcopper complexes are realized by the former two strategies.

9.2.1 Mono-alkenyl Organocopper Compounds with Intramolecular Coordination

Intramolecular coordination of heteroatoms was developed by van Koten and coworkers to stabilize reactive organometallic complexes, including alkenylcopper compounds, and aid their syntheses and structural characterizations [32–43]. The reaction between 1-aryl-propenyllithium **1** and excess CuBr *via* a ligand-exchange process produced a tetranuclear alkenylcopper aggregate **2** [37, 39, 44, 45], which can react with another aryllithium reagent to give a heterolytic organocopper aggregate **3** that is free of halides (Scheme 9.1). Copper atoms of both **2** and **3** formed a Cu_4 core structure with a rhombic configuration. The edges of the rhombus in **2** and **3** average to 2.48 (3) and 2.46 (2) Å, respectively. The dimethylamino groups of the 1-aryl-propenyl unit in **2** and **3** coordinate to copper atoms at diagonal positions of the rhombus, while those of the bridging aryl ligand in **3** are unbound. As a result, the di- and tri-coordinate copper atoms are located at alternate vertices of the rhombus.

Behrens and coworkers reported the formation of heterolytic mixed alkenyl–aryl organocopper aggregate **5** as crystals suitable for X-ray diffraction with a very low yield (Scheme 9.2) [46]. The authors proposed a mechanism where the addition of a phenylcopper unit, generated by initial transmetalation of phenyllithium and copper salts, to a C–C triple bond afforded vinylcopper units, which finally formed a tetranuclear mixed vinyl–phenyl–copper (I) complex **5**. The phenyl and vinyl moieties are approximately perpendicular (83.5 or 87.9°) to the Cu–Cu vector. Every anionic sp^2 carbon atom bridges two copper atoms *via* Cu–C–Cu three-center two-electron (3c–2e) bonds [36, 47–52], in which each copper atom contributes with an empty sp hybrid orbital and the carbon atom contributes with an occupied sp^2 hybrid orbital. In this case, the aryl groups might provide stabilization for the mixed alkenyl–aryl molecule.

Besides intramolecular heteroatoms, the alkyne groups were also used to stabilize the resulting alkenylcopper–alkyne π-complexes [53–55]. The precursor 1,4-diyne–Cu–R **9**, which can be synthesized *via* two methods (Scheme 9.3), transformed into the alkenyl 1,1-bimetallic complex **10** by intramolecular rearrangement [54]. The copper atom of **10** is trigonally planar, coordinated by π-bonding to a $C{\equiv}CSiMe_3$ fragment [Cu–C_α 1.972(16) Å, Cu–C_β 2.236(18) Å], σ-bonding to the bridging carbon atom (C_b) of the $C{=}C(SiMe_3)(R)$ entity [Cu–C_b 2.031(14) Å] together with the intramolecular nitrogen coordination. The bridging carbon atom bridges the Ti and Cu atoms with an acute Cu–C_b–Ti angle of 80.7(5)°. The Ti–C_b–C angle 160.6(11)° is highly distorted indicating that the Ti–C_b bond can be best described as a bent bond, in which the orbitals are positioned outside

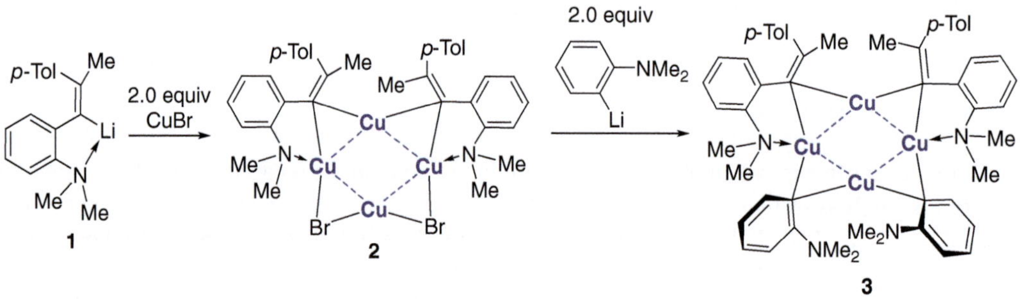

Scheme 9.1 Synthesis of 1-aryl-propenyl copper compounds with *N*-coordination.

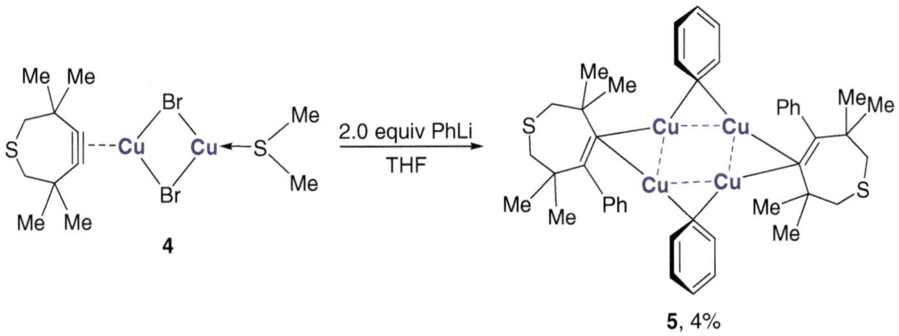

Scheme 9.2 Synthesis of a mixed alkenyl–aryl copper complex.

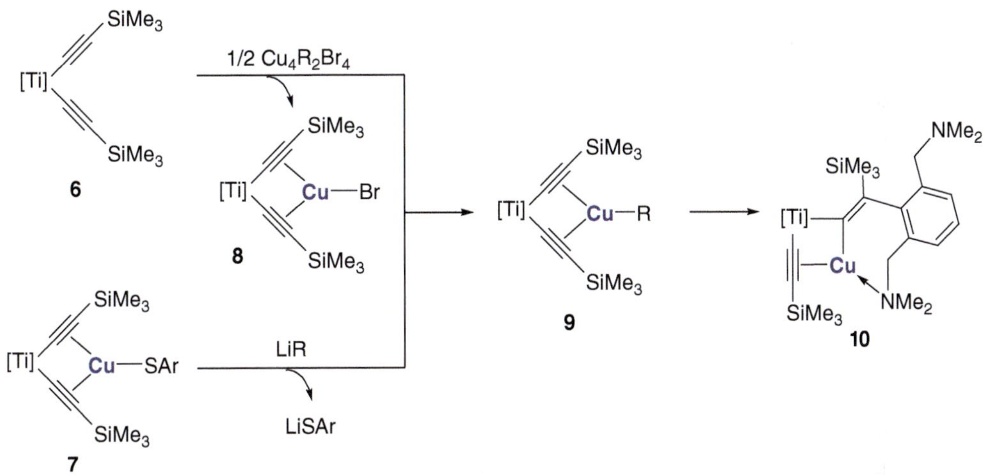

Scheme 9.3 Synthesis of an alkenylcopper–alkyne π-complex.

the Ti–C_b–Cu triangle. The coordination geometry of the Cu atom and the Cu–C_b–C angle 117.6(10)° suggest an above-mentioned Cu–C_b σ-bond.

9.2.2 Mono-alkenyl Organocopper Compounds Stabilized by *N*-heterocyclic Carbene

NHC, which can be readily prepared, behaves as a strong σ-donor to metal atoms, forming exceptionally stable metal–carbene bonds [56]. Cu–NHC complexes are ubiquitously found in catalytic reactions [57]. Several alkenylcopper–carbene compounds have been isolated from the *cis*-hydrocupration reaction of Cu–H species with carbon–carbon triple bonds.

Sadighi and coworkers reported the facile synthesis of dimeric (NHC)copper(I) hydride **12** from the reduction of carbene-ligated copper(I) *tert*-butoxide **11** with triethoxysilane (Scheme 9.4) [58]. **12** is unstable in solution at an ambient temperature, which indicates a reactive character. Treatment of **12** with 3-hexyne afforded a well-defined monomeric vinyl (NHC)copper complex **13** *via cis*-hydrocupration of a carbon–carbon triple bond [59]. The Cu–C_{vinyl} bond length [1.902(4) Å] is notably shorter than that of di-coordinate aryl–CuL (L = SR_2 or PR_3) complexes [60-67], but comparable to those of aryl organocuprates [68], which reveals the advantage of bulky NHCs for stabilization of organocopper complexes. Besides, ring-expanded NHCs were also used to stabilize alkenylcopper complexes [69, 70].

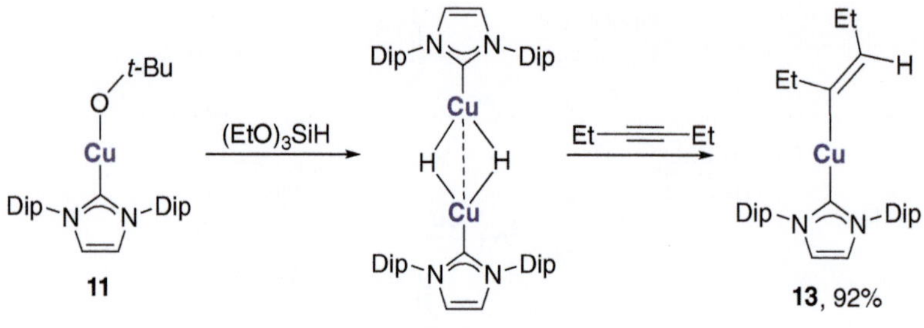

Scheme 9.4 Syntheses of mononuclear alkenylcopper–carbene complex.

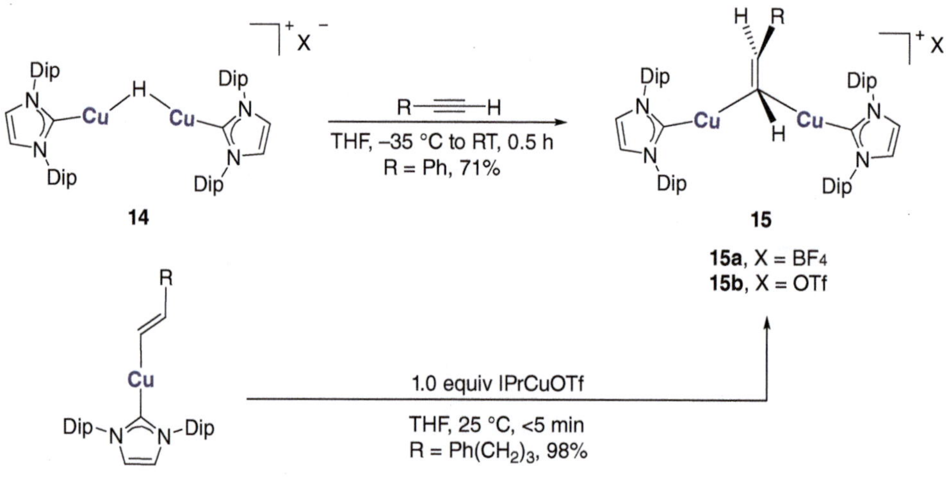

Scheme 9.5 Syntheses of dinuclear alkenylcopper–carbene complexes.

Later, Gray, Sadighi and coworkers found that the μ-hydrido dicopper cationic complex **14** reacted with phenylacetylene, forming a (*trans*-phenylvinyl)-bridged dicopper(I) complex **15a** (Scheme 9.5) [71]. The Cu–Cu distance in **15a** [2.6303(4) Å] is slightly longer than that of hydride-bridged **14** [2.5331(15) Å]. Density functional theory (DFT) calculations reveal that the core $[Cu_2C_{vinyl}]^+$ structure involves an open three-center interaction [72, 73], in which the metal–metal interaction occurs mainly through the bridging carbon atom. The alkenyl-bridged dicopper(I) complex **15b** could be alternatively obtained by adding 1.0 equiv. of IPrCuOTf to alkenyl monocopper(I) complex **16** [74].

These works facilitate the deep study into the mechanism of hydrogenation and hydrofunctionalization reactions of alkynes mediated by copper hydride complexes [59,74-76]. For instance, Lalic and coworkers provided experimental evidence on a revised mechanism **II** in which the vinyl dicopper cationic complexes [(NHC–Cu)$_2$(μ-X)(OTf) (X = F, H, alkenyl)] rather than the mononuclear alkenyl–Cu–NHC species of mechanism **I** behave as the key intermediate in the hydroalkylation reaction of alkynes with alkyl triflates (Scheme 9.6) [74]. As demonstrated by further mechanistic experiments, **14b** is responsible for the reaction selectivity, since it reacts with alkynes preferentially to give the hydrocupration product **15b** whereas it remains inert toward the highly electrophilic alkyl triflates, which excludes byproduct formation from reduction. Meanwhile,

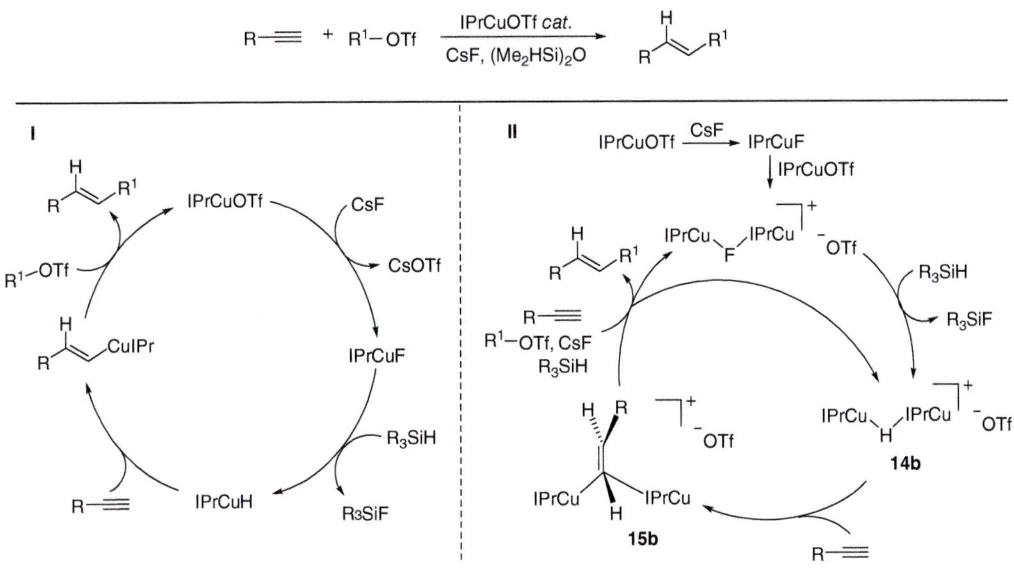

Scheme 9.6 Revised mechanism of hydroalkylation of alkynes involving a dicopper intermediate [IPr = 1,3-bis (2,6-diisopropylphenyl)imidazol-2-ylidene].

fluoride-bridged intermediate reacts with silanes much faster than with alkyl triflates, preventing the fluorination process during the course of the hydroalkylation.

9.2.3 Butadienyl Copper Compounds

Xi group has been investigating the syntheses, structural characterizations, and reaction chemistry of *cis*-1,3-butadienyl-based organometallic complexes [77–98]. The two coordinating sites on the 1,4-positions and the conjugate effect of *cis*-butadienyl ligands were found to cooperate synergistically to increase the stability of reactive organometallic complexes and to make butadienyl-based reagents exhibit unique, different reaction patterns from mono-alkenyl organometallic reagents [99–103].

Treatment of 1,4-dilithio-1,3-butadiene (dilithio reagent for short) **17** with 2.0 equiv. of CuCl in Et_2O afforded tetrameric 1,4-dicopper compound **18** in a 75% isolated yield (Scheme 9.7) [81]. When 2.5 equiv. of CuCl and 2.0 equiv. of *N,N,N′,N′*-tetramethylethylenediamine (TMEDA) were added into the solution of **17** in Et_2O, a different butadienyl copper compound **19** was obtained in a 63% isolated yield. **19** could be transformed into **18** when treated with another 1.0 equiv. of **17**. In addition, interestingly, solvents and extra lithium salts were found to have a significant influence on the formation of alkenylcopper compounds [104]. **17** reacted with 2.5 equiv. of CuCl in the polar solvent tetrahydrofuran (THF) instead of Et_2O, forming the organocopper aggregate **20**, linked by 1,3-butadienyl and 1,3,5,7-octatetraenyl units, in a 43% isolated yield. When dilithio reagent **17** was treated with 2.5 equiv. of CuCl and 4.0 equiv. of extra LiI in THF, 1,8-dicopper-1,3,5,7-octatetraene derivative **21** was obtained in a 63% isolated yield. The 1,3,5,7-octatetraenyl unit was formed from the dimerization of butadienyl copper complexes. All the above reactions were conducted at a low temperature of −78 °C to avoid the uncontrollable thermal decomposition of alkenylcopper compounds. When treated with extra LiI in THF at an elevated temperature of 0 °C, **20** disaggregates to **21** in an 82% yield along with a small amount of **18**.

Alkenylcopper aggregates **18–21** have very different structural configurations, however, they share some parameters in common. The Cu–Cu distances are all in close proximity (2.37–3.01 Å), which might suggest a weak d^{10}–d^{10} metal–metal interaction [105–107]. Cu–$C_{alkenyl}$ bond lengths

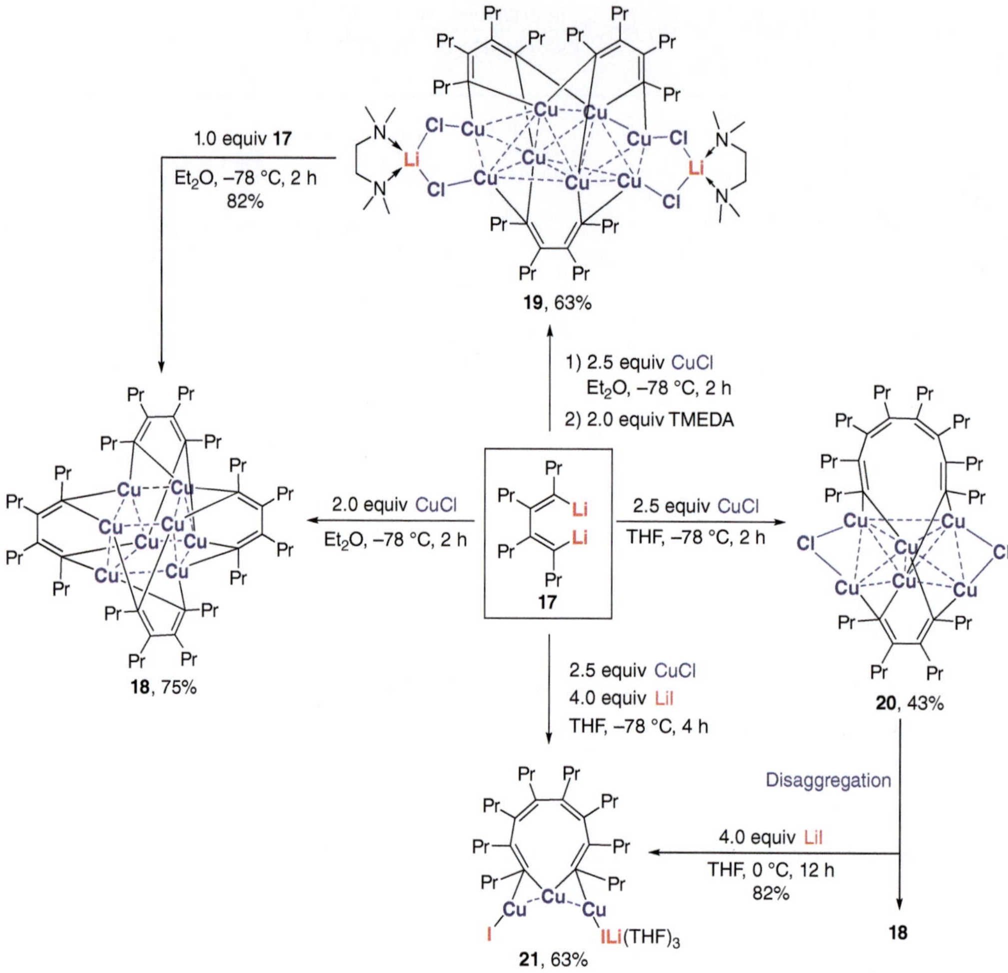

Scheme 9.7 Syntheses and transformations of butadienyl and octatetraenyl copper complexes.

and Cu–C–Cu bite angles are in the range 1.96–2.08 Å and 72.9–83.5°, respectively, which are comparable to the reported values of arylcopper aggregates [12, 13].

The crystals of **18–21** were all thermally stable under an inert atmosphere at room temperature (RT). When the temperature increased, **18** and **21** will decompose in THF to release homocoupling products **22** and **24** in high yield as well as copper mirror (Scheme 9.8). However, when heated in toluene, **21** afforded semibullvalene **25** as the major product in a moderate isolated yield along with a small amount of **24**. **21** might be the intermediate of CuCl-mediated reactions of dilithio reagents in the earlier work [108]. Lithium salts could improve the solubility and disaggregation of copper clusters, which might result in an increase in their reactivity. In the presence of LiI, **18** could react with aryl isocyanide, forming iminocyclopentadiene derivative **23** in high yield *via* 1,1-insertion followed by intermolecular coupling.

Copper-mediated reactions of zirconacyclopentadienes (or zirconaindenes) and electrophiles were developed by Takahashi *et al.* to construct a variety of useful organic molecules [11, 109–116]. In these reactions, the key step is proposed as the transmetalation process of the zirconium reagent

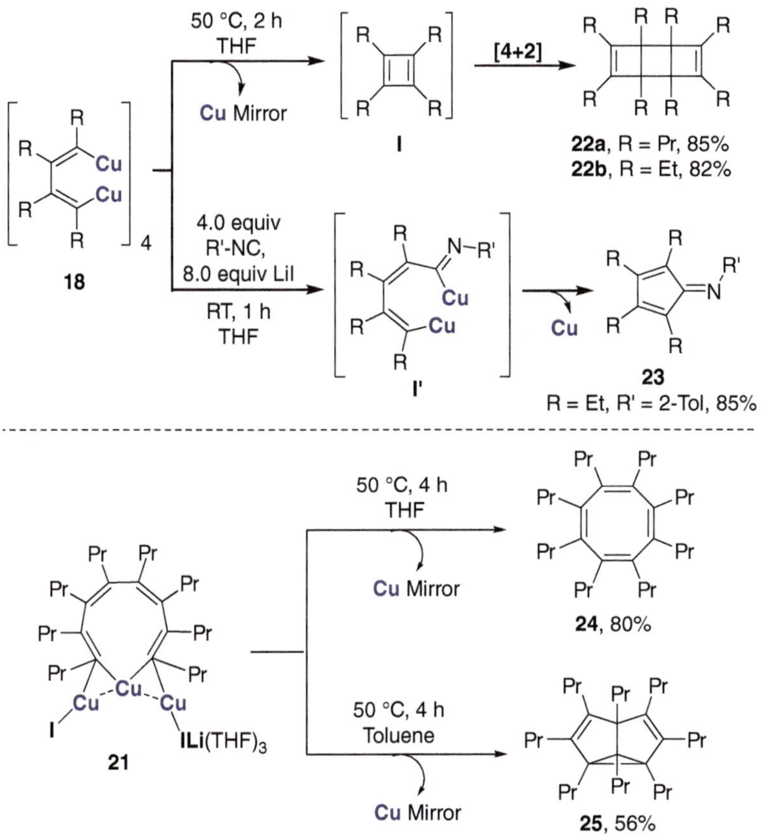

Scheme 9.8 Reaction of butadienyl 1,4-dicopper tetramer and octatetraenyl tricopper-LiI adduct.

and copper salts to generate reactive butadienyl (or styrenyl) dicopper or hybrid mono-copper mono-zirconium intermediates.

Inspired by these works, Xi group developed a new method to synthesize alkenyl organocopper aggregates using the readily prepared zirconium reagents as starting materials instead of dilithio reagents[83]. Treatment of zirconaindene **26** with 3.0 equiv. of CuCl and TMEDA in THF afforded the tetrameric styrenyl copper aggregate **28** in a 53% isolated yield (Scheme 9.9). When 2.5 equiv. of CuCl was added to the solution of **26** in THF and TMEDA, the hexameric styrenyl copper aggregate **29** was obtained in a 27% isolated yield after recrystallization in Et_2O and tetrahydrothiophene. Zirconacyclopentadienes **27a,b** with two Zr–$C_{alkenyl}$ bonds, which are much different from **26** with one Zr–$C_{alkenyl}$ and one Zr–C_{aryl} bond, reacted with 3.4 equiv. of CuCl in THF, affording trimeric butadienyl copper aggregates **30** and **31** in 72% and 51%, respectively. **30** is relatively stable in Et_2O below 10 °C, whereas **31** is more unstable and would thermally decompose in Et_2O to give copper mirror at 10 °C within 0.5 hours.

28–31 are all moisture- and oxygen-sensitive and have been determined by single-crystal X-ray analysis. A total of ten copper atoms in **28** formed two distorted tetrahedra and two butterfly-like quadrilaterals. The lengths of Cu–$C_{alkenyl}$ bonds [1.983(7)–2.010(7) Å] are comparable to those of Cu–C_{aryl} bonds [1.982(7)–2.015(8) Å] in **28**. **28** is notably different from **18** which has a cube-like core structure and is free of additional chlorides and TMEDA ligands. **29** consists of twelve copper atoms forming eight distorted tetrahedra. The Cu–Cu distances [2.4373(8)–3.0318(9) Å] are slightly

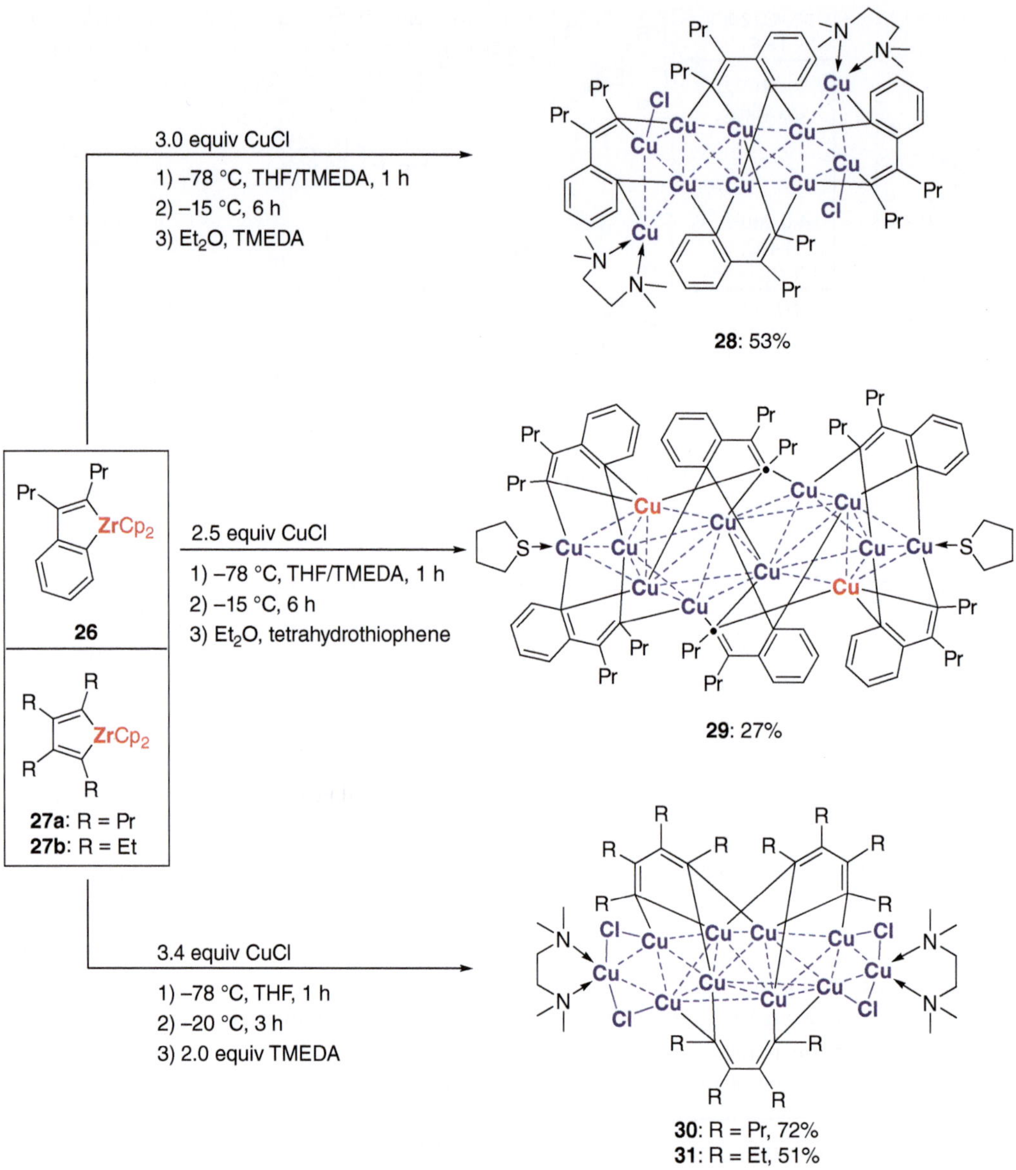

Scheme 9.9 Syntheses of styrenyl and butadienyl copper aggregates from zirconium reagents.

longer than those of **28** [2.3909(18)–2.8820(11) Å]. The bonding mode of **29** is more complicated yet unique since it has two unusual μ_3-alkenyl carbon atoms (black dots: C17, C17′) and, correspondingly, two tri-coordinate copper atoms (red: Cu2, Cu2′) [117]. The three Cu–C17 bond lengths are 2.019(4), 2.032(5), and 2.041(5) Å, respectively. **30** and **31** have almost identical configurations of the core structures. **30** and **31** can be regarded as variants of **19** by replacement of two lateral lithium atoms with copper atoms.

As shown above, the reaction of butadienyl-based polar organometallic (organolithium and organozirconium) reagents with excess copper salts afforded a series of structurally diverse organocopper aggregates. It is reasonable to think that organocuprates would be formed by the treatment of dilithio reagents with deficient copper salts.

It was previously reported by Xi group that the dilithio reagent reacted with low-valent transition-metal salts, e.g. $Ni(COD)_2$, $[RhCl(COD)]_2$, and $Pd(Pt\text{-}Bu_3)_2$, affording unprecedented aromatic metalloles [84, 85, 96], in which the 1,3-butadienyl ligand not only coordinates to the metal center but also accepts two electrons from d or p orbitals of the transition metal using the empty π^* orbital. In these reactions, the dianionic *cis*-butadienyl, which is isoelectronic to the α-diimine ligands, behaves as a novel type of noninnocent ligand and consequently an unusual formal oxidant as it accepts electrons from the metal center [118]. Later, Xi and coworkers found that the dilithio reagent **32a** could react with 0.5 equiv. of CuCl supported by nitrogen or phosphine ligands to produce aromatic dicupra[10]annulenes **33** along with a small amount of byproduct **34** (Scheme 9.10) [89]. The macrocyclic dibutadienyl cuprate **34a** can be alternatively prepared in high yield by the reaction between **32a** and 2.0 equiv. of mesitylcopper. **34a** converted into the corresponding aromatic dicupra[10]annulene **33a** when reduced by excess metal lithium.

33a–c and **34a** were all determined by single-crystal X-ray analysis. **33a** and **34a** are obviously different since **33a** has four lithium atoms, while **34a** has two lithium atoms. In comparison with the alternating bonds of butadienyl in **34a** [1.518(10), 1.367(10) Å], C–C bonds of the annulene moiety in **33a** are notably averaged [1.470(4), 1.421(3) Å], indicating a considerable π-conjugation. The Cu–C(sp^2) bonds of **33a** [1.917(2) Å] were much shorter that those of **34a** [1.957(8) Å]. The C–Cu–C bond angles of both **33a** [164.3(8)°] and **34a** [170.6(3)°] are nearly linear, suggesting sp hybridization of copper atoms. A weak interaction might exist between copper atoms of **33a** and **34a**, as supported by their short Cu–Cu distances (2.438 and 2.344 Å, respectively) and small Wiberg bond indices (0.59 and 0.45, respectively).

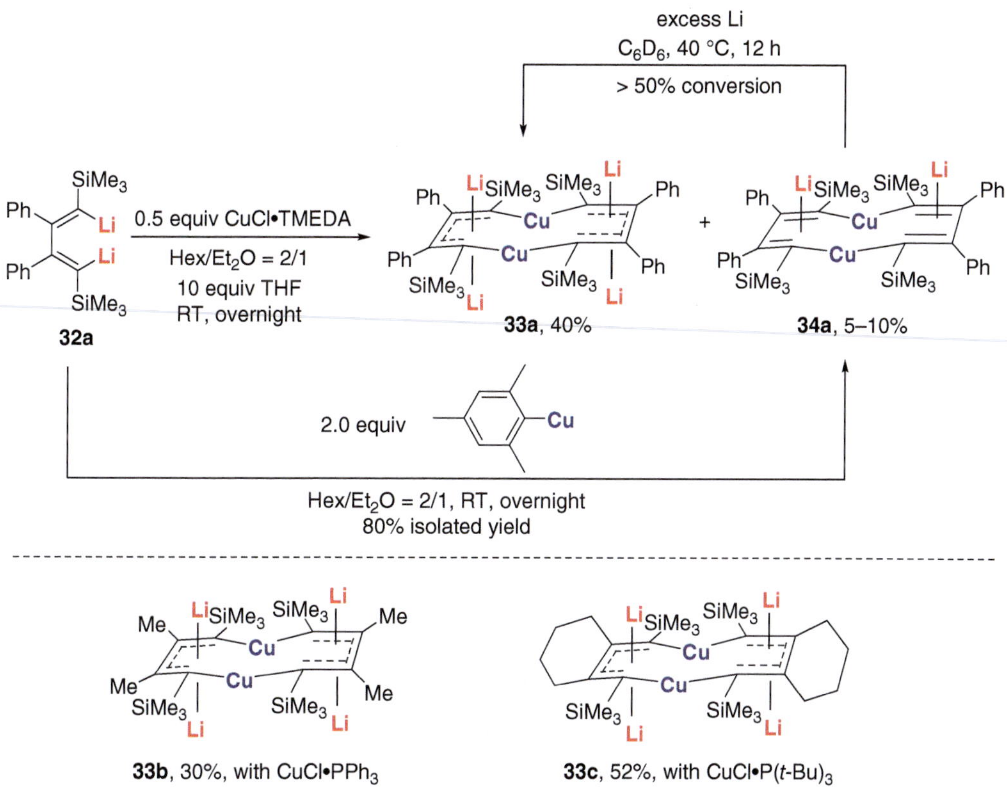

Scheme 9.10 Formation of aromatic dicupra[10]annulenes.

The abnormal coordination in **33**, especially of the four lithium cations, brings out an interesting question about the oxidation state of the copper atoms. The Cu $2p_{3/2}$ binding energy (932.9 eV) and Cu LMM kinetic energy (915.8 eV) of **33b**, as displayed by X-ray photoelectron spectroscopy (XPS) data, result in a conclusion that the copper atoms of **33b** are more likely to be copper(I). The corresponding energies of **34a** (933.2 and 914.6 eV) also suggest that it is a Cu(I) species.

Low-frequency resonance of ^{7}Li nuclear magnetic resonance (NMR) (−5.09 to −6.17 ppm), negative values of isomerization stabilization energy (ISE) (−11.7 and −21.5), and large negative nucleus-independent chemical shift (NICS) values (−9.0 to −12.8) all indicate the considerable aromaticity of dicupra[10]annulenes. Adaptive natural density partitioning (AdNDP) analysis suggest a 10π aromatic system for the dicupra[10]annulenes. Other calculations supported a naphthalene-like or 16e Möbius aromaticity for **33** [119, 120]. It was also calculated that the lithium atoms affect the electronic and molecular structures of the dicupra[10]annulenes [121].

The mechanism of the above-mentioned reaction is not clear yet. Based on the experiments and literature, a proposed mechanism is presented in Scheme 9.11. **34** and an alkenylcopper aggregate species **IM-11** were initially produced by the transmetalation process of **32** and copper salts. **IM-11** would release a Cu(0)-cluster together with octatetraene and cyclooctatetraene derivatives. The Cu(0)-cluster generated *in situ* finally reacts with **32** to give **33**, in which the butadienyl dianions behave as Z-type ligands. However, no target product was detected during the test of **32** with other copper sources (activated Cu powder, copper hydride cluster).

Biphenyl and styrenyl backbones are analogues to the butadienyl moiety, however, they might exhibit different properties resulting from the difference in conjugation. Biphenyl dilithio reagent **35a** reacted with 0.5 equiv. of CuBr·SMe_2 in THF at RT to produce copper(I) spiro complex **36a** in a quantitative NMR yield and a 75% isolated yield (Scheme 9.12) [92]. Beside CuBr·SMe_2, CuCl and CuI could also be applied to the above reactions, affording the copper(I) spiro complex in comparable yields. Treatment of **36a** with 1.0 equiv. of oxidant gave the corresponding copper(III) spiro complex **37a** in a more than 85% NMR yield and a 65% isolated yield. A range of oxidizing agents, such as *p*-benzoquinone, iodines, 1-fluropyridinium tetrafluoroborate, or molecular oxygen, worked well for this reaction. In return, **37a** transformed into **36a** by reduction of excess lithium. Biaryl spiro copper(I) and copper(III) compounds **36b** and **37b** can be generated in good yields. Styrenyl analogues **36c** and **37c** were obtained in low yields, probably due to their decomposition to benzosilolyl species or homocoupling products of styrenyl units. In these reactions, the dianionic ligands behaved as preorganized ligands to lock the copper atom in the center.

36a, **36c**, and **37a–c** were all obtained as moisture- and oxygen-sensitive crystals suitable for X-ray crystallographic analysis. The copper(I) atom of **36a** is linked with four carbon atoms with a distorted tetrahedral coordination, of which the dihedral angle between spiro-fused 5-membered cupracycles is 84.1(1)°. This unique bonding mode make **36** rare examples of high-order

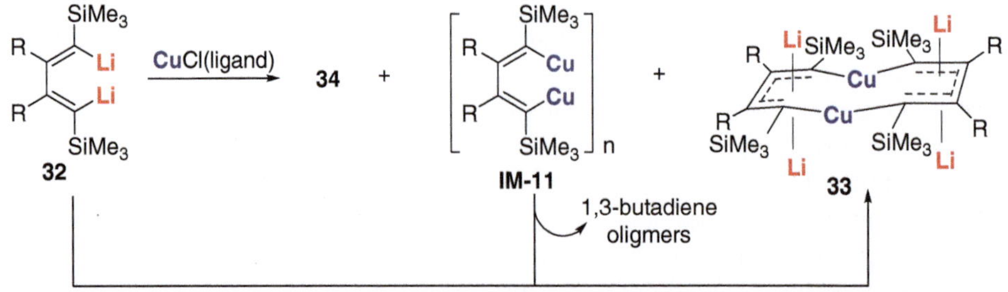

Scheme 9.11 Proposed mechanism for the formation of dicupra[10]annulenes.

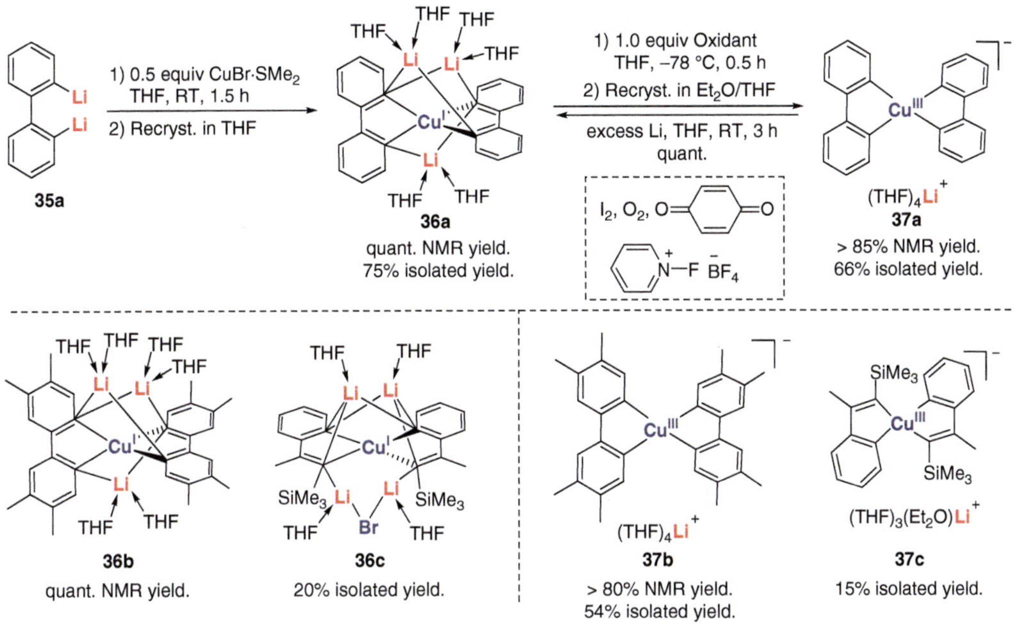

Scheme 9.12 Syntheses of spiro organocopper(I) compounds and organocopper(III) complexes.

organocuprates(I) [68]. The copper(III) atom of **37a** adopted a nearly square-planar geometry with a corresponding 27.2(5)° dihedral angle. The lengths of Cu(III)–C(sp^2) bonds [1.952(3)–1.968(3) Å] are comparable to those of previously reported Cu(III) compounds and slightly shorter than the Cu(I)–C(sp^2) bonds of **36a** [2.030(3)–2.040(3) Å], in accordance with the fact that the higher valence gets a smaller radius. The *cis*-bonding angle of C–Cu(III)–C [83.9(1)°] of **37a** is smaller than those of C–Cu(I)–C in **36a** [86.6(1)° and 86.8(1)°]. The dihedral angle of **37b** [29.2(1)°] and **37c** [37.9(4)°] remains almost the same and gets much larger, respectively, compared to that of **37a**. This observation might be ascribed to the steric effect of the substituents on the α-positions.

Owing to its *cis*-C–Cu(III)–C bonding configuration, **37a** was expected to undergo reductive elimination reactions. Treatment of **37a** with electrophiles afforded **38–40** in high yields (Scheme 9.13). The above reactions shared an intramolecular C(sp^2)–C(sp^2) bond-forming process, which was then investigated by reactions with methylation reagents. The reaction of **37a** with 1.0 equiv. of trimethylsulfonium fluoroborate mainly afforded the intermediate **41**, an arylcopper(I) species, along with a tiny amount of **40**. The solution of **41** generated *in situ* reacted with 1.0 equiv. of methyl iodide, producing **40** in about a 90% overall yield. CuI(PMe_3)$_3$ was isolated from this reaction and determined by single-crystal analysis. In addition, **43** and **44** were obtained in high yields, when 1.2 equiv. of iodine or benzoyl chloride was added into the *in situ* generated solution of **41**. The high-yield formation of **40**, **42**, **43**, and **44** gave strong evidence of the existence of intermediate **41**. Overall, the experimental results directly supported that the reductive elimination of the spiro copper(III) complex occurred, leading to the formation of C(sp^2)–C(sp^2) bonds and regeneration of copper(I) species. Notably, only quaterphenyl derivatives were formed and no biphenylene derivatives were detected in the process of reductive elimination, which showed an excellent selectivity.

Tetraalkyl-substituted butadienyl dilithio reagent **45** can also react with 0.5 equiv. of copper salts to give butadienyl copper(I) spiro complexes **46a** and **46b** in a 65% combined isolated yield

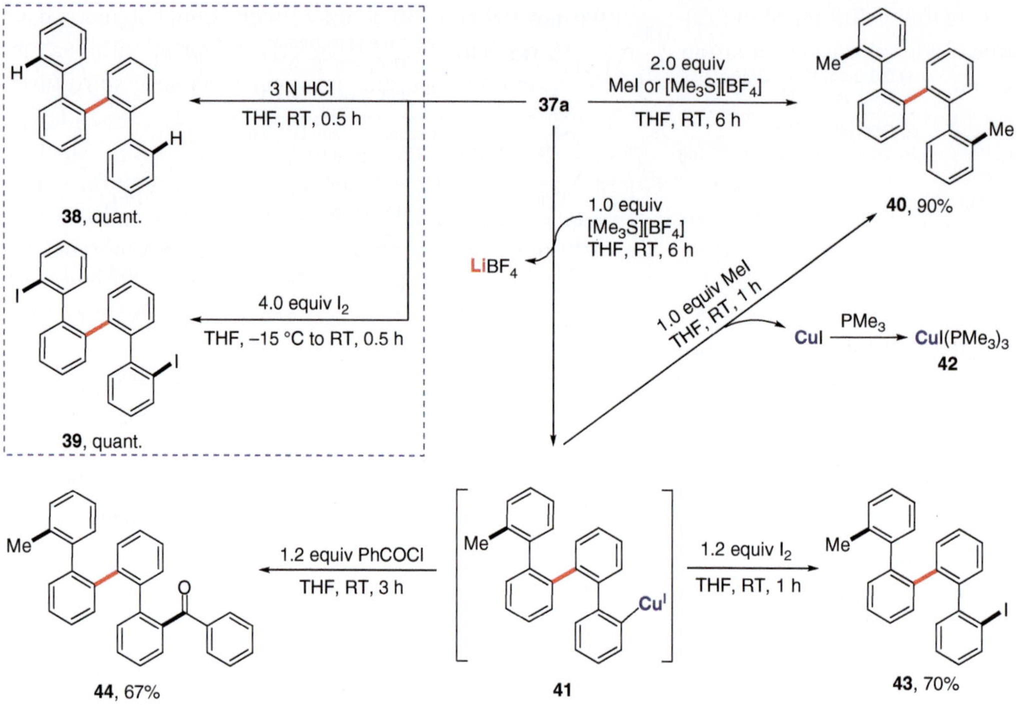

Scheme 9.13 Reductive elimination of tetra-carbon-linked copper(III) complexes.

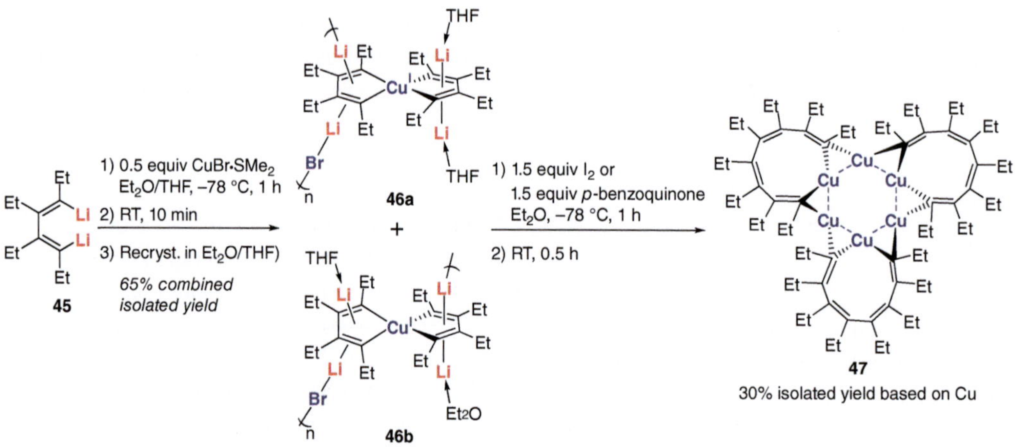

Scheme 9.14 Synthesis and oxidation of a butadienyl spiro copper complex.

(Scheme 9.14) [98]. However, when treated with I_2 or *p*-benzoquinone, **46a** and **46b** were oxidized to form a hexanuclear 1,3,5,7-octatetraenyl organocopper(I) aggregate **47**, rather than their corresponding spiro copper(III) complex. Cyclooctatetraenes were formed simultaneously in the reaction, and were detected by gas chromatography–mass spectrometry (GC–MS). The results revealed that, depending on the ligands of organocuprates(I), either the organic skeletons or the copper(I) atom could undergo an oxidation reaction. In this particular case, the more electron-rich butadienyl ligands instead of the copper(I) atom are oxidized preferentially.

Due to the difference in LiBr coordination, **46a** and **46b**, as one-dimensional polymers, have linear and zigzag coordination geometries, respectively. The copper(I) atoms of both **46a** and **46b** are tetra-coordinated with four carbon atoms. The dihedral angles between spiro-fused cupracyclopentadienes are 63 and 83°, respectively. The average lengths of Cu–$C_{alkenyl}$ bonds (2.011 Å) are longer than those of Cu–C(sp^2) bonds in diarylcuprates [e.g. $Ph_4Cu_2Li_2(Et_2O)_2$; 1.918 Å] [122]. Octatetraenyl copper(I) aggregate **47** features a core Cu_6 structure, which approaches a closely regular hexagon. The close Cu–Cu distances of each edge are in a narrow range [2.3903(5)–2.4092(5) Å]. The interior angles of the hexagon are in the range of 111.13(2)–129.96(3)°. Every bridging carbon atom binds to two copper(I) atoms with acute bite angles [72.80(10)–74.02(10)°]. The average Cu–C_b bond lengths and C–Cu–C bond angles are 2.009 Å and 168.2°, respectively.

A proposed mechanism depicting the transformation from **46** to **47** was provided based on experimental results and literature (Scheme 9.15). Take the iodine as an example of oxidant. Oxidation of **46** by iodine generated a butadienyl copper(III) spiro complex **IM-15-1** (path A), whose analogues **37a–c** were successfully isolated and characterized by X-ray crystallography before [92]. **IM-15-1** would readily transform to 9-membered cupracyclic species **IM-15-2**, which could also be formed by direct oxidative coupling of two butadienyl moieties (path B). Then, **IM-15-2** was further reacted with iodine to give cyclooctatetraene along with copper salts, which reacted with **IM-15-2** to finally produce **47**.

In sharp contrast to the lithium organocuprates [15, 68], magnesium organocuprates remain less explored, especially in terms of their well-defined structures. The Xi group recently reported a facile synthesis of multi-substituted magnesiacyclopentadienes **48** as the first example of alkaline earth metallocyclopentadienes [82]. The small bite angle of C–Mg–C (ca. 90°) and coplanar geometry of the 5-membered ring indicate a reactive character of magnesiacyclopentadienes. As shown in Scheme 9.16, **48a** with phenyl groups at *β*-positions could react with both inorganic and organic copper salts *via* formal cleavage of Cu–X bonds to give monomeric (Cu/Mg = 1/2) magnesium organocuprates **49** and **50[88]**. Additionally, Ag–Cl and Ag–C(sp) bonds can be cleaved and fused

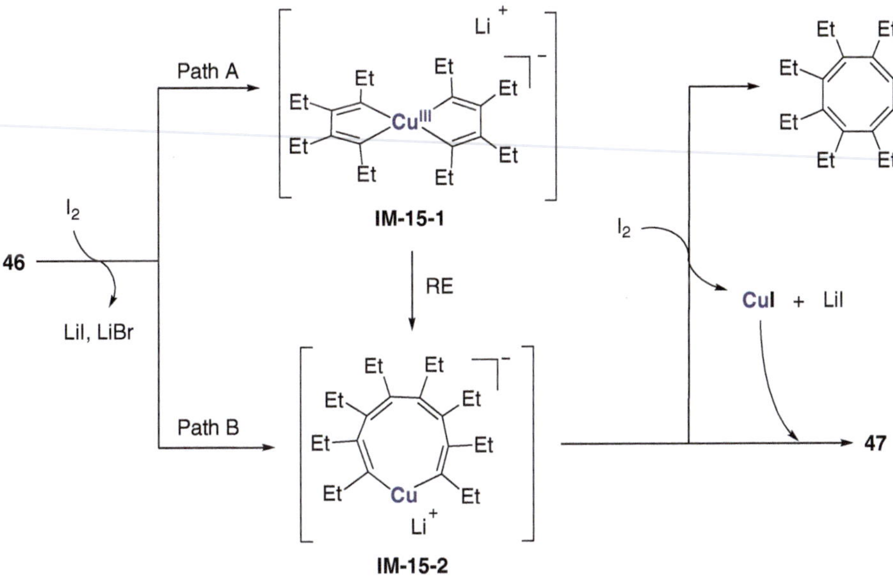

Scheme 9.15 Proposed mechanism of the formation of octatetraenyl copper aggregate **47**.

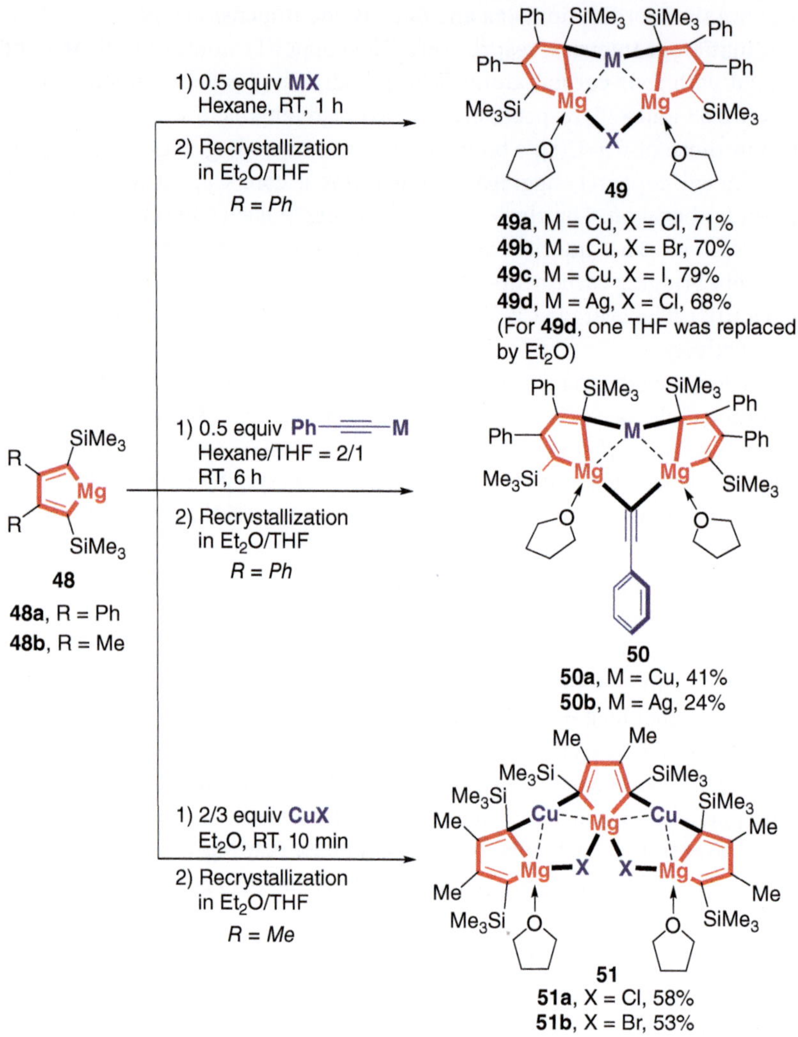

Scheme 9.16 Syntheses of rigid magnesium organocuprates and organoargentates.

in the final product. The reaction between **48b** with methyl groups at β-positions and copper salt afforded linearly linked dimeric (Cu/Mg = 2/3) magnesium organocuprates **51** as major products.

Compared to that of **48a**·TMEDA [2.129(3) Å], Mg–C bond lengths [2.113(7), 2.341(7) Å] of **49b** indicate a normal and a noticeably weakened Mg–C(sp^2) bond, respectively. Meanwhile, it means that **49b** results from an incomplete transmetalation event because the ligand-exchange process was partially completed and the bromine atom was kept in the product instead of releasing as $MgBr_2$. Accordingly, **49** can be regarded as a resting-state intermediate in the transmetalation reaction between organomagnesium reagents and coinage metal salts. Although the Cu–C bond lengths [1.945(6) Å] are in the normal range for Cu–C(sp^2) bond lengths, the C–Cu–C bond angle [156.8(4)°] is significantly smaller than those of diaryl lithium organocuprates (163–178°), indicating an increase of Cu–C–Mg 3c–2e bonding character as a result of a higher degree of covalency

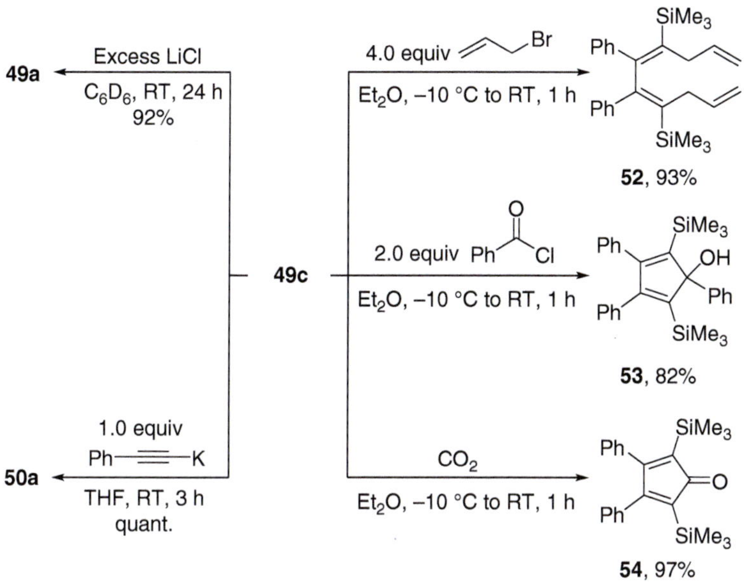

Scheme 9.17 Transformation and preliminary reactivity of magnesium organocuprates.

in Mg–C bonds. Natural bond orbital (NBO) and atom-in-molecule (AIM) analysis gave more evidence of the nonexistence of the Cu–Br interaction, which is in accordance with the long Cu–Br distance [3.5885(13) Å] in the molecular structure. Overall, the synergistic coordination of exposed terminal C(sp^2) atoms and unsaturated Mg atoms upon MX salts led to the heterolytic cleavage of M–X bonds by the transfer of anionic X units from M centers to Mg centers.

49c can be transformed into **49a** and **50a** in high yields (Scheme 9.17). The preliminary reactivity of **49c** with electrophiles displayed excellent reaction specificity, which might result from the close proximity of two nucleophilic anionic carbon atoms on the *cis*-butadienyl units.

9.3 Summary

A colourful world of molecules was reflected by the diverse structures of alkenylcopper compounds. The preference to form polynuclear aggregates is a nature of non-ate organocopper compounds. In these compounds, the bridging carbon atoms are typically attached to a couple of copper atoms, each of which binds with two carbonic ligands in a nearly linear geometry. Consequently, a cyclic polynuclear core structure in either two- or three-dimensional configurations forms. The close Cu-Cu distances indicate a d^{10}–d^{10} interaction.The resulting Cu–$C_{bridging}$–Cu motif forms an electron-deficient three-center two-electron bond. When bulky or strong donating ligands, e.g. NHC, or preorganized ligands, e.g. 1,4-diyne, are incorporated into the structures, low-nuclear neutral organocopper compounds are formed predominantly. In terms of the alkenyl cuprate complexes, they are mainly based on butadienyl skeletons. The copper atoms are di- and tetra-coordinated with the bidentate ligands, which never show *cis* to *trans* rotation of configuration in the solid structures. In these cases, the butadienyl ligands also provided preorganized geometries for the incoming copper atoms.

References

1 Surry, D.S. and Spring, D.R. (**2006**). *Chem. Soc. Rev. 35*: 218–225.
2 Breit, B. and Schmidt, Y. (**2008**). *Chem. Rev. 108*: 2928–2951.
3 Evano, G., Blanchard, N., and Toumi, M. (**2008**). *Chem. Rev. 108*: 3054–3131.
4 Meldal, M. and Tornøe, C.W. (**2008**). *Chem. Rev. 108*: 2952–3015.
5 Rijs, N.J. and O'Hair, R.A.J. (**2010**). *Organometallics 29*: 2282–2291.
6 van Koten, G., Perez, P., and Liebeskind, L. (**2012**). *Organometallics 31*: 7631–7633.
7 Beletskaya, I.P. and Cheprakov, A.V. (**2012**). *Organometallics 31*: 7753–7808.
8 Evano, G. and Blanchard, N. (eds.) (**2014**). *Copper-Mediated Cross-Coupling Reactions*. Hoboken: John Wiley & Sons
9 Allen, S.E., Walvoord, R.R., Padilla-Salinas, R., and Kozlowski, M.C. (**2013**). *Chem. Rev. 113*: 6234–6458.
10 McCann, S.D. and Stahl, S.S. (**2015**). *Acc. Chem. Res. 48*: 1756–1766.
11 Yan, X. and Xi, C. (**2015**). *Acc. Chem. Res. 48*: 935–946.
12 Krause, N. (ed.) (**2002**). *Modern Organocopper Chemistry*. Weinheim: Wiley-VCH.
13 Rappoport, Z. and Marek, I. (eds.) (**2009**). *The Chemistry of Organocopper Compounds*. Chichester: Wiley.
14 Sperotto, E., van Klink, G.P.M., van Koten, G., and de Vries, J.G. (**2010**). *Dalton Trans. 39*: 10338–10351.
15 Yoshikai, N. and Nakamura, E. (**2012**). *Chem. Rev. 112*: 2339–2372.
16 Casitas, A. and Ribas, X. (**2013**). *Chem. Sci. 4*: 2301–2318.
17 Sambiagio, C., Marsden, S.P., Blacker, A.J., and McGowan, P.C. (**2014**). *Chem. Soc. Rev. 43*: 3525–3550.
18 Rasika, D.H.V. and Jiang, W. (**2008**). *Eur. J. Inorg. Chem.*: 509–522.
19 Basheer, A. and Marek, I. (**2010**). *Beilstein J. Org. Chem. 6*: 77.
20 Yun, J. (**2013**). *Asian J. Org. Chem. 2*: 1016–1025.
21 Fujihara, T., Semba, K., Terao, J., and Tsuji, Y. (**2014**). *Catal. Sci. Tech. 4*: 1699–1709.
22 Piazza, C. and Knochel, P. (**2002**). *Angew. Chem. Int. Ed. 41*: 3263–3265.
23 Yang, X., Rotter, T., Piazza, C., and Knochel, P. (**2003**). *Org. Lett. 5*: 1229–1231.
24 Calaza, M.I., Yang, X., Soorukram, D., and Knochel, P. (**2004**). *Org. Lett. 6*: 529–531.
25 Yang, X., Althammer, A., and Knochel, P. (**2004**). *Org. Lett. 6*: 1665–1667.
26 Yang, X. and Knochel, P. (**2006**). *Chem. Commun.*: 2486–2488.
27 Yang, X. and Knochel, P. (**2006**). *Org. Lett. 8*: 1941–1943.
28 Yang, X. and Knochel, P. (**2006**). *Chem. Commun.*: 2170–2172.
29 Yang, X. and Knochel, P. (**2006**). *Synthesis*: 2167–2172.
30 Ebert, G. and Rieke, R.D. (**1984**). *J. Org. Chem. 49*: 5280–5282.
31 Ebert, G.W. and Rieke, R.D. (**1988**). *J. Org. Chem. 53*: 4482–4488.
32 Guss, J.M., Mason, R., Thomas, K.M. et al. (**1972**). *J. Organomet. Chem. 40*: C79–C80.
33 Guss, J.M., Mason, R., Sotofte, I. et al. (**1972**). *J. Chem. Soc. Chem. Commun.*: 446–447.
34 van Koten, G. and Noltes, J.G. (**1975**). *J. Organomet. Chem. 84*: 129–138.
35 van Koten, G. and Noltes, J.G. (**1975**). *J. Organomet. Chem. 102*: 551–563.
36 ten Hoedt, R.W.M., Noltes, J.G., van Koten, G., and Spek, A.L. (**1978**). *J. Chem. Soc., Dalton Trans.*: 1800–1806.
37 ten Hoedt, R.W.M., van Koten, G., and Noltes, J.G. (**1979**). *J. Organomet. Chem. 179*: 227–240.
38 Camus, A., Marsich, N., Nardin, G., and Randaccio, L. *J. Organomet. Chem 174*: 121–128.

39 Noltes, J.G., ten Hoedt, R.W.M., van Koten, G. et al. (**1979**). *J. Organomet. Chem. 225*: 365–376.
40 Wehman, E., van Koten, G., and Jastrzebski, J.T.H.B. (**1986**). *J. Organomet. Chem. 302*: C35–C39.
41 Wehman, E., van Koten, G., Knotter, M. et al. (**1987**). *J. Organomet. Chem. 25*: 293–309.
42 Wehman, E., van Koten, G., Erkamp, C.J.M. et al. (**1989**). *Organometallics 8*: 94–99.
43 Janssen, M.D., Corsten, M.A., Spek, A.L. et al. (**1996**). *Organometallics 15*: 2810–2820.
44 ten Hoedt, R.W.M., van Koten, G., and Noltes, J.G. (**1980**). *J. Organomet. Chem. 201*: 327–342.
45 Smeets, W.J.J. and Spek, A.L. (**1987**). *Acta Cryst. C 43*: 870–873.
46 Schulte, P., Behrens, U., and Olbrich, F. (**2000**). *Z. Anorg. Allg. Chem. 626*: 1692–1696.
47 Mason, R. and Mingos, D.M.P. (**1973**). *J. Organomet. Chem. 50*: 53–61.
48 van Koten, G. (**1990**). *J. Organomet. Chem. 400*: 283–301.
49 Antes, I. and Frenking, G. (**1995**). *Organometallics 14*: 4263–4268.
50 Belanzoni, P., Rosi, M., Sgamellotti, A. et al. (**1996**). *Chem. Phy. Lett. 257*: 41–48.
51 Davenport, T.C. and Tilley, T.D. (**2011**). *Angew. Chem. Int. Ed. 50*: 12205–12208.
52 Green, J.C., Green, M.L.H., and Parkin, G. (**2012**). *Chem. Commun. 48*: 11481–11503.
53 Janssen, M.D., Smeets, W.J.J., Spek, A.L. et al. (**1995**). *J. Organomet. Chem. 505*: 123–126.
54 Janssen, M.D., Köhler, K., Herres, M. et al. (**1996**). *J. Am. Chem. Soc. 118*: 4817–4829.
55 Bruce, M.I., Zaitseva, N.N., Skelton, B.W., and White, A.H. (**2008**). *J. Organomet. Chem. 693*: 1400–1404.
56 Hopkinson, M.N., Richter, C., Schedler, M., and Glorius, F. (**2014**). *Nature 510*: 485.
57 Lazreg, F., Nahra, F., and Cazin, C.S.J. (**2015**). *Coord. Chem. Rev. 293–294*: 48–79.
58 Mankad, N.P., Laitar, D.S., and Sadighi, J.P. (**2004**). *Organometallics 23*: 3369–3371.
59 Jordan, A.J., Lalic, G., and Sadighi, J.P. (**2016**). *Chem. Rev. 116*: 8318–8372.
60 Gambarotta, S., Strologo, S., Floriani, C. et al. (**1984**). *Organometallics 3*: 1444–1445.
61 He, X., Olmstead, M.M., and Power, P.P. (**1992**). *J. Am. Chem. Soc. 114*: 9668–9670.
62 Mark, N. (**2003**). *Z. Anorg. Allg. Chem. 629*: 1535–1540.
63 Schaper, F., Foley, S.R., and Jordan, R.F. (**2004**). *J. Am. Chem. Soc. 126*: 2114–2124.
64 Mankad, N.P., Gray, T.G., Laitar, D.S., and Sadighi, J.P. (**2004**). *Organometallics 23*: 1191–1193.
65 Laitar, D.S., Tsui, E.Y., and Sadighi, J.P. (**2006**). *J. Am. Chem. Soc. 128*: 11036–11037.
66 Dubinina, G.G., Furutachi, H., and Vicic, D.A. (**2008**). *J. Am. Chem. Soc. 130*: 8600–8601.
67 Takeshi, O., Masayoshi, N., and Zhaomin, H. (**2008**). *Angew. Chem. Int. Ed. 47*: 5792–5795.
68 Davies, R.P. (**2011**). *Coord. Chem. Rev. 255*: 1226–1251.
69 Collins, L.R., Riddlestone, I.M., Mahon, M.F., and Whittlesey, M.K. (**2015**). *Chem. Eur. J. 21*: 14075–14084.
70 Jordan, A.J., Wyss, C.M., Bacsa, J., and Sadighi, J.P. (**2016**). *Organometallics 35*: 613–616.
71 Wyss, C.M., Tate, B.K., Bacsa, J. et al. (**2013**). *Angew. Chem. Int. Ed. 52*: 12920–12923.
72 Churchill, M.R. and Ni, S.W.Y. (**1973**). *J. Am. Chem. Soc. 95*: 2150–2155.
73 Churchill, M.R., DeBoer, B.G., and Rotella, F.J. (**1976**). *Inorg. Chem. 15*: 1843–1853.
74 Suess, A.M., Uehling, M.R., Kaminsky, W., and Lalic, G. (**2015**). *J. Am. Chem. Soc. 137*: 7747–7753.
75 Uehling, M.R., Rucker, R.P., and Lalic, G. (**2014**). *J. Am. Chem. Soc. 136*: 8799–8803.
76 Uehling, M.R., Suess, A.M., and Lalic, G. (**2015**). *J. Am. Chem. Soc. 137*: 1424–1427.
77 Liu, L., Zhang, W.X., Luo, Q. et al. (**2010**). *Organometallics 29*: 278–281.
78 Luo, Q., Wang, C., Gu, L. et al. (**2010**). *Chem. Asian J. 5*: 1120–1128.
79 Zhou, Y., Zhang, W.X., and Xi, Z. (**2012**). *Organometallics 31*: 5546–5550.
80 Li, H., Wei, B., Xu, L. et al. (**2013**). *Angew. Chem. Int. Ed. 52*: 10822–10825.
81 Geng, W., Wei, J., Zhang, W.X., and Xi, Z. (**2014**). *J. Am. Chem. Soc. 136*: 610–613.
82 Wei, J., Liu, L., Zhan, M. et al. (**2014**). *Angew. Chem. Int. Ed. 53*: 5634–5638.

83 Liu, L., Geng, W., Yang, Q. et al. (**2015**). *Organometallics 34*: 4198–4201.
84 Wei, J., Zhang, W.X., and Xi, Z. (**2015**) *Angew. Chem. Int. Ed. 54*: 5999–6002.
85 Wei, J., Zhang, Y., Zhang, W.X., and Xi, Z. (**2015**). *Angew. Chem. Int. Ed. 54*: 9986–9990.
86 Xu, L., Wang, Y., Wei, J. et al. (**2015**). *Chem. Eur. J. 21*: 6686–6689.
87 Zhang, Y., Wei, J., Zhang, W.X., and Xi, Z. (**2015**). *Inorg. Chem. 54*: 10695–10700.
88 Liu, L., Wei, J., Chi, Y. et al. (**2016**). *Angew. Chem. Int. Ed. 55*: 14762–14765.
89 Wei, J., Zhang, Y., Chi, Y. et al. (**2016**). *J. Am. Chem. Soc. 138*: 60–63.
90 Xu, L., Chi, Y., Du, S. et al. (**2016**). *Angew. Chem. Int. Ed. 55*: 9187–9190.
91 Du, S., Yin, J., Chi, Y. et al. (**2017**). *Angew. Chem. Int. Ed. 56*: 15886–15890.
92 Liu, L., Zhu, M., Yu, H.T. et al. (**2017**). *J. Am. Chem. Soc. 139*: 13688–13691.
93 Ma, W., Yu, C., Chi, Y. et al. (**2017**). *Chem. Sci. 8*: 6852–6856.
94 Wei, B., Liu, L., Zhang, W.X., and Xi, Z. (**2017**). *Angew. Chem. Int. Ed. 56*: 9188–9192.
95 Zhang, Y., Chi, Y., Wei, J. et al. (**2017**). *Organometallics 36*: 2982–2986.
96 Zhang, Y., Wei, J., Chi, Y. et al. (**2017**). *J. Am. Chem. Soc. 139*: 5039–5042.
97 Zhu, M., Liu, L., Zhang, Y. et al. (**2017**). *Chem. Eur. J. 24*: 3186–3191.
98 Liu, L., Zhu, M., Yu, H.T. et al. (**2018**). *Organometallics 37*: 845–847.
99 Xi, Z. (**2010**). *Acc. Chem. Res. 43*: 1342–1351.
100 Zhang, S., Zhang, W.X., and Xi, Z. (**2014**). In: *Organo-di-Metallic Compounds (or Reagents): Synergistic Effects and Synthetic Applications* (ed. Z. Xi), 1–41. Berlin: Springer.
101 Zhang, W.X. and Xi, Z. (**2014**). *Org. Chem. Front. 1*: 1132–1139.
102 Ma, W., Yu, C., Chen, T. et al. (**2017**). *Chem. Soc. Rev. 46*: 1160–1192.
103 Wei, J., Zhang, W.X., and Xi, Z. (**2018**). *Chem. Sci. 9*: 560–568.
104 Yi, H., Yang, D., Xin, J. et al. (**2017**). *Nature Commun. 8*: 14794.
105 Harvey, P.D. and Gray, H.B. (**1988**). *J. Am. Chem. Soc. 110*: 2145–2147.
106 Ford, P.C., Cariati, E., and Bourassa, J. (**1999**). *Chem. Rev. 99*: 3625–3648.
107 Yam, V.W.W. and Lo, K.K.W. (**1999**). *Chem. Soc. Rev. 28*: 323–334.
108 Wang, C., Yuan, J., Li, G. et al. (**2006**). *J. Am. Chem. Soc. 128*: 4564–4565.
109 Takahashi, T., Kotora, M., and Xi, Z. (**1995**). *J. Chem. Soc. Chem. Commun.*: 361–362.
110 Takahashi, T., Xi, Z., Yamazaki, A. et al. (**1998**). *J. Am. Chem. Soc. 120*: 1672–1680.
111 Ura, Y., Li, Y., Xi, Z., and Takahashi, T. (**1998**). *Tetrahedron Lett. 39*: 2787–2790.
112 Takahashi, T. and Li, Y. (**2003**). In: *Titanium and Zirconium in Organic Synthesis* (ed. I. Marek), 50–83. Weinheim: Wiley-VCH.
113 Chen, C., Xi, C., Jiang, Y., and Hong, X. (**2005**). *J. Am. Chem. Soc. 127*: 8024–8025.
114 Dufková, L., Kotora, M., and Císařová, I. (**2005**). *Eur. J. Org. Chem.*: 2491–2499.
115 Song, Z., Li, Y., Liu, M. et al. (**2006**). *Organometallics 25*: 5035–5044.
116 Zhou, Y., Yan, X., Chen, C., and Xi, C. (**2013**). *Organometallics 32*: 6182–6185
117 Lang, H, Jakob, A., Milde, B. (**2012**). *Organometallics*, *31*, 7661–7693.
118 Berben, L.A., de Bruin, B., and Heyduk, A.F. (**2015**). *Chem. Commun. 51*: 1553–1554.
119 Grande-Aztatzi, R., Mercero, J.M., Matito, E. et al. (**2017**). *Phy. Chem. Chem. Phy. 19*: 9669–9675.
120 An, K., Shen, T., and Zhu, J. (**2017**). *Organometallics 36*: 3199–3204.
121 Dimitrova, M. and Sundholm, D. (**2018**). *Phys. Chem. Chem. Phys. 20*: 1337–1346.
122 Peter, L.N. and Erwin, W. (**1990**). *Angew. Chem. Int. Ed. Eng. 29*: 300–302.

Index

Polar Organometallic Reagents: Synthesis, Structure, Properties and Applications, First Edition. Edited by Andrew E. H. Wheatley and Masanobu Uchiyama.

m

n

o